Lecture Notes in Computer Science 6488

Commenced Publication in 1973
Founding and Former Series Editors:
Gerhard Goos, Juris Hartmanis, and J

Editorial Board

David Hutchison
Lancaster University, UK

Takeo Kanade
Carnegie Mellon University, Pittsburgh, PA, USA

Josef Kittler
University of Surrey, Guildford, UK

Jon M. Kleinberg
Cornell University, Ithaca, NY, USA

Alfred Kobsa
University of California, Irvine, CA, USA

Friedemann Mattern
ETH Zurich, Switzerland

John C. Mitchell
Stanford University, CA, USA

Moni Naor
Weizmann Institute of Science, Rehovot, Israel

Oscar Nierstrasz
University of Bern, Switzerland

C. Pandu Rangan
Indian Institute of Technology, Madras, India

Bernhard Steffen
TU Dortmund University, Germany

Madhu Sudan
Microsoft Research, Cambridge, MA, USA

Demetri Terzopoulos
University of California, Los Angeles, CA, USA

Doug Tygar
University of California, Berkeley, CA, USA

Gerhard Weikum
Max Planck Institute for Informatics, Saarbruecken, Germany

Lei Chen   Peter Triantafillou   Torsten Suel (Eds.)

Web Information Systems Engineering – WISE 2010

11th International Conference
Hong Kong, China, December 12-14, 2010
Proceedings

 Springer

Volume Editors

Lei Chen
Hong Kong University of Science and Technology
Clear Water Bay, Kowloon, Hong Kong, China
E-mail: leichen@cse.ust.hk

Peter Triantafillou
University of Patras
26504 Patras, Greece
E-mail: peter@ceid.upatras.gr

Torsten Suel
Polytechnic Institute of NYU
Brooklyn, NY 11201, USA
E-mail: suel@poly.edu

Library of Congress Control Number: 2010940105

CR Subject Classification (1998): H.3.3-5, H.4-5, K.3.2, J.1, C.2, E.5

LNCS Sublibrary: SL 3 – Information Systems and Application, incl. Internet/Web and HCI

ISSN	0302-9743
ISBN-10	3-642-17615-1 Springer Berlin Heidelberg New York
ISBN-13	978-3-642-17615-9 Springer Berlin Heidelberg New York

springer.com

© Springer-Verlag Berlin Heidelberg 2010
Printed in Germany

Typesetting: Camera-ready by author, data conversion by Scientific Publishing Services, Chennai, India
Printed on acid-free paper   06/3180

Preface

Welcome to the Proceedings of WISE 2010 — the 11th International Conference on Web Information Systems Engineering. This year, WISE returned to the place where the inaugural conference was held in 2000, Hong Kong. WISE has also been held in: 2001 Kyoto (Japan), 2002 Singapore, 2003 Rome (Italy), 2004 Brisbane (Australia), 2005 New York (USA), 2006 Wuhan (China), 2007 Nancy (France), 2008 Auckland (New Zealand), and 2009 Poznan (Poland).

Continuing its trend, this year's WISE provided a forum for engineers and scientists to present their latest findings in Web-related technologies and solutions. The submitted contributions address challenging issues in Web services, search, modeling, recommendation and data mining, as well as keyword search, social network analysis, query languages, and information retrieval and extraction.

This year, WISE received 170 submissions from 25 countries, including Argentina, Australia, Austria, Belgium, Canada, China, Czech Republic, France, Germany, Hong Kong, Greece, Iran, Ireland, Italy, Japan, The Netherlands, Norway, Singapore, South Korea, Spain, Sweden, Switzerland, Taiwan, UK, and the USA. After a thorough reviewing process, 32 papers were selected for presentation as full papers – the acceptance rate was 18.8%. In addition, 19 papers were selected for presentation as short papers, yielding an overall acceptance rate of 30%.

This conference was made possible through the efforts of many people. We wish to thank everyone involved, including those who worked diligently behind the scenes and without formal recognition. First, we would like to thank the WISE Steering Committee (WISE-SC), in particular Yanchun Zhang and Xiaohua Jia representing the WISE-SC who selected City University of Hong Kong to host the WISE 2010 conference. We would like to thank the Workshops Chairs, Dickson Chiu and Ladjel Bellatreche, for their efforts to organize workshops in association with WISE. We thank the Publicity Chairs Hua Wang, Stephane Jean, Raymond Wong, and Feifei Li, as well the Organization Chairs, Hong Va Leong and Howard Leung. Special thanks go to our Financial Chair, Xiaohua Jia, and Jiying Wang (Jean) of the Hong Kong Web Society serving as the treasurer for the conference. We would like to thank the Program Committee members and reviewers for a very rigorous and outstanding reviewing process. We wish to thank Beng Chin Ooi (National University of Singapore, Singapore), Edward Chang (Google Research Lab, China), Christian S. Jensen (Aarhus University, Denmark), and Kyu-Young Whang (KAIST, Korea) for graciously accepting our invitation to serve as keynote speakers. Last but not least, we are grateful to the hard work of our webmaster Haoran Xie, and to our sponsors including the Croucher Foundation and the K.C. Wong Education Foundation, who generously supported our conference.

September 2010

Lei Chen
Peter Triantafillou
Torsten Suel
Qing Li
Karl Aberer
Dennis McLeod

Organization

Conference Co-chairs

Qing Li	City University of Hong Kong
Karl Aberer	EPFL, Switzerland
Dennis McLeod	USC, USA

PC Co-chairs

Lei Chen	HKUST, Hong Kong
Peter Triantafillou	University of Patras, Greece
Torsten Suel	NYU Poly, USA

Organization Co-chairs

Hong Va Leong	Poly University, Hong Kong
Howard Leung	City University of Hong Kong

Workshop Co-chairs

Dickson Chiu	Dickson Computer Sys., Hong Kong
Ladjel Bellatreche	ENSMA Poitiers University, France

Publicity Co-chairs

Hua Wang	University of Southern Queensland, Australia
Stephane Jean	Poitiers University, France
Raymong Wong	HKUST, Hong Kong
Feifei Li	Florida State University, USA

Finance Chair

Xiaohua Jia	City University of Hong Kong

Steering Committee Representatives

Yanchun Zhang	Victoria University, Australia
Xiaohua Jia	City University of Hong Kong

Webmaster

Haoran Xie	City University of Hong Kong

Program Committee

Vo Ngoc Anh	University of Melbourne, Australia
Amitabha Bagchi	IIT Delhi, India
Wolf Tilo Balke	University of Hannover, Germany
Srikanta Bedathur	MPI Saarbruecken, Germany
Sourav Bhowmick	NTU, Singapore
Roi Blanco	Yahoo! Research, Spain
Jinli Cao	Latrobe University, Australia
Sven Casteleyn	Vrije Universiteit Brussel, Belgium
Wei Chen	Beijing Institute of Technology, China
Yueguo Chen	Renmin University of China, China
Vassilis Christophides	University of Crete, Greece
Bin Cui	Beijing University, China
Theodore Dalamagas	National Technical University of Athens, Greece
Arjen Devries	CWI, Amsterdam, The Netherlands
Fernando Diaz	Yahoo! Lab, Canada
Marcus F. Fontoura	Yahoo, USA
Yunjun Gao	Zhejiang University, China
Aristides Gionis	Yahoo! Spain
Vivekanand Gopalkrishnan	NTU Singapore
Utku Irmak	Yahoo! Labs, USA
Yoshiharu Ishikawa	Nagoya University, Japan
Panos Karras	NUS, Singapore
Jinho Kim	Kangwon National University, Korea
Markus Kirchberg	A*STAR, Singapore
Ramamohanarao Kotagiri	University of Melbourne, Australia
Manolis Koubarakis	University of Athens, Greece
Hady W. Lauw	Microsoft Research, USA
Wenyin Liu	City University Hong Kong, Hong Kong, China
An Liu	USTC, China
Claudio Lucchese	CNR, Italy
Ashwin Machanavajjha	Yahoo Research, USA
Alex Markowetz	University of Bonn, Germany
Weiyi Meng	Binghamton University , USA
Sebastian Michel	University of Saarland, Germany
Raymond Ng	University of British Columbia, Canada
Nikos Ntarmos	University of Ioannina, Greece
Alex Ntoulas	Microsoft Research, USA
Josiane Xavier Parreira	NUIG, Ireland
Evaggelia Pitoura	University of Ioannina,Greece
Dimitris Plexousakis	University of Crete, Greece
Alexandra Poulovassilis	Birbeck, UK

Tieyun Qian   Wuhan University, China
Michael Rabinovich   CWRU, USA
Matthias Renz   University of Munich, Germany
Ralf Schenkel   University of Saarbruecken, Germany
Hengtao Shen   University of Queensland, Australia
Shuming Shi   Microsoft Research, Asia, China
Gleb Skobeltsyn   Google, Zurich
Guojie Song   Peking University, China
Vincent Tseng   National Cheng Kung University, Taiwan
Ozgur Ulusoy   Bilkent University, Turkey
Liping Wang   Victoria University, Australia
Wei Wang   The University of New South Wales, Australia
Ingmar Weber   Yahoo!, Spain
Raymong Wong   HKUST, Hong Kong
Quanqing Xu   NUS, Singapore
Jianliang Xu   Hong Kong Bapatist University, Hong Kong
Jefferey Xu Yu   CUHK, Hong Kong, China
Philip Yu   UIC, USA
Pavel Zezula   Masaryk University, Czech Republic
Xiaofang Zhou   University of Queensland, Australia
Lei Zou   Beijing University, China

External Reviewers

Ismail Sengor Altingovde   Bilkent University, Turkey
Duygu Atilgan   Bilkent University, Turkey
Nikos Bikakis   NTU Athens, Greece
Berkant Barla Cambazoglu   Yahoo! Research
Michal Batko   Masaryk University, Czech Republic
George Baryannis   University of Crete, Greece
Alexander Bergmayer   University of Crete, Greece
Thomas Bernecker   Ludwig-Maximilians-Universität München,
                              Germany
Thomas-Ogbuji Chimezie   Case Western Reserve University and
                              Cleveland Clinic, USA
Bilal Choudry   La Trobe University, Australia
Theodore Dalamagas   IMIS Institute, Greece
Jeff Dalton   University of Massachusetts Amherst, USA
Aggeliki Dimitriou   NTU Athens, Greece
Martin Doerr   FORTH-ICS, Greece
Vlastislav Dohnal   Masaryk University, Czech Republic
Tobias Emrich   Ludwig-Maximilians-Universität München,
                              Germany
Yuzhang Feng   A*STAR, Singapore
Giorgos Giannopoulos   IMIS Institute, Greece
Rossi Gustavo   Universidad Nacional de La Plata, Argentina
Yu Jiang   Binghamton University, USA

Sadiye Kaptanoglu	Bilkent University, Turkey
Haris Kondylakis	University of Crete, Greece
Dimitris Kotzinos	FORTH-ICS, Greece
Wookey Lee	Inha University, Korea
Xian Li	Binghamton University, USA
Lian Liu	Hong Kong University of Science and Technology, Hong Kong
Peter Mika	Yahoo! Research
Yang-Sae Moon	Kangwon National University, Korea
Irene Ntoutsi	Ludwig-Maximilians-Universität München, Germany
Duy Ngan Le	A*STAR, Singapore
Khanh Nguyen	La Trobe University, Australia
David Novak	Masaryk University, Czech Republic
Rifat Ozcan	Bilkent University, Turkey
George Papastefanatos	IMIS Institute, Czech Republic
Yu Peng	Hong Kong University of Science and Technology, Hong Kong
Kanagasabai Rajaraman	A*STAR, Singapore
Erich Schubert	Ludwig-Maximilians-Universität München, Germany
Yannis Stavrakas	IMIS Institute, Greece
Loan Vo	La Trobe University, Australia
Ouyang Tu	Case Western Reserve University, USA
William Van Woensel	Vrije Universiteit Brussel, Belgium
Qian Wan	Hong Kong University of Science and Technology, Hong Kong
Raymond Wan	University of Tokyo, Japan
Clement Yu	University of Illinois at Chicago, USA
Chrysostomos Zeginis	University of Crete, Greece
Bin Zhang	Hong Kong University of Science and Technology, Hong Kong

Table of Contents

Keynotes

Providing Scalable Database Services on the Cloud   1
    Chun Chen, Gang Chen, Dawei Jiang, Beng Chin Ooi,
    Hoang Tam Vo, Sai Wu, and Quanqing Xu

Search and Social Integration (Abstract)   20
    Edward Y. Chang

Elements of a Spatial Web (Abstract)   21
    Christian S. Jensen

The Ubiquitous DBMS (Abstract)   22
    Kyu-Young Whang

Web Service

Building Web Services Middleware with Predictable Service
Execution   23
    Vidura Gamini Abhaya, Zahir Tari, and Peter Bertok

Event Driven Monitoring for Service Composition Infrastructures   38
    Oliver Moser, Florian Rosenberg, and Schahram Dustdar

On Identifying and Reducing Irrelevant Information in Service
Composition and Execution   52
    Hong-Linh Truong, Marco Comerio, Andrea Maurino,
    Schahram Dustdar, Flavio De Paoli, and Luca Panziera

Propagation of Data Protection Requirements in Multi-stakeholder
Web Services Systems   67
    Tan Phan, Jun Han, Garth Heward, and Steve Versteeg

Social Networks

Refining Graph Partitioning for Social Network Clustering   77
    Tieyun Qian, Yang Yang, and Shuo Wang

Fast Detection of Size-Constrained Communities in Large Networks   91
    Marek Ciglan and Kjetil Nørvåg

Evolutionary Taxonomy Construction from Dynamic Tag Space   105
    Bin Cui, Junjie Yao, Gao Cong, and Yuxin Huang

Co-clustering for Weblogs in Semantic Space 120
  Yu Zong, Guandong Xu, Peter Dolog, Yanchun Zhang, and
  Renjin Liu

Web Data Mining

A Linear-Chain CRF-Based Learning Approach for Web Opinion
Mining 128
  Luole Qi and Li Chen

An Unsupervised Sentiment Classifier on Summarized or Full
Reviews 142
  Maria Soledad Pera, Rani Qumsiyeh, and Yiu-Kai Ng

Neighborhood-Restricted Mining and Weighted Application of
Association Rules for Recommenders 157
  Fatih Gedikli and Dietmar Jannach

Semantically Enriched Event Based Model for Web Usage Mining 166
  Enis Söztutar, Ismail H. Toroslu, and Murat Ali Bayir

Keyword Search

Effective and Efficient Keyword Query Interpretation Using a Hybrid
Graph 175
  Junquan Chen, Kaifeng Xu, Haofen Wang, Wei Jin, and Yong Yu

From Keywords to Queries: Discovering the User's Intended Meaning ... 190
  Carlos Bobed, Raquel Trillo, Eduardo Mena, and Sergio Ilarri

Efficient Interactive Smart Keyword Search 204
  Shijun Li, Wei Yu, Xiaoyan Gu, Huifu Jiang, and Chuanyun Fang

Relevant Answers for XML Keyword Search: A Skyline Approach 216
  Khanh Nguyen and Jinli Cao

Web Search I

A Children-Oriented Re-ranking Method for Web Search Engines 225
  Mayu Iwata, Yuki Arase, Takahiro Hara, and Shojiro Nishio

TURank: Twitter User Ranking Based on User-Tweet Graph
Analysis 240
  Yuto Yamaguchi, Tsubasa Takahashi, Toshiyuki Amagasa, and
  Hiroyuki Kitagawa

Identifying and Ranking Possible Semantic and Common Usage
Categories of Search Engine Queries 254
  Reza Taghizadeh Hemayati, Weiyi Meng, and Clement Yu

Best-Effort Refresh Strategies for Content-Based RSS Feed
Aggregation 262
  Roxana Horincar, Bernd Amann, and Thierry Artières

Mashup-Aware Corporate Portals 271
  Sandy Pérez and Oscar Díaz

Web Data Modeling

When Conceptual Model Meets Grammar: A Formal Approach to
Semi-structured Data Modeling 279
  Martin Nečaský and Irena Mlýnková

Crowdsourced Web Augmentation: A Security Model 294
  Cristóbal Arellano, Oscar Díaz, and Jon Iturrioz

Design of Negotiation Agents Based on Behavior Models 308
  Kivanc Ozonat and Sharad Singhal

High Availability Data Model for P2P Storage Network 322
  BangYu Wu, Chi-Hung Chi, Cong Liu, ZhiHeng Xie, and Chen Ding

Recommender Systems

Modeling Multiple Users' Purchase over a Single Account for
Collaborative Filtering 328
  Yutaka Kabutoya, Tomoharu Iwata, and Ko Fujimura

Interaction-Based Collaborative Filtering Methods for Recommendation
in Online Dating 342
  *Alfred Krzywicki, Wayne Wobcke, Xiongcai Cai, Ashesh Mahidadia,
  Michael Bain, Paul Compton, and Yang Sok Kim*

Developing Trust Networks Based on User Tagging Information for
Recommendation Making 357
  *Touhid Bhuiyan, Yue Xu, Audun Jøsang, Huizhi Liang, and
  Clive Cox*

Towards Three-Stage Recommender Support for Online Consumers:
Implications from a User Study 365
  Li Chen

RDF and Web Data Processing

Query Relaxation for Star Queries on RDF   376
    Hai Huang and Chengfei Liu

Efficient and Adaptable Query Workload-Aware Management for RDF
Data   390
    Hooran MahmoudiNasab and Sherif Sakr

RaUL: RDFa User Interface Language – A Data Processing Model for
Web Applications   400
    Armin Haller, Jürgen Umbrich, and Michael Hausenblas

Synchronising Personal Data with Web 2.0 Data Sources   411
    *Stefania Leone, Michael Grossniklaus, Alexandre de Spindler, and
    Moira C. Norrie*

An Artifact-Centric Approach to Generating Web-Based Business
Process Driven User Interfaces   419
    Sira Yongchareon, Chengfei Liu, Xiaohui Zhao, and Jiajie Xu

XML and Query Languages

A Pattern-Based Temporal XML Query Language   428
    Xuhui Li, Mengchi Liu, Arif Ghafoor, and Philip C-Y. Sheu

A Data Mining Approach to XML Dissemination   442
    Xiaoling Wang, Martin Ester, Weining Qian, and Aoying Zhou

Semantic Transformation Approach with Schema Constraints for
XPath Query Axes   456
    *Dung Xuan Thi Le, Stephane Bressan, Eric Pardede,
    Wenny Rahayu, and David Taniar*

Domain-Specific Language for Context-Aware Web Applications   471
    *Michael Nebeling, Michael Grossniklaus, Stefania Leone, and
    Moira C. Norrie*

Web Search II

Enishi: Searching Knowledge about Relations by Complementarily
Utilizing Wikipedia and the Web   480
    Xinpeng Zhang, Yasuhito Asano, and Masatoshi Yoshikawa

Potential Role Based Entity Matching for Dataspaces Search   496
    Yue Kou, Derong Shen, Tiezheng Nie, and Ge Yu

Personalized Resource Search by Tag-Based User Profile and Resource
Profile   510
   Yi Cai, Qing Li, Haoran Xie, and Lijuan Yu

Incremental Structured Web Database Crawling via History Versions . . .   524
   Wei Liu and Jianguo Xiao

Web Information Systems

An Architectural Style for Process-Intensive Web Information
Systems   534
   Xiwei Xu, Liming Zhu, Udo Kannengiesser, and Yan Liu

Model-Driven Development of Adaptive Service-Based Systems with
Aspects and Rules   548
   Jian Yu, Quan Z. Sheng, and Joshua K.Y. Swee

An Incremental Approach for Building Accessible and Usable Web
Applications.   564
   Nuria Medina Medina, Juan Burella, Gustavo Rossi,
   Julián Grigera, and Esteban Robles Luna

CPH-VoD: A Novel CDN–P2P-hybrid Architecture Based
VoD Scheme   578
   Zhihui Lu, Jie Wu, Lijiang Chen, Sijia Huang, and Yi Huang

Information Retrieval and Extraction

A Combined Semi-pipelined Query Processing Architecture for
Distributed Full-Text Retrieval   587
   Simon Jonassen and Svein Erik Bratsberg

Towards Flexible Mashup of Web Applications Based on Information
Extraction and Transfer   602
   Junxia Guo, Hao Han, and Takehiro Tokuda

On Maximal Contained Rewriting of Tree Pattern Queries Using
Views   616
   Junhu Wang and Jeffrey Xu Yu

Implementing Automatic Error Recovery Support for Rich
Web Clients   630
   Manuel Quintela-Pumares, Daniel Fernández-Lanvin,
   Raúl Izquierdo, and Alberto Manuel Fernández-Álvarez

Author Index   639

Providing Scalable Database Services on the Cloud

Chun Chen[2], Gang Chen[2], Dawei Jiang[1], Beng Chin Ooi[1],
Hoang Tam Vo[1], Sai Wu[1], and Quanqing Xu[1]

[1] National University of Singapore
{jiangdw,ooibc,voht,wusai,xuqq}@comp.nus.edu.sg
[2] Zhejiang University, China
{chenc,cg}@cs.zju.edu.cn

Abstract. The Cloud is fast gaining popularity as a platform for deploying Software as a Service (SaaS) applications. In principle, the Cloud provides unlimited compute resources, enabling deployed services to scale seamlessly. Moreover, the pay-as-you-go model in the Cloud reduces the maintenance overhead of the applications. Given the advantages of the Cloud, it is attractive to migrate existing software to this new platform. However, challenges remain as most software applications need to be redesigned to embrace the Cloud.

In this paper, we present an overview of our current on-going work in developing epiC – an elastic and efficient power-aware data-intensive Cloud system. We discuss the design issues and the implementation of epiC's storage system and processing engine. The storage system and the processing engine are loosely coupled, and have been designed to handle two types of workload simultaneously, namely data-intensive analytical jobs and online transactions (commonly referred as OLAP and OLTP respectively). The processing of large-scale analytical jobs in epiC adopts a phase-based processing strategy, which provides a fine-grained fault tolerance, while the processing of queries adopts indexing and filter-and-refine strategies.

1 Introduction

Data has become an important commodity in modern business where data analysis facilitates better business decision making and strategizing. However, a substantial amount of hardware is required to ensure reasonable response time, in addition to data management and analytical software. As the company's business grows, its workload outgrows the hardware capacity and it needs to be upgraded to accommodate the increasing demand. This indeed presents many challenges both in terms of technical support and cost, and therefore the Cloud becomes a feasible solution that mitigates the pain.

The web applications, such as online shopping and social networking, are currently the majority of applications deployed in the Cloud. Excellent system scalability, low service response time and high service availability are required

L. Chen, P. Triantafillou, and T. Suel (Eds.): WISE 2010, LNCS 6488, pp. 1–19, 2010.

for such applications, as they are generating unprecedented massive amounts of data. Therefore, large-scale ad-hoc analytical processing of the data collected from those web services is becoming increasingly valuable to improving the quality and efficiency of existing services, and supporting new functional features. However, traditional online analytical processing (OLAP) solutions, such as parallel database systems and data warehouses, fall short of scaling dynamically with load and need.

Typically, OLTP (online transaction processing) and OLAP workloads are often handled separately by separate systems with different architectures – RDBMS for OLTP and data warehousing system for OLAP. To maintain the data freshness between these two systems, a data extraction process is periodically performed to migrate the data from the RDBMS into the data warehouse. This system-level separation, though provides flexibility and the required efficiency, introduces several limitations, such as the lack of up-to-date data freshness in OLAP, redundancy of data storage, as well as high startup and maintenance cost.

The need to dynamically provide for capacity both in terms of storage and computation, and to support online transactional processing (OLTP) and online analytical processing (OLAP) in the Cloud demands the re-examination of existing data servers and architecting possibly "new" elastic and efficient data servers for Cloud data management service. In this paper, we present epiC – a Cloud data management system which is being designed to support both functionalities (OLAP and OLTP) within the same storage and processing system. As discussed above, the approaches adopted by parallel databases cannot be directly applied to the Cloud data managements. The main issue is the elasticity. In the Cloud, thousands of compute nodes are deployed to process petabyte of data, and the demand for resources may vary drastically from time to time. To provide data management service in such environment, we need to consider the following issues.

1. Data are partitioned among multiple compute nodes. To facilitate different access patterns, various storage systems are developed for the Cloud. For example, GFS [20] and HDFS [1] are designed for efficient scan or batch access, while Dynamo [16], BigTable [12] and Cassandra [25] are optimized for key-based access. OLAP queries may involve scanning multiple tables, while OLTP queries only retrieve a small number of records. Since it is costly to maintain two storage systems, to support both workloads, a hybrid storage system is required by combining the features of existing solutions.

2. The efficiency and scalability of the Cloud is achieved via parallelism. The query engine must be tuned to exploit the parallelism among the compute nodes. For this purpose, we need to break down the relational operators into more general atomic operators. The atomic operators are tailored for the cluster settings, which are naturally parallelizable. This approach is adopted by MapReduce [15], Hive [35], Pig [30], SCOPE [11], HadoopDB [8] and our

proposed epiC system [6]. The query engine should be able to transform a SQL query into the atomic operators and optimize the plans based on cost models.

3. Machine failure is not an exception in the Cloud. In a large cluster, a specific node may fail at any time. Fault tolerance is a basic requirement for all Cloud services, especially the database service. When a node fails, to continue the database service, we need to find the replicas to recover the lost data and schedule the unfinished jobs of the failed node to others. Indeed, the fault tolerance issues affect the designs of both the storage layer and the processing engine of the Cloud data management system.

4. Last but not the least, the Cloud data management system should provide tools for users to immigrate from their local databases. It should support a similar interface as the conventional database systems, which enables the users to run their web services, office softwares and ERP systems without modification.

Due to the distinct characteristics of OLAP and OLTP workload, the query processing engine of epiC is loosely coupled with the underlying storage system and adopts different strategies to process queries from the two different workloads. This enables the query process and the storage process to be deployed independently. One cluster machine can host one or more query processes or storage processes, providing more space for load balancing.

In epiC, OLAP queries are processed via parallel scans, while OLTP queries are handled by indexing and localized query optimization. The OLAP execution engine breaks down conventional database operations such as join into some primitives, and enables them to run in MapReduce-like or filter-and-refine phases. The motivation for this design is that, although the widely adopted MapReduce computation model has been designed with built-in parallelism and fault tolerance, it does not provide data schema support, declarative query language and cost-based query optimization. To avoid the access contention between the two workloads, we relax the data consistency of OLAP queries by providing snapshot-based results, which are generally sufficient for decision making.

The rest of paper is organized as follows. Section 2 reviews the efforts of previous work and discusses the challenges of implementing a database service in the Cloud. Section 3 presents the design and implementation of our proposed epiC system for Cloud data management service. We conclude the paper in Section 4.

2   Challenges of Building Data Management Applications in the Cloud

In this section, we review related work and discuss how they affect the design of epiC system.

2.1   Lessons Learned from Previous Work

Parallel Database Systems. Database systems capable of performing data processing on shared-nothing architectures are called parallel database systems [1]. The systems mostly adopt relational data model and support SQL. To parallelize SQL query processing, the systems employ two key techniques pioneered by GRACE [19] and Gamma [18] projects: 1) horizontal partition of relational tables and 2) partitioned execution of SQL queries.

The key idea of horizontal partitioning is to distribute the tuples of relational tables among the nodes in the cluster based on certain rules or principles so that those tuples can be processed in parallel. A number of partitioning strategies have been proposed, including hash partitioning, range partitioning, and round-robin partitioning [17]. For example, to partition the tuples in a table T among n nodes under the hash-partitioning scheme, one must apply a universal hash function on one or more attributes of each tuple in T in order to determine the node that it will be stored.

To process SQL queries over partitioned tables, the partition based execution strategy is utilized. Suppose we want to retrieve tuples in T within a given date range (e.g., from '2010-04-01' to '2010-05-01'). The system first generates a query plan P for the whole table T, then partitions P into n subquery plans $\{P_1, \ldots, P_n\}$ such that each subquery plan P_i can be independently processed by node n_i. All the subquery plans apply the same principles by applying the filtering condition to the tuples stored on the local node. Finally, the intermediate results from each node are sent to a selected node where a merge operation is performed to produce the final results.

Parallel database systems are robust, high-performance data processing platforms. In the past two decades, many techniques have been developed to enhance the performance of the systems, including indexing, compression, materialized views, result caching and I/O sharing. These technologies are matured and well tested. While some earlier systems (e.g., Teradata [2]) have to be deployed on proprietary hardware, recent systems (e.g., Aster [3], Vertica [4] and Greenplum [5]), can be deployed on commodity machines. The ability of deploying on low-end machines makes parallel database systems Cloud-ready. To our knowledge, Vertica and Aster have in fact already released their Cloud editions.

However, despite the fact that some parallel database systems can be deployed on Cloud, these systems may not be able to take full advantage of the Cloud. Cloud allows users to elastically allocate resources from the Cloud and only pay for the resources that are actually utilized. This enables users to design their applications to scale their resource requirements up and down in a pay-as-you-go manner. For example, suppose we have to perform data analysis on two datasets with the size of 1TB and 100GB consecutively. Under the elastic

---

[1] In database context, database systems employing shared-memory and shared-disk architectures are also called parallel database systems. In this paper, we only cover shared-nothing parallel database systems since Cloud as of now is mostly deployed on a large shared-nothing cluster.

scale-up scheme, we can allocate a 100 node cluster from Cloud to analyze the 1TB dataset and shrink down the cluster to 10 nodes for processing the 100GB dataset. Suppose the data processing system is linearly scaled-up, the two tasks will be completed in roughly the same time. Thus, the elastic scale-up capability along with the pay-as-you-go business model results in a high performance/price ratio.

The main drawback of parallel database systems is that they are not able to exploit the built-in elasticity feature (which is deemed to be conducive for startups, small and medium sized businesses) of the Cloud. Parallel database systems are mainly designed and optimized for a cluster with a fixed or fairly static number of nodes. Growing up and shrinking down the cluster requires a deliberate plan (often conducted by a DBA) to migrate data from existing configuration to the new one. This data migration process is quite expensive as it often causes the service to be unavailable during migration and thus is avoided in most production systems. The inflexibility for growing up and shrinking down clusters on the fly affect parallel database systems' elasticity and their suitability for pay-as-you-go business model. Another problem of parallel database systems is their degree of fault tolerance. Historically, it is assumed that node failure is more of an exception than a common case, and therefore only transaction level fault tolerance is often provided. When a node fails during the execution of a query, the entire query must be restarted. As argued in [8], the restarting query strategy may cause parallel database systems not being able to process long running queries on clusters with thousands of nodes, since in these clusters hardware failures are common rather than exceptional. Based on this analysis, we argue that parallel database systems are best suitable for applications whose resource requirements are relatively static rather than dynamic. However, many design principles of parallel database systems could form the foundation for the design and optimization of systems to be deployed in the Cloud. In fact, we have started to witness the introduction of declarative query support and cost based query processing into MapReduce-based systems.

MapReduce-Based Systems. MapReduce [15], developed in Google, is a programming model and associated implementation for processing datasets on shared-nothing clusters. This system was designed as a purely data processing system with no built-in facilities to store data. This *"pure processing engine"* design is in contrast to parallel database systems which are equipped with both processing engine and storage engine.

Although MapReduce has been designed to be independent of the underlying data storage system, the system makes at least one assumption about the underlying storage system, namely the data stored in the storage system are already (or can be) horizontally partitioned and the analytical program can be independently launched on each data split [2].

---

[2] In this perspective, MapReduce actually shares the same key techniques with parallel database systems for parallel data processing: 1) horizontal data partitioning and 2) partitioned execution.

The MapReduce programming model consists of two user specified functions: `map()` and `reduce()`. Without loss of generality, we assume that the MapReduce system is used to analyze data stored in a distributed file system such as GFS and HDFS. To launch a MapReduce job, the MapReduce runtime system first collects partitioning information of the input dataset by querying the metadata of the input files from the distributed file system. The runtime system subsequently creates M map tasks for the input, one for each partition and assigned those map tasks to the available nodes. The node which is assigned a map task reads contents from the corresponding input partition and parses the contents into key/value pairs. Then, it applies the user-specified map function on each key/value pair and produces a list of key/value pairs. The intermediate key/value pairs are further partitioned into R regions in terms of the key and are materialized as local files on the node. Next, the runtime system collects all intermediate results and merges them into R files (intermediate results in the same region are merged together), and launches R reduce tasks to process the files, one for each file. Each reduce task invokes user specified reduce function on each key/value list and produces the final answer.

Compared to the parallel database systems, MapReduce system has a few advantages: 1) MapReduce is a pure data processing engine and is independent of the underlying storage system. This storage independence design enables MapReduce and the storage system to be scaled up independently and thus goes well with the pay-as-you-go business model. The nice property of the Cloud is that it offers different pricing schemes for different grades and types of services. For example, the storage service (e.g., Amazon S3) is charged by per GB per month usage while the computing service (e.g., Amazon EC2) is charged by per node per hour usage. By enabling independent scaling of the storage and processing engine, MapReduce allows the user to minimize the cost on IT infrastructure by choosing the most economical pricing schemes. 2) Map tasks and reduce tasks are assigned to available nodes on demand and the number of tasks (map or reduce) is independent of the number of nodes in the cluster. This runtime scheduling strategy makes MapReduce to fully unleash the power of elasticity of the Cloud. Users can dynamically increase and decrease the size of the cluster by allocating nodes from or releasing nodes to the Cloud. The runtime system will manage and schedule the available nodes to perform map or reduce tasks without interrupting the running jobs. 3) Map tasks and reduce tasks are independently executed from each other. There are however communications or dependencies between map tasks or reduce tasks. This design makes MapReduce to be highly resilient to node failures. When a single node fails during the data processing, only map tasks and reduce tasks on the failed node need to be restarted; the whole job needs not to be restarted.

Even though MapReduce has many advantages, it has been noted that achieving those capabilities comes with a potentially large performance penalty. Benchmarking in [31] shows that Hadoop, an open source MapReduce implementation, is slower than two state of the art parallel database systems by a factor of 3.1 to 6.5 on a variety of analytical tasks and that the large performance gap between

MapReduce and parallel databases may offset all the benefits that MapReduce provides. However, we showed that by properly choosing the implementation strategies for various key operations, the performance of MapReduce can be improved by a factor of 2.5 to 3.5 for the same benchmark [23]. The results show that there is a large space for MapReduce to improve its performance. Based on our study, we conclude that MapReduce system is best suitable for large-scale deployment (thousands of nodes) and data analytical applications which demand dynamic resource allocation. The system has not been designed to support real time updates and search as in conventional database systems.

Scalable Data Management Systems. Providing scalable data management has posed a grand challenge to database community for more than two decades. Distributed database systems were the first general solution that is able to deal with large datasets stored on distributed shared-nothing environment. However, these systems could not scale beyond a few machines as the performance degrades dramatically due to synchronization overhead and partial failures. Therefore it is not surprising that modern scalable Cloud storage systems, such as BigTable [12], Pnuts [13], Dynamo [16], and Cassandra [25], abandon most of the designs advocated by distributed database systems and adopt different solutions to achieve the desired scalability. The techniques widely adopted by these scalable storage systems are: 1) employ simple data model 2) separate meta data and application data 3) relax consistency requirements.

Simple Data Model. Different from distributed databases, most of current scalable Cloud storage systems adopt a much simpler data model in which each record is identified by a unique key and the atomicity guarantee is only supported at the single record level. Foreign key or other cross records relationship are not supported. Restricting data access to single records significantly enhance the scalability of system since all data manipulation will only occurred in a single machine, i.e., no distributed transaction overhead is introduced.

Separation of Meta Data and Application Data. A scalable Cloud data management system needs to maintain two types of information: meta data and application data. Meta data is the information that is required for system management. Examples of meta data are the mappings of a data partition to machine nodes in the cluster and to its replicas. Application data is the business data that users stored in the system. These systems make a separation between meta data and application data since each type of data has different consistency requirements. In order for the system to operate correctly, the meta data must always be consistent and up-to-date. However, the consistency requirement of application data is entirely dependent on the applications and varies from application to application. As a result, Cloud data management systems employ different solutions to manage meta data and application data in order to achieve scalability.

Relaxed Consistency. Cloud data management systems replicate application data for availability. This design, however, introduces non-negligible overhead of synchronization of all replicas during updating data. To reduce the synchronization overhead, relaxed consistency model like eventual consistency (e.g. in

Dynamo [16]) and timeline consistency (e.g. in Pnuts [13]) are widely adopted. The detailed comparison between current scalable cloud data serving systems on their supported consistency models and other aspects such as partitioning and replication strategies can be found in [32].

2.2   How to Manage the Cloud - The Essential of Cloud

In this section, we describe the desired properties of a Cloud data management system and our design consideration.

Scalability. There is a trend that the analytical data size is growing exponentially. As an example, Facebook reports that 15TB new data are inserted into their data warehouse every day, and a huge amount of the scientific data such as mechanobiological data and images is generated each day due to the advancement in X-ray technologies and data collection tools (as noted in the second keynote [28] at VLDB 2010). To process such huge amount of data within a reasonable time, a large number of compute nodes are required. Therefore, the data processing system must be able to deploy on very large clusters (hundreds or even thousands of nodes) without much problems.

Elasticity. As we argued previously, elasticity is an invaluable feature provided by Cloud. The ability of scaling resource requirements up and down on demand results in a huge cost saving and is extremely attractive to any operations when the cost is a concern. To unleash the power of Cloud, the data processing system should be able to transparently manage and utilize the elastic computing resources. The system should allow users to add and remove compute nodes on the fly. Ideally, to speed up the data processing, one can simply add more nodes to the cluster and the newly added nodes can be utilized by the data processing system immediately (i.e., the startup cost is negligible). Furthermore, when the workload is light, one can release some nodes back to the Cloud and the cluster shrinking process will not affect other running jobs such as causing them to abort.

Fault Tolerance. Cloud is often built on a large number of low-end, unreliable commodity machines. As a result, hardware failure is fairly common rather than exceptional. The Cloud data processing system should be able to highly resilient to node failures during data processing. Single or even a large number of node failures should not cause the data processing system to restart the running jobs.

Performance. A common consensus in MapReduce community is that scalability can compensate for the performance. In principle, one can allocate more nodes from the Cloud to speed up the data processing. However this solution is not cost efficient in a pay-as-you-go environment and may potentially offset the benefit of elasticity. To maximize the cost saving, the data processing system indeed needs to be efficient.

Flexibility. It is well accepted that the Cloud data processing system is required to support various kinds of data analytical tasks (e.g., relational queries, data mining, text processing). Therefore, the programming model of the Cloud data processing system must be flexible and yet expressive. It should enable users to easily express any kinds of data analytical logic. SQL is routinely criticized for its insufficient expressiveness and thus is not ideal for Cloud data processing systems. The MapReduce programming model is deemed much more flexible. The `map()` and `reduce()` functions can be used to express any kinds of logic. The problem of the programming model is that MapReduce has no built-in facilities to manage a MapReduce pipeline. Due to its simple programming model (only two functions are involved), all real world data analytical tasks must be expressed as a set of MapReduce jobs (called a MapReduce pipeline). Hence, the synchronization and management of jobs in the MapReduce pipeline poses a challenge to the user.

3   epiC, Elastic Power-Aware Data Intensive Cloud System

A typical web data management system has to process real-time updates and queries by individual users, and as well as periodical large scale analytical jobs. While such operations take place in the same domain, the transactional and periodical analytical processing have been handled differently using different systems or even hardware. Such a system-level separation naturally leads to the problems of data freshness and data storage redundancy. To alleviate such problems, we have designed and implemented epiC, an elastic power-aware data-intensive Cloud platform for supporting both online analytical processing (OLAP) and online transaction processing (OLTP).

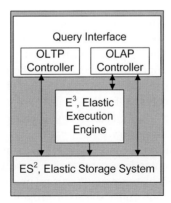

Fig. 1. Architecture of epiC

Figure 1 shows the architecture of epiC system, which is composed of three main modules, the **Query Interface**, the **Elastic Execution Engine** (E^3) and the **Elastic Storage System** (ES^2). The query interface provides a SQL-like language for up-level applications. It compiles the SQL query into a set of analytical jobs (for OLAP query) or a series of read and write operations (for OLTP query). E^3, a sub-system of epiC, is designed to efficiently perform large scale analytical jobs on the Cloud. ES^2 is the underlying storage system, which is designed to provide an always-on data service. In what follows, we shall briefly introduce the implementation of each module.

3.1   Query Interface and Optimization

Some systems, such as MapReduce [15] and its open source Hadoop [1], provide a flexible programming model by exposing some primitive functions (e.g. map and reduce). Users can implement their processing logic via customized functions. This design facilitates the development of applications, but does not provide an interface for the end-users, who are more familiar with SQL. In epiC, we provide a SQL-like language (in fact, a subset of SQL) as our query language. Currently, non-nested queries and nested queries in the where clause are supported. DBMS users can adapt to epiC without much difficulty.

Inside the query interface module, two controllers, namely OLAP controller and OLTP controller, are implemented to handle different types of queries and monitor the processing status. After a query is submitted to epiC, the query interface first checks whether the query is an analytical query or a simple select query. In the former case, the query is forwarded to the OLAP controller. It transforms the query into a set of E^3 jobs. The OLAP controller interacts with E^3 to process the jobs. Normally, the jobs are processed by E^3 one by one. A specific processing order of jobs is actually a unique query plan. The OLAP controller employs a cost based optimizer to generate a low-cost plan. Specifically, histograms are built and maintained in underlying ES^2 system by running some E^3 jobs periodically. The OLAP controller queries the metadata catalog of the ES^2 to retrieve the histograms, which can be used to estimate the cost of a specific E^3 job. We iteratively permute the processing order of the jobs and estimate the cost of each permutation. The one with lowest cost is then selected as the query plan. Based on the optimized query plan, the OLAP controller submits jobs to E^3. For each job, the OLAP controller defines the input and output (both are tables in ES^2). The processing functions of E^3 are auto-generated by the controller. After E^3 completes a job, the controller collects the result information and returns to the user if necessary.

If the query is a simple select query, the OLTP controller will take over the query. It first checks the metadata catalog of ES^2 to get histogram and index information. Based on the histograms, it can estimate the number of involved records. Then, the OLTP controller selects a proper access method (e.g. index lookup, random read or scan), which reduces the total number of network and disk I/Os. Finally, the OLTP controller calls the data access interface of ES^2 to perform the operations. For more complex queries, the OLTP controller will make use of histograms and other statistical information, available join strategies and system loads to generate an efficient query plan, and execute it as in OLAP query processing. Both OLAP and OLTP controller rely on the underlying ES^2 system to provide transactional support. Namely, we implement the transaction mechanism in the storage level. Detailed descriptions can be found in the Section 3.3.

Besides above optimization, we also consider multi-query optimizations in both OLAP and OLTP controller. When multiple queries involve the same table, instead of scanning it repeatedly, we group the queries together and process them in a batch. This strategy can significantly improve the performance. A similar approach on MapReduce is adopted in [29].

In addition, since MapReduce [15] has not been designed for generic data analytical workload, most cloud-based query processing systems, e.g. Hive [35], may translate a query into a long chain of MapReduce jobs without optimization, which incurs a significant overhead of startup latency and intermediate results I/Os. Further, this multi-stage process makes it more difficult to locate performance bottlenecks, limiting the potential use of self-tuning techniques. The OLAP controller can exploit data locality, as a result of offline low-cost data indexing, to efficiently perform complex relational operations such as n-way joins. The detailed implementation of these optimization techniques and experimental results are presented in our technical report [21].

3.2 E^3: Elastic Execution Engine

To perform the data analytical tasks (jobs), OLAP controller submits the jobs to the master node of E^3. The master node then distributes the jobs to all available nodes for parallel execution. Figure 2 describes the overall architecture of E^3.

Like MapReduce [15], E^3 is also a pure data processing system and is independent of the underlying storage systems it operates on. We only assume that the input data can be partitioned into even sized data chunks and the underlying storage system can provide necessary information for the data partitions. This design should enable users to take full advantage of the flexible pricing schemes offered by Cloud providers and thereby minimize their operational costs.

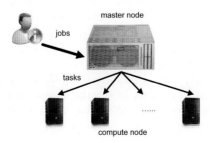

Fig. 2. Architecture of E^3

Users may choose to write data analytical tasks in Java [3]. The core abstractions of E^3 are implemented as Java classes and can be used in any Java programs. The goal of this design is twofolds: 1) By choosing a general programming language such as Java for writing data analytical tasks, E^3 enables users to express arbitrarily complex data processing logic; 2) As E^3 jobs are normal Java programs, users can utilize conventional IDE and profiling tools to debug and tune the jobs. Compared to the approach of embedding SQL to a general programming language such as Java for data processing (commonly adopted by major parallel database systems), this design should significantly increase the productivity of development as it is difficult to provide a powerful tool to debug and profile complex SQL. If queries are submitted via the epiC's query interface, the query interface will automatically generate the Java codes for each job based on predefined code templates.

E^3 provides three core abstractions for users to specify data processing logics. The first abstraction is Table<K,V> which represents the input dataset as a key/value data structure. By default, a table is unordered. There is another class OrderedTable<K,V> represents an ordered table. Tables can be loaded from any kinds of data sources (e.g., HDFS or databases). An example of loading an input dataset from HDFS looks like follows [4]:

```
Table<Integer,String> input = load("/root/data", splitter, reader)
```

In this example, the data are loaded from HDFS to the table object input. E^3 implements the load() function using deferred evaluation. When the code is executed, nothing happens. The system just initializes a Table object and populates the object with necessary metadata. The actual data retrieval only occurs when the data is being processed. The splitter and reader specify how to split and read data from a given data partition respectively.

The way to process the data in a Table is to call Table.do() function. The simplest version of this function requires two parameters: grouper and processor. The grouper specifies a grouping algorithm which groups records according to the key. The processor specifies the logic to process each group of records that the grouper produces. Typically, users only need to specify processor as E^3 provides many built-in groupers. However, if the built-in groupers are not sufficient, users can introduce new groupers by implementing the Grouper interface. The following code describes how to apply a filter operation on each record of the input table:

```
Table<Integer, String> result = input.do(SingleRecGrouper(), Filter())
```

In the above code, SingleRecGrouper() is a grouper which places each record of input in a separate group. Filter() denotes the filtering operation that will be applied to each group of records. In E^3, the grouper is the only primitive that will run in parallel. This design simplifies the implementation significantly. Furthermore, E^3 enforces groupers to be state-less. Therefore, parallelizing groupers

---

[3] For standard SQL query, users are suggested to use epiC's query interface, unless they try to apply special optimization techniques for the query.

[4] For clarity, the sample code is an abstraction of the real code

is straightforward. The runtime system just launches several copies of the same grouper code on the slave nodes and runs them in parallel. By default, the number of processes for launching groupers is automatically determined by the runtime system based on the input data size. Users, however, can override the default behavior by providing a specific task allocation strategy. This is accomplished by passing a `TaskAlloc` object as the final argument of `Table.do()`.

The `Table.do()` function returns a new table object as the result. Users can invoke `Table.do()` on the resulting table for further processing if necessary. There is no limitation on the invocation chain. User can even call `Table.do()` in a loop. This is a preferred way to implement iterative data mining algorithms (e.g., k-means) in E^3.

The execution of `Table.do()` also follows deferred evaluation strategy. In each invocation, the runtime system just records necessary information in internal data structures. The whole job is submitted to the master node for execution when the user calls `E3Job.run()`. The runtime system merges all the `Table.do()` calls, analyzes and optimizes the call graphs and finally produces the minimum number of groupers to execute.

Compared to MapReduce [15], E^3 provides built-in support for specifying and optimizing data processing flows. Our experiences show that specifying complex data processing flows using E^3 is significantly easier than specifying the same logic using a MapReduce chain.

3.3   ES²: Elastic Storage System

This section presents the design of ES^2, the Elastic Storage System of epiC. The architecture of ES^2 comprises of three major modules, as illustrated Figure 3, **Data Import Control**, **Data Access Control** and **Physical Storage**. Here we will provide a brief description of each module. For more implementation details of ES^2 and experimental results, please refer to our technical report [10].

In ES^2, as in conventional database systems, data can be fed into the system via the OLTP operations which insert or update specific data records or via the **data import control** module which supports efficient data bulk-loading from external data sources. The data could be loaded from various data sources such as databases stored in conventional DBMSs, plain or structured data files, and the intermediate data generated by other Cloud applications. The data import control module consists of two sub-components: *import manager* and *write cache*. The import manager implements different protocols to work with the various types of data sources. The write cache resides in memory and is used for buffering the imported data during the bulk-loading process. The data in the buffer will be eventually flushed to the physical storage when the write cache is full.

The **physical storage module** contains three main components: *distributed file system (DFS)*, *meta-data catalog* and *distributed indexing*. The DFS is where the imported data are actually stored. The meta-data catalog maintains both meta information about the tables in the storage and various fine-grained statistics information required by the data access control module.

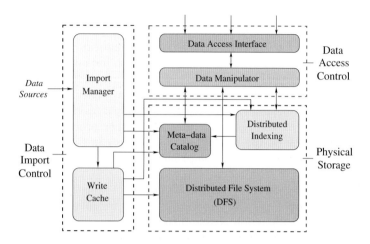

Fig. 3. Architecture of ES^2

The **data access control module** is responsible for performing data access requests from the OLTP/OLAP controller and the E3 engine. It has two sub-components: *data access interface* and *data manipulator*. The data access interface parses the data access requests into the corresponding internal representations that the data manipulator operates on and chooses a near optimal data access plan such as parallel sequential scan or index scan or hybrid for locating and operating on the target data stored in the physical storage module.

Now, we briefly introduce the implementation of ES^2 in various aspects including data model, data partitioning scheme, load-adaptive replication, transaction management and secondary indexes.

Data Model. We adopt the widely accepted *relational* data model. Although this model has been said to be an overkill in Cloud data management and is replaced by the more flexible *key-value* data model for systems such as BigTable [12], Dynamo [16] and Cassandra [25], we observe that all these systems are transaction-oriented with heavy emphasis on the handling of OLTP queries. On the other hand, systems that focus on ad-hoc analysis of massive data sets (i.e., OLAP queries), including Hive [35], Pig [30] and SCOPE [11], are sticking to the relational data model or its variants. Since we aim to provide effective and efficient supports for both OLTP and OLAP queries and multi-tenancy in the future, we choose the relational model for our storage system.

Data Partitioning. ES^2 has been designed to operate on a large cluster of shared-nothing commodity machines. Consequently, ES^2 employs both vertical and horizontal data partitioning schemes. In this hybrid scheme, columns in a table schema that are frequently accessed together in the query workload are grouped into a column group and stored in a separate physical table. This vertical partitioning strategy facilitates the processing of OLAP queries which often access only a subset of columns within a logical table schema. In addition, for each physical table corresponding to a column group, a horizontal partitioning scheme is carefully designed based on the database workload so that transactions which span multiple partitions are only necessary in the worst case.

Load-adaptive Replication. Cloud services need always-on (24×7) data provision, which can be achieved via data replication. A straightforward replication approach is to replicate all data records in the system with the same replication level. However, if the replication level is set to too high, the system storage and the overhead to keep them consistent can be considerably high. Additionally, the data access pattern in web applications is often skewed and changes frequently. Therefore, we develop a two-tier load-adaptive replication strategy to provide both data availability and load balancing function for ES^2.

In this replication scheme, each data record is associated with two types of replicas - namely secondary and slave replicas - in addition to its primary copy. The first tier of replication consists of totally K copies of data inclusive of the primary copy and its secondary replicas. The objective of this replication tier is to facilitate the data reliability requirement (K is typically set to small values). At the second tier, frequently accessed records are associated with additional replicas, called slave replicas, as a way to facilitate load balancing for the "hot" queried data. When a primary copy or secondary replica faces a flash crowd query, it will create slave replicas (which become associated with it) to help resolve the sudden surge in the workload. The two-tier load-adaptive replication strategy incurs much less replication costs, which include the storage cost and consistency maintenance cost, than the approach replicating all data records at high replication level, and it can efficiently facilitate load balancing at the same time.

For Cloud data management service, we have to meet the service level agreement (SLA) on various aspects such as service availability and response time. Therefore, synchronous replica consistency maintenance is not suitable due to the high latency for write operations, especially when there are storage node failures or when storage nodes are located in distributed clouds [9]. Instead, we employ the asynchronous replication method in ES^2. In particular, the primary copy is always updated immediately, while the update propagation to secondary replicas can be deferred until the storage node has spare network bandwidth. Although this optimistic replication approach guarantees low end-user latency, it might entail other problems such as "lost updates" due to different types of machine failures. For instance, when the primary copy crashes suddenly, there is a possibility that the modification to this copy gets lost because the update has not been propagated to other secondary copies. In ES^2, we adopt the write-ahead logging scheme and devise a recovery technique to handle the problem of "lost updates". In this manner, ES^2 guarantees that updates to the primary copy are durable and eventually propagated to the secondary copies.

Transaction Management. In the Cloud context, the management of transactions is a critical and broad research problem. MegaStore [7] and G-Store [14] have started to provide transactional semantics in Cloud storages which guarantee consistency for operations spanning multiple keys clustered in an entity group or a key group. In [26], it has been proposed that Cloud storages can be designed as a system comprising of loosely coupled transactional components and data components. The consistency rationing approach, which categorizes application data into three types and devises a different consistency treatment for each category, has been recently proposed in [24].

In ES^2, allowing OLAP and OLTP queries to operate within the same storage system further complicates the problem of transaction management. In our recent study [36], we examined transaction management in distributed environment where distributed data structures and replications are common. ES^2 uses replication mainly for load balancing and data reliability requirements, and multi-versioning transaction management technique to support both OLTP and OLAP workloads. Consequently, the OLTP operations access the latest version of the data, while the OLAP data analysis tasks execute on a recent consistent snapshot of the database.

Secondary Indexes. For OLTP queries and OLAP queries with high selectivities, it is not efficient to perform sequential or parallel scan on the whole table just to retrieve a few records. However, scanning the whole table is inevitable if query predicates do not contain attributes that have been used to horizontally partition the data. To handle this problem, we support various types of distributed secondary indexes over the data in ES^2. Recently, we have proposed two types of distributed secondary indexes for Cloud data: the distributed B^+-tree index which supports one-dimensional range queries [39] and the distributed multi-dimensional index which supports multi-dimensional range queries and nearest neighbor (NN) queries [38]. Both approaches share the common key idea of two-level indexing, whereby P2P routing overlays are used to index and guide the search based on the index information published by local storage nodes based on their local indexes.

Different P2P overlays are proposed to handle different types of queries. [37] gives a survey of existing P2P overlays. In reality, we cannot afford the cost of maintaining multiple overlays in the cluster for different types of distributed indexes. Consequently, we develop a *unified indexing framework*, which provides an abstract template overlay based on the Cayley graph model [27]. Based on this framework, the structure and behaviors of different overlays can be customized and mapped onto the template. In current implementation, we have integrated the structures of Chord [34], CAN [33] and BATON [22] in the framework. More P2P overlays will be included in the future.

4   Conclusion

Cloud computing is the next generation computation model. It hides the underlying implementation details and provides a resizable resource pool to users, which simplifies the deployment of large-scale applications and services. In this paper, we have reviewed some previous work on building scalable Cloud systems. We have briefly analyzed the advantages and disadvantages of each design. We have applied our background on parallel database systems and observations on other systems to an on-going project, epiC, which aims to provide a flexible framework for supporting various database applications. We have briefly introduced its design and implementations, and we will conduct extensive system benchmarking in the near future.

Acknowledgement. We would like to thank other team members from National University of Singapore and Zhejiang University for their valuable contributions, and we would also like to thank Divyakant Agrawal for his valuable comments and the numerous discussions during the course of the implementation of epiC. We would like to thank the conference chairs and PC chairs for their patience.

References

1. http://hadoop.apache.org/
2. http://www.teradata.com/
3. http://www.asterdata.com/
4. http://www.vertica.com/
5. http://www.greenplum.com/
6. epiC project, http://www.comp.nus.edu.sg/~epic/
7. Google MegaStore's Presentation at SIGMOD (2008),
   http://perspectives.mvdirona.com/2008/07/10/GoogleMegastore.aspx
8. Abouzeid, A., Bajda-Pawlikowski, K., Abadi, D., Silberschatz, A., Rasin, A.: Hadoopdb: an architectural hybrid of mapreduce and dbms technologies for analytical workloads. Proc. VLDB Endow. 2(1), 922–933 (2009)
9. Agarwal, S., Dunagan, J., Jain, N., Saroiu, S., Wolman, A., Bhogan, H.: Volley: automated data placement for geo-distributed cloud services. In: NSDI 2010: Proceedings of the 7th USENIX Conference on Networked Systems Design and Implementation, Berkeley, CA, USA, pp. 2–2. USENIX Association (2010)
10. Cao, Y., Chen, C., Guo, F., Jiang, D., Lin, Y., Ooi, B.C., Vo, H.T., Wu, S., Xu, Q.: A cloud data storage system for supporting both oltp and olap. Technical Report, National University of Singapore, School of Computing. TRA8/10 (2010)
11. Chaiken, R., Jenkins, B., Larson, P.-Å., Ramsey, B., Shakib, D., Weaver, S., Zhou, J.: Scope: easy and efficient parallel processing of massive data sets. PVLDB 1(2), 1265–1276 (2008)
12. Chang, F., Dean, J., Ghemawat, S., Hsieh, W.C., Wallach, D.A., Burrows, M., Chandra, T., Fikes, A., Gruber, R.E.: Bigtable: a distributed storage system for structured data. In: OSDI 2006: Proceedings of the 7th USENIX Symposium on Operating Systems Design and Implementation, Berkeley, CA, USA, pp. 15–15. USENIX Association (2006)
13. Cooper, B.F., Ramakrishnan, R., Srivastava, U., Silberstein, A., Bohannon, P., Jacobsen, H.-A., Puz, N., Weaver, D., Yerneni, R.: Pnuts: Yahoo!'s hosted data serving platform. Proc. VLDB Endow. 1(2), 1277–1288 (2008)
14. Das, S., Agrawal, D., Abbadi, A.E.: G-store: a scalable data store for transactional multi key access in the cloud. In: SoCC, pp. 163–174 (2010)
15. Dean, J., Ghemawat, S.: Mapreduce: simplified data processing on large clusters. ACM Commun. 51(1), 107–113 (2008)

16. DeCandia, G., Hastorun, D., Jampani, M., Kakulapati, G., Lakshman, A., Pilchin, A., Sivasubramanian, S., Vosshall, P., Vogels, W.: Dynamo: amazon's highly available key-value store. SIGOPS Oper. Syst. Rev. 41(6), 205–220 (2007)

17. DeWitt, D., Gray, J.: Parallel database systems: the future of high performance database systems. ACM Commun. 35(6), 85–98 (1992)

18. Dewitt, D.J., Ghandeharizadeh, S., Schneider, D.A., Bricker, A., Hsiao, H.I., Rasmussen, R.: The gamma database machine project. IEEE Trans. on Knowl. and Data Eng. 2(1), 44–62 (1990)

19. Fushimi, S., Kitsuregawa, M., Tanaka, H.: An overview of the system software of a parallel relational database machine grace. In: VLDB 1986: Proceedings of the 12th International Conference on Very Large Data Bases, pp. 209–219. Morgan Kaufmann Publishers Inc., San Francisco (1986)

20. Ghemawat, S., Gobioff, H., Leung, S.-T.: The google file system. In: SOSP, pp. 29–43 (2003)

21. Guo, F., Li, X., Ooi, B.C., Tan, K.-L.: Guinea: An efficient data processing framework on large clusters. Technical Report, National University of Singapore, School of Computing. TRA9/10 (2010)

22. Jagadish, H.V., Ooi, B.C., Vu, Q.H.: Baton: a balanced tree structure for peer-to-peer networks. In: VLDB, pp. 661–672 (2005)

23. Jiang, D., Ooi, B.C., Shi, L., Wu, S.: The performance of mapreduce: An in-depth study. Proc. VLDB Endow. 3(1), 472–483 (2010)

24. Kraska, T., Hentschel, M., Alonso, G., Kossmann, D.: Consistency rationing in the cloud: Pay only when it matters. PVLDB 2(1), 253–264 (2009)

25. Lakshman, A., Malik, P.: Cassandra: a decentralized structured storage system. SIGOPS Oper. Syst. Rev. 44(2), 35–40 (2010)

26. Lomet, D., Mokbel, M.F.: Locking key ranges with unbundled transaction services. Proc. VLDB Endow. 2(1), 265–276 (2009)

27. Lupu, M., Ooi, B.C., Tay, Y.C.: Paths to stardom: calibrating the potential of a peer-based data management system. In: SIGMOD 2008: Proceedings of the 2008 ACM SIGMOD International Conference on Management of Data, pp. 265–278. ACM, New York (2008)

28. Matsudaira, P.: High-end biological imaging generates very large 3d+ and dynamic datasets. Proc. VLDB Endow. (2010)

29. Nykiel, T., Potamias, M., Mishra, C., Kollios, G., Koudas, N.: Mrshare: Sharing across multiple queries in mapreduce. Proc. VLDB Endow. 3(1), 494–505 (2010)

30. Olston, C., Reed, B., Srivastava, U., Kumar, R., Tomkins, A.: Pig latin: a not-so-foreign language for data processing. In: SIGMOD 2008: Proceedings of the 2008 ACM SIGMOD International Conference on Management of Data, pp. 1099–1110. ACM, New York (2008)

31. Pavlo, A., Paulson, E., Rasin, A., Abadi, D.J., DeWitt, D.J., Madden, S., Stonebraker, M.: A comparison of approaches to large-scale data analysis. In: SIGMOD 2009: Proceedings of the 35th SIGMOD International Conference on Management of Data, pp. 165–178. ACM, New York (2009)

32. Ramakrishnan, R.: Data management challenges in the cloud. In: Proceedings of ACM SIGOPS LADIS (2009), http://www.cs.cornell.edu/projects/ladis2009/talks/ramakrishnan-key note-ladis2009.pdf

33. Ratnasamy, S., Francis, P., Handley, M., Karp, R., Shenker, S.: A scalable content-addressable network. In: SIGCOMM, pp. 161–172 (2001)

34. Stoica, I., Morris, R., Liben-Nowell, D., Karger, D.R., Kaashoek, M.F., Dabek, F., Balakrishnan, H.: Chord: a scalable peer-to-peer lookup protocol for internet applications. IEEE/ACM Trans. Netw. 11(1), 17–32 (2003)
35. Thusoo, A., Sarma, J.S., Jain, N., Shao, Z., Chakka, P., Zhang, N., Anthony, S., Liu, H., Murthy, R.: Hive - a petabyte scale data warehouse using hadoop. In: ICDE, pp. 996–1005 (2010)
36. Vo, H.T., Chen, C., Ooi, B.C.: Towards elastic transactional cloud storage with range query support. Proc. VLDB Endow. 3(1), 506–517 (2010)
37. Vu, Q., Lupu, M., Ooi, B.C.: Peer-to-peer computing: Principles and applications. Springer, Heidelberg (2009)
38. Wang, J., Wu, S., Gao, H., Li, J., Ooi, B.C.: Indexing multi-dimensional data in a cloud system. In: SIGMOD 2010: Proceedings of the 2010 International Conference on Management of Data, pp. 591–602. ACM, New York (2010)
39. Wu, S., Jiang, D., Ooi, B.C., Wu, K.-L.: Efficient b-tree based indexing for cloud data processing. Proc. VLDB Endow. 3(1), 1207–1218 (2010)

Search and Social Integration

Edward Y. Chang

Google Research China
edchang@google.com

Abstract. Search and Social have been widely considered to be two separate applications. Indeed, most people use search engines and visit social sites to conduct vastly different activities. This talk presents opportunities where search and social can be integrated synergistically. For instance, on the one hand, a search engine can mine search history data to facilitate uses to engage in social activities. On the other hand, user activities at social sites can provide information for search engines to improve personalized targeting. This talk uses Confucius, a Q&A system which Google develops and has launched in more than 60 countries, to illustrate how computer algorithms can assist synergistic integration between search and social. Algorithmic issues in data mining, information ranking, and system scalability are discussed.

L. Chen, P. Triantafillou, and T. Suel (Eds.): WISE 2010, LNCS 6488, p. 20, 2010.
© Springer-Verlag Berlin Heidelberg 2010

Elements of a Spatial Web

Christian S. Jensen

Department of Computer Science, Aarhus University, Denmark
csj@cs.au.dk

Abstract. Driven by factors such as the increasingly mobile use of the web and the proliferation of geo-positioning technologies, the web is rapidly acquiring a spatial aspect. Specifically, content and users are being geo-tagged, and services are being developed that exploit these tags. The research community is hard at work inventing means of efficiently supporting new spatial query functionality.

Points of interest with a web presence, called spatial web objects, have a location as well as a textual description. Spatio-textual queries return such objects that are near a location argument and are relevant to a text argument. An important element in enabling such queries is to be able to rank spatial web objects. Another is to be able to determine the relevance of an object to a query. Yet another is to enable the efficient processing of such queries. The talk covers recent results on spatial web object ranking and spatio-textual querying obtained by the speaker and his colleagues.

L. Chen, P. Triantafillou, and T. Suel (Eds.): WISE 2010, LNCS 6488, p. 21, 2010.
© Springer-Verlag Berlin Heidelberg 2010

The Ubiquitous DBMS

Kyu-Young Whang

Department of Computer Science, KAIST
kywhang@mozart.kaist.ac.kr

Abstract. Recent widespread use of mobile technologies and advancement in computing power prompted strong needs of database systems that can be used in small devices such as sensors, cellular phones, PDA, ultra PCs, and navigators. We call database systems that are customizable from small-scale applications for small devices to large-scale applications such as large-scale search engines ubiquitous database management systems (UDBMSs). In this talk, we first review requirements of UDBMSs. The requirements we identified include selective convergence (or "devicetization"), flash-optimized storage systems, data synchronization, supportability of unstructured/semi-structured data, and complex database operations. We then review existing systems and research prototypes. We first review the functionality of UDBMSs including the footprint size, support of standard SQL, supported data types, transactions, concurrency control, indexing, and recovery. We then review the supportability of requirements by those UDBMSs surveyed. We highlight ubiquitous features of a family of Odysseus systems that have been under development at KAIST for over 20 years. Functionalities of Odysseus can be "devicetized" or customized depending on the device types and applications as in Odysseus/Mobile for small devices, Odysseus/XML for unstructured/semistructured data, Odysseus/GIS for map data, and Odysseus/IR for large-scale search engines. We finally present research topics that are related to the UDBMSs.

L. Chen, P. Triantafillou, and T. Suel (Eds.): WISE 2010, LNCS 6488, p. 22, 2010.
© Springer-Verlag Berlin Heidelberg 2010

Building Web Services Middleware with Predictable Service Execution

Vidura Gamini Abhaya, Zahir Tari, and Peter Bertok

School of Computer Science and Information Technology
RMIT University, Melbourne, Australia
{vidura.abhaya,zahir.tari,peter.bertok}@rmit.edu.au

Abstract. This paper presents a set of guidelines, algorithms and techniques that enable web services middleware to achieve predictable execution times. Existing web service middleware execute requests in a *best-effort* manner. While this allows them to achieve a higher throughput, it results in highly unpredictable execution times, rendering them unsuitable for applications that require predictability in execution. The guidelines, algorithms and techniques presented are generic in nature and can be used, to enhance existing SOAP engines and application servers, or when newly being built. The proposed algorithms schedules requests for execution explicitly based on their deadlines and select requests for execution based on laxity. This ensures a high variance in laxities of the requests selected, and enables requests to be scheduled together by phasing out execution. These techniques need to be supported by specialised development platforms and operating systems that enable increased control over the execution of threads and high precision operations. Real-life implementation of these techniques on a single server and a cluster hosting web services are presented as a case study and with the resultant predictability of execution, they achieve more than 90% of the deadlines, compared to less than 10%, without these enhancements.

1 Introduction

Web Services technology has enabled the internet to be more than a large collection of information sources in a multitude of unforeseen ways [7]. Being the foundation of cloud-computing, web services have affirmed itself as the de-facto communication method in distributed systems [8]. Many applications are being developed for the cloud and many legacy applications are being moved onto the cloud. With applications, software, platforms and infrastructure being made available as a service, quality of service (QoS) aspects in web service execution has become increasingly important. Although being widely researched, only a few attempts have been made in achieving execution time QoS. None of them are on achieving consistent and predictable execution times.

Web Service (WS) middleware are optimised for throughput by design [2,10,14]. For instance, they accept requests unconditionally and execute them in a *best-effort* manner, using the *Thread-pool* design pattern [9]. Herein, multiple threads process requests in parallel, using the processor sharing approach. Although this approach yields a higher throughput, it increases the mean execution time of a web service invocation due to longer waiting times. Such a phenomenon adversely affects the predictability

L. Chen, P. Triantafillou, and T. Suel (Eds.): WISE 2010, LNCS 6488, pp. 23–37, 2010.

of service execution, thereby rendering web services unsuitable for applications with stringent QoS requirements. Moreover, the development (dev) platform used for such WS-middleware and the operating system (OS) used for deployment, do not support execution level predictability. For instance, the invocation of a web service may be interrupted by a process or a thread with higher priority, conducting house-keeping activities such as garbage collection [3]. Furthermore, thread priority levels on the dev platform may not be mapped directly to the same level on the operating system i.e. 'highest' priority on the dev platform may not be the highest available on the OS. Therefore, the execution of a thread with the highest priority on the dev platform can be interrupted by one outside of it.

Related work on QoS aspects of service execution could be found in literature. Some [17,20] assume that WS-middleware would meet expected levels of QoS and use the promised levels for matching clients to services or for service compositions. Some [4,6,11] use function level differentiation to achieve many levels of execution time, based on Service Level Agreements (SLAs). Requests belonging to different classes maybe accepted based on pre-defined ratios, to achieve levels of execution times promised in SLAs. Some [12] differentiate based on non-functional attributes such as the nature of application, the device being used, and nature of client. While these solutions either make the assumption that perceived QoS levels will be achieved by the middleware used or introduce some sort of differentiation to request processing, they fail to achieve predictability of execution for a couple of reasons. None of them consider a *deadline* to complete request execution as the main attribute for scheduling. Neither the middleware nor the software infrastructure used, have means of guaranteeing the consistency in meeting deadlines. Moreover, unconditional acceptance of requests may lead to overload conditions due to resource contentions.

Real-time systems consider predictability to be the most important aspect of request execution. They require the execution of a task to be completed within a given deadline, on every invocation. Even a valid result obtained with a deadline miss, is considered useless. Herein, the key aspect of task execution is the consistency in which deadlines are met. Such a feat is made possible by using specialised scheduling principles at design time of the system. In [5], we presented a mathematical model which uses real-time scheduling principles and an admission control mechanism based on it. The admission control mechanism selects web service requests for execution, at runtime under highly dynamic conditions. Moreover, the selected requests were scheduled to meet their respective deadlines.

As the contribution in this paper, we present software engineering aspects of achieving execution level predictability. First, we provide a set of guidelines for building or enhancing web service middleware, then provide a set of algorithms and techniques following these guidelines, to ensure predictable service execution. We demonstrate how these guidelines, techniques and algorithms can be used to enhance existing WS-Lmiddleware in the form of a case study. The guidelines we provide are generic, and valid for any SOAP engine or application server. They could be summarised into three important requirements. Firstly, the requests must be explicitly scheduled to meet their deadline requirement. Secondly, requests must be consciously selected for execution based on their laxity. Finally, the solution has to be supported by a software infrastructure (dev platform and OS) that provides features to achieve predictability of execution. The case study details on how an existing stand alone SOAP engine and a web service cluster can be enhanced to achieve these features. The uniqueness of our solution is the use of real-time scheduling principles (typically used at design time), at run-time in a highly dynamic environment. The proposed techniques work by selecting requests for execution based on their laxity. A schedulability check ensures that a request is selected only if its deadline could be met while not compromising the deadlines of the already accepted. The check results in a high variance of laxities at a server enabling requests competing for the same window of time, to be scheduled together. The selected requests are scheduled for execution using Earliest Deadline First (EDF) scheduling principle. In this research, network communication aspects are considered out of scope as we only concentrate on execution aspects of web services. In experiments, any delays in the network are not separately quantified and subsumed in the total execution time of a request.

Rest of this paper is organised as follows. In Sect. 2 we provide a set of guidelines to be used for achieving predictability in WS-middleware. It is followed by a case study of a real-life implementation of these guidelines in Sect. 3. Next we present the empirical evaluation of our solution in Sect. 4. We discuss a few related work (Sect. 5) and finally conclude in Sect. 6.

2   Guidelines

Execution of a web service request goes through many layers of the system. The middleware that handles the SOAP message typically makes use of many development libraries for message processing, transport and network communication. Ultimately, the OS handles the system level activities such has resource usage (memory, CPU, sockets, etc.) and the execution through processes and threads. At the OS level the preference of resource usage and execution is decided based on priority levels. Therefore, it is imperative that predictability of execution be achieved ground-up.

The following guidelines enable web service middleware to achieve predictability of service execution. These could be followed in building new middleware or to enhance existing ones.

Fig. 1. Software Stack

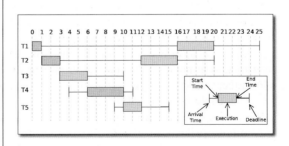

Fig. 2. Deadline based task schedule

G1. **Use a dev platform and an OS that guarantees predictability of execution.** Applications at upper levels of the software stack only achieves predictability of execution, if lower levels support such features. Most commonly used dev platforms and OSs lack such real-time features. For instance, the thread priority levels available in Java do not map to directly to the range available at the OS level and they are not explicitly enforced [15]. As a result, a Java thread running at the highest priority could be pre-empted by a non-Java thread running at a higher OS level priority. Moreover, the dev platform must ensure unhindered execution of requests, avoiding them being pre-empted by housekeeping activities such as garbage collection [3]. Figure 1 depicts the layers of software needed to achieve such a solution.

G2. **Schedule requests explicitly based on their deadlines.** For a request to finish execution within a deadline, it must be scheduled explicitly. Figure 2 contains a deadline based task schedule. The tasks have been scheduled in the order of increasing deadlines. As a result the requests finish execution in the order T3,T4,T5,T2,T1. Although T3,T4,T5 have arrived at the system after T2 and T1, they still manage to achieve their deadlines as a result of being explicitly scheduled to have the processor for execution, by pre-empting other tasks.

G3. **Select requests for execution based on laxity.** When requests are scheduled based on their deadlines, it becomes impossible to schedule all requests together on the same server. Typically, there will be requests vying for the same window of time. Therefore, requests needs to be selected based on whether their deadline could be met, while not compromising others already scheduled. The laxity of a request indicates how long its execution could be delayed without compromising its deadline. The higher the laxity of a request is, the more requests that could be scheduled within its lifespan. Thus, requests must be selected for execution to ensure that the variance of laxity at a server is high, thereby enabling more requests to be scheduled together with a guarantee of meeting their deadlines. Tasks T1 and T2 in Fig. 2 have their executions delayed and phased, enabling T3,T4 and T5 to be scheduled within their lifespan. This is possible due to the large difference of laxities between T1,T2 and T3,T4,T5. Similarly, if T5 had a smaller laxity, it cannot be scheduled with T3 and T4 as there is no slack time to delay its execution.

G4. **Introduce fine-grain differentiation in request processing.** The many stages of processing of a web service request is managed by one or more threads, throughout its lifetime in the system. The time spent on each stage is subsumed within the overall execution time of a request. Typically, web service middleware makes no differentiation of execution at thread level. With requests being scheduled for execution based on their deadlines, it is imperative that the system must have fine-grain control over their execution. This enables control over the execution of threads at the different functional units of the system thereby having control over their resource usage.

This could be achieved by introducing several levels of thread priorities and controlling the execution of threads by priority assignments.

G5. **Minimise priority inversions.** Priority inversion refers to the scenario where a lower priority task holds on to a resource needed by a higher priority task [13]. The higher priority task is not able to resume the execution until the lower priority task releases the resource. The lower priority task must run in order to release the resource. Such phenomenon could take place with I/O operations such as when displaying a message or writing to a file. The importance of this is more when debugging a solution where the normal practice is to trace the execution with messages. Such practices are unsuccessful in these type of applications due to resulting priority inversions not reflecting the actual execution scenario when the messages are not used. Moreover, such priority inversions introduced by the developer may result in unanticipated execution patterns that leads to unexpected behaviour in the application.

Following these guidelines will enable web service middleware of any type (SOAP based or REST based) to achieve predictability of execution. The application of these guidelines in the actual middleware may differ from one to another. In the case study that follows, two widely used WS-middleware are enhanced following these guidelines. Although these products are used as examples, the enhancements done are generic and can be applied for any other WS-middleware available.

3   Case Study

Dev Platform and OS

In this case study, we present enhancements done to Apache Axis2 [2] and Apache Synapse [1], two widely used open source WS-middleware products. These are implemented using Java which is considered not suitable for applications with stringent execution time requirements [18]. To support the implementation with a suitable dev platform and an OS, as per G1 we use Real-time Java Specification [15] (RTSJ) and Sun Solaris 10 (SunOS) as our software infrastructure. RTSJ introduces a set of strictly enforced priorities that are not available in standard Java. Moreover, it contains specialised thread implementations with higher precision execution. With the proper priority levels used, these threads cannot be interrupted by the garbage collector. RTSJ also provides timers with nanosecond accuracy when supported by a real-time OS. SunOS, provides these features for RTSJ by directly mapping RTSJ priorities to ones available at the OS level. RTSJ is able to directly use high precision clocks and other features of the OS with more control through specialised kernel modules installed.

3.1   Enhancements Made to Axis2

The enhancements made to Axis2 enables it to function as a stand-alone web services server providing consistent predictable execution times. Furthermore, instances of the enhanced version of Axis2 (RT-Axis2) can be used as executors in a web services cluster. For brevity, the enhancement made are summarised into major features introduced.

Schedulability Check. Following G3, requests are selected for execution subjected to a schedulability check. The check (Algorithm 1) ensures that a request is selected on the assurance of its deadline, without compromising the deadlines of already accepted. Firstly, the laxity of the new request (N) is checked with requests finishing within its lifespan, to ensure the deadlines of those requests can be met, if N is accepted (Lines 4-22). Next, the laxity of requests finishing after N is calculated to ensure N can be scheduled without compromising the others (Lines 23-36), as they make way for the execution of N (refer Fig. 2 for an example). The algorithm follows the mathematical model detailed in [5], but differs from the algorithm presented in it, with better execution time complexity of $O(n)$ where n is upper bound by the total number of accepted requests at the server. The schedulability check results in a large variance of laxity at a server, enabling more requests to be scheduled together to meet their deadlines.

Priority Model. Using the additional thread priority levels available on RTSJ, we introduced three priority levels to Axis2. The lowest level equals to the highest priority available for standard Java threads and is used for execution of metadata requests such as WSDL or schema requests. The *Mid* level is set to the mid level of RTSJ priorities and can be interrupted by the GC. This level is assigned to a thread when it needs to be pre-empted from execution. The *High* priority level is the highest available on RTSJ and assigned to a thread for it to gain the processor for execution. This level cannot be interrupted by the GC. Following G4, these priority levels are used by a newly introduced real-time scheduler component to achieve differentiation in request processing.

Real-Time Scheduler and Thread Pools. As seen in Fig. 4, RT-Axis2 contains multiple modules that carry out different stages of a service invocation. The execution of a request is carried out by two thread pools. These thread pools typically contain standard Java threads that execute requests in a best-effort manner. As part of the enhancements, they are replaced with custom made thread pools that use real-time threads, managed by a real-time scheduler component. Figure 3 contains the design of the real-time thread pool and the scheduler component. *RTWorkerThreads* based on RTSJ RealtimeThread class, enable the scheduler (*RTThreadPoolExecutor* and *RTEDFScheduler*) to have fine-grain control over their execution, using our priority model. Each request is internally represented by an instance of *RTTask*, which is assigned to a *RTWorkerThread*.

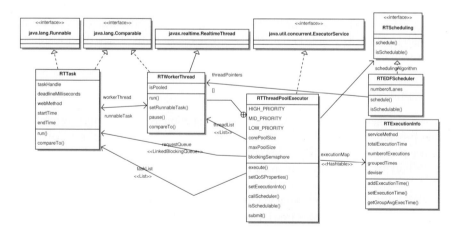

Fig. 3. Real-time Thread Pool Class Diagram

Algorithm 1. Schedulability Check

Require: New request N, List of Accepted Requests RL
Ensure: N is accepted or rejected
1. PDW ← 0; PDA ← 0
2. withinTasksChecked ← *false*
3. SortAscending(RL)
4. **while** RL has more and withinTasksChecked is *false* **do**
5.   nextReq ← RL.getNextReq
6.   **if** nextReq.startTime $\geq$ N.startTime and nextRequest.deadline $\leq$ N.deadline **then**
7.   **if** Exec. Info. for nextReq.Operation exists **then**
8.   PDW ← PDW + nextReq.getRemainingTime
9.   **else**
10.   PDW ← PDW + nextReq.getExecTime
11.   **end if**
12.   **else**
13.   **if** nextReq.deadline $\geq$ N.deadline **then**
14.   withinTasksChecked ← *true*
15.   **end if**
16.   **end if**
17. **end while**
18. PDW ← PDW + N.getExecTime
19. LoadingFactor ← $\frac{PDW}{N.deadline - N.startTime}$
20. **if** LoadingFactor > 1 **then**
21.   **return** *false*
22. **end if**
23. PDA ← PDW
24. **while** RL has more requests **do**
25.   nextReq ← RL.getNextReq
26.   **if** Exec. Info. for nextReq.Operation exists **then**
27.   PDA ← PDA + nextReq.getRemainingTime
28.   **else**
29.   PDA ← PDA + nextReq.getExecTime
30.   **end if**
31.   LoadingFactor ← $\frac{PDA}{nextReq.deadline - N.startTime}$
32.   **if** LoadingFactor > 1 **then**
33.   **return** *false*
34.   **end if**
35. **end while**
36. **return** *true*

Algorithm 2. LT.resetLatestThread method implementation

Require: Ordered active thread pointer list LT
Ensure: ptrLastThread points at the thread having the request with the latest deadline
1. **if** LT is not empty **then**
2.   ptrLastThread ← LT.first
3.   **for all** *thread* $\in$ LT **do**
4.   req ← *thread*.getRequest
5.   **if** req.deadline > ptrLastThread.deadline **then**
6.   ptrLastThread ← *thread*
7.   **end if**
8.   **end for**
9. **end if**

Algorithm 3. Scheduling of Threads

Require: Thread Queue TQ, Ordered active thread pointer list LT, New request N
Ensure: Execution of Threads assigned with earliest deadlines
1. found ← false
2. **while** found is false and LT.hasMore **do**
3.   **if** LT.ptrNextThread is not assigned **then**
4.   LT.ptrNextThread ← N.getThread
5.   N.getThread.priority ← High
6.   LT.resetLastThread
7.   found ← true
8.   **else**
9.   R ← LT.ptrNextThread.getRequest
10.   **if** N.deadline < R.deadline **then**
11.   LT.ptrLastThread.priority ← Mid
12.   TQ.queue(LT.ptrLastThread)
13.   LT.ptrLastThread ← N.getThread
14.   LT.resetLatestThread
15.   LT.ptrLastThread.priority ← High
16.   found ← true
17.   **end if**
18.   **end if**
19. **end while**
20. **if** found is false **then**
21.   N.getThread.priority ← Mid
22.   TQ.queue(N)
23. **end if**

Algorithm 4. RT-RoundRobin

Require: New request R, List of Executors E, Last Executor L
Ensure: R assigned to an executor or rejected
1. lastExecIndx ← L.getIndex
2. **if** lastExecIndx = E.size-1 **then**
3.   lastExecIndx = 0
4. **else**
5.   lastExecIndx ← lastExecIndx + 1
6. **end if**
7. nextExec ← E.getExec(lastExecIndx)
8. S ← IsSchedulable(R,nextExec)
9. **if** S is true **then**
10.   L ← nextExec
11.   Assign R to nextExec
12. **else**
13.   Reject R
14. **end if**

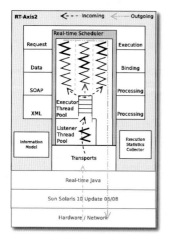

Fig. 4. RT-Axis2

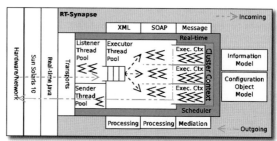

Fig. 5. RT-Synapse

Request Execution. A *Listener Thread Pool* (Fig. 4) listens to a specific socket for incoming requests with the aid of *Transports* module. The pool typically contains a single thread that fetches requests and immediately hands them over to an *Executor Thread Pool* using a queue (Fig. 3 - *requestQueue*). *RTWorkerThreads* pick requests off the queue and execute them through the different modules. Conforming with G2, the RT-scheduler, executes these requests in the order of their deadlines using EDF policy. The implementation makes use of multi-core and multi-processor hardware by having separate execution lanes for each core or processor. Each lane is treated as a separate execution time line when requests are assigned to them on acceptance. The RT-scheduler assigns the thread with the earliest deadline from each lane with the *High* priority for them to continue execution. It uses Algorithm 3 to reschedule threads whenever a new request (N) arrives at the system. RT-scheduler keeps track of all *High* priority threads through a list of references. The deadline of N is compared against requests that are being executed (Lines 9,10). If N has an earlier deadline than one of them, N is taken up for execution, pre-empting the thread with the latest deadline (Lines 11-15). To reduce the time complexity of this activity, the scheduler uses an explicit reference to the thread with the latest deadline. It gets reset when N is taken up for execution (Lines 6,14). This step detailed in Algorithm 2, results in a time complexity of $O(n)$ where n is the number of execution lanes. This results in Algorithm 3 having a similar overall time complexity of $O(n)$.

3.2 Enhancements Made to Synapse

Synapse is an Enterprise Service Bus that performs message mediation. Changes made to Synapse enables it to function as a dispatcher with predictable execution in a cluster hosting web services. The services are hosted on replica instances of RT-Axis2. The enhanced version of Synapse (RT-Synapse) dispatches a request to one of the executors based on the laxity of the request. Most of the enhancements made are common to both RT-Synapse and RT-Axis2.

Real-Time Scheduler and Thread Pools. As seen in Fig. 5, Synapse has three separate thread pools. These were replaced with custom built real-time thread pools. Moreover, a real-time scheduler component manages the execution of all thread pools using our priority model, supported by RTSJ and SunOS at the system level.

Schedulability Check. All requests directed to the cluster goes through the dispatcher. Prior to them being dispatched to an executor, they are subject to the schedulability check. The check has the same implementation as in RT-Axis2. However, the schedulability of a request is checked with already accepted and assigned requests at a particular executor.

Request Dispatching. RT-Synapse functions similar to RT-Axis2 in receiving requests and carrying out the processing. Rather than performing service invocation, RT-Synapse dispatches a request in the least possible time to an executor in the cluster. The state of the executors are kept at RT-Synapse using *Cluster Context* and *Executor Context* structures. Once a request is accepted and assigned to an executor, it is queued in the respective *Executor Context* and its execution thread pre-empted from execution. The time of dispatching is decided by the real-time scheduler, based on the completion of the request currently in execution. This prevents unnecessary processing at an executor, decreasing the time required to meet the deadline of a request. Lanes of execution is also introduced into RT-Synapse where each lane corresponds to an executor and a separate core or processor assigned for its use.

In a cluster, requests can be dispatched based on many algorithms. In this paper, we present the predictability gain achieved by modifying a simple round-robin algorithm. Algorithm 4, details the steps of RT-RoundRobin (RT-RR) algorithm. It works by keeping track of the last executor a request was assigned to (Line 1), and assigning the new request to the next executor in the list (Lines 2-7). Thereafter, the schedulability check is conducted on the assigned executor for the new request (Line 8). The round-robin nature of the algorithm makes it distribute requests evenly among the cluster members. The schedulability check ensures a high variance in the laxities at each executor. The RT-Axis2 instances, execute the requests using EDF policy, thereby enabling the cluster to achieve predictability of execution.

3.3   Minimising Priority Inversions

Priority inversions could impact the predictability of execution in two ways. Mainly, such a scenario would add an unwarranted delay to the execution of a request. Following G5, any output activities such as on screen reporting of operation statuses and output to log files, were changed to follow an in-memory model. Buffers in memory corresponding to such activities stores the output and directs it to the physical object at the end of a request execution. This prevents any inversion having an effect on the actual execution.

When debugging such an application, the typical debugging practices such as trace messages could lead to priority inversions and present a different view of the actual execution events. Specialised debugging tools such as the Thread Scheduling Visualiser [16], memory based logging techniques has to be used for such scenarios.

4   Empirical Evaluation

To evaluate the impact of the changes made to RT-Axis2 and RT-Synapse, they are compared with their unmodified versions. The systems are exposed to a request stream with all requests having hard deadline requirements and a high variability of request sizes. We use a web service that allows us to create different request sizes on the server using simple input parameters for the evaluation. Task sizes and inter-arrival times are generated using a uniform distribution. We conduct separate experiments for stand-alone SOAP engine and web service cluster scenarios. The metrics taken for the comparison are the percentage of requests accepted for execution and the percentage of deadlines met out of the requests accepted. Although the unmodified systems do not reject requests explicitly, all requests may not be accepted due to processing overloads. Test setup is deployed on PCs with Intel Core 2 Duo 3.4 GHz processors and 4 Gigabytes of RAM, running Solaris 10 update 05/08 with RTSJ version 2.2. RT-Axis2 is configured to have 3 execution lanes with 100 worker threads for the stand-alone deployment and 30 threads when deployed as an executor. RT-Synapse is also configured to have 3 execution lanes with 30 threads per lane.

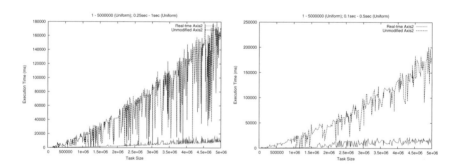

Fig. 6. Execution Time Comparisons - Axis2 and RT-Axis2

Table 1. Performance Comparison of Unmodified Axis2 vs. RT-Axis2

Inter-arrival time(sec)	Unmod. Axis2		RT-Axis2	
	% Acc.	% D. Met off % Acc.	% Acc.	% D. Met off % Acc.
0.25 - 2	100	36.2	96.7	100
0.25 - 1	62.4	18.3	58.6	100
0.1 - 0.5	55.1	9.1	30.7	99.7
0.1 - 0.25	28.7	8.8	18.1	96.7

Stand-Alone SOAP Engine Performance

Evaluating predictability of execution times gained by the enhancements made to RT-Axis2, it is compared with an unmodified version of Axis2 (Table 1). Unmodified Axis2 accepts more requests than RT-Axis2 as there is no explicit condition for request acceptance. As the inter-arrival times decrease, both systems accept less number of requests. However, it can be clearly seen that RT-Axis2 outperforms Axis2 in the number of deadlines met. Although Axis2 is able to meet all the deadlines when request arrival is at a very slow rate, it fails in scenarios with faster arrival rates, meeting less than 10% of the deadlines. Even in such conditions RT-Axis2 is able to meet at least 96% of the deadlines maintaining comparable acceptance rates with Axis2. Due to its best-effort nature, Axis2 results in very large execution times compared to RT-Axis2 (Fig. 6). This result in it missing most of the deadlines.

Cluster Based Web Services Performance

In the cluster based scenario, we compare a cluster setup with RT-Synapse and RT-Axis2 with one using unmodified versions of the products. With the unmodified setup, Synapse uses simple round-robin technique for dispatching requests and both Synapse and Axis2 execute requests in a best-effort manner. The experiments are done gradually increasing the number of executors in the cluster from 2 to 4 as seen in Table 2. The first graph of Fig. 7 visually summaries the results. Simple round-robin (RR) results in

Table 2. Performance Comparison of Round Robin vs. RT-RoundRobin

| | Round Robin (Non real-time) | | | | | | RT-RoundRobin | | | | | |
| | 2 Executors | | 3 Executors | | 4 Executors | | 2 Executors | | 3 Executors | | 4 Executors | |
Inter-arrival time(sec)	% Acc.	% D. Met off % Acc.	% Acc.	% D. Met off % Acc	% Acc.	% D. Met off % Acc	% Acc.	% D. Met off % Acc	% Acc.	% D. Met off % Acc	% Acc.	% D. Met off % Acc
0.25 - 1	99.5	28.8	99.8	37.2	99.9	51.5	88.0	99.0	99.0	100	99.9	100
0.1 - 0.5	62.3	20.3	89.0	28.4	98.0	39.7	52.0	96.4	74.0	99.0	99.4	99.9
0.1 - 0.25	49.0	15.0	67.3	20.0	74.1	33.2	28.0	96.0	47.0	97.6	78.0	99.0
0.05 - 0.1	38.8	6.3	52.6	9.1	68.0	13.6	20.5	90.0	37.5	95.0	46.3	99.0

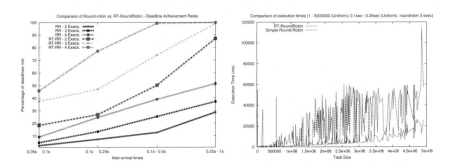

Fig. 7. Execution Time Comparisons - Round Robin and RT-RoundRobin

a better acceptance rate of requests as the cluster unconditionally accepts all incoming requests. However, RT-RR outperforms it consistently when the number of deadlines met is considered. Even with 4 executors in the cluster, RR is able to achieve only 51.5% of the deadlines with almost all requests being accepted, for the longest inter-arrival times. Comparatively, RT-RR accepts almost all requests and is able to meet all deadlines of the accepted requests. Although the dispatcher accepts all requests in the unmodified scenario, the requests are made to wait at the dispatcher as a result of the overload conditions at executors due to their best effort nature. As seen on the second graph of Fig. 7, this leads to longer execution times that result in higher a rate of deadline misses.

Discussion

The results achieved by the experiments, clearly demonstrate that predictability of execution could be achieved on stand-alone servers and clusters hosting web services, using the proper techniques, algorithms and software. Testing the implementation for worst case scenarios with all requests having hard deadlines, higher variability of requests and very high arrival rates gives an indication of the scalability of the solution.

With RT-Axis2, we demonstrated how existing stand-alone existing web service middleware could be enhanced to achieve predictability of execution. The implementation clearly outperformed the unmodified version securing at least 97% of the deadlines while achieving comparable request acceptance rates. Unmodified Axis2, accepts requests unconditionally and executes them in parallel on a best-effort basis. The number of requests executed in parallel is upper bound by the number of threads in the pool. This leads to a build-up of requests at the server and results in high average execution times as seen in Fig. 6. Many threads compete for resources such as memory and processor time, overloading the server. This leads to some of them timing out and being rejected. The remaining, execute through to completion however missing their deadlines due to long waiting times. The schedulability check in RT-Axis2 ensures that requests are selected based on their laxity. Although the main motive of this is to create a higher variance of laxities at the server, this ensures the stream of requests selected does not overload the server. Scheduling the selected requests based on their deadlines, ensures that they could be met successfully. RTSJ and SunOS ensures that the necessary resources are available for this an unexpected interruptions are avoided. As a result, RT-Axis2 is able to achieve almost all the deadlines of the accepted requests. With increased arrival rates, the number of requests competing for the same window of time increases. This leads to some of them being rejected by the schedulability check. These rejections can be minimised by having a cluster based deployment with more than one server hosting the services.

With the cluster based setup, the evaluation measures the performance of the entire cluster. Any delays at the dispatcher needs to be prevented and the RT-Synapse implementation guarantees this with better control over the execution of threads. The cluster is exposed to higher arrival rates than the stand-alone experiments. With RT-RR algorithm, we demonstrate the effectiveness of introducing real-time scheduling principles into a simple dispatching technique. The unmodified cluster, as expected accepts more requests for execution due to unconditional acceptance of requests and the best-effort nature of execution. Even though RR algorithm distributes requests evenly among the executors, the high variability of request sizes creates overloaded conditions at the executors. As a result the requests are made to wait at the dispatcher until executors become responsive. This leads to some of the requests timing out and being rejected while others are ultimately executed with very long waiting times (Fig. 7). These long waiting times leads more than 90% of the requests missing their deadlines. With the unmodified cluster, the request build-up happens at both the dispatcher and the executor. This could lead to request rejections and server overloads at multiple points of the cluster. With RT-RR, the additional step of the schedulability check ensures that the evenly distributed requests does not overload the servers. Request rejections could only happen up-front as a result of the schedulability check. Rejecting a request as early as possible enables it to be scheduled on another server. The high variability of requests created by the schedulability check at the executors ensures deadlines could easily be met. This is further guaranteed by scheduling of requests based on their deadlines at executors and the supporting software infrastructure. The results confirms the scalability of the solution, with higher acceptance rates being achieved and all deadlines being met, with increasing the number of executors in the cluster.

5   Related Work

Many of the research work on QoS on service execution, assume that web service middleware guarantees the perceived QoS levels. [17,20] match clients to servers based on QoS level promised, and assume that the servers guarantee the QoS levels. [19] takes it to a higher level where multiple services are composed depending on the QoS requirements. Attempts at introducing some method of service differentiation can also be found. [4,6,11] use functional differentiation to achieve different levels of execution times based on Service Level Agreements (SLAs). Requests belonging to different classes maybe accepted based on pre-defined ratios to achieve levels of execution times promised in SLAs. Similarly, [12] introduce differentiation by means of regulating the throughput of requests based on non-functional properties such as the nature of application, the device being used, and nature of client.

Although some of the work discussed above achieve some method of differentiation, they fail to guarantee predictability of execution for several reasons. Firstly, none of them accept a deadline to finish the execution of request, nor do they schedule the request for execution based on a deadline. Secondly, there is no acceptance criteria for requests and as a result, the execution of one request may impact on another. This could lead to high resource contention at servers and create overloaded conditions. None of these implementations make use of features from the underlying software infrastructure to ensure predictability of execution. Without support from the development platform and the operating system used, the implementations will not be effective in securing support for their operations at the system level.

6   Conclusion

In this paper, we presented a set of guidelines, software engineering techniques and algorithms that could be used to make web service middleware achieve predictable service execution times. Web services being the foundation of the cloud computing paradigm, stringent execution level QoS becomes a necessity with platforms and infrastructure being exposed as services. Moreover, this will also enable applications with real-time requirements to use web services as middleware and enjoy the inherent benefits the technology. The case study presented, demonstrates how to apply these guidelines and make use of the techniques and algorithms in enhancing existing widely used web services middleware. The validity and the performance gain by these guidelines and techniques are empirically evaluated by comparing with unmodified versions of the middleware used in the case study. The results prove that predictability in execution can indeed be achieved by using the proper techniques in these software.

The network communication aspects plays an important role in web services. Therefore, predictability in such aspects is vital to achieve overall predictability of a web service invocation. Although network aspects were considered out of scope for the research presented in this paper, this aspect will be researched in future. Moreover, specialised dispatching algorithms based on real-time scheduling principles are also being researched, to increase the acceptance rates in web service clusters.

Acknowledgements. This work is supported by the ARC (Australian Research Council) under the Linkage scheme (No. LP0667600, Titled "An Integrated Infrastructure for Dynamic and Large Scale Supply Chain").

References

1. Apache Software Foundation: Apache Synapse,
   http://synapse.apache.org/ (June 9, 2008)
2. Apache Software Foundation: Apache Axis2, http://ws.apache.org/axis2/ (June 8, 2009)
3. Arnold, K., Gosling, J., Holmes, D.: The Java programming language. Addison-Wesley Professional, Reading (2006)
4. Tien, C.-M., Cho-Jun Lee, P.: SOAP Request Scheduling for Differentiated Quality of Service. In: Web Information Systems Engineering - WISE Workshops, pp. 63–72. Springer, Heidelberg (October 2005)
5. Gamini Abhaya, V., Tari, Z., Bertok, P.: Achieving Predictability and Service Differentiation in Web Services. In: Baresi, L., Chi, C.-H., Suzuki, J. (eds.) ICSOC-ServiceWave 2009. LNCS, vol. 5900, pp. 364–372. Springer, Heidelberg (2009)
6. García, D.F., García, J., Entrialgo, J., García, M., Valledor, P., García, R., Campos, A.M.: A qos control mechanism to provide service differentiation and overload protection to internet scalable servers. IEEE Transactions on Services Computing 2(1), 3–16 (2009)
7. Gartner: SOA Is Evolving Beyond Its Traditional Roots,
   http://www.gartner.com/it/page.jsp?id=927612 (April 2, 2009)
8. Gartner, Forrester: Use of Web services skyrocketing,
   http://utilitycomputing.com/news/404.asp (September 30, 2003)

9. Graham, S., Davis, D., Simeonov, S., Daniels, G., Brittenham, P., Nakamura, Y., Fremantle, P., Konig, D., Zentner, C.: Building Web Services with Java: Making Sense of XML, SOAP, WSDL and UDDI, 2nd edn. Sams Publishing, Indianapolis (July 8, 2004)
10. Microsoft: Windows Communications Foundation,
    http://msdn.microsoft.com/library/ee958158.aspx
11. Pacifici, G., Spreitzer, M., Tantawi, A., Youssef, A.: Performance management for cluster-based web services. IEEE Journal on Selected Areas in Communications 23(12), 2333–2343 (2005)
12. Sharma, A., Adarkar, H., Sengupta, S.: Managing QoS through prioritization in web services. In: Proceedings of Web Information Systems Engineering Workshops, pp. 140–148 (December 2003)
13. Stankovic, J.A., Spuri, M., Ramamritham, K., Buttazzo, G.C.: Deadline scheduling for real-time systems: EDF and related algorithms. Kluwer Academic Publishers, Dordrecht (1998)
14. Sun Microsystems: Glassfish Application Server - Features (2009),
    http://www.oracle.com/us/products/middleware/application-server/oracle-glassfish-server/index.html
15. Sun Microsystems: Sun Java Real-time System (2009),
    http://java.sun.com/javase/technologies/realtime/
16. Sun Microsystems: Thread Scheduling Visualizer 2.0 - Sun Java RealTime Systems 2.2,
    http://java.sun.com/javase/technologies/realtime/reference/TSV/JavaRTS-TSV.html (August 21, 2009)
17. Tian, M., Gramm, A., Naumowicz, T., Ritter, H., Freie, J.: A concept for QoS integration in Web services. In: Proceedings of Web Information Systems Engineering Workshops, pp. 149–155 (2003)
18. Wang, A.J., Baglodi, V.: Evaluation of java virtual machines for real-time applications. Journal of Computing Sciences in Small Colleges (4), 164–178 (2002)
19. Zeng, L., Benatallah, B., Dumas, M., Kalagnanam, J., Sheng, Q.: Quality driven web services composition. In: Proceedings of the 12th International Conference on World Wide Web, pp. 411–421 (2003)
20. Zeng, L., Benatallah, B., Ngu, A., Dumas, M., Kalagnanam, J., Chang, H., Center, I., Yorktown Heights, N.: QoS-aware middleware for web services composition. IEEE Transactions on Software Engineering 30(5), 311–327 (2004)

Event Driven Monitoring for Service Composition Infrastructures

Oliver Moser[1], Florian Rosenberg[2], and Schahram Dustdar[1]

[1] Distributed Systems Group, Vienna University of Technology, Austria
[2] CSIRO ICT Centre, GPO Box 664, Canberra ACT 2601

Abstract. We present an event-based monitoring approach for service composition infrastructures. While existing approaches mostly monitor these infrastructures in isolation, we provide a holistic monitoring approach by leveraging Complex Event Processing (CEP) techniques. The goal is to avoid fragmentation of monitoring data across different subsystems in large enterprise environments by connecting various event producers. They provide monitoring data that might be relevant for composite service monitoring. Event queries over monitoring data allow to correlate different monitoring data to achieve more expressiveness. The proposed system has been implemented for a WS-BPEL composition infrastructure and the evaluation demonstrates the low overhead and feasibility of the system.

Keywords: Monitoring, Composition, Complex Event Processing.

1 Introduction

Service-Oriented Computing (SOC) methodologies and approaches [18] are becoming an established and predominant way to design and implement process-driven information systems [23]. Service composition approaches are often used to orchestrate various loosely coupled services to implement business processes within and across enterprise boundaries. In such loosely-coupled environments, it is crucial to monitor the execution of the overall system and the business processes in particular. This ensures that problems, instabilities or bottlenecks during system execution can be detected before, for example, customers become aware of the problem.

In many applications, the composition engine acts as a central coordinator by orchestrating numerous services that implement the business processes. Therefore, the composition infrastructure needs to provide effective mechanisms to monitor the executed business processes at runtime. These include technical aspects such as monitoring Quality of Service (QoS) attributes of the services [16,15], resource utilization as well as business-related information relevant for Key Performance Indicator (KPI) progression or Service Level Agreement (SLA) fulfillment [21].

A number of monitoring systems have been proposed, however they are mainly tailored for a particular composition or workflow runtime [22,19,12,3,6,5,13]. Thus, these systems are usually tightly coupled or directly integrated in the

L. Chen, P. Triantafillou, and T. Suel (Eds.): WISE 2010, LNCS 6488, pp. 38–51, 2010.
© Springer-Verlag Berlin Heidelberg 2010

composition engine without considering other subsystems that live outside the composition engine but still influence the composite service execution (such as message queues, databases or Web services). The lack of an integrated monitoring mechanism and system leads to a fragmentation of monitoring information across different subsystems. It does not provide a holistic view of all monitoring information that can be leveraged to get detailed information of the operational system at any point in time. Additionally, most monitoring approaches do not allow a so-called multi-process monitoring, where monitoring information from multiple composite service instances can be correlated. This calls for an integrated and flexible monitoring system for service composition environments. We summarize the main requirements below, and give concrete and real world examples in Section 2 when describing the case study.

– **Platform agnostic and unobtrusive**: The monitoring system should be *platform agnostic*, i.e., independent of any concrete composition technology and runtime. Additionally, it should be *unobtrusive* to the systems being monitored, i.e., there should be no modifications necessary to interact with the monitoring runtime.
– **Integration with other systems**: The monitoring system should be capable of integrating monitoring data from other subsystems outside the composition engine (such as databases, message queues or other applications). This enables a holistic view of all monitoring data in a system.
– **Multi-process monitoring**: Composite services often cannot be monitored in isolation because they influence each other (cf. Section 2). The monitoring system should enable monitoring across multiple composite services, and its instances. Correlating monitoring data across several independent service compositions can be achieved by matching key information of the payload of the underlying message exchange.
– **Detecting anomalies**: The monitoring system should be capable of unveiling potential anomalies in the system, e.g., problems in backend services due to a high load outside of usual peak hour times. Additionally, the system should support dependencies between monitoring events. This feature can be used to find fraudulent and malicious behavior based on certain event patterns, e.g., absent events in a series of expected events.

We propose an event-driven monitoring system to address these requirements. Our approach leverages Complex Event Processing (CEP) to build a flexible monitoring system [9] that supports temporal and causal dependencies between messages. The system provides a loosely coupled integration at the messaging layer of a composition engine (e.g., on the SOAP message layer in the context of WS-BPEL). A monitoring component intercepts the messages sent and received by the composition engine and emits various events to the monitoring runtime. Furthermore, the monitoring runtime allows to connect arbitrary subsystems by using event source adapters. These external events are also processed within the monitoring runtime to provide an integrated monitoring environment.

This paper is organized as follows. Section 2 presents an illustrative example from the telecommunication domain. Section 3 introduces the concepts of our event driven monitoring approach and presents a proof of concept implementation. A detailed evaluation of our system is presented in Section 4. Section 5 gives an overview of the related work with respect to the main contributions of this paper. Finally, Section 6 concludes this paper.

2   Illustrative Example

This section introduces a subscriber activation scenario in a telecommunications enterprise called *Phonyfone*. A subscriber in our example is an active Phonyfone customer. Figure 1a shows a high level view of the activation process. A *Customer* provides required information such as personal and payment data to a *Shop Assistant* (1a). The Shop Assistant creates a new contract by entering the related data into the *PhonyWeb Terminal* (1b). The PhonyWeb Terminal transmits the customer (2a) and service data (2b) to an on-premises *Business Gateway*. The Business Gateway dispatches the two requests into a *Message Queue* (3), which delivers the requests (4a and 4b) to Phonyfone's *Composition Engine*. The Composition Engine triggers the execution of both, the *AddSubscriber* (AddSub) and *AddService* (AddSvc) composite services (Figure 1b). They enable Phonyfone to automate the steps required to create a *subscriber* object for a new customer as well as a *service* object for an additional service, such as mobile Internet access. Upon normal completion of these service compositions, the subscriber becomes active, and the PhonyWeb Terminal provides feedback to the Shop Assistant (5a). Finally, the Shop Assistant can handover a printout of the contract to the customer (5b).

From an operational perspective, Phonyfone adopts composition engines from two different vendors. This situation stems from a recent merger between Phonyfone and its parent company. Phonyfone faces between 500 and 600 subscriber activations per day, with 90 percent of the related composite service invocations executing between 09:00 am and 05:00 pm. On average, the AddSub request requires 4000ms for processing, and the AddSvc request requires 2500ms. The

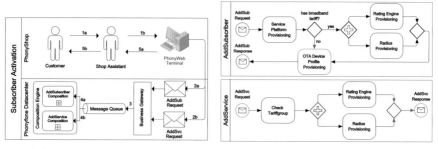

(a) Subscriber Activation Workflow   (b) Phonyfone Service Compositions

Fig. 1. Phonyfone Enterprise Scenario

combined processing time should not exceed 10 seconds. Deviations in both the throughput as well as the execution time can indicate potential problems, either in the backend systems or the network backbone itself. On the other hand, a rise in subscriber activations can prove customer acceptance for a promotion or special sale. Detecting such anomalies through an appropriate monitoring system *before* customers report them is of uttermost importance, both in terms of ensuring normal operation and deriving customer satisfaction with new or improved products.

2.1   The AddSubscriber and AddService Composition

Both the AddSub and AddSvc composite services are initiated by incoming XML requests (not shown for brevity). The following paragraphs discuss these services and their corresponding monitoring needs in more detail.

AddSubscriber. The AddSub composition (Figure 1b, top) is created by receiving an AddSub request. This request contains the Msisdn (i.e., the phone number of the new customer), the tariff information, the originator (i.e., shop assistant id) and an optional reduced recurring charge (RRC) field with a given start and end date. Provisioning of core subsystems starts with persisting relevant data in Phonyfone's *Service Delivery Platform* (SDP). The SDP provides access to subscriber and service relevant data. Then, if the tariff information provided in the request indicates a broadband product (for mobile Internet access), subscriber data is provisioned to two additional systems (*RADIUS* and the *Rating Engine* for dial-in services). Next, the *Over The Air* (OTA) device profile provisioning stores cellphone model and brand information in a device configuration server. Finally, an AddSub response, indicating a success or failure message, is sent back to the requester.

A potential risk emerges when reasoning about the RRC feature. Phonyfone allows its shop assistants to grant five RRCs per day, however, this limit is not enforced by any system and depends solely on the loyalty of the shop assistant. In the worst case scenario, this loophole can be exploited by a dishonest shop assistant. Monitoring of message exchange ensures compliance with enterprise policies and can be used to detect fraudulent behavior.

AddService. The AddSvc composition (Figure 1b, bottom) starts upon receiving a AddSvc request. It contains a subset of the request from the AddSub composition plus the name of the requested service (e.g., AntiVirus). After checking the subscriber's tariff to ensure that the subscriber is allowed to order the service, the Rating Engine as well as the RADIUS service is provided with service information. Finally, an AddSvc response is returned to the requester, indicating success or failure.

It is important to note that an AddSvc request will fail if the related AddSub request has not been successfully processed, i.e., the subscriber does not (yet) exist in Phonyfone's subscriber database. On the other hand, the AddSub request alone does not complete the subscriber activation – both the AddSub and the AddSvc request have to be processed timely and in the right order. This implies that a monitoring solution must be capable of (1) detecting situations where the temporal ordering between requests is wrong and (2) finding out which client issued the requests, e.g., to fix it. Finally, both service compositions are deployed on composition engines from different vendors, which hampers correlation of monitoring data due to non-uniform logging facilities. A holistic view on monitoring data from the various runtime environments greatly improves manageability and enforcement of monitoring requirements.

3   Event Driven Monitoring

Considering the monitoring needs of Phonyfone, it is rather obvious that a log file or basic performance reports cannot cover the requirements discussed in Section 1 and 2. A comprehensive monitoring solution that is capable of reflecting causal and temporal context within the composition engine's message exchange is needed. Moreover, the necessary platform independence requires that the monitoring solution has to operate on a layer that is common to most composition platforms. Operating on the *message level* makes the system we propose agnostic to implementation details of the monitored platform. Hence, it supports a large variety of current and future composition platforms. The temporal and causal requirements are covered by interpreting the message exchange as a stream of *events*. Each incoming and outgoing message is associated with an event object that has certain properties (cf. Section 3.1). Clearly, some messages are dependent on other messages, both from a causal as well as temporal point of view. These dependencies can be modeled by combining single message events into more complex composite events. This concept is generally known as *Complex Event Processing* (CEP) [14] and represents the fundamental building block of the *event-driven* monitoring system we will discuss below.

3.1   Event Model

As aforementioned, the monitoring approach we propose is built around the concept of events. From various available definitions [7], we follow a rather technical view of an event as a detectable condition that can trigger a notification. In this context, a notification is an event-triggered signal sent to a runtime-defined recipient [11]. In the Phonyfone scenario, each monitoring requirement breaks down to a detectable condition that can be represented by events of various types. An *event type* represents the formal description of a particular kind of event. It describes the data and structure associated with an event. As a simple example, detecting the anomalies described in Section 2 can be covered by an event type `AddSubCompositionInvocation`. This event type has several data items such as a timestamp or the execution duration. Using these data items, an *event processor* can filter the vast amount of events and separates events of interest, e.g., events of type `AddSubCompositionInvocation` where the timestamp is between 09:00 and 05:00pm and the execution time exceeds 4000ms.

To reason about and to provide a generic way to deal with the various event types, an *event model* is needed. Such an event model organizes the different event types and generalizes them by introducing an event hierarchy. Thus, it is possible to leverage simple base events to define more specific and complex events. Figure 2 displays a simplified version of our proposed event model. Various other types of events, such as service or composition life cycle events, are out of scope of this work.

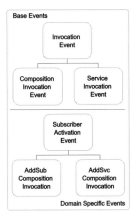

Fig. 2. Our event model distinguishes between *Base Events* and *Domain Specific Events*. Base Events are triggered upon invocation of either a service composition or a single service. Both `CompositionInvocationEvent` and `Service-InvocationEvent` inherit common attributes, such as a timestamp, the execution duration and a success indicator, from an abstract `InvocationEvent`. Domain Specific Events are user-defined and built upon Base Events. They are designed for a particular monitoring use case, e.g., our subscriber activation scenario. Additionally, they have access to the related message context to analyze the payload of the underlying message exchange. This feature enables the correlation of events based upon message payload, such as the `<Msisdn>` of the XML requests mentioned in Section 2.

3.2 Event Processing Language

In the previous section, we argued that an event processor is required to select specific event objects out of a stream of events, based on certain properties. Similar to SQL of relational databases, most event processors support an *Event Processing Language* (EPL). However, and contrary to relational databases, the processing model of event stream processors is *continuous* rather than only when queries are submitted. In particular, user defined queries are stored in memory and event data is pushed through the queries. This approach provides better performance and scalability than disk based database systems for large amounts of data, which makes it ideal for our application domain.

From our illustrative example, we can identify two requirements for an EPL. First, the EPL must provide capabilities to model complex *patterns* of events. Considering the coupling between invocations of the AddSub and AddSvc composite services, this sequence can be modeled as the following event pattern: `AddSub → AddSvc(msisdn = AddSub.msisdn)`. The arrow ($\rightarrow$) denotes that an AddSvc event follows an AddSub event. Moreover, we need to specify that the Msisdn included in both messages need to be the same. A *filter criteria* (msisdn = AddSub.msisdn) can be set to connect several events through a common attribute, such as the Msisdn. Besides detecting the presence of a particular event, the EPL should also support detecting the absence of an event, e.g., where the PhonyWeb Terminal issues an AddSub request but no related AddSvc request. Second, the EPL has to support event stream *queries* to cover online queries against the event stream, ideally using a syntax similar to existing query languages such as SQL. This feature can be used to analyze specific information about the business domain, e.g., the number of failed subscriber activations during a particular time window or even the related Msisdns: `SELECT msisdn FROM AddSub.win:time(15 minutes) WHERE success = false`. This query returns all Msisdns of AddSub requests that failed within the last 15 minutes. Furthermore, event stream queries can be leveraged to deal with marketing specific requirements such as finding the most wanted postpaid tariff of the last business day.

Reasoning about the fraudulent behavior of a shop assistant discussed in Section 2 leads us to yet another EPL requirement. Detecting an unusual high number of requests with a reduced recurring charge set can be handled by defining the following event pattern: `[5] (a = AddSub → b = AddSub(originator = a.originator and a.reducedrc = reducedrc and a.reducedrc > 0).win: time(1 day)`. The repeat operator ($[n]$, $n > 0$) triggers when the statement following the operator evaluates to true n times. However, we need to know which shop assistant triggered the activation. Thus, the final requirement for our EPL is the combination of both event pattern matching and event stream queries: `SELECT a.originator FROM PATTERN ([5] (a = AddSub → b = AddSub(originator = a.originator and a.reducedrc = reducedrc and a.reducedrc > 0).win:time(1 day)`. This query will unveil the dishonest shop assistant.

3.3   Architectural Approach

Figure 3 depicts a high-level overview of the proposed system architecture for our monitoring runtime. As discussed earlier, our approach is agnostic to the *Composition Runtime* under observation. The Composition Runtime is comprised of a *Composition Processor* and a *Messaging Stack*. The Composition Processor interprets and executes the service composition. It uses the Messaging Stack to process incoming and outgoing messages. The former represent process

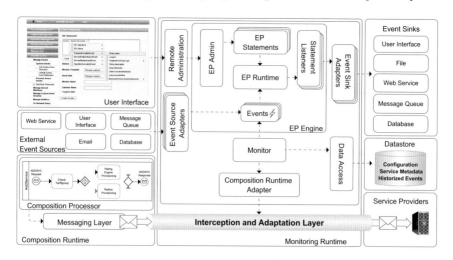

Fig. 3. Architectural Approach

instantiation messages (cf., BPEL receive elements) whereas the latter represent outgoing service invocations (cf., BPEL invoke elements).

The *Monitoring Runtime* introduces an *Interception and Adaption Layer* (IAL), which enables the inspection and, eventually, adaptation of message exchange [16]. The *Monitor* is the key component of our Monitoring Runtime. It serves two main purposes. First, it extracts service related metadata, such as endpoint URLs, and persists this information in a *Datastore* using the *Data Access* component. Second, it emits *Events*, such as `ServiceInvocationEvents` (cf. Section 3.1). The Monitor uses a *Composition Runtime Adapter* (CRA) to access the IAL. The CRA covers all implementation specific details of the monitored Composition Runtime. The Events emitted by the Monitor are handed over to the *Event Processing Runtime* (EP Runtime), which is part of the *Event Processing Engine* (EP Engine) incorporated in our system. The EP Runtime provides the capabilities discussed in Section 3.2. An *Event Processing Statement* (EP Statement) is registered with the EP Runtime and represents the choice of an event pattern, an event stream query or a combination of both. For each EP Statement, one or more *Statement Listeners* can be configured. Each Statement Listener is provided with a reference to the Event in case the EP Runtime produces a match for the underlying event pattern or query. To make use of this information, a Statement Listener has a list of registered *Event Sinks*. An Event Sink subscribes to the information represented by the Event objects, and will leverage this information to perform further analysis, aggregation or simply alert an operator. *Event Sink Adapters* encapsulate these details to abstract from the implementation and protocol specifics of the supported Event Sinks. Similarly, *Event Source Adapters* provide integration for *External Event Sources*, which can push external event data into our monitoring system. This allows to reflect data in EP statements that does not originate from the monitored service (composition), hence enabling a holistic view on monitoring data from various heterogeneous systems. Finally, a *User Interface* (UI) simplifies numerous administrative tasks. It provides a statement builder, auto completion features and syntax highlighting, and is decoupled from the Monitoring Runtime through the *Remote Administration Interface* to allow distributed deployment.

3.4   Implementation

This section briefly describes the Java based proof of concept implementation on top of our existing VieDAME framework. Figure 4 provides an overview of implementation technologies and highlights components related to event driven monitoring. For an in-depth discussion of VieDAME please refer to [16].

For our prototype implementation, WS-BPEL [17] is used as the underlying composition technology. *Engine Adapters* provide support for particular *BPEL Runtime* providers, such as ActiveBPEL or Apache ODE. The interception of the SOAP messages that are created by `<invoke>` or `<receive>` BPEL activities is done by the AOP based *Message Interception Layer*. *JBoss AOP* is leveraged to weave this processing layer in between the BPEL Runtime and the *SOAP Stack*, e.g., Apache Axis. Using AOP as a technology to extend a base system, i.e., the BPEL Runtime, stems from the fact that it minimizes the coupling

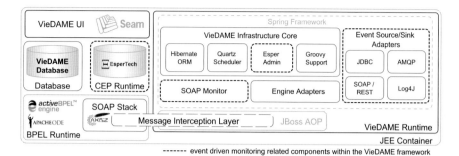

Fig. 4. Implementation Technologies

between the base system and our monitoring runtime. Thus, it effectively fulfills the requirements in terms of unobtrusiveness and platform independence from Section 1.

The *SOAP Monitor* taps the Message Interception Layer using an Engine Adapter suitable for the underlying BPEL runtime. It processes the message context and extracts service and process metadata, which is then stored in the *VieDAME Database*. Most important, it creates event objects that are passed to the *Esper CEP Runtime* [9] for event processing. Esper matches the processing language expressiveness and performance requirements postulated earlier and is easily embeddable into existing systems. Additionally, it provides XML DOM and StAX based support for XML payload processing of the message exchanges. Both processing modes use XPath expressions to access certain message elements, e.g., the <Msisdn> element from the AddSub and AddSvc messages from Section 2. For an in-depth discussion of the incorporated EPL please refer to the excellent Esper documentation [10].

The *Infrastructure Core* of our system provides common functionalities, such as Object/Relational Mapping (ORM), task scheduling or dynamic language support. To allow runtime integration and adaptation of event sinks such as a message queue, or external event sources, e.g., an SMTP host, Groovy is leveraged. It supports scripting and runtime deployment of templates that are used to create and modify the various supported *Event Sink Adapters* and *Event Source Adapters*.

4   Evaluation

For the evaluation of our event-driven monitoring system, we used the VieDAME based prototype discussed in Section 3.4. Additionally, we created an ActiveBPEL [1] implementation of the subscriber activation scenario from our illustrative example. Due to space restrictions, only results for the AddSub process are shown below[1].

---

[1] Full results of the load tests can be obtained on the project website
http://viedame.omoser.com

4.1   System Performance

We created a 50 minute load test using an industry grade tool (Mercury Loadrunner), measuring the performance of a default ActiveBPEL based (no viedame) process execution in comparison to a process execution where our monitoring system was enabled (viedame). The prototype was hosted on a quad core machine (3Ghz, 8GB RAM) running Linux. Implementations of the orchestrated Web services were created using JAX-WS and provide a randomized processing delay to mimic backend logic. Figure 5a shows the number of transactions per second (TPS) and the CPU utilization (CPU) for a 200 concurrent users scenario, with the viedame scenario having 100 EP statements registered in the system. The results show that performance in terms of TPS is almost identical, while the CPU utilization is only slightly higher in the viedame scenario. Both numbers confirm that the monitoring system we propose does not impose a real performance penalty on existing systems. Compared to our findings from previous work [16], we could effectively minimize the processing overhead by introducing a CEP engine for measuring QoS data and relying on a relational database only for historical data.

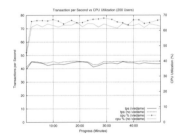
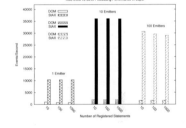

(a) Monitoring Overhead   (b) XML DOM vs StAX Performance

To drill down to the source of the processing overhead, we used a Java profiler (YourKit) to determine the component that uses most CPU cycles. The profiler showed clearly that the XML processing required by the domain specific events is rather time consuming. Besides the XML DOM processing mode, Esper contains a plugin that supports StAX (Streaming API for XML). We created another test scenario where we evaluated the event processing performance isolated from the BPEL engine to compare both processing modes. This scenario featured test runs for 1, 10 and 100 concurrent event emitters, i.e., Java threads concurrently issuing XML requests. The tests were executed with Esper having 10, 100 and 1000 statements registered. Figure 5b shows that the StAX based processing mode has far superior performance than the XML DOM based mode. Using the StAX mode, Esper is capable of processing more than 35000 events/second in the 10 emitter scenario, compared with 1900 events/second for the DOM processing mode in the same scenario. Service compositions with more than 2000 messages per second are rather rare. Hence, it is very unlikely that the CEP runtime will constrain the performance of a composition engine that uses our monitoring system, even in the most demanding environments.

4.2   Discussion

Stakeholders. The primary stakeholders of the monitoring system we propose are *domain experts*. Similar to a business analyst, a domain expert has solid knowledge of the business domain, e.g., the telecommunications domain. More-over, and different from the business analyst, a domain expert has more in depth know-how of the technical details of the underlying services and systems. Do-main experts only have basic programming skills, which separates them from developers or system engineers.

Defining Event Queries. When applying the proposed Monitoring Runtime to the Phonyfone scenario, the domain expert does not have to deal with a complex system or service setup. The inherent unobtrusiveness of our approach, i.e. the automatic capturing of services and the related service compositions deployed in VieDAME, enables the domain expert to focus on the definition of monitoring events and the analysis of monitoring results. The first step in defining monitoring events is to decide which event type satisfies the monitoring requirement. The domain experts leverages the UI to create EP statements using base events or domain specific events (cf. Section 3.1). Certainly, existing events can be combined into more complex events. For the following, we assume that the domain expert wants to determine the average response times of all available services (Listing 5a), and creates a domain specific event (Listing 5b) triggered in case the AddSvc request for an AddSub request is not received within 10 seconds.

```
1 select
2   median(executionTime), operation.name
3 from
4   ServiceInvocationEvent
5 group by
6   operation.name
7
```

(a) Querying Service Response Times

```
1 every AddSubAlert =
2   AddSub ->
3     (timer:interval(10 sec)
4 and not
5     AddSvcAlert = AddSvc(
6       msisdn = AddSubAlert.msisdn
7     )
```

(b) Detecting Absent Messages

Fig. 5. EP Statements for Base and Domain Specific Events

For the base event from Listing 5a, only the EP statement needs to be en-tered in the UI. Considering the domain specific event from Listing 5b, the EP Runtime needs to know how to interpret the msisdn attribute (line 6) of the AddSub and AddSvc events. The domain expert has two choices to communicate the meaning of the attribute. First, if a schema definition for the XML message exchange is available, the XPath expression required to reference the attribute can be inferred automatically. Second, in lack of a related schema definition, the XPath expressions have to be setup manually. Assuming that no schema definition is available, the <Msisdn> element from the AddSub request (cf. Section 2) is selected with the expression /soapenv:Envelope/soapenv:Body/ addSubscriber/subscriber/msisdn.

Once the EP statements have been defined, the domain expert chooses a EP statement listener template. These templates support the domain expert in integrating Event Sinks (cf. Section 3.3). Assuming that a remote Web service should be invoked for events matching the statements shown in Listing 5, the domain expert chooses the Web service statement listener template and adapts it to specific needs, e.g., defines the endpoint URL or authentication credentials. All required steps can be performed during runtime, which minimizes time-to-market and maximizes responsiveness to new and changing monitoring requirements. For more examples of domain specific events, please consult the project website[2].

5   Related Work

A large body of monitoring work exists that can be classified as assertion-based monitoring. Baresi and Guinea [3,4] propose WSCoL, a constraint language for the supervision of BPEL processes. Monitoring information is specified as assertions on the BPEL code. Their approach uses Aspect-Oriented Programming (AOP) to check the assertions at runtime. In [12], Baresi et al. propose SEC-MOL, a general monitoring language, that sits on top of concrete monitoring languages such as WSCoL. Thus, it is flexible by separating data collection, aggregation and computation from the actual analysis. Sun et al. [19] also propose a monitoring approach based on AOP. Their goal is to check business process conformance with the requirements that are expressed using WS-Policy. The properties (e.g., temporal or reliability) of a Web service are described as Extended Message Sequence Graph (EMSG) and Message Event Transferring Graph (METG). A runtime monitoring framework is then used to monitor the corresponding properties that are then analyzed and checked against the METG graphs. Wu et al. [22] propose an AOP-based approach for identifying patterns in BPEL processes. They use a stateful aspect extension allowing the definition of behavior patterns that should be identified. If identified, different actions can be triggered. It also allows to monitor certain patterns by using history-based pointcuts. However, monitoring is restricted to instances of a BPEL process.

In contrast to the aforementioned approaches, we do not use assertions to define what needs to be monitored using a proprietary language. We propose a holistic and flexible monitoring system based on CEP techniques. Our system measures various QoS statistics by using a non-intrusive AOP mechanism and has access to message payloads of the composition engine. Additionally, it allows to connect external event sources that might be of interest. CEP techniques are very powerful because they operate on event streams (i.e., the monitoring data). CEP operators, e.g., for temporal or causal event correlation, can then be used on event streams to define what monitoring information needs to be retrieved. Domain experts can then use the simple visual tool to define event queries without programming expertise.

Another group of work focuses on monitoring temporal properties of Web service compositions. Kallel et al. [13] propose an approach to specify and monitor temporal constraints based on a novel formal language called XTUS-Automata. Monitoring itself is based on AO4BPEL [8]. Similar to that, Carbon et al. [2] propose RTML (Runtime Monitoring Specification Language) to monitor temporal properties. Their approach translates monitoring information to Java code to monitor instances of services and process classes. Contrary to these approaches, we do

---

[2] http://viedame.omoser.com

not need to translate a formal language to specific monitoring code. However, our system provides an holistic view on all available monitoring information as event streams. An event query language can then be used retrieve temporal properties.

Suntinger et al. [20] provide a visualization approach using the notation of an Event Tunnel. It allows to visualize complex event streams of historical information and enables the detection and analysis of business patterns. This can be used, for example, to detect and prevent fraud. Finally, Beeri et al. [6,5] propose a query-based monitoring approach and system. It consists of a query language allowing to visually design monitoring tasks. The designer also allows to define customized reports based on those queries. Complementary to our work, we also allow a query based approach based on an event query language. However, we do not provide reporting functionality.

6 Conclusion

In this paper we introduced a generic monitoring system for message based systems. It interprets each message as an event of a particular type and leverages complex event processing technology to define and detect situations of interest. Our system can be used to correlate events across multiple service compositions and integrates with external event providers to minimize fragmentation of monitoring data. Moreover, the inherent unobtrusiveness of the approach makes our monitoring system applicable to most current and future message based systems. This minimizes integration and setup costs. The evaluation of the prototype shows only a very small performance penalty. We also emphasized on the practical usefulness of our approach.

For future work, we are planning to provide visual support for EP statement creation to assist users who are not familiar with the underlying event query language to define basic event patterns and queries. Another point of interest includes service life cycle events, which were out of scope of this work. We will investigate how life cycle events can be combined with other runtime events to extract information related to service evolution.

Acknowledgements. The research leading to these results has received funding from the European Community's Seventh Framework Programme [FP7/2007-2013] under grant agreement 215483 (S-Cube).

References

1. Active Endpoints: ActiveBPEL Engine (2007), http://www.active-endpoints.com/ (last accessed: September 07, 2010)
2. Barbon, F., Traverso, P., Pistore, M., Trainotti, M.: Run-Time Monitoring of Instances and Classes of Web Service Compositions. In: IEEE Intl. Conf. on Web Services (ICWS 2006), pp. 63–71. IEEE Computer Society, Los Alamitos (2006)
3. Baresi, L., Guinea, S.: Self-supervising BPEL Processes. IEEE Transactions on Software Engineering (2010) (forthcoming)
4. Baresi, L., Guinea, S.: Dynamo: Dynamic Monitoring of WS-BPEL Processes. In: Benatallah, B., Casati, F., Traverso, P. (eds.) ICSOC 2005. LNCS, vol. 3826, pp. 478–483. Springer, Heidelberg (2005)

 5. Beeri, C., Eyal, A., Milo, T., Pilberg, A.: Query-based monitoring of BPEL business processes. In: Proc. of the ACM SIGMOD Intl. Conf. on Management of Data (SIGMOD 2007), pp. 1122–1124. ACM, New York (2007)
 6. Beeri, C., Eyal, A., Milo, T., Pilberg, A.: BP-Mon: query-based monitoring of BPEL business processes. SIGMOD Rec. 37(1), 21–24 (2008)
 7. Chandy, K., Schulte, W.: Event Processing - Designing IT Systems for Agile Companies. McGraw Hill Professional, New York (2010)
 8. Charfi, A., Mezini, M.: AO4BPEL: An Aspect-oriented Extension to BPEL. World Wide Web 10(3), 309–344 (2007)
 9. EsperTech: Esper (2009), http://esper.codehaus.org (last accessed: October 25, 2009)
10. EsperTech: Esper EPL Documentation (2009), http://esper.codehaus.org/esper-3.5.0/doc/reference/en/html/index.html (last accessed: September 11, 2010)
11. Faison, T.: Event-Based Programming: Taking Events to the Limit. Apress (2006)
12. Guinea, S., Baresi, L., Spanoudakis, G., Nano, O.: Comprehensive Monitoring of BPEL Processes. IEEE Internet Computing (2009)
13. Kallel, S., Charfi, A., Dinkelaker, T., Mezini, M., Jmaiel, M.: Specifying and Monitoring Temporal Properties in Web Services Compositions. In: Proc. of the 7th IEEE European Conf. on Web Services (ECOWS 2009), pp. 148–157 (2009)
14. Luckham, D.C.: The Power of Events: An Introduction to Complex Event Processing in Distributed Enterprise Systems. Addison-Wesley Longman Publishing Co., Inc., Boston (2001)
15. Michlmayr, A., Rosenberg, F., Leitner, P., Dustdar, S.: Comprehensive QoS Monitoring of Web Services and Event-Based SLA Violation Detection. In: Proc. of the 4th Intl. Workshop on Middleware for Service Oriented Computing (MWSOC 2009), pp. 1–6. ACM, New York (2009)
16. Moser, O., Rosenberg, F., Dustdar, S.: Non-Intrusive Monitoring and Service Adaptation for WS-BPEL. In: Proc. of the 17th Intl. World Wide Web Conf. (WWW 2008), pp. 815–824. ACM, New York (2008)
17. OASIS: Web Service Business Process Execution Language 2.0 (2006), http://www.oasis-open.org/committees/tc_home.php?wg_abbrev=wsbpel (last accessed: April 17, 2007)
18. Papazoglou, M.P., Traverso, P., Dustdar, S., Leymann, F.: Service-Oriented Computing: State of the Art and Research Challenges. IEEE Computer 11 (2007)
19. Sun, M., Li, B., Zhang, P.: Monitoring BPEL-Based Web Service Composition Using AOP. In: Proc. of the 8th IEEE/ACIS Intl. Conf. on Computer and Information Science (ICIS 2009), pp. 1172–1177 (2009)
20. Suntinger, M., Schiefer, J., Obweger, H., Groller, M.: The Event Tunnel: Interactive Visualization of Complex Event Streams for Business Process Pattern Analysis. In: IEEE Pacific Visualization Symposium (PacificVIS 2008), pp. 111–118 (2008)
21. Wetzstein, B., Leitner, P., Rosenberg, F., Brandic, I., Leymann, F., Dustdar, S.: Monitoring and Analyzing Influential Factors of Business Process Performance. In: Proc. of the 13th IEEE Intl. Enterprise Distributed Object Computing Conf. (EDOC 2009), pp. 141–150. IEEE Computer Society, Los Alamitos (2009)
22. Wu, G., Wei, J., Huang, T.: Flexible Pattern Monitoring for WS-BPEL through Stateful Aspect Extension. In: Proc. of the IEEE Intl. Conf. on Web Services (ICWS 2008), pp. 577–584 (2008)
23. Zdun, U., Hentrich, C., Dustdar, S.: Modeling Process-Driven and Service-Oriented Architectures Using Patterns and Pattern Primitives. ACM Transactions on the Web (TWEB) 1(3), 14:1–14:14 (2007)

On Identifying and Reducing Irrelevant Information in Service Composition and Execution[*]

Hong-Linh Truong[1], Marco Comerio[2], Andrea Maurino[2],
Schahram Dustdar[1], Flavio De Paoli[2], and Luca Panziera[2]

[1] Distributed Systems Group, Vienna University of Technology, Austria
{truong,dustdar}@infosys.tuwien.ac.at
[2] Department of Informatics, Systems and Communication
University of Milano, Bicocca, Italy
{comerio,maurino,depaoli,panziera}@disco.unimib.it

Abstract. The increasing availability of massive information on the Web causes the need for information aggregation by filtering and ranking according to user's goals. In the last years both industrial and academic researchers have investigated the way in which quality of services can be described, matched, composed and monitored for service selection and composition. However, very few of them have considered the problem of evaluating and certifying the quality of the provided service information to reduce irrelevant information for service consumers, which is crucial to improve the efficiency and correctness of service composition and execution. This paper discusses several problems due to the lack of appropriate way to manage quality and context in service composition and execution, and proposes a research roadmap for reducing irrelevant service information based on context and quality aspects. We present a novel solution for dealing with irrelevant information about Web services by developing information quality metrics and by discussing experimental evaluations.

1 Introduction

Identifying and reducing irrelevant information have been always one of the main goals of Web information systems. Among other possibilities, information aggregation, filtering and ranking according to user's goals can be realized by means of (composite) Web services (either based on SOAP or REST). Due to the advantage of service-oriented computing models, data sources have been widely published using Web service technologies, and several service composition and execution engines and tools have been developed to foster the composition and execution of service-based information systems. The large number of available services and the easy-to-use composition tools leads to our questions of facing irrelevant information problems (i) during the service composition phase, and

---

[*] This work is partially supported by the Vienna Science and Technology Fund (WWTF), project ICT08-032 and by the SAS Institute srl (Grant Carlo Grandi).

L. Chen, P. Triantafillou, and T. Suel (Eds.): WISE 2010, LNCS 6488, pp. 52–66, 2010.

(ii) due to poor response of composite services. The former question is mostly related to developers and novice users who build composite services and the latter is mostly for service consumers who suffer from unwanted results delivered by composite service execution. These questions become more relevant when we consider that there are many types of data provided by data-intensive services, and, in the Internet and cloud environments, such data and services have different context and quality constraints which, if not handled properly, might cause severe decrease of perceived service quality. In this paper, we contribute with a detailed analysis of factors affecting information relevance in service composition and execution, a road-map for future research for overcoming irrelevant information using quality and context information associated with data and services, and a particular solution for filtering irrelevant service information using quality of data metrics developed for Web services information.

The rest of this paper is organized as follows: Section 2 presents motivations of this paper. Section 3 analyzes irrelevant service information problems. Section 4 presents a roadmap to solutions for these problems. Section 5 presents a novel approach for reducing irrelevant service information using information quality metrics. Section 6 presents experimental results. Related works are described in Section 7. We conclude the paper and outline our future work in Section 8.

2   Motivating Examples

Motivating Example 1 – Irrelevant Service Information in Service Composition: Let us consider the service discovery scenario in the logistic domain presented in the Semantic Web Service Challenge 2009[1]. Several logistic operators offer shipping services each one characterized by offered non-functional properties (NFPs) specified in *service contracts* (i.e., conditional joint offers of NFPs). Relevant NFPs in this domain are *payment method, payment deadline, insurance, base price,* and *hours to delivery.* Let us consider a developer looking for a shipping service to be included into a service composition. Assume 100 functional-equivalent services, each one associated with an average of 5 different service contracts. Therefore, a developer might have to select the best among a set of 500 service contracts without the quality of contract information. For example, a developer is unaware of timeliness (i.e., how current the service information are) and completeness (i.e., the number of available information respect to an expected minimum set) of NFPs in the service contracts. Given this lack of quality information, the developer has to perform a time-consuming task to filter information to detect irrelevant contracts, such as contracts outdated, incomplete or not applicable to user context. In our scenario, the developer must perform a manual process to evaluate the 500 service contracts and discover the incomplete ones. Moreover, the lack of automatic support to evaluate applicability conditions and the presence of outdated contracts might lead to potentially wrong selection (e.g., contracts not compliant with user context or expired contracts). We believe that quality of service information should be evaluated in

---

[1] http://sws-challenge.org/wiki/index.php/Scenario:_Logistics_Management

advance in order to reduce the presence of irrelevant information that causes waste of time and energy in the service discovery and selection.

Motivating Example 2 – Irrelevant Information between Service Composition and Execution: Let us consider a developer who wants to define a composite service out of services associated with contracts. The service composition process needs to produce a composite service contract by composing the contracts associated with the services involved into the composition [1]. At execution time the service users must follow the clauses specified in the composite service contract. Typically, a temporal distance exists between the composition (i.e., the time in which the composite service contract is created) and the execution (i.e., the time in which the contract must be enforced). A problem is that, in this time span, some information in the composite service contract might become outdated and, therefore, irrelevant when the composition needs to be executed. Therefore, it is important to detect such unsuitable information to rate the contract w.r.t. irrelevancy. Such a rate could be used to define when a service composition needs to be adapted or even discarded.

Motivating Example 3 – Irrelevant Information in Service Usage: Let us consider a novice user who wants to create a data-intensive composite Web service to manage contents from different RESTful services and to present top news headlines on current events associated with images and videos. The diversity and complexity of context constraints (e.g., indicating free for non commercial purpose, particular user country, and suitable user devices) and of quality of data and services (e.g., indicating accuracy and timeliness of images and response time of service requests) under which services and data can be used and displayed, together with the growth in development and deployment of services, have led to irrelevant information augmenting the complexity of service composition. For example, at the time of writing, `Flickr` proposes Creative Commons[2] to define licenses related to content usage that can be retrieved only after accessing the service, making this information useless for service selection. `YouTube` adopts copyright policy to cover all the published contents which is unstructured and therefore it cannot be used for automatic service selection. Quality, context and legal aspect information must be considered during service selection to fulfill user expectations. Currently, we lack the evaluation of quality and context information associated with the service and legal aspects related to service usage (e.g., data ownership), preventing us to reduce irrelevant information faced by the developers and novice users in selecting the right services.

3 Irrelevant Information Problems in Service Composition and Execution

In order to examine possible irrelevant information problems in the context of service composition and execution, we consider typical information flows inherent in service composition and execution. Figure 1 depicts actors, components and

---

[2] http://www.flickr.com/creativecommons/

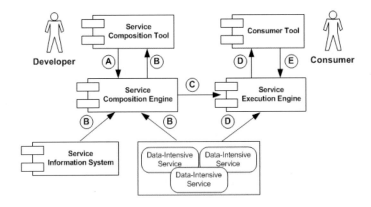

Fig. 1. Basic information flows in service composition and execution

interactions involved in service composition and execution in service-oriented computing environments, with a focus on data-intensive services, such as services providing company financial sheets and credits, images, and biodiversity data. In such environments, services offer well-defined interfaces (mostly implemented using SOAP and REST technologies) to consumers to access and store their data. In a typical service composition and execution lifecycle, the exchanged data are of:

- *Type A* - requirements about service and data schemas (functional information), NFPs, documentation, service contracts, and provenance information for the service composition.
- *Type B* - service and data schemas, NFPs, documentation, service contracts, and provenance information.
- *Type C* - information about the composite service and data provided by the composite service.
- *Type D* - data delivered by data-intensive services.
- *Type E* - information about data requested by the consumer.

These types of data are disseminated and manipulated through several abstract information flows in the service composition and execution lifecycle, as shown in Figure 1. In our analysis, these information flows are:

- *Flow 1 - developer → composition tool → composition engine*: the developer provides *Type A* information to the service composition tool and engine which support the composition process.
- *Flow 2 - services/service information system → composition engine → composition tool → developer*: the service and service information system, respectively, directly or indirectly, provide *Type B* information to the service composition engine and tool which filter, process, and propagate the information to the developer.
- *Flow 3 - composition engine → execution engine*: the composition engine passes *Type C* information to the execution engine.

- *Flow 4 - consumer → consumer tool → execution engine*: the consumer specifies *Type E* information through the consumer tool which invokes the service execution engine.
- *Flow 5 - service → execution engine → consumer tool → consumer*: the service provides *Type D* information to the execution engine which filters and processes them before returning the data to the consumer tool which presents the data to the consumer.

In all the above mentioned flows irrelevant information can exist. Irrelevant information during service composition is faced mostly by the composition developer, who relies on vast sources of information in order to construct composite services. A developer perceives this problem when information about services and data returned by the service composition engine/tool is incomparable, unsure, incomplete, or overwhelming. During the service execution, the service consumer faces irrelevant information problems when the data returned by data-intensive services are not comparable, inadequate or overwhelming. Table 1 gives some examples of irrelevant information. When we consider Internet-scale and cloud-based service composition and execution scenarios, in which services are provided by different providers, irrelevant information associated with these flows increases in many aspects due to the diversity and complexity of services and their descriptions.

Table 1. Examples of irrelevant information in service composition and execution

Problems	Examples
Relevant to context and quality information models	
Unstructured description of context, quality of service, and quality of data	Several data intensive services do not provide structured description of context, quality of service and of data information [2]. Mostly, they publish such information in HTML.
Different specifications and terminologies	Several specifications with similar terminologies are used. The semantics are not the same and the specifications are not interoperable [1].
Mismatching semantics of information about services and data	Similar services are classified in different classes. Similar metric names have different meanings and different data quality metrics represent the same thing [3].
Relevant to context and quality information access APIs	
No/Limited description of data and service usage	Information about service and data licensing is not associated with service description. Unclear/no service contract clauses (e.g., data ownership) exist.
No/Limited quality of data	Quality of data (e.g., the completeness and timeliness of information about a service) not available. Services and data are registered but they are no longer available.
No API for retrieving quality and context information	Impossibility to query context and quality information directly from services [4].
No quality and context information associated with requested data	Data returned by services have not been linked with information about their usage context and quality.
Relevant to context and quality evaluation techniques	
Missing evaluation of compatibility of context and quality of multiple services	Composition tools and engines cannot deal with multiple types of context and quality information specified in different languages [1].
Large/Irrelevant data quantity	Several services returned due to the impossibility to match the context and qualities of the requested information.

Table 2. Causes and effects of irrelevant information and information flows

Causes	Effects	Flow 1	Flow 2	Flow 3	Flow 4	Flow 5
Heterogeneous specs and terminologies	Not comparable service information	X	X	X	X	X
Untrusted or low-quality data	Unsure, noisy service information		X	X		
No/Limited context and quality specifications	Incomplete service information		X	X		
Large quantity of data	Overwhelming service information		X			X

Table 2 provides a mapping between causes and effects of irrelevant information problems and information flows described in Figure 1. The main causes and effects of irrelevant information problems that we have identified are:

- *Heterogeneous specifications and terminologies*: the management of information specified using different languages and terminologies causes semantic mismatching and the impossibility to apply automatic information processing in service composition and execution.
- *Untrusted or low-quality data*: the management of untrusted or low-quality information about service and data prevents from correct filtering activities, i.e., the selection of the best information according to consumer requests.
- *No/Limited context and quality specifications*: the lack of context and quality specifications about service and data prevents the implementation of information filtering because of incompleteness.
- *Large quantity of data*: the impossibility to filter information determines a large amount of data that overwhelms composition and execution tools.

4 Enhancing Context and Quality Support for Information about Services and Their Data

Existing irrelevant information problems in service composition and execution can be dealt by using several different techniques, for example, semantic matching, data mining, and similarity analysis [5,6,7]. In our work, we focus on how to evaluate and exchange the quality and the context of service information and utilize the quality and context for dealing with missing, ambiguous and inadequate service information used in the service composition and execution. In this paper, context information specifies situation under which services are constructed and used, the situation of the consumer, the situation under which the requested data can be used, whereas quality specifies the quality of service and quality of data provided by the service. We believe that if we are able to combine and utilize context and quality information in a unified way, several techniques could be developed to deal with irrelevant information problems. To this end, we propose the following research agenda to deal with information about services:

Topic 1 – Developing a meta-model and domain-dependent semantic representations for quality and context information specifications: by using such representations, the problem of unstructured descriptions of quality and context information and ambiguous semantics when using multiple specifications can be reduced. Researchers have shown the benefit of linked data models for Web information and we believe that such semantic representations can help to link context and quality information for data-intensive services. Currently, context and quality information for data-intensive services are not modeled and linked in an integrated manner. The use of a common meta-model and domain-dependent ontologies should provide a partial solution to the lack of standard models and terminologies for service information specification. Furthermore, we should develop techniques to map quality and context descriptions defined using different formalisms and terminologies [1]. With such techniques the problem to compare heterogeneous service information specifications can be partially solved.

Topic 2 – Developing context and quality information that can be accessed via open APIs: these APIs should be implemented by services and service information systems. In particular, current data-intensive services do not provide APIs for obtaining quality and context information associated with their data. We foresee several benefits if services provide such APIs [8]. The composition engine could utilize such information to perform service selection and compatibility checking. The service execution engine could use context and quality to filter and select the most relevant requested information. The composition engine could improve the selection of relevant resources and services by utilizing service context and quality description together with consumer's context and quality description. Similarly, the execution engine could utilize this information to filter information according to user requests.

Topic 3 – Developing techniques for context and quality evaluation: these techniques are for context and quality compatibility evaluation and composition. They should be implemented in composition and execution engines. In particular, composition engines can utilize these techniques (i) for matching context and quality information requested by service developers and offered by service providers, (ii) for checking the compatibility among context and quality information associated with services involved in the composition and (iii) for defining the context and quality information to be associated with the composite service. Viceversa, execution engines can utilize these techniques for matching context and quality information for data requested by service consumers and offered by services. While several techniques have been developed for quality of service or context matching in service composition and execution [9,10,11], techniques to deal with context and quality of data in an integrated way are missing. The use of such techniques could increase the relevance of information about services and of data provided by services.

5   Reducing Services and Resources by Qualifying Non-functional Information

In order to deal with untrusted/low-quality data and large quantity of data, in this section, we present a particular solution to reduce irrelevant information about service and data (*Type B* in Section 3). As stated before (see Figure 1), such information are exchanged between services/service information systems, composition engines, composition tools and developers.

The proposed solution consists in removing irrelevant information by evaluating the quality of information specified into service descriptions. Our solution is a concrete step in Topic 3 of the research agenda described in Section 4 since it could increase the relevance of information about services to be evaluated for service composition.

5.1   Quality of Data Metrics for Information about Service

Stimulated by information quality research, several metrics for evaluating the quality of information about services can be proposed. In this section, we propose only examples of these metrics that will be used to demonstrate the feasibility and the efficiency of our solution. An extended list of metrics is in [12].

Definition 1 (Interpretability). *Interpretability specifies the availability of documentation and metadata to correctly interpret the (functional and non-functional) properties of a service.*

This metric was originally described in [12] for data sources without concrete evaluation methods. In this paper, we extend it to service information using weighted factors and service document classification shown in Table 3.

Table 3. Types of information used for evaluating the Interpretability metric

Category	Service Information	Examples
schema	service and data schemas	WSDL, SAWSDL, pre/post conditions, data models
documentation	documents	APIs explanation, best practices
NFP	non-functional properties	categorization, location, QoS information
contract	service contracts and contract templates	service level agreements, policies, licenses
provenance	provenance information	versioning of schemas, NFPs, contracts

The *Interpretability* metric can be evaluated as follows:

$$Interpretability = \frac{\sum score(category_i) \times w_i}{\sum w_i} \qquad (1)$$

where $\forall category_i \in \{schema, documentation, NFP, contract, provenance\}$, w_i and $score(category_i) \in [0..1]$ are a weighted factor (i.e., its relevance for interpretability evaluation) and the degree of available information of $category_i$, respectively. In our assumption, $score(category_i)$ can be obtained automatically, e.g., from the utilization of (document) analysis tools or from service information systems which collect, manage and rank such information.

For what concerns the evaluation of information about NFPs of a service, we propose the following QoD metrics:

Definition 2 (Completeness). *Completeness specifies the ratio of missing values of provided NFP information (NFP_p) respect to the expected minimum set (NFP_{min}).*

NFP_{min} includes all the NFPs that are considered relevant for service selection by the service developer (e.g., $NFP_{min} = \{availability, reliability, response\ time\}$).

$$Completeness = 1 - \frac{\|NFP_p \cap NFP_{min}\|}{\|NFP_{min}\|} \tag{2}$$

Definition 3 (Timeliness). *Timeliness specifies how current a NFP description is.*

Timeliness is evaluated based on the age of the NFP description and expected validation. Let $ExpectedLifetime$ be the expected lifetime of a NFP description whose age is Age. The following formula can be used:

$$Timeliness = 1 - min(\frac{Age}{ExpectedLifetime}, 1) \tag{3}$$

5.2   Filtering Service Information Using Quality of Data Metrics

Currently, most service composition tools do not support service information filtering based on QoD metrics and therefore service selection algorithms assume to have complete and clean service information. Actually, information about services is incomplete and noisy, as like many other types of information on the Web. Based on the above-mentioned QoD metrics, we illustrate two service information filters (i) based on all service documents or (ii) based on NFP descriptions. To implement the former by using the *Interpretability* metric, we have to set weighted factors and determine scores of different categories of service documents. While scores can be evaluated based on service information provided by service information systems, weighted factors are request-specific. To implement the latter, the following steps are proposed:

- Step 1: Extract NFP_{min} and $ExpectedLifetime$ from requests;
- Step 2: Evaluate QoD metrics (e.g., *Completeness* and *Timeliness*);
- Step 3: Establish filtering thresholds based on QoD metrics;
- Step 4: Eliminate services whose information does not meet the thresholds.

The above-mentioned information filters should be used before service selection and composition, either conducted by the developer or automatic tools, in order to support functional and non-functional matching only on relevant and high-quality service information. Note that the way to use these filters and how to combine them with other service selection and composition features are tool- and goal-specific.

6   Experiments

6.1   Reducing Services by Using the Interpretability Metric

We illustrate how the developer can detect irrelevant services by using *Interpretability* metric. Using `seekda!` Web Services portal a service developer can discover more than 150 Weather services[3]. We assume that `seekda!` is a service information system in our scenario. But due to the lack of well-structured documents in `seekda!`, for our experiments, we manually prepared information about the first 50 services returned by `seekda!` including service interface (e.g., WSDL file), information about documentation, availability, user rating, etc. For every service, we considered $score(schema) = 1$ as their schemas are basically a WSDL file. `seekda!` classifies service documentation to $\{none, partially, good\}$ which are equivalent to $score(documentation) = \{0, 0.5, 1\}$. We assumed $NFP_{min} = \{availability, reliability, responsetime\}$ whereas `seekda!` provides only `availability` and `response time`. Provenance information and service contract are missing.

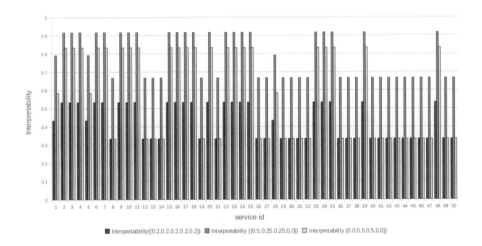

Fig. 2. Experimental values of the Interpretability metric for 50 weather services

Figure 2 describes different values of *Interpretability* metric based on different weighted factor sets: (i) all categories have the same weight (*Interpretability* $\{0.2,0.2,0.2,0.2,0.2\}$), (ii) only *schema*, *documentation* and NFP are considered and category *schema* is more important (*Interpretability* $\{0.5,0.25,0.25,0,0\}$), and (iii) only *documentation* and NFP are considered (*Interpretability* $\{0,0.5,0.5,0,0\}$). As shown in Figure 2 the values of *Interpretability* vary based on weighted factors reflecting different requirements, but in all three cases, less than a half of services have high *Interpretability* values (e.g., > 0.5). Consequently, the other half of services can be removed.

---

[3] Search `http://webservices.seekda.com` with the keyword `weather` on 20 June 2010.

6.2   Reducing Irrelevant Services by Qualifying NFPs

In this experiment, we show how to improve the service contract selection described in Section 2. We start from the experiment where 500 WSML service contracts are ranked to find the best service according to a user request using the PoliMaR framework[4].

The filtering conditions are established on the basis of the user request. In this experiment, the user is looking for a shipping service able to satisfy specified conditions on *payment method, payment deadline, insurance, base price* and *hours to delivery*. Moreover, we suppose that the user submits the request on 19 June 2010 and that she is interested in service information not older than 1 year. According to the user request, the *Completeness* and *Timeliness* metrics (see Section 5.1) are applied with the following parameters: (i) NFP_{min}={*payment method, payment deadline, insurance, base price, hours to delivery*} and (ii) $ExpectedLifetime = 1year$.

In order to perform the experiments, we implemented two filters: one selects/discards WSML contracts according to specified NFP_{min} and completeness threshold; the other selects/discards WSML contracts according to specified $ExpectedLifetime$ and timeliness threshold. We performed two different experiments[5] using an Intel(R) Core(TM)2 CPU T5500 1.66GHz with 2GB RAM and Linux kernel 2.6.33 64 bits. The first experiment analyzed the time required for ranking of 500 WSML contracts in the following cases: (i) without filters; (ii) applying a filtering phase on *Completeness*; (iii) applying a filtering phase on *Timeliness* and (iv) applying a filtering phase on *Completeness* and *Timeliness*. As an example, Table 4 reports the results of the experimentation for the following thresholds: *Completeness* $\geq$ 0.6 and *Timeliness* $>$ 0.2. Applying both the filters, we are able to halve the number of service contracts to be evaluated discarding irrelevant information with a reduction of the processing time equal to 52.8%.

Table 4. Results of applying Completeness (Filter 1) and Timeliness (Filter 2) filters

	Filter 1	Filter 2	Filtered Contracts	Filtering Time	Ranking Time	Total Time
Exp. 1	no	no	500	0 sec	37.5 sec	37.5 sec
Exp. 2	yes	no	309	2.7 sec	19.9 sec	22.6 sec
Exp. 3	no	yes	395	2.2 sec	25.9 sec	28.1 sec
Exp. 4	yes	yes	246	3.5 sec	14.2 sec	17.7 sec

The second experiment filtered and ranked 500 WSML contracts according to different *Completeness* and *Timeliness* thresholds. We consider the combination of the threshold values {0, 0.2, 0.4, 0.6, 0.8, 1} which are equivalent to {*not required, optional, preferred, strong preferred, required, strict required*}. Figure 3

---

[4] http://polimar.sourceforge.net/

[5] The dataset and the experimental results are available at http://siti-server01.siti.disco.unimib.it/itislab/research/experiments-for-wise10/

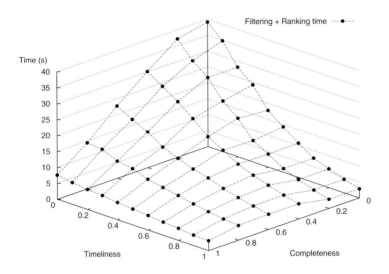

Fig. 3. Filtering and ranking time with different *Completeness* and *Timeliness* thresholds

shows the time required for filtering and ranking the 500 contracts. The following considerations emerge: (i) using equal threshold values for *Timeliness* and *Completeness*, the filter on timeliness results to be more selective; (ii) applying a filter on completeness with threshold equals to 0.2, no irrelevant contracts are discharged. This means that each contract in the dataset (i.e., 500 WSML contracts) contains at least one of the elements in NFP_{min}; (iii) applying a filter on *Timeliness* with threshold equals to 1, all the contracts are discarded.

7 Related Work

Information overloading problems can be considered as a particular case of irrelevant information problem that are mainly due to the sheer volume of information that one receives and processes. Ho and Tang [13] have studied information overloading problems in five industry cases and concluded that the three major causes are information quantity, information structure, and information quality. While certain solutions to deal with information overload are generic, such as filtering or eliminating sources of information given in [14], existing solutions are targeted to information overload in personal and business organizational environments. Other proposals (e.g., [15]) to reduce irrelevant information are based on the use of tag clouds. A tag cloud is a visual representation of user-generated tags, or simply the word content of a resource description. Tags are usually hyperlinks that are listed alphabetically and by popularity. In this case, the relevance of the information is not ensured since a tag cloud interpretation is based on user intuition. Moreover, it is not clear how tag clouds can be used to describing quality of service information.

The most common way to deal with irrelevant information problems in service-oriented computing is to apply data mining or semantic computing techniques. Chen and Cohen have discussed irrelevant service information [6] faced by developers and users in service composition by means of data mining techniques to classify and rank services. This approach, however, does not deal with dynamic and complex data sources, such as context and quality information. Semantic computing techniques, such as semantic annotation, similarity analysis, and concept analysis can be applied to improve service and data matching and selection processes [5,7,16]. Two limitations emerge in these techniques: (i) they concentrate on service interface and semantic meta-data about services, disregarding quality of data; (ii) they assume the availability of complete and clean service information so they cannot be used with unspecified/missing or incomplete information. This is very constrictive in particular for the approach in [16] which can require a wide set of information for the evaluation of applicability conditions associated with service contracts.

User context and quality of service (QoS) have been long considered as valuable source of information for supporting Web service design, discovery and composition [10,11,9,17]. Most related works using QoS and context to deal with irrelevant information in SOC can be divided into two classes: (i) techniques to increase the relevance of service discovery, such as QoS-based Web service discovery [9], and (ii) techniques to improve the relevance of information offered by the services based on users and their interaction with a service [10], user experiences [17], and personalized Web services [11]. However, in both classes existing works do not deal with QoD and they do not combine QoD, QoS and context together in order to improve the relevance of the service discovery results.

With respect to reducing irrelevant service information, one can argue that by using existing research efforts for service ranking [9] irrelevant services can be discarded since they are associated with a lower rank. However, most of existing efforts assume the existence of high-quality service information. Our work is different since we remove this assumption and we propose techniques to identify irrelevant information before service ranking in order to reduce the service information to be processed.

8   Conclusion and Future Work

The quality of information about services is crucial to identify and reduce irrelevant information and to support efficient service selection algorithms and service execution. To overcome the lack of appropriate way to manage quality and context in Web services, we proposed a research roadmap, centered around the development of context and quality models, APIs and evaluation techniques for information about services, for reducing irrelevant information during the composition and execution of (data-intensive) Web services. We have proposed a concrete solution based on some basic quality of information metrics and demonstrated the usefulness of these metrics in reducing irrelevant service information.

Currently, we focus on the extension of metrics and the development of techniques for context and quality compatibility evaluation and on the integration of our solutions into other service selection mechanisms and service composition and design tools.

References

1. Comerio, M., Truong, H.L., De Paoli, F., Dustdar, S.: Evaluating contract compatibility for service composition in the seco2 framework. In: Baresi, L., Chi, C.-H., Suzuki, J. (eds.) ICSOC-ServiceWave 2009. LNCS, vol. 5900, pp. 221–236. Springer, Heidelberg (2009)
2. Truong, H.L., Dustdar, S.: On Analyzing and Specifying Concerns for Data as a Service. In: Proceedings of The 4nd IEEE Asia-Pacific Services Computing Conference, APSCC 2009, Singapore, December 7-11. IEEE, Los Alamitos (2009)
3. Devillers, R., Jeansoulin, R.: Fundamentals of Spatial Data Quality. Geographical Information Systems series. ISTE (2006)
4. Al-Masri, E., Mahmoud, Q.H.: Discovering web services in search engines. IEEE Internet Computing 12, 74–77 (2008)
5. Paolucci, M., Kawamura, T., Payne, T.R., Sycara, K.P.: Semantic matching of web services capabilities. In: Horrocks, I., Hendler, J. (eds.) ISWC 2002. LNCS, vol. 2342, pp. 333–347. Springer, Heidelberg (2002)
6. Chen, Y., Cohen, B.: Data mining and service rating in service-oriented architectures to improve information sharing. In: IEEE Aerospace Conference, pp. 1–11 (2005)
7. Lamparter, S., Ankolekar, A., Studer, R., Grimm, S.: Preference-based selection of highly configurable web services. In: Proc. of the 16th International Conference on World Wide Web, WWW (2007)
8. Truong, H.L., Dustdar, S., Maurino, A., Comerio, M.: Context, quality and relevance: Dependencies and impact on restful web services design. In: Proc. of the Second International Workshop on Lightweight Integration on the Web, ComposableWeb 2010 (2010)
9. Ran, S.: A model for web services discovery with qos. ACM SIGecom Exchanges 4(1) (2003)
10. Pernici, B. (ed.): Mobile Information Systems: Infrastructure and Design for Adaptivity and Flexibility. Springer-Verlag New York, Inc., Secaucus (2006)
11. Maamar, Z., Mostefaoui, S.K., Mahmoud, Q.H.: Context for personalized web services. In: Proc. of the 38th Annual Hawaii International Conference on System Sciences (HICSS), Washington, DC, USA. IEEE Computer Society, Los Alamitos (2005)
12. Batini, C., Scannapieco, M.: Data Quality: Concepts, Methodologies and Techniques (Data-Centric Systems and Applications). Springer-Verlag New York, Inc., Secaucus (2006)
13. Ho, J., Tang, R.: Towards an optimal resolution to information overload: an infomediary approach. In: Proc. of International ACM SIGGROUP Conference on Supporting Group Work (GROUP), pp. 91–96. ACM, New York (2001)

14. Farhoomand, A.F., Drury, D.H.: Managerial information overload. ACM Commun. 45(10), 127–131 (2002)
15. Hassan-Montero, Y., Herrero-Solana, V.: Improving tag-clouds as visual information retrieval interfaces. In: InScit2006: International Conference on Multidisciplinary Information Sciences and Technologies (2006)
16. Garcia, J.M., Toma, I., Ruiz, D., Ruiz-Cortes, A.: A service ranker based on logic rules evaluation and constraint programming. In: Proc. of 2nd Non Functional Properties and Service Level Agreements in SOC Workshop (NFPSLASOC), Dublin, Ireland (2008)
17. Kokash, N., Birukou, A., D'Andrea, V.: Web service discovery based on past user experience. In: Abramowicz, W. (ed.) BIS 2007. LNCS, vol. 4439, pp. 95–107. Springer, Heidelberg (2007)

Propagation of Data Protection Requirements in Multi-stakeholder Web Services Systems[*]

Tan Phan[1], Jun Han[1], Garth Heward[1], and Steve Versteeg[2]

[1] Faculty of ICT, Swinburne University of Technology, Melbourne, Australia
{tphan,jhan,gheward}@swin.edu.au
[2] CA Labs, Melbourne, Australia
Steve.Versteeg@ca.com

Abstract. Protecting data in multi-stakeholder Web Services systems is a challenge. Using current Web Services security standards, data might be transmitted under-protected or blindly over-protected when the receivers of the data are not known directly to the sender. This paper presents an approach to aiding collaborative partner services to properly protect each other's data. Each partner derives an adequate protection mechanism for each message it sends based on those of the relevant messages it receives. Our approach improves the message handling mechanisms of Web Services engines to allow for dynamic data protection requirements derivation. A prototype is used to validate the approach and to show the performance gain.

Keywords: Security requirements conformance, performance-sensitive security provision, Web Services security, collaborative Web Services systems.

1 Introduction

In a multi-stakeholder Web Services system where services belong to different owners/stakeholders, one stakeholder often has only a limited view of the topology of the system. The party thus does not directly know every party that might have access to its data. However, each party in the system wants other (potentially unknown) parties to protect their data as the data flows through the system. Currently, WS-Security [1], which is the standard for security in Web Services systems, only allows for the end-to-end protection of data between known parties. Moreover, it is performance-wise expensive [2] due to the excessive overhead added by encryption and canonicalization operations performed on XML-based SOAP messages.

This paper proposes a mechanism to enable a given party to derive security requirements on data it receives and propagate such requirements to other parties with the flow of the data. This enables each party to apply a proper level of protection to data as required, without blindly over protecting the data. This paper is structured as follows. Section 2 presents a business scenario motivating the need for a mechanism

---

[*] This work is supported by the Australian Research Council and CA Labs.

L. Chen, P. Triantafillou, and T. Suel (Eds.): WISE 2010, LNCS 6488, pp. 67–76, 2010.

to propagate data protection requirements, which is followed by our proposed approach for addressing that need. The approach includes a method for modeling services and its data (Section 3), a set of techniques for ranking and combining different security protection mechanisms (Section 4), and a message processing algorithm for deriving data protection mechanism at each service (Section 5). Section 6 presents a prototype implementation and performance experiments. Related work is discussed in section 7, followed by conclusions.

2 A Motivating Business Scenario

The international bank **SwinBank (SB)** needs to do background checking for a newly employed staff member in Australia, who originally came from France. The bank contracts **BackgroundChecker (BC)**, based in Singapore, to do the background checking and forwards the employee's details *{firstName, lastName, email, passportNo}*. **BC** then contacts the target employee to collect his date of birth (*DOB*), a list of countries (*countries*) he has resided in and *scans* of his passport. There are two types of passport *scans*: *scan1* contains the main *passport* and *visa pages* which has information about the employee; *scan2* is the immigration clearance when the employee enters or leaves a country, which does not contain personal information. **BC** sub-contracts **PoliceChecker (PC)** to conduct a criminal history check for the employee and forward the employee's *{passportNo, fullNames, DOB, scan1, scan2}*. **PC** then contacts the national/federal police of the countries that the employee has resided, in this case France and Australia, to get the employee's police record. **BC** also sub-contracts another company **EmploymentChecker (EC)** for checking past employment and forwards *{passportNo, fullName, scan1, scan2}*.

SB is subject to the Australian Privacy Act which stipulates that a person's *name* and *passportNo* need to be kept confidential while **BC** is subject to the Singapore Privacy Act, which requires the completeness and accuracy (integrity) of personal information that it collects. In contrast, France prohibits the use of encryption that has a key length greater than or equal to 40 bits on message data, so that all transactions can be monitored[1]. While **BC** is required to respect all its clients' data protection requirements, **BC** also has an interest in reducing its system load and optimizing its processing for performance. When forwarding **SA**'s data to **PC** or **EC, BC** thus needs to know what protection mechanisms are required for such messages to optimize its performance while still respecting client's requirements. Therefore, for a service, there is the need for real time analysis and establishment of protection requirements for data based on individual messages and individual senders, as well as target receivers. The following sections present our approach to enabling each service to derive adequate protection mechanisms for output data based on those of the input data. The aim of our approach is to ensure that at each service, outgoing data will be protected with a security mechanism at least as secure as those used for protecting the related incoming data while not unnecessarily or blindly overprotecting the data.

---

[1] http://www.tc-forum.org/topicus/ru10tech.htm

3   Service Modeling

In this section, we present our models for data, messages and services in a service-based system. We model a service as a black box that is characterized by inputs, outputs and relationships between inputs and outputs. In particular, in our service model, we rely on a semantic mapping or annotation of input and output sets provided by each service component to be able to identify the corresponding input messages for a given output message. This is used for the purpose of determining protection requirements.

Definition 1: *Conceptually a service S is a tuple S = (I, O, MI, MO, R), where:*
$I = \{I_1,...,I_m\}$ is the set of input elements coming into S.
$O = \{O_1,...,O_n\}$ is the set of output elements S sends.
$MI = \{MI_1,...,MI_x\}$ is the set of *incoming messages* of S, containing data in I
$MO = \{MO_1,...,MO_y\}$ is the set of *outgoing messages* of S, containing data in O.
$R = \{R_1,...,R_z\} \subseteq O \times I$ is the set of dependency relations between O and I. Each relation R_i maps one output element $O_m \in O$ to a set of input elements $\{I_1,...,I_k\} \in I$. This implies O_m was derived (through some transformations) from a combination of $\{I_1,...,I_k\}$. The relationship is provided by developers and captures the *what* (i.e. which input elements are related to which output) but not the *how* (which mechanisms/techniques are used to transform input elements to output elements).

For correlating input and output messages related to a single request (e.g. all the background checking messages for a certain employee from **SB**), in our approach we use the standard content-based message correlation mechanism. In particular, based on a unique combination of some data elements (e.g. the *passportNo* of the employee in this case), different message related to a given service request are grouped together.

4   Deriving Data Protection Mechanisms

In this section we present our model for ranking and combining different data protection mechanisms (used in input messages) in deriving appropriate message protection mechanisms (for related output messages).

4.1   Message Confidentiality and Integrity

We model a protection mechanism PM as a combination of encryption (confidentiality) and digital signature mechanisms (integrity), $PM = enc \wedge sig$, where *enc* or *sig* can be empty. We define an encryption mechanism as a tuple $enc(b,ea,kl)$ where b is the encryption binding method, which can be symmetric (using a shared secret key) or asymmetric (using a key pair), *ea* is the encryption algorithm and *kl* is the length (number of bits) of the key used for encryption. For example, to ensure *confidentiality* for its customer's personal data, **SB** *encrypts* customer data with *asymmetric binding* using the *AES algorithm* and with a key length of 192 bits: $enc(asymmetric, AES,192)$. A digital signature mechanism is a tuple $sig(b,dga,dea,kl)$, where b is the binding method, which can be symmetric or asymmetric, *dga* is the digest (hash) generation algorithm, *dea* is the digest encryption algorithm, and *kl* is the length of the key used in *dea*. For example, to ensure the *integrity* of its data, **BC** *signs* its data following *asymmetric binding* using *SHA1* as the *dga algorithm, HMACSHA1* as the *dea algorithm,* and a key length of 192 bits: $sig(asymmetric, SHA1, HMACSHA1, 192)$.

The encryption and signature mechanisms are performed on sets of data elements D that need to be protected. These are represented as enc_D and sig_D respectively. The protection mechanism that includes enc_D and sig_D is represented as $PM_D = (enc \wedge sig)_D$. The absence of an encryption mechanism or a digital signature mechanism is denoted $\varnothing$. For example, **SB**'s mentioned encryption mechanism on its data is represented as $PM_{\{firstName, lastName, email, passportNo\}} = (enc(asymmetric, AES, 192) \wedge \varnothing)$.

4.2 Partial Ordering of Protection Mechanisms

We define the security protection mechanism partial ordering $(PM, \leq)$ in Definition 2 below to rank security protection mechanisms according to their strength.

Definition 2: *A protection mechanism* $PM_1 = (enc_1 \wedge sig_1)$ *is said to be as or more secure than another protection mechanism* $PM_2 = (enc_2 \wedge sig_2)$, $PM_1 \geq PM_2$ *if* $(enc_1 \geq enc_2) \wedge (sig_1 \geq sig_2)$. Our approach assumes the existence of a ranking function that can take two encryption mechanisms (or two digital signature mechanisms) and based on their binding mechanisms, their algorithms and the key lengths used, determine whether one is stronger than the other, or if they are not comparable. There has been some work that discusses the relative protection strength of different encryption algorithms and bindings such as [2] and [3], where different encryption mechanisms and key lengths are measured for the time and effort taken to break. However, at the moment, there is no universally agreed semantic for comparing the protection strength across multiple algorithms and binding methods.

4.3 Determining Data Protection Requirements for Output Messages

In our approach, we use the protection requirements for input messages to derive those of output messages. We give the following definition for combining different protection requirements from different sources.

Definition 3: *Given an output data element o, which is derived from the combination of a set of input elements* $I_1, \ldots, I_n$, *with each being protected by protection mechanisms* $PM_{I1}, \ldots, PM_{In}$, *respectively where* $PM_i = (Enc_i \wedge Sig_i), i = 1 \ldots n$, *the protection mechanism for o, denoted* PM_o, *is defined as*
$$PM_o = (Max(Enc_1, \ldots, Enc_n) \wedge Max(Sig_1, \ldots, Sig_n)).$$

This means that the combination of different protection mechanisms on a set of data is a new protection mechanism for which the encryption mechanism is the strongest of the elementary encryption mechanisms (i.e. those on input data) and the signature mechanism is the strongest of the elementary signature mechanisms. Here we assume that all elementary encryption mechanisms and digital signature mechanisms are comparable, applying Definition 3. If some mechanisms are not comparable, a maximum mechanism cannot be derived and thus an error is raised. This

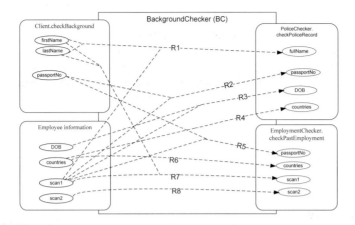

Fig. 1. BC's service model and protection requirements derivation

implies that to calculate the required protection mechanism for a given unit of data that is provided by different stakeholders, the strictest protection requirements, if determinable, from them will be selected.

Example: We apply the semantic of our protection mechanism derivation to the motivating example to illustrate our approach in different message transformation scenarios. When a set of messages comes to a service, data elements from them can be combined and mixed with data elements that the service sources internally, to create output messages.

- **Direct Mapping** is when an input element is directly carried forward to an output element without being combined with other elements or split (R8), the protection mechanism applied on the input element is combined with the default protection mechanism of the service, if any, to become the required protection mechanism on the corresponding output element. As can be seen in Figure 1, output element *checkPastEmployement.scan2* is directly forwarded through relationship R2 from input element *scan2* that **BC** sources from the employee directly. The outgoing *scan2* element will need to be digitally signed using a protection mechanism PM_{scan2} so that $PM_{scan2} >= (\varnothing \wedge sig(asymmetric, SHA1, HMACSHA1,192))$, applying the ranking in Definition 2.

- **Aggregation** is when multiple input elements are combined together to form a given output element (R1-R5, R7), we apply Definition 3 above to derive the required protection mechanism of the output element based on those of the corresponding input elements. For example, in Figure 1 the input element *scan1* is combined with input elements *firstName* and *lastname* to derive the output element *fullName*. Applying Definition 3, it can be seen that the resultant output element fullName has to be protected using a protection mechanism $PM_{fullName}$ so that it is stronger than or equal to the combined protection requirements of *firstName, lastName* (**SB**'s requirements) and *scan1* (**BC**'s requirements). That is: $PM_{fullName} >= (enc(asymmetric, AES,192) \wedge sig(asymmetric, SHA1, HMACSHA1,192))$. Note the difference between the eventual derived protection requirements for *scan1* (here under aggregation) and *scan2* (above under direct mapping) even though the two elements had the same original protection requirement from **SB**.

- **Splitting** is when an input element is carried over to multiple output elements *through* multiple relationships (e.g. *R*4, *R*6). In this case, we consider each resultant outgoing element independently and apply either the direct mapping or aggregation semantic above to derive the required protection mechanism for the output element.

5   Service Message Processing Algorithm

In this section, we present a message processing algorithm (Figure 2), utilizing the techniques discussed in previous sections, that can be used in a Web Service container (e.g. Axis2) and its security infrastructure (e.g. Apache Rampart) for data protection.

```
Inputs:
    I, service's set of input elements; O , service's set of output elements
    MI, service's set of input messages; MO, service's set of output messages
    R, the set of input, output correlation relations following Definition 2
    C, the correlation set for all messages in MI and MO
    M : ((c_p, i_q) ↦ pm_t), a map from a pair of correlation set value c_p ∈ C
        and an input element i_q ∈ I to a security protection mechanism pm_t
1  begin
       // process incoming messages
2      foreach incoming message (mi_k) ∈ MI do
3          identify the value for the correlation set c_mik for mi_k
4          identify the security policy and the policy alternative pa applied on mi_k
5          foreach input element i_l ∈ mi_k do
6              from pa, identify the protection mechanism pm applied on i_l
               // add an entry with key = pair of c_mik and i and value = pm into M
7              set M = M ⊕ {{c_mik, i_l}, pm}

       // process outgoing messages
8      foreach outgoing message mo_g ∈ MO do
9          create a protection mechanism pm_mog for mo_g which is initially = ∅
10         identify the value for the correlation set c_mog for mo_g
11         foreach output element o_h ∈ mo_g do
12             look up R to identify the set of input elements I_oh from which o_h was derived
               foreach input element i_r ∈ I_oh do
13                 using the pair c_mog and i_r, look up M to identify
14                     the protection mechanism pm_ir applied on M recorded before
                       // combine the existing protection mechanism of o_h with the
                       requirement of each of its input
15                 set pm_mog = pm_mog ⊎ pm_ir using Definition 3
                   // combine the protection requirements of inputs and the service's own
                   requirements
16             set pm_mog = pm_mog ⊎ service's protection requirements for mo_g,
17                 using Definition 3
18         foreach policy alternative for outgoing message pao do
19             compare the protection mechanism implied by pao and pm_mog, with respect to mo_g
20                 using Definition 2
21         if there exist mechanism(s) with adequate level of protection then
22             select the one with least over protection
23         else reject the sending of mo_g and raise an error
24     clear M
25 end
```

Fig. 2. Services' message processing algorithm

When an incoming message arrives at a service, the value of the (content-based) correlation id of the message of the service is first recorded. For example, when *checkBackground("John", "Smith", "jsmith@gmail.com, "N1000000")* from **SB** arrives at **BC***,* the correlation value (*"N1000000"*) is recorded (line 3). For each input element in the message, its protection mechanism is determined based on the security policy applied on the message (line 4, 5, 6). The identified protection mechanism is then stored in a map M, with a key being a pair of the message correlation value (*'N1000000'*) and the element itself (e.g. *firstName*) (line 7).

When a message is sent by the service, a protection mechanism for data in the message must be identified. First, an empty protection mechanism is initialized for the message (line 9). The value of the correlation id (in this case *'N1000000'*) for the message is then extracted. For each output element in the message, the set of input elements from which the output element was derived is determined by looking up the services' static dependency R (line 12). For example, when *checkPoliceRecord("N1000000", "John Smith", {"Aus", "FR"}, scan1, scan2)* is being sent by **BC**, the protection requirements for the element *"fullName"* are computed by looking up from M and combining the protection mechanisms for *"firstName"* and *"lastName"* and *"scan1"*. The protection requirement for each of these input elements is then identified from M using the message correlation value and the element name. These protection requirements are then combined (line 11-15). The protection requirement of the message is then computed by combining the requirements of each output element using Definition 3 (line 16). This computation results in $(enc(asymmetric, AES, 192) \wedge (enc(asymmetric, AES, 192))$ being the required protection mechanism for *"fullName"* as previously discussed in Section 4.3.

When the overall protection requirement is determined, an appropriate security channel for use is identified. This channel will be the one that satisfies all the protection requirements with the least level of over-protection (line 20). If there is no channel with an adequate level of protection available, the message will be rejected from being sent and an error will be reported. For example, using the presented algorithm, one of the derived protection requirements for the message *checkPoliceRecord("N1000000", "John Smith", "Jan 1 1980", scan1, scan2)* sent by **PC** to **FPC** to check for the police record of the employee in France is to apply an encryption mechanism at least as secure as $(enc(asymmetric, AES, 192) \wedge (enc(asymmetric, AES, 192))$ on *"passportNo"*, *"DOB"*, *"fullName"*, *"scan1"* and at least as secure as $(\varnothing \wedge sig(asymmetric, SHA1, HMACSHA, 192))$ on *"scan2"* . However, because **FPC** cannot support encryption mechanisms with key length >= 40, the message will be rejected from being sent and an error raised. This ensures that message data is never underprotected according to owner's requirements. The error raised is the result of a potential conflict in organization (and countries)'s security regulations and policies and might need to be resolved using an out of band negotiation mechanism, which is out of scope for this work.

6   Prototype Implementation and Performance Evaluation

In contrast to ensuring third-party message security by using maximum security all the time, our approach of dynamic security determination achieves higher efficiency.

We have implemented a prototype (Figure 3) for our approach based on Axis2 Web Service engine and the Apache Rampart WS-Security infrastructure. For this implementation, we adopt a conservative method of ranking protection mechanisms. In our implementation, encryption and signing mechanisms are compared on the basis of key length. That is, for two encryption mechanisms using the same algorithm and binding, one is considered stronger than the other when it uses a longer key length. We use the same semantic for digital signatures.

An Axis2 module called the *correlation handler* was implemented to handle correlation and message identification. The *correlation handler* allows for the storing and retrieval of relevant data elements based on some given correlation information. It stores them in a shared data pool accessible by inflow and outflow handlers so that outflow handlers can retrieve the protection requirements and message identification.

The protection mechanism ranking and combination techniques were deployed as an add-on to Rampart (we call it *Rampart++*). *Rampart++* uses the inflow and outflow message processing mechanism of Axis2 and the *correlation handler* to extract relevant details related to message protection mechanisms and correlation information from incoming messages. It then performs the necessary ranking and combinations of protection mechanism on data related to a given outgoing message to decide which security channel, if any, should be used for delivering the message. If no suitable channel is available, it prevents the message from being sent and raises an error. The message processing is transparent and independent to the service implementation.

We have evaluated the performance overhead and performance gain of using the prototype on a part of the motivating example system, which includes the three services **SB, BC**, and **PC**. During our experiment, we varied the size of the *scan1* and *scan2,* which **SB** sends **BC** and **BC** forwards to **PC**, from around 1kb to around 50 kb each to show the effect of different message sizes. There are three security channels for the connection between **BC** and **PC**, one with *full protection* (both encryption and signing are applied), one with only signing (*sign only*), and one with *no protection*. We changed the requirements of **SB** and **BC** during the experiment so that *Rampart++* will select one of the three channels, as appropriate.

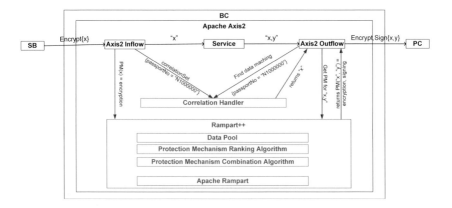

Fig. 3. Prototype architecture and deployment

The experiment results are shown in Figure 4, where "calculation" means the *correlation handler* and *Rampart++*are engaged while "no calculation" means the modules are disabled. In can be seen that the overhead incurred by our approach is relatively small (the columns for *"No calculation, Full security"* and *"Calculation, Full security"* are almost the same). It also shows that when our approach detects that full protection is not needed, but only a digital signature is needed (*"Calculation, Sign only"*), it can improve the response time by an average of around 33% compared with blind full protection. In the case when no security is needed (*"Calculation, No Security"*), the improvement is significant; the response time is on average only 17% of the response time for the case when full protection is blindly applied.

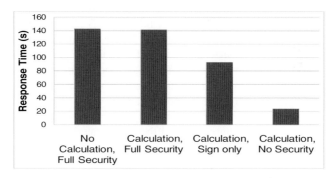

Fig. 4. Response time of **BC-PC** connection

7 Related Work

In traditional security modeling, work by Bell-LaPadula [4] and Biba [5] have set up the theoretical foundation for data protection. In this work, we implicitly apply the principles of "no-read- up" (from [4]) and "no-write-up" (from [5]) to provide data confidentiality and integrity. Existing work in the area of information flow security applied LaPadula and Bida principles for enforcing the secure data flow in programming languages, typically by introducing a typing system for data confidentiality and integrity. The major difference between our work and existing work in information flow security is that in such work data are explicitly labelled with "classifications" (clearance levels) while we do not explicitly model such clearance levels due to their complexity. Instead, we derive the implied clearance levels by analysing the protection mechanisms/levels that have been applied on the data. Moreover, existing work focuses on single-stakeholder systems whereby a unified set of principals and data classifications are assumed known. Our work supports multi-stakeholder distributed systems where each stakeholder only has a partial view of the system topology and it is not practical to construct a unified model of the system.

There have been a few research efforts on data protection in service-based systems. The SCIFC model in [6] is the most similar to our approach. SCIFC addresses the problem of access control for sensitive information when the information flows in a chain of Web Services. This is achieved by employing a combination of back-check procedures together with a mechanism of passing along digital certificates in themessage flow. The key difference between our work and SCIFC is that SCIFC focuses on preventing

76 T. Phan et al.

untrusted services from accessing sensitive data, while our work assumes collaborative partner services and focuses on preventing accidental under-protection of data by such partners. Moreover, the work of SCIFC is a theoretical model and no implementation was presented. The work of Hutter and Volkamer [7] provides a method for planning a service composition taking into consideration the restriction of data flow to elementary services of the composition. The approach in [7] relies on a known system topology with a centralized planner, which is different to our focus on multi-stakeholder systems.

8 Conclusions

With the assumption that parties collaborate in securing each other's data (but the communication channels between them can be attacked) our approach ensures that data will be adequately protected according to an accumulation of all stakeholders' requirements. Our processing mechanism ensures whenever possible, messages will not be over-protected. This provides a better balance between performance and security compared with blindly applying the strongest mechanism. Another benefit of our approach is that it does not require a known global system topology. It works in a peer-to-peer manner with each peer only needing to be aware of the peers it directly interacts with. The security requirements are then propagated recursively and dynamically together with the flow of data in the system.

The main contributions of our work presented in this paper include 1) a method for calculating the protection requirements for a service's output messages from those of input messages and the service's own requirements and 2) a mechanism for propagating the requirements together with the flow of the data in the system. In future work, we will investigate the application of our approach in service-based systems that are not implemented using Web Services.

References

1. Nadalin, A., Kaler, C., Monzillo, R., Hallam-Baker, P.: Web Services Security: SOAP Message Security (WS-Security) 1.1. OASIS Standard (2006)
2. Yau, S.S., Yin, Y., An, H.G.: An Adaptive Tradeoff Model for Service Performance and Security in Service-Based Systems. In: International Conference of Web Services 2009, ICWS 2009, pp. 287–294. IEEE Computer Society, Los Angeles (2009)
3. Lenstra, A.K., Verheul, E.R.: Selecting cryptographic key sizes. Journal of cryptology 14, 255–293 (2001)
4. Bell, D., LaPadula, L.: Secure computer systems: Mathematical foundations. Technical Report MTR-2547 (1973)
5. Biba, K.: Integrity considerations for secure computer systems. Storming Media (1977)
6. Cheng, W.W.-Y.: Information Flow for Secure Distributed Applications. Department of Electrical Engineering and Computer Science, Doctor of Philosophy, p. 179. Massachusetts Institute of Technology (2009)
7. She, W., Yen, I.-L., Thuraisingham, B., Bertino, E.: The SCIFC Model for Information Flow Control in Web Service Composition. In: International Conference of Web Services 2009, ICWS 2009, pp. 1–8. IEEE Computer Society, Los Angeles (2009)
8. Hutter, D., Volkamer, M.: Information flow control to secure dynamic web service composition. In: Clark, J.A., Paige, R.F., Polack, F.A.C., Brooke, P.J. (eds.) SPC 2006. LNCS, vol. 3934, pp. 196–210. Springer, Heidelberg (2006)

Refining Graph Partitioning for Social Network Clustering

Tieyun Qian[1], Yang Yang[2], and Shuo Wang[2]

[1] State Key Laboratory of Software Engineering, Wuhan University
[2] International School of Software, Wuhan University
16 Luojiashan Road, Wuhan, Hubei, 430072, China
qty@whu.edu.cn, sherlockbourne@gmail.com,
wangshuokevin@gmail.com

Abstract. Graph partitioning is a traditional problem with many applications and a number of high-quality algorithms have been developed. Recently, demand for social network analysis arouses the new research interest on graph clustering. Social networks differ from conventional graphs in that they exhibit some key properties which are largely neglected in popular partitioning algorithms. In this paper, we propose a novel framework for finding clusters in real social networks. The framework consists of several key features. Firstly, we define a new metric which measures the small world strength between two vertices. Secondly, we design a strategy using this metric to greedily, yet effectively, refine existing partitioning algorithms for common objective functions. We conduct an extensive performance study. The empirical results clearly show that the proposed framework significantly improve the results of state-of-the-art methods.

Keywords: graph partitioning, small world property, network clustering.

1 Introduction

Graph partitioning is a fundamental problem, with applications to many areas such as parallel computing and VLSI layout. The goal of graph partitioning is to split the nodes in graph into several disjoint parts such that a predefined objective function, for example, ratio-cut, or normalized cut, is minimal. The optimal graph partitioning is NP-complete and a variety of heuristic or approximation algorithms have been proposed to solve this problem [1], [2], [3], [4], [5].

In recent years, with the increasing demand of social network analysis, cluster detection, or community discovery, arouses new research interest [6], [7], [8], [9]. Social networks show some unique properties different from those in conventional graphs. Among them, small-world property is a key feature of real life social networks. For example, new members in a club tend to be introduced to other members via their mutual friends. Topology in graphs derived from underlying meshes in parallel systems is to some extent regular and the degree distribution is comparatively uniform. In contrast, the neighborhood of a node in social networks is often densely

L. Chen, P. Triantafillou, and T. Suel (Eds.): WISE 2010, LNCS 6488, pp. 77–90, 2010.

connected and the entire network exhibits a strong characteristic of small-world. The small-world feature suggests the presence of strong local community structure and will surely affect the performance of clustering algorithm. Unfortunately, this key property is not considered in most of existing partitioning methods.

In this paper, we focus on the problem of graph clustering in social networks. In order to leverage the small-world property, we present a novel framework for enhancing the community quality obtained by popular partitioning algorithms. This framework is independent of the original approach and can be integrated with various types of clustering algorithms. In this paper, we choose three typical partitioning techniques, i.e., Metis, Fast Modularity, and Kernel K-Means, as the baseline methods for comparison.

The main contributions of the paper are:

- A new metric is defined to measure the small world strength between two vertices.
- A greedy yet effective strategy is presented to refine current partitioning methods.
- A system evaluation on several real world datasets show that the proposed approach can significantly improve the objective functions.

The rest of this paper is organized as follows: Section 2 reviews related work. Section 3 describes the basic notions and notations. Section 4 defines a metric for measuring the small-world weight. Section 5 presents an algorithm for refining existing partitions of graph. Section 6 provides experimental results. Finally, Section 7 concludes the paper.

2   Related Work

A number of approaches have been developed for graph partitioning problem. The methods are categorized into three principle themes, i.e., geometric techniques, spectral methods, and multi-level methods. The geometric ones partition the graph based on coordinate information by putting vertices that are spatially near each other into one cluster. This kind of methods works only when coordinate information is available. The spectral methods are recently widely used in a number of fields including computer vision and speech segmentation [2], [5]. Spectral clustering relies on computing the first k eigenvectors corresponding to the Laplacian or some other matrix. Spectral approaches are in general computationally demanding and it is infeasible to compute the eigenvectors for large graphs. The third theme is based on multi-level schemes [3], [10], [11], which are considered as the state-of-the-art technique due to its high speed and quality. Multi-level partitioning methods consist of three main phases, i.e., coarsening phase, initial partitioning phase, and partition refinement phase. A sequence of successive approximations of the original graph is obtained in coarsening. And partitioning is performed on the smallest graph using recursive bisection methods. Finally, refinement on un-coarsened graphs is executed to gain the objective function by swapping pairs of nodes for Kernighan-Lin (K-L) method [12] or annotating each vertex along the partition for Fiduccia-Mattheyses (F-M) method [13]. As there is no strategy for guiding the search process, these two refinements are very time-consuming, i.e. $\Theta(n^2 \log n)$ for K-L and $\Theta(n*\log n + e)$ for F-M method, where n and e are the number of nodes and number of edges in the graph. Recently, A. Abou-Rjeili and G. Karypis [10] examine the power-law property of real networks develop new strategies for graph clustering.

For most graph clustering methods, it is necessary to specify the number of clusters as input. However, people know little about the graph structure in advance. One solution to this problem is hierarchically clustering the groups into multi-level structure by merging the sub-clusters with the largest similarity, and let the user choose the best representation, as what agglomerative algorithms do [14]. Fast modularity algorithm [6] is another kind of hierarchically clustering algorithms by merging two clusters that increases a quality function named modularity Q. The Q value in each iteration is recorded and the largest one is chosen for deciding the best partitioning. This algorithm thus has the privilege that no input parameter or human interaction is required. The optimization of modularity is also NP-complete. Many modularity-based approaches have been presented along this line, with various optimization or search strategies such as simulated annealing [7], [15], [16], genetic algorithms [17], external optimization [18]. By embedding graphs in Euclidean space [8] or replacing the Laplacian matrix with the modularity matrix [19], spectral methods can also be used for modularity optimization. An intrinsic limitation of modularity based algorithms is that they are not appropriate for finding partitions whose sizes are diverse.

In summary, the problem of community detection is typically NP-hard and a lot of heuristics or approximation algorithms have been presented to optimize the objective function. In contradiction to the large amount of clustering methods, there is less research interest on partition refinement. Kernighan-Lin [12] and Fiduccia-Mattheyses [13] are two basic methods for refinement phase. The main drawbacks of these methods are the high time complexity. Thus they are not appropriate for enhancing large-scale networks. Most importantly, they are not developed specifically for social networks with unique properties. Though the approach presented in [10] utilizes the power-low feature of networks, it is only designed for multilevel partitioning, and cannot be added on the top of other partitioning methods like modularity and spectral clustering.

3 Problem Statement

The problem of graph clustering is to assign the vertices in a graph G into k disjoint groups such that a predefined quality function is optimized. Formally, given a graph $G=(V, E)$, where V is the vertex set and E is the edge set. The problem consists of dividing the vertices in G into k disjointed partitions $P_1 = (V_1, E_1), P_2 = (V_2, E_2),..., P_k = (V_3, E_3)$ such that $V_i \cap V_j = \phi$, $\bigcup_{i=1}^{k} V_i = V$.

There have been a lot of metrics such as min-max cut [20] and ratio cut [21] proposed to measure the quality of graph clustering in literature. Among them, ratio cut, normalized cut [2], and modularity [6] are most commonly used.

Assuming a graph G is divided into k disjointed partitions $P_1, P_2,...,P_k$. P_i is one of clusters and $G \backslash P_i$ is the rest of graph. The normalized cut is often used in image segmentation. It is defined as following:

$$NC = \sum_{i=1}^{k} \frac{|cut(P_i, G \setminus P_i)|}{d_{Pi}}$$
(1)
,

Where d_{Pi} is the total degree of sub-graph P_i, $cut(P_i, G \setminus P_i)$ is the number of edges lying between P_i and the rest of graph.

The ratio cut is often used in the domain of circuit partitioning. It is defined as:

$$RC = \sum_{i=1}^{k} \frac{|cut(P_i, G \setminus P_i)|}{n_{Pi} n_{G \setminus P_i}},$$
(2)

Where n_{Pi} and $n_{G \setminus Pi}$ is the sum of degrees of vertices in P_i and the rest of graph, respectively.

Both of these two metrics are proportional to the number of cut edges. The difference lies in the denominator. The smaller value of normalized cut or ratio cut, the higher quality of graph partition.

Recently, M. Newman presents a new function named modularity to judge the goodness of a graph partition. It is defines as:

$$Q = \sum_{i=1}^{k} (\frac{|e_{P_i}|}{|e_G|} - (\frac{|d_{P_i}|}{|2e_G|})^2)$$
(3)

Where e_{Pi} is the number of internal edges in P_i and e_G the total number of edges in G. The larger modularity close to 1 means a good partition.

4 Measuring the Small World Weight between Two Nodes

Traditionally, for an un-weighted graph, the connection of two nodes is either 1 or 0. While for a weighted graph, the connection strength is represented as the edge weight. Formally, the edge weight of node pair (i, j) in G is defined as:

$$w_{ij} = \begin{cases} 1, \text{ if i and j are connected in an unweighted graph} \\ 0, \text{ otherwise} \end{cases}$$
(4)

$$w_{ij} = \begin{cases} e_{ij}, \text{ if i and j are connected in a weighted graph} \\ 0, \text{ otherwise} \end{cases}$$
(5)

The adjacency information has been fully utilized in many existing graph partitioning methods to group nodes which are highly connected. However, in many applications, especially in social networks, the states of neighbors also have influence on the node pair. For example, if all the neighbors of one node i are connected with another node j, then node i and node j will have tighter connections, as if the neighbors are pulling them together. In contrast, the connection between node i and node j will become looser if all the neighbors of i are disconnected with j, as if the neighbors are pushing them away.

Figure 1 illustrates such phenomena.

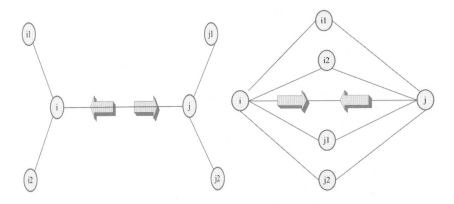

Fig. 1. Implications from nearby points

From Fig.1, we can see that besides the explicit relation between two nodes, there are also some implications from nearby points that will join or separate two vertices. Based on this observation, we propose a metric named small-world weight to measure such implications from neighborhood.

The small-world weight s_{ij} of node pair (i, j) is defined as follows:

$$s_{i->j} = \frac{\sum_{k \in \Gamma(i) \cap \Gamma(j)} w_{kj}}{\sum_{k \in \Gamma(i)} w_{ki}}, \ s_{j->i} = \frac{\sum_{k \in \Gamma(i) \cap \Gamma(j)} w_{ki}}{\sum_{k \in \Gamma(j)} w_{kj}}, \tag{6}$$

$$s_{ij} = s_{i->j} + s_{j->i} \tag{7}$$

Where Γ_i and Γ_j is the neighbour hood of node i and node j respectively.

5 Algorithm for Refining Network Clustering

5.1 A Total Order for Generating the Sorted List of Node Pairs

A general description of the goal of graph clustering is that the edges are dense within a partition but between which they are sparse [6]. To this end, most existing techniques aim at reducing the cut edges between different clusters. As we have illustrated in previous section, this strategy may not work well for complex networks like social networks. In this paper, we present a three-dimensional criterion to define the connection strength for node pair. As the edge and small-world weights reflect the direct and indirect relations existing between nodes, we set them as the first and second dimension accordingly. We also add the hub weight as the third dimension. The rationale is to promote the hub nodes. The hub value of node pair (i, j) is defined as follows:

$$h_{ij} = \max(d_i, d_j),\qquad(8)$$

Where d_i and d_j is the degree of node i and node j respectively.

Combining these three dimensions together, we now define a total order on the node pairs. This is used in measuring the strength of connection between node pair.

Definition: Given two pairs of nodes (i_1, j_1) and (i_2, j_2), we say that $(i_1, j_1) \succ (i_2, j_2)$ or (i_1, j_1) has a higher precedence than (i_2, j_2) if:

1. the edge weight of (i_1, j_1) is greater than that of (i_2, j_2), or

2. the edge weights are the same, but the small-world weights of (i_1, j_1) is greater than that of (i_2, j_2), or

3. both the edge weights and the small-world weights of (i_1, j_1) and (i_2, j_2) are the same, but the hub value of (i_1, j_1) is greater than that of (i_2, j_2), or

4. all the edge weights, the small-world weights, and the hub values of (i_1, j_1) and (i_2, j_2) are the same, but node id i_1 is greater than that of i_2.

5.2   The Refinement Algorithm

Let L be the order list of node pairs and π the partitioning by running any graph clustering methods. The basic idea of the algorithm is to sequentially merge the node pair in L into one cluster if the following two conditions are satisfied: (1) they are currently separated, and (2) the merge operation will enhance the quality function. The steps for refining network clustering are shown in Algorithm 1.

```
Algorithm  1  Algorithm  for  refining  network  clustering
(RENEC)
Input: a graph G=(V,E), π₀={P₁,P₂,...Pₖ}: the initial parti-
tion of G by a graph partitioning method, ω: a prede-
fined objective quality function, L the sorted list of
node pairs in G.
Output: πᵢ: the refined partition of G.
Method:
1.   Calculate ω₀ of π₀
2.   i = 0
3.   while L≠φ do
4.       Fetch the first node pair (x,y) in L
5.       Get the cluster labels lₓ and lᵧ in ωᵢ
6.       moved = false
```

```
7.      if  l_x ≠ l_y
8.          Calculate the temporary  ω_{t1}  by merging  x  to  P_{l_y}
9.          Calculate the temporary  ω_{t2}  by merging  y  to  P_{l_x}
10.         if  ω_{t1} > ω  or  ω_{t2} > ω
11.             ω_{i+1} = max(ω_{t1}, ω_{t2})
12.             π_{i+1} = π_{t1} if ω_{t1} > ω_{t2} ,  otherwise  π_{i+1} = π_{t2}
13.             moved = true
14.     if moved = false
15.         ω_{i+1} = ω_i
16.         π_{i+1} = π_i
17.     L = L − {(x, y)}
18.     i = i+1
19. Return  π_i
```

In Algorithm 1, line 1 calculates the quality function value of initial partition. Line 2 initiates iteration variable. Lines 4-5 fetch the first node pair (x, y) in L and get their cluster labels. Line 6 initiates a boolean variable to judge whether a moving operation actually happens. If the two vertices of the node pair are in separated clusters, then calculate the temporary quality function values $ω_{t1}$ and $ω_{t2}$ by merging x to y's cluster or merging y to x's cluster in lines 8-9. If either of these changes makes improvement, then lines 11-12 save the merging operation with larger value of objective functions, and line 13 sets the boolean variable. If the two nodes are in the same cluster, or the moving operation can not make any enhancement, lines 15-16 remain current partitions. Line 17 removes the handled pair from the ordered list L. Line 18 update iteration variable. This procedure repeats until all pairs have been tested. Line 19 returns the final refined partitioning.

The objective function $ω$ in our algorithm can be of any quality metrics for graph partitioning. Specifically, in this paper, we adopt three kinds of functions, i.e., normalized cut, ratio cut, and modularity. In addition, the initial partition of G can be made by *any* clustering methods such as spectral clustering and multilevel partitioning.

6 Experimental Results

6.1 Experimental Datasets

6.1.1 Datasets

We have conducted experiments on 4 typical social network data sets. The first one is the subset of Cora citation dataset [22] in the field of machine learning research. The second one is the CA-hepth collaboration network [23]. The third one is the Enron email communication network [24]. The forth one is the Epinions online interaction network [25].

Table 1 shows the number of nodes and edges of the datasets after pre-processing.

Table 1. Summary of data sets

dataset	|nodes|	|edges|
Cora-7ml	27891	65755
CA-hepth	9877	25975
Enron	36692	183831
Epinions	75879	405740

6.1.2 The Network Properties

Small-world networks have a small diameter and a high clustering coefficient. These properties as well as the density of the studied datasets are shown in Table 2. The Enron network has the smallest diameter and largest clustering coefficient, exhibiting a significant small-world feature. The Cora-7ml network has the smallest value of clustering coefficient and the lowest density. The clustering coefficient of Epinions network is also relatively small. However, the density of this graph is very large. The diameter of CA-hepth network is large.

Table 2. Properties of the studied networks

dataset	diameter	clustering coefficient	density
Cora-7ml	7.413	0.1261	2.36
CA-hepth	8.866	0.4714	2.63
Enron	5.978	0.4970	5.01
Epinions	6.329	0.1378	5.35

6.2 Improvements over FM Method

Fast modularity (FM) approach [6] is one of the most popular algorithms for community discovery. It does not require any hypothesis on the number of clusters. The objective in FM is modularity Q.

We first run FM algorithm on three datasets, and then apply the proposed RENEC algorithm. Figure 2 shows the improvements of Q values. The enhancements of modularity on Cora-7ml, Ca-hepth and Enron datasets are not very significant. Modularity value Q is often used to test whether a particular division is meaningful [6]. Q values greater than about 0.3 appear to show significant community structure. Initial Q values for these three datasets indicate that all these networks have already split quite cleanly into groups, and hence smaller enhancements can be made. The proposed method improves Q value for Epinions dataset from 0.358 to 0.409, reaching an improvement ratio of 14.25%.

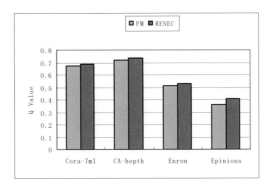

Fig. 2. Improvements of *Q Value* of RENEC over FM

6.3   Improvements over Metis

We choose Metis algorithm [3] as the second baseline clustering method, because it provides high quality partitions better than those produces by spectral partitioning algorithms, also because it is fast and requires less memory. The objective of Metis is to minimize the edgecut. Hence we set Normalized Cut (NC) (RC) or Ratio Cut as the objective functions to verify whether they can be further enhanced by our RENEC algorithm.

It is necessary to pre-assign the number of clusters, say k. A commonly used setting for this parameter is the square root of the number of nodes. We will investigate the sensitivity to the parameter k later.

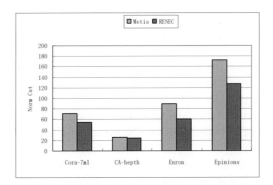

Fig. 3. Decrease of *Normalized Cut* of RENEC over Metis

Figure 3 and Figure 4 show the improvements of our RENEC over Metis in terms of Normalized Cut value and Ratio Cut value, respectively. Note that for these two evaluation metrics, the smaller, the better.

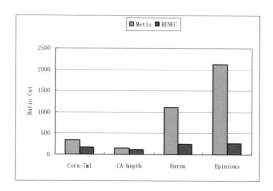

Fig. 4. Decrease of *Ratio Cut* of RENEC over Metis

The reduction of normalized cut on Enron is the biggest. It dramatically decreases from 89.02 to 61.27, a 31.17% improvement. The result on Epinions is also very attractive, reaching 26.92%. Despite of the diverse absolute ratio cut values for Metis, namely, 2120.75 for Epinions and 142.79 on Ca-hepth graph, the reductions of ratio cut are very similar to those of normalized cut. The reduction of ratio cut on Enron and Epinion reaches 77.40% and 87.43, respectively. The decrease on Ca-hepth and Cora-7ml datasets in these figures seems less significant. Please note that this is because of the different scale. For example, in Figure 4, the ratio cut for Cora-7ml dataset drops from 336.94 to 173.76, reaching a change ratio of 48.43%.

From Fig. 3 and Fig. 4, we can clearly see that, in all 4 networks, the proposed RENEC approach can significantly decrease the normalized cut and ratio cut obtained by Metis algorithm.

6.4 Improvements over Kernel K-Means Method

The third baseline clustering method is Kernel K-Means approach [26], which performs well for the task of linked-based clustering. The normalized cut and ratio cut are adopted as evaluation metrics.

Figure 5 and Figure 6 show the improvements of our RENEC over K-K-Means in terms of Normalized Cut value and Ratio Cut value, respectively. Again, the proposed RENEC approach can significantly decrease the objective functions obtained by K-K-Means algorithm. Take the Epinions online interaction network as an example, the decrease of Ratio Cut of RENEC over K-K-Means is very impressive. It drops from 5687.42 to 241.37, reaching a change ratio of 95.76%.

We notice that Ca-hepth dataset has the least reduction of normalized cut and ratio cut. The main reason can be that both Metis and K-K-Means methods work well for this kind of network, given the fact that the original function values on Ca-hepth are already the smallest among 4 graphs.

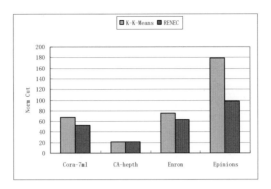

Fig. 5. Decrease of *Normalized Cut* of RENEC over K-K-Means

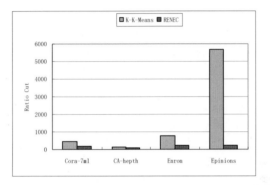

Fig. 6. Decrease of *Ratio Cut* of RENEC over K-K-Means

6.5 Sensitivity to the Number of Clusters

In this subsection, we evaluate the effect of the parameter k on the performance of the refinement approach by varying the input number of clusters. We set k to 5, 10, 15, 20, and 25, respectively.

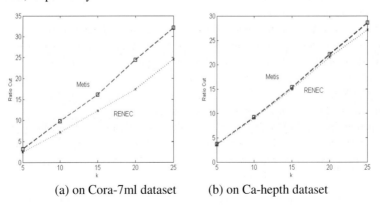

(a) on Cora-7ml dataset   (b) on Ca-hepth dataset

Fig. 7. Sensitivity of *Ratio Cut* to the number of clusters on Cora-7ml and Ca-hepth datasets

From the results in Figure 7 and Figure 8, both the values of ratio cut and the effects of the refinement algorithm increase with the growth of the parameter k on 4 data sets. This is reasonable as larger k will bring larger number of cut edges, as well as more chance for changing the destinations of the nodes.

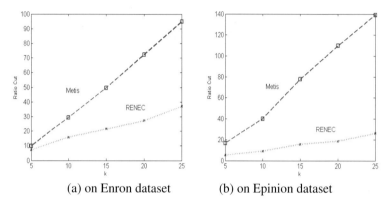

(a) on Enron dataset   (b) on Epinion dataset

Fig. 8. Sensitivity of *Ratio Cut* to the number of clusters on Enron and Epinion datasets

The benefit of using the proposed framework is clearly demonstrated in this experiment. Note that we also observe the similar improvements in terms of normalized cut and modularity on all other datasets. We do not present these results here only due to space limits.

6.6 Time Efficiency

Table 3 shows the timing results for the proposed method. It can be seen that the algorithm is very fast. All experiments were performed on a PC with Core 2.4 GHz CPU and 4G byte of memory running the Windows Vista Operating system. Please note that due to the memory limit, we implement the experiments using the data structure of adjacent list. If other data structure such as matrix is used, the time efficiency for our method can be further improved.

Table 3. Time efficiency of the proposed RENEC approach

dataset	k	CPU Execution Time in Seconds	
		time for sorting	time for refining
Cora-7ml	167	17	2
CA-hepth	99	3	0.2
Enron	192	118	12
Epinion	275	608	40

7 Conclusion

The small-world feature is one of the most important properties for social networks. Existing algorithms for graph partitioning do not take this key property into consideration. This paper presents a refinement approach for adjusting the current graph partitioning methods. By incorporating the small-world and hub weight with the edge weight, we successfully define a three dimensional total order on edges. The ordered list is further used to merge the node pairs into one cluster which are currently separated by the original partitioning method. The extensive evaluation on several real-world social network datasets demonstrates the effectiveness of the proposed approach over other state-of-the-art techniques.

Acknowledgments. This research was supported in part by the 111 Project (B07037), the NSFC Project (60873007), the NSF of Hubei Province (2008CDB340), Specialized Research Fund for the Doctoral Program of Higher Education, China (20090141120050).

References

1. Bui, T., Jones, C.: A heuristic for reducing fill in sparse matrix factorization. In: 6th SIAM Conf. Parallel Processing for Scientific Computing, pp. 445–452 (1993)
2. Shi, J., Malik, J.: Normalized Cuts and Image Segmentation. In: Proc. of CVPR, pp. 731–737 (1997)
3. Karypis, G., Kumar, V.: A fast and highly quality multilevel scheme for partitioning irregular graphs. SIAM Journal on Scientific Computing 20(1), 359–392 (1998)
4. Leighton, F., Rao, S.: Multi-commodity max-flow min-cut theorems and their use in designing approximation algorithms. J. ACM 46(6), 787–832 (1999)
5. Ng, A., Jordan, M., Weiss, Y.: On spectral clustering: analysis and an algorithm. In: Advances in Neural Information Processing Systems, vol. 14, pp. 849–856. MIT Press, Cambridge (2001)
6. Newman, M.: Fast algorithm for detecting community structure in networks. Phys. Rev. E 69 art. (066133) (2004)
7. Guimerà, R., Sales-Pardo, M., Amaral, L.A.N.: Modularity from fluctuations in random graphs and complex networks. Phys. Rev. E 70(2), 025101 (R) (2004)
8. White, S., Smyth, P.: A spectral clustering approach to finding communities in graphs. In: Proc. of SIAM International Conference on Data Mining, pp. 76–84 (2005)
9. Tang, L., Wang, X., Liu, H.: Uncovering Groups via Heterogeneous Interaction Analysis. In: Proc. of ICDM, pp. 503–512 (2009)
10. Abou-rjeili, A., Karypis, G.: Multilevel Algorithms for Partitioning Power-Law Graphs. Technical Report, TR 05-034 (2005)
11. Hauck, S., Borriello, G.: An evaluation of bipartitioning technique. In: Proc. Chapel Hill Conference on Advanced Research in VLSI (1995)
12. Kernighan, B.W., Lin, S.: An efficient heuristic procedure for partitioning graphs. The Bell System Technical Journal 49(2), 291–307 (1970)
13. Fiduccia, C.M., Mattheyses, R.M.: A linear time heuristic for improving network partitions. In: Proc. 19th IEEE Design Automation Conference, pp. 175–181 (1982)
14. Hastie, T., Tibshirani, R., Friedman, J.H.: The Elements of Statistical Learning. Springer, Berlin (2001)
15. Massen, C.P., Doye, J.P.K.: Identifying communities within energy landscapes. Phys. Rev. E 71(4), 46101 (2005)
16. Medus, A., Acuña, G., Dorso, C.O.: Detection of community structures in networks via global optimization. Physica A 358, 593–604 (2005)

17. Tasgin, M., Herdagdelen, A., Bingol, H.: Community detection in complex networks using genetic algorithms, eprint arXiv: 0711.0491
18. Duch, J., Arenas, A.: Community detection in complex networks using extremal optimization. Phys. Rev. E 72(2), 27104 (2005)
19. Newman, M.E.J.: From the cover: Modularity and community structure in networks. Proc. Natl. Acad. Sci. USA 103, 8577–8582 (2006)
20. Ding, C.H.Q., He, X., et al.: A min-max cut algorithm for graph partitioning and data clustering. In: Proc. of ICDM, pp. 107–114 (2001)
21. Wei, Y.-C., Cheng, C.-K.: Towards efficient hierarchical designs by ratio cut partitioning. In: Proc. of Intl. Conf. on Computer Aided Design, pp. 298–301. Institute of Electrical and Electronics Engineers, New York (1989)
22. McCallum, A., Nigam, K., Rennie, J., Seymore, K.: Automating the Construction of Internet Portals with Machine Learning. Information Retrieval Journal 3, 127–163 (2000)
23. Leskovec, J., Kleinberg, J., Faloutsos, C.: Graph Evolution: densification and Shrinking Diameters. ACM Transactions on Knowledge Discovery from Data (ACM TKDD) 1(1) (2007)
24. Leskovec, J., Kleinberg, J., Faloutsos, C.: Graphs over Time: Densification Laws, Shrinking Diameters and Possible Explanations. In: ACM SIGKDD International Conference on Knowledge Discovery and Data Mining (2005)
25. Richardson, M., Agrawal, R., Domingos, P.: Trust Management for the Semantic Web. In: Fensel, D., Sycara, K., Mylopoulos, J. (eds.) ISWC 2003. LNCS, vol. 2870, pp. 351–368. Springer, Heidelberg (2003)
26. Dhillon, I., Guan, Y., Kulis, B.: Weighted graph cuts without eigenvectors: a multilevel approach. IEEE. Transactions on PAMI 29(11), 1944–1957 (2007)

Fast Detection of Size-Constrained Communities in Large Networks

Marek Ciglan and Kjetil Nørvåg

Dept. of Computer and Information Science, NTNU, Trondheim, Norway
marek.ciglan@idi.ntnu.no

Abstract. The community detection in networks is a prominent task in the graph data mining, because of the rapid emergence of the graph data; e.g., information networks or social networks. In this paper, we propose a new algorithm for detecting communities in networks. Our approach differs from others in the ability of constraining the size of communities being generated, a property important for a class of applications. In addition, the algorithm is greedy in nature and belongs to a small family of community detection algorithms with the pseudo-linear time complexity, making it applicable also to large networks. The algorithm is able to detect small-sized clusters independently of the network size. It can be viewed as complementary approach to methods optimizing modularity, which tend to increase the size of generated communities with the increase of the network size. Extensive evaluation of the algorithm on synthetic benchmark graphs for community detection showed that the proposed approach is very competitive with state-of-the-art methods, outperforming other approaches in some of the settings.

1 Introduction

Many real-world data sets have the form of graphs. The most straightforward examples are those of various types of networks; e.g., information networks, citation networks, social networks, communication networks, transportation networks or biological networks. This wide variety of existing graph data has triggered an increased attention to the graph analysis and the graph mining within the research community. One of the prominent tasks in graph mining is that of the community detection. The community detection aims to discover the community structure of a graph by dividing nodes of graph to clusters, where nodes of the same cluster are densely linked among themselves and less densely linked to the nodes of other communities.

Real-world networks exhibit the community structures, along with other properties like power law degree distribution and small-world property. Although the notion of a graph community structure has not been formally defined and is still intuitive for the most part (a number of diverse definition can be found through the literature), research on the community detection has gained a momentum in recent years. The most widely used approach in community detection algorithms is to maximize the modularity measure of the graph partition [16]. The problem of the modularity based algorithms is the resolution limit [8] – identified communities get bigger as the size of the network increases.

L. Chen, P. Triantafillou, and T. Suel (Eds.): WISE 2010, LNCS 6488, pp. 91–104, 2010.

In this paper we first formalize the problem; we define affinity partition of the network, based on the notion of community as defined by Hu et al in [12]. We then propose a new greedy algorithm for community detection in graphs. It builds on the label propagation algorithm [19] and overcome its drawback, which is the collapse of the result into a single community when the community structure of the given network is not very clear. Our algorithm is pseudo-linear in the execution time and it is designed to allow a user to specify the upper size limit of the communities being produced. As we will illustrate later on a real-world example, the communities identified in large networks are often of large sizes. This can be very impractical for certain classes of applications; in case the identified community structure is used by human expert (e.g. identification of a social group for a user of a social network), the community of several hundred thousands of nodes is not very useful. Another example of usefulness of size-constrained community detection is the task of partitioning the graph data for a distributed graph database into k balanced components.

The ability to constrain the size of communities is a feature that distinguish our approach from other methods. It is especially suited to identify small-scale communities in large networks and can be viewed as a complementary approach to the methods maximizing the modularity. The main contribution of the paper is the new algorithm for community detection, which has following distinctive features:

- Allows user to specify upper size limit of the produced communities
- Fast in execution time; as it is pseudo-linear in time complexity, it can be used also for clustering of large graphs
- Achieves very good quality in unveiling community structure for networks with communities of small sizes and has results competitive with state-of-the-art methods (often with higher time complexity)

We first present related work in Section 2. In the Section 3, we describe measures useful for the evaluation of the goodness of the identified graph partition into communities and we discuss other existing fast algorithms for community detection. In Section 4 we propose a new algorithm for community detection. Section 5 presents the evaluation of our approach. We conclude the paper in Section 6 where we also provide directions for future research.

2   Related Work

The problem of the community detection has been quite popular in recent years and a large number of different algorithms addressing the problem were proposed. The interest in the problem boomed after introduction of divisive methods based on betweenness centrality measures, mainly the Girvan-Newman algorithm [9]. Same authors have later proposed modularity measure [16] – a quality function that can be used to evaluate how good a partition of the network is. The underlying idea is to compare how community-like the given partition is compared to a random graph with the same node degrees. This is perhaps the most influential work in the community detection research. A number of proposed methods rely on maximizing the modularity when identifying the community structure; e.g., greedy techniques [1] [4], simulated annealing based method [11] or extremal optimization [3]. As was shown in [8], the modularity function has a res-olution limit that prevents communities of small sizes (compared to the graph size) to be identified, even if they are clearly defined. Variants of basic community detection problem were studied in literature, e.g. clustering of bipartite graphs [17], identification of overlapping communities [10, 23] or clustering based on graph structure combined with additional content associated with nodes [22]. The most related to our work are fast, greedy algorithms for community detection [19, 1, 20]. We discus those works in more detail in Section 3. A number of other approaches to the community detection task has been proposed; an extensive overview of the methods and graph clustering related problems can be found in surveys on the topic [7, 18].

3   Preliminaries

In this section, we discuss the measures used for the evaluation of the goodness of iden-tified community structures, when the community structure is known, e.g. is artificially planted in the network by a benchmark graph generator. We than discuss in more detail three algorithms closely related to our algorithm proposed in Section 4.

Community detection formalization. Although there is no consensus on the for-mal definition of the community detection task, we try to explain the problem at hand in a more formal manner, following the approach in [7]. Let the G, be a graph $G = \{V, E, f\}$, where V is a set of nodes, E is a set of edges, $f : V \times V \rightarrow E$. The partition of the node set is $P(V) = \{C_1, C_2, \ldots, C_k\}$ where $\bigcup_{\forall C_i \in P(V)} C_i = V$ and $\bigcap_{\forall C_i \in P(V)} C_i = \emptyset$. Let $E_{C_i,in}$ be a set of intra-cluster edges of cluster C_i : $E_{C_i,in} = \{e_{k,l} \in E : k \in C_i \wedge l \in C_i\}$ and let $E_{C_i,out}$ be a set of inter cluster edges: $E_{C_i,out} = \{e_{k,l} \in E : (k \in V \setminus C_i \wedge l \in C_i) \vee (k \in C_i \wedge l \in V \; C_i)\}$. Let $|V| = n$ and $|C_i| = n_{C_i}$; we can define

$$\delta_{in}(C_i) = \frac{|E_{C_i,in}|}{n_{C_i} \times (n_{C_i} - 1)/2} \quad \text{and} \quad \delta_{out}(C_i) = \frac{|E_{C_i,out}|}{n_{C_i} \times (n - n_{C_i})}$$

The goal of community detection is then finding a partition with a good balance be-tween large $\delta_{in}(C_i)$ and small $\delta_{out}(C_i)$. We formalize out perception of the community structure in 4.1.

Comparing partitions of a network. Benchmark graph generators produce a defi-nition of network and a division of nodes to communities – the 'ground truth' partition of the given network. The community detection algorithm also produces a partition of the given network. The problem is how to evaluate how good the partition provided by a community detection algorithm approximates the original partition. In the community detection literature, an established similarity measure for comparing network partitions is *Normalized Mutual Information* (NMI). This measure originates from the informa-tion theory and was first adopted for the community detection by Danon et al. in [6]. Lancichinetti et al. proposed in [15] a modification of the NMI measure able to compare graph partitions with overlapping communities. As argued by authors, although it does not reproduce exactly the same values as NMI, it is close. We will refer to this mea-sure as cNMI. The cNMI was used as the similarity measure in comparative analysis of community detection algorithms in [14]. In order to be able to perform head-to-head comparison of our approach with those evaluated in [14], we use NMI as well as cNMI in our experiments.

Greedy approaches to community detection. In the following, we discuss greedy algorithms for community detection with the pseudo-linear execution time. We discuss three algorithms, the Label Propagation by Raghavan et al. [19], heuristic method for modularity optimization by Bondel et al. [1] and multi-resolution community detection algorithm using Potts model proposed by Ronhovde and Nussinov [20] (we will refer to this algorithm as RN). Algorithms are similar in their basic operational principle, where the label propagation is a basic approach, the two other can be regarded as the extension and modification of the label propagation mechanism. Label Propagation algorithm [19] is based on the greedy assignment of a node to the community which contains the most of its neighbors. If several communities contain the same highest number of n's neighbors, the ties are broken uniformly at random. The algorithm is initialized by assigning unique labels to all nodes in the network. Labels are propagated through the network in iterations in which the community membership of all the nodes are updated in random order. The process should continue until no node changes its label during iteration. As the convergence of such a greedy approach might be hard to prove, one may use the constraint on the iterations number to ensure that the algorithm will stop. Resulting communities are created from the nodes with the same labels. This approach, as the result of randomizations, does not have a unique solution. The algorithm might reach the stop criterion for multiple different partitions of the network. Authors also propose to aggregate multiple different partition of the network by creating new labels for nodes based on the labels they received in different runs and re-run the algorithm on that initial setting. The problem of label propagation approach is that it's solution often collapses into one single community, in case the boundaries between communities are not clearly defined. Processing similar to label propagation is used by Bondel et al. in [1], where the computation is done in two phases. The first phase is similar to label propagation, with the difference in the greedy step, where the authors choose the community to join based on the gain of modularity. The second phase of their algorithm consist of contracting partition into a new network. Those two phases are repeated iteratively until no gain in modularity can be achieved. Ronhovde and Nussinov in [20] use the processing of the label propagation style, with different decision function for changing the node's community membership. Their decision function, referred to as absolute Potts model (APM), can be parameterized to produce communities at different resolutions. Their proposed multi-resolution algorithm computes partition at different resolutions and they compute correlation among multiple partitions, identifying significant structures by strong correlations.

4   Size Constrained Greedy Community Detection

In this section, we propose a new community detection algorithm that allows a user to constrain the size of the communities being generated. It is named Size Constrained Greedy Community Detection algorithm (SizConCD). First, we describe the motivation for the work. We then provide the formalization of the problem, which correspond to our perception of the communities and the community structure (Subsection 4.1). and we provide the description of the proposed greedy algorithm.

Our work was motivated by the limitations of existing approaches to the problem of the community detection. In our view, there are the two major limitations. The first is the computational complexity of majority of methods which prevents them to be applied on large networks. The second is that algorithms with pseudo-linear computational time, which can be used on large networks, often produce partitions with very large communities. This is not very useful for the detailed analysis of a node (e.g. a community of 100 000 nodes is not particularly useful when one wants to identify social group of a user in a social network). Let us provide an illustrative example. We have tried to detect the community structure of the Wikipedia link graph. Our expectation was that semantically similar topics should be grouped together in communities. The size of the link graph extracted from Wikipedia XML dump was 3.1 million nodes and 91 million edges. The large size of the network limited our choice of a community detection algorithm only to those running in linear time. We have analyzed the link graph using Label Propagation [19] algorithm and the greedy modularity optimization [1]. The community structure produced by label propagation algorithm had the largest community with over 2.96 million of nodes. The size of the largest community in the partition identified by the greedy modularity optimization method was smaller (containing around 400 000); however, 20 largest communities of this partition comprised more than 95% of the nodes. In general, producing the partition with communities of limited sizes is useful for certain applications; e.g. when the community structure is analyzed by a human expert (inspecting social community of a user in social network, inspecting related concepts in semantic network). Another example of usefulness of size-constrained communities is the task of splitting the network into k groups, minimizing the number of edges connecting them. This task is useful for example for partition of the graph data for distributed graph database. This motivated us to develop a new algorithm for the community detection, which would allow us to constrain the community sizes.

4.1 Problem Formalization

Currently, there is no consensus on the formalization of the community detection problem. Several definitions can be found in the literature; often, the problem is not formalized at all and the definition of the task is provided as an informal, intuitive description.

In our work we adopt the definition of the community proposed by Hu et al., in [12]. Informally, every node of a community C should have higher or equal number of edges connecting it with other nodes of C, than number of edges connecting it with other communities. Let $G = (V, E, f, w)$ be a graph, with nodes V, edges E and function $f : V \times V \to E$ defining the mapping between nodes and edges and $w : E \to \mathbb{R}$ be the function defining the weights of the edges.

We will use the term *affinity* of a node n towards a cluster C to denote the sum of weights of edges connecting n with nodes of the cluster C:

$$aff(n, C) = \sum_{i \in C} w(e_{n,i}) : e_{n,i} \in E$$

Let us consider an example graph in Figure 1 and let all edges have the weight of 1. The affinity of n towards cluster A is 1 ($aff(n, A) = 1$); $aff(n, B) = 2$ and $aff(n, C) = 3$.

Based on work in [12], we define *affinity partition* of a graph to be $\gamma = C_1, C_2, .., C_m$, such that

$$\bigcup_{k \in \langle 1,...,m \rangle} C_k = V \quad \text{and} \quad \bigcap_{k \in \langle 1,...,m \rangle} C_k = \emptyset$$

and

$$\forall j \in C_k, aff(j, C_k) \geq max\{aff(j, C_l), C_l \in \gamma\}$$

It is obvious that there is more than one *affinity partition* of a graph. In fact, trivial partitions (single community containing all the nodes and partition where each node is a member of different community) also comply with the definition of the *affinity partition*. The authors in [12] propose to favor the partition which minimize number of intra communities edges, and provide mathematic formulation of the criterion. We do not adopt this criterion, as it favor partition with small number of large communities. (E.g., it can be shown that for GN-benchmark graphs [9], using this criterion, we would favor the partition of two communities instead the canoni-

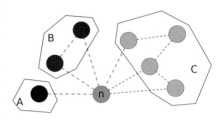

Fig. 1. Example of node's affinity towards clusters

cal four communities partition.) We propose a new criterion to compare *affinity partitions*. It is based on the following: if a node has the highest value of affinity towards multiple communities it should be assigned to the community where the ratio of the affinity towards the community and the number of community members is highest. This means that it should be assigned to the smallest community (of the candidate communities). We can define the *compactness* of the community as: $compactness(C) = \sum_{n \in C} \frac{aff(n,C)}{|C|^2}$ and we can define average compactness of a community partition as: $avg\_copactness(\gamma) = \frac{\sum_{C \in \gamma} compactenss(C)}{|\gamma|}$. Thus, the goal of our work is to approximate the affinity partition γ of a given network with highest value of $avg\_compactness(\gamma)$.

4.2   SizConCD Algorithm

The label propagation algorithm uses the function identical to our definition of *affinity* of a node towards a community to resolve node's community membership in its greedy step. However, when the community structure is not very clear, the label propagation algorithm often fails, and produces a single community comprising all the nodes as a result. The reason is the following: let there be a canonical affinity partition of a network; if we modify this partition, e.g. by margining two communities into one, the balance of the affinity partition can be broken and applying the greedy label propagation algorithm on this setting can lead to collapsing, node after node, the solution into a single community. We want to keep low computational complexity of the greedy approach, but we want to prevent the collapse of the community structure being produced into a single community. To achieve that, we first build seed groups containing small number of nodes.

Algorithm 1 SizConCD function: ComputeOneLevel

Require: int *iteration_limit* > 0
Require: Graph *G*
Require: int *upper* {Upper limit for community size}
 1: int *current_iteration* = 0
 2: int *moves* = −1
    {create partition where each node is in separate community}
 3: Set of Set of Node *C* = *initialize_partition(G)*
 4: **while** (*current_iteration* < *iteration_limit*) ∨ (*moves* = 0) **do**
 5:    List *l* = *randomizeNumbering(G)* {create randomized list of nodes}
 6:    int *moves* = 0
 7:    **for** ∀n ∈ *l* **do**
 8:       *moves* += *move(n, G, C, upper)* {greedy step - choose community for n}
 9:    **end for**
10: **end while**
11: **return** *C*

Building seed groups. We first introduce the basic mechanics of the algorithm and than describe it in a step by step manner. Basic procedure of the algorithm is the iteration over all nodes of the graph; for every node we consider all of its neighboring communities and select the best one to join. To identify the best cluster for node n, we define a *gain* function *seed_gain*(n, C). The gain function in seed groups building phase is:

$$seed\_gain(n, C) = aff(n, C) \times log\left(\frac{UpperLimit}{|C|}\right) \qquad (1)$$

where *UpperLimit* is the user provided upper limit for community sizes, this parameter is set in seed groups building phase to N (number of nodes in the graph); Once we compute the gain function for all neighboring clusters, we choose to assign node n with the community with highest gain value. In case when multiple clusters have the same (highest) gain value, we choose the smallest one. If there is still multiple candidates, we break the ties uniformly at random. For the example depicted in Figure 1, given the *UpperLimit* = 100, the gain values in SizConCD algorithm are: *seed_gain*(n, A) = 1.699, *seed_gain*(n, B) = 3.045 and *seed_gain*(n, C) = 3.903. Thus, the node n would be assigned to community C in this particular iteration.

We now discuss the whole algorithm step by step. The algorithm is initialized by crating communities for all the nodes – i.e., each node is initially placed in a separate cluster. The processing is done in iterations (Algorithm 1).Each iteration begins by creating a list of nodes l and randomize their order in the list (line 5). We than traverse nodes in the list l and process each node separately (lines 7–9). The node processing is as follows (pseudocode is presented in Algorithm 2): we remove the node from it's original community (line 5) and compute the gain function for all the neighboring communities (lines 7–17). We assign the node to the community with the highest gain value (lines 18–19). If the community the node is assigned to is different from it's original community, we say that the node moves between communities (return value of Algorithm 2 indicates whether the node has moved). After iteration over the list l is finished, we decide whether we continue with another iteration or finish the computation.

Algorithm 2 SizConCD function: move

Require: Graph G
Require: Node $n \in G$
Require: Set of Set of Node C {graph partition}
Require: int *upper* {Upper limit for community size}
1: double *max_gain* = 0
2: Set of Node *orig_comm* = getCommunityForNode(n, C) {get community n belongs to}
3: int *orig_comm_id* = getCommunityId(orig_comm)
4: Set of Set of Node *candidate_comms* = ∅
5: *orig_comm* = *orig_comm* \ n {First, remove n from original community}
6: Set of Set of Node *neighbor_comms* = getNeighboringCommunities(n, G, C) {get communities adjacent to n}
7: **for** $\forall comm \in neighbor\_comms$ **do**
8:     double *affinity* = getAffinity(n, comm) {compute affinity of n towards comm}
9:     double *gain* = *affinity* $\times log\left(\frac{upper}{|comm|}\right)$ {compute gain}
10:     **if** *gain* = *max_gain* **then**
11:         *candidate_comms* = *candidate_comms* ∪ *comm*
12:     **end if**
13:     **if** *gain* > *max_gain* **then**
14:         *max_gain* = *gain*
15:         *candidate_comms* = {*comm*}
16:     **end if**
17: **end for**
18: Set of Node *best_comm* = selectCommunity(candidate_comms) {select smallest community from the candidates}
19: *best_comm* = *best_comm* ∪ n {add node to the best fitted community}
20: **if** getCommunityId(best_comm) = *orig_comm_id* **then**
21:     **return** 0
22: **end if**
23: **return** 1

The stop criteria are: a) no node has moved in the whole iteration; b) user specified maximum number of iteration has been reached. The condition b) is introduced to ensure the convergence of the process and avoid the oscillation of nodes between communities with equal gain value for the node (we remind that we break ties by randomly picking one of the communities with the highest score). In our experiments, we have set the maximum number of iteration to 25. The Algorithm 1 handles iterations and terminal conditions, while Algorithm 2 is the greedy step, where we select the community the given node will join, based on current state of the intermediate partition of the network. We omit the definitions of functions used only for manipulation with data structures, as we consider them to be quite simple and unnecessary for the comprehension of algorithm's mechanism. Considering the definition of the gain function, our expectation is to receive communities of smaller sizes and rather balanced in sizes. This expectation is based on modification of affinity introduced in *seed_gain* function.

Approximation of affinity partition and size constrained community detection.
The algorithm for building seed groups is very accurate itself for identifying good approximations of the affinity partition of a network with small communities. However, it fails when the range in community sizes is high. The algorithm, in this case, identifies large number of small communities. On the other hand, if we use $aff(n, C)$ as a gain function instead of gain function 1, the algorithm collapses into a single community, when the community structure is not very clear. The following function was proposed as a gain function for size constrained community detection, which allows a user to impose constraint on the community sizes. Let $UpperLimit$ be the desired size limit (user provided parameter).

$$sizcon\_gain(n, C) = \frac{aff(n, c)}{\lfloor \frac{|C|}{UpperLimit} \rfloor + 1}$$

The function returns value equal to $aff(n, c)$ for the groups smaller then $UpperLimit$; the gain of joining communities larger then $UpperLimit$ is purposely lowered. When the user does not wish to constrain the size of the communities, he/she uses $UpperLimit$ equal to N (number of node in the network).

Experimentation with the use of different gain functions lead us to the following solution, the SizConCD algorithm: we first build the seed groups as described in Subsection 4.2; we then continue in the iterations with altering gain functions, switching $sizcon\_gain(n, C)$ and $seed\_gain(n, C)$ as the gain function. The intuition is that use of $sizcon\_gain(n, C)$ as a gain function pushes the intermediate result towards the state of *affinity partition*, while the use of the $seed\_gain(n, C)$ prevents the procedure from collapsing the result into a single large community. This greedy heuristics leads to very good results on synthetic benchmark graphs.

Time complexity. We first express time complexity of a single iteration of the algorithm. Let n be the number of nodes and m be the number of edges. The randomization of the order in which the nodes are processed takes $O(n)$ steps. In the subsequent loop we process all the edges when computing the gain function for each node, taking $O(m)$ steps. The time complexity of an iteration is then $O(n + m)$. To ensure convergence of the algorithm, we use the limit on the number of iterations, a constant; we thus perform at most k iterations. This means that the resulting time complexity of the algorithm stays $O(n + m)$.

5 Evaluation

In this section, we report on our experiments with the proposed SizConCD algorithm. We first describe the benchmark used for the evaluation. Our experiments include evaluation on artificial benchmark graphs and experiments on the Wikipedia link graph as a real-world network. To be able to take advantage of the comparative analysis of community detection algorithms conducted by Lancichinetti et al. in [14], we have redone their experiments with the use of our method. This allows us a head-to-head comparison with a number of popular community detection algorithms. In our experiments, we have used NMI similarity measure, which is dominant in the literature. We have used the cNMI measure as well , as it has been used in the comparative analysis paper. We perform thorough evaluation of our approach on networks with various properties.

5.1   Benchmarks

For research purposes, it is practical to compare results of community detection algorithms with a ground truth, analyzing networks with the known community structure. For those purposes, community detection benchmarks were proposed. Benchmarks generate artificial networks containing communities.

The most widely used approach for generating artificial networks with communities is *planted l-partition* model [5]. In this model, we generate defined number of groups of nodes; nodes are connected with probability of p_{in} to the other members of their group and with probability p_{out} to the nodes in other groups. Girvan and Newman used in their work [9] planted l-partition graphs with 128 nodes (each node having degree 16) divided into 4 disjoint communities, each having 32 nodes (GN-benchmark). This class of graphs becomes quite popular within community detection research, we refer to this type of benchmark graphs as GN-benchmark.

The criticism of the GN-benchmark is that all the nodes have the same degree and communities are of the same size, which makes them dissimilar to real-world networks where power-law distribution of node degrees and community sizes has been observed [2]. Lancichinetti et al. [13] have proposed LFR-benchmark that overcomes the drawbacks of GN-benchmark and generates networks with more realistic properties. LFR-benchmark is based on planted l-partition model; degrees of nodes and sizes of communities are assigned from a power law distribution, instead of probabilities p_{in}, p_{out} the mixing parameter μ is used. The value of μ defines the percentage of node's edges that connects it to the nodes outside it's own community. We believe that, to date, LFR-benchmark provides the most reliable way to test and compare community detection algorithms. Therefor, we have used the LFR-benchmark to evaluate the algorithm presented in this paper. Moreover, Lancichinetti et al. [14] have conducted a comparative analysis of several popular community detection algorithms on their benchmark. Thus, performing the evaluation on the graph with the same properties as the graphs used in [14] allows us the head-to-head comparison with numerous existing approaches to community detection.

5.2   Evaluation on the LFR Benchmark, on Small Undirected Graphs

We have performed experiments, using LFR-benchmark, with the proposed method on small undirected networks (1000 and 5000 nodes), with the same settings as in [14]. This benchmark test evaluates community detection methods on small undirected networks with small communities (10-50 nodes) and big communities (20-100 nodes). The results are depicted in Figure 2, each results represent the average of 100 trials, using SizConCD algorithm without constraining community sizes (upper limit: N).

SizConCD algorithm has clearly better performance on the networks with smaller communities. Surprising are consistently high values of NMI measure, even in case of network with high value of the mixing parameter ($\mu = 0.9$). Those high values of NMI

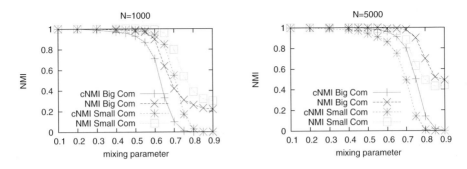

Fig. 2. LFR benchmark on undirected small graphs (1000 and 5000 nodes); small communities are of size 10–50, size of big communities ranges between 20–100

are caused by the bias of NMI measure towards partition with very small communities and do not indicate the accuracy of the proposed approach. SizConCD algorithm produced a large number of very small communities (3-5 nodes) when the community structure is very unclear (high values of μ); this is the cause of high NMI values. By comparing achieved results with the performance of those tested in [14], we can conclude that for LFR-benchmark on small undirected networks, the SizConCD has performance similar to the algorithm by Bondel et al. [1] and it is outperformed by algorithm introduced in [21] and RN [20] algorithms.

5.3   Evaluation on Small Directed Graphs

The next set of test was performed on small directed networks of LFR-benchmark. The experiment settings were identical to [14]. The results are depicted in Figure 3, using the SizConCD algorithm without size constraint (upper: N). We can observe better performance on networks with smaller communities. Comparison with benchmark results presented in [14] is rather favorable; in the comparative analysis paper, two algorithms were tested on benchmark for directed graphs – Infomap [21] and the modularity optimization via simulated annealing [11]. We observe higher values of cNMI for the partition produces by SizConCD than partitions by Infomap in all cases. Simulated annealing method has a slightly higher values for the case of 1000 nodes network with big communities, approach proposed in this paper is better in other settings. As the simulated annealing method optimizes the modularity, its results are rather poor for the case of 5000 node network with small communities due to the resolution limit of modularity. We can thus conclude that SizConCD method achieved the best results on this setting.

5.4   Evaluation on Large Undirected Graphs with Wide Range of Communities

In this experiment, we verify the performance of the algorithm on large networks of 50k and 100k nodes (community sizes: 20-1000, node degrees: 20-200). The results are shown in Figure 4(a). Again, the comparison of the proposed SizConCD algorithm with

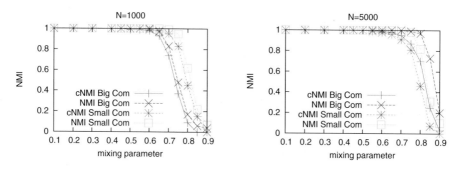

Fig. 3. LFR benchmark on directed small graphs (1000 and 5000 nodes); small communities are of size 10–50, size of big communities ranges between 20–100.

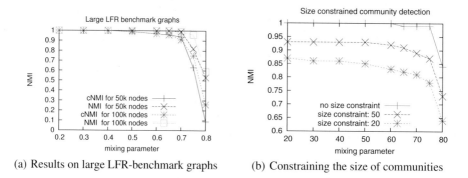

(a) Results on large LFR-benchmark graphs   (b) Constraining the size of communities

Fig. 4. Left: On large LFR-benchmark graphs; Right: Effect of constraining the community size

the algorithms tested in [14] is rather favorable. Our approach achieves better cNMI values than all the algorithms tested in the comparative analysis study.

5.5 Applying Size Constraint

To evaluate the results of the SizConCD algorithm when constraining the community size, we have performed following experiment. Using LFR-benchmark, we have generated directed graphs of 10000 nodes, containing 10 communities, each of 100 nodes; the average node degree was set to 20, maximal degree set to 50. We have then run the SizConCD algorithm a) without the size constraint (as a baseline), b) with the size constraint set to 50 nodes and c) with the size constraint set to 25 nodes. The results are depicted in Figure 4(b). High values of NMI indicates that algorithm constructs meaningful groupings when the size of generated communities is constraint.

5.6 Detecting Communities in Wikipedia

With promising results achieved on artificial networks, we wanted to apply the SizConCD algorithm on a real-world network. We have used proposed algorithm to cluster the link graph of Wikipedia, where each node represents an article and edges represent hyperlinks between articles. We obtain the link graph by processing Wikipedia XML dump (from November 2009) by a custom script. Links to redirect pages were replaced by links to targets of the redirects. Resulting link graph contains 3.1 million nodes and 91 million edges. Thanks to the low computational complexity of SizConCD algorithm, we were able to analyze the graph of this size. We have first run the SizConCD algorithm on Wikipedia link graph without constraining the size of the communities. The resulting partition had 51 communities with more than 10000 nodes, with the largest community containing 170000 nodes. Nodes had on average 34% of edges going out of the community. We have rerun the experiments, setting upper size constraint to 100. The resulting partition had the largest community of 1900 nodes, nodes had on average 56% of edges going out of the community. After clustering Wikipedia link graph, one would expect nodes in communities to be somehow semantically related. As an example we provide the listing of titles of article sharing the community with article on 'Community structure'. We consider this clustering to be semantically correct, containing semantically close concepts. The members of the community are the following: *Duncan J. Watts, Six Degrees: The Science of a Connected Age, Simon model, Clustering coefficient, Modularity (networks), Random regular graph, Random graph, Preferential attachment, Luciano Pietronero, Watts and Strogatz model, Generalized scale-free model, Mixing patterns, Shlomo Havlin, Sexual network, Community structure, Small world experiment, Social-circles network model, Assortativity, Steven Strogatz, Average path length, Copying mechanism, Countability, Adilson E. Motter, Scale-free network, Degree distribution, Complex network zeta function, Fitness model (network theory), Reciprocity in network, Mark Newman, Giant component, Guido Caldarelli, Fitness model, Assortative mixing, Shortcut model, Triadic closure, Derek J. de Solla Price, Fractal dimension on networks, Erdős-Rényi model, Complex network, Small-world network, Barabási-Albert model, Eli Upfal.*

6   Conclusion

We have proposed a new algorithm for the community detection in networks. A notable feature of the proposed algorithm is that a user can constrain the size of the communities being generated, which might be a practical feature for a number of applications. The algorithm has a pseudo-linear time complexity which makes it applicable also to large networks. Recent work on the LFR-benchmark for community detection algorithms allowed us to perform thorough evaluation of the performance of the proposed approach. The comparative analysis of community detection algorithms on the LFR-benchmark enabled direct, head-to-head comparison of our approach with other community detection algorithms. Evaluation showed very competitive performance of the proposed algorithm, which outperformed other approaches in several of the benchmarks.

As the algorithm has the potential to identify communities at different size resolutions (by varying size limit), the direction for the future work is to extend the algorithm for hierarchical clustering. Another direction for the future work is to identify multi-community membership of the nodes as a post-processing step of the algorithm.

References

1. Blondel, V.D., Guillaume, J.-L., Lambiotte, R., Lefebvre, E.: Fast unfolding of communities in large networks. Journal of Statistical Mechanics: Theory and Experiment (10) (2008)
2. Boccaletti, S., Latora, V., Moreno, Y., Chavez, M., Hwang, D.-U.: Complex networks: Structure and dynamics. Physics Reports 424(4-5), 175–308 (2006)
3. Boettcher, S., Percus, A.G.: Optimization with extremal dynamics. Complex Adaptive systems: Part I 8(2), 57–62 (2002)
4. Clauset, A., Newman, M.E.J., Moore, C.: Finding community structure in very large networks. Phys. Rev. E 70(6), 66111 (2004)
5. Condon, A., Karp, R.M.: Algorithms for graph partitioning on the planted partition model. Random Struct. Algorithms 18(2), 116–140 (2001)
6. Danon, L., Duch, J., Diaz-Guilera, A., Arenas, A.: Comparing community structure identification. Journal of Statistical Mechanics: Theory and Experiment (October 2005)
7. Fortunato, S.: Community detection in graphs. Physics Reports 486(3-5), 75–174 (2010)
8. Fortunato, S., Barthélemy, M.: Resolution limit in community detection. Proceedings of the National Academy of Sciences of the United States of America 104(1), 36–41 (2007)
9. Girvan, M., Newman, M.E.: Community structure in social and biological networks. Proc. Natl. Acad. Sci. USA 99(12), 7821–7826 (2002)
10. Gregory, S.: A fast algorithm to find overlapping communities in networks. In: Daelemans, W., Goethals, B., Morik, K. (eds.) PKDD 2008, Part I. LNCS (LNAI), vol. 5211, pp. 408–423. Springer, Heidelberg (2008)
11. Guimera, R., Amaral, L.A.N.: Functional cartography of complex metabolic networks. Nature 433(7028), 895–900 (2005)
12. Hu, Y., Chen, H., Zhang, P., Li, M., Di, Z., Fan, Y.: Comparative definition of community and corresponding identifying algorithm. Phys. Rev. E 78(2), 26121 (2008)
13. Lancichinetti, A., Fortunato, S.: Benchmarks for testing community detection algorithms on directed and weighted graphs with overlapping communities. Phys. Rev. E 80(1) (2009)
14. Lancichinetti, A., Fortunato, S.: Community detection algorithms: A comparative analysis. Phys. Rev. E 80(5), 56117 (2009)
15. Lancichinetti, A., Fortunato, S., Kertész, J.: Detecting the overlapping and hierarchical community structure in complex networks. New Journal of Physics 11(3), 33015 (2009)
16. Newman, M.E.J., Girvan, M.: Finding and evaluating community structure in networks. Phys. Rev. E 69(2), 26113 (2004)
17. Papadimitriou, S., Sun, J., Faloutsos, C., Yu, P.S.: Hierarchical, parameter-free community discovery. In: Daelemans, W., Goethals, B., Morik, K. (eds.) PKDD 2008, Part II. LNCS (LNAI), vol. 5212, pp. 170–187. Springer, Heidelberg (2008)
18. Porter, M.A., Onnela, J.-P., Mucha, P.J.: Communities in networks. CoRR, abs/0902.3788 (2009)
19. Raghavan, U.N., Albert, R., Kumara, S.: Near linear time algorithm to detect community structures in large-scale networks. Phys. Rev. E 76(3), 36106 (2007)
20. Ronhovde, P., Nussinov, Z.: Multiresolution community detection for megascale networks by information-based replica correlations. Phys. Rev. E 80(1), 16109 (2009)
21. Rosvall, M., Bergstrom, C.T.: Maps of random walks on complex networks reveal community structure. Proceedings of the National Academy of Sciences 105(4), 1118–1123 (2008)
22. Yang, T., Jin, R., Chi, Y., Zhu, S.: Combining link and content for community detection: a discriminative approach. In: KDD 2009: Proceedings of the 15th ACM SIGKDD, pp. 927–936. ACM, New York (2009)
23. Zhang, Y., Wang, J., Wang, Y., Zhou, L.: Parallel community detection on large networks with propinquity dynamics. In: KDD 2009: Proceedings of the 15th ACM SIGKDD, pp. 997–1006. ACM, New York (2009)

Evolutionary Taxonomy Construction from Dynamic Tag Space

Bin Cui[1], Junjie Yao[1], Gao Cong[2], and Yuxin Huang[1]

[1] Department of Computer Science
Key Lab of High Confidence Software Technologies (Ministry of Education)
Peking University
{bin.cui,junjie.yao,yuxin}@pku.edu.cn
[2] Nanyang Technological University
gaocong@ntu.edu.sg

Abstract. Collaborative tagging allows users to tag online resources. We refer to the large database of tags and their relationships as a tag space. In a tag space, the popularity and correlation amongst tags capture the current social interests, and taxonomy is a useful way to organize these tags. As tags change over time, it is imperative to incorporate the temporal tag evolution into the taxonomies. In this paper, we formalize the problem of evolutionary taxonomy generation over a large database of tags. The proposed evolutionary taxonomy framework consists of two key features. Firstly, we develop a novel context-aware edge selection algorithm for taxonomy extraction. Secondly, we propose several algorithms for evolutionary taxonomy fusion. We conduct an extensive performance study using a very large real-life dataset (i.e., Del.ici.ous). The empirical results clearly show that our approach is effective and efficient.

Keywords: social media, collaborative tagging, hierarchical taxonomy, temporal evolution.

1 Introduction

Social tagging systems have emerged recently as a powerful way to organize large amount of online data (e.g., web page, image, video). Notably, sites such as Flickr (http://www.flickr.com), Del.icio.us (http://www.delicious.com) and CiteULike (http://www.citeulike.org) have been used as the platform for the sharing of photos, bookmarks and publications respectively. Social tagging systems allow communities of users to sieve through large amount of online data by allowing users to tag the online data. We refer to a large database of tags in a collaborative tagging system as a tag space. In such a tag space, tags are not independent and often have strong semantic correlations, such as homonymy, synonymy, subsumption [5]. Additionally, the popularity of tags generally provides an intuitive way to understand the interests of communities of users.

More recently, the taxonomy, which is a hierarchical classification structure, has attracted more and more interests from research community to organize

L. Chen, P. Triantafillou, and T. Suel (Eds.): WISE 2010, LNCS 6488, pp. 105–119, 2010.
© Springer-Verlag Berlin Heidelberg 2010

tags [11,14]. The application of taxonomy is especially valuable for online resource organization and interactive user navigation [8]. It is important to note that taxonomies for tag space are different from formal predefined ontologies by domain experts, which are not suitable to dynamic tag space in collaborative tagging systems, as tags are not restricted by a controlled vocabulary and cover a variety of domains [9].

Another important characteristic of tag space in collaborative systems is that a tag space is a dynamic database, with millions of tagging actions being performed each week [10]. By watching the behavior of users in tag space over time, we can explore the evolution of community focus, and thus understand the evolution of users' interests. Traditional tag taxonomy construction methods [11,14] treat the tag space as a static corpus. However, a stagnant tag taxonomy cannot conform with the underlying dynamic tag space.

To capture such time-dependent correlations between tags, we propose *evolutionary taxonomy* to describing the dynamic trends of a tag space over time. Taxonomy only provides a static view of online data at a certain period. However, the evolutionary taxonomy will take into account the temporal dimension of tagging actions, and balance the effect of historical tagging actions and current tagging actions. Thus, the evolutionary taxonomy can represent the temporally evolving views of online tag data. This taxonomy construction technique is particularly applicable in burgeoning collaborative environments with increasing online data. Collaborative tagging systems can benefit from this evolutionary construction for better organization of tagged data, and utilize the evolutionary taxonomy for object ranking, similar content grouping, and hot resource recommendation. Users can also better interact with collaborative systems using this concise and updating topic taxonomy, which facilitates the navigation and browsing.

In this paper, we propose a solution for evolutionary taxonomy generation to organize a large corpus of evolving tags. The solution consists of the following: (a) We model the tag space with a graph and propose a new taxonomy extraction algorithm to performing taxonomy extraction from a large tag space, which employs a new context-aware tag correlation selection mechanism. We also design a quality measurement to quantitatively evaluate how well the extracted taxonomy hierarchy represents the tag relations in tag space. (b) Taxonomy fusion forms the basis for generating evolutionary taxonomy, and can be perfectly integrated into the proposed taxonomy extraction algorithm. We propose several algorithms to performing taxonomy fusion using historical graphs (to model the historical tag space), historical taxonomies, and current graph. To the best of our knowledge, this work is the first attempt to investigate the evolutionary taxonomy problem in a tag space.

We conduct an extensive experimental evaluation on a large real data set containing 270 million tagging actions from Del.cio.us. The results show that our approach can generate taxonomy of high quality, and algorithms of building evolutionary taxonomy can balance the current and historical tagging actions.

The rest of this paper is organized as follows. In Section 2, we briefly review the related work. Section 3 presents the proposed taxonomy extraction algorithm. In Section 4, we introduce the approach for evolutionary taxonomy generation. Section 5 describes a performance study and gives a detailed analysis of results. Finally, we conclude this paper in Section 6.

2   Related Work

Social tagging systems have grown in popularity on the web, which allow users to various types of online resources, e.g., tag pages, photos, videos and tweets. The huge amount of user generated tags reflect the interest of millions of users, and have been applied in many fields, such as web search[12], information retrieval [15,20] and community recommendation [18].

Taxonomy Extraction: Tagging is usually treated as a categorization process, in contrast to the pre-optimized classification as exemplified by expert-created Web ontology [10]. Tagging systems allow greater malleability and adaptability in organizing information than classification systems. The co-occurrence relations of tags often indicate the semantic aspects of tags, e.g., homonymy, synonymy, subsumption, etc. [5]. Association rules are able to capture the co-occurrence relations of tags and thus the semantic relations among tags [13]. Compared to flat tags, hierarchical concepts in a taxonomy is useful for browsing and organization. Automatically extracting taxonomy from tag spaces has received interests in recent years, e.g., [11,16,14] It is shown in [11] that k-means or agglomerative hierarchical clustering cannot get satisfactory results for taxonomy extraction. Schwarzkopf presented an association rule based approach to extract subsumption relations from tag space [16]. In [14], user provided subsumption relationships are utilized and combined.

Tag Evolution: Existing taxonomy extraction algorithms only consider static tag space. However, the tag space keeps evolving due to the huge amount of new tagging activities everyday. Hence, the stagnant taxonomies will fail to capture the evolving nature of tag space. Halpin et al [10] examined the dynamics of collaborative tagging systems. They introduced a generative model of collaborative tagging and showed that the distribution of the frequencies of tags for "popular" sites over a long history tend to stabilize into power law distributions. Dubinko et al [7] studied the problem of visualizing the evolution of tags within Flickr. They presented an approach based on a characterization of the most salient tags associated with a sliding interval of time. All aforementioned methods on detecting tag trend and evolution only consider individual tags.

3   Taxonomy Extraction

In this section, we present our context-aware taxonomy extraction algorithm that is able to cope with a huge tag space. We first introduce the data model for tags in a collaborative tagging system, and then present the proposed context aware correlation selection mechanism followed by a heuristic algorithm for context-aware taxonomy extraction. Our taxonomy extraction algorithm is the foundation of the proposed evolutionary taxonomy generation algorithms to be presented in the next section.

3.1   Data Model in Tag Space

In a collaborative tagging system, a tagging action involves user, tag, resource, and time [4]. We model a tagging action with a quadruple $< Usr_i, Res_j, Tag_k,$ $Tim_t >$ which says that user i tagged resource j with a set of tags k at time t. Note that, users may use multiple tags in a single tagging action, and these tags together describe the resource. We denote the set of quadruples in a tag space as D and the set of distinct tags in the tag space as T. For ease of presentation, we do not take the time dimension into consideration in the following discussion.

Tag Co-occurrence Graph: We say that there exists a "co-occurrence" relationship between two tags if the two tags are used by the same user to annotate an object in a single tagging action. Note that we can relax the "co-occurrence" relationship definition such that tags tag_1 and tag_2 have "co-occurrence" relationship if there exists an object labeled with tags tag_1 and tag_2 no matter whether they are from the same or different users, or are tagged at different time. In this paper, we use the first definition in our implementation, but our approach is equally applicable to the latter definition.

By establishing a link between two tags if there exists co-occurrence relationship between them, we generate an implicit tag relation graph. We project the quadruples in a tag space into a basic tag co-occurrence graph, which is an undirected graph, denoted as $G = (V, E)$ where V is the set of vertices and E is the set of undirected edges. Each distinct tag in T corresponds to a vertex in V. There exists an edge between two nodes u and v if and only if there exists "co-occurrence" relationship between them.

Tag Association Rule Graph: The tag co-occurrence graph could show the mutual relationship between tags, but is still not sufficient to capture the rich semantic of tag space. The association rules discovered in tag space are more valuable and can provide richer semantic of tag relations than co-occurrence relation [13].

Frequency: it represents the number of tag_i appearances in the tag space.

$$Freq(tag_i) \quad = \quad |\{(Usr_u, Res_r, Tag_k)|tag_i \in Tag_k\}| \quad . \tag{1}$$

Support: it represents the number of tagging actions which contain both tag_i and tag_j.

$$Supp(tag_i, tag_j) \quad = \quad |\{(Usr_u, Res_r, Tag_k)|tag_i, tag_j \in Tag_k\}| \quad . \tag{2}$$

Confidence: it represents how likely tag_j is given tag_i, where $Freq(tag_i) < Freq(tag_j)$ and the edge is directed to tag_j.

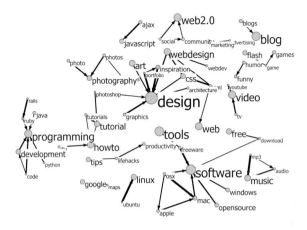

Fig. 1. An example of association rule graph (Del.icio.us)

$$Conf(tag_i \quad \rightarrow \quad tag_j) \quad = \quad Supp(tag_i, tag_j)/Freq(tag_i) \quad . \tag{3}$$

Note that the definitions of frequency and support are the same as those in traditional association rule mining [1]. However, the definition of confidence is different in that we impose an extra constraint $Freq(tag_i) < Freq(tag_j)$. This constraint means that given a pair of tags $\{tag_i, tag_j\}$, we at most generate one association rule for this pair where the tag with lower frequency will be the antecedent and the other tag will be consequent. The reason for imposing the constraint is that the statistical weight $Frequency$ typically represents the generality of tag [11], and highly frequent tags are expected to be located in the higher levels in the taxonomy tree.

We next proceed to define an **association rule graph** $ARG := (V, E)$, where V is the set of vertices, each of which represents a distinct tag which is also attached with a weight v_w computed by the tag frequency in the vertex; E is the set of directed edges representing the co-occurrence of tag pairs, each of which is attached with the weights of edge e_w and the weights are the *support* and *confidence* of the association rule represented by the edge. Figure 1 shows an example of association rule graph of a tag space. For the clarity of presentation, we set the frequency and confidence thresholds when plotting the graph, so it only shows highly frequent and correlated tags. However, we can still find lots of interesting information from the visualized graph.

3.2   Association Rule Based Taxonomy Extraction

The association rule graph is a concise representation of tag correlation in a tag space. We propose a context aware taxonomy extraction algorithm and introduce a novel quality measure which attempts to quantitatively evaluate taxonomy quality in Section 3.2.

Determining the Context-aware Confidence. As we introduced previously, the taxonomy generated for tagging system is to help organize the online content and provide users an hierarchical view of the tag space. Users in the tagging system usually browse the taxonomy to broaden or narrow down the areas of interests [4], which means it is desirable that the high level concepts can accommodate the semantics of their children concepts.

The method used in previous work [16] is to set a global confidence threshold and all rules whose confidence is below the threshold will not be considered for taxonomy generation. But, such an approach may fail to exclude noisy edges in tagging scenario. This is a weakness of [16] by simply applying confidence. The problem with confidence is that it is sensitive to the frequency of the antecedent. Caused by the way confidence is calculated, consequents with higher support will be more likely to produce higher confidence values even if there exists no tight association between the tags.

Since tags with high frequencies are more likely to be involved in noisy association rules, and thus a desirable strategy would be to set higher threshold for frequent tags and lower threshold for less frequent tags to eliminate those less meaningful association rules. However it is infeasible to manually specify a threshold for each tag. Therefore, we propose the context aware edge significance to measure the "real confidence" of the edge. Here, the context refers to some statistical information of underlying tag space, such as frequency of tags with parent-child relationship which will be used to calculate the $confidence$ between two tags.

Suppose we have a weighting parameter w for confidence, if the antecedent and consequent of an edge have the same frequency, w should be 1; on the other hand, if the frequency of consequent is approaching ∞, w will be approaching 0, where the confidence of the edge is nearly meaningless. So the frequency difference of antecedent and consequent, $Freq(B) - Freq(A)$, could be ranged from 0 ($Freq(B) = Freq(A)$) to ∞ ($Freq(B) \to \infty$).

Intuitively, we can find the small value of $Freq(B) - Freq(A)$ have higher effect on the utility of confidence. To deal with the "infinity" problem, we use the $MaxFreq$ to replace the ∞ in the formula, where $MaxFreq$ is the highest value of frequency in the tag space. The formula to calculate the context aware confidence $CConf(tag_i \to tag_j)$ is presented as follows:

$$CConf(tag_i \to tag_j) = Conf(tag_i \to tag_j) \times (1 - \frac{Freq(tag_j) - Freq(tag_i)}{MaxFreq - Freq(tag_i)}) . \quad (4)$$

Our strategy is a better choice than **confidence**, as it is enhanced with the implication from tag_j to tag_i, i.e., from the higher level to lower level, which is consistent with the philosophy of taxonomy. Another benefit of our strategy is that it is able to provide smaller confidence threshold in small cliques of the graph, which makes the taxonomy more balanced. We also notice that a number of association rule measures have been proposed in literature [3,1]. We actually tested some well-known association rule measures to represent the correlation between tags, but their performances in tag space are not satisfactory.

Algorithm for Taxonomy Extraction. In the preprocessing stage, we generate the association rule graph for tag dataset. Algorithm 1 takes in tag association rule graph G and a tuning parameter confidence pruning threshold Thr to extract taxonomy of a tag space. For each tag, we first compute a context aware confidence for its out edges (Line 3). By comparing with threshold Thr, we determine whether an edge is qualified to be inserted into the tree T (Line 4). From the candidate edge pool, we find the out edge with maximum context aware confidence for a certain tag, and this edge is inserted into the taxonomy (Lines 6-7). The associated two tags of this edge have father-child relationship in the taxonomy, i.e., the tag with higher frequency serves as father node in the tree structure. Finally, all the tags in T without out edges are collected as top layer concepts in the taxonomy. In the experimental evaluation, we will demonstrate our algorithm based on dynamic context-aware confidence strategy is more adaptive.

Algorithm 1: Adaptive Taxonomy Extraction from Tag Space

Input: Association rule graph G; Confidence threshold, Thr
Output: G's taxonomy tree, T

1 Get the $maxFreq$ from G ;
2 **forall** tag_i *in* G **do**
3   Compute the $CConf(tag_i \rightarrow tag_j)$;
4      $candidate\_edges \leftarrow \{< tag_i \rightarrow tag_j > | CConf(tag_i \rightarrow tag_j) > Thr\}$;
5   **if** $candidate\_edges$ *not empty* **then**
6   Select $< tag_i \rightarrow tag_{up} > \in candidate\_edges$ with $max(CConf)$;
7   Add $< tag_i \rightarrow tag_{up} >$ into T ;
8   **end**
9 **end**
10 The tags in T without out edges are assigned to *root* node ;
11 **return** T;

Measurement of Taxonomy Quality. It is usually difficult to measure the quality of tag taxonomies, since the collaborative tagging behaviors are unpredictable, and tags are typically free texts which do not rely on a controlled vocabulary. Human judgement is very natural, however performing unbiased assessment on the taxonomy for a large tag space is an extremely time-consuming task, which is inapplicable in reality. Existing taxonomy extraction algorithms did not conduct quantitative evaluation [11,16]. A possible alternative is to compare how "similar the induced taxonomies are to hand-crafted taxonomies, such as ODP or WordNet. However, tagged objects cover various domains and there exist no general standard taxonomies in social tagging systems. Even if there exists a standard taxonomy, tags' dynamic evolution invalidates this kind of evaluation. Here we introduce some statistical metrics to measure the quality of induced taxonomy.

The general user browsing behaviors in a tagging system are to find the objects tagged with a given tag or more tags, to find a list of the most popular tags, or to find tags with tight correlation with user interest. Therefore, the tag taxonomy should be highly navigable and informative. First, the hierarchical taxonomy structure should have informative "portal" concepts (tags), which are expected to be popular and capture the overview of the tag space. Here, we refer the "portal" tags to the first layer tags under the root of taxonomy. Second, the tags in the subtree rooted a certain "portal" tag should be highly correlated to provide users satisfactory browsing experience. If we extract a taxonomy with few portal concepts, these portal concepts are generally popular and with large subtrees, while the tags inside the subtree may not be highly correlated; vice versa.

The quality of taxonomy is the tradeoff of these two factors, and various evaluation criteria can be used to measure the taxonomy quality, e.g., the number of "portal" tags, the sizes of subtrees, the correlation (dependency) between tags, etc. Taking these issues into consideration, we design two novel quality measures to evaluate the overall quality of taxonomy in this paper.

Global quality of taxonomy: Let $top_1, top_2, \ldots, top_m$ be the "portal" tags in the taxonomy. $Freq(top_i)$ could represent this tag's concept generality [13], thus frequent tags are preferred to be "portal" concepts in the taxonomy. Furthermore, we choose the *geometric average*, which is more effective summary for skewed data to measure the global quality, instead of *arithmetic average*, because we also want the "portal" concepts to be more balanced to improve the experience of taxonomy navigation.

$$Global\_Quality(T) = (\prod_{i=1}^{m} Freq(top_i))^{\frac{1}{m}} . \tag{5}$$

Local quality of taxonomy: It can be also considered as the degree of co-relationship of the tags within the subtrees of the portal tags. Most conceptual clustering methods are capable of generating hierarchical category structures, and category utility is used to measure the "category goodness" of concept hierarchy [6]. We meliorate probability-theoretic definition of category utility given in [17] to facilitate our scenario.

$$Local\_Quality(T) = \sum_{C_i \in C} \sum_{j=1}^{m} (Conf(tag_j \rightarrow C_i))^2 . \tag{6}$$

where C_i refers to a non-leaf tag in taxonomy, $Conf(tag_j \rightarrow C_i)$ is the product of the confidence of the edges in the path from tag_j to C_i.

4   Evolutionary Taxonomy

The taxonomy extraction algorithm in Section 3 is able to generate hierarchical concept structure of a static tag space. However, it cannot capture the evolving and dynamic natures of the tag space. In this section, we formalize the problem of evolutionary taxonomy over time in a tag space, and propose the evolutionary taxonomy fusion mechanism to integrate the tag evolution.

4.1   Measuring the Evolutionary Taxonomy

Intuitively, evolutionary taxonomy should meet the following two different requirements: the generated taxonomy should represent the current status of the underlying tag space and should not deviate too much from the taxonomy in preceding interval.

Formally, let $t_1, t_2, \ldots t_m$ be m consecutive temporal intervals, and $G_1, G_2, \ldots G_m$ be the association rule graph of tags identified for each of the intervals $t_1, t_2, \ldots t_m$. Evolutionary taxonomy generation is to build a sequence of taxonomies, $T_1, T_2, \ldots T_m$, where T_i is the generated evolutionary taxonomy corresponding to t_i, such that T_i satisfies the current tag characteristics represented by G_i and at the same time, does not deviate too much from the previous tag taxonomy, T_{i-1}. The constructed evolutionary taxonomy is a tradeoff of two factors which can be determined according to user/system preference.

The difference between two taxonomies can be measured with the tree edit distance, which is defined as the minimum cost sequence of edit operations (node insertion, node deletion, and label change) that transforms one tree into another [2,19]. In our experimental evaluation, we utilize this criteria to measure the difference of two taxonomy trees T and T', and the smaller distance score indicates higher similarity between taxonomies.

4.2   Algorithms for Constructing Evolutionary Taxonomy

We first introduce two baselines, and then present our new evolutionary taxononmy algorithms which integrate the fusion mechanism.

Approach 1: Local Taxonomy Evolution. To capture the evolution of temporal taxonomies, the naïve approach generates taxonomies for each consecutive time interval, and computes the differences with preceding taxonomy to capture the taxonomy evolution. Given the successive temporal graph series $G_1, G_2, \ldots G_m$, this approach extracts a taxonomy for each time interval using the taxonomy extraction algorithm of Section 3. The generated taxonomy at time interval i is independent of tagging history in preceding intervals.

Approach 2: Historical Taxonomy Evolution. Another straightforward approach is to compute the taxonomy for a time interval from a "smoothed" graph, which is obtained by summing up the underlying tag association rule graph at the time interval and all the previous historical graphs with a decaying parameter. Consider the successive temporal graph series $G_1, G_2, \ldots G_m$. In this approach, at each time interval t_i, we accumulate the tagging history in a new graph G_i'. This graph is a combination of G_i based on the current data and all the previous graphs $G_j|j < i$. It is important to define appropriate weighting for current tag/correlation and the historical ones. Here we define the tag frequency and correlation decaying formulas:

$$Freq(tag)'_{t_i} = freq(tag)_{t_i} + \lambda \times Freq(tag)'_{t_{i-1}} . \tag{7}$$

$$weight(edge)'_{t_i} = weight(edge)_{t_i} + \lambda \times weight(edge)'_{t_{i-1}} . \tag{8}$$

where parameter λ is to determine the degree of information decay. With $Freq(tag)'_{t_i}$ and $weight(edge)'_{t_i}$, we can build a new association rule graph G'_i with the historical information, from which the historical taxonomy T'_i is extracted. The global historical graphs may be indistinguishable between historical tag correlations and the newly emerged links, which makes the taxonomies for two consecutive intervals nearly consistent as the historical tag information might be over-weighted.

Approach 3: Taxonomy Evolution with One-step Fusion. To model the evolutionary taxonomy smoothly, the historical tag information can help digest the noisy or burst tag. In the first fusion strategy, our approach takes the graph and taxonomy of preceding time interval into consideration. Figure 2 (a) shows the procedure of evolutionary taxonomy construction with Approach 3. The first taxonomy T_1 is simply generated using the underlying graph. While for the subsequent taxonomies, we need to fuse the information of local graph and preceding taxonomy and graph, e.g., fusing the T_{i-1}, G_{i-1} and G_i to generate T_i.

In this approach, the tags' frequencies and correlations at interval t_i are fused with T_{i-1} and G_{i-1}. If the current chosen edge for T_i is reflected consistently in T_{i-1}, we just add the edge into T_i. When it is different from T_{i-1}, we exploit the information in G_{i-1} and G_i, to select a suitable edge by edge weight accumulation. For all the tags in the tag space, we first compute context aware confidence of out edges from tag_i and find the candidate edges from the underlying graph. For each tag_i, find the out edge $< tag_i \rightarrow tag_{up} >$ with the largest context aware confidence from the candidate edges. If the edge appears in the preceding taxonomy T_{t-1}, we insert it into the current taxonomy. Otherwise, we find edge $< tag_i \rightarrow tag_{up'} >$ from T_{t-1}, and compare the accumulated confidence of edge $< tag_i \rightarrow tag_{up} >$ and $< tag_i \rightarrow tag_{up'} >$. After that, the edge with higher

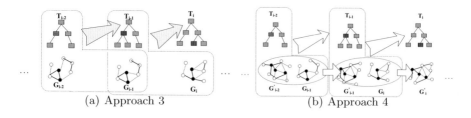

(a) Approach 3   (b) Approach 4

Fig. 2. Evolutionary Taxonomy Construction

confidence will be included in T_t. Finally, all the tags in T without out edges serve as "portal" concepts in the generated taxonomy.

Approach 4: Taxonomy Evolution with Historical Fusion. In Approach 3, we only consider the tags of the preceding time interval to generate the current taxonomy. Although we could infer some hints from the earlier taxonomy chain $T_1, T_2, \ldots T_{t-1}$, some earlier tag information of graph may be lost. Therefore, we exploit the historical tag association rule graph to better maintain the evolutionary taxonomy. When we select the correlation for taxonomy in time interval t, Approach 4 considers the historical graph G'_{t-1} (defined in Approach 2) in contrast to Approach 3 that uses local graph of preceding time interval G_{t-1}. Additionally, we generate G'_t by integrating G'_{t-1} and G_t for future usage. Note that, the decaying parameter λ will be applied for graph integration.

The detailed algorithm of Approach 4 is similar to that of Approach 3, while differs from Approach 3 in twofold: 1) at each time interval, Approach 4 generates the historical graph by integrating the decayed information of preceding historical graph; 2) Approach 4 fuses the information of local graph, preceding taxonomy and preceding historical graph for evolutionary taxonomy generation. Figure 2 (b) shows the procedure of evolutionary taxonomy generation with Approach 4. The first taxonomy T_1 is generated using the underlying graph G_1, which is same as in Approach 3; while we fuse the G_i, T_{i-1} and G_{i-1}' to generate T_i, and integrate G_{i-1}' and G_i into the historical graph G_i' for later usage.

5   Empirical Study

We conduct extensive experiments to evaluate the performance of taxonomy extraction and evolutionary taxonomy. All the experiments are conducted on an 8 core 2.0 GHz processor Server with 12GB RAM and running Redhat. For our experiment, we collect a large scale dataset from Del.icio.us. The dataset consists of 270,502,498 tagging actions, and we set one month as basic time interval, and evaluate the taxonomy evolution from the dataset range from 2007.01 to 2008.10.

5.1   Performance on Taxonomy Extraction

We first evaluate the performance of our proposed taxonomy extraction algorithm by comparing with existing taxonomy extraction approaches. As reported in [16], the association rule based method performs better than other statistical based approaches, such as co-occurrence based method [11]. Hence, we only demonstrate the comparison with [16] in this paper.

To evaluate the quality of extracted taxonomy, we use the proposed evaluation quality measures, i.e., **Global quality** and **Local quality**. We do not report efficiency for taxonomy extraction here, since both approaches take similar time, and generating the association rule graph of the tag space dominates the runtime.

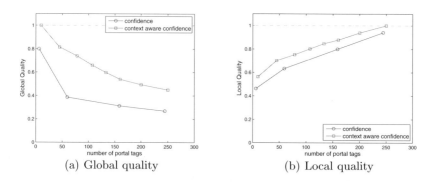

(a) Global quality   (b) Local quality

Fig. 3. Quality of extracted taxonomies

Results of Taxonomy Quality. We first filter less significant edges, which are probable noise, from association rule graph using a minimum confidence threshold 5% and minimum support threshold 10 as the empirical study in [16]. We then generate a series of taxonomies, each of which has different sizes, using different confidence thresholds. Here, we use *the number of portal tags* in a taxonomy to represent the *taxonomy size*.

As the association rule based approaches actually rely on statistical information of the tag space. It is fair to compare the quality of taxonomies generated by different approaches but with similar size. Hence, we need different thresholds to make the taxonomies generated by two approaches comparable. The comparison between [16] and our approach in terms of both "Global quality" and "Local quality" is shown in Figure 3, where our approach is represented by "context aware confidence" and the approach [16] is represented by "confidence". The size of taxonomies ranges from 10 to 250.

Figure 3 shows that our approach outperforms [16] for both quality measures. This is due to two reasons. First, our approach is enhanced with the implication from the father concept to child concept, and thus provides better subsumption capability when evaluating the correlation of father-child tag pairs. Second, our context aware threshold strategy is more robust to noise by actually providing large threshold for large cliques and low threshold for small cliques in terms of "original confidence". Additionally, the figure clearly shows the effect of *taxonomy size* on two quality measures. Generally, when we set higher threshold to select the edge of the taxonomy, it will generate a flatter taxonomy, i.e., more portal tags in the taxonomy. While we decrease the threshold, the number of portal tags decreases as well, as more correlations between tags are included.

5.2   Performance on Evolutionary Taxonomy

We proceed to evaluate the four approaches discussed in Section 4, which adopt the proposed taxonomy extraction algorithm as the basis. Based on the monthly tagging actions, we are able to generate a series of temporal tag graphs.

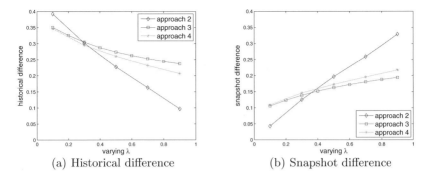

(a) Historical difference   (b) Snapshot difference

Fig. 4. Effect of tuning parameter λ

We evaluate how different approaches can trade off the difference between snapshot and historical taxonomies by showing the *historical difference*, i.e., the difference between evolutionary taxonomy and historical taxonomy, and *snapshot difference*, i.e., the difference between evolutionary taxonomy and snapshot taxonomy. In our implementation, we follow an effective approximation *pq-grams* of the well-known tree edit distance [2] to compute the difference between taxonomies, and the smaller distance indicated higher similarity.

Effect of λ. In this experiment, we evaluate the effect of λ on the performance of evolutionary taxonomies. λ is a decaying parameter to tune the weight of historical tag correlations, which ranges from 0 to 1. If we set λ as 0, actually approaches 2-4 degenerate to approach 1, as all the historical information is ignored. Figure 4 shows the performance of various approaches when λ is varied from 0.1 to 0.9.

From Figure 4, we can identify the tendency of evolutionary taxonomies with different λ. When λ increases, the current taxonomies generated by approaches 2-4 become closer to the historical taxonomies.

Figure 4 shows that the difference between local taxonomy and the historical taxonomy, i.e., the taxonomy of preceding month, of approach 1 is the largest among 4 approaches, which is around 0.44, while the snapshot difference is 0.

The approach 2 is the most sensitive to the value of λ. The fusion based approaches, i.e., approaches 3 and 4, are more stable, as we fuse the historical taxonomy to generate the current taxonomy. Additionally, approach 4 takes the underlying historical graph into account, its historical difference is smaller than that of approach 3. From Figure 4 (b), we observe that the difference from snapshot taxonomy is opposite to historical difference. If an approach integrates more historical information, the difference from the snapshot will be bigger as expected.

Effect of Temporal Evolution. We also illustrate the evolutionary taxonomy to show the effect of evolution. Since the taxonomies are generally with large size, we only shows a subset of taxonomy tree. The first taxonomy a is

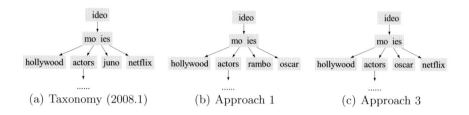

(a) Taxonomy (2008.1)   (b) Approach 1   (c) Approach 3

Fig. 5. An Example of Evolutionary taxonomies

simply generated from the tags in Jan 2008. Under "movies", there are some common topics, e.g., "hollywood", "actors", "netflix" (a DVD rental site) and several temporal hot tags, e.g., newly released movie "juno". The taxonomy b is the snapshot of Feb 2008, where the newly released movie "rambo" (Rambo 4) and "oscar" appear. We generate the taxonomy c with fusion of the historical information. By integrating the previous information, the tag "netflix" remains in taxonomy c as its accumulated frequency is higher than "rambo". "oscar" remains in c as it is a very hot topic in Feb due to the annual OSCARS movie event.

Users or tagging systems may have different perspectives from different views. Our work provides multiple choices by varying degrees of taxonomy evolution, i.e., Approach 1 generates latest snapshot taxonomy, Approach 2 constructs taxonomy with historical tagging information, and Approaches 3 & 4 balance both historical and current information and provide more smooth evolutionary taxonomy.

6   Conclusions

In this paper, we have addressed the problem of evolutionary taxonomy construction in collaborative tagging systems. A novel context aware mechanism is introduced to generate hierarchical taxonomy of a large tag corpus. Afterward we investigate how the temporal properties of tag correlation can be used to model the evolving hierarchical taxonomy, and propose several algorithms to construct evolutionary taxonomy. We conduct an extensive study using a large real-life dataset which shows the superiority of proposed method. Our approach makes an interesting attempt to incorporate temporal dimension into taxonomy summarization which improves the current tag organization in collaborative tagging sites.

Acknowledgements. This research was supported by the National Natural Science foundation of China under Grant No. 60933004 and 60811120098.

References

1. Agrawal, R., Imielinski, T.: Mining association rules between sets of items in large databases. In: Proc. of SIGMOD, pp. 207–216 (1993)
2. Augsten, N., Böhlen, M., Gamper, J.: Approximate matching of hierarchical data using pq-grams. In: Proc. of VLDB, pp. 301–312 (2005)
3. Brin, S., Motwani, R., Ullman, J.D., Tsur, S.: Dynamic itemset counting and implication rules for market basket data. In: Proc. of SIGMOD, pp. 255–264 (1997)
4. Cattuto, C., Benz, D., Hotho, A., Stumme, G.: Semantic grounding of tag relatedness in social bookmarking systems. In: Sheth, A.P., Staab, S., Dean, M., Paolucci, M., Maynard, D., Finin, T., Thirunarayan, K. (eds.) ISWC 2008. LNCS, vol. 5318, pp. 615–631. Springer, Heidelberg (2008)
5. Cattuto, C., Schmitz, C., Baldassarri, A., Servedio, V., Loreto, V., Hotho, A., Grahl, M., Stumme, G.: Network properties of folksonomies. AI Communications 20(4), 245–262 (2007)
6. Corter, J.E., Gluck, M.A.: Explaining basic categories: Feature predictability and information. Psychological Bulletin 111(2), 291–303 (1992)
7. Dubinko, M., Kumar, R., Magnani, J., Novak, J., Raghavan, P., Tomkins, A.: Visualizing tags over time. In: Proc. of WWW, pp. 193–202 (2006)
8. Fontoura, M., Josifovski, V., Kumar, R., Olston, C., Tomkins, A., Vassilvitskii, S.: Relaxation in text search using taxonomies. In: Proc. of VLDB, pp. 672–683 (2008)
9. Golder, S., Huberman, B.: Usage patterns of collaborative tagging systems. Journal of Information Science 32(2), 198–208 (2006)
10. Halpin, H., Robu, V., Shepherd, H.: The complex dynamics of collaborative tagging. In: Proc. of WWW, pp. 211–220 (2007)
11. Heymann, P., Garcia-Molina, H.: Collaborative creation of communal hierarchical taxonomies in social tagging systems. Technical Report 2006-10, Stanford University (2006)
12. Heymann, P., Koutrika, G., Garcia-Molina, H.: Can social bookmarking improve web search? In: Proc. of WSDM, pp. 195–206 (2008)
13. Heymann, P., Ramage, D., Garcia-Molina, H.: Social tag prediction. In: Proc. of ACM SIGIR, pp. 531–538 (2008)
14. Plangprasopchok, A., Lerman, K., Getoor, L.: Growing a tree in the forest: constructing folksonomies by integrating structured metadata. In: Proc. of KDD, pp. 949–958 (2010)
15. Schenkel, R., Crecelius, T., Kacimi, M., Michel, S., Neumann, T., Parreira, J., Weikum, G.: Efficient top-k querying over social-tagging networks. In: Proc. of SIGIR, pp. 523–530 (2008)
16. Schwarzkopf, E., Heckmann, D., Dengler, D., Kroner, A.: Mining the structure of tag spaces for user modeling. In: Proc. of the Workshop on Data Mining for User Modeling, pp. 63–75 (2007)
17. Witten, I.H., Frank, E.: Data mining: Practical machine learning tools and techniques. Morgan Kaufmann, San Francisco (2005)
18. Yahia, S.A., Benedikt, M., Lakshmanan, L.V.S., Stoyanovich, J.: Efficient network aware search in collaborative tagging sites. In: Proc. of VLDB, pp. 710–721 (2008)
19. Zhang, K., Shasha, D.: Simple fast algorithms for the editing distance between trees and related problems. SIAM Journal on Computing 18(6), 1245–1262 (1989)
20. Zhou, D., Bian, J., Zheng, S., Zha, H., Giles, C.: Exploring Social Annotations for Information Retrieval. In: Proc. of WWW, pp. 715–724 (2008)

Co-clustering for Weblogs in Semantic Space

Yu Zong[1], Guandong Xu[2,3,*], Peter Dolog[2],
Yanchun Zhang[3], and Renjin Liu[1]

[1] Department of Information and Engineering, West Anhui University, China
[2] IWIS - Intelligent Web and Information Systems, Aalborg University, Computer
Science Department Selma Lagerlofs Vej 300 DK-9220 Aalborg, Denmark
[3] Center for Applied Informatics, School of Engineering & Science, Victoria
University, PO Box 14428, Vic 8001, Australia

Abstract. Web clustering is an approach for aggregating web objects
into various groups according to underlying relationships among them.
Finding co-clusters of web objects in semantic space is an interesting
topic in the context of web usage mining, which is able to capture the
underlying user navigational interest and content preference simultaneously. In this paper we will present a novel web co-clustering algorithm
named Co-Clustering in Semantic space (COCS) to simultaneously partition web users and pages via a latent semantic analysis approach. In
COCS, we first, train the latent semantic space of weblog data by using
Probabilistic Latent Semantic Analysis (PLSA) model, and then, project
all weblog data objects into this semantic space with probability distribution to capture the relationship among web pages and web users, at last,
propose a clustering algorithm to generate the co-cluster corresponding
to each semantic factor in the latent semantic space via probability inference. The proposed approach is evaluated by experiments performed
on real datasets in terms of precision and recall metrics. Experimental
results have demonstrated the proposed method can effectively reveal
the co-aggregates of web users and pages which are closely related.

1 Introduction

Recently, the Internet becomes an important and popular platform for distributing and acquiring information and knowledge due to its rapid evolution of web
technology and the influx of data sources available over the Internet in last
decades [1]. Web clustering is emerging as an effective and efficient approach
to organize the data circulated over the web into groups/collections in order to
re-structure the data into more meaningful blocks and to facilitate information
retrieval and representation [2,3,4]. The proposed clustering on web usage mining is mainly manipulated on one dimension/attribute of the web usage data
standalone. However, in most cases, the web object clusters do often exist in the
forms of co-occurrence of pages and users - the users from the same group are
particularly interested in one subset of total web pages in a e-commerce site. As

---

* Corresponding author.

L. Chen, P. Triantafillou, and T. Suel (Eds.): WISE 2010, LNCS 6488, pp. 120–127, 2010.
© Springer-Verlag Berlin Heidelberg 2010

a result, a simultaneous grouping of both sets of users and pages is more appropriate and meaningful in modeling user navigational behavior and adapting the web design, in some contexts [5,6,7].

Although a considerable amount of researches of co-clustering on weblogs have been done, the major approaches are developed on a basis of graph partition theory[6,7,8,5]. One commonly encountered difficulty of such kind of approaches is how to interpret the cluster results and capture the underlying reason of such clustering. In contrast, latent semantic analysis is one of the effective means to capturing the latent semantic factor hidden in the co-occurrence observations. As such combining the latent semantic analysis with clustering motivates the idea presented in this paper. In this paper, we aim to propose a novel approach addressing the co-clustering of web users and pages by leveraging the latent semantic factors hidden in weblogs. With the proposed approach, we can simultaneously partition the web users and pages into different groups which are corresponding to the semantic factors derived by employing the PLSA model. In particular, we propose a new co-clustering framework, named Co-Clustering in Semantic Space (COCS) to deal with simultaneously finding web object co-clusters in a semantic space. Upon COCS, we first capture the semantic space by introducing the PLSA model; and then, we project the web objects into this semantic space via referring to the probability estimates; at last, we generate the web object co-clusters for all semantic factors in the semantic space. We conduct experimental evaluations on real world datasets.

The rest of the paper is organized as follows. In section 2, we introduce the framework of COCS. In section 3, we first introduce the PLSA model and describe the process of capturing the semantic space, and then, we discuss how to project the data objects into the captured semantic space, at last, a co-clustering clustering algorithm is proposed. To validate the proposed approach, we demonstrate experiment and comparison results conducted on two real world datasets in section 4, and conclude the paper in section 5.

2   The Framework of COCS

Prior to introduce the framework of COCS, we discuss briefly the issue with respect to sessionization process of web usage data. The user session data can be formed as web usage data represented by a session-page matrix $SP = \{a_{ij}\}$. The entry in the session-page matrix, a_{ij}, is the weight associated with the page p_j in the user session s_i, which is usually determined by the number of hit or the amount time spent on the specific page.

The framework of COCS is composed of three parts and is described as follow.

1. Learning the semantic space of web usage data;
2. Projecting the original web pages and web sessions into the semantic space respectively, i.e. each web session and page is expressed by a probability distribution over the semantic space;
3. Abstracting the co-clusters of web pages and web sessions corresponding to various semantic factors in the semantic space.

The first part is the base of COCS. For a given session-page matrix $SP = \{a_{ij}\}$, we can learn the latent semantic space Z by using the PLSA model, which is detailed in section 3.1. The main component of the second part in COCS is based on the result of first part of COCS. Two probabilistic relationship matrices of web pages and web sessions with semantic space are obtained by employing the PLSA model. Following the semantic analysis, we then project the web objects into the captured semantic space by linking these probability matrices. We further discuss this part in section 3.2. In the last part of COCS, we propose a clustering algorithm to find out the co-clusters in the semantic space by filtering out the web pages and sessions based on the probability cutting value. The detail will be described in section 3.3.

3   The Details of COCS

In this section, we will discuss the organization of each part of COCS in detail.

3.1   Capturing Semantic Space by Using PLSA

The PLSA model is based on a statistic model called aspect model, which can be utilized to identify the hidden semantic relationships among general co-occurrence activities. Similarly, we can conceptually view the user sessions over web page space as co-occurrence activities in the context of web usage mining to discover the latent usage pattern. For the given aspect model, suppose that there is a latent factor space $Z = \{z_1, z_2, \cdots z_K\}$ and each co-occurrence observation data (s_i, p_j) is associated with the factor $z_k \in Z$ by a varying degree to z_k.

From the viewpoint of aspect model, thus, it can be inferred that there are existing different relationships among web users or pages related to different factors, and the factors can be considered to represent the user access patterns. In this manner, each observation data (s_i, p_j) can convey the user navigational interests over the K-dimensional latent factor space. The degrees to which such relationships are explained by each factor derived from the factor-conditional probabilities. Our goal is to discover the underlying factors and characterize associated factor-conditional probabilities accordingly.

By combining probability definition and Bayesian formula, we can model the probability of an observation data (s_i, p_j) by adopting the latent factor variable z as:

$$P(s_i, p_j) = \sum_{z_k \in Z} P(z_k) \cdot P(s_i|z_k) \cdot P(P_j|z_k) \tag{1}$$

Furthermore, the total likelihood of the observation is determined as

$$L_i = \sum_{s_i \in S, p_j \in P} m(s_i, p_j) \cdot \log P(s_i, p_j) \tag{2}$$

where $m(s_i, p_j)$ is the element of the session-page matrix corresponding to session s_i and page p_j.

We utilize Expectation Maximization (EM) algorithm to perform the maximum likelihood estimation in latent variable model [9], i.e. the probability distributions of $P\left(s_i|z_k\right)$, $P\left(p_j|z_k\right)$ and $P(z_k)$.

3.2   Projecting Data Objects into Semantic Space

For each user session s_i, we can compute a set of probabilities $P\left(z_k|s_i\right)$ corresponding to different semantic factors via the Bayesian formula.

With the learned latent semantic factor space, it is noted that for any semantic factor $z_k \in Z$, each user session has a probability distribution with it. And the web page has the similar relationship with each semantic factor as well. For brevity we use matrix SZ to represent the relationship between all the user sessions and the semantic space Z, and PZ denotes the relationship between all the pages and the semantic space Z. Fig.1 shows these relationships in schematic structure of matrix. Fig.1(a) represents the matrix SZ and Fig.1(b) shows the matrix PZ respectively. In each matrix, the element represents the probabilistic relatedness of each session or page with each semantic factor. In summary after the semantic space learning, the original user session and web page are simultaneously projected onto a same semantic factor space with various probabilistic weights.

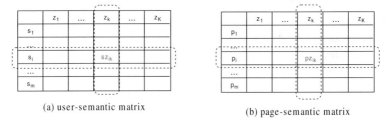

(a) user-semantic matrix   (b) page-semantic matrix

Fig. 1. Relationship between user, page and semantic factor

Since we aim to find out the co-cluster of pages and sessions corresponding to each semantic factor, so we join the matrix SZ and PZ together into a single $SPZ = \begin{bmatrix} SZ \\ PZ \end{bmatrix}$. Then, we apply a co-clustering algorithm, which will be presented in section 3.3 to find out the co-clusters in the semantic space by referring to the calculated matrices .

3.3   Abstracting Co-clustering in Semantic Space

In this section, a clustering algorithm is proposed to find out co-clusters embedded in semantic space. For each semantic $z_k \in Z$, there is a corresponding co-cluster C_k. Since web pages and web sessions have a latent semantic relationship with each semantic factor in probability model, so we use Definition 1 to define a co-cluster C_k.

Definition 1. *Given a semantic $z_k \in Z$, and a probability threshold μ, the co-cluster C_k corresponding to the semantic factor z_k is defined as: $C_k = (S_k, P_k)$*

where $S_k = \{s_i | sz_{ik} \geq \mu, i = 1, \ldots, m\}$ and $P_k = \{p_j | pz_{jk} \geq \mu, j = 1, \ldots, n\}$.

We use hard partition method to generate co-clusters in semantic space. In the semantic space Z, we presume there are K co-clusters with respect to K semantic factors. Definition 2 shows the process of obtaining the co-clustering results in the semantic space Z.

Definition 2. *Given a co-cluster C_k corresponding to a semantic factor $z_k \in Z$, $k = 1, \ldots, K$. The co-clustering result C in the semantic space Z is defined as: $C = \{C_1, C_2, ..., C_K\} = \{(S_1, P_1), \cdots, (S_k, P_k)\}$*

Algorithm 1 shows the details of finding co-clusters. At first, we set $C \leftarrow \emptyset$ as the initialization after the semantic factor space is learned via the PLSA model, and then, for each semantic factor z_k, we simultaneously filter out the web sessions and web pages which have the occurrence probability to the factor z_k greater than a probability threshold μ, and the co-cluster C_k corresponding to z_k is generated by aggregating these filtered sessions and pages. At last, we insert C_k into C.

Algorithm 1. Discovering co-clusters in the semantic space.

Input: the web usage matrix SP, the probability threshold μ
Output: C
1: employ the PLSA model to capture the latent semantic space Z, and obtain the projection matrices of web sessions and pages SZ and PZ over the semantic space and join them together;
2: $C \leftarrow \emptyset$;
3: for $k = 1, \ldots, K$
    3.1 find out the co-cluster C_k corresponding to the semantic factor Z_k according to Definition 1;
    3.2 $C \leftarrow C \cup C_k$;
4: end
5: return $C = \{C_1, C_2, ..., C_K\}$.

For each semantic z_k in Z, we must check the probability matrix SZ and PZ to decide whether their probabilities are over the probability threshold μ. So it needs $O(m + n)$ time. Algorithm 1 needs to check the K semantic factors in Z, so the time cost is $O(K(m + n))$.

4   Experiments and Evaluations

In order to evaluate the effectiveness of our proposed method, we have conducted preliminary experiments on two real world data sets. The first data set we used is downloaded from KDDCUP website [1]. After data preparation, we have setup an evaluation data set including 9308 user sessions and 69 pages and we refer this data set to "KDDCUP data". The second data set is from a university website log file [10]. The data is based on a 2-week weblog file during April of 2002 and the filtered data contains 13745 sessions and 683 pages. For brevity we refer this data as "CTI data". By considering the number of web pages and the content of the web site carefully and referring the selection criteria of factors in [11,10], we choose 15 and 20 factors for KDDCUP and CTI dataset for experiment, respectively.

Our aim is to validate how strong the correlation between the user sessions and pages within the co-cluster is. Here we assume a better co-clustering representing the fact that most of the user sessions visited the pages which are from the same co-cluster whereas the pages were largely clicked by the user sessions within the same subset of sessions and pages. In particular we use precision, recall and F-score measures to show how likely the users with similar visiting preference are grouped together with the related web pages within the same cluster. For each $C_k = (S_k, P_k)$, its precision and recall measures are defined as below.

Definition 3. *Give a co-cluster $C_k = (S_k, P_k)$, its precision and recall measures are defined as the linear combination of the row (i.e.session) precision and column (i.e. page) precision, and the linear combination of the row and column recall.*

$$precision(C_k) = \alpha * precision(R_{C_k}) + (1 - \alpha) * precision(C_{C_k})$$
$$recall(C_k) = \alpha * recall(R_{C_k}) + (1 - \alpha) * recall(C_{C_k})$$

where the row and column precision, and the row and column recall are defined as follows, respectively

$$precision(R_{C_k}) = \frac{\sum_{i \in S_k} \sum_{j \in P_k} a'_{ij}/|P_k|}{|S_k|}, \quad precision(C_{C_k}) = \frac{\sum_{j \in P_k} \sum_{i \in S_k} a'_{ji}/|S_k|}{|P_k|},$$

$$recall(R_{C_k}) = \sum_{i \in S_k} \frac{\sum_{j \in P_k} a'_{ij}}{\sum_{l=1}^{n} a'_{il}}/|S_k|, \quad recall(C_{C_k}) = \sum_{j \in P_k} \frac{\sum_{i \in S_k} a'_{ji}}{\sum_{l=1}^{n} a'_{jl}}/|P_k|,$$

where α is the combination factor, a'_{ji} is the binary representation of usage data, that is, $a'_{ji} = 1$, if $a_{ij} \geq 1$; is 0, otherwise.

Definition 4. *Given the co-clustering of $C = \{C_1, ..., C_K\}$, the precision and recall of C is defined as follows:*

$precision = \frac{1}{K} \sum_{k=1}^{K} precision(C_k)$, $recall = \frac{1}{K} \sum_{k=1}^{K} recall(C_k)$

Because we consider that sessions and pages have the equal contribution on the cluster conformation, in the experiment, the combination factor α is set as 0.5. It is clear that, the higher value of precision and recall denotes a better clustering being executed. We conducted evaluations on these two datasets in terms of precision, recall. We run 30 times of the proposed algorithm COCS and the compared algorithm: Spectral Co-Clustering (SCC) on CTI and KDDCUP datasets first, and then we denote the process as COCS_CTI, COCS_KDD, SCC_CTI and SCC_KDD, respectively. The results are presented in Fig.2-3.

---

[1] http://www.ecn.purdue.edu/KDDCUP/

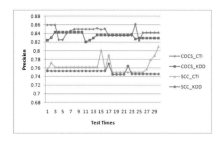
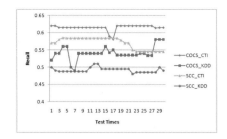

Fig. 2. Precision Comparison Results   **Fig. 3.** Recall Comparison Results

From Fig.2, it is seen that the precisions of COCS on CTI and KDDCUP data-sets are between 0.82 to 0.86 while the precisions of SCC on these two datasets are ranging from 0.74 to 0.81, resulting in by 7.7% worse than COCS. In COCS, sessions and pages are organized according to different semantic factors, that is, each semantic factor in the semantic space has a corresponding co-cluster. However, in SSC, the co-clusters are generated by calculating the cosine similarity between sessions and pages on a projected spectrum (i.e attribute) space. The co-cluster of sessions and pages in same co-cluster have no straight relationship with the semantic factor. According to the co-clustering result of COCS and SCC, we conclude the fact that co-clusters of COCS are grouped more semantic-related than those of SSC, so the precision values of COCS on these two datasets are both higher than that of SCC.

Fig.3 shows the recall comparison of COCS and SCC on CTI and KDDCUP datasets. From this figure, we can know that, in addition to the finding that COCS always outperforms SSC by around 15%, the recall values of COCS and SCC on the CTI dataset are much higher than those on the KDDCUP dataset respectively - recall values ranging in (0.58-0.62) vs. (0.49-0.55). This is likely because that the KDD dataset is relative sparse due to the bigger granularity level.

5   Conclusion and Future Work

Web clustering is an approach for aggregating web objects into various categories according to underlying relationships among them. In this paper, we address the question of discovering the co-clusters of web users and web pages via latent semantic analysis. A framework named COCS is proposed to deal with the simultaneous grouping of web objects. Experiments results have shown that the proposed method largely outweighs the existing co-clustering approaches in terms of precision and recall measures due to the capability of latent semantic analysis. Our future work will focus on the following issues: we intend to conduct more experiments to validate the scalability of our approach, and investigate the impact of the selection of different semantic factor number on the co-clustering performance.

Acknowledgment

This work has been partially supported by FP7 ICT project "M-Eco: Medical Ecosystem Personalized Event-Based Surveillance" under grant No. 247829 and "KiWi - Knowledge in a Wiki" under grant agreement No. 211932, and the Nature Science Foundation of Anhui Education Department under the Grant No KJ2009A54, KJ2010A325 and KJ2009B126.

References

1. Zhang, Y., Yu, J.X., Hou, J.: Web Communities: Analysis and Construction. Springer, Heidelberg (2006)
2. Wang, X., Zhai, C.: Learn from web search logs to organize search results. In: SIGIR 2007: Proceedings of the 30th Annual International ACM SIGIR Conference on Research and Development in Information Retrieval, pp. 87–94. ACM, New York (2007)
3. Kummamuru, K., Lotlikar, R., Roy, S., Singal, K., Krishnapuram, R.: A hierarchical monothetic document clustering algorithm for summarization and browsing search results. In: WWW 2004: Proceedings of the 13th International Conference on World Wide Web, pp. 658–665. ACM, New York (2004)
4. Flesca, S., Greco, S., Tagarelli, A., Zumpano, E.: Mining user preferences, page content and usage to personalize website navigation. World Wide Web Journal 8(3), 317–345 (2005)
5. Zeng, H.J., Chen, Z., Ma, W.Y.: A unified framework for clustering heterogeneous web objects. In: WISE 2002: Proceedings of the 3rd International Conference on Web Information Systems Engineering, Washington, DC, USA, pp. 161–172. IEEE Computer Society, Los Alamitos (2002)
6. Xu, G., Zong, Y., Dolog, P., Zhang, Y.: Co-clustering analysis of weblogs using bipartite spectral projection approach. In: Proceedings of 14th International Conference on Knowledge-Based and Intelligent Information & Engineering Systems (2010)
7. Koutsonikola, V.A., Vakali, A.: A fuzzy bi-clustering approach to correlate web users and pages. I. J. Knowledge and Web Intelligence 1(1/2), 3–23 (2009)
8. Newman, M.E.J., Girvan, M.: Finding and evaluating community structure in networks. Physical Review 69, 26113 (2004)
9. Dempster, A.P., Laird, N.M., Rubin, D.B.: Maximum likelihood from incomplete data via the em algorithm. Journal Of The Royal Statistical Society, Series B 39(1), 1–38 (1977)
10. Jin, X., Zhou, Y., Mobasher, B.: A maximum entropy web recommendation system: Combining collaborative and content features. In: Proceedings of the ACM SIGKDD Conference on Knowledge Discovery and Data Mining (KDD 2005), Chicago, pp. 612–617 (2005)
11. Hofmann, T.: Unsupervised learning by probabilistic latent semantic analysis. Machine Learning Journal 42(1), 177–196 (2001)

A Linear-Chain CRF-Based Learning Approach for Web Opinion Mining

Luole Qi and Li Chen

Department of Computer Science,
Hong Kong Baptist University,
Hong Kong, China
{llqi,lichen}@comp.hkbu.edu.hk

Abstract. The task of opinion mining from product reviews is to extract the product entities and determine whether the opinions on the entities are positive, negative or neutral. Reasonable performance on this task has been achieved by employing rule-based, statistical approaches or generative learning models such as hidden Markov model (HMMs). In this paper, we proposed a discriminative model using linear-chain Conditional Random Field (CRFs) for opinion mining. CRFs can naturally incorporate arbitrary, non-independent features of the input without making conditional independence assumptions among the features. This can be particularly important for opinion mining on product reviews. We evaluated our approach base on three criteria: recall, precision and F-score for extracted entities, opinions and their polarities. Compared to other methods, our approach was proven more effective for accomplishing opinion mining tasks.

Keywords: Conditional Random Field (CRFs), Web Opinion Mining, Feature Function.

1 Introduction

User-generated reviews have been increasingly regarded useful in business, education, and e-commerce, since they contain valuable opinions originated from the user's experiences. For instance, in e-commerce sites, customers can assess a product's quality by reading other customers' reviews to the product, which will help them to decide whether to purchase the product or not. Nowadays, many e-commerce websites, such as Amazon.com, Yahoo shopping, Epinions.com, allow users to post their opinions freely. The reviews' number has in fact reached to more than thousands in these large websites, which hence poses a challenge for a potential customer to go over all of them.

To resolve the problem, researchers have done some work on Web opinion mining which aims to discover the essential information from reviews and then present to users. Previous works have mainly adopted rule-based techniques [2] and statistic methods [10]. Lately, a new learning approach based on a sequence model named Hidden Markov model (HMMs) was proposed and proved more

L. Chen, P. Triantafillou, and T. Suel (Eds.): WISE 2010, LNCS 6488, pp. 128–141, 2010.

effective than previous works. However, the HMMs-based method is still limited because it is difficult to model arbitrary, dependent features of the input word sequence.

To address the limitation, in this paper, we have particularly studied the Conditional Random Field (CRFs) [11], because it is a discriminative, undirected graphical model that can potentially model the overlapping, dependent features. In prior work on natural language processing, CRFs have been demonstrated to out-perform HMMs [12][13]. Thus, motivated by the earlier findings, we propose a linear-chain CRF-based learning approach to mine and extract opinions from product reviews on the Web. Specifically, our objective is to answer the following questions: (1) how to define feature functions to construct and restrict our linear-chain CRFs model? (2) How to choose criteria for training a specific model from the manually labeled data? (3) How to automatically extract product entities and identify their associated opinion polarities with our trained model? In the experiment, we evaluated the model on three evaluation metrics: recall, precision and F-score, on extracted entities and opinions. The experimental results proved the higher accuracy of our proposed approach in accomplishing the web opinion mining task, in comparison with the related rule-based and HMMs-based approaches.

To highlight our contributions, we have demonstrated that the linear-chain CRF-based learning method can perform better than L-HMMs approach, in terms of integrating linguistic features for opinion mining. The feature functions defined in our work have been also indicated robust and effective for the model construction.

The rest of this paper is therefore organized as follows: we will discuss related work in Section 2 and describe in detail our proposed CRF-based opinion mining learning method in Section 3. In Section 4, we will present experiment design and results. Finally, we will conclude our work and give its future directions.

2   Related Work

Thus far, many researchers have attempted to extract opinion values from user reviews (or called documents in some literatures). At the document level, Turney et al. [2] used pointwise mutual information (PMI) to calculate the average semantic orientation (SO) of extracted phrases for determining the document's polarity. Pang et al [4] examined the effectiveness of applying machine learning techniques to address the sentiment classification problem for movie review data. Hatzivassiloglou and Wiebe [6] studied the effect of dynamic adjectives, semantically oriented adjectives, and gradable adjectives on a simple subjectivity classifier, and proposed a trainable method that statistically combines two indicators of grad ability. Wiebe and Riloff [7] proposed a system called OpinionFinder that automatically identifies when opinions, sentiments, speculations, and other private states are present in text, via the subjectivity analysis. Das and Chen [9] studied sentimental classification for financial documents. However, although the above works are all related to sentiment classification, they just use the sentiment to represent a reviewer's overall opinion and do not find what features the reviewer actually liked and disliked. For example, an overall negative sentiment on an object does not mean that the reviewer dislikes every aspect of the object, which can indeed only indicate that the average opinion as summarized from this review is negative.

To in-depth discover a reviewer's opinions on almost every aspect that she mentioned in the text, some researchers have tried to mine and extract opinions at the feature level. Hu and Liu [10] proposed a feature-based opinion summarization system that captures highly frequent feature words by using association rules under a statistical framework. It extracts the features of a product that customers have expressed their opinions on, and then concludes with an opinion score for each frequent feature (one of the top ranked features) while ignoring infrequent features. Popescu and Etzioni [3] improved Hu and Liu's work by removing frequent noun phrases that may not be real features. Their method can identify part-of relationship and achieve a better precision, but a small drop in recall. Scaffidi et al [8] presented a new search system called Red Opal that examined prior customer reviews, identified product features, and then scored each product on each feature. Red Opal used these scores to determine which products to be returned when a user specifies a desired product feature. However, the limitation of these works is that they failed to identify infrequent entities effectively.

The work that is most similar to our focus is a supervised learning system called OpinionMiner [1]. It was built under the framework of lexicalized HMMs that integrates multiple important linguistic features into an automatic learning process. The difference between our method and it is that we employ CRFs in order to avoid some limitations that are inherent in HMMs. Indeed, it can not represent distributed hidden states and complex interactions among labels. It can neither involve rich, overlapping feature sets. Miao et al [15] have recently also attempted to adopt CRFs to extract product features and opinions, but they did not identify sentiment orientation using CRFs and did not make a comparison across different methods.

3   The Proposed Approach

Fig. 1 gives the architecture overview of our approach. It can be divided into four major steps: (1) pre-processing which includes crawling raw review data and cleaning; (2) review data tagging; (3) training the liner-chain CRFs model; and (4) applying the model to obtain opinions from new review data.

3.1   CRFs Learning Model

Conditional random fields (CRFs) are conditional probability distributions on an undirected graph model [11]. It can be defined as follows: considering a graph $G = (V, E)$ for which V indicates the nodes and E indicates the edges. Let $Y = (Y_v)_{v \in V}$, and (X, Y) is a CRF, where X is the set of variables over the

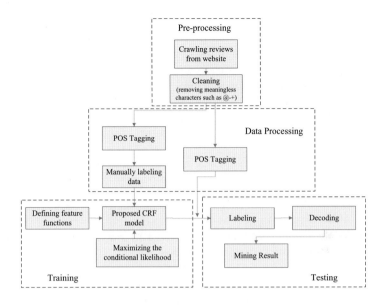

Fig. 1. The architecture of our proposed system

observation sequences to be labeled (e.g., a sequence of textual words that form a sentence), and Y is the set of random variables over the corresponding sequence. The (X, Y) obeys the Markov property with respect to the graph(e.g., part-of-speech tags for the words' sequence). Formally, the model defines $p(y|x)$ which is globally conditioned on the observation of X:

$$p(y|x) = \frac{1}{Z(x)} \prod_{i \in N} \phi_i(y_i, x_i) \tag{1}$$

where $Z(x) = \sum_y \prod_{i \in N} \phi_i(y_i, x_i)$ is a normalization factor over all state sequences for the sequence x. The potentials are normally factorized on a set of features f_k, as

$$\phi_i(y_i, x_i) = \exp(\sum_k \lambda_k f_k(y_i, x_i)) \tag{2}$$

Given the model defined in equation (1), the most probable labeling sequence for an input x is

$$\widehat{Y} = \arg\max_y p(y|x) \tag{3}$$

3.2 Problem Statement

Our goal was to extract product entities from reviews which also include the opinion polarities. The product entities can be divided into four categories according to [1]: Components, Functions, Features and Opinions. Please notice that the features mentioned here refer to the product's features (e.g., the camera's size, weight) which are different from the meaning of features in the feature functions for constructing CRFs models (see the definition of feature functions later). Table 1 shows the four categories of entities and their examples. In our work, we follow this classification scheme.

Table 1. Four types of product entities and their examples (referred to [1])

Components	Physical objects of a product, such as cellphone's LCD
Functions	Capabilities provided by a product, e.g., movie playback, zoom, automatic fillflash, auto focus
Features	Properties of components or functions, e.g., color, speed, size, weight
Opinions	Ideas and thoughts expressed by reviewers on product features, components or functions.

We employ three types of tags to define each word: entity tag, position tag and opinion tag. We use the category name of a product entity to be the entity tag. As for a word which is not an entity, we use the character 'B' to represent it. Usually, an entity could be a single word or a phrase. For the phrase entity, we assign a position to each word in the phrase. Any word of a phrase has three possible positions: the beginning of the phrase, the middle of the phrase and the end of phrase. We use characters 'B', 'M' and 'E' as position tags to respectively indicate the three positions. As for "opinion" entity, we further use characters 'P' and 'N' to respectively represent Positive opinion and Negative opinion polarity, and use "Exp" and "Imp" to respectively indicate Explicit opinion and Implicit opinion. Here, explicit opinion means the user expresses opinion in the review explicitly and implicit opinion means the opinion needs to be induced from the review. These tags are called opinion tags. Thus with all of above defined tags, we can tag any word and its role in a sentence. For example, the sentence "The image is good and its ease of use is satisfying" from a camera review is labeled as:

The(B) image(Featue-B) is(B) good(Opinion-B-P-Exp) and(B) its(B) ease(Feature-B) of(Feature-M) use(Feature-E) is(B) satisfying(Opinion-B-P-Exp) .

In this sentence, 'image' and 'ease of use' are both features of the camera and 'ease of use' is a phrase, so we add '-B', '-M' and '-E' to specify the position of each word in the phrase. 'Good' is a positive, explicit opinion expressed on the feature 'image', so its tag is 'Opinion-B-P-Exp' (such tag combination is also called hybrid tags in [1]). Other words which do not belong to any entity categories are given the tag 'B'.

Therefore, when we get each word's tag(s), we could obtain the product entity it refers to and identify the opinion orientation if it is an "opinion" entity. In this way, the task of opining mining can be transformed to an automatic labeling task. The problem can be then formualized as: given a sequence of words $W = w_1 w_2 w_3 ... w_n$ and its corresponding parts of speech $S = s_1 s_2 s_3 ... s_n$, the objective is to find an appropriate sequence of tags which can maximize the conditional likelihood according to equation (3).

$$\widehat{T} = \arg \max_T p(T|W, S) = \arg \max_T \prod_{i=1}^{N} p(t_i|W, S, T^{(-i)}) \qquad (4)$$

In equation (4), $T^{(-i)} = \{t_1 t_2 ... t_{i-1} t_{i+1} ... t_N\}$ (which are tags in our case, and called hidden states in the general concept). From this equation, we can see that the tag of word at position i depends on all the words $W = w_{1:N}$, part-of-speech $S = s_{1:N}$ and tags. Unfortunately it is very hard to compute with this equation as it involves too many parameters. To reduce the complexity, we employ linear-chain CRFs as an approximation to restrict the relationship among tags. It is a graphic structure as shown in Figure 2 (in the figure, Y forms a simple first-order chain). In the linear-chain CRF, all the nodes in the graph form a linear chain and each feature involves only two consecutive hidden states. Equation (4) can be hence rewritten as

$$\widehat{T} = \arg \max_T p(T|W, S) = \arg \max_T \prod_{i=1}^{N} p(t_i|W, S, t_{i-1}) \qquad (5)$$

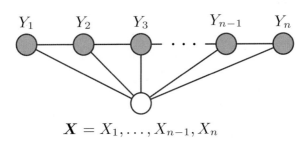

Fig. 2. Liner-CRFs graphic structure

3.3   Feature Functions

From the model above, we can see there are still many parameters to be processed. To make the model more computable, we need to define the relationships among the observation states $W = w_{1:N}$, $S = s_{1:N}$ and hidden states $T = t_{1:N}$ so as to reduce the unnecessary calculations. Thus, behaving as the important constructs of CRFs, feature functions are crucial to resolve our problem. Let $w_{1:N}$, $s_{1:N}$ be the observations (i.e., words' sequence and their corresponding parts of speech), $t_{1:N}$ the hidden labels (i.e., tags). In our case of linear-chain CRF (see equation (5)), the general form of a feature function is $f_i(t_{j-1}, t_j, w_{1:N}, s_{1:N}, n)$, which looks at a pair of adjacent states t_{j-1}, t_j, the whole input sequence $w_{1:N}$ as well as $s_{1:N}$ and the current word's position. For example, we can define a simple feature function which produces binary value: the returned value is 1 if the current word w_j is "good", the corresponding part-of-speech s_j is "JJ" (which means single adjective word) and the current state t_j is "Opinion":

$$f_i(t_{j-1}, t_j, w_{1:N}, s_{1:N}, j) = \begin{cases} 1 \; if \; w_j = good, \; s_j = JJ \; and \; t_j = Opinion \\ 0 \; otherwise \end{cases} \quad (6)$$

Combining feature function with equation (1) and equation (2), we have:

$$p(t_{1:N}|w_{1:N}, s_{1:N}) = \frac{1}{Z} \exp(\sum_{j=1}^{N} \sum_{i=1}^{F} \lambda_i f_i(t_{j-1}, t_j, w_{1:N}, s_{1:N}, j)) \quad (7)$$

According to equation (7), the feature function f_i depends on its corresponding weight λ_i. That is if $\lambda_i > 0$, and f_i is active (i.e., $f_i = 1$), it will increase the probability of the tag sequence $t_{1:N}$ and if $\lambda_i < 0$, and f_i is inactive (i.e., $f_i = 0$), it will decrease the probability of the tag sequence $t_{1:N}$.

Another example of feature function can be like:

$$f_i(t_{j-1}, t_j, w_{1:N}, s_{1:N}, j) = \begin{cases} 1 \; if \; w_j = good, \; s_{j+1} = NN \; and \; t_j = Opinion \\ 0 \; otherwise \end{cases}$$
$$(8)$$

In this case, if the current word is "good" such as in the phrase "good image", the feature function in equation (6) and (8) will be both active. This is an example of overlapping features which HMMs can not address. In fact, HMMs cannot consider the next word, nor can they use overlapping features.

In addition to employing linear-chain CRFs to simplify the relations within hidden states T, we also define several different types of feature functions to specify state-transition structures among W, S and T. The different state transition features are based on different markov orders for different classes of features. Here we define the first order features:

1. The assignment of current tag t_j is supposed to only depend on the current word. The feature function is represented as $f(t_j, w_j)$.
2. The assignment of current tag t_j is supposed to only depend on the current part-of-speech. The feature function is represented as $f(t_j, s_j)$.
3. The assignment of current tag t_j is supposed to depend on both the current word and the current part-of-speech. The feature function is represented as $f(t_j, s_j, w_j)$.

All the three types of feature functions are first-order, with which the inputs are examined in the context of the current state only. We also define first-order+transition features and second-order features which are examined in the context of both the current state and previous states. We do not define third-order or higher-order features because they create data sparse problem and require more memory during training. Table 2 shows all the feature functions we have defined in our model.

Table 2. The feature function types and their expressions

Feature type	Expressions
First-oder	$f(t_i, w_i)$, $f(t_i, s_i)$, $f(t_i, s_i, w_i)$
First-order+transitions	$f(t_i, w_i)f(t_i, t_{i-1})$, $f(t_i, s_i)f(t_i, t_{i-1})$, $f(t_i, s_i, w_i)f(t_i, t_{i-1})$
Second-order	$f(t_i, t_{i-1}, w_i,)$, $f(t_i, t_{i-1}, s_i)$, $f(t_i, t_{i-1}, s_i, w_i)$

3.4 CRFs Training

After the graph and feature functions are defined, the model is fixed. The purpose of training is then to identify all the values of $\lambda_{1:N}$. Usually one may set $\lambda_{1:N}$ according to the domain knowledge. However, in our case, we learn $\lambda_{1:N}$ from training data. The fully labeled review data is $\{(w^{(1)}, s^{(1)}, t^{(1)}), ..., (w^{(M)}, s^{(M)}, t^{(M)})\}$, where $w^{(i)} = w^{(i)}_{1:N_i}$ (the ith words sequence), $s^{(i)} = s^{(i)}_{1:N_i}$ (the ith part-of-speech sequence), $t^{(i)} = t^{(i)}_{1:N_i}$ (the ith tags sequence) respectively. Given that in CRFs, we defined the conditional probability $p(t|w, s)$, the aim of parameter learning is to maximize the conditional likelihood based on the training data:

$$\sum_{j=1}^{M} \log p(\mathbf{t}^{(j)} | \mathbf{w}^{(j)}, \mathbf{s}^{(j)}) \tag{9}$$

To avoid over-fitting, log-likelihood is usually penalized by some prior distributions over the parameters. A commonly used distribution is a zero-mean Gaussian. If $\lambda \sim N(0, \sigma^2)$, the equation of (9) becomes

$$\sum_{j=1}^{M} \log p(t^{(j)} | w^{(j)}, s^{(j)}) - \sum_{i}^{F} \frac{\lambda_i^2}{2\sigma^2} \tag{10}$$

The equation is concave, so λ has a unique set of global optimal values. We learn parameters by computing the gradient of the objective function, and use the gradient in an optimization algorithm called Limited memory BFGS (L-BFGS).

The gradient of the objective function is formally computed as follows:

$$
\begin{aligned}
&\frac{\partial}{\partial \lambda_k} \sum_{j=1}^{m} \log p(t^{(j)}|w^{(j)}, s^{(j)}) - \sum_{i}^{F} \frac{\lambda_i^2}{2\sigma^2} \\
&= \frac{\partial}{\partial \lambda_k} \sum_{j=1}^{m} (\sum_{n}\sum_{i} \lambda_i f_i(t_{n-1}, t_n, w_{1:N}, s_{1:N}, n) - \log T^{(j)}) - \sum_{i}^{F} \frac{\lambda_i^2}{2\sigma^2} \\
&= \sum_{j=1}^{m}\sum_{n} f_k(t_{n-1}, t_n, w_{1:N}, s_{1:N}, n) \\
&\quad - \sum_{j-1}^{m}\sum_{n} E_{t'_{n-1}, t'_n}[f_k(t'_{n-1}, t'_n, w_{1:N}, s_{1:N}, n)] - \frac{\lambda_k}{\sigma^2}
\end{aligned}
\tag{11}
$$

In equation (11), the first term is the empirical count of feature i in the training data, the second term is the expected count of this feature under the current trained model and the third term is generated by the prior distribution. Hence, the derivative measures the difference between the empirical count and the expected count of a feature under the current model. Suppose that in the training data a feature f_k appears A times, while under the current model, the expected count of f_k is B: when $|A| = |B|$, the derivative is zero. Therefore, the training process is to find λs that match the two counts.

4   Experiment

In this section, we present the measurements of recall, precision and F-score with our approach. Recall is $\frac{|C \cap P|}{|C|}$ and Precision is $\frac{|C \cap P|}{|P|}$, where C and P are the sets of correct and predicted tags, respectively. F score is the harmonic mean of precision and recall, $\frac{2RP}{R+P}$. In the experiment, two related methods were compared to our method: the rule-based method as the baseline and the L-HMM model as described [1] We used two datasets of product reviews: one was crawled from Yahoo shopping, and another was the sharable corpus from Liu and Hu's work [10]. For example, Figure 3 gives one digital camera's user review (in XML format) crawled from Yahoo Shopping site, for which we mainly focused on the "Posting" part as it provides the user-generated textual comments.

We finally collected 476 reviews in total for three cameras and one cellphone, and manually labeled them by using the tags as defined in Section 3.2. After the pre-processing to remove meaningless characters (e.g., @-+), we applied the LBJPOS tool [17] to produce the part-of-speech tag for each word in every review. All tagged data were then divided into 4 four sets to perform 4-fold cross-validation: one set was retained as the validation data for testing, and the other 3 sets were used as training data. The cross-validation process was repeated four times, and every time one set was used as the validation data. Afterwards, the results were averaged to produce precision, recall and F score.

4.1   Compared Methods in the Experiment

Rule-base Method. Motivated by [2], we designed a rule-based method as the baseline system for comparison. The first step was performing Part-of-Speech (POS) task. One example of POS result is:

```
<Review>
  <Title>Great Camera</Title>
  <Reviewer>I_infante69</Reviewer>
  <CreateTime>1133976475</CreateTime>
  <HelpfulRecommendations>3</HelpfulRecommendations>
  <TotalRecommendations>4</TotalRecommendations>
- <Ratings>
    <Rating ratingType="Features">5</Rating>
    <Rating ratingType="Overall">5</Rating>
    <Rating ratingType="Quality">5</Rating>
    <Rating ratingType="Support">5</Rating>
    <Rating ratingType="Value">5</Rating>
  </Ratings>
  <OverallRating>5</OverallRating>
  <Pro>Light weight, great battery power</Pro>
  <Con>PC Picture Software and Users Guide</Con>
  <Posting>This is a great camera. I shopped around and got a great
    price. This is my first digital camera. No problems with the
    pictures or the screen. The battery power is fantastic, the size is
    great, and the pictures and photo options are really nice. <br>
    <br>The user guide isn&#39;t very user friendly. If you are not
    electronic savy, it may take some time to figure out this camera.
    <br> <br>The software to load the pictures on my PC is also not
    very user friendly. The only way I can crop and edit pictures is by
    loading into a different application (such as HP photo
    director).</Posting>
</Review>
```

Fig. 3. An example of one digital camera's user review in XML format. The <Posting>... </Posting>part gives the textual comment that was emphasized in our algorithm

(PRP I) (VBD used) (NNP Olympus) (IN before) (, ,) (VBG comparing) (TO to) (NN canon) (, ,) (PRP it) (VBD was) (DT a) (NN toy) (, ,) (NNP S3) (VBZ IS) (VBZ is) (RB not) (DT a) (JJ professional) (NN camera) (, ,) (CC but) (RB almost) (VBZ has) (NN everything) (PRP you) (VBP need) (..)

In the example, each word gets a tag of POS such as NN (noun word), JJ (adjective word) etc. We then applied several basic rules to extract objective product entities (i.e., components, functions and features as defined in Table 1).

1) One rule is that a single noun that follows an adjective word or consecutive adjective words will be regarded as a product entity, such as JJ + NN or JJ.

2) Any single noun word that connects an adjective word to a verb will be taken as a product entity, such as NN + VBZ +JJ.

3) Any consecutive noun words that appear at the position described in 1) or 2) will be taken as a product entity phrase.

The three rules were actually derived from observations obtained in [10][1].

As for opinion words, the adjective words that appear in rules 1 and 2 will be opinion entities, and their sentimental orientation was determined by a lexicon with polarities for over 8000 adjective words [16].

L-HMMs Method. The work in [1] integrated linguistic features such as part-of-speech results and lexical patterns into HMMs. Their aim was to maximize the conditional probability as defined in:

$$\widehat{T} = \arg\max_{T} axp(W, S|T)p(T) = \arg\max_{T} axp(S|T)p(W|T, S)p(T)$$

$$= \arg\max_{T} \prod_{i=1}^{N} \left\{ \begin{array}{l} p(s_i|w_1...w_{i-1}, s_1...s_{i-1}, t_1...t_{i-1}t_i) \times \\ p(w_i|w_1...w_{i-1}, s_1...s_{i-1}s_i, t_1...t_{i-1}t_i) \times \\ p(t_i|w_1...w_{i-1}, s_1...s_{i-1}, t_1...t_{i-1}) \end{array} \right\}$$

Three assumptions were made for simplifying the problem: (1) the assignment of the current tag is supposed to depend not only on its previous tag but also on the previous J words. (2) The appearance of the current word is assumed to depend not only on the current tag, the current POS, but also on the previous K words. (3) The appearance of the current POS is supposed to depend both on the current tag and previous L words (J=K=L). Then the objective was to maximize

$$\arg\max_{T} \prod_{i=1}^{N} \left\{ \begin{array}{l} p(s_i|w_{i-1}, t_i) \times \\ p(w_i|w_{i-1}, s_i, t_i) \times \\ p(t_i|w_{i-1}, t_{i-1}) \end{array} \right\}$$

Maximum Likelihood Estimation (MLE) was used here to estimate the parameters. Other techniques were also used in this approach, including information propagation using entity synonyms, antonyms and related words, and token transformations, in order to in order to lead to an accurate extraction.

4.2 Experimental Results and Discussion

Table 3 shows the experimental results of recall, precision and F-score through the 4-fold cross-validation, from which we can see that the CRF-based learning method (henceforth CRF) increases the accuracy regarding almost all the four types of entities, except the slightly lower recall land F-score than L-HMMs method (henceforth L-HMM) in respect of component entity. Please note that the empty value of the baseline approach in terms of the four entity types is because the baseline method does not support to extract the four types of entities, so it only has the results averaged on all entities.

More specifically, CRF improved the precision from 83.9% to 90.0% on the average value and the F-score from 77.1% to 84.3% in comparison with L-HMM. Two major reasons can lead to this result. Firstly, L-HMM assumes that each feature is generated independently of hidden processes. That is, only tags can affect each other and the underlying relationships between tags and words/POS-tags are ignored. Secondly, HMMs does not model the overlapping features. As for recall, it was also averagely improved (from 72.0% by L-HMM to 79.8% by CRF). This is promising because the recall can be easily affected by errors in tagging. For example, if a sentence's correct tags should be "Opinion-B-P-Exp, Opinion-M-P-Exp, Opinion-M-P-Exp, Opinion- E-P-Exp" and it was labled

Table 3. Experimental results from the comparison of the three approaches: Baseline - the rule-based opinion mining method, L-HMM - the Lexicalized-Hidden Markov model based learning method, CRF - the Conditional Random Field based learning method (R: recall, P: precision, F: F-score)

Methods		Baseline	L-HMM	CRF
Feature Entities(%)	R	-	78.6	81.8
	P	-	82.2	93.5
	F	-	80.4	87.2
Component Entities(%)	R	-	96.5	91.8
	P	-	95.3	98.7
	F	-	96.0	95.1
Function Entities(%)	R	-	58.9	80.4
	P	-	81.1	83.7
	F	-	68.2	82.0
Opinion Entities(%)	R	-	53.7	65.3
	P	-	76.9	84.2
	F	-	63.2	73.5
All Entities(%)	R	27.2	72.0	79.8
	P	24.3	83.9	90.0
	F	25.7	77.1	84.3

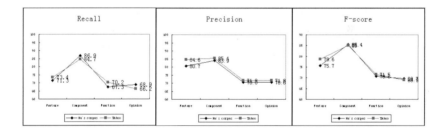

Fig. 4. Recall, Precision and F-score resulting from training CRF respectively with the two datasets

as "Opinion-B-P-Exp, Opinion-E-P-Exp, Opinion-M-P-Exp, Opinion-E-P-Exp", the labeling accuracy is 75%, but recall is 0.

We also conducted a comparison between the two datasets: 238 documents from [10] and 238 documents from Yahoo Shopping. The procedure is that we first trained the CRF on the first dataset and then tested it with the second one, and vice versa. Fig. 4 shows the precision, recall and F-scores averaged over four entities. It can be seen that there is no big difference between the two datasets. The largest distance occurs with the precision on feature entity, but it is only 3.9%. The result infers that our approach can achieve a stable performance with different datasets and can be hence easily scalable to mine review data from various sources (e.g., from other sites like Amazon), without the need of training examples for each site.

It is also worth noting that the overlapping feature functions as defined in our CRF model can likely increase the discovery of infrequent entities, which however were largely ignored in related approaches. For example, although the entity "ISO" only appears once in our data, functions $f(t_i, s_i, w_i)$ and $f(t_i, w_i)$ can be still active in finding this feature. Moreover, the uneasily discoverable entities, such as non-noun product entities (e.g., "flashing") and non-adjective opinions (e.g., "strongly recommended"), can be also identified by our approach.

5   Conclusion

Thus, in conclusion, we proposed a novel linear-chain CRF-based learning approach for Web opinion mining. Contrasting to L-HMM-baesd method which assumes that each feature is generated being independent of hidden states, CRF-based approach can more effectively handle with dependent input features . The experiment results indeed demonstrated the effectiveness of our proposed approach in comparison with the rule-based and L-HMM-based ones. In the future, we will target to incorporate more feature functions to improve our model. Due to the complicacy of natural languages, some long-distance features are beyond our assumption, but if some appropriate feature functions can be included, we could possibly employ automatic feature induction techniques to find non-obvious conjunction of features, which may hence further improve the algorithm's performance. Moreover, we will apply the opinion mining result from product reviews to enhance existing recommender systems, such as to address the rating sparsity limitation and infer users' preferences on products from their expressed opinions.

Acknowledgement

Great thanks to my colleague Victor Cheng for his insightful suggestions and comments on our work.

References

1. Wei, J., Hung, H., Rohini, S.K.: OpinionMiner: a novel machine learning system for web opinion mining and extraction. In: 15th ACM SIGKDD International Conference on Knowledge Discovery and Data Mining, pp. 1195–1204 (2009)
2. Turney, P.D.: Thumbs up or thumbs down?: semantic orientation applied to unsupervised classification of reviews. In: 15th ACM SIGKDD International Conference on Knowledge Discovery and Data Mining, pp. 417–424 (2002)
3. Popescu, A., Etzioni, O.: Extracting product features and opinions from Reviews. In: Conference on Empirical Methods in Natural Language Processing, pp. 339–346 (2005)
4. Pang, B., Lee, L., Vaithyanathan, S.: Thumbs up?: sentiment classification using machine learning techniques. In: The ACL 2002 Conference on Empirical methods in Natural Language Processing, pp. 79–86 (2002)

5. Dave, K., Lawrence, S., Pennock, D.M.: Mining the peanut gallery: opinion extraction and semantic classification of product reviews. In: 12th International Conference on World Wide Web, pp. 519–528 (2002)
6. Hatzivassiloglou, V., Wiebe, J.M.: Effects of adjective orientation and gradability on sentence subjectivity. In: 18th Conference on Computational linguistics, pp. 299–305 (2000)
7. Wilson, T., Hoffmann, P., Somasundaran, S., Kessler, J., Wiebe, J., Choi, Y., Cardie, C., Riloff, E., Patwardhan, S.: OpinionFinder: a system for subjectivity analysis. In: HLT/EMNLP on Interactive Demonstrations, pp. 34–35 (2005)
8. Scaffidi, C., Bierhoff, K., Chang, E., Felker, M., Ng, H., Jin, C.: Red Opal: product-feature scoring from reviews. In: 8th ACM Conference on Electronic Commerce, pp. 182–191 (2007)
9. Das, S., Mike, C.: Yahoo! for Amazon: Extracting market sentiment from stock message boards. In: Asia Pacific Finance Association Annual Conference (2001)
10. Hu, M., Liu, B.: Mining and summarizing customer reviews. In: 10th ACM SIGKDD International Conference on Knowledge Discovery and Data Mining, pp. 168–177 (2004)
11. John, L., Andrew, M., Fernando, P.: Conditional random fields: probabilistic models for segmenting and labeling sequence data. In: International Conference on Machine Learning, pp. 282–289 (2001)
12. Fuchun, P., Andrew, M.: Accurate information extraction from research papers using conditional random fields. In: Human Language Technology Conference and North American Chapter of the Association for Computational Linguistics (2004)
13. Fei, S., Fernando, P.: Shallow parsing with conditional random fields. In: The 2003 Conference of the North American Chapter of the Association for Computational Linguistics on Human Language Technology, pp. 134–141 (2003)
14. McCallum, A.: Efficiently inducing features of conditional random fields. In: Conference on Uncertainty in Artificial Intelligence (2003)
15. Miao, Q., Li, Q., Zeng, D.: Mining fine grained opinions by using probabilistic models and domain knowledge. In: 2010 IEEE/WIC/ACM International Conference on Web Intelligence and Intelligent Agent Technology (2010)
16. http://www.cs.cornell.edu/People/pabo/movie-review-data/review_polarity.tar.gz
17. http://l2r.cs.uiuc.edu/~cogcomp/software.php

An Unsupervised Sentiment Classifier on Summarized or Full Reviews

Maria Soledad Pera, Rani Qumsiyeh, and Yiu-Kai Ng

Computer Science Department, Brigham Young University, Provo, Utah, U.S.A.

Abstract. These days web users searching for opinions expressed by others on a particular product or service PS can turn to review repositories, such as Epinions.com or Imdb.com. While these repositories often provide a high quantity of reviews on PS, browsing through archived reviews to locate different opinions expressed on PS is a time-consuming and tedious task, and in most cases, a very labor-intensive process. To simplify the task of identifying reviews expressing positive, negative, and neutral opinions on PS, we introduce a simple, yet effective sentiment classifier, denoted $SentiClass$, which categorizes reviews on PS using the semantic, syntactic, and sentiment content of the reviews. To speed up the classification process, $SentiClass$ summarizes each review to be classified using $eSummar$, a single-document, extractive, sentiment summarizer proposed in this paper, based on various sentence scores and anaphora resolution. $SentiClass$ ($eSummar$, respectively) is domain and structure independent and does not require any training for performing the classification (summarization, respectively) task. Empirical studies conducted on two widely-used datasets, Movie Reviews and Game Reviews, in addition to a collection of Epinions.com reviews, show that $SentiClass$ (i) is highly accurate in classifying summarized or full reviews and (ii) outperforms well-known classifiers in categorizing reviews.

1 Introduction

The rapid growth of social search websites, such as Imdb.com and Epinions.com, which allow users to express their opinions on products and services, yields large review repositories. As a side effect of the growth, finding diverse sentiment information on a particular product or service PS from these review repositories is a real challenge for web users, as well as web search engine designers. To facilitate the task of identifying reviews on PS that share the same polarity, i.e., positive, negative, or neutral, we introduce a simple, yet effective sentiment classifier, denoted $SentiClass$. Given a review R on PS, which can be extracted from existing review repositories using a simple keyword-based query, $SentiClass$ relies on the $SentiWordNet$ (sentiwordnet.isti.cnr.it) scores of each non-stopword[1] in R, which are numerical values that quantify the positive, negative, and neutral

---

[1] *Stopwords* are commonly-occurred words, such as articles, prepositions, and conjunctions, which carry little meaning. From now on, unless stated otherwise, whenever we mention (key)word(s), we mean non-stopword(s).

L. Chen, P. Triantafillou, and T. Suel (Eds.): WISE 2010, LNCS 6488, pp. 142–156, 2010.

connotation of a word, and considers the presence of *intensifiers* (e.g., "very", "extremely", "least"), *connectors* (e.g., "although","but", "however"), *reported speech* (which report what someone has said), and *negation terms* (e.g., "not", "except", "without") in R to precisely capture the sentiment on PS expressed in R and categorize R according to its polarity.

To reduce the overall classification time of *SentiClass* and shorten the length of the classified reviews a user is expected to examine, we summarize the reviews to be classified using *eSummar*, which is a single-document, extractive, sentiment summarizer introduced in this paper. *eSummar* pre-processes a review R using *anaphora resolution*, which identifies successive references of the same discourse entity in R to eliminate ambiguity when interpreting the content of R. Hereafter, *eSummar* computes for each sentence S in R the (i) *similarity* of the words in S and in the remaining sentences of R to determine how representative S is in capturing the content of R, (ii) *word significance factor*, which quantifies the significance of each word in S in representing the content of R, and (iii) *sentiment score* that reflects the degree of sentiment on PS expressed in S and is calculated using the linguistic type (such as adjective, adverb, or noun) and the *SentiWordNet* score of each word in S. A number of sentences in R with high combined scores yield the summary of R.

SentiClass (*eSummer*, respectively) does not require any training for performing the categorization (summarization, respectively) task, which simplifies and shortens the classification (summarization, respectively) process. In addition, *SentiClass* (*eSummar*, respectively) is domain independent and thus can classify (summarize, respectively) reviews with diverse structures and contents.

We proceed to present our work as follows. In Section 2, we discuss existing (sentiment) classification and summarization approaches. In Section 3, we introduce *SentiClass* and *eSummar*. In Section 4, we present the performance evaluation of *SentiClass* based on widely-used benchmark datasets and metrics, in addition to verifying the correctness of using *eSummar*, as compared with other (sentiment) summarizers, for summarizing reviews to be classified by *SentiClass*. In Section 5, we give a concluding remark.

2   Related Work

Pang et al. [18] evaluate three different machine learning approaches for (sentiment) classification: Naive Bayes, Support Vector Machines (SVM), and Maximum Entropy. Kennedy and Inkpen [6] compare two sentiment classification approaches, one of which identifies positive and negative terms in a document and labels the document positive (negative, respectively) if it contains more positive (negative, respectively) than negative (positive, respectively) terms, and the other approach trains an SVM (using single words as features) to determine the sentiment expressed in a document. As opposed to [6], the sentiment classifier in [4] trains an SVM using only a subset of the words in a document, which indicate a positive or negative intent and are determined using a maximum entropy model. Zhao et al. [22] categorize movie reviews using a probabilistic approach based on Conditional Random Field that captures the context of a sentence to infer its sentiment. Unlike *SentiClass*, all of these classification methods rely on training data and do not consider the semantics, such as negation terms, reported speech, and words in different contexts, of a sentence, which affects the polarity of words and thus their sentiment in a review.

Given an opinion-rich document D, Ku et al. [9] first employ a manually-created set of seed words with pre-determined orientation to generate a list L of positive and negative words using synsets from *WordNet*, a widely-used English lexical database. Thereafter, the authors label each sentence S in D as positive (negative, respectively) if the majority of words in S are positive (negative, respectively) according to the sentiment information of the words in L. D is classified as positive (negative, respectively) if the majority of sentences in D are labeled as positive (negative, respectively). While *SentiClass* uses *SentiWordNet* to determine the polarity of a word, the approach in [9] relies on the synsets extracted from *WordNet*, which are purely based on the original choice of seed words and thus is restricted in word usage.

The polarity of a review is determined in [21] by identifying fixed sequences of words (stems) which, when they appear together, tend to have a polarity (i.e., negative or positive orientation). Unlike *SentiClass*, this method excludes word types, such as connectors, i.e., "but", "although", "however", etc., which could eliminate or reverse the polarity of a sentiment in a review.

For single-document summarization, Radev et al. [20] claim that the most promising approach is *extraction*, which identifies and retains *sentences* that capture the content of a text as its summary. Techniques commonly used for generating extractive, single-document summaries include the Hidden Markov Model (HMM), Latent Semantic Analysis (LSA), Word Significance, and Support Vector Machines (SVM). Unlike *eSummar*, SVM is a semi-supervised method that relies on training data for generating a summary, whereas HMM, LSA, and Word Significance are unsupervised methods that (i) fail to capture the sentiments expressed in a document because they are not sentiment summarizers and (ii) do not consider the relative degree of significance of a sentence in capturing the content of a document D when selecting sentences for the summary of D. Given a review R on a product P, Hu and Wu [3] adopt a score algorithm that considers the positive or negative orientation of each word in a sentence S (in R) to determine the sentiment of S. Thereafter, the summarizer extracts key phrases in each sentence that represent positive and negative opinions on P and groups the phrases according to their polarity to generate the summary of R. While the summarizer in [3] simply lists positive and negative phrases describing P, *eSummar* actually creates a complete summary of R that reflect the sentiment on P expressed by the author of R. Zhuang et al. [23] rely on *WordNet*, statistical analysis, and movie knowledge to generate extractive, feature-based summaries of movie reviews. Unlike *eSummar*, the summarizer in [23] is domain dependent, which employs a pre-defined set of movie features, such as screenplay, and cannot be generalized. (See in-depth discussions on existing sentiment summarizers and classifiers in [17].)

3   Sentiment-based Classification

In this section, we present *SentiClass* whose overall process is shown in Figure 1.

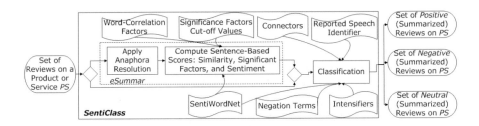

Fig. 1. Processing steps of the proposed classifier, *SentiClass*

3.1   Sentiment Classification

In classifying a review R, *SentiClass*, the proposed sentiment classifier, first determines the polarity of each word w in R such that w is positive (negative, respectively) if its positive (negative, respectively) *SentiWordNet* score is higher than its negative (positive, respectively) counterpart. (The *SentiWordNet* scores of three sample words are shown in Figure 2.) Thereafter, *SentiClass* calculates the overall sentiment score of R, denoted $SentiScore(R)$, by *subtracting* the *sum* of its *negative* words' scores from the *sum* of its *positive* words' scores, which reflects the sentiment orientation, i.e., positive, negative, or neutral, of R. Since the longer R is, the more sentiment words are in R and thus the higher its sentiment score is, we normalize the score by dividing it with the *number* of *sentiment words* in R, which yields $SentiScore(R)$. If the normalized $SentiScore(R)$ is higher (lower, respectively) than a pre-defined range (as given in Section 3.2), then R is labeled as *positive* (*negative*, respectively). Otherwise, R is *neutral*.

We support the claim made in [19] which asserts that the sentiment expressed in a review R cannot be properly captured by analyzing *solely* individual words in R. Thus, prior to computing the *SentiScore* of R, we consider *intensifiers*,

Fig. 2. The *SentiWordNet* scores of words "Excellent", "Movie", and "Poor", where P, O, and N denote Positive, Objective (i.e., neutral), and Negative, respectively

connectors, *negation terms*, and *reported speech* in R, if there are any, which affect the polarity of the keywords in R.

Intensifiers, which can be extracted from wjh.harvard.edu/~inquirer/home cat.htms, are words, such as "extremely" and "barely", that either *weaken* or *strengthen* the polarity of their adjacent words in R. *SentiClass* doubles (*divides* by half, respectively) the *SentiWordNet* score of a word w in R if an intensifier strengthens (weakens, respectively) the polarity of w. In addition, we consider **connectors**, such as "however" and "but", which can mitigate the polarity of words in a sentence S and can be compiled using *WordNet*. *SentiClass* assigns zero as the *SentiWordNet* score of any word following a connector in S.

In another initial step, *SentiClass* identifies words in R that are preceded by **negation terms**, such as "not" and "never", which are listed in [15], and labels them as "Negated". During the classification process, *SentiClass* first inverts the polarity of a word w (in R) affected by a preceding negation term in the same sentence so that if w has a *positive* (*negative*, respectively) polarity (based on the *SentiWordNet* score of w) and is labeled as "Negated" in the initial step, *SentiClass* treats w as a *negative* (*positive*, respectively) word in R and assigns to w the corresponding *negative* (*positive*, respectively) *SentiWordNet* score.

Polanyi et al. [19] further suggest that **reported speech** sentences can have a detrimental effect in adequately determining the polarity of a text, and thus it is a common computational linguistic treatment to *ignore* them. Reported speech, also known as *indirect speech*, refers to sentences reporting what someone else has said. Consider the sentence S, "My friend said that he did not enjoy the movie", in a review R. Since S does not reflect the opinion expressed by the author of R, it should not be considered when determining the polarity of R. *SentiClass* removes from R reported speech sentences (using the algorithm in [8]) prior to detecting intensifiers, connectors, and negation terms in R.

3.2   Classification Ranges

In establishing the pre-defined, sentiment range for determining whether a review R should be treated as *positive*, *negative*, or *neutral*, we conducted an empirical study using a set of 200 *neutral* reviews extracted from Gamespot.com, along with a set of 800 (randomly selected) reviews on Books, Cars, Computers, and Hotels, which were extracted from Epinions.com such that each review comes with a three-star (out of five) rating and is treated as *neutral*. (None of these reviews was used in Section 4 for assessing the performance of *SentiClass*.) Using the 1,000 reviews, we first computed the *SentiScore* (see Section 3.1) of each review. Hereafter, we calculated the *mean* and *standard deviation* of the *SentiScores* of the reviews, which are 0.02695 and 0.01445, respectively, that yield the range [0.0125, 0.0414]. If the *SentiScore* of R falls into the range, then *SentiClass* classifies R as *neutral*.

Since, as previously stated, *SentiScore* is a normalized value that is not affected by the *length* of a review, the range established for classifying R can also be used by *SentiClass* for classifying the summarized version of R.

3.3   *SentiClass* Using Sentiment Summarization

To further enhance the efficiency of *SentiClass* in terms of minimizing its over-all processing time, we introduce our single-document, sentiment summarizer, called *eSummar*. Instead of classifying an entire review R, *SentiClass* can apply *eSummar* on R to reduce the length of R for categorization while preserving the main content and polarity of R, which as a side-effect reduces the overall classification time of *SentiClass*.

In designing *eSummar*, we rely on sentence similarity, significance factor, and sentiment scores, in addition to applying anaphora resolution on R, to (i) first determine the expressiveness of each sentence in R in capturing the content of R and (ii) then choose sentences in R to be included in the summary of R.

Sentence Similarity. The *sentence-similarity* score of sentence S_i in a review R, denoted $Sim_R(S_i)$, indicates the relative degree of significance of S_i in reflecting the overall *content* of R. $Sim_R(S_i)$ is computed using the *word-correlation factors*[2] [7] of words in S_i and in each remaining sentence S_j in R to determine the *degree of resemblance* of S_i and S_j. The higher $Sim_R(S_i)$ is with respect to the remaining sentences in R, the more promising S_i is in capturing the content of R to a certain degree. To compute $Sim_R(S_i)$, we adopt the *Odds ratio* $= \frac{p}{1-p}$ [5], where p denotes the *strength* of an association between a sentence S_i and the remaining sentences in R, and 1-p reflects its complement.

$$Sim_R(S_i) = \frac{\sum_{j=1,i\neq j}^{|S|} \sum_{k=1}^{n} \sum_{l=1}^{m} wcf(w_k, w_l)}{1 - \sum_{j=1,i\neq j}^{|S|} \sum_{k=1}^{n} \sum_{l=1}^{m} wcf(w_k, w_l)} \qquad (1)$$

where $|S|$ is the number of sentences in R, n (m, respectively) is the number of words in S_i (S_j, respectively), w_k (w_l, respectively) is a word in S_i (S_j, respectively), and $wcf(w_k, w_l)$ is the word-correlation factor of w_k and w_l.

Sentence Significance Factor. As a summary of a review R reflects the *content* of R, it should contain sentences that include *significant words* in R, i.e., words that capture the main content in R. We compute the *significance factor* [12] of each sentence S in R, denoted $SF_R(S)$, based on the number of *significant words* in S.

$$SF_R(S) = \frac{|significant\_words|^2}{|S|} \qquad (2)$$

where $|S|$ is the number of words in S and $|significant\_words|$ is the number of significant words in S such that their frequency of occurrence in R is between

---

[2] Word-correlation factors indicate the degree of similarity of any two words calculated using the *frequency* of *word co-occurrence* and *word distances* in a Wikipedia dump of 880,000 documents on various subjects written by more than 89,000 authors.

pre-defined high- and low-frequency cutoff values. A word w in R is *significant* in R if

$$f_{R,w} \geq \begin{cases} 7 - 0.1 \times (25 - Z) & \text{if } Z < 25 \\ 7 & \text{if } 25 \leq Z \leq 40 \\ 7 + 0.1 \times (Z - 40) & \text{otherwise} \end{cases} \tag{3}$$

where $f_{R,w}$ is the frequency of w in R, Z is the number of sentences in R, and 25 (40, respectively) is the low- (high-, respectively) frequency cutoff value.

Sentence Sentiment. Since the *eSummar*-generated summary of R captures the *sentiment* expressed in R, we measure to what extent a sentence S in R reflects its sentiment, i.e., polarity, during the summarization process. In defining the *sentiment score* of S, denoted $Senti_R(S)$, we determine the *polarity* of each word w in S by multiplying its (sentiment) *weight* (i.e., the *SentiWordNet* score of w) with its *linguistic score*, denoted $LiScore$, as defined in [3], which reflects the *strength* of the *sentiment* expressed by w based on its type, i.e., adverb, adjective, verb, or conjunction, the commonly-used types to express sentiment.

$$Senti_R(S) = \sum_{i=1}^{n} StScore(w_i) \times LiScore(w_i) \tag{4}$$

where n is the number of words in S, w_i is a word in S, $StScore(w_i)$ is the highest *SentiWordNet* score of w_i (among the three possible *SentiWordNet* scores for w_i), and $LiScore(w_i)$ is the *linguistic score* of w_i, which is the multiplicative factor assigned to the linguistic type of w_i. As defined in [3], the *linguistic scores* of adjectives and adverbs, verbs, and conjunctions are 8, 4, and 2, respectively.

Anaphora Resolution. Prior to computing the sentence scores previously introduced, we adopt anaphora resolution to eliminate the *ambiguity* that arises in interpreting the content of R by assigning a *discourse entity* to proper nouns, acronyms, or pronouns in R [10]. Consider the sentence S, "Shrek was great and its animation was very well-done." By applying anaphora resolution on S, "it" is transformed into the referenced entity, i.e., "Shrek". We identify anaphoric chains, which are entities and their various discourse referents in R, using the anaphora resolution system *Guitar* (cswww.essex.ac.uk/Research/nle/GuiTAR/) and replace all discourse referents by their corresponding entities in R.

Ranking Sentences and Summary Size. *eSummar* computes a *ranking score* for each sentence S in R, denoted $Rank_R(S)$, which reflects the relative degree of *content* and *sentiment* in R captured by S using $Sim_R(S)$, $SF_R(S)$, and $Senti_R(S)$. Hereafter, the sentences in R with the highest *Rank* scores are included in the summary of R. *Rank* applies the *Stanford Certainty Factor* [11][3] on $Sim_R(S)$, $SF_R(S)$, and $Senti_R(S)$, which is a measure that integrates

---

[3] Since $Sim_R(S)$, $SF_R(S)$, and $Senti_R(S)$ are in different numerical scales, prior to computing *Rank*, we normalize the range of the scores using a logarithmic scale.

different assessments, i.e., sentence scores in our case, to determine the *strength* of a hypothesis, i.e., the content in R captured by S in our case.

$$Rank_R(S) = \frac{Sim_R(S) + SF_R(S) + Senti_R(S)}{1 - Min\{Sim_R(S), SF_R(S), Senti_R(S)\}} \tag{5}$$

In choosing the *size* of *eSummar*-generated summaries, we adopt the length of 100 words determined by DUC (www-nlpir.nist.gov/projects/duc/index.html) and TAC (www.nist.gov/tac/data/forms/index.html), which provide benchmark datasets for assessing the performance of a summarizer. Since *eSummar* is an extractive summarization approach, it creates the summary of R by including the maximum number of sentences in R in the order of their *Rank* values, from the highest to lowest, excluding the ranked sentence (and successive ones) in R that causes the total word count to exceed 100.

Weights of Sentence Scores. Since the $Sim_R(S)$, $SF_R(S)$, and $Senti_R(S)$ values of sentence S in R provide different measures in determining the *Rank* value of S, their *weights* in computing *Rank* should be different. To adequately determine their weights in *Rank*, we apply the *multi-class perceptron* algorithm [14] which establishes the weights through an iterative process[4]. We constructed a training set TS with 1,023 sentences in 100 blog posts (on an average of 10 sentences per post) extracted from the TAC-2008[5] dataset (nist.gov/tac/data/forms/index.html). Each sentence S (in TS) is represented as an input vector with four different values associated with S, i.e., $Sim_R(S)$, $SF_R(S)$, $Senti_R(S)$, and *Rouge*-1. While $Sim_R(S)$, $SF_R(S)$, and $Senti_R(S)$ are as defined earlier, *Rouge*-1 quantifies the degree of a sentence S in R in capturing the content and sentiment of the corresponding expert-created summary, *Expert-Sum*, of R. *Rouge*-1 is determined by the overlapping of unigrams between S and *Expert-Sum*. The higher the *Rouge*-1 score is, the more representative S is in the content of *Expert-Sum* (and thus R). Based on the conducted experiment, the weights of $Senti_R(S)$, $Sim_R(S)$, and $SF_R(S)$ are set to be 0.55, 0.25, and 0.20, respectively, and Equation 5 is modified to include the weights as follows: $Enhanced\_Rank_R(S)$

$$= \frac{0.25 \times Sim_R(S) + 0.20 \times SF_R(S) + 0.55 \times Senti_R(S)}{1 - Min\{0.25 \times Sim_R(S), 0.20 \times SF_R(S), 0.55 \times Senti_R(S)\}} \tag{6}$$

4   Experimental Results

In this section, we first introduce the datasets and metrics used for assessing the performance of *SentiClass* (in Sections 4.1 and 4.2, respectively). Thereafter, we present the accuracy of *SentiClass* in sentiment classification (in Section 4.3)

---

[4] The training of the multi-class perceptron occurs only once and as a pre-processing step prior to performing the summarization task by *eSummar*.

[5] TAC-2008 includes posts and their respective expert-created reference summaries.

and the effectiveness of using *eSummar* (compared with other summarizers) to generate summaries of reviews which enhance the classification process of *SentiClass* (in Section 4.4).

4.1   The Datasets

To evaluate the effectiveness of *SentiClass*, we have chosen two widely-used datasets, Movie Reviews (cs.cornell.edu/people/pabo/movie-review-data/) and Game Reviews (cswiki.cs.byu.edu/cs679/index.php/Game_Spot). The former includes 2,000 reviews extracted from Imdb.com in which 1,000 reviews are *positive* and the others are *negative*, whereas the latter consists of reviews that were downloaded from Gamespot.com between April 2005 and January 2007. Each game review includes (i) a score (between 1.0 and 10.0 inclusively, rounded to one decimal point), which denotes the *rating* of a reviewed video game given by an author, (ii) the author's name, and (iii) the text of the review. Reviews with a score up to 6, between 6 and 8 exclusively, and 8 or higher are labeled as *negative*, *neutral*, and *positive*, respectively, which yields a set of 2,044 game reviews: 548 *negative*, 1067 *neutral*, and 429 *positive*.

Besides movie and game reviews, we created a new dataset, denoted *Epinions-DS*, with 1,811 reviews extracted from Epinions.com[6], out of which 940 reviews are *positive* and the remaining ones are *negative*. The reviews in *Epinions-DS* are uniformly distributed into four different subject areas, Books, Cars, Computers, and Hotels, which are diverse in contents and structures.

4.2   Evaluation Metrics

To evaluate the performance of *SentiClass* in sentiment classification on either entire reviews or *eSummar*-generated summaries, we use the *classification accuracy measure* defined below.

$$Accuracy = \frac{Number\ of\ Correctly\ Classified\ (Summarized)\ Reviews}{Total\ Number\ of\ (Summarized)\ Reviews\ in\ a\ Collection} \quad (7)$$

4.3   Performance Evaluation of *SentiClass*

In this section, we analyze the performance of *SentiClass* on classifying (*eSummar*-generated summaries of) reviews in Movie Reviews, Game Reviews, and *Epinions-DS*. Thereafter, we compare the effectiveness of *SentiClass* with other existing sentiment classifiers in categorizing reviews.

Classification Accuracy. *SentiClass* is highly accurate in classifying (*eSummar*-generated summarized) reviews. As shown in Figure 3, the average accuracy achieved by *SentiClass* using entire reviews in the three test datasets

---

[6] Epinions.com is a well-known public and free product review source from where test datasets can be created for sentiment classification studies [17].

is 93%, as opposed to 88% using *eSummar*-generated summaries. Classification accuracy on entire and summarized reviews in Game Reviews yields the largest difference due to the inclusion of the third class, *neutral*. Identifying *neutral* reviews is more difficult than categorizing positive or negative ones, since a mixed number of positive and negative terms often co-occur in neutral reviews.

A significant impact of using *eSummar*-generated summaries, instead of the entire reviews, is that the overall average classification time of *SentiClass* is dramatically reduced by close to 50% in classifying *eSummar*-generated summaries, instead of full reviews, from an average of 112 minutes to 59 minutes, as shown in Figure 3.

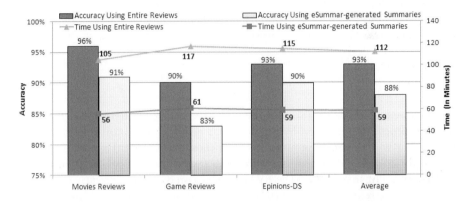

Fig. 3. (Average) Accuracy and overall processing time of *SentiClass* in classifying *eSummar*-generated summaries and their entire reviews in the test datasets

SentiClass **and Other Classifiers.** To further assess the effectiveness of *SentiClass*, we compare its classification accuracy with other well-known (sentiment) classifiers, as presented in [18], which include Multinomial Naive Bayes (MNB), Maximum Entropy (ME), Support Vector Machines (SVM), and Linear/Log Pooling [13], in addition to two other sentiment classifiers by design, i.e., Hybrid and Extract-SVM, on Movie Reviews and *Epinions-DS*. We do not consider Game Reviews, since most of the sentiment classifiers to be compared are designed for two-class, i.e., positive and negative, categorization.

MNB is a simple and efficient probabilistic classifier that relies on a conditional word independence assumption to compute the *probability* of word occurrence in a pre-defined class, which dictates to which class a review R should be assigned according to the occurrences of its words. ME, which is a classification technique that has been shown effective in several natural language processing applications, estimates the conditional distribution of the pre-defined class labels. The classifier represents a review R as a set of word-frequency counts and relies on labeled training data to estimate the expected values of the word counts in R on a class-by-class basis to assign R to its class.

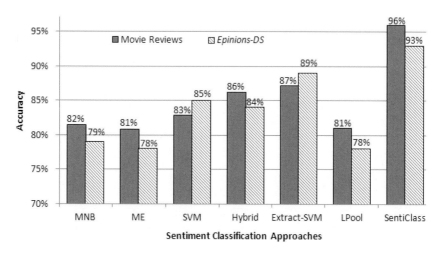

Fig. 4. Sentiment classification accuracy achieved by various sentiment classifiers

The premise of SVM [18], a highly effective text classifier, is to conduct a training procedure to find a hyperplane that accurately separates reviews (represented as word vectors) in one class from those in another. Extract-SVM [16], another SVM-based classifier, first categorizes sentences in a review as subjective or objective and then applies the SVM classifier on only the subjective sentences in performing sentiment classification, since removing objective sentences prevents the classifier from considering irrelevant or even potentially misleading text [16]. Hybrid [6], on the other hand, is a weighted, voted classifier that combines the classification score on a review R computed by a support vector machine with the score on R generated by a term-counting approach, which considers context valence shifters, such as negations, intensifiers, and diminishers, to classify reviews based on their sentiments.

The authors in [13] propose a *Linear/Log Pooling* method, denoted *LPool*, which combines lexical knowledge and supervised learning for text classification. *LPool* might lead to overfitting (i.e., over-training) its model, which increases training time, lowers its accuracy, and as a result the classification performance on unseen data becomes worse.

As shown in Figure 4, *SentiClass* outperforms MNB, ME, SVM, Hybrid, Extract-SVM, and *LPool* on categorizing reviews in Movie Reviews from 9% to 15% and from 4% to 15% on *Epinions-DS*.

4.4 Assessment of *SentiClass* Using Summarization

Having demonstrated the effectiveness of *SentiClass* in Section 4.3, we proceed to assess and compare the classification accuracies achieved by *SentiClass* using summarized reviews generated by *eSummar* and other summarizers on the test datasets introduced in Section 4.1.

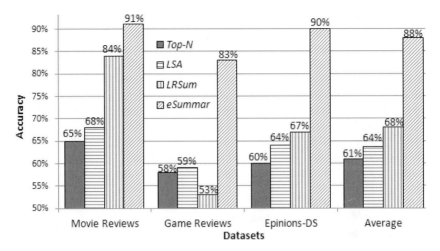

Fig. 5. (Average) Accuracy achieved by *SentiClass* using summaries created by *Top-N*, *LSA*, *LRSum*, and *eSummar* on test datasets, respectively

The *Top-N* summarizer, a naive summarization approach, assumes that introductory sentences in a review R contain the overall gist of R and extracts the first N (≥ 1) sentences in R as its summary. We treat *Top-N* as a baseline measure for summarization. Gong [2], on the contrary, applies Latent Semantic Analysis (*LSA*) for creating single-document summaries. *LSA* analyzes relationships among documents and their terms to compile a set of *topics* related to the documents, which are described by word-combination patterns recurring in the documents. *LSA* selects sentences with high frequency of recurring word-combination patterns in a document D as the summary of D.

LRSum [1] creates summaries of reviews by first training a logistic regression model using sentences in reviews represented by a set of features, which include the position of a sentence S within a paragraph, its location in a review, and the frequency of occurrence of words in S. Using the trained model, *LRSum* determines *sentiment* sentences based on their features and selects the *one* in a review R with the *maximal conditional likelihood* as the summary of R.

As shown in Figure 5, performing the classification task in *SentiClass* using *eSummar*-generated summaries is more accurate, by an average of at least 20% higher, than using summaries generated by *Top-N*, *LSA*, or *LRSum*. Note that we set the length of the summaries generated by *Top-N*, *LSA*, *LRSum*, and *eSummar* to be 100 words, as previously discussed in Section 3.3.

As opposed to *eSummar* which computes sentence sentiment scores, neither *Top-N* nor *LSA* considers the sentiment expressed in a review in creating its summary, even though *LRSum* does. Unlike *eSummar*, *LRSum* requires (i) training, which increases the complexity and processing time of the summarizer, and (ii) labeled data, which may not always be available nor is easy to compile.

Figure 6 shows the (average) processing time of *SentiClass* in creating and classifying the summarized versions of the reviews in the test datasets

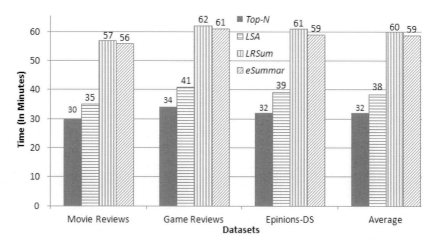

Fig. 6. (Average) Processing time of *SentiClass* in creating summaries using *Top-N*, *LSA*, *LRSum*, or *eSummar* on test datasets and classifying the summaries

(introduced in Section 4.1) using *Top-N*, *LSA*, *LRSum*, and *eSummar*, respectively. Not only the processing time of *SentiClass* in classifying *eSummar*-generated summaries is (slightly) faster than the one achieved by using *LRSum*-generated summaries, its classification accuracy is higher than the one using *LRSum*-generated summaries by 7%, 30%, and 23%, on Movie Reviews, Game Reviews, and *Epinions-DS*, respectively, as shown in Figure 5. Although the average processing time using *Top-N*- and *LSA*-generated summaries is on an average 40% faster than the one achieved by using *eSummar*-generated summaries, the average classification accuracy using summaries created by *eSummar* is significantly higher than the ones using summaries created by using *Top-N* or *LSA* by an average of 27% and 24%, respectively (see Figure 5).

Since *SentiClass* (i) achieves high accuracy and (ii) does not significantly prolong its classification processing time by using summaries created by *eSummar* as compared with others created using alternative summarizers, *eSummar* is an ideal choice for summarizing reviews to be classified by *SentiClass*.

5   Conclusions

With the increasing number of reviews posted under social search websites, such as Epinions.com, a heavier burden is imposed on web users to browse through archived reviews to locate different opinions expressed on a product or service *PS*. To assist web users in identifying reviews on *PS* that share the same polarity, i.e., positive, negative, or neutral, we have developed a simple, yet effective sentiment classifier, denoted *SentiClass*. *SentiClass* considers the *SentiWordNet* score of words, intensifiers, connectors, reported speech, and negation terms in reviews to accurately categorize the reviews according to the sentiment reflected on PS. To (i) reduce the processing time required by *SentiClass* for categorizing reviews and (ii) shorten the length of the (classified) reviews a user is expected to browse through, we have introduced *eSummar*, a single-document, extractive, sentiment summarizer, which considers word-correlation factors, sentiment words, and significance factors to capture the opinions on PS expressed in a review R and generate the summary of R. Empirical studies conducted using well-known datasets, Movie Reviews and Game Reviews, along with a set of reviews extracted from Epinions.com, denoted *Epinions-DS*, show that *SentiClass* is highly effective in categorizing (*eSummar*-generated summaries of) reviews according to their polarity. Using Movie Reviews and *Epinions-DS*, we have verified that *SentiClass* outperforms other classifiers in accomplishing the sentiment classification task. We have also demonstrated that *eSummar* is an ideal summarizer for *SentiClass* by comparing *eSummar* with other existing (sentiment) summarizers for creating summaries to be classified.

References

1. Beineke, P., Hastie, T., Manning, C., Vaithyanathan, S.: An Exploration of Sentiment Summarization. In: Proc. of AAAI, pp. 12–15 (2003)
2. Gong, Y.: Generic Text Summarization Using Relevance Measure and Latent Semantic Analysis. In: Proc. of ACM SIGIR, pp. 19–25 (2001)
3. Hu, X., Wu, B.: Classification and Summarization of Pros and Cons for Customer Reviews. In: Proc. of IEEE/WIC/ACM WI-IAT, pp. 73–76 (2009)
4. Jie, S., Xin, F., Wen, S., Quan-Xun, D.: BBS Sentiment Classification Based on Word Polarity. In: Proc. of ICCET, vol. 1, pp. 352–356 (2009)
5. Judea, P.: Probabilistic Reasoning in the Intelligent Systems: Networks of Plausible Inference. Morgan Kaufmann, San Francisco (1988)
6. Kennedy, A., Inkpen, D.: Sentiment Classification of Movie Reviews Using Contextual Valence Shifters. Computational Intelligence 22(2), 110–125 (2006)
7. Koberstein, J., Ng, Y.-K.: Using Word Clusters to Detect Similar Web Documents. In: Lang, J., Lin, F., Wang, J. (eds.) KSEM 2006. LNCS (LNAI), vol. 4092, pp. 215–228. Springer, Heidelberg (2006)
8. Krestel, R., Bergler, S., Witte, R.: Minding the Source: Automatic Tagging of Reported Speech in Newspaper Articles. In: Proc. of LREC, pp. 2823–2828 (2008)
9. Ku, L., Liang, Y., Chen, H.: Opinion Extraction, Summarization and Tracking in News and Blog Corpora. In: Proc. of AAAI 2006 Spring Symposium on Computational Approaches to Analyzing Weblogs, pp. 100–107 (2006)
10. Lappin, S., Leass, H.: An Algorithm for Pronominal Anaphora Resolution. Computational Linguistics 20(4), 535–561 (1994)
11. Luger, G.: Artificial Intelligence: Structures and Strategies for Complex Problem Solving, 6th edn. Addison-Wesley, Reading (2009)
12. Luhn, H.: The Automatic Creation of Literature Abstracts. IBM Journal of Research and Development 2(2), 159–165 (1958)
13. Melville, P., Gryc, W., Lawrence, R.: Sentiment Analysis of Blogs by Combining Lexical Knowledge with Text Classification. In: Proc. of KDD, pp. 1275–1284 (2009)

14. Minsky, M., Papert, S.: Perceptrons: An Introduction to Computational Geometry. The MIT Press, Cambridge (1972)
15. Na, J., Khoo, C., Wu, P.: Use of Negation Phrases in Automatic Sentiment Classification of Product Reviews. Library Collections, Acquisitions, and Technical Services 29(2), 180–191 (2005)
16. Pang, B., Lee, L.: A Sentimental Education: Sentiment Analysis Using Subjectivity Summarization Based on Minimum Cuts. In: Proc. of ACL, pp. 271–278 (2004)
17. Pang, B., Lee, L.: Opinion Mining and Sentiment Analysis. Foundations and Trends in Information Retrieval 2(1-2), 1–135 (2008)
18. Pang, B., Lee, L., Vaithyanathan, S.: Thumbs up? Sentiment Classification using Machine Learning Techniques. In: Proc. of EMNLP, pp. 79–86 (2002)
19. Polanyi, L., Zaenen, A.: Contextual Valence Shifters. In: Computing Attitude and Affect in Text: Theory and Applications, pp. 1–10. Springer, Heidelberg (2006)
20. Radev, D., Hovy, E., McKeown, K.: Introduction to the Special Issue on Summarization. Computational Linguistics 28(4), 399–408 (2002)
21. Wiebe, J., Wilson, T., Bruce, R., Bell, M., Martin, M.: Learning Subjective Language. Computational Linguistics 30, 277–308 (2004)
22. Zhao, J., Liu, K., Wang, G.: Adding Redundant Features for CRFs-based Sentence Sentiment Classification. In: Proc. of EMNLP, pp. 117–126 (2008)
23. Zhuang, L., Jing, F., Zhu, X.: Movie Review Mining and Summarization. In: Proc. of ACM CIKM, pp. 43–50 (2006)

Neighborhood-Restricted Mining and Weighted Application of Association Rules for Recommenders

Fatih Gedikli and Dietmar Jannach

Technische Universität Dortmund,
44221 Dortmund, Germany
{firstname.lastname}@tu-dortmund.de

Abstract. Association rule mining algorithms such as Apriori were originally developed to automatically detect patterns in sales transactions
and were later on also successfully applied to build collaborative filtering
recommender systems (RS). Such rule mining-based RS not only share
the advantages of other model-based systems such as scalability or robustness against different attack models, but also have the advantages
that their recommendations are based on a set of comprehensible rules.
In recent years, several improvements to the original Apriori rule mining
scheme have been proposed that, for example, address the problem of
finding rules for rare items. In this paper, we first evaluate the accuracy
of predictions when using the recent IMSApriori algorithm that relies
on multiple minimum-support values instead of one global threshold. In
addition, we propose a new recommendation method that determines
personalized rule sets for each user based on his neighborhood using IM-
SApriori and at recommendation time combines these personalized rule
sets with the neighbors' rule sets to generate item proposals. The evaluation of the new method on common collaborative filtering data sets shows
that our method outperforms both the IMSApriori recommender as well
as a nearest-neighbor baseline method. The observed improvements in
predictive accuracy are particularly strong for sparse data sets.

1 Introduction

Association rule mining is a popular knowledge discovery technique which was
designed as a method to automatically identify buying patterns in sales transactions, or, in a more broader view, to detect relations between variables in
databases. One of the earliest efficient techniques to find such rules is the Apriori algorithm proposed by Agrawal and Srikant in [1]. A common example of an
association rule that could be found in the sales transactions in a supermarket
could be [2]: $cheese \Rightarrow beer$ [$support = 10\%$, $confidence = 80\%$] which can be
interpreted that in 10% of all transactions beer and cheese were bought together
(support of the rule) and that in 80% of the transactions, in which cheese was
bought, also beer was in the shopping basket (confidence of the rule). Confidence and support are thus statistical measures that indicate the "strength" of
the pattern or rule.

L. Chen, P. Triantafillou, and T. Suel (Eds.): WISE 2010, LNCS 6488, pp. 157–165, 2010.

Quite obviously, the knowledge encoded in such automatically detected association rules (or frequent itemsets) can be exploited to build recommender systems (RS). Since the rules can be mined in an offline model-learning phase, rule mining-based approaches do not suffer from scalability problems like memory-based algorithms [3]. A further advantage of these approaches lies in the fact that association rules are suitable for explaining recommendations.

Regarding the predictive accuracy of rule mining-based approaches, previous research has shown that the accuracy of rule mining-based recommenders is comparable to nearest-neighbor (kNN) collaborative filtering approaches. However, using the original Apriori algorithm can lead to the problem of reduced coverage as shown in [3]. This phenomenon can be caused by the usage of a global minimum support threshold in the mining process, which leads to the effect that no rules for rare items can be found. Lin et al. [4] therefore propose an "adaptive-support" method, in which the minimum support value is determined individually for each user or item (depending on whether item associations or user associations are used). Their experiments show a slight increase in accuracy when compared with the baseline kNN-method.

More recently, Kiran and Reddy [5] proposed a new method called IMSApriori that uses a particular metric to determine appropriate minimum support values per item (see also [2]) in order to mine rare itemsets; their experiments indicate that this method is better suited to mine rare itemsets than previous methods. An evaluation of the approach for recommendation purposes has, however, not been done so far.

In this work, we evaluate the predictive accuracy of a recommender system based on the IMSApriori algorithm and describe our extension to the *Frequent Itemset Graph* used in [6] for enabling a fast recommendation process. In addition, we propose a new scheme for association rule-based recommendation called NRR (*Neighborhood-Restricted Rule-Based Recommender*), which is based on the idea to learn a personalized set of rules for each user based on his nearest neighbors and not based on the whole database of transactions. Similar to kNN-approaches, the underlying idea of this is that close neighbors will be better predictors than others. After the model-building phase, the user's personalized knowledge base is at recommendation time combined with the rule sets of his nearest neighbors to generate recommendation lists.

2   Algorithms

In the following we will shortly summarize the ideas of the IMSApriori algorithm in order to give the reader a quick overview of the algorithm parameters that were varied in the experimental evaluation. In addition, we will describe how the *Frequent Itemset Graph* proposed, e.g., in [6], has to be extended for a recommender based on IMSApriori.

2.1   IMSApriori

In order to deal with the problem of "missing rules" for rare, but interesting itemsets, different proposals have been made in literature. IMSApriori [5], which is used in this work, is a very recent one that builds on the idea of having several minimum support thresholds, an idea also proposed earlier as MSapriori in [2]. The general idea is to calculate a minimum item support (MIS) value for each item with the goal to use a lower support threshold for rare itemsets. In [2] a user-specified value β (between 0 and 1) is used to calculate a MIS value based on the item's support and a lower support threshold value LS as $MIS(item) = max(\beta \times support(item), LS)$. In order to be counted as a *frequent* itemset, itemsets containing only frequent items have to pass a higher minimum support threshold than itemsets consisting of frequent and rare or only rare items. Thus, rare itemsets are found when using a low value for LS while at the same time not too many uninteresting, but more frequent rules are accepted.

Recently, in [5], a different approach to calculate the MIS values was proposed because MSapriori fails to detect rare itemsets in situations with largely varying item support values. This phenomenon can be attributed to the fact that due to the constant proportional factor β the difference between the item support and the MIS value decreases when we move from frequent to rare items. The main idea of the *improved* MSapriori (IMSApriori) is therefore the use of the concept of "support difference" (SD) to calculate MIS values as $MIS(item) = max(support(item) - SD, LS)$. SD is calculated as $SD = \lambda(1 - \alpha)$, where λ is a parameter "like mean, median, mode, maximum support of the item supports" and α is a parameter between 0 and 1. The net effect of the support difference concept is that the difference between item support values and the MIS values remains constant so that rare items can also be found in data sets with strongly varying item supports. Finally, in this approach, an itemset is considered to be frequent if its support is higher than the minimum of the MIS values of its components. Regarding the generation of candidates, it has to be noted that the Apriori assumption that all subsets of frequent itemsets are also frequent does not hold and that a different algorithm for finding frequent itemsets has to be used.

2.2   The Neighborhood-Restricted Rule-Based Recommender (NRR)

The idea of the herein proposed NRR algorithm is to learn personalized rule sets for each user in an offline phase and to exploit these rule sets in combination with the neighbor's rule sets to generate more accurate predictions. The algorithm is summarized in Algorithm 1. The parameters of the algorithm include – beside the IMSApriori parameters – two neighborhood sizes (for rule learning and for the prediction phase). In the online phase, the calculated user-specific frequent itemsets (UserFISs) of the target user and of the neighbors of the target user are used to calculate predictions using the *Extended Frequent Itemset Graph* (EFIG) which is introduced in the next section. The resulting confidence scores

---

Algorithm 1 NRR (sketch)

In: user, ratingDB, learnNeighborSize, predictNeighborSize, λ, α
Out: recommendedItems
(Offline:) UserFISs = CalcUserFISsIMSApriori(ratingDB, learnNeighborSize, λ, α)
neighborhood = user $\cup$ findNeighbors(user, predictNeighborSize, ratingDB)
recommendedItems = $\emptyset$
for all u $\in$ neighborhood **do**
    userRecs = Recommend(u, buildEFIG(UserFISs(u)))
    weightedUserRecs = adjustConfidenceScoresBySimilarity(userRecs, user, u)
    recommendedItems = recommendedItems $\cup$ weightedUserRecs
end for
recommendedItems = sortItemsByAdjustedScores(recommendedItems)

---

are weighted according to the similarity of the target user and the neighbor (using Pearson correlation as a metric). These user-specific predictions are finally combined and sorted by the weighted confidence scores.

2.3   The Extended Frequent Itemset Graph

The *Frequent Itemset Graph* (FIG) is a data structure (a directed acyclic graph) proposed in [6], which can be used by a recommendation engine to make real-time recommendations without learning explicit association rules first. Figure 1(a) shows such a graph in which the elements of the frequent itemsets are lexico-graphically sorted and organized in a tree structure where the size of the item-sets are increased on each level. Note that the numbers in brackets stand for the support values of the itemsets. Given, for example, a set of past transactions $T = \{A, D\}$ of user u, recommendations can be produced by traversing the tree in depth-first order and looking for (single-element) supersets of $\{A, D\}$ in the graph. In the example, given the superset $\{A, D\}$, C could be recommended to u if the recommendation score of item C is high enough. The recommendation score of this item corresponds to the confidence value $\frac{support(\{A,D\}\cup\{C\})}{support(\{A,D\})} = \frac{2}{5}$ which

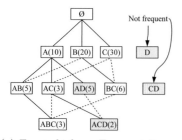

(a) Example for a Frequent Item-set Graph.

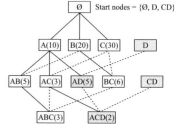

(b) *Extended* Frequent Itemset Graph example.

Fig. 1. Extended Frequent Itemset Graph approach

is at the same time the confidence value of the association rule $\{A, D\} \Rightarrow \{C\}$. The solid arrows in the figure indicate how the graph would be traversed in depth-first order.

Since the assumption that "any subset of a frequent itemset must be frequent" does not hold when using multiple minimum-support values, the standard FIG-based method has to be extended. Let us assume that in the example Figure 1(a) the itemsets $\{D\}$ and $\{C, D\}$ are not frequent, although they are subsets of the frequent itemset $\{A, D\}$ and $\{A, C, D\}$ respectively. We could therefore not recommend $\{A\}$ to users who purchased $\{C, D\}$ or $\{D\}$ alone although this would be plausible.

We propose to solve this problem by extending the FIG in a way that it also contains all subsets of the frequent itemsets and connect these additional nodes with their supersets as shown in Figure 1(b). In order to find frequent itemsets like $\{A, C, D\}$ from $\{C, D\}$ we re-start the depth-first search on the not-yet-visited parts of the subgraph beginning from the additional nodes.

Algorithm 2 shows the steps that are necessary for building the *Extended Frequent Itemset Graph* (EFIG).

Algorithm 2 Build EFIG.

1: **for** $k = n$ to 1 **do**
2:     **for all** frequent k-itemsets f **do**
3:         **for all** $k - 1$ subsets s of f **do**
4:             add new edge (s, f);
5:             **if** $s \notin L_{k-1}$ **then**
6:                 add new **start node** s;
7:             **end if**
8:         **end for**
9:     **end for**
10: **end for**

The algorithm works as follows. The EFIG is constructed bottom-up, starting with all frequent itemsets of size k, beginning with the maximum size n of the frequent itemsets (lines 1-2). Lines 3 and 4 connect each $k - 1$ subset s of a k-frequent itemset with its superset f. If subset s is not a frequent itemset, i.e., $s \notin L_{k-1}$ where L_{k-1} contains all frequent itemsets of length $k - 1$, s will be added to the set of *start nodes* which contains all possible entry-points to the graph. This set consists of the root node ($\emptyset$, the original start node) and all not frequent itemsets that are subsets of frequent itemsets. In the example Figure 1(b), the nodes D and CD are finally also included in the start nodes and can be considered in the recommendation process.

3   Experimental Evaluation

The proposed NRR algorithm has been evaluated in an experimental study on different data sets. In particular, the predictive accuracy was measured using different sparsity levels and compared to (a) a recommender based on IMSApriori and a classical prediction scheme and (b) the user-based k nearest neighbor method using the Pearson correlation coefficient (kNN). In the following, we will summarize the findings of this evaluation.

3.1 Experimental Setup / Evaluation Metrics

Data sets. As data sets for the evaluation, we used the MovieLens rating database consisting of 100,000 ratings provided by 943 users on 1,682 items and a snapshot of the Yahoo!Movies data set containing 211,231 ratings provided by 7,642 users on 11,915 items[1]. In order to test our NRR scheme also in settings with low data density, we varied the density level of the original data sets by using subsamples of different sizes of the original data set. Four-fold cross-validation was performed for each data set; in each round, the data sets were split into a 75% training set and a 25% test set.

Accuracy metrics. In the study, we aim to compare the predictive accuracy of two rule mining-based methods and the kNN-method. We proceed as follows. First, we determine the set of existing "like" statements (ELS) in the 25% test set and retrieve a top-N recommendation list with each method based on the data in the training set[2]. In the kNN-case, the rating predictions are converted into "like" statements as described in [3], where ratings above the user's mean rating are interpreted as "like" statements. The set of predicted like statements returned by a recommender shall be denoted as *Predicted Like Statements (PLS)*.

We use standard information retrieval accuracy metrics in our evaluation. *Precision* is defined as $\frac{|PLS \cap ELS|}{|PLS|}$ and measures the number of correct predictions in PLS. *Recall* is measured as $\frac{|PLS \cap ELS|}{|ELS|}$ and describes how many of the existing "like" statements were found by the recommender.

In the evaluation procedure, we use "top-10", that is, the list of the top ten movies for a test user with predicted rating values above the user's mean rating, and calculate the corresponding precision and recall values for all users in the test data set. The averaged precision and recall values are then combined in the usual F-score, where $F = 2 * \frac{precision * recall}{precision + recall}$.

Algorithm details and parameters. For the neighborhood-based algorithms, we used Pearson correlation as a similarity metric both for the kNN-baseline method and for determining the neighborhood in the NRR algorithm. For the kNN-method, we additionally applied *default voting* and used a neighborhood-size of 30, which was determined as an optimal choice in literature.

The IMSApriori implementation used in the experiments corresponds to above-described algorithm and learns the rules from the whole database of

---

[1] http://www.grouplens.org/node/73, http://webscope.sandbox.yahoo.com

[2] The top-N recommendation lists are created either based on the confidence of the producing rule or based on the prediction score of the kNN-method.

transactions. Recommendations are generated by using the Extended Frequent Itemset Graph structure.

For the NRR method, two further parameters can be varied: *neighborhood-size-learn* is the number of neighbors used to learn association rules; *neighborhood-size-predict* determines on how many neighbors the predictions should be based.

The sensitivity of these parameters was analyzed through multiple experiments on the MovieLens and Yahoo! data set with different density levels. The parameter values for both data sets and all density levels were empirically determined to be 900 and 60 for *neighborhood-size-predict* and *-learn* respectively.

In order to establish fair conditions in our study, we have used individual, empirically-determined *LS* values for each rule learning algorithm (IMSApriori: 3%; NRR: 9%), which we have then used for all density levels and data sets.

3.2   Results

Figure 2 summarizes the evaluation results for the three algorithms kNN, IMSApriori and NRR. The results show that our NRR algorithm consistently outperforms the IMSApriori method on the F1-measure and is better than the kNN algorithm in nearly all settings for both data sets. The observed accuracy improvements are particularly high for low density levels, i.e., for sparse data sets. With higher density levels, the relative improvements become smaller for both data sets.

As a side-observation, we can see that the pure IMSApriori version does not always reach the accuracy level of the kNN-method, especially in settings with lower and medium density levels. However, the herein proposed NRR algorithm,

		Density →	10%	20%	30%	40%	50%	60%	70%	80%	90%
MovieLens	F1	kNN	30,76	41,48	48,06	51,25	55,39	57,13	58,89	59,99	61,32
		IMSApriori	3,75	32,80	50,00	55,94	59,34	60,97	62,66	62,84	62,82
		NRR	37,14	44,09	50,80	56,54	59,10	61,38	63,10	63,51	63,57
	Precision	kNN	39,85%	53,62%	59,76%	61,73%	64,44%	65,29%	66,18%	66,61%	67,07%
		IMSApriori	5,92%	48,43%	63,03%	64,91%	65,31%	65,35%	65,70%	65,24%	64,48%
		NRR	47,25%	57,73%	63,62%	65,74%	65,50%	65,49%	65,69%	65,30%	64,75%
	Recall	kNN	25,05%	33,83%	40,21%	43,81%	48,57%	50,78%	53,05%	54,57%	56,47%
		IMSApriori	2,76%	24,81%	41,43%	49,15%	54,38%	57,15%	59,90%	60,61%	61,24%
		NRR	30,60%	35,67%	42,28%	49,60%	53,84%	57,75%	60,71%	61,81%	62,42%
Yahoo!Movies	F1	kNN	9,24	15,93	21,91	27,95	33,30	37,53	41,32	44,02	46,29
		IMSApriori	6,95	17,06	24,96	31,95	37,86	41,76	43,84	45,61	47,05
		NRR	15,10	20,70	26,81	31,50	36,53	40,30	43,05	45,00	47,01
	Precision	kNN	10,93%	19,75%	27,31%	35,28%	41,95%	47,35%	52,08%	55,47%	58,23%
		IMSApriori	7,65%	20,20%	30,11%	39,37%	46,07%	51,11%	53,81%	56,28%	57,70%
		NRR	17,38%	24,76%	32,37%	38,71%	44,65%	49,46%	52,73%	54,94%	57,36%
	Recall	kNN	8,01%	13,35%	18,30%	23,15%	27,61%	31,09%	34,24%	36,49%	38,41%
		IMSApriori	6,37%	14,77%	21,32%	26,89%	32,13%	35,31%	36,99%	38,35%	39,73%
		NRR	13,36%	17,79%	22,88%	26,55%	30,90%	34,00%	36,37%	38,11%	39,83%

Fig. 2. Top-10 F1, precision and recall values for different density levels

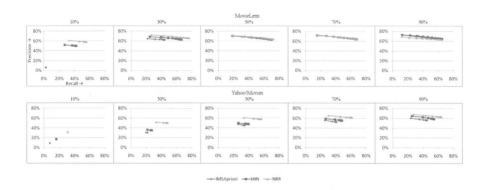

Fig. 3. Precision and recall values for varying list lengths and densities

which is based on the IMSApriori algorithm, consistently outperforms both the IMSApriori method and the kNN algorithm.

In a further experiment, we analyzed the accuracy of our method for different data sets, density levels and recommendation set sizes. Figure 3 shows the recall (x-axis) and precision (y-axis) values for different density levels. Each data point in a plot corresponds to the precision and recall values for given recommendation set size; the sizes were varied from 1 to 10 (i.e., top-1 to top-10 lists). The recall values naturally improve when the recommendation lists are longer.

In Figure 3, we can observe that the improvement of our NRR algorithm is stronger on the Yahoo! data set. A possible explanation for this observation could be the different sparsity levels of the two data sets, i.e., assuming that NRR works particularly well for sparse settings, it is intuitive that even better results can be achieved on the sparser Yahoo! data set (0.9976 sparsity) than on the MovieLens data set (0.9369 sparsity). As already observed in Figure 2, the performance improvements of NRR are higher for lower density levels.

4   Summary

Association rule mining is a powerful method that has been successfully used for various personalization and recommendation tasks in the past.

In this paper we have shown how the personalization of the learned model in rule mining-based approaches to recommendation can help to increase the accuracy of the system's prediction while at the same time the advantages of model-based approaches such as robustness against attacks and the possibility to generate explanations can be preserved. Data structures such as the Extended Frequent Itemset Graph can be used to efficiently generate recommendations online. Furthermore, given the explicit and comprehensible nature of the frequent itemsets, these (personalized) frequent itemsets can be easily manually extended with additional manually-engineered domain rules.

References

1. Agrawal, R., Srikant, R.: Fast algorithms for mining association rules in large databases. In: VLDB 1994, Chile, pp. 487–499 (1994)
2. Liu, B., Hsu, W., Ma, Y.: Mining association rules with multiple minimum supports. In: SIGKDD 1999, USA, pp. 337–341 (1999)
3. Sandvig, J.J., Mobasher, B., Burke, R.: Robustness of collaborative recommendation based on association rule mining. In: RecSys 2007, USA, pp. 105–112 (2007)
4. Lin, W., Alvarez, S., Ruiz, C.: Efficient adaptive-support association rule mining for recommender systems. Data Mining and Knowledge Discovery 6, 83–105 (2002)
5. Kiran, R.U., Reddy, P.K.: An improved multiple minimum support based approach to mine rare association rules. In: CIDM 2009, USA, pp. 340–347 (2009)
6. Mobasher, B., Dai, H., Luo, T., Nakagawa, M.: Effective personalization based on association rule discovery from web usage data. In: WIDM 2001, USA, pp. 9–15 (2001)

Semantically Enriched Event Based Model for Web Usage Mining[*]

Enis Söztutar[1], Ismail H. Toroslu[1], and Murat Ali Bayir[2]

[1] Computer Engineering Dept., Middle East Technical University
Ankara, Turkey
[2] Computer Science & Eng. Dept., University at Buffalo, SUNY
14260, Buffalo, NY, USA

Abstract. With the increasing use of dynamic page generation, asynchronous page loading (AJAX) and rich user interaction in the Web, it is possible to capture more information for web usage analysis. While these advances seem a great opportunity to collect more information about web user, the complexity of the usage data also increases. As a result, traditional page-view based web usage mining methods have become insufficient to fully understand web usage behavior. In order to solve the problems with current approaches our framework incorporates semantic knowledge in the usage mining process and produces semantic event patterns from web usage logs. In order to model web usage behavior at a more abstract level, we define the concept of semantic events, event based sessions and frequent event patterns.

Keywords: Web Usage Mining, Semantic Web Usage Mining, Semantic Events, Event Mining , Apriori.

1 Introduction

Recent trends in WWW (Web 2.0, AJAX, and Asynchronous page loading) paved the way to more interactive web applications that are similar to traditional desktop applications. While this new style of web application development enables developers to collect more information about users, the complexity of the usage data also increases. As a result, traditional page-view based usage mining methods [6,7,16] becomes insufficient to capture and analyze complex data. During the site visit, a user hardly thinks of the web site as a set of pages, but rather she uses applications in web pages to achieve higher level goals, such as finding content, searching for information, listening to a song etc. As an example, consider the following scenario.

Scenario: Bob is planning to go to a dinner with his girlfriend. He opens his favorite 'restaurants web site' and sends a query with 'Mediterranean Restaurants' keywords and his zip code. He finds an overwhelming number of results for his query and decides to narrow his search by writing 'Italian Restaurants'. Then, he selects 2 of them, and by clicking blog icons on the map he reads the reviews for the restaurants. Finally, Bob picks one of them, and he visits the web site of the place.

---

[*] The project is supported by the The Scientific and Technological Research Council of Turkey (TÜBİTAK) with industrial project grants TEYDEB 7070405 and TUBITAK 109E239.

L. Chen, P. Triantafillou, and T. Suel (Eds.): WISE 2010, LNCS 6488, pp. 166–174, 2010.

In similar scenarios as discussed above, it is difficult to interpret users' behaviors in terms of sessions including only page-view sequences. Rather, semantically enriched event series like ['Search Mediterranean Restaurants', 'search Italian Restaurants', 'view reviews for Restaurant A', 'view reviews for Restaurant B', 'click web site link of Restaurant A'] provides more information about site usage and the higher level goals of the user. By viewing sessions as sequences of events, the site owner can improve the user experience.

A natural direction to handle similar scenarios as above is incorporating semantic knowledge [11,12] in the web usage mining process. However, designing a semantically enriched session model is not an easy task due to the following requirements. First, a method should be devised to capture the user intent more robustly than page-views. Second, usage data should be mapped to the semantic space to model user behaviors in a more structured way. Lastly, an algorithm exploiting semantic relations should be devised in the knowledge discovery phase. In order to meet these requirements we propose a semantically enriched event based approach to the Web Usage Mining problem.

Web usage mining can be enhanced by injecting semantic information into various phases. Especially with increasing popularity of Semantic Web, integrating semantics to the web mining process is a natural direction for research. In [4], the authors elaborate different ways for how the fields of Semantic Web and Web Mining can co-operate. In web usage mining perspective, the concept of semantic application events is proposed by Berendt et al. [5].

The concept of domain level usage profiles is introduced in [8] which is a weighted set of instances in the domain. In this work each instance corresponds to an object in the ontology rather than semantic events. In a recent work [14], Nasraoui et al. proposed a semantic web usage mining method for mining changing user profiles of dynamic web sites. Unlike our session model, these works do not include semantically enriched event based model which can capture detailed user actions during the site visit. To the best of our knowledge, this is the first work to encode user browsing behavior into rich semantic events. The problem of mining multi-level association rules has been studied in [10,15]. However, our algorithm extends these methods by the use of richer ontology concepts rather than a taxonomy.

2 Semantic Event Based Session Concept

Similar to [5], our event concept corresponds to the higher level tasks specific to the domain of the site. Events are conceptual actions that the user performs to achieve a certain affect. Events are used to capture business actions that are defined in the site's domain. Some examples of events are 'Play' event for a video on the page, 'Add to shopping cart' event for and E-Commerce site, or 'add friend' action for a social site. Most of the time, events have associated properties that define the event. For example a 'search event' has the search query as a property, and 'play a song' event has a property showing which song is played.

In our model, every event is defined as an object which is represented as an atom-tree, a special type of object graph. Objects can have two types of properties. In the first type, a property of an object relates the object to another object, and in the second type, a property relates the object to a data literal. In this simple event model, capturing arbitrary detail about an object becomes possible since all the properties of an event can be modelled with properties and related objects. The total information about an event is the event object graph (atom tree), including the event object itself, and its related objects recursively. Events in this work are tracked and logged from the JavaScript client, which stores session identification information in session cookies. This information is then used in the session construction step. It should be noted that proposed approach is indifferent to the session construction strategy [3].

Events are considered to capture all the aspects of user's interaction with the site, including page-views. When events are mapped to semantic space, by means of a predefined ontology, we can call them *Semantic Events*. The graph of event objects can be considered as a tree structure since the event object dependency graph is non-cyclic, there is a root, and objects can only be linked to one object. Three types of nodes are defined: nodes containing an individual, a datatype property or an object property . The properties of an individual are stored as children. A datatype property node is always a leaf node and contains the property name and value. An object property node contains the property name and has a single child, which is another individual node.

An illustration of the tree structure is given in Figure 1(a) for a hypothetical video streaming site. When a user views a video with the title "mykitten", and event is created with an individual of class `PlayVideoEvent`, as the root of the tree. The individual has a single object property child, video, which relates the individual to an individual of type `Video`. The `Video` individual has several datatype properties such as the `title`, ID and `Length` of the video. Also it has several `Tags`. The category object property relates this video to an individual of type `KittenVideos`. And the `Submitter` property relates a `User` individual with `username` and `isSubscriber` properties. Note that, even when the user views the same video, different individuals for `Video`, `User`, and `KittenV ideos` are created, since there is no way to refer to the definitions in the ABox[13,2]

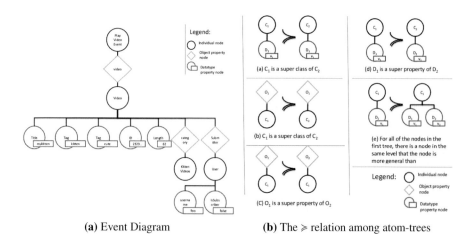

(a) Event Diagram   (b) The ⩾ relation among atom-trees

Fig. 1. Event Diagram and ⩾ relation among atom-trees

from the event tracking client. However, similar events will be aggregated appropriately using anonymous objects in the algorithm.

2.1 Definitions

For all the definitions in this paper, we denote knowledge base, KB, interpretation function $\mathcal{I}$, domain of interpretation $\Delta^{\mathcal{I}}$, domain of data types, $\Delta_D{}^{\mathcal{I}}$, Classes C_i, Object Properties R_i, Individuals o_i, Data Values v_i, and Datatype Properties U_i, where $U^{\mathcal{I}} \subseteq \Delta^{\mathcal{I}} \times \Delta_D{}^{\mathcal{I}}$.

Definition 1 (Session). *A session,* $S = < a_1, a_2, a_3, \ldots, a_n >$, *is an ordered set of atom-trees that corresponds to events for a single user in a certain browsing activity, where* a_i *is an atom-tree.*

Definition 2 (Atom-tree (Event)). *An atom-tree, denoted with a, is a semantic structure that is used to represent a domain event in the web site's ontology. Each node in the atom-tree is an atom(defined below), denoted with* α_i. α_1 *is the root of the tree.*

Definition 3 (Atom). *An atom is either an individual of class C, a datatype property assertion U, or an object property assertion R. An atom is denoted with* α. *Atoms correspond to nodes of the atom-tree.*

Atoms are connected in a tree structure in atom-trees, where property assertion nodes have their objects as their parent. The object property assertion nodes have the child atom's individual as the subject of the statement. Individuals in the Knowledge Base can be unique or anonymous. The unique individuals have a unique URI, whereas the anonymous individuals do not. The anonymous individuals are used as representative samples of their class.

Definition 4 (isMoreGeneralThan, $\geqslant$ relation)

1. *An individual* o_1, *is more general than* o_2, *iff the class of* o_1 *is a superclass of the class of* o_2, *and* o_1 *is anonymous. If* o_1 *is a named individual, than it is not more general than another named individual.*
2. *An assertion* $R_1(o_1, o_2)$ *is more general that* $R_2(o'_1, o'_2)$, *iff first property is a super-property of the second, and the subject/object of the first is more general than the subject/object of the second*
3. *A datatype assertion* $U_1(o_1, v_1)$ *is more general than* $U_2(o_2, v_1)$, *iff the object of the first one is more general than the second, they have the same subject and* U_1 *is a super-property of* U_2
4. *An atom* α_1 *is more general than* α_2 *iff the atoms are both individual assertions, datatype property assertions, or object property assertions, and the first assertion is more general than the second according to rules 1–4.*
5. *An atom-tree* a_1 *is more general than* a_2, *iff for each atom in the first tree, there is a less general atom in the second tree in the same level, all of whose ancestors are less general than their corresponding pairs in the first tree.*

To illustrate the $\geqslant$ relation among atom-trees, some atom-trees are compared in Figure 1(b). The knowledge base contains classes C_1, C_2, datatype properties D_1, D_2, and object properties O_1, O_2. C_1 is a super class of C_2, D_1 is a super property of D_2, and O_1 is a super property of O_2.

The $\geqslant$ relation is the key point of injecting semantics to the web usage mining process. It is defined over four sets, which are the set of individuals in the ABox[13,2], the set of datatype and object property assertions in the ABox, and the set of atom-trees. The relation leverages the class and property hierarchy semantic constructs of the terminological box, and usage of anonymous objects for representing the class. The domain and range restrictions for properties are used to limit the search space.

3   Frequent Pattern Discovery from Semantic Event Logs

The frequent semantic event mining task is to find the frequent sequence of events, from the set of sessions. A frequent pattern, shows the sequence of events that users perform frequently. Patterns are similar to the sessions in the data set. However, the atom-trees in the patterns are constructed from the ones in the data set, using the $\geqslant$ relation. A session S contains, or supports, the pattern Q iff Q is a subsequence of S, where the subsequence relation uses $\geqslant$(rather than equality) for determining inclusion.

With the extended definition of *is a subsequence of* at hand, the *frequency* of a pattern is defined as the number of sessions that this pattern is a subsequence of, and the *support* of an atom-tree or pattern is the frequency of the pattern divided by the total number of sessions. Then for given Dataset, D, the problem is to find the set of patterns, with support greater than the threshold value, *minSupport*.

The algorithm to find frequent sequences of atom-trees from the set of sessions is inspired by the level-wise search in the GSP[15] algorithm. The algorithm consists of two phases. In the first phase the frequent atom-trees are found, and in the second phase, frequent atom-tree sequences are searched.

The algorithm is similar to the Apriori[1] algorithm, in that, it iterates by generating candidates at each level, counting frequency of the candidates and generating new candidates from the previous ones. A close examination of the definitions of support and $\geqslant$ from the previous section reveal that, there is an *apriori* property among atom-trees. If an atom-tree is more general than the other, then its support is greater. Then by the following lemma, the algorithm can prune the search space by eliminating atom-trees that are less general than some infrequent atom-tree.

Lemma 1. *Given atom-trees* a_1, a_2, *if* $a_1 \geqslant a_2 \Rightarrow support(a_1) \geq support(a_2)$

The proof for lemma 1 is straightforward from the definition of subsequence relation and support. All the sessions supporting a_2 supports a_1 by the transitivity of $\geqslant$.

To eliminate infrequent candidates in the early iterations, we count the most general forms of the atom-trees and gradually refine the frequent ones. To obtain the initial set of atom-trees, a *getMostGeneralForms*, ψ, operation is defined over atom-trees.

Definition 5 (*getMostGeneralForms*, ψ)**.** *Given an atom-tree, a, getMostGeneralForms, ψ, operation returns the set of trees, that are more general than the given atom-tree and that are not less general than any other tree in the set of atom-trees.*

getMostGeneralForms function takes an atom-tree as input and returns a set of atom-trees. For each top-level super class of the root's ontology class, we construct a new tree with an anonymous individual atom of super class. Once the most general forms of the atom trees are found, the frequent trees are refined iteratively until no more candidates can be generated. A one-step refinement operator, ϕ is defined over the sets of individual atoms, object and datatype property assertion atoms.

Definition 6 (one step refinement, ϕ). *Given individual atoms o_1, o_2, object properties R_1, R_2, datatype properties U_1, U_2, the ϕ operator is defined over pairs where the first one is more general than the other as :*

(1) if $o_1 \succcurlyeq o_2$, then $\phi(o_1, o_2)$ is the set of individual atoms that are individuals of direct subclasses of the class of o_1, and have the o_2's class as a subclass.
(2) if $R_1(o_1, o_2) \succcurlyeq R_2(o_1, o_2)$, then $\phi(R_1(o_1, o_2), R_2(o_1, o_2))$ is the set of property assertions, that have a direct subproperty of R_1 and have R_2 as a subproperty.
(3) if $R_1(o, o_1) \succcurlyeq R_1(o, o_2)$, then $\phi(R_1(o, o_1), R_1(o, o_2))$ is the set of property assertions, that have the subject belonging to the set $\phi(o_1, o_2)$
(4) if $U_1(o_1, v_1) \succcurlyeq U_2(o_1, v_1)$, then $\phi(U_1(o_1, v_1), U_2(o_1, v_1))$ is the set of datatype property assertions, that have a direct subproperty of U_1 and have U_2 as a subproperty.

The one-step refinement operator over the set of atom-trees is defined using the above definitions. We use the uppercase symbol, Φ for the operator.

Definition 7. *Given two atom-trees, a_1, a_2, and $a_1 \succcurlyeq a_2$, $\Phi(a_1, a_2)$ returns a set of atom-trees that is constructed by either (i) refining a single node in the tree (by applying ϕ), or (ii) adding the mostGeneralForm of a child (by applying ψ) in a_2, that does not exists in a_1, to the corresponding parent node of a_1.*

Algorithm that finds frequent atom-trees is as follows. We first find the initial candidate set, and iterate until no more candidates can be generated. At each iteration the atom-trees in the candidate set(C_i) is compared towards the dataset and the frequency of the candidates is incremented for each atom-tree in the dataset that is less general than the candidate. The next set of candidates is computed in this iteration by adding all the refinements of the candidate ($\Phi(a_c, a)$) with the atom-tree in the data set. Once all the candidates are counted, those with support at least *minSupport* are kept in pattern set(L_i), as the frequent atom-trees. The candidates in C_{i+1} which are generated by refining some infrequent candidate are removed from the set. The algorithm returns the set of frequent events in the web site.

In the next stage of the algorithm the set of frequent atom-tree sequences are found. This stage of the algorithm is much similar to GSP [15]. In GSP, item hierarchies (taxonomies) can be introduced in the mining process. To mine frequent patterns with taxonomy, each data item is converted to an extended sequence, where each transaction contains all the items, and all the ancestors of the items in the hierarchy.

In the second phase of finding frequent event patterns, we use the $\succcurlyeq$ relation, which defines a partial order over the set of atom-trees. This order is used to introduce a taxonomy over the atom-trees, so that GSP is used with little modification. With the $\succcurlyeq$ operator, and the subsequence definition, the task of finding frequent event patterns is reduced to finding ordered frequent sequences. We use an algorithm that is similar to GSP[15], however other sequential pattern mining algorithms such as SPADE [17] can also be adapted. We have chosen an Apriori[1] like algorithm since it offers straightforward parallelization via MapReduce [9](Planned as future work). Since GSP is well known, the second phase is not further discussed here due to limited space.

4   Experimental Results

For testing the usability and performance of the proposed algorithm and framework for discovering frequent semantic events, we have deployed two experimental setups.

The first experiments were performed in the music streaming site, called BulDinle[1]. A moderate amount of traffic of average 2700 visits and 8000 page-views per day is measured on the site. We have defined 7 type of events that capture the user's interaction with the site. We used the event logs for June 2009 which contain 263,076 page-views and 284,819 events in 75,890 sessions, for an average of 3.47 page-views and 3.75 events per session. 15959 sessions contains events, which gives an average number of 17.85 events per session. The algorithm was run with $minSupport = 0.03$.

In the first phase, the frequent events of most general form showed that in 38.9% of the sessions, the user made a search, in 9.3%, the user removed a song from her playlist, and in 95.5%, the user made an action about a song. In the next iteration of the algorithm, it is calculated that in 92.6% of the sessions, the user played a song, and in 27.5% the user added a song to playlist, which confirms the expected user behavior. For the popular songs, the number of times each song is played or added to a playlist are found at the last level of refined events. In the first phase, a total of 55 events are found to be frequent with *support* greater than 0.03. 26 of the events are found to be maximal, where a maximal patterns are defined as the least general patterns.

In the second phase of the algorithm, a total of 139 frequent patterns are found. An interesting pattern of length 2 appearing in 673 sessions indicates a search is performed after playing a particular song. Another pattern indicates that a specific song which is played 2287 times in total(according to length-1 event pattern), is played after 5 searches, in 697 sessions, thus in 30% of the time, the song may be searched up to 5 times. Another pattern of length 6, shows that sequential removal of songs from playlist is frequent, which is clearly due to a lack of 'clear playlist' button in the interface.

The second set of experiments was performed on the logs of a large mobile network operator. The web site offers various set of features ranging from browsing static content to sending SMS online. Two days worth of logs, on 28-29 June 2008, is used in the analysis. A total of more than 1 million page-views occurred in 175K sessions. In the given time interval, 138,058 distinct URL's were accessed.

We have run the proposed algorithm over the converted logs with min support 0.03. A total of 35 events are counted to be frequent in the first phase. 27 of the events are found to be maximal. We found that nearly 10% of the sessions contain at least on search action, nearly all of the sessions contain browsing static content, and in 38% of the sessions, a page not categorized in the ontology is visited. After the second iteration, page-view events for specific URL's are output so that the visit counts for individual

---

[1] Buldinle Web Site, http://www.buldinle.com/

URLs are listed. An interesting pattern states that in 71% of the sessions the user visits the home page. The first phase finishes after 7 iterations. In the second phase of the algorithm, a total of 173 frequent patterns of length more than one are found. Some subjectively interesting patterns include user's browsing behaviors between subclasses of content class, patterns in which users visited home page then jumped to some specific content and patterns in which users searched and moved on to specific category. Due to space limitations, no further discussion is given for this experiment.

5   Conclusion and Future Work

In this work, we have introduced semantic events and an algorithm for mining frequent semantic event sequences. The proposed system has several advantages over traditional web usage mining systems and some of the earlier approaches to integrating semantics to the process. First, the definition of semantic events is demonstrated as a valid model for capturing web surfing behavior in the experiments. Second, the ordering relation among event atom trees is intuitive and sound. By using this relation, we can employ an ordering among possible events, and use the Apriori property to avoid counting infrequent candidates. Third, semantics is richly injected to the process, which unlike previous systems, exploits classes, properties, sub-class and sub-property relations, individuals and property statements of individuals. Moreover, possible future work for this project include using other language constructs, grouping or aggregating datatype literals, etc. Further, it is clear from the examples that, the approach can be applicable to web sites with very different characteristics, usage patterns and site structures.

References

1. Agrawal, R., Imielinski, T., Swami, A.: Mining Association Rules between Sets of Items in Large Databases, pp. 207–216 (1993)
2. Baader, F., Horrocks, I., Sattler, U.: Description Logics as Ontology Languages for the Semantic Web. In: Festschrift in honor of Jörg Siekmann. LNCS (LNAI), pp. 228–248. Springer, Heidelberg (2003)
3. Bayir, M.A., Toroslu, I.H., Cosar, A., Fidan, G.: Smart miner: a new framework for mining large scale web usage data. In: WWW 2009: Proceedings of the 18th International Conference on World Wide Web, pp. 161–170. ACM Press, New York (2009)
4. Berendt, B., Hotho, A., Stumme, G.: Towards Semantic Web Mining. In: Horrocks, I., Hendler, J. (eds.) ISWC 2002. LNCS, vol. 2342, pp. 264–278. Springer, Heidelberg (2002)
5. Berendt, B., Stumme, G., Hotho, A.: Usage Mining for and on the Semantic Web. In: Data Mining: Next Generation Challenges and Future Directions, pp. 461–480. AAAI/MIT Press (2004)
6. Chen, M.-S., Park, J.S., Yu, P.S.: Efficient data mining for path traversal patterns. IEEE Trans. Knowl. Data Eng. 10(2), 209–221 (1998)
7. Cooley, R.: Web Usage Mining: Discovery and Application of Interesting Patterns from Web Data. PhD thesis, Dept. of Computer Science, Univ. of Minnesota (2000)
8. Dai, H.K., Mobasher, B.: Using Ontologies to Discover Domain-Level Web Usage Profiles (2002)
9. Dean, J., Ghemawat, S.: MapReduce: simplified data processing on large clusters. In: OSDI 2004, Berkeley, CA, USA. USENIX Association (2004)

10. Han, J., Fu, Y.: Discovery of multiple-level association rules from large databases. In: Proc. 1995 Int. Conf. Very Large Data Bases, pp. 420–431 (1995)
11. Jin, X., Zhou, Y., Mobasher, B.: Web usage mining based on probabilistic latent semantic analysis. In: KDD, pp. 197–205 (2004)
12. Józefowska, J., Lawrynowicz, A., Lukaszewski, T.: Intelligent Information Processing and Web Mining, pp. 121–130 (2006)
13. Nardi, D., Brachman, R.J.: An Introduction to Description Logics, pp. 1–40 (2003)
14. Nasraoui, O., et al.: A web usage mining framework for mining evolving user profiles in dynamic web sites. IEEE Trans. Knowl. Data Eng. 20(2), 202–215 (2008)
15. Srikant, R., Agrawal, R.: Mining Sequential Patterns: Generalizations and Performance Improvements. In: Apers, P.M.G., Bouzeghoub, M., Gardarin, G. (eds.) EDBT 1996. LNCS, vol. 1057, pp. 3–17. Springer, Heidelberg (1996)
16. Xiao, Y., Dunham, M.H.: Efficient mining of traversal patterns. Data Knowl. Eng. 39(2), 191–214 (2001)
17. Zaki, M.J.: SPADE: an efficient algorithm for mining frequent sequences. Machine Learning 42(1/2), 31–60 (2001)

Effective and Efficient Keyword Query Interpretation Using a Hybrid Graph

Junquan Chen[1], Kaifeng Xu[1], Haofen Wang[1], Wei Jin[2], and Yong Yu[1]

[1] Apex Data and Knowledge Management Lab
Shanghai Jiao Tong University, Shanghai, 200240, China
{jqchen,kaifengxu,whfcarter,yyu}@apex.sjtu.edu.cn
[2] Department of Computer Science
North Dakota State University, Fargo, ND 58108
wei.jin@ndsu.edu

Abstract. Empowering users to access RDF data using keywords can relieve them from the steep learning curve of mastering a structured query language and understanding complex and possibly fast evolving data schemas. In recent years, translating keywords into SPARQL queries has been widely studied. Approaches relying on the original RDF graph (*instance-based approaches*) usually generate precise query interpretations at the cost of a long processing time while those relying on the summary graph extracted from RDF data (*schema-based approaches*) significantly speed up query interpretation disregarding the loss of accuracy. In this paper, we propose a novel approach based on a hybrid graph, for the trade-off between interpretation accuracy and efficiency. The hybrid graph can preserve most of the connectivity information of the corresponding instance graph in a small size. We conduct experiments on three widely-used data sets of different sizes. The results show that our approach can achieve significant efficiency improvement with a limited accuracy drop compared with instance-based approaches, and meanwhile, can achieve promising accuracy gain at an affordable time cost compared with schema-based approaches.

1 Introduction

On the way to Semantic Web, Resource Description Framework (RDF) is a language for representing information about resources in the World Wide Web. An example data repository available in RDF format is semanticweb.org[1]. A snippet of RDF data from semanticweb.org is shown in Figure 1(a), where the entity "Statue of Liberty" (a statue) is described by its attribute values and relations to other entities.

The ever growing semantic data in RDF format provides fertile soil for semantic search, and formal query languages (e.g. SPARQL) are adopted by most current semantic search systems[1,2] to accurately express complex information needs. An example query in SPARQL is shown in Figure 1(b) which is used

---

[1] http://www.semanticweb.org/

L. Chen, P. Triantafillou, and T. Suel (Eds.): WISE 2010, LNCS 6488, pp. 175–189, 2010.
© Springer-Verlag Berlin Heidelberg 2010

```
xmlns:sw="http://semanticweb.org/id/"          PREFIX  sw:<http://semanticweb.org/id/>
                                                SELECT  ?stat
<rdf:Description rdf:about= "Statue_of_Liberty">  WHERE  {
  <rdfs:label> Statue of Liberty </rdfs:label>          ?stat      sw:Built_by      ?person.
  <sw:Instance_of rdf:resource= "Statue"/>              ?stat      sw:Instance_of   "Statue".
  <sw:Built_by rdf:resource= "Gustave_Eiffel"/>         ?person    sw:name          "Eiffel".
</rdf:Description>                              }
```

Fig. 1. (a) RDF snippet, and (b) SPARQL query from semanticweb.org

to find out all the statues built by Eiffel. However, the disadvantages of formal queries are: (1) *Complex Syntax*: It is hard to learn and remember complex syntax of formal queries for ordinary users. (2) *Priori Knowledge*: Users have to know the schema of the underlying semantic data beforehand. In contrast, keyword queries cater to user habits since keywords (or known as keyword phrases) are easier to be understood and convenient to use. An approach that can leverage the advantages of both query types is to provide a keyword user interface and then translate keyword queries into formal queries.

In XML and database communities [3,4,5,6,7,8], bridging the gap between keyword queries and formal queries has been widely studied. However, there exists a limited amount of work on how to answer keyword queries on semantic data (mainly RDF graph). As an early attempt to build a semantic search system, SemSearch [9] employed a template-based approach to capture the restricted interpretations of given keywords. Later, improved approaches [10,11] have been proposed to address the problem of finding all possible interpretations. In particular, Thanh et al. [11] employed the RDF graph (*instance-based approaches*) to discover the connections between nodes matching the input keywords, through which the interpretation accuracy can be ensured, but at the cost of a longer processing time. This problem has been recently tackled by [12,13], where keyword queries are translated using a summary graph extracted from the RDF data (*schema-based approaches*). Although schema-based approaches significantly speed up the processing, they cannot always guarantee the interpretation accuracy.

Our major contributions in this research can be summarized as follows.

(1) Instead of using only an instance graph or a schema graph, a hybrid graph is proposed for the effective and efficient keyword query interpretation, which combines the advantages of an instance graph capable of preserving most of connectivity information and a schema graph with a relatively small size requirement.

(2) A score function is defined which can reflect the trade-off between expected interpretation accuracy and efficiency. Based on which, we propose a novel approach to construct the hybrid graph to fit the requirements of the expected interpretation accuracy and efficiency.

(3) Experiments are conducted on three widely-used data sets (semanticweb.org, DBLP[2] and DBpedia[3]) of different sizes. The results show that our

---

[2] http://www.informatik.uni-trier.de/%7Eley/db/
[3] http://dbpedia.org/

approach has achieved a 61.66% efficiency improvement with a 6.17% accuracy drop over instance-based approaches on average. Meanwhile, our approach achieves 132.30% accuracy gain with a 20.08% time increase compared with schema-based approaches.

The rest of this paper is organized as follows. Section 2 gives the overview of our solution. Section 3 elaborates on the construction of the hybrid graph and Section 4 describes the interpretation process together with the ranking mechanisms. The experimental results are presented in Section 5 followed by the related work in Section 6 and the conclusion in Section 7.

2   Overview

We aim at translating keyword queries to their corresponding formal queries (i.e. conjunctive SPARQL queries). For example, in order to get the SPARQL query mentioned in Figure 1(b), the user just needs to input the keywords "Statue" and "Eiffel". In general, a keyword query Q_K is composed of keyword phrases $\{k_1, k_2, \ldots, k_n\}$ where k_i has correspondence (i.e. can be mapped) to literals contained in the underlying RDF graph. A formal query Q_F can be represented as a tree of the form $\langle r, \{p_1, p_2, \ldots, p_n\} \rangle$, where r is the root node of Q_F and p_i is a path in Q_F which starts from r and ends at the leaf node that corresponds to k_i. The root node of Q_F represents the target variable of a tree-shaped conjunctive query corresponding to the sub-tree semantics, which is mentioned in [14]. In the example of Figure 1(b), k_1 = "Statue" and k_2 = "Eiffel", $Q_F = \langle r, \{p_1, p_2\}\rangle$, where $r = stat$, $p_1 = \langle stat, Build\_by, person, name, Eiffel \rangle$, and $p_2 = \langle stat, Instance\_of, Statue \rangle$.

The online process to construct such a tree-shaped conjunctive query on a hybrid graph (abbreviated to HG) includes two main phases: (1) *Keyword Mapping*, which retrieves the nodes on HG (a graph) matching the given keywords. (2) *Query Construction and Ranking*, which searches the HG to construct top-k potential tree tree-shaped conjunctive queries (formal queries). The details will be discussed in Section 4.

The offline construction of HG is an iterative process which extracts and refines a graph from the "crude" RDF graph (i.e. the original data graph) by means of a score function. The score function is used to define the overall interpretation performance, which plays an important role from the starting point to the ending point of each interaction in the whole construction process. This is similar to that of oil refinery where the crude oil is refined into more useful petroleum products separated by different boiling points. In this case, the score function serves as the role these boiling points play. The details will be discussed in Section 3.1. The workflow of HG construction is illustrated in Figure 2(a) which takes the crude RDF graph as input. A *construction unit* (CU) is employed to carry out the refinement on the given RDF graph to generate the HG subgraph. The CUs of the same type are additionally used multiple times for further refinement on the remaining crude RDF graphs. As a result, several HG subgraphs are returned, and combined together to form an overall HG. The

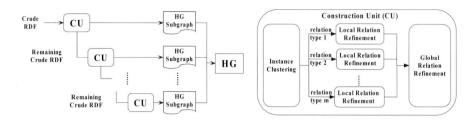

Fig. 2. (a) The workflow of HG construction. (b) The components inside a CU.

CU process will be described in Section 3.2 followed by the merging method of HG subgraphs to be introduced in Section 3.3.

Figure 2(b) shows the components inside a CU. Given the crude RDF graph G and the specific performance preference p, CU will work through the following streamline: (1) *Instance Clustering*: Each instance in G is represented by the feature set of its relations and regarded as a trivial cluster initially. After that, the cluster pairs with the highest similarity are continuously clustered until a cluster tree T is derived. The nodes of the HG are constructed in this phase. (2) *Local Relation Refinement*: The relations of a single type are linked upon T with the guide of the score function w.r.t. the given p. The relations of HG are constructed in this phase. (3) *Global Relation Refinement*: The quality of the resulting relations returned by *Local Relation Refinement* is assessed using the score function within the context of all relations. The high-quality relations are picked out to form a subgraph, while the others are fed into the remaining crude graph for further refinements.

3   Hybrid Graph Construction

3.1   Score Function

Taking both the advantages and disadvantages of the instance-based and schema-based approaches into mind, we treat the issue of interpretation performance as a trade-off between *interpretation accuracy* (IA) and *interpretation efficiency* (IE). IA is used to estimate the probability of a translated query to return non-empty answers from the underlying RDF data, while IE is to help assess how efficient it might take for the query to be translated.

IA score of the graph used for the keyword query interpretation is high when the graph can preserve the connectivity information of the instance graph as much as possible. In other words, most of the son instance nodes should have the same connectivity as their father cluster nodes. Suppose that c_1 and c_2 are two nodes in the schema graph which are connected by a relation r, and then the probability that any two instances i and j which are child nodes (directly or indirectly) of c_1 and c_2 respectively can connect to each other's cluster ($c1$ or $c2$) with relation r is:

$$P_r(c_1, c_2) = \frac{|\{i | i \rightarrow c_1 \wedge \exists j \rightarrow c_2, r_{ij} = r\}|}{|\{i | i \rightarrow c_1\}|} \cdot \frac{|\{j | j \rightarrow c_2 \wedge \exists i \rightarrow c_1, r_{ji} = r\}|}{|\{j | j \rightarrow c_2\}|} \quad (1)$$

where $i \rightarrow c_1$ means i is a child node (directly or indirectly) of c_1. Then the IA score of the graph is defined as the arithmetic average of all the probabilities.

$$IA_R = \frac{1}{|R|} \cdot \sum_{<c_m, r_k, c_n>} P_{r_k}(c_m, c_n) \quad (2)$$

where $< c_m, r_k, c_n >$ is a part of the graph in which the node c_m and the node c_n are connected by r_k, while R is the set of all relations in the graph and $r_k \in R$.

The average interpretation time is proportional to the quantity of edges in the graph. In the construction of formal query, more edges mean larger search space. So the IE score is defined as the reciprocal of the quantity of edges:

$$IE_R = \frac{1}{|R|} \quad (3)$$

In order to let the user express his/her preference on the trade-off of IA and IE, we further define a configurable score function for an $|R|$-size graph as follows:

$$Score_R = p \cdot IA_R + (1 - p) \cdot IE_R \quad (4)$$

where p is a preference value given by the user. It reflects the user's preference for interpretation accuracy or efficiency. Note that the range of p is $[0, 1]$. If p is set to 1, it indicates that the user concerns about the interpretation accuracy only. On the other hand, if p is set to 0, the user is most interested in the interpretation efficiency. We will discuss the selection of p and its impact on the construction of HG in the following sections.

3.2   Construction Unit

The Construction Unit (CU) takes the responsibility of extracting an HG subgraph from the RDF graph by means of a score function. It then separates the subgraph from the remaining part of the RDF graph which will be further refined at the next iteration. Given an RDF graph G and performance preference p as inputs, the whole process inside CU is composed of three main sub-processes: *Instance Clustering*, *Local Relation Refinement* and *Global Relation Refinement*.

Instance Clustering clusters the underlying instances of G into a cluster tree. During the hierarchical clustering, each instance i in the RDF graph is first mapped into a feature set where each element corresponds to one connected relation with its direction to the instance as the additional information. Note that we strictly distinguish outgoing relations from incoming relations connected to an instance even if these relations are of the same type. The feature set of i as well as that of a cluster $c = \{i_1, i_2, \ldots, i_n\}$ are defined as follows:

$$F(i) = \{r_1{}^{+/-}, r_2{}^{+/-}, \ldots, r_j{}^{+/-}, \ldots, r_m{}^{+/-}\}, \; and \; F(c) = \bigcup_{k=1}^{n} F(i_k) \qquad (5)$$

where $r_j{}^{+/-}$ is a relation connected to i, $+/-$ denotes the direction of r_j (either incoming or outgoing) w.r.t. the instance i; m represents the number of connected relations for i, and n is the instance count in the cluster c.

The clustering process starts from a set of trivial clusters in which each contains one instance, and then similar clusters are merged into a single cluster iteratively until we reach a cluster tree. At each iteration, the cluster pair with the highest similarity is selected for merging. The similarity between two clusters is calculated by means of a Jarcard Index which can be expressed by the following equation:

$$sim(c_1, c_2) = \frac{|F(c_1) \cap F(c_2)|}{|F(c_1) \cup F(c_2)|} \qquad (6)$$

The straightforward implementation of the clustering process is inefficient with time complexity $O(|I|^3)$ where I denotes the entire instance set. Inspired by the community detection algorithm proposed by [15], we introduce the global and local max heaps to effectively cut down the total category pairs for similarity calculation. As a result, the time complexity of instances clustering is reduced to $O(|I|^2 \log |I|)$.

Local Relation Refinement tries to substitute high-level clusters for those instances (leaf nodes) with link relations between them, which results in a hybrid graph having relations that connect (trivial) clusters. The size of the resulting hybrid graph can be much smaller compared with that of the crude RDF graph (having only trivial clusters connected with relations in our context). The relation refinement process aims to achieve the required interpretation accuracy, while at the same time, improve the efficiency on the resulting graph.

Since relations of the same type share the same label or identifier, their connected instances always behave similarly. We call an $|R|$-size graph containing the same relation type k an R_k-style graph. Then, we define the IA score of an R_k-style graph as:

$$IA_{R_k} = \frac{1}{|R_k|} \sum_{<c_m, r_i, c_n> \wedge r_i \in R_k} Pr_i(c_m, c_n) \qquad (7)$$

Similarly, we define the IE score of an R_k-style graph as follows:

$$IE_{R_k} = \frac{1}{|R_k|} \qquad (8)$$

Then the score function for the R_k-style graph is defined as:

$$Score_{R_k} = p \cdot IA_{R_k} + (1 - p) \cdot IE_{R_k} \qquad (9)$$

After instance clustering, a cluster tree is obtained. We first simply link the instances in the cluster tree with the same relation type k according to the original instance graph to derive a naive hybrid graph HG_{R_k}. Obviously, IA_{R_K} of HG_{R_k} might be high but IE_{R_k} might be low. Intuitively, we can migrate some relation edges of instances up to the corresponding clusters to increase IE_{R_k} because some lower nodes can share a relation which decreases $|R_k|$, and at the same time IA_{R_k} might decrease. Therefore our goal is now to identify a graph with the maximum $Score_{R_k}$ in terms of IA_{R_K} and IE_{R_K}.

Instead of enumerating all the possible relations linking between clusters to find out the graph with maximum $Score_{R_k}$, a heuristic relation refinement order from top to bottom is adopted to derive a hybrid graph with local maximum $Score_{R_k}$. The basic idea is as follows. Two relatively high level clusters are selected and linked by one relation edge. In the following step, in order to increase the performance, the subject and object clusters w.r.t. the relation edge are to be replaced with some lower clusters in turn. The replacement will continue until the $Score_{R_k}$ does not increase any more. Then a hybrid graph with local maximum $Score_{R_k}$ is generated.

Here is the detailed process. To construct a hybrid graph HG_{R_k}, we first select two lowest level clusters c_1 and c_2 which are ancestor nodes of all the subject instances and object instances respectively. Note that an instance which has at least one out-going (in-coming) relation is called a subject (object) instance. A relation edge is assigned from c_1 to c_2. Now the IE_{R_k} of HG_{R_k} is 1 because there is only one relation edge in HG_{R_k}. However, the IA_{R_k} is likely low. To increase IA_{R_k}, we would replace the subject and object clusters with lower level clusters with the result that after the replacement, the IA_{R_k} is supposed to increase at the cost of a IE_{R_k} drop. The main cycle is to execute the two replacement operations $replace(obj)$ and $replace(subj)$ iteratively until both replacements cannot increase the $Score_{R_k}$ any more. The replacement operation $replace(obj)$ is to execute $replace(r, obj)$ for each relation edge r in HG_{R_k}, where obj is the object node of r. Likewise, $subj$ is the subject node of r. The operation $replace(r, obj)$ will first select a set A of some children (direct or indirect) of obj, and then assign relation edges between the subject node and each node in A and remove r finally. The selection of set A is fulfilled by a function $select(r, obj)$.

The basic idea of $select(r, obj)$ is as follows. The result set of $select(r, obj)$ first consists of all the object instances which are the children of obj. In order to increase the $Score_{R_k}$ of HG_{R_k}, higher level clusters are used to replace lower level clusters or instances until $Score_{R_k}$ cannot increase any more. In practice, a recursive algorithm is developed where $select(r, obj)$ is implemented as a recursive function. obj can be an instance or a non-trivial cluster.

If obj is an instance, then

$$select(r, obj) = \begin{cases} \{obj\} \ If \ obj \ is \ an \ object \ instance \\ \emptyset \ otherwise \end{cases}$$

And if obj is a non-trivial cluster, then

$$A = select(r, obj) = \bigcup_{\forall \, child\_node \, of \, obj} select(r, child\_node)$$

Two relation linking strategies are adopted. One is linking a relation between *subj* and *obj*, and the other is linking relations between *subj* and each node in the set A. The strategy which can achieve a better $Score_{R_k}$ will be chosen, that is, if the first strategy wins, then the set with one element *obj* will be returned, otherwise A will be returned.

Figure 3 gives an example of local relation refinement process. In the graphs [a-e], nodes [A-H] are non-trivial clusters and nodes [I-Q] are instances. Graph (a) is the crude graph in which there are only relation edges between instances derived after instance clustering. Step 1, the lowest clusters B and C which are ancestor nodes of all the subject instances and object instances respectively are first linked by a relation edge. Note that the number in the edge represents the step in which this edge is added. Step 2, the object node C is replaced by lower level clusters F and P to improve the $Score_{R_k}$ while the subject node is still B. Step 3, it's the subject node B's turn, and it is replaced by D and K and the corresponding subjects and objects are connected. Step 4, the object node F is replaced by H and O. After step 4, graph (e) is derived in which the $Score_{R_k}$ cannot be improved through local relation refinement any more. Finally, graph (e) is the output of local relation refinement.

Global Relation Refinement denotes the task of selecting the refined subgraph from the linked cluster graph to output an HG subgraph and leaving the unrefined crude RDF subgraph for further refinement.

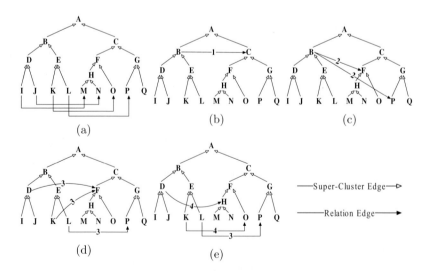

Fig. 3. An example of Local Relation Refinement

Note that the extraction of the refined subgraph is necessary since clusters in the hierarchical cluster tree are always disjoint. More clusters are necessary to enable high quality of HG. One solution is to adopt more than one tree.

The detailed process is as follows. The score function is first applied on the derived graph to calculate its global $Score$ value. We directly output the derived graph as the HG subgraph as long as the graph's $Score$ exceeds a predefined threshold. Otherwise, the best subgraph whose $Score$ can satisfy the threshold will be calculated and extracted. To find such a subgraph, a list of $Score_{R_k}$ computed by the process of relation linking is ranked in descending order. We check the $Scores$ of those kinds of relations which are ranked in the top half of the list and return the best subgraph whose $Score_{R_k}$ is above the threshold, otherwise those ranked in top $1/4, 1/8, \ldots, 1/2^n$ will be subsequently checked until the best subgraph is found. The worst condition is that only the top 1 in the list is left, in this case, we directly recognize the subgraph corresponding to it without checking the $Score$. After the HG subgraph generation, the remaining unrefined crude RDF subgraph which has less types of relations will be further clustered and refined at next iteration.

3.3 Hybrid Subgraph Merging

Having generated several refined HG subgraphs from the crude RDF graph, the final step is to merge them into one. The basic idea is to integrate clusters coming from different subgraphs with inclusion relations linked by a super- or sub-cluster edge.

4 Query Interpretation using Hybrid Graph

4.1 Keyword Mapping

The input keywords (keyword phrases) by end users are first mapped to the related nodes in HG via a pre-computed inverted index, in which the literals of each node are stored as terms and the reference of the node is treated as the document. The literal set $L(i)$ of an instance i is defined as:

$$L(i) = \{l_{ca_1}, l_{ca_2}, \ldots, l_{ca_{m_1}}, l_i, v_{a_1}, v_{a_2}, \ldots, v_{a_{m_2}}\}, \tag{10}$$

where l_{ca_j} is the label of the jth category of i, l_i is the label of i and v_{a_k} is the kth attribute value of i. Correspondingly, the literal set $L(c)$ of a cluster c is defined as:

$$L(c) = \bigcup_{j=1}^{n} L(i_j), \ c = \{i_1, i_2, \ldots, i_n\}. \tag{11}$$

For each literal l in $L(c)$, we shall count its occurrence frequency N_l in the instances of c. The calculated frequency is used in the definition of $\sigma_{lc} = N_l/|c|$

which represents the importance of the literal l in cluster c. The intuition behind the definition is that those literals that have high co-occurrence among the instances of the cluster have priority of being descriptors of such a cluster.

Each cluster or instance is a document whose terms are its literals in the inverted index. The frequency of a term is σ_{lc} and 1 for cluster and instance respectively. As a result the matching score of a keyword related node is its $tf \cdot idf$ value after searching the inverted index.

4.2   Query Construction and Ranking

Query construction tries to construct top-k minimum cost subtrees by which keywords related nodes can be connected in HG.

We implement a similar query construction and ranking method as mentioned in Q2Semantic [12]. Due to space limitations, we omit details here. Note that only the path length and keyword matching score are taken into account to calculate the cost of a subtree here. For a subtree $\langle r, \{p_1, p_2, \ldots, p_n\}\rangle$, the ranking schema R is defined as:

$$R = \sum_{i=1}^{n}(\frac{1}{D_i}\sum_{e\in p_i}1), \tag{12}$$

where D_i is the matching score corresponding to the ith keyword-related node. In this case, R prefers shorter queries with higher matching score of keyword phrases. This is also supported in [11] with the conclusion that shorter queries tend to capture stronger connections between keyword phrases.

It is mentioned in Q2Semantic [12] that nodes and edges are assigned scores which can be used for ranking. In particular, the scores of nodes and edges will guide the expansion process, which can increase the accuracy of keyword query interpretation. However, this ranking schema is not necessary in our approach since the interpretation accuracy has been guaranteed by the score function and the quality of derived HG.

5   Experiments

5.1   Experiment Setup

Data sets semanticweb.org, DBLP and DBpedia are used for the experiments, and the statistics of the three data sets are listed in Table 1. We manually construct 42 scenarios (17 from semanticweb.org, 10 from DBpedia, and 5 from DBLP). To compare impacts of the interpretation efficiency and effectiveness on three different graph-based approaches (i.e., schema, hybrid and instance), we ensure that the keyword mapping in the 42 scenarios is correct and consistent for three representations. Due to space limitations, only three example keywords queries and their corresponding potential information needs are shown in Table 2 (one query for each data set). For the complete set of queries, please refer to the website[4]. The experiments are conducted on a PC with 2.3GHz AMD Athlon 64bit processor and 8GB memory.

---

[4] http://hermes.apexlab.org/keyword_queries.htm

Table 1. Statistics of semanticweb.org, DBLP and DBpedia

Data set	♯Category	♯Instance	♯Relation	♯Inst.degree	♯Rel.kind	♯Rel/kind
semanticweb.org	5.06×10^2	7.483×10^3	1.628×10^4	2.18	4.77×10^2	3.413×10^1
DBLP	1.0×10^1	1.640×10^6	3.176×10^6	1.94	1.0×10^1	3.176×10^5
DBpedia	2.694×10^5	2.520×10^6	6.868×10^6	2.73	1.128×10^4	6.088×10^2

Table 2. Sample queries on semanticweb.org, DBLP and DBpedia

Keywords Query	Potential information needs	Data Set
"Eiffel", "United States"	What's the relationship between Eiffel and United States?	semanticweb.org
"Object SQL", "William Kent"	Which papers are written by William Kent on Object SQL?	DBLP
"Computer Games", "Windows"	Which computer games can be played on Windows?	DBpedia

5.2   Evaluation of Interpretation

In order to evaluate the overall interpretation performance, the metrics are defined on interpretation accuracy and efficiency as follows. (1) For interpretation accuracy, precision and recall as applied to information retrieval cannot be directly used because only one of the query results matches the meaning of the keywords intended by the user. Hence, the interpretation accuracy on a data set is measured by the ratio of correctly-translated keyword queries to the selected ones from such a data set. The correctly-translated keyword query is defined as the query from which the intended formal query is derived and can be found in the top 5 results (we assume users are only interested in the top 5 results). (2) For interpretation efficiency, we evaluate the running time of different interpretation models. Generally, the longer the running time the interpretation requires, the lower the interpretation efficiency is. The impacts of the three graph representations (Schema, Instance and Hybrid) on query interpretation accuracy and efficiency have been detailedly examined and the comparison is conducted with other subcomponents (i.e. keyword matching, query construction, and ranking) remaining the same for all graph representations.

Figure 4 illustrates the variation trends of interpretation accuracy and efficiency when translating keywords on semanticweb.org with different approaches. In particular, the vertical axis represents interpretation accuracy and horizontal axis represents running time. The performance of instance-based approaches is mapped to the upper right corner of the figure with the coordinate $(1, 1)$ indicating the best interpretation accuracy but the longest running time achieved by instance-based approaches. The interpretation accuracy and running time of the schema-based approach and HG-based approach with different preference settings can be assigned a value in range $[0, 1]$ respectively.After normalization with reference to performance of instance-based approach, the interpretation performances of three approaches are mapped into their corresponding coordinates as in Figure 4. It is obvious that the optimal performance corresponds

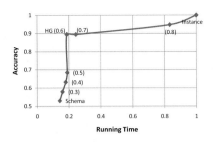
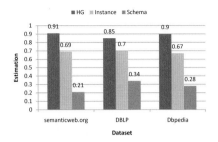

Fig. 4. The trends of the interpretation performance on semanticweb.org

Fig. 5. The estimation of interpretation performance on different data sets

to point $(0,1)$, which means queries can be correctly translated with the least time. However, this ideal case can never be achieved in practice. Fortunately, the HG-based approach with preference value 0.6 achieves the best performance among all three approaches, which is closest to the optimal value (i.e.,has the shortest distance to the optimal point $(0,1)$). In the following experiments on semanticweb.org, we consider this point the best balanced performance point achieved by the HG-based keyword query interpretation in terms of accuracy and efficiency, and the preference value 0.6 is chosen as the default preference setting for the HG-based evaluation.

Before the evaluation of keyword query interpretations on different experimental data sets, we first examine the effectiveness of the proposed score function, which is used to estimate the overall interpretation performance. Figure 5 demonstrates the estimation of interpretation performance on semanticweb.org, DBLP and DBpedia using three different approaches. On all data sets, the HG-based approach achieves the best estimated performance than the others according to the defined score function.

Having estimated the interpretation performance by means of the score function, Figure 6 illustrates the practical interpretation accuracy and efficiency on the given data sets with different approaches. On the whole, the practical evaluation is consistent with the theoretical performance estimation. In particular, the HG-based approach improves the accuracy by an average of 132.30% compared

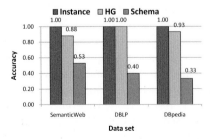
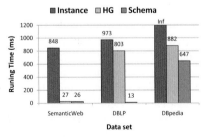

Fig. 6. The practical interpretation accuracy and efficiency on different data sets

with schema-based approaches and there is only at an average 6.17% accuracy drop compared with instance-based approaches on the three datasets. On the other hand, the running time is reduced by 61.66% compared with instance-based approaches and increased by 20.08% compared with schema-based approaches on average. Note that the keyword query interpretation on DBpedia cannot be handled by instance-based approaches as the size of the instance graph is rather large. Our method achieves a high precision (93%) in the limited time on DBpedia. Another bias is that the interpretation runs slower on DBLP than the other two datasets. The reason may be the fact that much fewer relations are contained in DBLP which increases the running time of distributedly constructing HGs.

6   Related Work

In recent years, a lot of work on keyword search has been made in both XML [3,4] and database fields [5,6,7,8]. However, these approaches could not be directly applied to semantic search on RDF data since the underlying data model is a graph rather than tree-shaped XML or relational data.

As an early attempt on semantic search systems, SemSearch [9] used pre-defined triple patterns and query templates to make resulting queries with a fixed structure. As a result, a limited number of the interpretations derived from the underlying RDF graph w.r.t. keyword queries can be computed. Improved approaches [10,11] were proposed to address the problem of generating all possible interpretations. In [11], a more generic graph-based approach (*instance-based approach*) was proposed to explore all possible connections between keyword related nodes, through which the interpretation accuracy could be ensured. However, instance-based approaches are too costly to be acceptable in terms of their interpretation processing time when large-scale RDF data is involved. Such a problem has been tackled recently by [12,13]. In [12], a summary graph is extracted from the original RDF data by merging the properties and literals (*schema-based approach*). In addition, several ranking mechanisms are adopted on the summary graph to help find the top-k tree shaped formal queries. Extended work was presented in [13] by proposing a concrete algorithm for top-k exploration of query graph candidates. Although schema-based approaches achieve significant efficiency improvement, the accuracy of the interpretation has largely been decreased.

Compared with instance-based and schema-based approaches, our approach is distinct in two aspects. First, a score function is proposed to accurately assess the overall interpretation performance on a given RDF graph. Second, a novel process of constructing hybrid subgraphs from the original RDF graph under the guidance of the score function is devised which involves three major steps: Instance Clustering, Local Relation Refinement and Global Relation Refinement. The proposed framework has been experimentally proven effective in reaching the Best Balanced point between the interpretation accuracy and efficiency.

7   Conclusions and Future Work

In this paper, we propose a new solution to the problem of translating keyword queries to formal queries. A hybrid graph based approach is developed which combines the advantages of both schema-based and instance-based methods. The solution also offers the capability of finding the best balanced point in terms of interpretation accuracy and efficiency. We believe we are among the first to reach the above mentioned goals. A score function is first introduced to assess the overall expected interpretation performance on a given RDF graph, which is then employed to construct hybrid graphs (HG) for keyword query interpretation. The proposed framework has been implemented and compared with instance-based and schema-based approaches on three different datasets. The experimental results show that our approach has achieved significant efficiency improvement over instance-based approaches with a limited accuracy drop, and meanwhile, achieved promising accuracy gain at an affordable time cost compared to schema-based approaches.

Future directions include the exploration of alternative clustering methods and score functions for relation refinement. We are also researching improvements of the local relation refinement process for generating a hybrid graph with maximum $Score_{R_k}$. Different methods for approximate relation refinement will be examined and compared with the currently employed top-down heuristic relation refinement order.

References

1. Broekstra, J., Kampman, A., Van Harmelen, F.: Sesame: A generic architecture for storing and querying rdf and rdf schema. In: Horrocks, I., Hendler, J. (eds.) ISWC 2002. LNCS, vol. 2342, pp. 54–68. Springer, Heidelberg (2002)
2. Lu, J., Ma, L., Zhang, L., Brunner, J., Wang, C., Pan, Y., Yu, Y.: SOR: a practical system for ontology storage, reasoning and search. In: VLDB, pp. 1402–1405 (2007)
3. Hristidis, V., Koudas, N., Papakonstantinou, Y., Srivastava, D.: Keyword proximity search in XML trees. In: TKDE, pp. 525–539 (2006)
4. Guo, L., Shao, F., Botev, C., Shanmugasundaram, J.: XRANK: Ranked keyword search over XML documents. In: ACM SIGMOD, pp. 16–27 (2003)
5. Hristidis, V., Papakonstantinou, Y.: DISCOVER: Keyword search in relational databases. In: VLDB, pp. 670–681 (2002)
6. Bhalotia, G., Hulgeri, A., Nakhe, C., Chakrabarti, S., Sudarshan, S.: Keyword searching and browsing in databases using BANKS. In: ICDE, pp. 431–440 (2002)
7. Balmin, A., Hristidis, V., Papakonstantinou, Y.: Objectrank: Authority-based keyword search in databases. In: VLDB, pp. 564–575 (2004)
8. Qin, L., Yu, J.X., Chang, L., Tao, Y.: Querying communities in relational databases. In: ICDE, pp. 724–735 (2009)
9. Lei, Y., Uren, V., Motta, E.: Semsearch: A search engine for the semantic web. In: Staab, S., Svátek, V. (eds.) EKAW 2006. LNCS (LNAI), vol. 4248, p. 238. Springer, Heidelberg (2006)

10. Zhou, Q., Wang, C., Xiong, M., Wang, H., Yu, Y.: Spark: Adapting keyword query to semantic search. In: Aberer, K., Choi, K.-S., Noy, N., Allemang, D., Lee, K.-I., Nixon, L.J.B., Golbeck, J., Mika, P., Maynard, D., Mizoguchi, R., Schreiber, G., Cudré-Mauroux, P. (eds.) ASWC 2007 and ISWC 2007. LNCS, vol. 4825, p. 694. Springer, Heidelberg (2007)
11. Tran, T., Cimiano, P., Rudolph, S., Studer, R.: Ontology-based interpretation of keywords for semantic search. In: Aberer, K., Choi, K.-S., Noy, N., Allemang, D., Lee, K.-I., Nixon, L.J.B., Golbeck, J., Mika, P., Maynard, D., Mizoguchi, R., Schreiber, G., Cudré-Mauroux, P. (eds.) ASWC 2007 and ISWC 2007. LNCS, vol. 4825, p. 523. Springer, Heidelberg (2007)
12. Wang, H., Zhang, K., Liu, Q., Tran, T., Yu, Y.: Q2Semantic: A lightweight keyword interface to semantic search. In: Bechhofer, S., Hauswirth, M., Hoffmann, J., Koubarakis, M. (eds.) ESWC 2008. LNCS, vol. 5021, p. 584. Springer, Heidelberg (2008)
13. Tran, T., Wang, H., Rudolph, S., Cimiano, P.: Top-k exploration of query candidates for efficient keyword search on graph-shaped (RDF) data. In: ICDE, pp. 405–416 (2009)
14. Horrocks, I., Tessaris, S.: Querying the semantic web: a formal approach. In: Horrocks, I., Hendler, J. (eds.) ISWC 2002. LNCS, vol. 2342, pp. 177–191. Springer, Heidelberg (2002)
15. Cheng, J., Ke, Y., Ng, W., Yu, J.X.: Context-aware object connection discovery in large graphs. In: ICDE, pp. 856–867 (2009)

From Keywords to Queries: Discovering the User's Intended Meaning

Carlos Bobed, Raquel Trillo, Eduardo Mena, and Sergio Ilarri

IIS Department
University of Zaragoza
50018 Zaragoza, Spain
{cbobed,raqueltl,emena,silarri}@unizar.es

Abstract. Regarding web searches, users have become used to keyword-based search interfaces due to their ease of use. However, this implies a semantic gap between the user's information need and the input of search engines, as keywords are a simplification of the real user query. Thus, the same set of keywords can be used to search different information. Besides, retrieval approaches based only on syntactic matches with user keywords are not accurate enough when users look for information not so popular on the Web. So, there is a growing interest in developing semantic search engines that overcome these limitations.

   In this paper, we focus on the front-end of semantic search systems and propose an approach to translate a list of user keywords into an unambiguous query, expressed in a formal language, that represents the *exact* semantics intended by the user. We aim at not sacrificing any possible interpretation while avoiding generating semantically equivalent queries. To do so, we apply several semantic techniques that consider the properties of the operators and the semantics behind the keywords. Moreover, our approach also allows us to present the queries to the user in a compact representation. Experimental results show the feasibility of our approach and its effectiveness in facilitating the users to express their intended query.

1   Introduction

The Web has made a huge and ever-growing amount of information available to its users. To handle and take advantage of this information, users have found in web search engines their best allies. Most search engines have a keyword-based interface to allow users to express their information needs, as this is an easy way for users to define their searches. However, this implies that search engines have to work with an input that is a simplification of the user's information need. Besides, most search engines follow an approach based mainly on syntactic matching techniques, without taking into account the semantics of the user keywords. Thus, when users are not looking for information very popular on the Web, they usually need to browse and check many hits [18].

   Under these circumstances, Semantic Search [7] can help to overcome some current problems of the searching process and provide users with more precise

L. Chen, P. Triantafillou, and T. Suel (Eds.): WISE 2010, LNCS 6488, pp. 190–203, 2010.

results. In any search system, and particularly in a semantic search engine, the main steps performed are: query construction, data retrieval, and presentation of results [1]. In this paper we focus on the first step, as capturing what users have in mind when they write keywords is a key issue to retrieve what they are looking for. Although keyword-based interfaces create a semantic gap, we focus on this type of systems due to their massive adoption [9]. Moreover, it is not clear that current natural language (NL) interfaces perform better than keyword-based interfaces in the context of Web search, as systems based on NL interfaces require performing more complex tasks, as indicated in [19].

Several approaches (e.g., [15,18]) advocate finding out first the intended meaning of each keyword among the different possible interpretations. For instance, the keyword "book" could mean "a kind of publication" or "to reserve a hotel room". These approaches consult a pool of ontologies (which offer a formal, explicit specification of a shared conceptualization [5]) and use disambiguation techniques [13] to discover the intended meaning of each user keyword. So, plain keywords can be mapped to ontological terms (concepts, roles, or instances).

Identifying the meaning of each input keyword is a step forward. However, several queries might be behind a set of keywords, even when their semantics have been properly established. For example, given the keywords "fish" and "person" meaning "a creature that lives and can breathe in water" and "a human being", respectively, the user might be asking for information about either biologists, fishermen, or even other possible interpretations based on those meanings. Therefore, it is interesting to find queries that represent the possible semantics intended by the user, to allow her/him choose the appropriate one and perform a more precise search [17]. Moreover, the queries generated should be expressed in a formal language to avoid the ambiguity of NL and express the user information need in a more precise way. The formal queries generated can be used for different purposes, such as semantic information retrieval or any other task that needs to find out the intended meaning behind a list of user keywords.

In this paper, we present a system that translates a set of keywords with well defined semantics (*semantic keywords*, they are mapped to ontology concepts, roles or instances) into a set of formal queries (expressed in different selectable query languages) which attempt to define the user's information need. Any query that the user may have in mind will be generated as a candidate query as long as an expressive enough output query language has been selected. Our system applies several semantic techniques to reduce the number of possible queries while avoiding the generation of inconsistent and semantically equivalent ones. Finally, it uses a compact representation of the queries to present them to the user. Experimental results show the feasibility of our approach and its effectiveness in reducing the number of queries generated, achieving the goal of not overwhelming users with a huge list of queries.

The rest of this paper is as follows. Firstly, in Section 2, we explain our approach to query generation. The semantic techniques used to reduce the number of generated queries are explained in Section 3. Then, the way in which our system obtains compact representations of the final queries is shown in Section 4. Experimental results that prove the interest of our proposal are presented in Section 5. We discuss some related work in Section 6. Finally, the conclusions and future work are drawn in Section 7.

2 Semantic Query Generation

In this section, we present our approach to query generation [3]. Our system takes as input a list of keywords with well-defined semantics (*semantic keywords*), chosen from the set of terms of one or several ontologies or mapped automatically to ontology terms by using the semantic techniques described in [18]. Then, considering the semantics of the output query language selected, it outputs a list of possible queries that could comprise the user's information need. The main steps are shown in Figure 1. Part of the generation process is performed by considering only the type of each keyword (concept, role, or instance). This allows the system to cache results and improve its performance in future iterations.

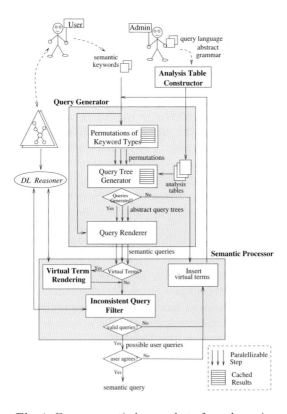

Fig. 1. From semantic keywords to formal queries

To illustrate the different steps along the paper, let us assume that a user enters "person fish" to find information about people devoured by fishes. They are mapped to the homonym terms in the ontology Animals[1]. Simplified versions of the language BACK [14] and the language DIG [16] will be used to show the flexibility of the approach[2]. The following steps are executed:

- *Permutations of keyword types*: To decouple the generation process of any specific language, its first stage is syntax-based. So, the system firstly obtains all the possible permutations of the types of terms (concepts, roles, and instances) corresponding to the input keywords, to discover any syntactically possible query in latter steps. In this case, *person* and *fish* are concepts, so the output of this step would be <C,C>, as no more permutations are possible.
- *Generation of abstract query trees*: For each permutation obtained in the previous step, the system generates all syntactically possible combinations according to the syntax of the selected output query language. We call these combinations *abstract queries* because they have *gaps* that will be filled later with specific concepts, roles, or instances. These abstract queries are represented as trees, where the nodes are operators and the leaves are *typed gaps* (concept, role, or instance gaps). Following the previous example, with the input <C,C> and simplified BACK as output language, And(C,C) would be built as an abstract query.

  In this process, the system uses bottom-up parsing techniques and analysis tables where the semantics of the selected output query language, but not its syntactic sugar, is stored. These tables are built only once, using a special annotated grammar, when a new output query language is made available to the system (*Analysis Table Construction* step). The semantics of the operators are considered to avoid generating equivalent queries (see Section 3.1).
- *Query rendering*: For each abstract query tree generated, the gaps in the leaves are filled with the user keywords matching the corresponding gap type. Then, the query trees are rendered into a full query in the selected target language. Several properties of the operators are also considered in this step to avoid generating equivalent queries (see Section 3.1). The result of this step for the running example would be And(Person, Fish).
- *Inconsistent query filtering*: The result of the previous steps is a set of syntactically correct queries. However, some of these queries might not be semantically correct according to the used ontologies. So, the system filters out the inconsistent queries with the help of a Description Logics (DL) reasoner [2] compatible with the chosen output query language. In our example, And(Person, Fish) would be removed in this step as it is classified as being inconsistent (*person* and *fish* are defined as disjoint classes).

---

[1] http://www.cs.man.ac.uk/~rector/tutorials/Biomedical-Tutorial/ Tutorial-Ontologies/Animals/Animals-tutorial-complete.owl

[2] The simplified version of BACK consists of four operators (*And, Some, All, Fill*). Simplified DIG is equivalent to simplified BACK plus the *Or* operator.

The performance of this step is greatly boosted by the fact that the set of generated queries forms a *conservative extension* [4] of the original ontology. Once an ontology has been classified, this property makes it possible to evaluate the satisfiability of the queries in polynomial time, as each query does not assert new knowledge into that ontology.

- *Insert virtual terms*: When no query is generated or satisfies the user, the system is able to consider new ontology terms. This is performed by adding *virtual terms* (*VTs*) to the original list of user keywords. VTs represent possible terms that the user may have omitted as part of her/his query. Then, the previous steps are executed again. In the example, the new inputs considered would be "person fish concept" and "person fish role". This allows the system to build, among others, And(Person (Some (role, Fish)).

- *Virtual term rendering*: If a built query includes VTs, this step replaces them by appropriate terms extracted from the background knowledge ontologies considered, in order to build a possible query. To build only relevant queries, the system narrows the set of candidate terms by using the ontology modularization techniques described in [10] (see Section 3.2). So, in the example, the previous enriched abstract query is rendered into And(Person (Some (is_eaten_by, Fish)), which was the initial user's intended query.

Note that a query generation that involves filling a semantic gap will generate many queries. Even when the meaning of each keyword has been perfectly established (and thus, its polysemy avoided), the number of possible interpretations will grow as the number of user keywords increases and the output query language allows to combine them in more ways. Besides, we do not want to limit our approach to provide the most popular results (syntactic search engines do that already), so we aim at not missing any possible interpretation (according to the accessible knowledge). We are aware that this may lead to the generation of a high number of queries (as there are many possible interpretations), and therefore in the rest of the paper we propose different semantic techniques to deal with this problem.

3   Avoiding the Generation of Useless Queries

As stated previously, trying to guess the user's information need is a hard task due to the high number of possible interpretations. In our example, for the input "person fish" and one extra VT, there exist 780 syntactically possible queries for simplified BACK and 2832 for simplified DIG. So, it is critical to reduce the number of generated queries. Apart from the number of input keywords, there are two main elements that lead to this high number of possible queries: the expressivity of the output query language, and the semantic enrichment step with VTs performed to fill the semantic gap.

On the one hand, expressive query languages are very valuable, as they are more likely to be able to represent the user's intended query. However, the higher the number of operators, the higher the number of possible queries (the operands can be combined in more different ways). On the other hand, adding new terms to the user's input can help to discover the intention of the user. However, this will also increase the number of possible queries, mainly for two reasons: 1) new possible interpretations appear, and 2) there may be a high number of candidates to replace a given VT (some of them probably irrelevant).

Along this section, we show how the system deals with these two issues. Then, in Section 4, we present a semantic technique that is applied to further simplify the output of the query generation.

3.1   Considering the Expressivity of the Query Language

For expressivity of a query language we understand its set of operators and the possible ways in which they can be combined to form a proper query. The more operators the language has and the more ways to combine them exist, the more queries will be possible. For example, if you add the *Or* operator to a language that had the *And* operator, the number of possible queries for a user input considering both operators will be larger than the double. There is apparently no way to reduce this number because the different options express different queries. However, we can avoid building equivalent queries along the generation process by considering the semantic properties of the operators.

With each new output query language that is made available to the system, a special grammar is provided as stated in Section 2. In this grammar, the properties of each operator are asserted. In particular:

- *associativity*: It is used by the *Query Tree Generator* to avoid, for example, building *and(and(c1, c2),c3)* if it has already generated *and(c1,and(c2,c3))*.
- *involution*: It is used by the *Query Tree Generator*, for example, to avoid building *not(not(c))*, which is equal to *c*.
- *symmetry*: It is used by the *Query Renderer* to avoid, for example, building *and(c2,c1)* if it has already generated *and(c1,c2)*.

Apart from those well-known properties, we consider two other properties of the operators: *restrictiveness* and *inclusiveness*. They are defined as follows:

Definition 1. *A binary operator* op *is restrictive if* $\exists f : K \to C$, $K = \{R, C\}$ | $f(x) \sqsubseteq y \Rightarrow op(x, y) \equiv op(x, f(x))$.

Definition 2. *A binary operator* op *is inclusive if* $\exists f : K \to C$, $K = \{R, C\}$ | $f(x) \sqsupseteq y \Rightarrow op(x, y) \equiv op(x, f(x))$.

From these definitions, and according to the semantics of the operators in BACK and DIG, it directly follows that the *And*, *Some* and *All* operators are restrictive and *Or* is inclusive. These properties are used in the *Virtual Term Rendering* step to avoid substituting the VTs by terms that would result in equivalent queries. For example, if we have *and(c1, concept)*, all the candidates that subsume c1 can be avoided as any of them *and* c1 would result in c1.

Following with the running example in Section 2, let us suppose that a user enters "person fish" to find information about people devoured by fishes.

Adding one VT to that input, the system generated 780 queries using simplified BACK and 2832 queries using simplified DIG, which are reduced to 72 queries (90, 77% reduction) and 364 (87, 15% reduction), respectively, by considering the semantics of the operators. In both cases the intended query *person and some(is_eaten_by, fish))* was among the final results.

3.2   Reducing the Number of Candidate Terms for Query Enrichment

Usually, users simplify the expression of their information needs when they write keyword queries, which contributes to the semantic gap. Thus, a user may omit terms that form part of the actual information need. To deal with the problem of possible information loss, some VTs can be added to the input, in order to generate the possible queries as if these VTs were proper terms introduced by the user. Then, the system performs a substitution of those VTs with actual terms extracted from the ontologies considered.

Following this idea, one can think about using each term of the same type as the inserted virtual one. For example, if the VT was a concept, the system could substitute it with all the concepts of the input ontologies (and generate one query for each different candidate concept). However, as the number of queries with VTs could also be high, considering all the terms of the input ontologies to render each VT for each query is too expensive.

In order to reduce the number of candidates for rendering a VT while avoiding losing any possible related term, we apply the modularization and re-using techniques explained in [10]. More specifically, the system uses ProSÉ, which, given a set of terms of an ontology (*signature*), allows to extract different *modules* that can be used for different purposes. Our system uses the user input terms as signature and extracts a module such that the same information can be inferred from the extracted module as from the whole ontology, as shown in [10]. This allows the discovery process to focus only on what is related to the input terms.

After applying the modularization techniques for the user query "person fish", the system generates 32 queries using BACK (15 after filtering, 98.07% less than the 780 original possible ones) and 148 queries using DIG (73 after filtering, 97.42% less than the 2832 original possible ones). The intended query is still among the final results, as the system does not miss any possible interpretation.

4   Extraction of Relevant Query Patterns

Besides reducing the number of generated queries, the way in which they are presented to the user also makes a difference. Users' attention is a capital resource and they can get easily tired if they are forced to browse over too many options to select their intended query.

In this stage of the search process, recall seems to be crucial as only one interpretation will fit the user's information need. Ranking the generated queries according to their probability of reflecting the user's interest can be an approach to minimize the interaction with the user. However, it is not clear how to identify the query that a specific user had in mind when writing a set of keywords, even though the meanings of the individual keywords in the input have been identified previously. For example, with the input "person fish" the user might be looking for information about people devoured by fishes, but also about people who work with fishes (e.g., biologists, fishermen, ...), among other interpretations. Besides, a statistics-based approach may be not suitable for ranking queries, as it would hide possible interpretations that are no popular in the community. Approaches based on semantic and graph distances would also hide possible meanings.

Due to these reasons, we advocate a different and orthogonal approach to ranking. Our approach tries to minimize the amount of information that will be presented to the user by identifying query patterns, and could be combined with any query ranking approach if it is available. The semantic technique that we propose takes advantage of the syntactic similarity of the generated queries and of the ontological classification of the terms that compose the queries. A small example is shown in Table 1, where we can see that several queries may have a similar syntactic structure.

Table 1. Queries and patterns generated for "person bus" using BACK

Queries	Patterns
all(drives, person) and bus all(drives, bus) and person	all(Role, Concept) and Concept
...	
some(drives, person) and bus some(drives, bus) and person	some(Role, Concept) and Concept
...	
all(drives, bus and person)	all(Role, Concept and Concept)
...	
some(drives, bus and person)	some(Role, Concept and Concept)
...	

Thus, the system analyzes the structure of the queries to extract common query patterns which lead to a compact representation of the queries. This is especially useful when the system tries to find out the user's intention by adding VTs. A query pattern shows an expression with the VTs not substituted (i.e., with *gaps*) and the system maintains a list of potential candidates for each gap.

At this point, a new challenge arises: how can the candidates for a gap be organized to facilitate their selection by the user? We advocate the use of a DL reasoner to show the candidates and allow users to navigate through their taxonomy. The interface shows a list of candidate terms and three buttons for each gap in each pattern (see Fig. 2):

– *Fix*: performs the substitution with the candidate term selected.
– *Subsumers*: enables the user to generalize the candidate term selected.
– *Subsumees*: enables the user to refine the candidate term selected.

Fig. 2. Example of query patterns for "person drives" and "person fish"

This allows to show a high number of queries in a really compact way and provide a navigation in a top-down style through the terms of the ontology. Thus, the more general terms are initially presented to the user, who can then move through the taxonomy by accessing each time a direct subsumers/subsumees level. Users are allowed to select only terms that are relevant for the corresponding gap, i.e., their selections will never lead to an inconsistent query.

Thus, for example, for the input "person fish", the system is able to show the 15 final queries obtained using BACK under 7 patterns and the 73 obtained using DIG under 20 patterns. Moreover, this representation allows the system to establish an upper bound on the number of user clicks needed to select their query, which is equal to the depth of the taxonomy of the ontology multiplied by the number of substitutions to be performed (number of gaps).

5   Experimental Results

We developed a prototype to evaluate our approach in a detailed and systematic way. It was developed in Java 1.6 and using Pellet[3] 1.5 as background DL reasoner, and was executed on a Sunfire X2200 (2 x AMD Dual Core 2600 MHz, 8GB RAM). As knowledge bases, we selected two well-known ontologies: *People+Pets*[4] and *Koala*[5], which are two popular ontologies of similar size to those used in benchmarks such as the OAEI[6]. Due to space limitations, we only show the experimental results obtained with simplified BACK as output query language because most search approaches are based only on conjunctive queries. Nevertheless, we have also performed the experiments with simplified DIG, and we obtained similar conclusions (e.g., a similar percentage of query reduction).

For the experiments, we considered different sample sets of input terms and measured average values grouped by the number of terms in the set. It is important to emphasize that the terms in each set were not selected randomly but based on actual queries proposed by students of different degrees with skills in Computer Science. The sets were chosen according to the following distribution: 10 sets with a single term (5 roles and 5 concepts), 15 sets with two terms (5 with

---

[3] http://clarkparsia.com/pellet

[4] http://www.cs.man.ac.uk/~horrocks/ISWC2003/Tutorial/people+pets.owl.rdf

[5] http://protege.stanford.edu/plugins/owl/owl-library/koala.owl

[6] http://oaei.ontologymatching.org/

2 roles, 5 with 2 concepts, and 5 with 1 concept and 1 role), 20 sets with three terms (5 with 2 concepts and 1 role, 5 with 1 concept and 2 roles, 5 with 3 concepts, and 5 with 3 roles) and, following the same idea, 25 sets with four terms and 30 sets with five terms. Notice that, even though our approach can effectively deal with instances as well, we do not consider sets with instances because the selected ontologies do not have instances (as it happens frequently [20]). We set the maximum number of keywords to 5, as the average number of keywords used in keyword-based search engines "is somewhere between 2 and 3" [12]. This way we can see how our system performs with inputs below and above this value.

Our experiments measured the effectiveness in reducing the amount of queries generated and the query pattern extraction, and the overall performance.

5.1   Reducing the Number of Queries Generated

Firstly, we analyze the impact of the techniques explained in Section 3. For each set of terms, we executed the prototype considering no VTs and one VT in the enrichment step. The results of the experiment are shown in Figs. 3.a and 3.b, where the Y-axis represents the number of generated queries in log scale and the X-axis is the number of user keywords. We observe in both cases (Figs. 3.a and 3.b) that the number of queries generated increases with the number of input terms, as the more operands there are the more queries can be built. Moreover, performing the semantic enrichment leads also to a significant increase in the number of queries because many new interpretations appear. However, the percentages of reduction are satisfactory as they vary from the 33% when using 1 keyword up to the 90% when using 5 keywords, and, what is more important, the user's intended query is always in the generated set.

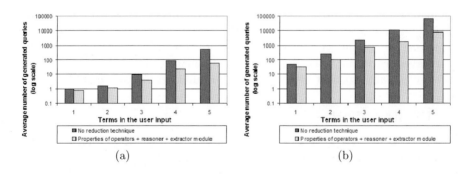

Fig. 3. Number of queries with: a) no VT, b) one VT

5.2   Extracting Relevant Query Patterns

Having all the reduction techniques turned on, we now focus on the detection of query patterns (explained in Section 4). So, in this experiment we measure the number of query patterns that are extracted by the system to group the queries generated. Figure 4, where the Y-axis is in log scale, shows the benefits of providing query patterns to the user, instead of the plain list of queries generated. With few input terms and no VTs considered, the number of queries generated is quite low. However, as the number of input terms increases or when VTs are considered for query enrichment, the list of query patterns is very useful. Thus, it is about 50% smaller than the list of queries when no VTs are considered and about 92% when one extra VT is added. So, thanks to the use of the patterns identified with our semantic approach, the user can select her/his intended query much more easily than by browsing a plain list of possible choices.

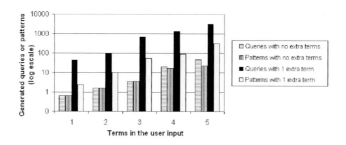

Fig. 4. Reduction obtained by extracting relevant query patterns

5.3 Performance Evaluation

Finally, we focus on the performance of the system as a whole. The average times that the generation process takes are shown in Figure 5.a. They include the generation and the semantic filtering times. Being the low they are makes the system suitable to be a responsive front-end. Figure 5.b shows the maximum number of clicks required to reach any possible query allowed by the output query language considered in the experiments. With one VT, more effort may be needed as the number of possible queries grows considerably. However, as the user navigates through the taxonomy of candidate terms, s/he could consider different possibilities that s/he overlooked to fulfil her/his information needs.

6 Related Work

One of the first systems whose goal is building queries from keywords in the area of the Semantic Web was SemSearch [11]. In this system, a set of predefined templates is used to define the queries in the language SeRQL. However, not all the possible queries are considered in those templates, and therefore the system could fail in generating the user's intended query. Besides, it requires that at least one of the user keywords matches an ontology concept. On the contrary, our system considers all the possible queries syntactically correct in a query language selectable by the user and it works with any list of keywords.

Other relevant systems in the area of semantic search are [17] by Tran et al., Q2Semantic [8], SPARK [22], and QUICK [21]. These systems find all the paths

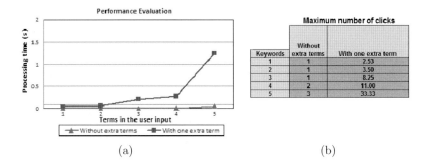

Keywords	Maximum number of clicks	
	Without extra terms	With one extra term
1	1	2.53
2	1	3.50
3	1	8.25
4	2	11.00
5	3	33.33

(a)   (b)

Fig. 5. Performance evaluation: processing time and maximum number of clicks

that can be derived from a RDF graph, until a predefined depth, to generate the queries. However, these approaches can only generate conjunctive queries, whereas our system is able to generate any query that can be expressed in the selected output query language. Besides, they do not support reasoning capabilities for query reduction, as opposed to our system. In Q2Semantic, three possible ranking schemes are proposed, but they are too attached to their query generation model and are not applicable to our approach. Regarding these approaches, our approach works at a higher semantics level, as it exploits the background knowledge not only to build new queries but also to infer which ones are satisfiable and to avoid the generation of irrelevant ones.

There are also some works in the area of databases to provide a keyword-based interface for databases, i.e., translating a set of keywords into SQL queries. However, as emphasized in [6], most of these works rely only on extensional knowledge and neglect the intensional one (the structural knowledge). Besides, most of these approaches build only conjunctive queries and require the use of a specific language to express some constraints (e.g., *less than*). As opposed to these proposals, our approach considers both the intensional and the extensional knowledge, it is not tied to a specific query language, and it considers all the possible constraints of the selected query language.

Summing up, to our knowledge, our system is the only one with the goal of building formal queries from a set of keywords that uses a DL reasoner to infer new information and remove semantically inconsistent queries. Besides, no other approach is as independent as ours of the specific output query language. Finally, despite the potentially high number of queries, our system keeps it manageable by using several semantic techniques.

7   Conclusions and Future Work

In this paper, we have proposed an approach to discover the intended meaning behind a set of semantic keywords. To do so, our system translates the set of input keywords into unambiguous formal queries. Our approach is flexible enough to deal with any output query language (once the annotated grammar is provided), and so it can be used as a front-end for different semantic search engines. The proposed system has the following features:

- It exploits semantic properties of the operators (*associativity, involution, symmetry, restrictiveness,* and *inclusiveness*) of the output query language to avoid generating semantically equivalent queries.
- It filters inconsistent queries according to the used ontologies.
- It performs a semantic enrichment to fill the gap between the user keywords and the user's intended query. In this process, it uses modularization techniques to avoid generating irrelevant queries.
- It considers the semantics and structure of the generated queries to discover patterns that allow the user to locate her/his intended query iteratively with a small number of clicks.

Our experimental results show that, on average, our approach avoids generating an 80% of the possible queries without losing the user's intended meaning. Besides, the extraction of query patterns allows the user to reach any generated query with little effort, even if the user's query is not a popular interpretation.

As future work, we plan to integrate in a complete semantic search system, taking as starting point plain keywords [18]. We will also study ranking schemas and natural language translations for the query patterns in order to ease the query selection. Further experiments would be driven to evaluate the overall performance of our system and evaluate the user's satisfaction levels. Finally, we will move to the next stage of search by using the queries to access data once the exact meaning intended by the user is known.

Acknowledgments

This work has been supported by the CICYT project TIN2007-68091-C02-02.

References

1. Proceedings of the Intl. Workshop on TRECVID Video Summarization (TVS 2007), Germany (2007)
2. Baader, F., Calvanese, D., McGuinness, D., Nardi, D., Pastel-Scheneider, P.: The Description Logic Handbook. Theory, Implementation and Applications. Cambridge University Press, Cambridge (2003)
3. Bobed, C., Trillo, R., Mena, E., Bernad, J.: Semantic discovery of the user intended query in a selectable target query language. In: 7th Intl. Conference on Web Intelligence (WI 2008), Australia. IEEE, Los Alamitos (2008)
4. Cuenca, B., Horrocks, I., Kazakov, Y., Sattler, U.: Modular reuse of ontologies: Theory and practice. J. of Artificial Intelligence Research 31, 273–318 (2008)
5. Gruber, T.R.: Towards principles for the design of ontologies used for knowledge sharing. In: Guarino, N., Poli, R. (eds.) Formal Ontology in Conceptual Analysis and Knowledge Representation. Kluwer Academic Publishers, Dordrecht (1993)

6. Guerra, F., Bergamaschi, S., Orsini, M., Sala, A., Sartori, C.: Keymantic: A keyword based search engine using structural knowledge. In: 11th Intl. Conference on Enterprise Information Systems (ICEIS 2009), Italy. Springer, Heidelberg (2009)
7. Guha, R.V., McCool, R., Miller, E.: Semantic search. In: 12th Intl. Conference on World Wide Web (WWW 2003), Hungary. ACM, New York (2003)
8. Liu, Q., Wang, H., Zhang, K., Tran, T., Yu, Y.: Q2semantic: A lightweight keyword interface to semantic search. In: Bechhofer, S., Hauswirth, M., Hoffmann, J., Koubarakis, M. (eds.) ESWC 2008. LNCS, vol. 5021, pp. 584–598. Springer, Heidelberg (2008)
9. Yu, L.Q.J.X., Chang, L.: Keyword Search in Databases. Morgan & Claypool (2010)
10. Jimenez, E., Cuenca, B., Sattler, U., Schneider, T., Berlanga, R.: Safe and economic re-use of ontologies: A logic-based methodology and tool support. In: Bechhofer, S., Hauswirth, M., Hoffmann, J., Koubarakis, M. (eds.) ESWC 2008. LNCS, vol. 5021, pp. 185–199. Springer, Heidelberg (2008)
11. Lei, Y., Uren, V.S., Motta, E.: SemSearch: A search engine for the semantic web. In: Staab, S., Svátek, V. (eds.) EKAW 2006. LNCS (LNAI), vol. 4248, pp. 238–245. Springer, Heidelberg (2006)
12. Manning, C.D., Raghavan, P., Schütze, H.: Introduction to Information Retrieval. Cambridge University Press, Cambridge (2008)
13. Navigli, R.: Word sense disambiguation: A survey. ACM Computing Surveys 41(2), 1–69 (2009)
14. Peltason, C.: The BACK system – an overview. ACM SIGART Bulletin 2(3), 114–119 (1991)
15. Rungworawut, W., Senivongse, T.: Using ontology search in the design of class diagram from business process model. Transactions on Engineering, Computing and Technology 12 (March 2006)
16. Sean Bechhofer, P.C., Möller, R.: The DIG description logic interface: DIG/1.1. In: Intl. Workshop on Description Logic (DL 2003), Italy (2003)
17. Tran, D.T., Wang, H., Rudolph, S., Cimiano, P.: Top-k exploration of query candidates for efficient keyword search on graph-shaped (RDF) data. In: 25th Intl. Conference on Data Engineering (ICDE 2009), China. IEEE, Los Alamitos (2009)
18. Trillo, R., Gracia, J., Espinoza, M., Mena, E.: Discovering the semantics of user keywords. Journal on Universal Computer Science. Special Issue: Ontologies and their Applications, 1908–1935 (November 2007)
19. Wang, Q., Nass, C., Hu, J.: Natural language query vs. keyword search: Effects of task complexity on search performance, participant perceptions, and preferences. In: Costabile, M.F., Paternó, F. (eds.) INTERACT 2005. LNCS, vol. 3585, pp. 106–116. Springer, Heidelberg (2005)
20. Wang, T.D., Parsia, B., Hendler, J.A.: A survey of the web ontology landscape. In: Cruz, I., Decker, S., Allemang, D., Preist, C., Schwabe, D., Mika, P., Uschold, M., Aroyo, L.M. (eds.) ISWC 2006. LNCS, vol. 4273, pp. 682–694. Springer, Heidelberg (2006)
21. Zenz, G., Zhou, X., Minack, E., Siberski, W., Nejdl, W.: From keywords to semantic queries–incremental query construction on the semantic web. Web Semantics: Science, Services and Agents on the World Wide Web 7(3), 166–176 (2009)
22. Zhou, Q., Wang, C., Xiong, M., Wang, H., Yu, Y.: SPARK: Adapting keyword query to semantic search. In: Aberer, K., Choi, K.-S., Noy, N., Allemang, D., Lee, K.-I., Nixon, L.J.B., Golbeck, J., Mika, P., Maynard, D., Mizoguchi, R., Schreiber, G., Cudré-Mauroux, P. (eds.) ASWC 2007 and ISWC 2007. LNCS, vol. 4825, pp. 694–707. Springer, Heidelberg (2007)

Efficient Interactive Smart Keyword Search

Shijun Li, Wei Yu, Xiaoyan Gu, Huifu Jiang, and Chuanyun Fang

School of Computer, Wuhan University, Wuhan, China, 430072
shjli@whu.edu.cn,
{whu_wyu,qiushui881,huifujiang,chyfang88}@163.com

Abstract. Traditional information systems usually return few answers if a user submits an incomplete query. Users often feel "left in the dark" when they have limited knowledge about the underlying data. They have to use a try-and-see approach to modify queries and find answers. In this paper we propose a novel approach to keyword search which can provide predicted keywords when a user submits a few characters of the underlying data in order. We study research challenges in this framework for large amounts of data. Since each keystroke of the user could invoke a query on the backend, we need efficient algorithms to process each query within milliseconds. We develop an incremental-search algorithm using previously computed and cached results in order to achieve an interactive speed. Some experiments have been conducted to prove the practicality of this new computing paradigm.

Keywords: Interactive Search, Smart Search, Autocomplete, Fuzzy Search.

1 Introduction

In a traditional keyword search system, users compose a complete query and submit it to the system to find relevant answers. This information-access paradigm requires users to formulate a complete query to find interesting answers. If users have limited knowledge about the underline data, they often feels"left in the dark"when issuing queries. They have to use a try-and-see approach to modify the query and find information. For example, users want to search the football star *"Maradona"* without knowing the exact spelling. Because sometimes we only know the pronunciation of some word and can just figure out some characters of the word from the pronunciation. In this case, it is not convenient for users to find the star.

Many systems are introducing various features to solve this problem. One of the commonly used methods is *autocomplete*, which predicts a word or phrase that users may type based on the partial query they have entered. Search engines nowadays automatically suggest possible keyword queries when users type in a partial query.

In the paper, we study how to return answers efficiently and interactively when a user inputs only some parts of query orderly which he/she thinks are right. For instance, suppose the user wants to search the football star "maradona", while he/she only remembers that it is somewhat like "m...r...d...n...". **Table 1** shows the process of our system searching it. It has two features: (1) Interactive: each keystroke of the user could invoke a query on the backend, update the list of predicted keywords and

L. Chen, P. Triantafillou, and T. Suel (Eds.): WISE 2010, LNCS 6488, pp. 204–215, 2010.
© Springer-Verlag Berlin Heidelberg 2010

Table 1. Process of searching

Query	Display	Result
m	maalouf, marabini	not found
mr	maradei, marabini	not found
mrd	maradei,maradona	found

show them to users in real time; (2) Smart and no prefix-preference: the system can search records based on the partial input of the query. The fuzzy keyword search proposed by S. Ji. has the problem of prefix-preference[1]. If errors exist in the front part of the query, usually it will not be able to find the relevant keywords.

Our work is similar to the search in database using the wildcard '%'. They both can search data on partial input that needn't be continuous. For example, if the user inputs query "mrd", we can expand it to "%m%r%d%", and then submit it to the database. But in essence and in effect, they are different in the following aspects. (1) Efficiency. In traditional database, it will find and cache all the relevant keywords, which will cost much memory and time. This is unnecessary, because users will just look through the front parts of the keywords. We proposed an incremental/cached algorithm to address this problem, which is efficient and costs little memory. We have proved it through experiments. (2) Interactive. The efficient interactive smart keyword search (*EISKS*) is interactive, which means as the user type in each character of the query, EISKS can return predicted keywords which would be of interest to the user according to the current partial query. This feature requires the algorithm to be efficient. In traditional database, the efficiency cannot meet this requirement, so it won't be able to be interactive.

A main requirement of our smart search on large amounts of data is the need of high interactive speed. Each keystroke from a user could invoke a query to the system, which needs to compute the answers within milliseconds. This high-speed requirement is technically challenging especially due to the following reason. Users only selectively give some letters of their interesting keyword and it would cause many answers matching the query.

In this paper, we develop a novel technique to address the challenge and an on-demand caching technique for incremental search. Its idea is to cache only part of the results of a query. As the user types in letters, unfinished computation will be resumed if the previously cached results are not sufficient. In this way, we can efficiently compute and cache a small amount of results. We conducted a thorough experimental evaluation of the developed techniques on a DBLP dataset with about 1 million publication records, and show the practicality of this new computing paradigm. All experiments were done using a single desktop machine, which can still achieve response time of milliseconds on millions of records.

1.1 Our Contribution

In this paper, we introduce a novel query model where the user doesn't have to submit the complete query. This feature is especially important in number search such as telephone number search. For example, if we want to know someone's telephone number. We don't remember the exact number, but we know some part of it. The number is like "*13476****58*", where '*' denotes the information we don't remember exactly. In existing search system, we can only input the query "*13476*", and try whether we can find the relevant information. However, we didn't take full use of the information "*58*". In EISKS, users can input the query"*1347658*", and with more information provided, EISKS is more likely to return the relevant information.

We develop techniques to minimize resource requirements and make our system EISKS interactive and efficient. In EISKS, each query will invoke many results. Naïve algorithm will cost lots of memory and time, because it will return all the results which are often unnecessary. Such methods will not be able to work, especially when a lot of user accesses the system at the same time. We improve this algorithm and propose an algorithm using incremental and cache technique (the algorithm will be described in Section 4). The experiments prove the efficiency and convenience of our EISKS.

2 Related Work

Prediction and Autocomplete: There have been many studies on predicting queries and user actions [2, 3, 4, 5, 6]. Using these techniques, a system predicts a word or a phrase that the user may next type in based on the sequence of partial input the user has already typed. The main difference between these techniques and our work is as follows. The existed methods do prediction using sequential data to construct their model, but the partial sequence should be a complete prefix of the underlying data. For instance, supposing a user wants to search '*Maradona*', he/she inputs 'Mara' and EISKS may offer '*Maradona*' as a reference, while inputs '*Mrad*', it would never return '*Maradona*' to the user.

CompleteSearch: Bast et al proposed technique to support "CompleteSearch", in which a user types in keywords letter by letter, and it finds records that include these keywords (possibly at different places) [7, 8, 9, 10]. Our work differs from theirs as follows. (1) CompleteSearch mainly focused on compression of index structures, especially in disk-based settings. Our work focuses on efficient query processing using in-memory indexes in order to achieve a high interactive speed. (2) Our work allows fuzzy search, making the computation more challenging. (3) EISKS can return answers based on user's partial input while CompleteSearch needs a fully complete input.

Efficient interactive fuzzy keyword search (EIFKS): There has been recent study to support efficient interactive fuzzy keyword search [1]. The technique finds relevant records that have keywords matching query keywords approximately. EISKS also has the features of efficient, interactive and fuzzy. The differences are as follows. (1) EIFKS has prefix-preference. If the first several letters are not input correctly, it can hardly return relevant records. In our EISKS, the errors can exist in any part of the query; it is able to return the relevant records. (2) They have different tolerance. Generally in EIFKS, the edit distance threshold is often 2, which means there can be 2 errors at most. If there are more than 2 errors, EIFKS is not able to return relevant records. In our system EISKS, the maximum number of tolerated errors is not specified; the user can just ignore the uncertain part of the query. Most of the fuzzy search algorithms are based on *edit distance* [1, 12].

3 Preliminaries

Ordered Substring: For a string $S1(s_1s_2...s_m)$, we refer to string $S2(s_is_j...s_k)$ as an *ordered substring* of S1, where $1 \leq i \leq j \leq k \leq m$. That is, S1 has an ordered substring S2. We define that S1 is a *complete string* of S2. Given two strings, "$S2 \prec S1$" denotes S2 is an ordered substring of S1. For example, string "clr" is an ordered substring of "scholarship" and "**clear**".

Indexing: We use a Trie to index the words in the relational table. Each word ω in the table corresponds to a unique path from the root of the Trie to a leaf node. Each node on the path has a label of a character in ω. For simplicity a node is mentioned interchangeably with its corresponding string in the remainder of the paper. Each leaf node has an inverted list of IDs of records that contain the corresponding word, with additional information such as the attribute in which the keyword appears and its position. For instance, given the word set W= {"beckham", "maalouf", "marabini", "maradei", "maradona", "ronald"}, **Fig. 1** shows the corresponding Trie. For simplicity, the figure doesn't give the additional information of each node. The node with underlined character in it denotes a keyword. The node with ID of 7 denotes the keyword "beckham".

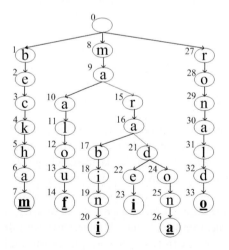

Fig. 1. Structure of Trie

Table 2. Inverted list of character

Character	IDs
A	6,9,10,16,26,30
b	1,17
c	3
d	21,32
e	2,22
f	14
h	5
i	18,20,23
k	4
l	11,31
m	7,8
n	19,25,29
o	12,24,28,33
r	15,27
u	13

Inverted List of character: In order to get the position of characters in the Trie, we build an inverted list for each character, count the appearance of each character after the Trie is built, and allocate corresponding space for the character, which stores the ID of the node that has a label of the character. For example, the inverted list of character corresponding to **Fig. 1** is shown in **Table 2**.

4   Algorithm

4.1   Single Query Based on Cursor

In this section we will introduce the *single query algorithm based on cursor*. After a query search, the node ID of all the returned keywords are greater than a node ID called *cursor*. Since the size of the results for each query may be very large, in order to solve this problem, EISKS returns the front part of the results. In section 4.2, we will introduce an algorithm using cache to store the partial results.

Suppose the query is q. Let $q[i]$ denote the i-th character of the query q. The length of the query q is s. *WP* denotes *inverted list of character* of the dataset. *WP(c)* denotes the list of the nodes which has the character c in Trie. Let $top(n)$ denote the maximum node ID of all the descendant nodes of n. If the prefix string of n is a complete string of q, the node n is called an active node.

Our algorithm uses an auxiliary array structure called *thruster*. *Thruster[i]* denotes the i-th unit of it and its content is $<c, n>$, where c is $q[i]$ and n is the node ID of a node whose value is $q[i]$. Specially, the value of n can be infinity and in this case, the thruster is called *invalid thruster*. Otherwise, it is called *valid thruster*. The procedure *GT(c, n)* demonstrates how to get next n from its current value. Since the length of the query q is s, *thruster[s]* is the last unit of the thruster. Supposing R is a *valid thruster*. For $1 < i <= s$, if all the $R[i].n$ is the descendant node of $R[i-1].n$, we call R is an *active thruster*. Otherwise R is called an *inactive thruster*. It's obvious that if R is an *active thruster*, the string that $R[s].n$ denotes is a complete string of the query q. So $R[s].n$ is an *active node*.

In the algorithm, the parameter *cache* denotes the memory cache, and the size of it is §. R is a thruster. Line 1 calls the procedure *InitThruster()* to initialize the thruster. While the cache is not full and R is still an active thruster, we keep searching keywords from the dataset to fill the cache (Line 3-Line9). We have demonstrated that if R is active, then $R[s].n$ is an active node. So we can call procedure *Load(R[s].n, cache)* to put the relevant keywords into the cache (Line4-Line8). If the thruster is active, we call the procedure *Next()* to get next thruster (Line 7). Otherwise, we call the procedure *Transform()* to transform the thruster from inactive to active or invalid (If there's no more active thruster, the thruster will be transformed to invalid).

We take an example to demonstrate this algorithm. **Table 3** shows the process of the algorithm.

ALGORITHM 4.1 Single Query Based on Cursor

INPUT: cache, Query q and cursor
OUTPUT: cache that has § keywords at most
FUNCTION: SearchBaseCursor

```
 1 R←InitThruster(q, cursor)
 2 s←the length of q
 3 WHILE R is a valid thruster
 4   IF R is an active thruster
 5     IF Load(R[s].n, cache) is FALSE
 6       RETURN cache
 7     R[s].n←Next(R[s].c, R[s].n)
 8     continue
 9   Transform(R)
10RETURN cache
```

PROCEDURE InitThruster(q, cursor)

```
 1 s←the length of q
 2 FOR i=1...s
 3   R[i].c←q[i]
 4   R[i].n←the minimum of node in array WP(q[i])
 5 IF EXIST node n in array WP(q[i]) that n> cursor
 6   R[s].n←minimum of all the n
 7 ELSE R[s].n←infinity
 8 RETURN R
```

PROCEDURE Transform(& R)

```
 1 WHILE R is valid thruster
 2   IF EXIST i that i<s and R[i].n•R[i+1].n
 3     R[i+1].n←GT(R[i+1].c, R[i+1].n)
 4   ELSE IF EXIST i that i<s and top(R[i].n)<
R[i+1].n
 5     R[i].n←GT(R[i].c, R[i].n)
 6   ELSE RETURN
```

PROCEDURE GT(c, n)

```
 1 IF EXIST u in array WP(c) that u>n
 2   RETURN minimum of all the u
 3 ELSE RETURN infinite
```

PROCEDURE Load(n, & cache)

```
 1 FOR each child (include n) of node n
 2   IF the size of cache is less than §
 3     put the child into cache
 4   ELSE RETURN FALSE
 5 RETURN TRUE
```

PROCEDURE Next(c, n)

```
 1 vd←GT(c, n)
 2 WHILE vd is not infinite and vd is a child of n
 3   vd←GT(c,vd)
 4 RETURN vd
```

Table 3. Process of the algorithm *Single Query Based on Cursor*

Step	Thruster R			Type	Cache
	R[1]	R[2]	R[3]		
1	<m, 7>	<r, 15>	<d, 21>	inactive	
2	<m, 8>	<r, 15>	<d, 21>	active	23, 26
3	<m, 8>	<r, 15>	<d, 32>	inactive	23, 26
4	<m, 8>	<r, 27>	<d, 32>	inactive	23, 26
	<m, ∞>	<r, 15>	<d, 32>	invalid	

Supposing the query is "mrd", and the size of the cache § is 3. Cursor is initialized to 0 and the cache is flushed empty. We can take 4 steps to complete our algorithm as follows:

Step 1: Since the first node IDs in $WP('m')$, $WP('r')$ and $WP('d')$ are 7, 15 and 21 separately, thruster R is initialized to {<m, 7>, <r, 15>, <d, 21>}. From **Fig. 1**, we can see that the node in $R[2]$ is not the child of the node in $R[1]$⟦ so R is an inactive thruster.

Step 2: We call the procedure *Transform()* to transform it to active. Since *top(7)* is 7 (larger than 15), so the node in $R[2]$ is not the child of $R[1]$. We should change $R[1]$ into a larger node. From **Table 2** we can see that it should be changed to 8. After that, we can find that R becomes an active thruster, and then we call the procedure *Load()* to load all the child nodes (of course it should represent a keyword) of node 21 into cache.

Step 3: Since the cache is not full, we call the procedure *Next()* to transform the thruster into another active one. From the $WP('d')$ in **Table 2**, we can see $R[3]$ is changed to 32. But from **Fig. 1** we find the node in $R[3]$ is not a child of the node in $R[2]$, so R is an inactive thruster.

Step 4: Since R is inactive, the procedure *Transform()* is called. We find that the node in $R[2]$ is 26 and *top(26)* is 26, which is smaller than the node in $R[3]$. So we should transform the node in $R[2]$ into a larger node. From $WP('r')$ in **Table 2** we can see it should be changed to node 27. However, the node in $R[1]$ is 8, and the value of *top(8)* is 26, which is smaller than the node in $R[2]$ (node 27). We should change $R[1]$ into a larger node, but we can't find any node in $WP('m')$ that is larger than 8. So $R[1]$ is set infinity. At this point, R becomes an invalid thruster and the loop stops here.

In the end, there are two nodes {23, 26} in the cache, which indicate the two keywords {"maradei", "maradona"}.

4.2 Single Query Based on Incremental/Cache Algorithm

In this section, we will introduce an algorithm named single query based on incremental/cache algorithm. The algorithm returns the top k relevant keywords from the cache and dataset. The cursor is first initialized 0 and will be updated after each query. The algorithm named Searchsingle is as follows:

ALGORITHM 4.2 Searchsingle

INPUT: &cache, query qi and &cursor
OUTPUT: top k keywords

```
1 FOR each keyword p in cache
2   IF p is not the complete string of q
3     delete p from cache
4 cache←SearchBaseCursor(cache, q, start)
5 IF the maximum of node in cache is larger than cur-
sor
6   cursor←maximum of node in cache
7 ELSE cache←infinite
8 rank cache
9 RETURN top k keyword in cache
```

For each query q, we first verify the keywords in cache and remove the keywords that are not the complete string of new query q. If the cache is not full, we can call the algorithm *SearchBaseCursor(cache, q, cursor)* to get new keywords to fill the cache. If the maximum node ID in the cache is greater than the cursor, then cursor is assigned to the maximum node ID. Otherwise, the cursor is assigned infinity ∞. At last we rank the keywords in the cache and return the top k keywords to the user.

Here we give an example to demonstrate the algorithm. Supposing the queries are {"m", "mr", "mrd", "mrdn"}, the number of keywords returned to the user k is set to 2, and the size of the cache is set to 3. **Table 4** shows the process of the search.

Before any character is input, the cache is empty and the cursor is set to 0 (the beginning). We can find three keywords "bechham", "maalouf" and "marabini" after we input "m". They are input into the cache and the cursor is changed to 20 (the id of keyword "marabini"). We return the top 2 keywords "maalouf" and "marabini" to the user.

Then the user inputs character 'r' and the query is "mr" now. We first verify the cache and judge which keywords should be removed. It's obvious that only the keyword "marabini" should be kept. Since there is only one keyword in the cache, we find two more keywords "maradei" and "maradona" from the rest dataset. And the cursor is now changed to 26 (the id of keyword "maradona"). The top 2 keywords "maradei" and "marabini" are returned to the user.

Table 4. Process of the algorithm *Searchsingle*

Query	Cache		Display	Cursor
	Verify	Search		
m		bechham, maalouf, marabini	maalouf, marabini	0
mr	marabini	maradei, maradona	maradei, marabini	20
mrd	maradei, maradona		maradei, maradona	26
mrdn	maradona		maradona	∞

The user now inputs character 'd', and the query is "mrd". After the cache is verified, only "maradei" and "maradona" are kept. We should find one more keyword from the rest of the dataset, but there are no more relevant keywords. So the cursor is set to ∞, which means the dataset has been searched through. The top 2 keywords "maradei" and "maradona" are returned to the user.

At last, the user inputs character 'n', and the query is "mrdn" now. After the cache is verified, only "maradona" is kept. Moreover, there is no more keyword to be searched. So we return the keyword "maradona" to the user.

4.3 Ranking

Each query will retrieve lots of predicted keywords, and it's important how to rank them. A ranking function considers various factors to compute an overall relevance score of a keyword to a query. The followings are several important factors.

(1) Length of best matching prefixes (LBMP): suppose the input query is q and keyword k_q is relevant to q. the LBMP is defined as follows:

$LBMP(k_q, q) = \min\{len\ of\ s \mid s\ is\ complete$ string $of\ k_q\ and\ s\ is\ substring\ of\ q\}$;

For example, if the query is "go" and "google" is a keyword relevant to query "go". Strings "go", "goo", "goog", "googl", and "google" are substring of "google" and *complete string* of "go", and the lengths are 2,3,4,5, and 6 separately. So the value of LBMP("google", "go") is 2. The smaller the LBMP is, the higher the keyword should be ranked.

(2) Keyword length: The length of the keyword should also be considered. For example, for the query "go", both "goal" and "google" are the relevant keywords, and the LBMP are both 2. However, since "goal" is shorter than "google", "goal" should be ranked higher.

(3) Lexicographic order: If the front two factors are the same, then the keywords are ranked in lexicographic order.

(4) Keyword weights [11]: different keywords could have different weights. For example, a famous author could be ranked higher than a less famous author. This factor can also be considered, but in our experiments, we didn't take this into consideration.

5 Experiments

This section presents our experimental results to evaluate the developed techniques. Here we report the results on the data sets DBLP mainly because of its relative large sizes. DBLP includes about one million computer science publication records, with six attributes: authors, title, conference or journal name, year, page numbers, and URL. The number of record and distinct keyword are 1,062,361 and 378,551 separately.

We implemented EISKS using JAVA. All the experiments were run on a PC, with a 2.5GHz Pentium® Dual-Core CPU and 204MB memory. The operating system was Windows XP.

5.1 Efficiency of Computing Complete String

We evaluated the efficiency of computing the complete string of a query keyword on the Trie. We generated 1,000 sing-keyword queries by randomly selecting keywords in the data set, and applying some delete operations on each keyword. The number of the operation is one third of the length of the query. The average length (number of letters) of keywords was 9.9 for the DBLP data set. For each query[1], we measured the time of finding the complete string of the query with the same length.

We implemented two methods to compute similar prefixes. (1) Incremental/Cache: We computed the active nodes of a query using the cached active nodes of previous complete string, using the incremental algorithm presented in Section 4.2. This algorithm is applicable when the user types a query letter by letter. (2) Incremental/NoCache: We used the incremental algorithm, but assumed no earlier active nodes have been cached, and the computation started from scratch. This case happens when a user copies and pastes a long query, and none of the active nodes of any complete string has been computed. It also corresponds to the traditional non-interactive-search case, when a user submits a query and clicks the "Search" button.

Fig. 2 shows the performance results of these two methods. The method Incremental/Cache was more efficient. As the user types in letters, its response time first increased slightly (less than 5 ms), and then started decreasing quickly after the fourth letter. The main reason is that the number of active nodes decreased, and the cached results made the computation efficient. The method Incremental/NoCache required longer time since each query needed to be answered from scratch, without using any cached active nodes.

5.2 Success Rate of Search

As mentioned above, each query can retrieve a lot of results. We hope the expected result can be placed at the top k (k is usually smaller than 40) position, so the user can find it quickly. We compare the success rate of search between EISKS and EIFKS described in [1]. We change the value of k, and compute the success rate of search in both two systems for each k. For each query, if the expected result appeared at the top k position of the results, we score it 1, else 0.

In this section, we generated 1,000 sing-keyword queries (called altered query) by randomly selecting keywords (called original result) in the data set, and applying some delete operations on each keyword. We input the altered query into the two systems, test whether the original result appeared at the top k position of the results, and give it the corresponding score. The success rate of search (SRS) is computed as follows:

$$SRS = \frac{total\ score}{1000} \times 100\% \tag{1}$$

Fig. 3 shows the success rate of search in two systems for each k. With the increase of k, the SRS of both systems increased. For the same value of k, our system EISKS has a higher SRS than the EIFKS.

---

[1] In fact, since the search is interactive, each keystroke will generate a query.

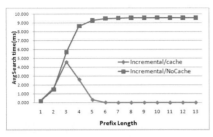

Fig. 2. Efficiency of computing complete string

Fig. 3. Success rate of search

5.3 Saved Typing Effort

Interactive search can also save user typing efforts, since results can be found before the user types in complete query. In this section we try to evaluate the convenience of the interactive feature. We constructed six queries on the DBLP data set as shown in **Table 5**. Each query was typed in letter by letter, until the system found the expected records. We measured how much letter-typing effort the system can save for the user. For each query Q_i, let $N(Q_i)$ be the number of letters the user typed before the relevant answers are found. We use $1 - N(Q_i)/Q_i$ to quantify the relative saved effort. For example, for query 6, the user could find relevant answers right after typing in "sim". The saved effort of it is $1 - 3/10 = 70\%$, as the user only needed to type in 3 letters, instead of 10 letters in the full query. **Table 5** shows that in general, the longer the query is, the more typing effort can be saved.

5.4 Comparing with Oracle

We have mentioned the EISKS is somewhat similar with the search with wildcard '%' in database. We also have introduced the differences between them. In this section, we compared them in efficiency.

Table 5. Queries and saved typing effort

ID	Query	Saved effort
1	suntan	50%
2	sarawgi	57%
3	nick	25%
4	approximate	82%
5	icde	25%
6	similarity	70%

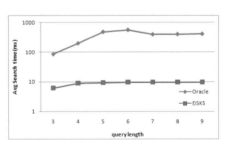

Fig. 4. Efficiency compared with Oracle 10g (log scale)

For each query length, we generated 1000 query randomly. We selected characters ranging from 'a' to 'z' randomly for each position of the query. For each query, we retrieve the top 1,000 results from the dataset. From **Fig. 4** we can see that EISKS has great advantage over the Oracle 10g in efficiency.

6   Conclusions

We presented a new information-access paradigm that supports interactive, smart search. We proposed efficient algorithms for incrementally computing answers to queries by using cached results of previous partial queries in order to achieve an interactive speed on large data sets. We also discussed the useful feature – ranking answers. We have conducted some experiments, and the results proved the practicality of this new computing paradigm.

Acknowledgments. This work was supported by a grant from the National Natural Science Foundation of China (No. 60970018), and the Fundamental Research Funds for the Central Universities.

References

1. Ji, S., Li, G., Li, C., Feng, J.: Efficient interactive fuzzy keyword search. In: WWW, pp. 371–380 (2009)
2. Grabski, K., Scheffer, T.: Sentence completion. In: SIGIR, pp. 433–439 (2004)
3. Kukich, K., Li, C.: Selectivity estimation for fuzzy string predicates in large data sets. In: VLDB, pp. 793–804 (2005)
4. Motoda, H., Yoshida, K.: Machine learning techniques to make computers easier to use. Artif. Intell. 103(1-2), 295–321 (1998)
5. Nandi, A., Jagadish, H.V.: Effective phrase prediction. In: VLDB, pp. 219–230 (2007)
6. Williams, H.E., Zobel, J., Bahle, D.: Fast phrase querying with combined indexes. ACM Trans. Inf. Syst. 22(4), 573–594 (2004)
7. Bast, H., Weber, I.: Type less, find more: fast autocompletion search with a succinct index. In: SIGIR, pp. 364–371 (2006)
8. Bast, H., Weber, I.: The completesearch engine: Interactive, efficient, and towards ir & db integration. In: CIDR, pp. 88–95 (2007)
9. Bast, H., Chitea, A., Suchanek, F.M., Weber, I.: Ester: efficient search on text, entities, and relation. In: SIGIR, pp. 671–678 (2007)
10. Bast, H., Mortensen, C.W., Weber, I.: Output-sensitive autocompletion search. In: Crestani, F., Ferragina, P., Sanderson, M. (eds.) SPIRE 2006. LNCS, vol. 4209, pp. 150–162. Springer, Heidelberg (2006)
11. Vernica, R., Li, C.: Efficient top-k algorithms for fuzzy search in string collection. In: KEYS, pp. 9–14 (2009)
12. Chaudhuri, S., Kaushik, R.: Extending autocompletion to tolerate errors. In: SIGMOD, pp. 707–718 (2009)

Relevant Answers for XML Keyword Search: A Skyline Approach

Khanh Nguyen and Jinli Cao

Department of Computer Science and Computer Engineering
La Trobe University, Melbourne, Australia
{tuan.nguyen,j.cao}@latrobe.edu.au

Abstract. Identifying relevant results is a key task in XML keyword search (XKS). Although many approaches have been proposed for this task, effectively identifying results for XKS is still an open problem. In this paper, we propose a novel approach for identifying relevant results for XKS by adopting the concept of Mutual Information and skyline semantics. Specifically, we introduce a measurement to effectively quantify the relevance of a candidate by using the concept of Mutual Information and provide an effective mechanism to identify the most relevant results amongst a large number of candidates by using skyline semantics. Extensive experimental studies show that in overall our approach is more effective than existing approaches and can identify relevant results and top k results in acceptable computational costs.

1 Introduction

XML is rapidly emerging as a standard for representing, publishing and exchanging data over the Internet. With the great success of keyword search over flat documents, keyword search over XML data has recently attracted lots of attentions of researchers from both database and information retrieval. Keyword search provides a friendly mechanism to access XML data without requiring the knowledge of the structured query languages and possibly complex data schemas. However, the limited expressiveness and the ambiguity of keyword queries cause identifying relevant results a very challenging task of XML keyword search.

A candidate of keyword search over XML databases is a subtree covering all query keywords. The baseline approach uses Lowest Common Ancestor (LCA) semantics from graph theory [2] to identify the result of a given keyword query. This approach returns all candidates, thus it has high recall but very low precision. Recently, many proposals [15,9,7,11,12] have been made to boost precision of the baseline approach. The common ideas of these work are (i) *Relevant evaluation*: defining heuristic-based rules that a relevant result has to satisfy; (ii) *Pruning*: eliminating all LCA nodes which do not satisfy the defined rules. It has been experimentally proved by [13] that these approaches not only miss relevant results but also return irrelevant results.

To improve the quality of results for keyword search, we need to deal with all of the following requirements: (R_1) effectively measuring relevant degree of a

L. Chen, P. Triantafillou, and T. Suel (Eds.): WISE 2010, LNCS 6488, pp. 216–224, 2010.

candidate; (R_2) providing an effective mechanism to identify the most relevant results amongst a large number of candidates.

In this paper, we investigate the challenging problems for fulfilling aforementioned requirements. Specifically, we introduce a measurement to quantify the relevance of a candidate by using the concept of Mutual Information for fulfilling requirement (R_1) . The requirement (R_2) is solved by using skyline semantics [4,5] which is proven as an effective mechanism to select the most relevant results (skyline answers).

Mutual Information is a central concept of information theory [8]. It is a quantitative measure of the dependency of two random variables. In other words, the Mutual Information of X and Y measures how much information X can tell us about Y and vice versa. In the context of an XML tree we see that the more information two nodes u and v can tell about each other, the more meaningful relationship between them is holding. More generally, the more information each node in a subtree tells about other nodes, the more relevancy the node associates to the query. From that observation, we adapt the *Mutual Information* to measure the relevancy degree of a candidate answer in this paper.

Skyline query can provide a set of relevant answers, even though those answers may not be the satisfactory ones in all criteria. Skyline queries have been well studied over relational databases [3,6,1,14]. However, applying skyline queries into the context of XML keyword search has not received many attentions yet. In this paper, we propose a novel approach for identifying results of interest for XKS using skyline semantics. We also introduce three different ranking criteria, based on the dominance relationship of skylines to retrieve the top k results. The contributions of this paper are described as follows. The contributions of this paper are described as follows.

- Proposed a novel approach to evaluate the relevancy degree of a candidate answer using the concept of Mutual Information from information theory.
- Introduced an approach to identify most relevant results amongst a large number of candidates using skyline semantics.
- Conducted extensive experiments on real data sets to prove the effectiveness and efficiency of our algorithms.

The remainder of this paper is organized as follows. Section 2 presents some preliminaries. In section 3, we briefly introduce Mutual Information (MI) and its related concepts. Then, we propose normalized MI to quantify the relevant degree of a candidate. Our approach for identifying relevant results using skyline semantics is presented in section 4. Algorithms are developed in section 5. Section 6 discusses experimental results and finally, conclusions are given in section 7.

2 Preliminaries

2.1 Data Model and Query

An XML database is modeled as a rooted labeled tree $T = (r, V, E, L, C, D)$, where V is the set of nodes, $r \in V$ is the root, E is the set of parent-child edges between nodes in V, $C \subset V$ is a subset of the leaf nodes of the tree called content nodes, L assigns a label to each member of $V \setminus C$, and D assigns a data value (e.g., a string) to each content node. We assume no node has both leaf and non-leaf children, and each node has at most one leaf child. Each subtree $S = (r', V', E', L', C', D')$ of T is a tree such that $V' \subseteq V, E' \subseteq E, L' \subseteq L$, and $C' \subseteq C$.

A keyword query Q is a sequence $w_1, \ldots w_n$ of words. A subtree S is a *candidate answer* to Q if its content nodes contain at least one instance of each keyword in Q. If there is more than one subtree of S containing the same instances of search keywords, we only choose the smallest subtree.

2.2   Related Work

In this section, we will discuss related work of XML keyword search. The concept of Mutual Information and skylines will be also briefly introduced.

Result Identification. The baseline algorithm returns all candidate answers as the result. This approach has perfect recall but very low precision. Recently, several attempts as Smallest LCA (SLCA) [15], Exclusive LCA(ELCA) [9], XSEarch [7], Compact Valuable LCA (CVLCA) [11] and Meaningful LCA (MLCA) [12] have been made to boost the precision of the baseline approach. The common idea of these approaches is to evaluate the relevance of a candidate in a boolean way. It means that a candidate either is evaluated as a relevant candidate or is not at all depending on whether it satisfies a set of pre-defined heuristics rules. However, given the ambiguity of keyword query in terms of search intentions, it is difficult (sometimes impossible) to exactly conclude a candidate as a relevant answer or otherwise. To more effectively measure the correlation between content of nodes in a subtree, we adopt the concept of *Mutual Information* from information theory [8] which has been widely used in mining the meaningful correlation between attributes in a relation. The details will be introduced in next section.

Skylines. Skyline queries have received a lot of attentions over the recent years, and several algorithms have been proposed [3,6,1,14,10]. Given a set of points in a d-dimension space. The skyline is defined as the subset containing those points that are not dominated by any other point, whereas a point p dominates p' if p is better than or equal to p' in all the dimensions and strictly better in at least one dimension. Thus, the best answer for such query exists in the skyline.

BNL [3], SFS [6] and SaLSa [6] are generic, in the sense that they do not require any specialized access structure to compute the skyline and can therefore be applied even when the points are the results of some other operations. Other works [14,10] rely on the existence of appropriate indexes, such as B^+-tree or R-tree to speed-up skyline computations. Note that these approaches only apply on static data, where the over-head for building the indexes is amortized across multiple queries. In our setting, the underlying data (candidates) are depended on the query. In this case, building indexes at query time is very expensive, thus it is not suitable.

3   Normalized Mutual Information

In this section, we review the concept of Mutual Information (MI) and its related concepts, and then make it applicable in measuring the meaningful relationship between two nodes.

Entropy and *Mutual Information (MI)* are two central concepts in information theory [8]. Entropy is a measure of the uncertainty of a random variable, while MI quantifies the mutual dependence of two random variables.

Definition 1 (Entropy). *Let X be a discrete random variable that takes on values from the set $\mathcal{X}$ with a probability distribution function $p(x)$. The entropy of X is defined as*

$$H(X) = - \sum_{x \in \mathcal{X}} p(x) \log p(x)$$

Definition 2 (Mutual Information). *Mutual information of two random variables is a quantity that measures the mutual dependence of the two variables. Given two discrete random variables X and Y, their mutual information can be defined as:*

$$I(X;Y) = \sum_{y \in \mathcal{Y}} \sum_{x \in \mathcal{X}} p(x,y) \log \frac{p(x,y)}{p(x)p(y)}$$

where $p(x,y)$ is is the joint probability distribution function of X and Y, and $p(x)$ and $p(y)$ are the marginal probability distribution functions of X and Y respectively.

Property 1 $I(X;Y) \leq H(X)$ *and* $I(X;Y) \leq H(Y)$.

Property 1 indicates that the MI of two nodes is bounded by the minimum of their entropy. The proof of this property can be found in [8]. Since the entropy of different nodes varies greatly, the value of MI also varies from different pairs of nodes. To make MI a good measure to quantify the closely relativeness of two nodes in a candidate, we require the MI of two nodes independent from their entropy. For this purpose, we propose normalized MI as follows.

Definition 3 (Normalized Mutual Information). *The Normalized Mutual Information (NMI) of two random variables X and Y is defined as:*

$$\widetilde{I}(X;Y) = \frac{I(X;Y)}{max\{H(X), H(Y)\}}$$

In next section, we will adopt this concept to measure the relationship between two nodes in a subtree. Then, skyline semantics is used to identify a set of relevant results.

4   Identifying Relevant Results Using Skyline Semantics

Let S be a candidate answer of Q in XML database T, we measure its relevance by calculating the NMI of every pair of content nodes. The question is how we can identify the candidate S relevant to Q. Normally, an aggregation function (*i.e., sum, average*) can be obtained to get the total NMI of all content nodes in S as $score(S) = \sum_{c_i, c_j \in C_S} \widetilde{I}(c_i; c_j)$, where C_S is a set of content nodes in candidate S. The relevance of S can be decided based on a pre-defined threshold α. For instance, if $score(S) \geq \alpha$, S can be considered as a desired result; otherwise it is an irrelevant result. However, selecting a suitable threshold α is not easy, because it is varied from query to query. A low threshold causes returning many answers (including less or not desired ones). In contrast, a high threshold may miss some desired answers. Even though we can let user select the threshold at query time, this is not a good option because users may need to query different times with different thresholds to get their desired results.

We apply skyline semantics which is an effective approach to select the most desired answers amongst numerous candidates. For every keyword $w_i \in Q$, we calculate the set $M_i = \{m | m$ is a leaf node in candidate T of Q and m contains $w_i\}$. The NMI of two keywords w_i and w_j in a candidate subtree S is defined as $max\{\widetilde{I}(m_i; m_j)\}$, where $m_i \in M_i$ and $m_j \in M_j$. Given a keyword query $Q = \{w_1, \ldots, w_n\}$ we measure the NMI of each pair of keywords in a candidate subtree S and store them in vector $D_S = [\widetilde{I}_k(w_i; w_j) | w_i, w_j \in Q \wedge (i < j)]$, where $k = 1, \ldots, C_n^2$ while n is the number of keywords in Q. To choose the most relevant result set to Q, we apply skyline semantics over the candidate set. The vector D_S plays a role as skyline dimensions of the candidate S. Because the high MNI of two nodes indicates their high relativeness, we refer those candidates with high values in their skyline dimensions. More formally, we define the *dominance* relationship between candidates in our context as follows.

Definition 4 (Dominance). *Let S and S' are two candidate answers of Q over an XML database T. S' dominates S, denoted as $S' \succ S$ if,*

- $\forall i (1 \leq i \leq d) D_S[i] \leq D_{S'}[i]$,
- *and* $\exists j (1 \leq j \leq d) D_S[j] < D_{S'}[j]$

where d is where d is the number of values in vector D_S and $d = C_n^2$. The $D_S[i]$ is the i-th value in D_S.

In words, $S' \succ S$ means that the relationship between every pair of keywords in S' is at least as meaningful as the relationship between the corresponding pairs in S. Consequently, S' is more relevant than S is to query Q if $S' \succ S$. Therefore, the problem of identifying relevant results of Q becomes finding a set of non-dominated results by adapting the skyline semantics.

Definition 5 (Relevant Results). *Given a keyword query Q and an XML database T. R is a set of relevant results for query Q over T if for each $S \in R$, there does not exist any other candidate S' of Q that S' dominates S.*

5   Algorithms

In this section, we introduce the algorithms for identifying relevant results based on skyline semantics To accelerate the query processing time, indexes are built off-line at the time we parse the XML database. The efficiency of the algorithms will be analyzed in details in next section.

5.1   Candidate Generation

Generating the candidates is the first step of XKS. The efficient computation has been well study in previous literature [15,16,17]. In this paper, we adapt the algorithm proposed in [17] which is experimentally proved to be the fastest one. Due to the limited space, the details of the algorithm is omitted here. The difference of our approach from others is that we concurrently measure the Normalized Mutual Information (NMI) between each pair of keywords in every candidate during the generating of candidates. The resultant candidates are stored in a sorted list by values of corresponding NMI vectors.

5.2   Skyline Answers

We integrate the skyline computation into the process of candidate generation. More specifically, at the time of generating a candidate, we also check whether it is a skyline answer. By doing this, the skyline computation is simplified to be a

---

Algorithm 1. *Skyline Answers*

Input: keyword query Q, XML databases T.
Output: a set of skyline answers $\mathcal{R}$.

1: $\mathcal{R} = \emptyset$;
2: **while** there are more candidates of query Q in T **do**
3:      $nextCan$ = select a candidate;
4:      $isDominated$ = false;
5:      **for** for each $(R \in \mathcal{R})$ **do**
6:          **if** $R \succ nextCan$ **then**
7:              $isDominated$ = true;
8:          **else**
9:              **if** $nextCan \succ R$ **then**
10:                 remove R from $\mathcal{R}$;
11:             **end if**
12:         **end if**
13:     **end for**
14:     **if** isDominated = false **then**
15:         $\mathcal{R} \leftarrow$ insert $nextCan$;
16:     **end if**
17: **end while**
18: **return** $\mathcal{R}$

---

dominance check procedure(as shown by Algorithm 1). The algorithm works as follows. *(i) Initialization (line 1)*: the result set is set to empty set. *(ii) Repeatedly generating a new candidate (line 2)*: it can be adapted from [17] with some minor modifications. We omit the details here due to space limit. *(iii) Dominance check (lines 3 - 13)*: for each new generated candidate, we apply skyline semantics to see whether it is a relevant result. *(iv) Updating results (line 14-16)*: if the new candidate is not dominated by any candidates so far, it is added to the result set.

6   Experimental Analysis

The experiments were conducted on a 3.2GHz P4 CPU running Windows XP Professional with 1GB of RAM. The algorithms were implemented in Java. We used Oracle Berkeley DB[1] as a tool for creating indexes. We have tested on two real data sets, including: DBLP[2] and IMDB[3].

6.1   Result Quality

We compare the search quality of our relevant results identification using skyline semantics (*referred as SkyAns from here*) with other state-of-art approaches (i.e., SLCA [15], XSEarch [7], CVLCA) [11] and MLCA [12]) The quality is measured in three popular metrics in IR literature, including *Precision (P), Recall (R) and F-measure*. The *F-measure* shows the trade-off between precision and recall and is computed as:

$$F - measure = \frac{(1 + \beta^2)PR}{\beta^2 P + R}$$

where $\beta = 1$ weights precision and recall equally; $\beta < 1$ emphasizes precision, while $\beta > 1$ focuses on recall.

We tested 20 queries on each data set. The correct answers for those queries are obtained by running the corresponding schema-aware XQuery, and the correctness of the answers is verified manually. We recorded the precision and recall for each query and take the average as the precision and recall on each data set. The results are summarized in Fig. 1. The results show that our approach is more effective than all other counterparts. However, The recall is lightly lower due to strict semantics of skylines. To further evaluate the overall of result quality, we take F-measure with different values of β (see Fig. 3). The result from Fig. 3 indicates the overall quality of our approach is higher than all their counterparts.

6.2   Computational Costs

The computational cost is tested on both extracted data sets with size of 200 MB. For each data set, we test a set of 10 queries with the average of 5 keywords.

---

[1] http://www.oracle.com/technology/products/berkeley-db/index.html
[2] http://dblp.uni-trier.de/xml/
[3] http://www.imdb.com/

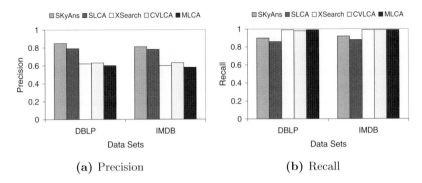

(a) Precision   (b) Recall

Fig. 1. Result quality

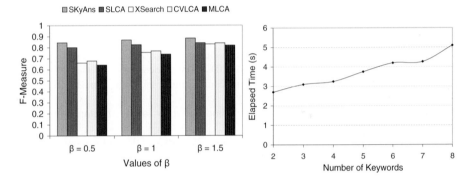

Fig. 2. Overall of result quality   **Fig. 3.** Computational costs

The response time is average time of the corresponding 10 queries, as shown in Fig. 3. The result shows that our approach responds in acceptable time (only few seconds).

7   Conclusions

In this paper, we have addressed some crucial requirements towards effective XML keyword search, including: measuring relevancy degree of a candidate and identifying desired results amongst numerous candidates. To fulfil those requirements, we have proposed an approach for relevancy measurement using Mutual Information concept from information theory. The skyline semantics are obtained for desired result identification.Finally, extensive experiments have been conducted and the results show that our work is very promising in terms of effectiveness and efficiency.

References

1. Bartolini, I., Ciaccia, P., Patella, M.: Efficient sort-based skyline evaluation. ACM Trans. Database Syst. 33(4), 1–49 (2008)
2. Bender, M.A., Farach-Colton, M., Pemmasani, G., Skiena, S., Sumazin, P.: Lowest common ancestors in trees and directed acyclic graphs. Journal of Algorithms 57, 75–94 (2005)
3. Börzsönyi, S., Kossmann, D., Stocker, K.: The skyline operator. In: Proceedings of the 17th International Conference on Data Engineering, Washington, DC, USA, pp. 421–430. IEEE Computer Society, Los Alamitos (2001)
4. Borzsonyi, S., Stocker, K., Kossmann, D.: The skyline operator. In: International Conference on Data Engineering, vol. 0, p. 421 (2001)
5. Chomicki, J., Godfrey, P., Gryz, J., Liang, D.: Skyline with presorting. In: ICDE, pp. 717–816 (2003)
6. Chomicki, J., Godfrey, P., Gryz, J., Liang, D.: Skyline with presorting. In: International Conference on Data Engineering, vol. 0, p. 717 (2003)
7. Cohen, S., Mamou, J., Kanza, Y., Sagiv, Y.: XSEarch: a semantic search engine for XML. VLDB Endowment, 45–56 (2003)
8. Cover, T.M., Thomas, J.A.: Elements of information theory. Wiley Interscience, New York (1991)
9. Guo, L., Shao, F., Botev, C., Shanmugasundaram, J.: XRANK: ranked keyword search over xml documents. In: SIGMOD 2003: Proceedings of the 2003 ACM SIGMOD International Conference on Management of Data, pp. 16–27. ACM, New York (2003)
10. Kossmann, D., Ramsak, F., Rost, S.: Shooting stars in the sky: an online algorithm for skyline queries. In: VLDB 2002: Proceedings of the 28th International Conference on Very Large Data Bases, pp. 275–286. VLDB Endowment (2002)
11. Li, G., Feng, J., Wang, J., Zhou, L.: Effective keyword search for valuable lcas over xml documents. In: CIKM 2007: Proceedings of the Sixteenth ACM Conference on Conference on Information and Knowledge Management, pp. 31–40. ACM, New York (2007)
12. Li, Y., Yu, C., Jagadish, H.V.: Enabling schema-free xquery with meaningful query focus. The VLDB Journal 17(3), 355–377 (2008)
13. Liu, Z., Chen, Y.: Reasoning and identifying relevant matches for xml keyword search. In: VLDB 2008: Proceedings of the 34th International Conference on Very Large Data Bases, pp. 921–932 (2008)
14. Tan, K.-L., Eng, P.-K., Ooi, B.C.: Efficient progressive skyline computation. In: VLDB 2001: Proceedings of the 28th International Conference on Very Large Data Bases, pp. 301–310 (2001)
15. Xu, Y., Papakonstantinou, Y.: Efficient keyword search for smallest lcas in xml databases. In: SIGMOD 2005: Proceedings of the 2005 ACM SIGMOD International Conference on Management of Data, pp. 527–538. ACM, New York (2005)
16. Xu, Y., Papakonstantinou, Y.: Efficient lca based keyword search in xml data. In: EDBT 2008: Proceedings of the 11th International Conference on Extending Database Technology, pp. 535–546. ACM, New York (2008)
17. Zhou, R., Liu, C., Li, J.: Fast elca computation for keyword queries on xml data. In: EDBT 2010: Proceedings of the 13th International Conference on Extending Database Technology, pp. 549–560. ACM, New York (2010)

A Children-Oriented Re-ranking Method
for Web Search Engines

Mayu Iwata[1], Yuki Arase[2], Takahiro Hara[1], and Shojiro Nishio[1]

[1] Dept. of Multimedia Engineering,
Grad. Sch. of Information Science and Tech.,
Osaka Univ.
1-5 Yamadaoka, Suita, Osaka, Japan
{iwata.mayu,hara,nishio}@ist.osaka-u.ac.jp
[2] Microsoft Research Asia 4F, Sigma Building, No.49, Zhichun Road,
Haidian District, Beijing 100190, P.R. China
yukiar@microsoft.com

Abstract. Due to the explosive growth of the Internet technology, children commonly search information using a Web search engine for their homework and satisfy their curiosity. However, there are few Web search engines considering children's inherent characteristics, e.g., children prefer to view images on a Web page rather than difficult texts. Therefore, general search results are neither friendly nor satisfactory to children. In this paper, to support children to obtain suitable information for them, we propose a method to re-rank a general search engine's ranking according to the children-friendly score. Our method determines the score based on the structure of a Web page and its text. We conduct an experiment to verify the re-ranked results match children's preferences. As a ground-truth, we chose 300 Web pages and asked 34 elementary school students whether these Web pages are preferable for them. The result shows that our method can re-rank children-friendly pages highly.

Keywords: Web search, Children, Re-ranking.

1 Introduction

According to the popularization of the Internet, children are commonly using the Internet for searching information, e.g., using a Web search engine. A survey conducted in 2009 has reported that the number of Internet users of elementary school students has been increasing every year at a rapid pace, and about 90% of 6-12 years old children access the Internet in Japan [11]. Another report investigated the motivations of Japanese children to access the Internet: 53.9% of children use the Internet to search information related to their course works and 53.3% of children search information related to entertainments, e.g., games and sports [5]. These data show that the Internet is a common tool for children to obtain information as the same with adults. This trend brings children opportunities to learn new knowledge much more easily and broadly than past, without being restricted by physical constrains.

L. Chen, P. Triantafillou, and T. Suel (Eds.): WISE 2010, LNCS 6488, pp. 225–239, 2010.
© Springer-Verlag Berlin Heidelberg 2010

To find information of interest from an enormous pool of information on the Web, it is essential to use search engines. However, there are few Web search engines that considers children's characteristics, e.g., children prefer images and animations to view on a Web page rather than difficult texts. Therefore, many Web pages highly ranked in a search result are difficult to understand or boring for children. For example, general search engines rank Wikipedia (http://en.wikipedia.org/wiki/) pages highly in the search results. While Wikipedia pages are useful for adults since they show a variety of information relating to a (queried) concept in a well organized way, they might not be preferable for children who have difficulty to understand long text with difficult expressions and fewer images. This mismatch between children's characteristics and search engines may discourage children to keep searching information and learn new knowledge. Another remarkable characteristic of children is that they generally browse only top five pages in the search engine's ranking [3]. Therefore, it is important to rank children-friendly pages higher to avoid getting children bored or discouraged.

In this paper, we propose a method that re-ranks a general search engine's result for children. Our target users are elementary school students, who would search information for their homework and satisfying their curiosity. We define children-friendly pages as ones being easy to understand and visually appealing for children. Based on this definition, our method calculates the children-friendly score for each page. More specifically, the children-friendly score is determined based on the structure of a Web page and the text. Then, our method re-ranks a search engine's ranking by sorting the pages according to their children-friendly scores so that children-friendly pages get ranked higher.

To evaluate our method, we invited 34 elementary school children to obtain the ground-truth of children-friendly pages. We chose 300 Web pages and asked the children to judge each of the Web pages based on the three criteria: a) Whether they want to read the page, b) Whether the page is visually appealing to them, and c) Whether the page is easily understandable for them. Based on this ground-truth, we compare our method and commercial search engines.

The contributions of this paper are summarized as follows:

- To our best knowledge, this is the first study that considers children's characteristics on Web search. We thoroughly investigate field studies of children's characteristics on information acquisition and design series of criteria to decide the children-friendly score of a Web page.
- We conducted an experiment to evaluate our re-ranking method with 34 elementary school children. This is one of the largest scale studies with real children. The obtained data would be an insightful reference for researchers working for children oriented works.

The remainder of this paper is organized as follows. Section 2 describes related work. Section 3 explains our re-ranking method in detail; describing features to decide the children-friendly score of a Web page. Then, Section 4 presents the evaluation result. Finally, Section 5 concludes the paper.

2   Related Work

Mima et al. [10] proposed a Web application to support children to search information which is for their study. This application expands a search query based on the ontology that is constructed based on text books for elementary school classes. For example, when a user queries "apple", this application expands the query to include some related words such as "fiber," a nutrition of apple. Nakaoka et al. [12] constructed a Web information retrieval system using the ontology for children based on their lifestyles. The children's lifestyle ontology defines concepts related to events for children. For example, it generates a concept of "popular Christmas presents" for an event "Christmas". This system can guess children's intention on Web search and helps children discover other keywords related to the query.

These applications aim to expand a search query, which is helpful for children to bridge the gap between their intention and the query. However, there is little effort to generate a search result that matches children's characteristics.

In Japan, there have been some search engines targeting children. Yahoo! KIDS [16], which is one of the most popular search engines for children, pushes pre-registered Web pages at the top of a result. These pages are registered by owners (generally commercial companies and organizations) and only those approved through the internal check by the Yahoo! KIDS administrator are presented. The rest of the ranking result is the same as that of Yahoo! Japan [15]. Kids goo [7] filters out Web pages that are judged as harmful for children. The ranking method is similar to Yahoo! KIDS, as it puts registered pages on the top of the search result and other pages follows the original ranking of goo [4].

The main purpose of these search engines is filtering harmful information out so that children can safely search information. Therefore, these search engines' rankings are almost same as that of general search engines. In addition, since the registrations of Web pages, which are recommended to be on the top of the search result, are basically done by adults, these pages are not always children-friendly. Considering that children tend to browse only top five pages of search result's ranking [3], we should highly rank children-friendly pages, which is the main focus of this work.

3   Re-ranking Method for Children

3.1   Definition of Children-Friendly Page

We have thoroughly investigated some conventional field studies on children's information acquisition. Base on the investigation result, we define children-friendly pages as ones that are easily understandable and visually appealing to children. Specifically, children-friendliness is judged from the following factors.

First, we consider whether the structure of a Web page is children-friendly or not based on [1] [3] [6] [8] [13]. While images and animations are appealing to children and support them to understand the contents of a page, children tend to lose their motivation to continue reading when browsing a page that contains densely lined characters. Additionally, when the size of a page is huge and has a lot of information, children get confused since they cannot decide where they should focus. Therefore, a children-friendly page should satisfy the following criteria:

- The page contains images and animations.
- The amount of text is small.
- The number of links is small.
- No need of scrolling.
- Colorful.

Additionally, we consider whether the expression of the text in a Web page is children-friendly or not based on [1] [2] [6] [8] [13]. It is important that the texts in the page are written in a friendly expression for children and easy to understand. Therefore, a children-friendly page should satisfy the following criteria:

- The length of a sentence is short.
- The text doesn't contain difficult expressions.
- The text contains colloquial expressions for children.

3.2 Design of Features

Based on the above definition of children-friendly pages, we set the following ten features on a trial basis. These features are used to calculate the children-friendly score of a Web page in our proposed page re-ranking method. Each feature is normalized to range from -1 to 1.

3.2.1 Structure Based Features

It is important that the structure of a Web page is easy to grasp the information (contents) for children and also visually appealing to children. Therefore, we set the following six features from the viewpoint of the page structure.

- *Size*

Since children don't prefer to scroll a Web page, a large sized page is not children-friendly. Thus, we set *Size* as a feature which is determined based on the area size of a Web page. $Size_i$ of page i is defined by the following equation, where smaller *Size* means more children-friendly:

$$Size_i = \begin{cases} -1 & (pagesize_i \geq maxsize) \\ -\dfrac{pagesize_i}{maxsize} & (pagesize_i < maxsize) \end{cases} \tag{1}$$

$pagesize_i$ is the area size of page i and $maxsize$ is the maximum area size of existing Web pages. Here, we use $1{,}000 \times 5{,}000$ [pix^2] as $maxsize$ based on our preliminary investigation on $1{,}000$ Web pages.

- *Image Rate*

Images and animations are appealing to children and support them to understand the contents. Therefore, we set *Image Rate* as a feature which is determined based on the amount of images and animations in a page. $Image\ Rate_i$ of page i is defined by the following equation, where larger *Image Rate* means more children-friendly:

$$Image\ Rate_i = \frac{\sum_{j=1}^{N} imagesize_j^i}{pagesize_i} \quad . \tag{2}$$

$imagesize_j^i$ is the area size of jth image or animation in page i, N is the number of images and animations in page i, and $pagesize_i$ is the area size of page i.

- *Text Rate*

Children tend to get bored when browsing pages containing a large amount of text. Therefore, we set *Text Rate* as a feature which is determined based on the amount of text in a page. *Text Rate$_i$* of page i is defined by the following equation, where smaller *Text Rate* means more children-friendly:

$$Text\ Rate_i = \frac{-\sum_{j=1}^{N} textlen_j^i \cdot fontsize}{pagesize_i} \quad . \tag{3}$$

$textlen_j^i$ is the number of characters in jth sentence in page i, N is the number of sentences in page i, *fontsize* is the size of a character, and $pagesize_i$ is the area size of page i. Here, for simplicity, we use 16 [point] as *fontstze* for all pages.

- *Link Rate*

Pages that contain a lot of links get children confused since they cannot decide which link they should select. Therefore, we set *Link Rate* as a feature which is determined based on the number of links in a page. *Link Rate$_i$* of page i is defined by the following equation, where smaller *Link Rate* means more children-friendly:

$$Link\ Rate_i = -\frac{num\ of\ link_i}{maxnum\ of\ link} \quad . \tag{4}$$

num of link$_i$ is the number of links in page i and *maxnum of link* is the maximum number of links for all existing Web pages. Here, we use 300 as *maxnum of link* based on our preliminary investigation on 1,000 Web pages.

- *Component*

A component is an information block of the relevant information in a page. Children tend to get confused when a Web page contains a lot of components, i.e., the page contains a large amount of information. Therefore, we set *Component* as a feature which is determined based on the number of components in a page. We assume that components are extracted using the method proposed in our previous work [9]. *Component$_i$* of page i is defined by the following equation, where smaller *Component* means more children-friendly:

$$Component_i = \begin{cases} -1 & (num\ of\ comp_i \geq maxnum\ of\ comp) \\ -\dfrac{num\ of\ comp_i}{maxnum\ of\ comp} & (num\ of\ comp_i < maxnum\ of\ comp) \end{cases} \quad . \tag{5}$$

num of comp$_i$ is the number of components in page i and *maxnum of comp* is the maximum number of components in a page for all existing Web pages. Here, we use 20 as *maxnum of comp* based on our preliminary investigation on 1,000 Web pages.

● *Color*

Children tend to prefer colorful pages. Therefore, we set *Color* as a feature which is determined based on the number of different colors in a page. *Color$_i$* of page i is defined by the following equation, where larger *Color* means more children-friendly:

$$Color_i = \frac{num\,of\,color_i}{maxnum\,of\,color} \qquad (6)$$

num of color$_i$ is the number of different colors in page i and *maxnum of color* is the maximum number of different colors. Here, we set $1{,}670 \times 10^4$ as *maxnum of color*, which is the maximum number of different colors available for JPEG images.

3.2.2 Text Based Features

As features based on text of a Web page, we set the following four features.

● *Children Expression*

Children are more familiar with colloquial expressions rather than formal ones. Therefore, we set *Children Expression* as a feature which is determined based on the number of colloquial expressions in the text of a page. For this aim, we have constructed a dictionary of colloquial expressions for children by extracting frequent terms from Web pages targeting children, such as Yahoo! KIDS and Kids goo. *Children Expression$_i$* of page i is defined by the following equation, where larger *Children Expression* means more children-friendly:

$$Children\,Expression_i = \frac{\sum_{j=1}^{N} num\,of\,childexpr_j}{num\,of\,term_i} \qquad (7)$$

num of childexpr$_j$ is the number of occurrences of jth colloquial expression in the text of page i, *num of term$_i$* is the total number of terms in the text of page i, and N is the number of occurrences of unique colloquial expressions in the text of page i, i.e., the number of unique terms that match with colloquial expressions in our dictionary.

● *Difficult Expression*

Children cannot understand difficult expressions. Thus, we set *Difficult Expression* as a feature which is determined based on the number of difficult expressions in the text of a page. For this aim, we have constructed a dictionary of difficult expressions by extracting frequent terms from Web pages featuring news and technical contents. *Difficult Expression$_i$* of page i is defined by the following equation, where smaller *Difficult Expression* means more children-friendly:

$$Difficult\,Expression_i = -\frac{\sum_{j=1}^{N} num\,of\,diffexpr_j}{num\,of\,term_i} \qquad (8)$$

num of diffexpr$_j$ is the number of occurrences of jth difficult expression in the text of page i, *num of term$_i$* is the total number of terms in the text of page i, and N is the number of occurrences of unique difficult expressions in the text of page i, i.e., the number of unique terms that match with difficult expressions in our dictionary.

- *Easy*

The difficulty of the entire text (not only difficult terms) is also an important factor that contributes to the children-friendliness. The lower the text's difficulty, the more easily children can understand it. Therefore, we set *Easy* as a feature which is determined based on the difficulty of the text in a page. The difficulty of a text is evaluated by the tool proposed in [14] that estimates the difficulty level by using texts extracted from textbooks for elementary school, high school, and college course works. $Easy_i$ of page i is defined by the following equation, where smaller *Easy* means more children-friendly:

$$Easy_i = -\frac{level_i}{maxlevel}$$
(9)

$level_i$ is the difficulty level of the text in page i. We set 13 as *maxlevel*, which is the maximum difficulty level in the tool [14].

- *Sentence Length*

A long sentence is difficult for children, since its grammatical structure is more complex. Therefore, we set *Sentence Length* as a feature which is determined based on the average length of sentences in a page. $Sentence\ Length_i$ of page i is defined by the following equation, where smaller *Sentence Length* means more children-friendly:

$$Sentence\ Length_i = \begin{cases} -1 & (average\ len_i \geq maxlen) \\ -\dfrac{average\ len_i}{maxlen} & (average\ len_i < maxlen) \end{cases}$$
(10)

$average\ len_i$ is the average length of all sentences in page i and *maxlen* is the maximum length of a sentence for all existing pages. Here, we use 100 as *maxlen* based on our preliminary investigation on 1,000 Web pages.

3.3 Steps of Re-rank

Based on the features described in Section 3.2, our proposed re-ranking method calculates the children-friendly score of a Web page as following steps.

(0) A user issues a query.
(1) Obtain, the search results from a general search engine.
(2) Discard, harmful Web pages for children from the result.
(3) Calculate, the value of each feature for each page in the result.
(4) Decide, the children-friendly score by summing the values of features.
(5) Re-rank, Web pages in the result by sorting them according to their children-friendly scores in descending order.

Since our proposed features are independent from the query, the steps (2) to (4) can be performed offline when a search engine crawls and indexes Web pages.

4 Evaluation

In this section, we present an evaluation we conducted to verify how search results re-ranked by our method match with actual children's perspectives towards Web pages. We examine the effectiveness of each of all features as well as their combinations.

4.1 Ground-Truth Dataset

To evaluate our method, we need a ground-truth dataset that shows actual children's perspectives towards Web pages. Therefore, we have constructed the ground-truth dataset as following steps.

First, we chose six popular queries for children (global warming, service dog, game, horoscope, winter solstice, and Karuta (a traditional Japanese card game)) to collect Web pages to re-rank. These are chosen from top ten popular queries on Yahoo! KIDS during the period between December 2009 and March 2010.

Next, on each query, we collected top 25 Web pages separately from the search result rankings by Yahoo! Japan and Yahoo! KIDS. Since Yahoo! KIDS ranks pre-registered pages on top of its ranking, resulted Web pages are different from Yahoo! Japan and Yahoo! KIDS.

We asked 34 children of six to twelve years old to judge whether collected Web pages are children-friendly or not. To investigate multiple aspects of children-friendly pages, we prepared three different questions for the children. We carefully determined these questions so that children can easily and intuitively answer the questions, which is important to achieve consistency among answers. Specifically, we asked children to browse each Web page with the corresponding query for about 30 seconds, and then select Yes/No to the three questions; "Do you want to read this page?", "Do you think this page is visually appealing?," and "Do you think this page is easy to understand?".

As a result, we collected 1,634 answers in total, i.e., each page was judged by six children on average. We calculated the average vote for each Web page to obtain ground truth ranking of Web pages by regarding 'Yes' as a vote to the page. Here, we define "average vote" as the ratio of number of votes for a page to the total number of children who judged the page. We ranked Web pages separately for above three questions by sorting pages according to their average votes in descending order. If two pages tie, we keep the original ranking order by Yahoo! Japan or Yahoo! KIDS. As a whole, we have 36 ground truth rankings (6 queries and 3 judgment criteria on Web pages obtained by 2 search engines).

4.2 Performance Metric

We used *NDCG (Normalized Discounted Cumulative Gain)* as a performance metric. *NDCG* measures the quality of the search result ranking when the ground truth ranking has cumulative score on a query. *NDCG* at rank k on a query is defined by the following equation:

$$NDCG @ k = \frac{1}{IDCG}\left(rel_1 + \sum_{i=2}^{k} \frac{rel_i}{\log_2 i} \right) \qquad (11)$$

where rel_i is the relevant score of ith page (in our case, it corresponds to the average vote) and k is the rank. We set k to 5 because children tend to browse only top five ranked pages. *IDCG* is the ideal score of *NDCG*, i.e., the *NDCG* value when sorting pages according to their average votes.

4.3   Results

In this section, we compare our re-ranking method with rankings of Yahoo! Japan and Yahoo! KIDS. We should note that Yahoo! KIDS tends to have a higher NDCG value than Yahoo! Japan, since it ranks pre-registered pages (which are regarded to be worth presenting to children) on the top of the ranking, although the rest of the ranking is the same as that of Yahoo! Japan.

4.3.1   Re-ranking Using a Single Feature

Fig. 1 shows the average NDCG value of our re-ranking results when using a single feature on the Yahoo! Japan dataset and that of Yahoo! Japan ranking. Fig. 2 shows the result on the Yahoo! KIDS dataset together with the average NDCG value of Yahoo! KIDS ranking.

Regarding question (a) (Do you want to read this page?), as Fig. 1(a) and 2(a) show, although the NDCG value on *Size* was 2% lower than that of Yahoo! Japan and Yahoo! KIDS, the NDCG values on the other features were 1% to 14% higher than that of Yahoo! Japan and Yahoo! KIDS. Especially, *Text Rate*, *Color*, and *Children Expression* archived a big improvement, i.e., 7% to 18% higher than the NDCG values of Yahoo! Japan and Yahoo! KIDS. This result shows that children tend to prefer

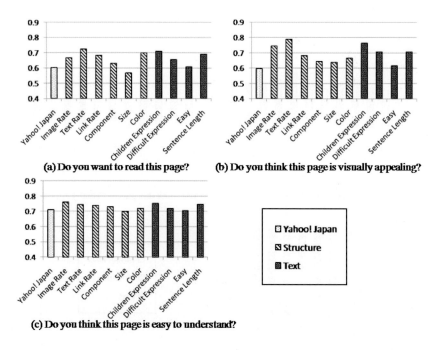

Fig. 1. NDCG of Yahoo! Japan ranking and our method using a single feature

pages that have colorful appearance and contain a smaller amount of text which is full of colloquial expressions. Regarding *Size*, its poor performance comes from the fact that children didn't want to browse too small pages in some cases, i.e., too less contents do not attract children. On the other hand, we observed that a lot of children didn't want to browse pages that are several times larger than the screen size (e.g. Wikipedia). These facts suggest us to appropriately set *maxsize* and extend our method to filter out both too small and too large pages.

Regarding question (b) (Do you think this page is visually appealing?), as Fig. 1(b) and 2(b) show, the NDCG values of all features were 2% to 20% higher than that of Yahoo! Japan and Yahoo! KIDS. *Image Rate* and *Text Rate* especially achieved a big improvement, as 5% to 18% higher than that of Yahoo! Japan and Yahoo! KIDS. This result also supports the result for question (a), as children tend to prefer pages containing a lot of images and lesser amount of text.

Regarding question (c) (Do you think this page is easy to understand?), as Fig. 1(c) and 2(c) show, the NDCG values of all features were 0% to 5% higher than that of Yahoo! Japan and 1% to 6% lower than that of Yahoo! KIDS. This result comes from the fact that pages that explain the contents relating to the query in detail by using both images and text were easy to understand for elder children, but only images are not enough for them. Therefore, we should extend our method to determine the score of each feature based on children's age, e.g., larger amount of text is not always worse for elder children.

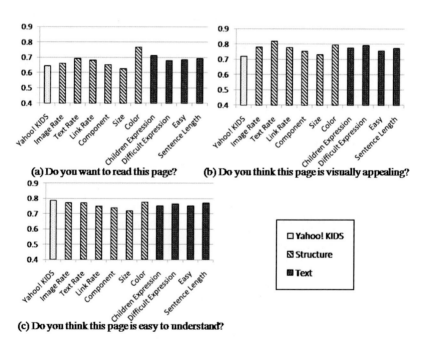

(a) Do you want to read this page? (b) Do you think this page is visually appealing?

(c) Do you think this page is easy to understand?

☐ Yahoo! KIDS
▨ Structure
▣ Text

Fig. 2. NDCG of Yahoo! KIDS ranking and our method using a single feature

Additionally, we can see that features that are effective for re-ranking are different among questions. The structure of pages such as the amount of images and text should be taken into account for all questions. On the other hand, the text of a page should be taken into account for questions (a) and (c). As for question (c), we should take into account not only difficulty of the text but also the amount of text according to children's age.

4.3.2   Re-ranking Using a Feature Combination

Next, for each question, we picked the top four features that achieved the highest NDCG values in the results described in Section 4.3.1 and examined the effectiveness of using the combination of the four features. Here, the score of the combination was defined as the sum of the scores of the four features. Table 1 shows the combinations of the four features for each question; *mix1* for question (a), *mix2* for question (b), and *mix3* for question (c). Fig. 3 shows the average NDCG values of re-ranked results using

Table 1. Combination of features.

Type	Equation	Explanation
mix1	*Text Rate + Color +* *Children Expression + Sentence Length*	For question (a) (Do you want to read this page?)
mix2	*Image Rate + Text Rate +* *Color + Children Expression*	For question (b) (Do you think this page is visually appealing?)
mix3	*Image Rate + Color +* *Children Expression + Sentence Length*	For question (c) (Do you think this page is easy to understand?)

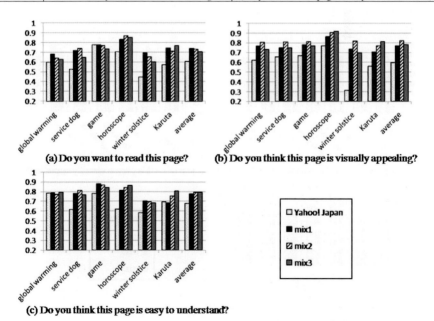

(a) Do you want to read this page?

(b) Do you think this page is visually appealing?

(c) Do you think this page is easy to understand?

☐ Yahoo! Japan
■ mix1
▨ mix2
▨ mix3

Fig. 3. NDCG of Yahoo! Japan ranking and our method using feature combinations

feature combinations and that of Yahoo! Japan ranking. Fig. 4 similarly shows the average NDCG values compared to Yahoo! KIDS.

These results show that the NDCG values using feature combinations outperform that of Yahoo! Japan and Yahoo! KIDS in most cases. Thus, we can confirm that combined features leverage with each other and improve the ranking quality.

Table 2 shows examples of re-ranked results using feature combinations when the query is "winter solstice". We can see that children-friendly pages get ranked higher than Yahoo! Japan ranking.

Fig. 3 and 4 also show that queries affect the re-ranking quality. We should take into account the characteristics of queries when deciding the combination of features:

- Queries related to entertainment, such as "game" and "horoscope", tend to be difficult to commonly provide a good feature combination for all children since individual child has a strong and different preference to such entertainment contents. For example, in our experiments, most boys did not like Web pages related to fancy characters and they judged these pages as low. Therefore, for queries related to entertainment, we should combine features based on individual preferences as well as children's age and gender.
- As for queries unfamiliar with children, such as "service dog", Fig. 3(a) and 4(a) show that the NDCG values of *mix2* were highest among all combinations. This is because *mix2* prioritize on the amount of images, and thus pages with full of visual contents were ranked higher, which helped children to learn new knowledge. This confirms us that we should use features relating to Web page visuals for unfamiliar queries to children.

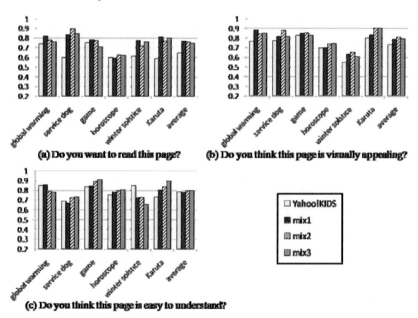

(a) Do you want to read this page?   (b) Do you think this page is visually appealing?

(c) Do you think this page is easy to understand?

Fig. 4. NDCG of Yahoo! KIDS ranking and our method using feature combinations

- As for query "winter solstice", Fig. 3(c) and 4(c) show that the NDCG values of all combinations were lower than that of Yahoo! Japan and Yahoo! KIDS. Web pages collected by querying "winter solstice" contain both children-oriented educational pages and general pages explaining about "winter solstice" (easy for adults). However, elder children regarded both types of pages as easy to understand, which is the reason of the bad performance in our method. According to our observation, this is because the logical structure of these pages was simple and children regard them easy to understand. Therefore, we should take into account whether the logical structure of the page is simple or not, even for general adult-oriented pages.

- As for queries familiar with children, such as "global warming" and "winter solstice", Fig. 3(c) and 4(c) show that the NDCG values of all combinations were lower compared with other queries. This is because they learn information relating to these queries in their classes, thus, presenting too simple and easy Web pages was not effective. This is confirmed by the result that *mix1*, which does not consider the amount of images, had higher NDCG value than other combinations. Web pages containing large amount of images and animations but not enough text explaining the contents are not preferable for children. These results confirm us that we should use features relating to text of a Web page for queries familiar to children in their classes.

Table 2. Examples of *mix1* to *mix3*'s re-ranking and Yahoo! Japan ranking on "winter solstice"

(a)   Re-ranked result by *mix1* on question (a) (Do you want to read this page?).

Rank	Original rank by Yahoo! Japan	Score (improvement from original score of the page ranked by Yahoo! Japan)
1	25	0.73 (+0.35)
2	18	0.23 (− 0.15)
3	23	0.73 (+0.50)

(b)   Re-ranked result by *mix2* on question (b) (Do you think this page is visually appealing?).

Rank	Original rank by Yahoo! Japan	Score (improvement from original score of the page ranked by Yahoo! Japan)
1	25	1.00 (+0.82)
2	23	0.73 (+0.35)
3	18	0.73 (+0.50)

(c)   Re-ranked result *mix3* on question (c) (Do you think this page is easy to understand?).

Rank	Original rank by Yahoo! Japan	Score (improvement from original score of the page ranked by Yahoo! Japan)
1	25	0.48 (+0.10)
2	23	0.48 (− 0.10)
3	9	0.58 (+0.35)

In summary, our experimental results show that features should be combined based on the characteristics of each query (e.g., a query related to entertainment, a query that children learn in their class, and a query unfamiliar for children) to rank Web pages more children-friendly.

Although we decided the feature combinations of *mix1* to *mix3* based on the dataset obtained from our experiment, we expect these combinations to be effective for general cases when children search Web pages. To rank Web pages satisfying "children want to browse" higher, *mix1* can prioritize Web pages structurally and textually familiar with children. To rank Web pages satisfying "visually appealing to children" higher, *mix2* can prioritize children-oriented Web pages in terms of visual aspects, e.g., image and color. To rank Web pages satisfying "easy to understand for children" higher, *mix3* can prioritize Web pages that contain images and easy-to-understand text for children. Therefore, each of *mix1*, *mix2*, and *mix3* takes into account the characteristics of each aspect of children-friendly pages, and thus, we believe that these combinations are effective for general cases, i.e., searching children-oriented pages. However, to further improve the ranking quality, we should extend our approach so that features can be combined based on the characteristics of queries and children, e.g., age and gender, as our evaluation results show.

5   Conclusion

To support children to use a search engine and find a children-friendly Web page, we proposed a method to re-rank a general search engine's ranking according to the children-friendly score. We thoroughly investigated some conventional field studies on children's acquisition and defined features of a Web page to decide the children-friendly score of the page. Specifically, our method takes into account the structure of the page, amount of images and animations, and its text, to decide the score.

We conducted an experiment to evaluate how a re-ranked result matches with the actual children's perspective to Web pages. As a ground-truth, we asked 34 elementary school students to judge whether a Web page is children-friendly or not. As a result, NDCG of our re-ranked result was 5% to 20% higher than original rankings of commercial search engines. We found that the amount of text, the number of colors, and the number of colloquial expressions for children are the most important features to detect children-friendly pages.

As our future work, we plan to examine an appropriate amount of text and images according to children's age. Additionally, we further examine the best combination of features according to characteristics of queries and children.

Acknowledgement

This research was partially supported by The Global Center of Excellence Program of the Ministry of Education, Culture, Sports, Science and Technology, Japan.

References

1. Bilal, D., Kirby, J.: Differences and Similarities in Information Seeking: Children and adults as Web Users. Information Processing and Management 38(5), 649–670 (2002)
2. Dale, E., Chall, J.: Readability Revisited: The New Dale-Chall Readability Formula. Brookline Books/Lumen edn. (1995)
3. Druin, A., Foss, E., Hatley, L., Golub, E., Guha, M.L., Fails, J., Hutchinson, H.: How Children Search the Internet with Keyword Interfaces. In: Proc. IDC 2009, pp. 89–96 (June 2009)

 4. Goo, http://www.goo.ne.jp/ (in Japanese)
 5. Goo research, http://research.goo.ne.jp (in Japanese)
 6. John, W.: Design Criteria for Children's Web Portals: the Users Speak Out. Journal of the American Society for Information Science and Technology 53(2), 79–94 (2002)
 7. Kids goo, http://kids.goo.ne.jp/ (in Japanese)
 8. Kikuchi, H., Kato, H., Akahori, K.: Analysis of Children's Web Browsing Process ICT Education in Elementary Schools. In: Proc. ICCE 2002, pp. 253–254 (2002)
 9. Maekawa, T., Hara, T., Nishio, S.: Two Approaches to Browse Large Web Pages Using Mobile Devices. In: Proc. MDM 2006 (2006)
10. Mima, H., Yoon, T.: Design and Implementation of a Web Information Retrieval Aid System for Children. In: Proc. Program Symposium of IPSJ, pp. 17–23 (August 2003) (in Japanese)
11. Ministry of Internal Affairs and Communications,
    http://www.johotsusintokei.soumu.go.jp/index.html (in Japanese)
12. Nakaoka, M., Shirota, Y., Tanaka, K.: Web Information Retrieval Using Ontology for Children based on Their Lifestyles. In: Proc. ICDEW 2005, p. 1260 (April 2005)
13. Nielsen, J.: Teenagers on the Web: 61 Usability Guidelines for Creating Compelling Websites for Teens, Nielsen Norman Group Report (January 2005)
14. Sato, S., Matsuyoshi, S., Kondoh, Y.: Automatic Assessment of Japanese Text Readability Based on a Textbook Corpus. In: Proc. LREC 2008, pp. 28–30 (May 2008)
15. Yahoo!Japan, http://www.yahoo.co.jp/ (in Japanese)
16. Yahoo!KIDS, http://kids.yahoo.co.jp/ (in Japanese)

TURank: Twitter User Ranking Based on User-Tweet Graph Analysis

Yuto Yamaguchi[1,*], Tsubasa Takahashi[1,*],
Toshiyuki Amagasa[1,2], and Hiroyuki Kitagawa[1,2]

[1] Graduate School of Systems and Information
Engineering, University of Tsukuba, Japan
{yuto_ymgc,tsubasa}@kde.cs.tsukuba.ac.jp
[2] Center for Computational Sciences, University of Tsukuba, Japan
{amagasa,kitagawa}@cs.tsukuba.ac.jp

Abstract. In this paper, we address the problem of finding authoritative users in a micro-blogging service, Twitter, which is one of the most popular micro-blogging services [1]. Twitter has been gaining a public attention as a new type of information resource, because an enormous number of users transmit diverse information in real time. In particular, authoritative users who frequently submit useful information are considered to play an important role, because useful information is disseminated quickly and widely. To identify authoritative users, it is important to consider actual information flow in Twitter. However, existing approaches only deal with relationships among users. In this paper, we propose TURank (Twitter User Rank), which is an algorithm for evaluating users' authority scores in Twitter based on link analysis. In TURank, users and tweets are represented in a *user-tweet graph* which models information flow, and ObjectRank is applied to evaluate users' authority scores. Experimental results show that the proposed algorithm outperforms existing algorithms.

1 Introduction

In recent years, micro-blogging services, where users exchange short messages, have attracted considerable attention as a new type of web services. Micro-blogging services are interesting in that they offer features similar to both blogs and SNS (Social Network Services), while they limit the length of a message that a user can send. Due to the limitation, a user can casually send messages even if messages being exchanged are not so meaningful or informative for other users. As a consequence, messages exchanged in those services are mixtures of useful and unuseful information, such as users' current status, news stories, reviews of a product or a service, and other interests. Another important feature is that a micro-blogging service allows users to communicate with each other.

Twitter [1] is one of the most famous micro-blogging services, and has shown an explosive growth in the past several years. A message posted by a user is called

---

[*] The current affiliation is Service Platforms Research Laboratories, NEC Corporation.

L. Chen, P. Triantafillou, and T. Suel (Eds.): WISE 2010, LNCS 6488, pp. 240–253, 2010.

a *tweet*, and a user can *follow* any user accounts if he/she finds those accounts interesting and/or useful. Conversely, a user account may have some *followers* depending on his/her popularity or usefulness. The approximate number of user accounts in Twitter is estimated to be 75 million as of the end of 2009.

User accounts in Twitter are of various types, such as ordinary users, companies, politicians, and news sites. Java et al. [7] classified Twitter accounts into three categories. *Information source* is the category of user accounts who posts useful information. Consequently, accounts in this category tends to collect many followers; *Friends* is the category of user accounts who are friends, families, and co-workers in the real world. *Information seeker* is the category where users who rarely post messages, but follow an enormous number of user accounts to obtain information from other users' tweets. Therefore, noticing the difference among the user categories is very important, because the way information is exchanged in Twitter inherently depends on users and their behaviors.

For this reason, identifying authoritative user accounts in Twitter is a challenging and an important task for obtaining useful information. To this end, some researchers have proposed algorithms for measuring users' authority scores by analyzing the link structure consisting of *follow* relationships [7] [12]. However, in Twitter, most of users follow back their followers in accordance with mere formal courtesy. Besides, only a few percent of users in *follow* relationships communicate with each other [6]. Therefore, the algorithm using only a Twitter social graph which consists of only *follow* relationships is not sufficient.

In this paper, we propose TURank (Twitter User Rank), which is an algorithm that measures the Twitter users' authority scores considering both a Twitter social graph and how tweets actually flow among users. To address this problem, we give our focus on *retweet* (RT hereafter). RT is originated from one of the user conventions, and it allows users to resend other user's tweets to his/her followers. Generally, a user retweets a tweet if it appears to contain useful information, because he/she wants to share it with his/her followers. As a consequence, a user is considered to be authoritative if his/her tweets contain useful information, and if so, those tweets tend to be retweeted by other users. To model this, we introduce the *user-tweet graph*. A user-tweet graph consists of nodes, corresponding to user accounts and tweets, and edges, corresponding to *follow* and retweet relationships. Unlike the Twitter social graph which is relatively static, the user-tweet graph is dynamic and reconstructed whenever a new retweet is observed. We perform link structure analysis on this graph based on ObjectRank [3] to evaluate the users' authority scores reflecting the actual information flow and the dynamic property of Twitter. We show the feasibility of our approach by some experiments.

The rest of this paper is organized as follows. Section 2 describes preliminaries including an overview of Twitter and the concept of ObjectRank. Our algorithm is presented in Section 3, and is experimentally evaluated and compared to other conventional algorithms in Section 4. Related works are discussed in Section 5. Finally, Section 6 concludes this paper and discuss future works.

2   Preliminaries

In this section, we overview Twitter and its related concepts, followed by an overview of ObjectRank [3], which is a link analysis scheme for linked objects.

2.1   An Overview of Twitter

Basics of Twitter. Twitter [1] is one of the most notable micro-blogging services founded in 2006. Messages exchanged in Twitter are called *tweets*, and the maximum length of them is limited at most 140 characters. For each user account, there is a profile page with its permanent link (permalink) for displaying posted tweets. Moreover, even a tweet has its own permalink, thereby making it possible to browse or link tweets regardless of Twitter account ownership. Note that protected user accounts do not allow non-approved users to browse their tweets; only approved users are allowed to refer to the tweets.

A user may *follow* other user accounts to subscribe their tweets. Once user accounts are followed by a user, called *follower*, all tweets posted by the accounts are then displayed in the follower's *timeline* instantly. Unlike most of the other SNS services, where mutual social networking model is applied, Twitter allows users to follow others without any permission. This feature is one of the most crucial reasons why Twitter won an enormous number of users.

An RT (retweet) is a tweet that quotes a past tweet in order to disseminate the past tweet to the followers. Because RTs are originated in user convention, there are multiple formats to represent RTs. However, most of them follow the following format:

```
[additional text] RT @[account]: [original tweet]
```

It represents that *original tweet* by *account* is retweeted with some *additional text*.

Recently, Twitter has adopted this user convention as a part of official functionality, because RT is commonly used as a de facto standard. The official version of RT is called *official RTs*, while conventional RTs are called *unofficial RTs*. The official RT differs from the unofficial one in several ways: 1) for official RTs, users are not allowed to add additional text, and 2) sending an official RT is easy, because it just requires pressing a dedicated button, whereas users must copy the original text to their tweets manually in unofficial RTs.

Semantics of RT. In Twitter, information contained in tweets spreads across users via RTs. It is important to understand this dissemination process, when we try to measure a user's authority score. Boyd et al. reported that RTs are used for different purposes, such as circulating of tweets to the followers and commenting on a tweet for initiating exchange of opinions [4]. Specifically, they made analysis on RTs used for conversations. In addition, they investigated how users retweet, why users retweet, and what users retweet, and reported that there are multiple formats to represent RTs. However, recently mots RTs are of the format described above due to the emergence of official RTs and popularization of Twitter client that support RTs.

A tweet may be retweeted by not only one user, but also two or more users, which result in a chain of RTs. For an RT chain, it can be regarded as a flow of information originated from a user to his/her followers and their descendant followers. When we look into the types of retweets, there is a tendency that a tweet is widely disseminated if the objective of the RT is to circulate useful information. On the other hand, it may not be widely disseminated if the objective is the conversation among a small number of users.

There have been several proposals that rank users based on the number of RTs. However, they have some problems. First, as mentioned above, RTs have different characteristics depending on their objectives, such as conversation and information circulation. For this reason, it is important to consider how RTs spread, that is, conversational RT spreads only for participants involved in the conversation. In contrast, recommending RTs spread widely. Hence, only counting the number of RTs will not be enough to measure the user's authority score, since RTs for conversation are less relevant to the user's authority score. Second, these methods do not consider the authority score of the user who retweets a tweet. It is natural that tweets retweeted by authoritative users are likely to be more useful than those that retweeted by unauthoritative users.

To address these problems, our algorithm considers the link structure of the tweet flow by RTs. In fact, an information dissemination process in Twitter can be modeled using a graph structure called *user-tweet graph*, which we shall introduce later. We then analyze the link structure of such graphs to evaluate users' authority scores.

2.2   ObjectRank

ObjectRank [3] is an extention of PageRank [11]. It performs link structure analysis over linked objects for measuring the importance of objects in the database. Unlike PageRank, ObjectRank takes account of edge types and node types in order to deal with multiple kinds of edges and nodes. Specifically, in ObjectRank, we differentiate each edge by setting an appropriate weight to control the flow of scores going through the edge.

At first, we construct a graph called an *authority transfer schema graph*, which is an intentional graph that models the domain of discourse. Figure 1 shows an example of an authority transfer schema graph. It illustrates the structure of the graph as well as the weight of each edge. The graph consists of the node set V containing all types of target objects, and the edge set E containing all types of edges existing between nodes in V. Notice that, for a pair of linked nodes, there always exist two edges with different directions, i.e., forward and backward edges. This means that the scores should flow both forward and backward directions. For example, the score of a paper should flow to its authors, and from the authors to the paper as well. Notice that the edge weights can sometimes be zero depending on the domain of discourse. For example, a paper which is cited by an important paper is also important, but a paper which cites an important

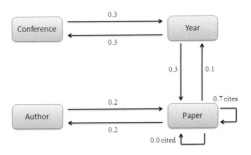

Fig. 1. An Authority Transfer Schema Graph

paper is not always important. Weights of all edges can be set by hand so that they reflect the semantics of relations between all objects. Note that the sum of weights of all edges starting from certain node must be less than or equal to 1.

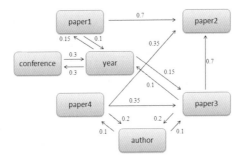

Fig. 2. An Authority Transfer Data Graph

Based on the authority transfer schema graph, we can create a *authority transfer data graph*. Figure 2 shows an example. We can see that various kinds of relationships among objects are represented. Link structure analysis is applied to this graph in such a way that, for each node, the score is propagated along the outgoing edges, and the propagated scores are computed by the weighted average based on the edge weights defined by the authority transfer schema graph. Specifically, scores are calculated by applying the Equation 1 to the authority transfer data graph constructed as above. Equation 1 is the same as the one employed by PageRank, which uses the random surfer model

$$\mathbf{r} = d\mathbf{A}\mathbf{r} + \frac{(1-d)}{|V|}\mathbf{e} \qquad (1)$$

where $\mathbf{r}$ is the ObjectRank score vector, d is the probability of random jump, and $\mathbf{A}$ is the transition matrix. Note that the element a_{ij} of the transition matrix $\mathbf{A}$ is the weight of the edge from node i to node j if it exists, otherwise, a_{ij} is 0.

3   Twitter User Rank (TURank)

In this section, we describe a scheme to evaluate Twitter users' authority scores based on the link structure analysis on the *user-tweet graph*. It reflects both Twitter social graph and actual information flow among users. We call this scheme Twitter User Rank (TURank).

3.1   Basic Idea

In Twitter, there are several relationships between users and tweets, such as post, follow, and RT. We take these relationships into account for measuring the authority score. Our scheme is based on the following observations:

 – A user followed by many authoritative users is likely to be an authoritative user.
 – A tweet retweeted by many authoritative users is likely to be a useful tweet.
 – A user who posts many useful tweets is likely to be an authoritative user.

Based on these observations, we construct the user-tweet graph in which user nodes and tweet nodes are interconnected by edges corresponding to the relationships among users and tweets. A user-tweet graph allows us to understand how information spreads among users by RTs. Then, the link structure analysis is applied to this graph to calculate authority scores based on ObjectRank. The remainder of this section describes these steps in detail.

3.2   User-Tweet Graph

To construct a user-tweet graph, we define a *user-tweet schema graph*, as illustrated in Figure 3. It corresponds to the authority transfer schema graph in ObjectRank. A user-tweet schema graph $UTG_S = (V_S, E_S)$ defines the structure and edge weights of a user-tweet graph. Here, V_S is the node set consisting of user nodes and tweet nodes, and E_S is the edge set consisting of post, posted, follow, followed, RT, and RTed edges. A post edge is from a user u to a tweet posted by u. A follow edge is from a user u to a user followed by u. An RT edge is from a tweet t to a tweet retweeted by t. Posted, followed, and RTed edges are the reverse edges corresponding to post, follow, and RT edges. The weight $w(e_S)$

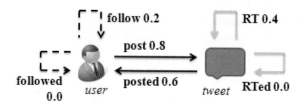

Fig. 3. A User-Tweet Schema Graph

is set to edge $e_S \in E_S$ by hand in order to reflect the semantics of each edge. Scores flow in accordance with the amount of weights from a node to another. The weight is shown beside the edge type in Figure 3.

A user-tweet graph $UTG = (V, E)$ is a graph corresponding to authority transfer data grpah in ObjectRank, and is derived from its corresponding user-tweet schema graph using actual data obtained from Twitter (Figure 4). Here, V is the node set which contains all tweet nodes and user nodes in obtained data. E is the edge set which contains all existing edges in obtained data, namely, post, posted, follow, followed, RT, and RTed edges. The weight $w(e)$ set to the edge $e \in E$ from node $u \in V$ is calculated by Equation 2.

$$w(e) = \frac{w(e_S)}{OutDeg(u, e_S)} \tag{2}$$

where $e_S \in E_S$ is the edge of the same type as e, and $OutDeg(u, e_S)$ is the number of outgoing edges of type e_S from node u.

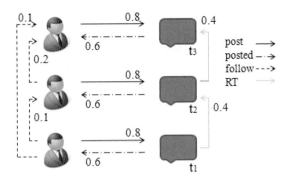

Fig. 4. A User-Tweet Graph

Although it looks quite simple, a user-tweet graph successfully represents important relationships in Twitter. If the graph consists of user nodes and follow edges, or alternatively tweet nodes and RT edges, it is almost the same as the graph for PageRank. However, in the case of a user-tweet graph, it has the edges between users and tweets, thereby making it possible to reflect the score flow from a user to its related tweets and vice versa. Specifically, scores of tweet nodes aggregated from RT edges are delivered to their authors through posted edges. Likewise, scores of user nodes aggregated from follow edges are delivered to their tweets through post edges. Therefore, we can calculate scores of both users and tweets concurrently. Moreover, a user-tweet graph is successful in capturing the RT chains appropriately. For example, as illustrated in Figure 4, when tweet t_1 retweets t_2, and t_2 retweets t_3, the score of t_1 affects t_3 through t_2. Of course the score of t_2 also affects t_3. In this way, scores of all tweets participating the RT chain affect the original tweet of the chain.

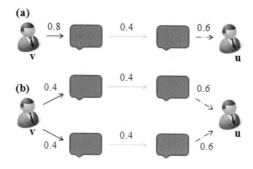

Fig. 5. Multiple RTs by the Same User

Even if user v retweets user u's tweets multiple times, u does not get severalfold scores from v. Because scores travel in accordance with the normalized weight as illustrated in Figure 5. In the case of (b), user u is retweeted by v twice. However, the same amount of score from v transfers to u as the case of (a), because the weights of the RT edges are normalized. This means that the user whose tweets are retweeted multiple times by only one user does not get a high score, while the user whose tweets are retweeted multiple times by different users gets a high score.

3.3   Calculating Scores

Having obtained a user-tweet graph, we then try to compute the scores of each user for subsequent ranking. The score calcuration is based on the Equation 1. Here, if edge e from node i to node j exists, element a_{ij} of **A** is $w(e)$, otherwise, a_{ij} is 0.

TURank
  $r^0 \leftarrow [1, \cdots, 1]$
  $\alpha \leftarrow 0$
  Repeat
    $\alpha \leftarrow \alpha + 1$
    foreach $r_i^\alpha \in r^\alpha$
      $r_i^\alpha \leftarrow \sum_{e=(j,i) \in E} w(e) r_j^{\alpha-1} + (1-d)/|V|$
    end
    $r^\alpha \leftarrow r^\alpha / ||r^\alpha||_1$
  until $||r^\alpha - r^{\alpha-1}||_1 < \epsilon$
  return r^α
end

Fig. 6. TURank Algorithm

Figure 6 shows the algorithm to calculate TURank scores in detail. The score of node i in step α is calculated by summing up scores of all nodes which have the edge to i in step $\alpha - 1$ and scores by random jump. This calculation is iterated until all scores are converged, where convergence threshold ϵ is set to be sufficiently small. Using this algorithm, we can measure the authority scores based on the link structure of the relationships among users and tweets.

4   Experimental Evaluation

This section shows experimental evaluation of the proposed TURank to show that the proposed algorithm outperforms other conventional algorithms. The dataset used in this experiment is explained in Section 4.1, and Section 4.2 presents the methodology of this experiment. Lastly, we discuss experimental results in Section 4.3.

4.1   Dataset

The dataset used in our experiment is crawled from Twitter using Twitter API [2] from January 26 to 28 in 2010. From the raw data, we extracted the following data: $D = (T, U, P, F, R)$, where T is the tweet set consisting of Japanese tweets which retweet other tweets or are retweeted by other tweets; U is the user set containing users who post tweet $t \in T$; P is the post edge set containing post edges from user $u \in U$ to a tweet posted by u; F is the follow edge set containing follow edges from user $u \in U$ to a user followed by u; and R is the RT edge set containing RT edges from tweet $t \in T$ to a tweet retweeted by t. The details of the dataset is shown in Table 1.

Table 1. Dataset details

	size
# of tweet nodes $|T|$	605,968
# of user nodes $|U|$	112,035
# of post edges $|P|$	605,968
# of RT edges $|R|$	369,383
# of follow edges $|F|$	14,631,014

Unfortunately, Twitter API does not provide information about RT edges between a retweet and its original tweet. For this reason, we extracted the correspondence information in the following way. At first, we extract the original user id which corresponds to the user who posted the original tweet, and the original text from a retweet, which is formed as shown in Section 2.1. Next, to identify the original tweet, we look into the latest 50 tweets posted by the original user. Actually, we compute the Levenshtein (edit) distance [10] between the original text and the latest 50 tweets, and find the tweet with the least distance which is less than the predefined threshold.

Here, the Levenshtein distance [10] is a metric that measures the similarity between two strings. It is defined as the minimum number of edit operations needed to transform one string into the other. Edits include deletion, insertion, and substitution of a character. For example, the Levenshtein distance between *apple* and *play* is 4:

1. *pple* (delete *a*)
2. *ple* (delete *p*)
3. *pla* (substitution of *a* for *e*)
4. *play* (insert *y* at the end)

It is possible to set the different cost to each edit. In the example above, the costs of all edits are 1. When cost 1 is set to insertion and deletion and cost 2 is set to substitution, the Levenshtein distance between *apple* and *play* is 5.

In the procedure of identifying the correspondence between retweets and their original tweets, we assign a larger cost for insertion than that of deletion, because users tend to retweet shortened messages by deleting some of the original texts due to the strict limitation of message length (140 characters).

4.2   Methodology

We compare 8 ranking schemes, including FollowNum, RTNum, PageRank, HITS [8], and 4 variants of TURank. FollowNum ranks users according to the number of followers. RTNum ranks users by the number of RTs. For PageRank and HITS, we apply each algorithm to the Twitter social graph, and rank the users according to the scores. For TURanks, we use the user-tweet graph weighted as shown in Table 2. The weights of followed edges in all TURanks are set to 0, because the authority score should not flow from users to their followers. Weights of other types of edges are varied in order to compare these TURanks and analyze how the weights work.

Table 2. TURank Weights

	follow	followed	post	posted	RT	RTed
TURank1	0.4	0.0	0.6	0.6	0.4	0.0
TURank2	0.2	0.0	0.8	0.6	0.4	0.0
TURank3	0.2	0.0	0.8	0.4	0.6	0.0
TURank4	0.2	0.0	0.8	0.6	0.2	0.2

The obtained rankings are evaluated by 34 examinees. We show top 25 authoritative users, and examinees evaluate the adequacy of the authority score of each user by scoring 1 to 5 by browsing his/her latest 100 tweets. In this experiment, an authoritative user is defined as a user who posts either some breaking news stories attracting many users, interesting or humorous tweets, or ideas about some topics which are regarded as useful by many users.

Average adequacy of evaluated the top k authoritative users is shown in Figures 7 and 8. Figure 7 compares all algorithms, and Figure 8 compares 4 TU-Ranks using the graph weighted as shown in Table 2.

4.3   Discussions

Figure 7 shows that our algorithm successfully suggests relatively high adequacy for the entire range. TURank outperforms both PageRank and HITS in that only *follow* relations are considered. The result suggests the importance of considering tweet/retweet relations, thereby making it possible to model information flow in Twitter. On the other hand, FollowNum and RTNum indicate low adequacy, probably because the number of followers or RTs alone is not adequate.

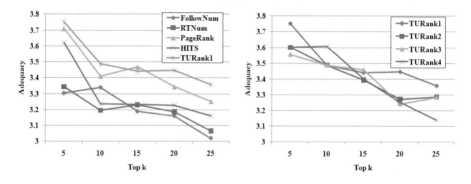

Fig. 7. Average Adequacy of Top k Users of All Algorithms   **Fig. 8.** Average Adequacy of Top k Users of TURanks

TURank shows more effectiveness than RTNum which only uses the number of RTs. Specifically, RTNum tends to extract users who use RTs for conversation in higher ranks. These users often use replies of the form @[*user id*] in order to specify the user to which he/she replies. In these cases, tweets do not convey useful information, and actually examinees evaluated most of them as useless.

The total number of replies obtained by the top 100 users of RTNum is 1753, while the one of TURank is 1241. For this reason, we observe that only counting the number of RTs is not sufficient to capture the semantics of RTs; RTNum do not discriminate the difference between conversational RTs and recommending RTs. In contrast, TURank is successful in discriminating the difference. TURank uses the user-tweet graph which represents how information is disseminated via RT chains. In this case, recommending RTs, which spread widely, gain larger scores, while conversational RTs, which spread across limited users, gain smaller scores.

The reason why TURank outperforms PageRank and HITS is as follows. TURank ranks users, whose tweets are retweeted many times, in higher ranks, and such users are regarded as authoritative users. If we only use *follow* relations, we may fail to rank these users higher. In contrast, assume a user who has many followers, but his/her tweets are not retweeted many times. Such users do not get higher ranks by TURank, and categorized as unauthoritative users. In Twitter, one wants to follow a user, if the user is likely to transmit useful information even without guarantee. In many cases followers do not stop following even if he/she turns out not to transmit useful information any more. This is partly because one does not completely remember all users whom he/she is following, and because the number of following user accounts can be very large (100 to more than 1,000) in many cases. Even if a user has a lot of followers in this way, his/her tweets may be rarely retweeted, which result in lower rankings.

The results also show the problem of TURank. In fact, there are some authoritative users, who have a lot of followers, but their tweets are rarely retweeted. Even in that case, TURank lowers such users' scores. For example, although the weather forecasting bot[1] is evaluated as very useful by examinees, TURank ranked it in lower ranks. This bot acquires a lot of followers, but the followers do not want to retweet its tweets. Actually, the followers of this bot is satisfied by receiving weather forecasts, but do not have a strong motivation to disseminate the forecasts to their followers. Therefore, proper evaluation of such user accounts is an important future issue.

Figure 8 shows that TURank1 is the best among all TURanks. In Twitter, as shown in Table 1, the number of RT edges is much less than follow edges. For this reason, if the weight of RT edges is larger than necessary, the sparse part of the user-tweet graph is weighed heavily, and consequently, TURank is unsuccessful in evaluating the users.

TURank1 shows a better results, because it is based on a proper setting of edge weights. Comparing TURank2 with TURank3, there are some differences between them. Setting larger weights to RT edges, a tweet which is retweeted many times obtains a higher score. For this reason, a user who posts such a tweet ranks higher position of TURank2 than its position of TURank3. Moreover, setting smaller weights to posted edges, smaller scores transfer from tweets to users through posted edges. Hence, scores of users totally decrease, and as a consequence, follow edges convey smaller amount of scores. However, evaluated effectiveness of these two rankings are almost the same in this experiment, because the difference of weights between TURank2 and TURank3 is not large enough. The evaluated result of TURank4 shows a little lower effectiveness than others. Because the weight of the RTed edge is not 0, it is expected that the user who often retweets useful tweets moves up on the list of TURank4. However, as a result, only two useless bots[2] entered the upper level in TURank4.

---

[1] A *bot* is a user who posts automatically by a program.

[2] Two bots are @yumemitter and @soysaucebot. The former is the bot which randomly retweets the tweet about dreams posted by other users, and the latter is the bot which randomly retweets the tweet about the soy sauce posted by other users.

5   Related Work

In recent years, many researchers have investigated Twitter. Java et al. [7] studied the topological and geographical properties of Twitter analyzing the social graph composed of *follow* relationships. In addition, they investigated the community formed by Twitter users at large, and then divided users into three categories. Honeycutt et al. [5] analyzed the conversational practice in Twitter, such as replies and RTs. Huberman et al. [6] reported that the social graph of Twitter does not describe relationships among users well. Besides, they indicated that there is the non-dense social graph composed of *friends* relationships, which cannot be illustrated by *follow* relationships. Boyd et al. [4] conducted the usage survey of RTs. They interviewed Twitter users in order to analyze the syntaxes of RTs, the purposes of RTs, and the targets of RTs.

Weng et al. [12] and Leavitt et al. [9] proposed the algorithm which ranks Twitter users. TwitterRank [12] measures the users' influence considering the link structure of *follow* relationships, the similarity between users, and the number of posts. TURank differs from TwitterRank in two ways: TURank takes RTs into accounts which is not considered by TwitterRank, and is the graphically-based algorithm which does not use the content of tweets. Leavitt et al. insisted that measuring the influence by only *follow* relationships is inadequate and take the communication such as RTs and replies into consideration. Although this research is close to our research, this research does not consider the link structure of the graph composed in Twitter.

6   Conclusions

This paper presented a scheme for identifying authoritative users in Twitter. Considering the actual information flow in Twitter, we proposed the TURank which is an algorithm to measure the authority scores and rank users applying the link structure analysis to the *user-tweet graph*. The user-tweet graph is constructed from user nodes and tweet nodes linked by post, follow, and RT edges in order to model how information flows and spreads among users. In spite of its simple structure, this graph describes these relationships well enough. Moreover, this graph appropriately represents RT chains and multiple RTs by the same user.

In this paper, we demonstrated the effectiveness of the proposed TURank. As a result, we showed that TURank can extract users, who are not followed by many users, but his/her tweets are retweeted for many times, with higher position in the ranking, when all edge weights in the user-tweet schema graph are appropriately defined. Second, users whose tweets are not retweeted many times tend to be regarded relatively useless by examinees, even if they have a large number of followers. Our algorithm can lower such users' authority scores. Lastly, users, most of their tweets are for conversation, are evaluated as completely useless by examinees. Our algorithm can successfully lower their authority scores.

Authoritativeness is differentiated by topics. For instance, a user who knows a lot about tennis and less about baseball is authoritative in the area of tennis but not in the area of baseball. We did not consider this topic-wise authoritativeness. However, the proposed algorithm can incorporate this type of authoritativeness easily, because the user-tweet graph includes tweet nodes. Each tweet has its topic, so we can measure each user's topic-wise authoritativeness analyzing the topic of each tweet and delivering it to its authors through the posted edges.

Tweets are posted at a furious speed as a stream of data, and hence the user-tweet graph grows dynamically. To evaluate the authority scores in such environments, we should consider this dynamic property of Twitter in the future.

Acknowledgement

This research has been supported in part by the Grant-in-Aid for Scientific Research from MEXT (#21013004).

References

1. Twitter, http://twitter.com
2. Twitter API, http://apiwiki.twitter.com/Twitter-API-Documentation
3. Balmin, A., Hristidis, V., Papakonstantinou, Y.: Objectrank: Authority-based keyword search in databases. In: VLDB (2004)
4. Boyd, D., Golder, S., Lotan, G.: Tweet, tweet, retweet: Conversational aspects of retweeting on twitter. In: HICSS-43, January 6. IEEE, Kauai (2010)
5. Honeycutt, C., Herring, S.C.: Beyond microblogging: Conversation and collaboration in twitter. In: Proc. 42nd HICSS. IEEE Press, Los Alamitos (2009)
6. Huberman, B.A., Romero, D.M., Wu, F.: Social networks that matter: Twitter under the microscope. First Monday 14(1) (January 5, 2009)
7. Java, A., Song, X., Finn, T., Tseng, B.: Why we twitter: Understanding microblogging usage and communities. In: Joint 9th WEBKDD and 1st SNA-KDD Workshop, San Jose, CA (2007)
8. Kleinberg, J.: Authoritative Sources in a Hyperlinked Environment. In: Proc. of the 9th ACM SIAM Symposium on Discrete Algorithms (SODA 1998), pp. 668–677 (1998)
9. Leavitt, A., Burchard, E., Fisher, D., Gilbert, S.: The influentials: New approaches for analyzing influence on twitter. A Publication of the Web Ecology Project (2009)
10. Levenshtein, I.V.: Binary codes capable of correcting deletions, insertions, and reversals. Cybernetics and Control Theory 10(8), 707–710 (1966)
11. Page, L., Brin, S., Motwani, R., Winograd, T.: The pagerank citation ranking: Bringing order to the web. Technical report, Stanford Digital Library Technologies Project (1998)
12. Weng, J., Lim, E., Jiang, J., He, Q.: Twitterrank: Finding topic-sensitive influential twitterers. In: WSDM (2010)

Identifying and Ranking Possible Semantic and Common Usage Categories of Search Engine Queries

Reza Taghizadeh Hemayati[1], Weiyi Meng[1], and Clement Yu[2]

[1] Department of Computer Science, Binghamton university, Binghamton, NY 13902, USA
{hemayati,meng}@binghamton.edu
[2] Department of Computer Science, University of Illinois at Chicago, Chicago IL 60607, USA
yu@cs.uic.edu

Abstract. In this paper, we propose a method for identifying and ranking possible categories of any user query based on the meanings and common usages of the terms and phrases within the query. Our solution utilizes WordNet and Wikipedia to recognize phrases and to determine the basic meanings and usages of each term or phrase in a query. The categories are ranked based on their likelihood in capturing the query's intention. Experimental results show that our method can achieve high accuracy.

Keywords: Query categorization, Search engine, Wikipedia, WordNet.

1 Introduction

Current search engines often return too many useless results for users' queries. One way to tackle this problem is to group search results into multiple categories such that all results in the same category correspond to the same meaning of the query. This makes it much easier for users to identify useful results. Most current result clustering techniques are based on word-match similarity. Although a few techniques have used semantic similarity [4, 5], they have various weaknesses. E.g., current techniques do not explicitly and systematically take *usages* of query terms into consideration. Here a term's usage means its use beyond its common meanings in dictionaries. Not considering these usages would lower the quality of search result clustering.

In this paper, we propose a new solution to identify and rank all possible categories of any user query based on both the possible meanings and the possible usages of the terms and/or phrases in the query. Our solution has the following unique features. First, our method utilizes both WordNet and Wikipedia. Second, we apply rule-based techniques to merge the meanings of individual query terms/phrases to increase the clarity of produced categories. Third, we generate candidate categories for a query by considering all combinations that can be formed from different meanings/usages of the terms/phrases in the query. Finally, our method ranks the generated categories by taking into consideration both the importance of each term/phrase in the query and the strength of the relationships between the terms and phrases in the query.

For the rest of the paper, Section 2 reviews related work, Section 3 provides an overview of our approach, Sections 4, 5 and 6 present the main steps of our approach, Section 7 concludes the paper.

L. Chen, P. Triantafillou, and T. Suel (Eds.): WISE 2010, LNCS 6488, pp. 254–261, 2010.
© Springer-Verlag Berlin Heidelberg 2010

2 Related Work

Our work is related to word sense disambiguation (WSD). While we aim to find all possible senses of a term, WSD tries to find the most likely sense only. [15] suggested an unsupervised knowledge-based WSD algorithm. [12, 14] proposed several WSD algorithms based on semantic relatedness using WordNet. In [13], Wikipedia was used for WSD. [1] presented techniques for clustering WordNet word senses and they could be used to merge WordNet senses. In our work, we identify all possible meanings/usages of each query term/phrase using both WordNet/Wikipedia. We also perform sense merging for meanings/usages obtained from both WordNet/Wikipedia. Furthermore, we keep all meanings and usages and rank them.

Some researchers used web directories like Yahoo directory or ODP to categorize/classify user queries. Mapping user queries to hierarchical sequences of topic categories was studied in [8, 14]. [11] proposed to map user queries to categories using a user profile. Our method does not use existing category hierarchies. We provide a systematic method to generate all possible categories for each query.

Some category ranking methods were studied in [6] and the best method is based on the similarity between a query and the clusters/categories. In contrast, our category ranking method is based on the importance of each term/phrase in the query and the strength of the relationships between the terms and phrases in the query.

3 Method Overview

Our method has the following three main steps:

Alternative query generation. For each user query Q, this step generates a set of *alternative queries* (AQs). All AQs contain the same set of query terms in Q but different phrases. The key task is to identify different possible phrases comprised of the terms in Q. For convenience, we call both query terms and phrases as ***concepts***.

Definition category generation. In this paper, a *definition category* (DC) is a combination of *meanings* or *usages* derived from the concepts of an AQ. This step is to generate all possible DCs for each AQ.

Definition category ranking. This step ranks the DCs generated in Step 2.

4 Alternate Query (AQ) Generation

Identify Valid Phrases:
We use Wikipedia and WordNet to recognize phrases. From these sources we can find all dictionary phrases and most well known proper nouns. The order of terms inside each phrase is significant. In this work, terms that are not part of a dictionary phrase or proper names will be considered individually. Given an n-word query $Q(w_1, w_2, ..., w_n)$, we send Q to Wikipedia and WordNet to check whether the n words form a phrase. If they do, we stop; else we search for possible $(n-1)$-word phrases, i.e., $(w_1, ..., w_{n-1})$ and $(w_2, ..., w_n)$. This process is repeated until reaching the two-word candidates. We do not look for sub-phrases inside any valid phrase already found.

To determine whether a sequence of words p forms a valid phrase using Wikipedia, we submit p as a query to Wikipedia and consider the following three cases:

Case 1: A definition page entitled by p is returned. This indicates that p is a well known phrase so we consider p as a valid phrase.

Case 2: A page saying something like "p may refer to the following definitions" is returned. This means that p refers to different definitions. In this case, by following the link for each definition, we will be directed to the definition page for that phrase. If one of the linked pages is entitled by p, we consider p as a valid phrase.

Case 3: If none of the above cases is true, we don't consider p as a valid phrase.

To determine whether p is a valid phrase using WordNet, we submit p to WordNet and consider the following two cases:

Case 1: A page containing some definitions (called synsets in WordNet) is returned, i.e., p is defined in the dictionary. In this case, we consider p as a valid phrase.

Case 2: If no entry for p in WordNet is found, we don't consider p as a valid phrase.

Building the Set of Alternative Queries:
In this step we form all possible combinations consisting of the phrases and the terms not appearing in any of the phrases in a query Q. Each combination forms an AQ. The original query consisting of individual terms (i.e., no phrase is used) also forms an AQ. Each AQ must satisfy the following: (1) it contains all the terms in Q; (2) its phrases do not overlap; (3) each AQ has a unique set of phrases relative to other AQs.

5 Definition Category (DC) Generation

For each AQ, we break this step into three tasks: (1) *Meaning/Usage Generation*, (2) *Meaning/Usage Merging*, and (3) *DC Generation*. Task 1 uses WordNet/Wikipedia to identify all possible meanings/usages of each concept in AQ. Task 2 first processes each concept without considering other concepts in the AQ. The possible meanings or usages retrieved from Wikipedia/WordNet for each concept are compared and the similar ones are merged. To generate DCs for each AQ in Task 3, we use the meanings/usages (including the merged ones) for different concepts to form different combinations. Each combination contains one meaning/usage from each concept of the AQ and forms one DC. Since Task 3 is straightforward, we will focus on the first two tasks only. Query terms having no entries in either WordNet or Wikipedia are assumed to have *unknown* meanings/usages and will be included in each DC of Q.

Merging WordNet meanings: Our synset-merging algorithm consists of six merging rules, each of which gives one condition under which two synsets should be merged. These rules have been reported in [9] and will not be repeated here.

Merging Wikipedia meanings/usages: For *meaning/usage generation* from Wikipedia, we send a concept C (phrase/term) to Wikipedia and obtain a returned page P. We consider the following cases based on the type of page that is returned.

Case 1: Concept C has a unique meaning/usage, i.e., there is no disambiguation link at the top of P and there are no multiple meaning/usage entries. In this case, we represent the meaning/usage of this concept as a vector of terms with weights calculated based on *tf*idf*. The first n (say 20) terms appearing P are used to generate the vector because they usually include the main meaning/usage of a concept.

Case 2: Concept *C* has multiple meanings/usages. Two sub-cases: (1) There exists a disambiguation link at the top of P saying "For other uses, see *C* (disambiguation)." By following that link we can see the possible meanings/usages of the concept. (2) There is no disambiguation link on P and there are multiple definition entries on P. This case can be identified by noticing that P has "*C* can refer to the following …". In both subcases, for each meaning/usage of the concept, a short definition is provided by Wikipedia. Sometimes the definitions are organized by categories. Each definition and its corresponding category (if exists) are used to generate the vector for the meaning/usage. Higher weights are given to terms that appear in the category labels.

For each *C*, the vector representations of different meanings/usages generated above are compared to see if some vectors should be merged using the following rule:

Rule 1: If the vector representations of two meanings/usages of *C* have common noun words (excluding *C*), then merge them.

Merging WordNet-Wikipedia meanings/usages: In many cases the same meaning or usage may be retrieved from both Wikipedia and WordNet for a given concept even though their vector representations may be different. In these cases, we merge their vector representations for the meaning/usage using the following rules:

Rule 2: If a synonym or a hyponym of a synset *S* in WordNet appears in the definition *D* of the meaning/usage in Wikipedia, then merge *S* and *D* (i.e., they will be merged into one document; same below).

Rule 3: If a definition *D* in Wikipedia and a synset *S* in WordNet have common content words (not counting the concept itself), then merge *D* and *S*.

6 Ranking the Definition Categories

We use two major weight formulas to rank the DCs, one calculates the importance of each AQ and the other estimates the importance of each DC within each AQ.

<u>Alternative Query Weighting:</u>

Importance of each phrase in AQ: We consider three factors:

<u>Phrase frequency:</u> We give higher weights to phrases that appear in more search result records (SRRs), i.e., have higher frequencies.

<u>Well-knownness:</u> We give higher values of *well-knownness* to phrases that are better known and have less ambiguity. In Section 4 we introduced different cases when we send a phrase to Wikipedia and WordNet. In our experiments, the *well-knownness* of a phrase in Case 1 (WordNet or Wikipedia) = 1, that of a phrase in Case 2 (Wikipedia) = 0.5, and those in Case 3 (Wikipedia) and Case 2 (WordNet) = 0.

<u>Phrase length:</u> This is the number of words in a phrase. We use number of words in the query to normalize this weight. Longer phrases usually have less ambiguity.

To summarize, we use the formula below to compute the weight of a phrase *p*:

$$W_p(p) = (df(p)/max\_df + well\text{-}knownness(p) + |p|/|Q|) / 3$$

where $df(p)$ is the phrase frequency of *p* in SRRs, max_df is the largest phrase frequency in SRRs among all phrases that appear in the set of AQs, $|p|$ and $|Q|$ are the lengths of *p* and user query *Q*, respectively.

Importance of each term *t* in AQ: We consider two factors in assigning weight to *t*.

Co-occurrence: We utilize the co-occurrences of *t* with all phrases in AQ among the SRRs. If AQ does not have phrases then we consider the co-occurrences of *t* with all proper names in AQ. We compute the co-occurrence based weight for term *t* as follows: $cow(t) = \sum_{j=1}^{m} nco(t, p_j)$, where $nco(t, p)$ is the number of co-occurrences of term *t* with phrase *p* among the SRRs and *n* is the number of valid phrases in AQ.

Well-knownness: We give higher values of *well-knownness* to terms that are better known and have less ambiguity. This is similar to assigning a well-knownness value to a phrase as discussed earlier.

We use $W_t(t) = (cow(t)/max\_cow + well\text{-}knownness(t)) / 2$ to compute the weight of *t* in AQ, where *max_cow* is maximum co-occurrence weight for all terms in the SRRs.

Importance of each AQ: Let *AQ* be a given alternative query. We add the weights of the terms and phrases in *AQ* and normalize the sum by |Q| to obtain the final weight of *AQ* and denote it as $W_{AQ}(AQ)$. Note that the terms here are only those terms that appear in the original query but not in any of the phrases in *AQ*.

Definition Category (DC) Weighting:

For ranking DCs, we consider two types of weights: *meaning/usage weight* (MUW) and *relationship weight* (RW). Recall that each DC is a combination of concepts with each concept bound to a specific meaning/usage. The MUW of a meaning/usage *u* of a concept *C* reflects the likelihood that *u* is the correct meaning/usage for *C* without considering other concepts in the same DC. The RW is used to capture the impact of different relationships between concepts or their usages in the same DC on the likelihood of the DC to be the correct DC for the original query. By giving higher weights to those DCs whose concepts are more closely related, those DCs that make little sense (i.e., whose concepts are not related) will be ranked low.

Meaning/Usage Weighting: Let $u(C, DC)$ denote the specific meaning/usage of concept *C* in a definition category *DC*. In WordNet, each $u(C,DC)$ is represented by up to four components: synonyms, definition, example(s), and domain. In Wikipedia, each $u(C, DC)$ is represented by a set of words, which is called its *definition*. In summary, each $u(C, DC)$ has a *representation*. Let $muw(u(C, DC))$ denote the MUW of $u(C, DC)$. This weight indicates the relative importance of $u(C, DC)$ among all possible meanings/usages of *C*.

For a meaning/usage $u(C, DC)$ from Wikipedia, we compute its weight by:

$$muw(u(C, DC)) = \frac{N - Rank(u(C, DC)) + 1}{N}$$

where *N* is the number of meanings/usages for concept *C* from Wikipedia, $Rank(u(C,DC))$ is the rank (order) of the meaning/usage of *C* in *DC* in the list of these *N* meanings/usages.

For a meaning (i.e., synset) $u(C, DC)$ from WordNet, we use the ratio of the *frequenc-of-use* of $u(C, DC)$ (denoted as $f(u(C, DC))$) to the sum of the frequencies-of-use of all synsets of *C* (denoted as $F(C)$) to compute $muw(u(C, DC))$.

Also, concepts with an *unknown* meaning will be given zero weight.

Relationship Weighting: Consider a given *DC* and let $u_1 = u(C_1, DC)$ and $u_2 = u(C_2, DC)$ be the (possibly merged) meanings/usages of concepts C_1 and C_2 in *DC*, respectively. The following three cases are possible: Case 1: Both u_1 and u_2 are from WordNet; Case 2: u_1 is from WordNet and u_2 is from Wikipedia; and Case 3: Both u_1 and u_2 are from Wikipedia.

Furthermore, there are two types of basic relationships: **Type 1** is between two specific meanings/usages such as u_1 and u_2; and **Type 2** is between a specific meaning/usage (e.g., u_1) and a concept (e.g., C_2). Since Type 1 relationships are more specific than Type 2 relationships, we assign a higher weight to the former than to the latter. Currently, all Type 1 relationships have the same weight (denoted $rwt1$) and all Type 2 relationships also have the same weight (denoted $rwt2$). In our experiments, $rwt1 = 2$ and $rwt2 = 1$ are used. We have identified 6 **Type 1** relationships and 4 **Type 2** relationships. Due to space limitation, they cannot be included here but can be found in [10].

It is possible that more than one basic relationship exists between two specific meanings/usages u_1 and u_2 of any two concepts C_1 and C_2 in a given DC. Since satisfying more basic relationships usually indicates a stronger overall relationship between u_1 and u_2, we use the sum of all weights of the basic relationships that exist between u_1 and u_2 as the *overall relationship weight* (ORW) between u_1 and u_2, and denote it as $ORW(u_1, u_2)$.

We now discuss how to compute the weight for a *definition category DC* within the AQ from which the *DC* is derived. In general, a *DC* may contain multiple concepts $C_1, ..., C_k$, $k \bullet 1$, with each having a specific meaning/usage in the *DC*. If $k = 1$, there will be no basic relationship. If $k > 1$, we need to consider all pairs of meanings/usages. In summary, we use the formula below to compute the weight of defintion category *DC*:

$$W_{DC}(DC) = \begin{cases} muw(u_1), & if \quad k = 1 \\ \sum_{1 \le i < j \le k} ORW(u_i, u_j) * (muw(u_i) + muw(u_j)), & if \quad k > 1 \end{cases}$$

To normalize the weight between 0 and 1, we divide $W_{DC}(DC)$ by the maximum $W_{DC}(DC)$ among all DCs in the corresponding AQ.

Final Definition Category Ranking:
Recall that each DC comes from a particular alternative query (AQ). In our DC ranking model, the importance of the AQ from which a DC is derived can impact the final ranking of the DC. We consider the following two general ways to combine $W_{AQ}(AQ)$ and $W_{DC}(DC)$ to obtain a final ranking for all DCs:

AQ-DC Ordering: First, all AQs are ordered in non-ascending value of $W_{AQ}(AQ)$; next, within each AQ, its DCs are ordered in non-ascending value of $W_{DC}(DC)$. For AQs with the same $W_{AQ}(AQ)$, their DCs are considered together and are ordered in non-ascending value of $W_{DC}(DC)$.

Weighted-Sum: The final weight of *DC*, denoted as $FW_{DC}(DC)$, is computed by:
$$FW_{DC}(DC) = c_1 * W_{AQ}(AQ) + c_2 * W_{DC}(DC) \qquad (1)$$
where c_1 and c_2 are non-negative weight parameters satisfying $c_1 + c_2 = 1$, and each DC in (1) is derived from the AQ in (1). Finally the DCs are ordered in non-ascending value of $FW_{DC}(DC)$.

7 Evaluation

Our dataset contains 50 queries. Among these queries, the numbers of queries having 1, 2, 3 and 4 terms are 9, 28, 10 and 3, respectively. 37 of the 50 queries have at least one phrase and 13 have no phrase; 44 queries are ambiguous (at least one term or phrase has more than one meaning/usage) and 6 are not. We submit each test query to the Yahoo search engine to collect the top 50 search result records.

We report two evaluation tests here: (1) evaluate our method for generating DCs; and (2) evaluate our algorithm for ranking the DCs. In each test, the results generated by the proposed methods are compared against the ideal results (the golden standard) judged by human expert based on the intention and meaning of each test query.

DC Generation Accuracy:

A DC is considered to be correct if it does not contain unrelated meanings/usages and does not miss related meanings/usages from the concepts in this DC. Our results can be summarized as follows: Our algorithm has an average *precision, recall,* and *F1-measure* of 0.97, 0.96 and 0.96, respectively. Our results show that our algorithm can provide very accurate DCs for the test queries. When phrases in a query can be recognized, the average *precision, recall,* and *F1-measure are 0.98, 0.99* and *0.99,* showing the importance of recognizing phrase(s) in a query.

DC Ranking Accuracy:

To evaluate the quality of the ranking of DCs produced by our ranking method, the list of ranked DCs by our method $_1$ is compared against the ideal list generated by human expert $_2$. We use *scaled/normalized Spearman's footrule distance* (denoted by *NFr*) [7] to compare two lists of rankings of the same set S. It is the sum (over all elements i from S) of the absolute difference between the ranks of i in the two lists divided by some normalization value.

Table 1. NFr Performance of DC Ranking

	AQ-DC Ordering	$c_1=0$, $c_2=1$	$c_1=0.2$ $c_2=0.8$	$c_1=0.5$ $c_2=0.5$	$c_1=0.8$ $c_2=0.2$
no phrase*	0.01	0.01	0.01	0.01	0.01
phrase	0.05	0.096	0.078	0.055	0.051
1 term	0.02	0.02	0.02	0.02	0.02
2 terms	0.01	0.01	0.01	0.01	0.01
3 terms	0.01	0.138	0.094	0.025	0.01
4 terms	0.1	0.26	0.19	0.13	0.106
overall	0.02	0.056	0.04349	0.0258	0.022

* This row does not contain results for 1-term queries.

In Table 1, the second column shows the DC ranking result when *AQ-DC Ordering* is used, and columns 3-6 show the results when different weights of c_1 and c_2 (See Formula 1) are used. Note that $c_1 = 0$ means that the ranking is based on DC weight only while the impact of AQ is ignored. It can be observed that the overall accuracy improves when more emphasis is given to the AQ weight, i.e., when c_1 increases. *AQ-DC Ordering* gives the best result. We can see that the accuracy for 4-term queries is noticeably lower than those for other queries; this is due to the lack of enough basic relationships among some of the 4-term queries used. The lower accuracy for 4-term queries also has a negative impact on the accuracy for queries with phrases (see the 3rd row in Table 1). Note that for queries without phrases (e.g., 1-term queries), only one AQ will be generated and thus there is no difference in performance for different ranking methods.

8 Conclusion

In this paper we studied the problem of generating all possible categories of any user query and ranking these categories according to their match with the intention of the user. These categories can be used to categorize search result records returned from search engines in response to user queries. Our approach focuses on leveraging the meanings and usages of terms/phrases in each query and their relationships. We utilize both WordNet and Wikipedia to identify phrases in queries, the basic meanings/usages of each term/phrase, and the basic relationships between each pair of terms/phrases and their meanings/usages.

References

1. Agirre, E., Alfonseca, E., Lopez, O.: Approximating Hierarchy-based Similarity for Word-Net Nominal Synsets Using Topic Signatures. In: Global WordNet Conf. (2004)
2. Al-Kamha, R., Embley, D.W.: Grouping search-engine returned citations for person-name queries. In: WIDM 2004 (2004)
3. Beitzel, S.: Varying Approaches to Topical Web Query Classification. In: ACM SIGIR (2007)
4. de Luca, E., Nürnberger, A., von Guericke, O.: Ontology-Based Semantic Online Classification of Documents: Supporting Users in Searching the Web. University of Magdeburg, Germany (2004)
5. de Simone, T., Kazakov, D.: Using WordNet Similarity and Antonymy Relations to Aid Document Retrieval. Recent Advances in Natural Language Processing, RANLP (2005)
6. Demartini, G., Chirita, P.-A., Brunkhorst, I., Nejdl, W.: Ranking Categories for Web Search. In: Macdonald, C., Ounis, I., Plachouras, V., Ruthven, I., White, R.W. (eds.) ECIR 2008. LNCS, vol. 4956, pp. 564–569. Springer, Heidelberg (2008)
7. Diaconis, P., Graham, R.L.: Spearman's Footrule as a Measure of Disarray. Journal of the Royal Statistical Society, Series B (Methodological) 39, 262–268 (1977)
8. He, M., Cutler, M., Wu, K.: Categorizing Queries by Topic Directory. In: WAIM 2008 (2008)
9. Hemayati, R., Meng, W.: Semantic-Based Grouping of Search Engine Results. In: Fung, G. (ed.) Introduction to the Semantic Web: Concepts, Technologies and Applications. iConcept Press (2010)
10. Hemayati, R., Meng, W., Yu, C.: Identifying and Ranking Possible Semantic and Common Usage Categories of Search Engine Queries. Technical report (2010)
11. Liu, F., Yu, C., Meng, W.: Personalize Web Search by Mapping User Queries to Categories. In: ACM CIKM (2002)
12. Liu, S., Yu, C., Meng, W.: Word Sense Disambiguation in Queries. In: ACM CIKM (2005)
13. Mihalcea, R.: Using Wikipedia for Automatic Word Sense Disambiguation. In: NAACL HLT 2007, pp. 196–203 (2007)
14. Sanderson, M.: Word Sense Disambiguation and Information Retrieval. In: ACM SIGIR (1994)
15. Patwardhan, S., Banerjee, S., Pedersen, T.: Using Measures of Semantic Relatedness for Word Sense Disambiguation. In: Gelbukh, A. (ed.) CICLing 2003. LNCS, vol. 2588, pp. 241–257. Springer, Heidelberg (2003)

Best-Effort Refresh Strategies for Content-Based RSS Feed Aggregation*

Roxana Horincar, Bernd Amann, and Thierry Artières

LIP6 - University Pierre et Marie Curie, Paris, France
{roxana.horincar,bernd.amann,thierry.artieres}@lip6.fr

Abstract. During the past several years RSS-based content syndication has become a standard technique for efficiently and timely disseminating information on the web. From a data processing perspective RSS feeds are standard XML resources which are periodically refreshed by feed aggregators for generating continuous streams of items. In this article, we study the problem of information loss in the context of a content-based feed aggregation system and we propose a new best-effort refresh strategy for RSS feeds under limited bandwidth. This strategy is evaluated experimentally and compared to other state-of-the-art crawling strategies for web pages.

1 Introduction

RSS [14] (and Atom [7]) have become the de-facto standards for efficient information dissemination and web sites providing RSS-based services face an increasing usage of their bandwidth for delivering RSS information to their clients. Web syndication standards are mainly concerned with the document format. From the point of view of a web server, there is no distinction between a RSS feed and a regular web resource, both of them being fetched by using the standard http protocol. Thereby web syndication can be considered as a particular publish-subscribe application where subscribers must regularly refresh the web feed sources in order to access the latest updates.

In this article we are interested in the issue of reducing bandwidth usage in the context of RoSeS (Really Open Simple and Efficient Syndication) [13], a content-based RSS feed aggregation system which allows individual users to create personalized feeds by defining content-based aggregation queries on selected collections of RSS feeds. Compared to centralized server-based feed aggregators like GoogleReader [6] and YahooPipes [17], RoSeS advocates a distributed client-based aggregation infrastructure which allows user to install and personalize their local feed aggregator. RoSeS has been designed as a pull-based system. This choice is first based on the fact that the great majority of RSS services and applications use the standard pull-based web protocol http where subscribers are the ones that regularly contact the feed sources to retrieve new postings. Alternative solutions based on push-based or hybrid pull/push-based protocols are presented in [1], [16].

---

* The authors acknowledge the support of the French Agence Nationale de la Recherche (ANR), under grant ROSES (ANR-07-MDCO-011) "Really Open, Simple and Efficient Syndication".

L. Chen, P. Triantafillou, and T. Suel (Eds.): WISE 2010, LNCS 6488, pp. 262–270, 2010.

Whereas RoSeS allows to formulate complex queries with joins and windows, in this article we consider only stateless continuous queries computing a filtered union of source feeds. We study the problem of information loss and we propose a new best-effort refresh strategy for RSS feeds under limited bandwidth.

In particular, we make the following contributions:

- a declarative feed aggregation model with a precise semantics for feeds and aggregation queries (Section 2),
- appropriate metrics (feed completeness) designed to measure the quality of newly produced aggregation feeds (Section 2),
- a best-effort refresh strategy to retrieve new postings from different feed sources which maintains an optimal level of aggregation quality (Section 3) and
- an experimental evaluation of our strategy (Section 5).

Additionally, we review related work in Section 4 and conclude in Section 6.

2   Content-Based Feed Aggregation and Feed Completeness

Feed aggregation queries are defined in a declarative RSS query language. The result of each query is a new feed that can be accessed locally and, if necessary, be published for other users. For example, a user wants to create a feed with news about volcano eruptions in Iceland fetched from "The Guardian" and images published by "The Big Picture" on the same topic. This can easily be translated into the following aggregation query which applies a simple disjunctive filtering condition on the union of the corresponding feeds:

```
CREATE FEED IcelandVolcanoEruption AS
RETURN "http://feeds.guardian.co.uk/theguardian/rss" |
       "http://www.boston.com/bigpicture/index.xml"
WHERE ITEM CONTAINS "iceland" OR "volcano" OR "eruption"
```

Aggregation Queries: Let $S = \{s_1, ..., s_m\}$ be a set of RSS feeds. An *aggregator* node n is defined by a set of *aggregation queries* $Q(n) = \{q_1, ...q_k\}$ on S. A query q applied on a source feed introduces a selectivity factor $sel(q) \in [0, 1]$: $sel(q) = 1$ if the aggregator query keeps all items published by the input feeds (simple union) and $sel(q) = 0$ if it filters out all items (empty result). The aggregator node n generates for each query $q \in Q(n)$ a query feed $f(q, n)$ and we will denote by $feeds(n)$ the set of feeds generated by n. We assume that S and $feeds(n)$ are disjoint.

Windows and Streams: There are mainly two different interpretations of a feed produced by some query q at some time instant t. In the first case, which corresponds to the standard document-based interpretation of RSS feeds, a feed is represented by a limited number of items available in a XML document at some time instant t. We will call this document a *publication window* of size W_s and denote it by $A(f, t)$. The publication process guarantees that for all time instants t and t' where $t < t'$ all items in $A(f, t') - A(f, t)$ have been published after all items in $A(f, t)$. This observation allows us to define the second interpretation which considers a feed as a continuous stream of items published until time instant t. This stream is called the *publication stream* of f and denoted $F(f, t)$.

Feed Completeness: There are two main reasons for losing items during the aggregation process that influence $F(f, t)$. On the back-end, the crawler might miss some items that have been published by the sources and never been read by the aggregator. This read loss depends on the publishing frequency of the sources, the available bandwidth b and the polling strategy. On the front-end, the query processing and publication threads might receive more items than they can consume. This is a well known data stream processing problem which can be solved in different ways (approximation, load shedding). This issue is considered to be outside the scope of this article.

The feed completeness of a query feed f at some time instant t is denoted $\mathcal{C}_F(f, t) = |F(f, t)|/|I(f, t)|$ and estimates the relative read loss of the aggregation process generating the feed by comparing the number of published items in $F(f, t)$ with the number of items in $I(f, t)$, where $I(f, t)$ represents an ideal stream generated by the aggregator with unlimited bandwidth and refresh frequency.

3   A Best-Effort Refresh Strategy for Feed Completeness

Window Divergence: We call *divergence* the degree to which a source feed differs from the feed published by an aggregator subscribed to that source. In our case, we define *window divergence* Div_A as the number of new items generated by the source feed s during time period $[t_0, t]$, available at time instant t (computed by $new(A(s, t))$) and that are relevant to query q, where t_0 denotes the time of the last refresh of source s: $Div_A(q, s, t) = |q(new(A(s, t)))|$. Observe that the divergence value depends on the publication frequency of s and $sel(q)$, the selectivity of q.

Since the number of relevant items present in the source publication window can decrease if they are replaced by new irrelevant ones, Div_A is a *non monotonic* function. This is illustrated in figure 1 representing the evolution of Div_A compared to the total number of new items published since the last refresh by some "ideal" source s publishing on average λ items per cycle with a uniform distribution of relevant items.

Both curves start at time t_0 when s is refreshed and evolve identically (*monotonically increasing*) until some time t' near $t = t_0 + W_s/\lambda$ where the number of new items reaches the capacity of the source publication window W_s. We will say that source s has *saturated* at time t'. At this moment, both curve values correspond to some value near $sel(q) \cdot W_s$ and from this moment on, window divergence Div_A can decrease since new irrelevant items might replace relevant items that have not been read yet (read loss).

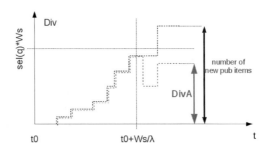

Fig. 1. Window divergence

Best-Effort Refresh Strategy for Unsaturated Sources: We adapt the web page approach presented in [9], where the authors define the utility gained by refreshing a web page based on a monotonic divergence function. In our case, we use the window divergence Div_A defined in Section 3, guaranteed to be monotonic for *unsaturated sources* to define $Uti_A(q_i, s_j, t)$, the utility of refreshing source s_j for the quality of query feed $f(q_i)$ at time instant t:

$$Uti_A(q_i, s_j, t) = (t - t_k) \cdot Div_A(q_i, s_j, t) - \int_{t_k}^{t} Div_A(q_i, s_j, x)dx \qquad (1)$$

Utility function $Uti_A(q_i, s_j, t)$ aggregates all divergence scores over the period $[t_k, t]$ where t_k is the last time when source s_j was refreshed ($Div(q_i, s_j, t_k) = Uti(q_i, s_j, t_k) = 0$).

We then define the following refresh strategy for *unsaturated sources*. This strategy is *best-effort* in the sense of [9], since there exists no other strategy which obtains a better result within the same cost:

Let τ be any positive constant and Uti_A be the utility function defined in equation 1. Then, at each time t and for each query q_i, refresh those source feeds s_j that have:

$$Uti_A(q_i, s_j, t) \geq \tau \qquad (2)$$

τ is a refresh threshold which controls the global polling frequency of sources with respect to their divergence scores. For a high value of τ, the aggregator will refresh its sources at a rather low rate and for low values of τ, it will refresh at higher rates. The actual value of τ depends on the maximum number of sources that the aggregator is allowed to refresh at time t and on the update frequencies of the sources. τ converges to a non-negative value in case all these remain constant (see our experiments in Section 5).

Best-Effort Refresh Strategy for Saturated and Unsaturated Sources: Since window divergence monotonicity is guaranteed only for unsaturated sources, all sources which already have reached their saturation point must be handled separately. Therefore we define a two-step refresh strategy as described in the algorithm 1 conceived for the special case of RSS feeds. The first step will refresh the top-b *saturated* sources that have maximum positive window divergence Div_A. This criteria ensures maximum feed completeness for saturated sources. The following step handles the rest of the sources (the *unsaturated* ones) and refreshes only those that meet the condition in equation 2. Since the refresh threshold τ in equation 2 corresponds to an average limited bandwidth of b refreshed sources, threshold τ must be adjusted to the fraction of the bandwidth available after the first step.

Estimating the Refresh Threshold τ: Since in the context of content-based feed aggregation the update frequency of items relevant to query q and the available resources b of the node might change, there is no single value for τ that is optimal all the time. If the system parameters change, the algorithm dynamically adjusts the value of τ so it converges to a new best value.

We consider a τ value specific to each aggregator n that can be set to any random initial value. We suppose that if n has selected b' sources to refresh, it will (i) increase

Algorithm 1. Refresh strategy in 2 steps

Input: $b, \tau, Div_A(n, s_j, t), Uti_A(n, s_j, t)$
  $R(n, t) := \emptyset, Sat(n, t) := \emptyset, b_{sat} = 0$
  {**First step: refresh up to** b **saturated sources ordered by divergence**}
  for all $s_j \in S(n)$ **do**
    if s_j is at saturation point and $Div_A(n, s_j, t) > 0$ **then**
      $Sat(n, t) := Sat(n, t) \cup \{s_j\}$

  $R(n, t) := top_b(Sat(n, t))$
  {**Second step: refresh unsaturated sources using utility threshold**}
  $b_{sat} = size(R(n, t))$
  if $b_{sat} \leq b$ **then**
    for all $s_j \in S(n)$ **do**
      if $Uti_A(n, s_j, t) \geq \frac{b}{b - b_{sat}} \cdot \tau$ **then**
        $R(n, t) := R(n, t) \cup \{s_j\}$

  estimate new threshold τ
  refresh sources in $R(n, t)$

τ with a factor β if more resources than available were used ($b' > b$) and it will (ii) decrease it with α if the available bandwidth has been exploited below some percentage p ($b' < p\%b$). α controls how aggressively n gives priority for more feed sources to be refreshed and β reflects how quickly it slows down the sources refresh rate. τ doesn't change otherwise ($p\%b \leq b' \leq b$). During experiments, we made τ vary with 5% since big variations may prevent τ from converging.

4 Related Work

The problem of efficient polling policies for web resources was largely studied in the context of web page crawling [2,3,4,5,8,9,10,11]. For example, Garcia-Molina and Cho [3,2,4] develop synchronization strategies to optimize web page freshness and age. In our paper, we study a measure more appropriate to our context: feed completeness. Another approach proposed in [10] introduces more elaborated quality measures: web page freshness perceived by users. This metric is optimized based on query load and click-through data. [11] proposes CAM (Continuous Adaptive Monitoring), a resource allocation algorithm which allocates limited monitoring resources across pages so as to minimize the information loss compared to an ideal monitoring algorithm which can monitor every change of the page. Whereas this setting is similar to ours, we explore the problem in a different direction by adapting existing best-effort refresh strategies for web pages.

Reference [9] proposes different types of divergence metrics, lag being similar to our window divergence. The authors introduce a best-effort synchronization scheduling policy that exploits cooperation between data sources and the cache in a push architecture, where data sources actively notify the cache of any changes. In our case, we assume a pull context with passive data sources that are periodically contacted by aggregator nodes. A best-effort strategy which focuses on the lifetime of content fragments that appear and disappear from the web pages over time (information longevity) is presented in [8].

Closer to our problem setting, reference [15] proposes a pull based aggregator architecture that monitors RSS data sources and quickly retrieves new postings, minimizing the delay between the appearance of a posting at the source and its retrieval by the aggregator. They propose a periodic inhomogeneous Poisson process to model the generation of RSS feed items. Whereas their approach is efficient for refreshing RSS feeds, we believe that content-based filtering destroys periodicity and we follow the idea of [9] which dynamically adjusts τ to the real workload.

5   Experiments

Parameter Settings: The simulation environment used to perform our experiments is PeerSim [12], a Java-based engine for testing peer-to-peer protocols. We constructed a cycle-based environment with an aggregator node that applies a query q computing a filtered union on a set of feed sources that publish following a Poisson process of rate parameter λ_i. The values chosen for the input parameters (summarized in table 1) are representative for the obtained experimental results.

Table 1. Input parameter values

No. of aggregator nodes	1	No. of cycles in a simulation	100
No. of source feeds	100	λ_i uniformly distributed in	[0, 6.5]
W_s	10	b	(0,100]
α	0.95	β	1.05
$p\%$	90%		

Comparing Strategies: In the following sections we compare the optimal refresh strategy presented in Section 3 (that we will identify as the $2steps$ strategy) with five other strategies. The considered strategies are all offline, based on a-priori knowledge of the divergence function and the average publication frequency λ_i of source feed s_i:

- $onlySat$ represents the first step of our optimal $2steps$ refresh strategy, taken separately. It refreshes only those b feed sources that have maximum window divergence Div_A among the saturated sources.
- $onlyTau$ represents the second step of the $2steps$ refresh strategy, considered separately. It refreshes only those feed sources s_i for which $Uti_A(n, s_i, t) \geq \tau$.
- $topKUtility$ refreshes b sources that have maximum utility Uti_A.
- $uniform$ refreshes every source feed at the same frequency b/m, where m represents the total number of feed sources.

Strategy Effectiveness: Besides measuring the feed completeness $\mathcal{C}_F$ introduced in Section 2, we also consider the *window freshness* $\mathcal{F}_W$, that compares the number of relevant items available both on the source feeds and on the query feed $f(q)$ in the same time. $\mathcal{F}_W$ is defined as a weighted average of the single-sourced window freshness

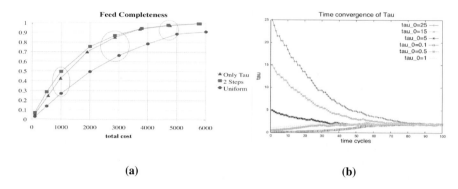

Fig. 2. Feed Completeness and Tau Convergence

$\mathcal{F}_W(f/s,t) = |A(f,t) \cap q(A(s,t))|/|q(A(s,t))|$, where $s \in S(q)$. Both $\mathcal{C}_F$ and $\mathcal{F}_W$ estimate the quality of the used refresh strategy in terms of item loss, where the former represents a refresh quality metric for long-term feed aggregation tasks (archiving) and the latter reflects the refresh quality for short-term feed aggregation tasks (news reader).

In figure 2a we plotted the feed completeness obtained for different values of b (on the abscissa we plotted the total cost, measured as the total number of refresh requests done during a simulation). In order to make the figure more clear, we present only the $2steps$, $onlyTau$ and $uniform$ strategies. The results for the complete set of strategies are presented in table 2, where we give the experimental values both for feed completeness $\mathcal{C}_F$ and window freshness $\mathcal{F}_W$, for three representative values of b. All the presented results are obtained after a warm-up period when the value of τ has converged.

When the aggregator is restricted to refresh few sources (for relatively small values of $b \in (0,22]$), very many sources that were not refreshed in time become quickly saturated. And thus, giving priority to refreshing the saturated sources (as the optimal $2steps$ or $onlySat$ strategies do) generates significantly better feed completeness values than all the other strategies, as shown in table 2 for $b = 10$. However, refreshing the saturated sources does not guarantee fetching many relevant items and thus, the strategies that refresh based on their utility ($onlyTau$, $topKUtility$) obtain better results for the window freshness.

Table 2. Feed completeness and Window freshness

	$b = 10$			$b = 30$			$b = 50$		
	C_F	$\mathcal{F}_W$	$Cost$	C_F	$\mathcal{F}_W$	$Cost$	C_F	$\mathcal{F}_W$	$Cost$
2steps	**0.4954**	0.3671	1000	**0.8691**	**0.7279**	2854	**0.9781**	**0.8762**	4700
onlyTau	0.4252	**0.4197**	982	0.8524	0.7048	2860	0.9759	0.8731	4762
topKUtility	0.4160	**0.4220**	1000	**0.8671**	**0.7200**	3000	0.9793	0.8868	5000
onlySat	**0.4934**	0.3671	1000	0.8082	0.6590	2516	0.8091	0.6625	2524
uniform	0.2682	0.3620	1000	0.6625	0.6398	3000	0.8871	0.7863	5000

If the aggregator is allowed to refresh more sources (b takes values in $[22, 40]$), there are fewer sources that reach the saturation point. Since the optimal $2steps$ strategy takes advantage both of refreshing saturated sources (as $onlySat$) and sources with best utility (as $onlyTau$), it obtains the best results for a lower cost, as presented in table 2 for $b = 30$.

For more permissive constraints ($b > 40$), the optimal $2steps$ strategy manages to refresh all the sources based on their utility before they reach saturation, just as the $onlyTau$ and $topKUtility$ strategies do, but at a lower cost.

Strategy Robustness: Using a self adaptive threshold-setting algorithm for finding the optimal value of τ in the second step of our $2steps$ refresh strategy makes our policy highly adjustable to the changes that may occur in real world. As discussed in Section 3, not only the average source publication frequency λ may vary in time, but also the update frequency of items relevant to some query q. Since τ depends on the available bandwidth and the divergence rates of the news feeds, there is no single best value that works well all the time. In order to show the high adaptability of the τ threshold (and consequently, of our optimal $2steps$ refresh policy), we show a study on the convergence of τ in figure 2b.

We plotted the evolution of τ for the optimal $2steps$ refresh strategy when the average number of refresh sources by time cycle is set to $b = 30$, for different τ initial values. From figure 2b, we can see that τ always converges to the optimal value. This convergence assures the robustness of our refresh strategy to different changes that may occur in real world.

6   Conclusion

In this article we have studied the problem of refreshing RSS feeds in a content-based feed aggregation context. We proposed a declarative feed aggregation model with precise semantics for feeds and aggregation queries. We defined a quality metric (feed completeness) and a two-step polling strategy optimizing this metric in time. This strategy has been evaluated experimentally by comparing its behavior to different other strategies on a synthetically generated setting.

As future work, we want to test our refresh strategy in a real world setting with real data sources and online estimation functions for the source publication frequencies and the divergence values. Besides that, we want to adapt our feed aggregation approach to a distributed context where users create personalized feeds aggregating feeds generated by other users.

References

1. Acharya, S., Franklin, M., Zdonik, S.: Balancing push and pull for data broadcast. In: SIGMOD 1997: Proceedings of the 1997 ACM SIGMOD international conference on Management of data, Tucson, Arizona, United States, pp. 183–194. ACM, New York (1997)
2. Cho, J., Garcia-Molina, H.: Synchronizing a database to improve freshness. SIGMOD Rec. 29(2), 117–128 (2000)

3. Cho, J., Garcia-Molina, H.: Effective page refresh policies for web crawlers. ACM Trans. Database Syst. 28(4), 390–426 (2003)
4. Cho, J., Garcia-Molina, H.: Estimating frequency of change. ACM Trans. Interent Techonol. 3(3), 256–290 (2003)
5. Cho, J., Ntoulas, A.: Effective change detection using sampling. In: VLDB 2002: Proceedings of the 28th international conference on Very Large Data Bases, Hong Kong, China, pp. 514–525. ACM, New York (2002)
6. google_reader, http://www.google.com/reader
7. Network Working Group: The atom publishing protocol, http://tools.ietf.org/html/rfc5023
8. Olston, C., Pandey, S.: Recrawl scheduling based on information longevity. In: WWW 2008: Proceeding of the 17th international conference on World Wide Web, Beijing, China, pp. 437–446. ACM, New York (2008)
9. Olston, C., Widom, J.: Best-effort cache synchronization with source cooperation. In: SIGMOD 2002: Proceedings of the 2002 ACM SIGMOD international conference on Management of data, Madison, Wisconsin, pp. 73–84. ACM, New York (2002)
10. Pandey, S., Olston, C.: User-centric Web Crawling. In: WWW 2005: Proceedings of the 14th international conference on World Wide Web, Chiba, Japan, pp. 401–411. ACM, New York (2005)
11. Pandey, S., Ramamritham, K., Chakrabarti, S.: Monitoring the dynamic web to respond to continuous queries. In: WWW 2003: Proceedings of the 12th international conference on World Wide Web, Budapest, Hungary, pp. 659–668. ACM, New York (2003)
12. peersim, http://peersim.sourceforge.net/
13. roses, http://www-bd.lip6.fr/roses/doku.php
14. rss, http://www.rssboard.org/
15. Sia, K.C., Cho, J., Cho, H.-K.: Efficient Monitoring Algorithm for Fast News Alerts. IEEE Trans. on Knowl. and Data Eng. 19(7), 950–961 (2007)
16. Silberstein, A., Terrace, J., Cooper, B.F., Ramakrishnan, R.: Feeding frenzy: selectively materializing users' event feeds. In: SIGMOD 2010: Proceedings of the 2010 international conference on Management of data, Indianapolis, Indiana, USA, pp. 831–842. ACM, New York (2010)
17. yahoo_pipes, http://pipes.yahoo.com/pipes/

Mashup-Aware Corporate Portals

Sandy Pérez and Oscar Díaz

ONEKIN Research Group, University of the Basque Country,
San Sebastián, Spain
{sandy.perez,oscar.diaz}@ehu.es

Abstract. Unlike other Web applications, corporate portals reckon to provide an integration space for corporate services. Mashups contribute to this goal by bringing a relevant customization technique whereby portal users can supplement portal services with their own data needs. The challenge is to find a balance between portal reliability and mashup freedom. Our approach is to split responsibilities between service providers and portal users. Providers decide on how services can be mashuped, portal users determine the supplemented content, and finally, the portal engine mediates between the two. This permits portal services to be reliably customized through user mashups. The approach is realized for *Liferay* as the portal engine, portlets as the realization of portal services, and XBL as the integration technology.

Keywords: portal, portlet, mashup, customization, presentation integration.

1 Introduction

Traditional mashuping distinguishes two scenarios w.r.t. source applications. In the first scenario, the mashup is a separate application from source applications (e.g. *Yahoo! Pipes*). In the second scenario, the mashup is an enhancement upon the source application (e.g. *MashMaker [4], MARGMASH* [3]). This normally requires the installation of a browser plugin for the mashup to be woven with the application markup. In both cases, source applications ignore they are being subject to mashup ("unaware mashuping"). Corporate portals are Web applications. Hence, unaware mashuping can also be applied to portals. However, being "unaware", this approach cannot capitalize on portal utilities (e.g. single-sign on, access control, customization, etc.). You can use *MashMaker* on portals but the role of the portal is totally passive. However, and unlike other Web applications, portals reckon to provide an integration space for corporate services. By mashuping at the back of the portal, mashuping misses the opportunity to benefit from this integration space. This paper introduces a collaborative approach where portals actively support mashuping ("mashup-aware portals").

This departs from traditional scenarios. First, and unlike *Yahoo! Pipes*-like approaches, the mashup is offered without leaving the portal. Second, and unlike *Mash-Maker*-like approaches, now the portal takes an active role on facilitating mashups on portal services. The implications are three-fold: (1) no additional plugin is necessary since mashup weaving is already engineered into the portal, (2) the portal "guides" users throughout the mashup process, and (3), the portal provides the context for mashups to be seamlessly integrated into portal services.

L. Chen, P. Triantafillou, and T. Suel (Eds.): WISE 2010, LNCS 6488, pp. 271–278, 2010.

From the portal perspective, mashuping becomes an additional approach to cus-tomize portal offerings. Customization helps portal services (e.g. booking flight tickets) to be adapted to the users' roles. Both content and services can be adapted to the current user (e.g. *flightBooking* is only available for senior engineers). Personalization goes one step further by permitting users themselves to set some configuration options (e.g. the *destinationAirport* parameter is set to *"New York"* by *John Douglas*). This work intro-duces mashups as an additional personalization mechanism. For instance, *John Douglas* is very apprehensive to weather conditions so that he looks at the weather forecast be-fore setting the trip date. This just applies to *Mr. Douglas*, and it is not contemplated by *flightBooking*. Hence, *Mr. Douglas* is forced to move outside the portal realm to sat-isfy this data need (e.g. through a *weatherForecast* widget), and to bridge himself the passing of data from the portal to the widget. By contrast, mashup-aware portals would assist *Mr. Douglas* in weaving the *weatherForecast* widget to the *flightBooking* service.

The paper addresses the challenges that this approach poses for both the portal and the providers of portal services. The approach is borne out for *Liferay*[1] as the portal engine, portlets as the service realization technology [2], and *XBL* as the binding tech-nology [8]. The paper begins by introducing the challenges.

2   Introducing the Challenges

We regard portal mashups as enhancements provided *by* users but accomplished *through* the portal. This introduces a distinction among the tasks offered through the portal: **main tasks** (e.g. *flightBooking*) and **mashup tasks** (e.g. *weatherForecast*).

Main tasks are set by the portal administrator. They support the functional backbone of the portal. Portlets are the standardized approach for supporting these presentation-oriented services. Portlets strive to play at the front end the same role that Web ser-vices currently enjoy at the back end, namely, enablers of application assembly through reusable services. Portlets are user-facing (i.e. return markup fragments rather than data-oriented XML) and multi-step (i.e. they encapsulate a chain of steps rather than a one-shot delivering) [2]. The latter is worth noticing: service fulfilment is rarely achieved through a single shot but a *set* of steps are needed (e.g. date setting, site booking, en-tering billing data and so on). Each step is supported through a markup (a.k.a. a portlet fragment). Backed by the WSRP and JSR286 standards, portlets are currently supported by major portal vendors.

On the other hand, **mashup tasks** are subordinated to main tasks. Normally, they consult and provide additional data rather than updating the service state. Unlike main tasks, mashup tasks are set by portal users. We choose *widgets* as the realization technol-ogy for mashup tasks. Widgets are full-fledged client-side applications that are authored using Web standards and packaged for distribution [7].

Traditionally, portal tasks (whether being supported through portlets or widgets) are readily presented as you enter the portal page. This is based on the premise that tasks are all equal. In our scenario, this premise however, is not longer valid: mashup tasks should not be readily available but only when they are needed. Otherwise you will end up with cluttered portal pages full of widgets with no obvious purpose. The purpose of

---

[1] http://www.liferay.com/

Fig. 1. Rendering of two portlets (i.e. *flightBooking* and *librarySearch*), and one widget (i.e. *weatherForecast*). The widget is user-specific and supports the selection of the flight date. Once *flightBooking* moves to the next stage, *weatherForecast* is not longer available.

a mashup task should be sought in the context of a main task. In our previous example, *weatherForecast* only makes sense when *flightBooking* reaches the point of prompting for the trip date. Once *flightBooking* moves to the next stage, *weatherForecast* is of no use. Additionally, the *weatherForecast* widget should be located the closest to the entry form for the destination airport. However, traditional side-by-side composition would assign *flightBooking* and *weatherForecast* distinct (although co-located) cells. It is then possible for the departure airport to appear at the bottom of the cell while forecast data is rendered upper on the page.

Mashup-aware portals permit main tasks to inlay mashup tasks (rather than being co-located). Figures 1 provides an example. First, *flightBooking* and *library* are presented side-by-side as realization of main portal tasks. By contrast, mashup tasks, such as *weatherForecast*, can now be inlayed within the rendering space of *flightBooking*.

This scenario introduces three actors, namely

- **the portal user**. *Once the portal is deployed*, users can require additional data to better accomplish main tasks. Akin to the DIY approach, it is up to users to find the appropriate sources for mashup tasks (e.g. widgets). *Requirement: hot-deployment of mashup tasks.*
- **the portlet provider**. Portlet providers ensure the quality of main tasks (data integrity, service throughput, etc.). Now, they are also responsible to decide how the portlet can be mashuped. Portlet designers should foresee placeholders to inlay mashup tasks on accomplishing the portlet task (e.g. on selecting the destination airport). *Requirement: placeholder specification for main tasks.*
- **the portal** (i.e. the portlet consumer). Akin to the portal-as-an-integration-space, portals should offer weaving mechanisms that permit data to seamlessly flow between main tasks and mashup tasks. *WeatherForecast* provides an example: its parameter *"location"* is to be obtained from the *flightBooking* destination airport, so that every time the airport is changed, the inlayed widget is refreshed. Requirement: *data-flow portal facilities.*

3   Realizing the Portlet Provider Perspective

Portlets support well-focus functionality: booking a flight seat, handling a bank transfer, and so on. On the other hand, portlets are born to be reused. The same portlet can

274   S. Pérez and O. Díaz

be offered through different portals. As in any other component technology, this implies an
attempt to foresee requirements
for distinct potential consumers.
However, traditional component
development already advises that
*"no design can provide informa-
tion for every situation, and no
designer can include personalized
information for every user"* [5].
This is when mashups come into
play. Mashups permit portal users
to complement portlet functionality. It is most important to notice
that we *do not mashup the portlet as such but the offering of this
portlet through this portal.* The
very same portlet can have a different mashup when offered through a distinct portal.

```
<div id="search_form_view" class="view">
  <div id="top-mashcell" class="mashcell"></div>
  <form method="post" action="<portlet:actionURL>...</portlet:actionURL>">
    <table width="100%" border="0" cellpadding="0" cellspacing="0"><tbody>
      <tr><!-- ROUNDTRIP FORM FIELD --></tr>
      <tr><!-- ORIGIN FORM FIELD --></tr>
      <tr>
        <td width="50%" align="right">Destination:</td>
        <td width="50%" align="left">
          <select name="destination">
            <c:forEach var="airport" items="${airports}">
              <option value="${airport.IATACode}">${airport.name}</option>
            </c:forEach>
          </select>
        </td>
      </tr>
      <!-- MORE FORM MARKUP -->
    </tbody></table>
  </form>
  <div id="bottom-mashcell" class="mashcell"></div>
</div>
```

Fig. 2. The *"search_form_view"* fragment: two *mash-cells* pingpoint mashup placeholders

A similar situation arises in *XML Schema*. Schema standards are set by international bodies. Since the specificities of each sector/country can be difficult or inappropriate to be directly captured by the general schema, extension points are defined for consumers to adapt the schema to their own contexts. Likewise, portlet designers need to find a balance when supporting the portlet functionality, i.e. the portlet should be general enough to be appropriate for a large set of consumers while including "mashup placeholders" to cater for mashup specifics (hereafter referred to as *"mashcells"*). Their role is similar to the *<any>* element in XML schemas.

Mashcells do not have any presentation impact other than pinpointing where portlet markups can be extended. Implementation wise, mashcells are supported as the *CSS* class *"mashcell"*. Figure 2 shows a snippet for the *"search_form_view"* fragment for *flightBooking*. The designer decides to provide two mashcells right before and after the entry form: *"top-mashcell"* and *"bottom-mashcell"*, respectively. Portal tools can then light up these mashcells, pinpointing mashup placeholders for users to fill up with desired widgets. This moves us to the portal perspective.

4   Realizing the Portal Perspective

Now, portlets become "the canvas" where widgets can be placed. The design space is then set on a portlet basis. Broadly, this space comprises three dimensions (see Fig. 3):

- *what* to include (i.e. widget selection). Values stand for the widgets available at the portal. Permission can be granted for portal users to add their own widgets, hence personalizing their own data purveyors,
- *where* is to be included. Values correspond to mashcells for the portlet at hand,
- *how* is to be included. Values stand for potential "data feeds" to be obtained from the portlet markup.

Figure 3 shows the design space for the *flightBooking* portlet. This space frames the setting for deciding *what-where* (hereafter, referred to as *"composition coordinate"*), and *where-how* (referred to as *"orchestration coordinate"*). For the sample problem, the *weatherForecast* gadget[2] is to be inlayed into the *"top-mashcell"* (composition coordinate). Additionally, this gadget is to be fed after the *destination* airport entry form (orchestration coordinate). Next paragraphs introduce two requirements to be fulfilled by

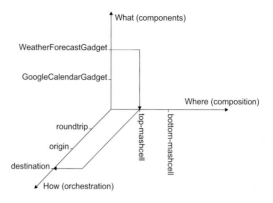

Fig. 3. Design space for the *flighBooking* portlet

the implementation technology: dynamic binding and presentation-based orchestration.

Dynamic binding: coordinates can be set once the portal is already deployed. Behaving as a kind of portal preferences for decision taken, widgets and the associated coordinates can be added by portal users at any time. This hot-deployment of widgets implies the ability to dynamically define coordinates.

Presentation-centred orchestration: widget parameters can be obtained from portlet markup. Being both portlets and widgets presentation components, it is just natural to use events for this purpose[3]. The fact of *HTML* being the standard for delivering rendering through the Web makes *HTML* the *lingua franca* for portlet-widget communication. That is, the *DOM API* is used [6]. This *API* provides a set of low-level UI events (e.g. load, mouse over, etc.) to operate on low-level *UI* components (e.g. menus, button, etc.). Therefore, we rely on *DOM* events to specify the orchestration model. This forces composition to take place at the client[4]. Based on these two requirements (i.e. dynamic binding, and client-based, *DOM* event-based orchestration), *XML Binding Language (XBL)* [8] is selected.

XBL is a *W3C* candidate recommendation for describing bindings that can be attached to elements in other documents. It is currently supported by Firefox. The element that the binding is attached to, called the *bound element*, acquires the new behaviour specified by the binding [8]. An XBL document has *<bindings>* as a root. The *<bindings>* tag contains a collection of *<binding>* elements that describe element behaviour. Each *<binding>* can include the *<content>* tag, which is used to describe anonymous content that can be inserted around a *bound element*, an *<implementation>* tag that captures new properties and methods, and a *<handlers>* tag to account for event handlers on the *bound element*. Next, bindings can be attached to elements through *CSS*

---

[2] Gadgets are a concrete technology for widgets.

[3] Portlets already have a standard that permits portlets to synchronize their lifecycles through events [1]. But this leaves widgets out.

[4] Actually, portlets produce the markup at the server while widgets are scripts to be run at the client.

```
<bindings     xmlns="http://www.mozilla.org/xbl"
              xmlns:html="http://www.w3.org/1999/xhtml">
  <binding id="gadget_21">
        <content>
            <html:iframe anonid="gadget_21" scrolling="no" style="display:block;" frameborder="no"></html:iframe>
        </content>
        <implementation>
            <field name="iframe"><![CDATA[ document.getAnonymousElementByAttribute(this,"anonid","gadget_21"); ]]></field>
            <field name="gadgetURL"><![CDATA[ "http://localhost:8080/shindig/.../WeatherForecastGadget.xml" ]]></field>
            <field name="moduleId">21</field>
            <field name="language"></field>
            <field name="country"></field>
            <field name="gadgetMetadata"></field>
            <constructor><!-- Initializes gadget's preferences & calls "refresh" method. --></constructor>
            <property name="location">
                <getter><![CDATA[ return this.getAttribute('location'); ]]></getter>
                <setter><![CDATA[ this.setAttribute('location',val); ]]></setter>
            </property>
            <property name="units">
                <getter><![CDATA[ return this.getAttribute('units'); ]]></getter>
                <setter><![CDATA[ this.setAttribute('units',val); ]]></setter>
            </property>
            <method name="store"><!-- Makes gadget's preferences persistent & calls "refresh" method. --></method>
            <method name="refresh"><!-- Asks the server for gadget's metadata & re-renders the gadget's markup. --></method>
        </implementation>
  </binding>
  ...
</bindings>
```

CSS
```
div[id="portlet-wrapper-FlightBookingPortlet2_WAR_FlightBookingPortlet2"]
*[id="search_form_view"]
*[id="top-mashcell"] {
    -moz-binding: url('http://localhost:8080/xbl/wrappers.xml#gadget_21');
}
```

Fig. 4. *XBL* support for the composition coordinate *(WeatherForecastGadget,top-mashcell)*

using the *-moz-binding* property. Next sections look at how *XBL* can be used to realize both *composition coordinates* and *orchestration coordinates*. *XBL* files can be bound dynamically, hence, changing the coordinates based on either the profile of the current user or the portal configuration parameters.

5   Setting Composition Coordinates

A composition coordinate *(what, where)* specifies *what* widget is to be place in *which* mashcell. *XBL* wise, the binding provides the *what* as a *<content>* element, while the *CSS* provides the *bound element*, i.e. the *where*. Figure 4 provides an example for the coordinate *(WeatherForecastGadget, top-mashcell)*.

What. The binding *gadget_21* describes this widget as anonymous content. Widget preferences are supported as properties. In this case, getter and setter methods are provided for the two widget preferences: *location* and *units*. This file is kept at *"localhost:8080/xbl/wrappers.xml"*.

Where. Widgets are placed in mashcells. The *CSS* file bounds the *gadget_21* binding to the *"top-mashcell"* to be found in the markup of the *flightBooking* portlet. This markup is shown in Figure 2: the *"top-mashcell"* div pertains to the *"search_form_view"* (notice that a portlet can deliver a set of markups) which in turn is wrapped by a portal decorator (a.k.a. portlet wrappers)[5]. The *CSS* addresses the whole portal page; hence the *CSS* style must identify a single cell within the portal page. This explains the three *ID* conditions in a raw found in the *CSS* sample.

---

[5] The latter identification is needed since it is possible for the same portlet to be instantiated more than once in the very same portal page. That is, they can be distinct *flightBooking portlet* instances in the very same portal page.

```
<bindings    xmlns="http://www.mozilla.org/xbl"
             xmlns:html="http://www.w3.org/1999/xhtml">
    <binding id="onDestinationChange">
        <handlers>
            <handler id="handler_1" event="change"><![CDATA[
                // Call the extractors.
                var extractor1_result = jQuery(this).val();

                // Call the converter
                jQuery.get("http://localhost:8080/emml/IATACode2YahooCode_Converter", {iataCode:extractor1_result}, callback);

                function callback(yahooCode){
                // Get the bound element.
                jQuery("    *[id=portlet-wrapper-FlightBookingPortlet2_WAR_FlightBookingPortlet2]
                            *[id=search_form_view]
                            *[id=top-mashcell]").each(function(index, boundElement){
                            boundElement.location = yahooCode;
                            boundElement.store();
                    });
                };
            ]]></handler>
        </handlers>
    </binding>
</bindings>
```

CSS

```
div[id="portlet-wrapper-FlightBookingPortlet2_WAR_FlightBookingPortlet2"]
*[id="search_form_view"]
> form > table > tbody > tr:nth-child(3) > td:nth-child(2) > select {
    -moz-binding: url('http://localhost:8080/xbl/extensions.xml#onDestinationChange');
}
```

Fig. 5. *XBL* support for the orchestration coordinate *(top-mashcell, destination)*

6   Setting Orchestration Coordinates

An orchestration coordinate *(where, how)* specifies how data flows from the portlet to the widget. For instance, the *weatherForecast's location* parameter is to be obtained through subscription to changes in the *destination* form field of the *flightBooking* portlet. This stands for the orchestration coordinate *(top-mashcell, destination)*. Figure 5 illustrates its XBL realization.

How. The binding *onDestinationChange* specifies how to extract the desired data (e.g. through a handler on the *DOM* event *"change"*). The handler proceeds along three steps: (1) calling the data extractors, (2) enacting the converters for data transformation, if required, and finally (3), localize the *bound element* where the widget resides. Data conversion is needed since data formats can differ between the provider and the consumer of data. In this case, a converter is used to map *IATA* airport identification used by the portlet to *Yahoo!* city code utilized in the gadget.

Where. The *CSS* file associates the previous *onDestinationChange* binding to the form field that collects the *destination* airport. The *CSS* locates the *bound element* by first identifying the portlet decorator, next the portlet view, and finally the DOM node that holds the data. At enactment time, this binding will cause changed in *destination* to be propagated to the *weatherForecast* markup.

7   Related Work

For the purpose of this work, mashup approaches can be classified based on the role taken by the source application: passive ("unaware mashuping") versus active ("aware mashuping"). Unaware mashuping is illustrated by *MashMaker* [4] and *MARGMASH* [3]. Both tools allow end users to augment the content of an *existing* Web application by inserting mashups' markup throughout the source application's pages. The source application is completely unaware of being mashup so that screen-scraping techniques are used "to glue" the mashup to the existing code. Browser plugins can be available to facilitate such data extraction. By contrast, corporate portals offer a more controlled environment for mashuping. Here, the applications to be mashuped (i.e. the portlets) are already integrated into the portal. Such integration implies that portlet providers need first to be certified, and this agreement can now include the existence of mashup place-holders that facilitate mashuping from portal users. The portal ensures access control, presentation guidelines, and data extraction. No plugin is required. The mashup is set through the portal as an additional utility.

8   Conclusions

Corporate portals achieve front-end integration for presentation-oriented Web Services (a.k.a. portlets). As any other Web application, portlets can also be subject to mashup. However, their special characteristics (i.e. reusable components being offered by third parties) make portlet mashup a combined endeavour of portlet consumers and portlet providers. We have presented an architecture where portlet providers facilitate mashup placeholders (i.e. *mashcells*) to add companion widgets. As for portlet consumers, *XBL* bindings are used to dynamically bound user mashups to *mashcells*. The approach strives to find a balance between portal reliability and mashup freedom. This architecture is borne out for *Liferay*. Next follow-on includes capitalizing on the portal utilities for leveraging mashup. Single-sign on, access control, customization mechanisms are now at our disposal to adapt mashup techniques when achieved through a corporate portal.

Acknowledgments. This work is co-supported by the Spanish Ministry of Education, and the European Social Fund under contract TIN2008-06507-C02-01/TIN (MODE-LINE). Perez enjoys a doctoral grant from the Basque Government under the "Researchers Training Program".

References

1. JSR 286: Portlet Specification Version 2.0, http://jcp.org/en/jsr/summary?id=286
2. Díaz, O., Rodríguez, J.J.: Portlets as Web Components: an Introduction. Journal of Universal Computer Science, 454–472 (2004)
3. Díaz, O., Pérez, S., Paz, I.: Providing Personalized Mashups Within the Context of Existing Web Applications. In: Benatallah, B., Casati, F., Georgakopoulos, D., Bartolini, C., Sadiq, W., Godart, C. (eds.) WISE 2007. LNCS, vol. 4831, pp. 493–502. Springer, Heidelberg (2007)
4. Ennals, R.J., Garofalakis, M.N.: MashMaker: mashups for the masses. In: Proceedings of the 2007 ACM SIGMOD International Conference on Management of Data (2007)
5. Rhodes, B.J.: Margin Notes: Building a Contextually Aware Associative Memory. In: Proceedings of the International Conference on Intelligent User Interfaces, IUI 2000 (2000)
6. W3C. Document Object Model (DOM) Level 2 Events Specification, http://www.w3.org/TR/DOM-Level-2-Events/
7. W3C. Widgets Family of Specifications, http://www.w3.org/2008/webapps/wiki/WidgetSpecs
8. W3C. XML Binding Language (XBL) 2.0 (2007), http://www.w3.org/TR/xbl/

When Conceptual Model Meets Grammar: A Formal Approach to Semi-structured Data Modeling⋆

Martin Nečaský and Irena Mlýnková

Department of Software Engineering, Charles University in Prague, Czech Republic
{necasky,mlynkova}@ksi.mff.cuni.cz

Abstract. Currently, XML is a standard for information exchange. An important task in XML management is designing particular XML formats suitable for particular kinds of information exchange. There exist two kinds of approaches to this problem. Firstly, there exist XML schema languages and their formalization – regular tree grammars. Secondly, there are approaches based on conceptual modeling and automatic derivation of an XML schema from a conceptual schema.

In this paper, we provide a unified formalism for both kind of approaches. It is based on formal specification of XML schemas, conceptual schemas, and mappings between both kinds of schemas. The formalism gives necessary conditions on the mappings. The mapping may then be applied in practice not only for unified process of designing XML schemas on both levels, i.e. conceptual and grammatical, but also for integration and evolution of XML schemas.

1 Introduction

Currently, XML is a de-facto standard meta-format for information exchange on the Web. An important task in XML management is designing particular XML formats suitable for particular kinds of information exchange. This means creating of XML schemas describing the required structure of XML documents. For this reason, there have recently appeared various works on this topic.

Related work. In one direction, there exist various languages for expressing XML schemas (XML schema languages), e.g. XML Schema [14] or Relax NG [2]. These languages are practical but lack a formal background. This background was recently introduced by Murata et al. in [9] and is called *regular tree grammars*. The introduced formalism unifies a family of XML schema languages based on specification of grammar rules. It allows for a formal comparison of the languages and also enables one to formally describe algorithms for validation of XML documents against XML schemas.

In the other direction, there is a bunch of works which look at the problem of designing XML schemas from the conceptual point of view. These approaches usually apply the ER model (such as [3,8,7,1,10]) or UML class model (such as [13,4]). They suppose designing a conceptual diagram of the problem domain first. After that, a representation in an XML schema language is derived automatically from the conceptual diagram. We provide an extensive survey of these approaches in [10].

⋆ Supported the Czech Science Foundation (GAČR), grants number P202/10/0573 and 201/09/P364.

L. Chen, P. Triantafillou, and T. Suel (Eds.): WISE 2010, LNCS 6488, pp. 279–293, 2010.

Contributions. The contribution of this paper is a formalism which unifies both directions. This is a first attempt in this area to our best knowledge. The introduced formalism is based on mappings between a regular tree grammar and conceptual schema. These mappings bring a lot of advantages when integrating XML schemas and managing their evolution as we described in [5,12,6]. We have implemented the results presented in this work in a case tool called XCase [11].

Outline. The outline of the paper is as follows. In Section 2, we introduce an extension to regular tree grammars as proposed in [9]. In Section 3, we introduce a formal conceptual model for XML data based on UML class diagrams. In Section 4, we introduce a mapping between both levels, i.e. a regular tree grammar and conceptual schema. We conclude in Section 5.

2   Describing Structure with Regular Tree Grammar

In this section, we formally introduce the model of XML documents and schemas. As it is usual in the recent literature, we will represent an XML document as an XML tree. An XML schema will be expressed in a form of a regular tree grammar.

For the purposes of the following text, we introduce some notation. Let $\mathcal{C} = \{m..n : m \in \mathbb{N}_0 \land n \in (\mathbb{N}_0 \cup *) \land (m \leq n \lor n = *)\}$ denote an infinite set of cardinality constraints where $\mathbb{N}_0$ denotes the set of natural numbers including 0. Further, let L denote an infinite set of labels (words over a finite alphabet Σ, i.e. $L \subseteq \Sigma^*$). Let D denote a finite set of supported data types. A data type $d \in D$ is a possibly infinite set of data values. For example, a data type `integer` is an infinite set of all integers. Moreover, let $\mathcal{D} = \bigcup_{d \in D} d$, i.e. $\mathcal{D}$ denotes the set of all values of all considered data types. Finally, let $\mathcal{X}$ be a set. We will use $\mathrm{SUBSETS}(\mathcal{X})$ and $\mathrm{SUBLISTS}(\mathcal{X})$ to denote the set of all subsets and ordered subsequences of $\mathcal{X}$, respectively.

Definition 1. *An* XML tree τ *is an expression described by the following grammar:*

$$\tau ::= l\,[\,\{f_a\}\,,\,(f_e)\,] \quad | \quad l\,[\,v\,]; \qquad f_a ::= f_a\,,\,f_a \quad | \quad @\,l\,[\,v\,] \quad | \quad ();$$
$$f_e ::= f_e\,,\,f_e \quad | \quad \tau \quad | \quad ()$$

where $()$ denotes an empty expression, v stands for a value from $\mathcal{D}$ and l stands for a label from L. An expression $@\,l\,[\,v\,]$ is an *XML attribute* with a name l and value v. An expression $l\,[\,\{a_1, \ldots, a_m\}, (e_1, \ldots, e_n)\,]$ is an *XML element with a complex content* with name l, XML attributes $a_1, \ldots, a_m$ and *child* XML elements $e_1, \ldots, e_n$. The XML attributes must have distinct names from each other. An expression $l\,[\,v\,]$ is an *XML element with a simple content* with a name l and simple content v.

Example 1. A sample XML tree is depicted in Fig. 1 (b). Fig. 1 (a) shows the corresponding XML document.

Let us note that Definition 1 unifies the term XML tree and XML element – an XML tree τ is in fact an expression which is an XML element at the same time. This XML element is called *root XML element*.

```
<purchase code="o123" date="26.5.10" version="1">
  <shipto>
    <street>Malostranska</street><city>Prague</city>
    <gps>505'18N,1424'13E</gps>
  </shipto>
  <billto>
    <street>Jasminova</street><city>Prague</city>
  </billto>
  <cust login="martin">
    <name>Martin Necasky</name>
    <phone>+420111222333</phone>
  </cust>
  <items>
    <item code="p1">
      <price>1</price><amount>1</amount>
    </item>
    <item code="p2" tester="true">
      <amount>5</amount>
    </item>
    <item code="p3">
      <price>2</price><amount>2</amount>
    </item>
  </items>
</purchase>
```

(a) XML document

$$N = \{Pur, Code1, Date, Ver, SAd, BAd, Ad, GPS,$$
$$\quad Str, City, Cust, Login, Nm, Cont, Email, Phone$$
$$\quad Items, Item, Code2, Tester, Price, Amount\}$$

$$T = \{purchase, shipto, billto, street, city, gps, cust,$$
$$\quad name, email, phone, items, item, price, amount$$
$$\quad @code, @date, @version, @login, @tester\}$$

$$S = \{Pur\}$$

$$P = \{Pur \rightarrow purchase[Code1, Date, Ver, SAd,$$
$$\quad\quad\quad\quad\quad\quad\quad\quad BAd, Cust, Items],$$

$$Code1 \rightarrow @code[string],$$
$$Date \rightarrow @date[string],$$
$$Ver \rightarrow @version[string],$$
$$SAd \rightarrow shipto[Ad, GPS],$$
$$BAd \rightarrow billto[Ad],$$
$$Ad \rightarrow [Str, City],$$
$$GPS \rightarrow gps[string],$$
$$Str \rightarrow street[string],$$
$$City \rightarrow city[string],$$
$$Cust \rightarrow cust[Login, Nm, Cont],$$
$$Login \rightarrow @login[string],$$
$$Nm \rightarrow name[string],$$
$$Cont \rightarrow [Phone\,1..*, Email\,0..1],$$
$$Email \rightarrow email[string],$$
$$Phone \rightarrow phone[string],$$
$$Items \rightarrow items[Item\,1..*],$$
$$Item \rightarrow item[Code2, (Tester|Price), Amount],$$
$$Code2 \rightarrow @code[string],$$
$$Tester \rightarrow @tester[string],$$
$$Price \rightarrow price[string],$$
$$Amount \rightarrow amount[string]\}$$

$purchase[\{@code['o123'], @date['26.5.10'], @version['1']\}, ($
$\quad shipto[\{\}, (street['Malostranska'], city['Prague'],$
$\quad\quad\quad\quad gps['50^o5'18N, 14^o24'13E'])],$
$\quad billto[\{\}, (street['Jasminova'], city['Prague'])],$
$\quad cust[\{@login['martin']\}, (name['MartinNecasky'],$
$\quad\quad\quad\quad\quad\quad\quad\quad phone[' + 420111222333'])],$
$\quad items[\{\}, ($
$\quad\quad item[\{@code['p1']\}, (price['1'], amount['1'])],$
$\quad\quad item[\{@code['p2'], @tester['true']\}, (amount['5'])],$
$\quad\quad item[\{@code['p3']\}, (price['2'], amount['2'])]$
$\quad)]$
$)]$

(b) XML tree   (c) Regular tree grammar

Fig. 1. Sample XML document (a), its XML tree (b), and corresponding regular tree grammar (c)

Definition 2. *A regular tree grammar is a 4-tuple* $G = (N, T, S, P)$, *where:*

- N *is a finite set of non-terminals.*
- $T \subseteq L$ *is a finite set of terminals.*
- P *is a set of production rules of one of the following forms:*
  - $Z \rightarrow @\,t\,[d]$, *where* $Z \in N$, $t \in T$ *and* $d \in D$. Z *is called* XML attribute declaration,
  - $Z \rightarrow t\,[d]$, *where* $Z \in N$, $t \in T$ *and* $d \in D$. Z *is called* XML element declaration with a simple type,
  - $Z \rightarrow t\,[re]$, *where* $Z \in N$, $t \in T$, *and* re *is a regular expression.* Z *is called* XML element declaration with a complex type, *and*
  - $Z \rightarrow [re]$, *where* $Z \in N$ *and* re *is a regular expression.* Z *is called* complex type definition.
- $S \subseteq N$ *is a set of XML element declarations called* start symbols.

For the rest of the paper, let N_c, N_a, and N_e denote the set of all complex type definitions, XML attribute declarations, and XML element declarations in N, respectively.

Definition 3. *Let* $G = (N, T, S, P)$ *be a regular tree grammar. A regular expression* re
is an expression described by the following grammar:

$$re ::= \quad Z \quad | \quad re\, m..n \quad | \quad (\, re\,) \quad | \quad re\, ,\, re \quad | \quad re\, |\, re \quad | \quad \{re\}$$

where Z *stands for a non-terminal from* N *and* $m..n$ *stands for a cardinality from* $\mathcal{C}$.

A regular expression of the form re , $re, re \mid re$ or $\{re\}$ is called *sequence model*, *choice model* or *set model*, respectively.

Example 2. A sample regular tree grammar is depicted in Fig. 1 (c). Non-terminal Pur is an XML element declaration with a complex type. $City$ is an XML element declaration with a simple type. $Code1$ is an XML attribute declaration. Ad is an example of a complex type definition.

We are now ready to specify the semantics of the introduced regular tree grammars. We start with the semantics of regular expressions. In the theory of regular expressions, a regular expression re models a language. A word belongs to this language if it *matches* re which is formally introduced by Definition 4.

Definition 4. *Let* Z *be a non-terminal. Let* re_1 *and* re_2 *be two regular expressions. Let* $\mathcal{Z}$ *be an ordered sequence of non-terminals. We say that* $\mathcal{Z}$ matches

- $Z \Leftrightarrow \mathcal{Z} = Z \vee (\exists(Z \to re) \in P)(\mathcal{Z}\ matches\ re)$
- $(re_1) \Leftrightarrow \mathcal{Z}\ matches\ re_1$
- $re_1\ m..n \Leftrightarrow \mathcal{Z} = \mathcal{Z}_1 \ldots \mathcal{Z}_k \wedge m \leq k \leq n \wedge (\forall 1 \leq i \leq k)\ (\mathcal{Z}_i\ is\ an\ ordered$ *sequence of non-terminals which matches* re_1)
- $re_1, re_2 \Leftrightarrow \mathcal{Z} = \mathcal{Z}_1\ \mathcal{Z}_2 \wedge \mathcal{Z}_1\ is\ an\ ordered\ sequence\ of\ non-terminals\ which$ *matches* $re_1 \wedge \mathcal{Z}_2$ *is an ordered sequence of non-terminals which matches* re_2
- $re_1 \mid re_2 \Leftrightarrow \mathcal{Z}\ matches\ re_1 \vee \mathcal{Z}\ matches\ re_2$
- $\{re_1\} \Leftrightarrow \mathcal{Z} = \mathcal{Z}_1 \ldots \mathcal{Z}_k \wedge (\forall 1 \leq i \leq k)\ (\mathcal{Z}_i \in N) \wedge (\exists\ \pi : \{1, \ldots, k\} \to$ $\{1, \ldots, k\})\ (\mathcal{Z}_{\pi(1)} \ldots \mathcal{Z}_{\pi(k)}\ matches\ re_1)$

Example 3. To demonstrate Definition 4, suppose the regular tree grammar from Fig. 1, an ordered sequence of non-terminals $\mathcal{Z} = Login\ Nm\ Phone\ Phone\ Email$, and regular expression $re = Login, Nm, Cont$. We show that $\mathcal{Z}$ matches re. Trivially, the sequence $Login\ Nm$ matches the regular expression $Login, Nm$. The sequence $Phone\ Phone\ Email$ matches the regular expression $Cont$ because of the production rule $Cont \to [Phone\ 1..*, Email\ 0..1]$ where $Phone\ Phone$ matches $Phone\ (1, *)$ and $Email$ matches $Email\ 0..1$.

Now we slightly extend the notion of *interpretation* of an XML tree against a regular tree grammar introduced in [9].

Definition 5. *An* interpretation I *of an XML tree* τ *against a regular tree grammar* G $= (N, T, S, P)$ *is a mapping from each XML element and attribute* x *in* τ *to a non-terminal, denoted* $I(x)$, *such that:*

- *If* x *is the root element then* $I(x) \in S$.
- *If* $x = l\,[v]$ *or* $x = @\,l\,[v]$ *where* $l \in L$ *and* $v \in \mathcal{D}$ *then* $(I(x) \to l\,[d]) \in P$ *where* d $\in D \wedge v \in d,$

– *if* $x = l \, [\, \{x_1, \ldots, x_{m_1}\}, \, (x_{m_1+1}, \ldots, x_{m_1+m_2}) \,] \,$ *where* $l \in L$ *then* $(I(x) \rightarrow l \, [re]) \in P$ *where* $(\exists \pi : \{1, \ldots, m_1 + m_2\} \rightarrow \{1, \ldots, m_1 + m_2\})((\forall 1 \leq i < m_2)(\pi(m_1 + i) \leq \pi(m_1 + i + 1)) \wedge I(x_{\pi(1)}) \ldots I(x_{\pi(m_1+m_2)})$ *matches* re)

Example 4. For example, the XML tree depicted in Fig. 1 (b) has an interpretation against the regular tree grammar depicted in Fig. 1 (c). The interpretation maps, e.g., the XML element *purchase* to the non-terminal Pur, *shipto* to SAd and each XML element *item* to $Item$.

Now, we can borrow all the results introduced in [9]. For clarity, we repeat the definitions of the language generated by a regular tree grammar and the notion of validity.

Definition 6. *Let* $G = (N, T, S, P)$ *be a regular tree grammar. The* language generated *by* G *is the set* $L(G) = \{\tau \, : \, \tau$ *is an XML tree with an interpretation against* $G\}$. *An XML tree* τ *is* valid *against a regular tree grammar* G *if* $\tau \in L(G)$.

3   Modeling Semantics with Platform-Independent Model (PIM)

A schema in a platform-independent model (*PIM schema*) models a problem domain at the conceptual level independently of its representation in the logical XML data model. For example, we would like to model the domain of purchasing and supplying products independently of its representation in XML documents valid against the regular tree grammar depicted in Fig. 1. Today, the most natural and popular way for modeling PIM schemas is to use UML class diagrams. We follow this common trend as well.

Definition 7. *A PIM schema is a 8-tuple* $\mathcal{S} = (\mathcal{S}_c, \, \mathcal{S}_a, \, \mathcal{S}_r,$ *name, type, class, acard, rcard*) *where:*

– $\mathcal{S}_c$ *and* $\mathcal{S}_a$ *denote sets of* classes *and* attributes *in* $\mathcal{S}$, *respectively.*
– $\mathcal{S}_r$ *denotes a multiset of* binary associations *in* $\mathcal{S}$. *A binary association is a multiset* $R = \{C_1, C_2\}$ *where* $C_1, C_2 \in \mathcal{S}_c$ *are called* participants *of* R.
– *name* $: \mathcal{S}_c \cup \mathcal{S}_a \cup \mathcal{S}_r \rightarrow L$ *assigns a* name *to each class, attribute or association. The name of an association may be empty.*
– *type* $: \mathcal{S}_a \rightarrow D$ *assigns a data type to each attribute.*
– *class* $: \mathcal{S}_a \rightarrow \mathcal{S}_c$ *assigns a class to each attribute.*
– *acard* $: \mathcal{S}_a \rightarrow C$ *assigns a* cardinality *to each attribute.*
– *rcard* $: \mathcal{S}_r \times \mathcal{S}_c \rightarrow C$ *assigns a cardinality to each participant of each association.*

We will call the members of $\mathcal{S}_c$, $\mathcal{S}_a$, and $\mathcal{S}_r$ *components* of $\mathcal{S}$. A PIM schema is displayed as a classical UML class diagram.

Example 5. A sample PIM schema is depicted in Figure 2. It models the domain of purchasing and supplying goods. A purchase (modeled by class $Purchase$[1]) has one or more items (modeled by class $Item$). An item is associated with a purchased product

---

[1] When referring to components of a PIM schema we will use their name. If an association has an empty name, we will use the pair of the names of its participants.

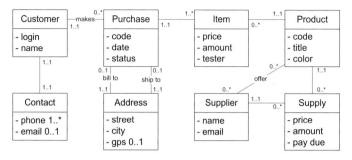

Fig. 2. Sample PIM schema

(class $Product$). A purchase is also associated with a billing and shipping address (class $Address$). A customer is associated with her contact information (class $Contact$). We also consider suppliers (class $Supplier$). A supplier may offer (association $offer$) and supply (class $Supply$) products. Classes have attributes. For example a purchase has its code, date and status modeled by attributes $code$, $date$, and $status$, respectively.

Let us shortly discuss relationships. The definition allows for an existence of two different associations with the same participants, i.e. $R_1 = \{C_1, C_2\} = R_2$. Similarly, participants of an association may be the same class, i.e. $R = \{C_3, C_4\}$ where $C_3 = C = C_4$. We therefore need a mechanism which enables to distinguish R_1 from R_2 and C_3 from C_4. However, we do not suppose concrete distinguishing mechanisms in this paper. We only require that some mechanisms are applied so that we can distinguish R_1 and R_2 as well as C_3 and C_4 in any case.

Finally, to simplify the further formalism, we will also use auxiliary functions $attributes : \mathcal{S}_c \rightarrow \text{SUBSETS}(\mathcal{S}_a)$ and $associations : \mathcal{S}_c \rightarrow \text{SUBSETS}(\mathcal{S}_r)$ defined for a given class $C \in \mathcal{S}_c$ as follows:

$$attributes(C) = \{A \in \mathcal{S}_a : class(A) = C\}$$
$$associations(C) = \{R \in \mathcal{S}_r : (\exists \overline{C} \in \mathcal{S}_c)(R = \{C, \overline{C}\})\}$$

Function $attributes(C)$ returns the set of all attributes assigned to a class C. Function $Associations(C)$ returns the set of all associations with C as a participant.

4   Binding via Platform-Specific Model (PSM)

We showed how structure of XML documents can be described by a regular tree grammar and how a problem domain can be modeled at the conceptual level. In this section, we formally introduce a binding between both levels. This binding is a schema in a platform-specific model (*PSM schema*). A PSM schema is a UML class diagram with some extensions necessary for modeling regular tree grammars. We will look at the PSM schema from two perspectives called *grammatical* and *conceptual*. From the *grammatical perspective*, the PSM schema models a particular regular tree grammar. From the *conceptual perspective*, the PSM schema specifies the semantics of the regular tree grammar in the terms of a given PIM schema. The semantics is given by a mapping from components of the PSM schema to components of the PIM schema.

Having a PIM schema, a designer can easily derive one or more PSM schemas required by her application. The PSM schemas present the XML schemas in a graphical user-friendly way and are mapped to the same PIM schema. Therefore, she can easily see what parts of her application domain are represented in a particular XML format and how. She can also use the mapping for evolution of her XML schemas. When the domain changes, these changes can be propagated to the XML schemas via the PSM schema. And, it is possible to start with already existing XML schemas, create corresponding PSM schemas and map them to the PIM schema. The PIM schema can be designed manually by the designer or derived from the XML schemas. All these features are experimentally demonstrated in our XCase tool [11] and described in detail in our recent papers [6,12].

In this section, we do not describe the mentioned methods. We introduce the formalism of mappings between the PIM and PSM schemas and between PSM schemas and corresponding regular tree grammars. The formalism also provides conditions which must be satisfied when deriving an XML schema from a PSM schema and vice versa.

Definition 8. *A PSM schema is a 12-tuple* $S' = (S'_c, S'_a, S'_r, S'_m, C'_{S'}, name', type',$ *class', xattr', card', content', repr').*

- S'_c, S'_a, *and* S'_m *denote sets of* classes, attributes, *and* content models *in* S', *respectively. We distinguish three types of content models:* sequence, choice *and* set.
- $S'_r \subseteq S'_c \times S'_c$ *denotes a set of oriented binary* associations *in* S'. *For* $(C'_1, C'_2) \in S'_r$, *we call* C'_1 *and* C'_2 parent *and* child *of* R'. *We will also sometimes call both* C'_1 *and* C'_2 participants *of* R' *and say that* C'_1 *is the* parent *of* C'_2 *and* C'_2 *is a* child *of* C'_1.
- $C'_{S'} \in S'_c$ *is a class called* schema class *of* S'.
- $name' : S'_c \cup S'_a \cup S'_r \to L$ *assigns a* name *to each class, attribute and association.*
- $type' : S'_a \to T$ *assigns a data type to each attribute.*
- $class' : S'_a \to S'_c$ *assigns a class to each attribute.*
- $xattr' : S'_a \to \{$true, false$\}$ *assigns a boolean value to each attribute.*
- $card' : S'_a \cup S'_r \to C$ *assigns a* cardinality *to each attribute and association.*
- $content' : S'_c \cup S'_m \to$ SUBLISTS(S'_r) *assigns a sequence of distinct associations to each class or content model* X', *s.t.* $(R' \in content'(X') \Leftrightarrow X'$*is the parent of* $R')$.
- $repr' : S'_c \to S'_c$ *assigns a class* C *to another class* $\overline{C}$. $\overline{C}$ *is called* structural representative *of* C. *Let* $C_1, ..., C_n$ *be classes such that* C_{i+1} *is a structural representative of* C_i. *Then* $C_1 \neq C_n$.

The graph $(S'_c \cup S'_m, S'_r)$ *must be a directed forest with one of its trees rooted in the schema class* $C'_{S'}$. *We will call each class* C' *which is a child of* $C'_{S'}$ top class.

Similarly to PIM schemas, we will call the members of S'_c, S'_a, S'_r, and S'_m *components* of S'. A PSM schema is displayed as a UML class diagram in a tree layout. Classes and attributes are displayed in a classical way with some extensions. The name of an attribute A' is prefixed with @ if $xattr'(A') = true$. Moreover, for a structural representative $\overline{C'}$ of a class C' we display the name of C' above the name of $\overline{C'}$. Associations are displayed oriented from their parent to child. Content models are displayed as small rounded rectangles. They have no semantic equivalent. They are relevant only for the

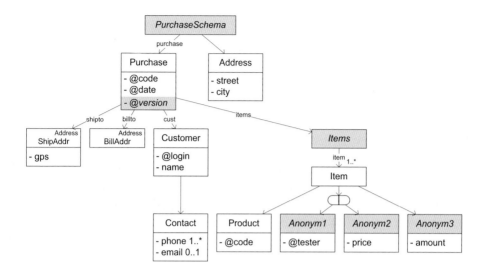

Fig. 3. Sample PSM schema

grammatical perspective. A sequence models a sequence of associations and allows to repeat the sequence in XML documents. A choice models that only one option of the included associations may appear in XML documents. And, finally, a set models that the associations may appear in any order in XML documents. To distinguish sequence, choice and set we display . . . , |, or {} in the box, respectively.

Example 6. A sample PSM schema is depicted in Figure 3. It binds the PIM schema from Fig. 2 and regular tree grammar from Fig. 1. We will discuss the binding later in a more detail. When referring to The figure displays the schema class $PurchaseSchema'$[2] and two top classes $Purchase'$ and $Address'$. There are two structural representatives of $Address'$: $ShipAddr'$ and $BillAddr'$. The schema also contains a choice content model containing classes $Anonym1'$ and $Anonym2'$.

We will use auxiliary functions $attributes' : \mathcal{S}'_c \rightarrow \text{SUBSETS}(\mathcal{S}'_a)$ and $parent'$, $child' : \mathcal{S}'_r \rightarrow \mathcal{S}'_c \cup \mathcal{S}'_m$ defined as follows:

$$attributes'(C') = \{A' \in \mathcal{S}'_a : class'(A') = C'\}$$
$$(\forall R' \in \mathcal{S}'_r)(R' = (X'_1, X'_2) \Leftrightarrow parent'(R') = X'_1 \wedge child'(R') = X'_2)$$

Function $attributes'(C')$ returns the set of all attributes assigned to class C'. Functions $parent'(R')$ and $child'(R')$ return the parent and child of association R'.

For the purposes of the following text, we will need a notion of *path* in a PSM schema. A path is a sequence of components of the PSM schema separated by the '.' symbol which meets certain conditions introduced by Definition 9.

---

[2] When referring to components of a PSM schema we will use their name suffixed with $'$ to distinguish them from components of a PIM schema.

Definition 9. *Let $\mathcal{S}'$ be a PSM schema and X' be a component of $\mathcal{S}'$. A* path *to X' is an expression defined as follows:*

$$path'(X') = \begin{cases} X', & \text{if } X' = C'_{\mathcal{S}'} \\ path'(parent'(X')).X', & \text{if } X' \in \mathcal{S}'_r \\ path'(class'(X')).X', & \text{if } X' \in \mathcal{S}'_a \\ path'(R').X', & \text{if } X' \in (\mathcal{S}'_c \setminus \{C'_{\mathcal{S}'}\}) \cup \mathcal{S}'_m, \text{ where} \\ & R' \in \mathcal{S}'_r \wedge child'(R') = X' \end{cases}$$

On the base of the notion of path, we define another auxiliary function $classcontext'$: $\mathcal{S}'_c \cup \mathcal{S}'_a \cup \mathcal{S}'_r \cup \mathcal{S}'_m \to \mathcal{S}'_c$ which assigns a given component X' with the last class C' on the path to C' except X' (if X' is a class). Formally, let $X'_1. \cdots .X'_k. \cdots .X_n$ be a path from $C'_{\mathcal{S}'}$ to C' ($X'_1 = C'_{\mathcal{S}'}$ and $X'_n = X'$). Let $X'_k \neq C'$ be a class such that $\forall$ $k < i < n$, X'_i is not a class. Then $classcontext'(X') = X'_k$. If there is no such X'_k then $classcontext'(X') = \lambda$.

Example 7. Suppose the PSM schema in Fig. 3. Suppose the attribute $tester'$ of the class $Anonym1'$. We have $classcontext'(tester') = Anonym1'$ and $classcontext'$ $(Anonym1') = Item'$.

4.1 Conceptual Perspective

From the conceptual perspective, PSM schema is mapped to a part of a PIM schema. This mapping is formalized as so called *interpretation* of the PSM schema against the PIM schema. Informally, each class or attribute in the PSM schema is mapped to a class or attribute in the PIM schema, respectively. Similarly, each association in the PSM schema is mapped to an association in the PIM schema. The only problem is that PSM associations are ordered while PIM associations are not. Therefore, we firstly define the notion of *ordered image* of a PIM association. The resulting interpretation gives us the semantics of the PSM schema in the sense of the PIM schema.

Definition 10. *Let $\mathcal{S}$ be a PIM schema with a set of associations $\mathcal{S}_r$. Let $R = \{C_1, C_2\}$ $\in \mathcal{S}_r$ be an association. An* ordered image *of R is $R^{C_1} = (C_1, C_2)$ or $R^{C_2} = (C_2, C_1)$. We also define the set of ordered images $\overrightarrow{\mathcal{S}_r} = \{R^{C_1}, R^{C_2} : R = \{C_1, C_2\} \in \mathcal{S}_r\}$.*

Definition 11. *An* interpretation *of a PSM schema $\mathcal{S}'$ against a PIM schema $\mathcal{S}$ is a partial function $I_{pim} : (\mathcal{S}'_c \cup \mathcal{S}'_a \cup \mathcal{S}'_r) \to (\mathcal{S}_c \cup \mathcal{S}_a \cup \overrightarrow{\mathcal{S}_r})$ which maps a class, attribute or association from $\mathcal{S}'$ to a class, attribute or ordered image of an association from $\mathcal{S}$, respectively. For $X' \in (\mathcal{S}'_c \cup \mathcal{S}'_a \cup \mathcal{S}'_r)$, we call $I_{pim}(X')$ interpretation of X'. $I_{pim}(X') = \lambda$ denotes that X' does not have an interpretation. The following must be satisfied:*

$$(\forall C' \in \mathcal{S}'_c \text{ s.t. } repr'(C') \text{ is defined })(I_{pim}(C') = I_{pim}(repr'(C'))) \tag{1}$$

$$(\forall A' \in \mathcal{S}'_a \text{ s.t. } I_{pim}(A') \neq \lambda)(classcontext'_{I_{pim}}(A') \neq \lambda \wedge$$
$$class(I_{pim}(A')) = I_{pim}(classcontext'_{I_{pim}}(A'))) \tag{2}$$

$$(\forall R' \in \mathcal{S}'_r \text{ s.t. } I_{pim}(child'(R')) \neq \lambda)(classcontext'_{I_{pim}}(R') \neq \lambda \wedge$$
$$I_{pim}(R') = (I_{pim}(classcontext'_{I_{pim}}(R')), I_{pim}(child'(R')))) \tag{3}$$

$$(\forall R' \in \mathcal{S}'_r \text{ s.t. } I_{pim}(child'(R')) = \lambda)(I_{pim}(R') = \lambda) \tag{4}$$

where the function $classcontext'_{I_{pim}} : \mathcal{S}'_c \cup \mathcal{S}'_a \cup \mathcal{S}'_r \cup \mathcal{S}'_m \to \mathcal{S}'_c$ is defined similarly to $classcontext'$. The difference is that $classcontext'_{I_{pim}}$ returns the closest class with a defined interpretation I_{pim} instead of simply the closest class.

The conditions (1) - (4) ensure consistency between the PIM and PSM schema. We demonstrate them on the following example.

Example 8. We will demonstrate an interpretation I_{pim} of the PSM schema in Fig. 3 against the PIM schema in Fig. 2. Firstly, there are classes and attributes in the PSM schema which do not have an interpretation defined. These are distinguished by a grey background. For example, the attribute $Purchase'.version'$ does not have an interpretation. The same if for classes $Items'$, $Anonym1'$, $Anonym2'$, and $Anonym3'$.

The classes and attributes with an empty background have an interpretation I_{pim}. However, we do not list the whole I_{pim}. Some sample interpretations of classes in the PSM schema are following: $I_{pim}(Purchase') = Purchase'$, $I_{pim}(Address') = Address'$, or $I_{pim}(ShipAddr') = Address$. The condition (1) ensures that a structural representative has the same interpretation as the represented class. Therefore, it must be that $I_{pim}(ShipAddr') = Address$ since $repr'(ShipAddr') = Address'$.

Sample interpretations of attributes in the PSM schema are following: $I_{pim}(Purchase'.code')=Purchase.code$, $I_{pim}(Contact'.email') = Contact.email$, or $I_{pim}(Anonym1'.tester') = Item.tester$. Condition (2) ensures that the interpretation of an attribute A' belongs to the class which is the interpretation of the closest interpreted class to A'. For example, we have that the closest interpreted class of attribute $Anonym1'.tester'$ is class $Item'$. We also have that $I_{pim}(Item') = Item$. Therefore, condition (2) requires that $I_{pim}(Anonym1'.tester')$ is an attribute of $Item$ which is satisfied in our case.

Finally, let us demonstrate interpretations of associations. First of all, conditions (3) and (4) require that only associations with an interpreted child may have their own interpretations. Therefore, the association with child class $Items'$ in our example does not have an interpretation. If the child is interpreted than the interpretation of an association R' must be an ordered image of an association R from the PIM schema such that one of its participants is the interpretation of the child. The other participant must be an interpretation of the closest interpreted class on the path to R'. For example, the interpretation of association R going from $Items'$ to $Item'$ in our example must be an association in the PIM schema connecting classes $Purchase$ and $Item$. We therefore have $I_{pim}(Items', Item') = (Purchase, Item)$.

4.2   Grammatical Perspective

From the grammatical perspective, a PSM schema models a regular tree grammar. A given PSM schema may be translated to a regular tree grammar. Vice versa, a regular tree grammar may be translated to a PSM schema. In both cases, there must exist a mapping of the grammar to the PSM schema. This mapping is called *interpretation* of the regular tree grammar against a PSM schema and is formally defined by Def. 12.

Definition 12. *Let G be a regular tree grammar and $\mathcal{S}'$ be a PSM schema. An interpretation of G against $\mathcal{S}'$ is a total injective function $I_{psm} : N \to \mathcal{S}'_c \cup \mathcal{S}'_a$. Let $Z \in N$,*

$t \in T$, $d \in D$, and re be a regular expression. The following conditions must be satisfied:

$$Z \in S \Rightarrow I_{psm}(Z) \text{ is a top class} \tag{5}$$
$$(Z \to t\,[d]) \in P \Rightarrow I_{psm}(Z) \in \mathcal{S}'_a \wedge type'(I_{psm}(Z)) = d \wedge \tag{6}$$
$$name'(I_{psm}(Z)) = t \wedge xattr'(I_{psm}(Z)) = \texttt{false}$$
$$(Z \to @\,t\,[d]) \in P \Rightarrow I_{psm}(Z) \in \mathcal{S}'_a \wedge type'(I_{psm}(Z)) = d \wedge \tag{7}$$
$$name'(I_{psm}(Z)) = t \wedge xattr'(I_{psm}(Z)) = \texttt{true}$$

$$(Z \to re) \in P \Rightarrow I_{psm}(Z) \in \mathcal{S}'_c \wedge \tag{8}$$
$$(\exists R' \in \mathcal{S}'_r)(child'(R') = I_{psm}(Z) \wedge$$
$$name'(R') = \lambda \wedge re \text{ corresponds to } I_{psm}(Z))$$
$$(Z \to t\,[re]) \in P \Rightarrow I_{psm}(Z) \in \mathcal{S}'_c \wedge \tag{9}$$
$$(\exists R' \in \mathcal{S}'_r)(child'(R') = I_{psm}(Z) \wedge$$
$$name'(R') = t \wedge re \text{ corresponds to } I_{psm}(Z))$$

We will say that $\mathcal{S}'$ *models* G. For a non-terminal Z from G and component X' of $\mathcal{S}'$ s.t. $I_{psm}(Z) = X'$, we will say that X' *models* Z.

Definition 12 specifies the regular tree grammar modeled by a PSM schema. A top class of the PSM schema models a start symbol of the corresponding grammar (rule (5)). An attribute A' models an XML attribute declaration or XML element declaration with a simple content depending on the value of $xattr'(A')$ (rules (6) and (7)). A complex type definition models a class and an association going to the class with an empty name (rule (8)). Finally, an XML element declaration with a complex content models a class and an association going to the class with a non-empty name (rule (9)). In both cases, the attributes and content of the class models the regular expression of the corresponding production rule.

Example 9. Our sample regular tree grammar from Fig. 1 has an interpretation I_{psm} against the PSM schema displayed at Fig. 3. We will demonstrate only a part of the interpretation.

- $I_{psm}(Pur) = Purchase'$
  Rule (5) is satisfied since we have that Pur is a start symbol and $Purchase'$ is a top class.
- $I_{psm}(Nm) = Customer'.name'$
  We have the production rule $Nm \to name\,[string]$. In other words, the attribute $Customer'.name'$ models an XML element declaration with a simple content. Rule (6) is satisfied because we have $type'(Customer'.name') = string$, $name'(Customer'.name') = name$, and $xattr'(Customer'.name') = \texttt{false}$.
- $I_{psm}(Login) = Customer'.login'$
  We have the production rule $Login \to @\,login\,[string]$. In other words, the attribute $Customer'.login'$ models an XML attribute declaration. Rule (7) is satisfied since we have $type'(Customer'.login') = string$, $name'(Customer'.login') = login$, and $xattr'(Customer'.login') = \texttt{true}$.

- $I_{psm}(Cont) = Contact'$
  We have the production rule $Cont \rightarrow [Phone\ 1..*, Email\ 0..1]$. I.e. the class $Contact'$ models a complex type definition. Rule (8) is satisfied. This is because we have an association R' with $child'(R') = Contact'$ and $name'(R') = \lambda$. Moreover, the regular expression $Phone\ 1..*, Email\ 0..1$ corresponds to $Contact'$ as we show later.
- $I_{psm}(Item) = Item'$
  We have the production rule $Item \rightarrow item\ [Code2, (Tester|Price), Amount]$. I.e. the class $Item'$ models an XML element declaration with a complex content. Rule (9) is satisfied. We have an association R' with $child'(R') = Item'$ and $name'(R') = item$. Moreover, the regular expression $Code2, (Tester|Price), Amount$ corresponds to $Item'$ as we show later.

The connection between a class and regular expression is given by the relation *correspond to* used in Def. 12. It is formally introduced by Def. 13.

Definition 13. *Let G be a regular tree grammar and S' be a PSM schema. We say that a regular expression re corresponds to a component $X' \in S'_c \cup S'_a \cup S'_r \cup S'_m$ when the following conditions are satisfied:*

$$X' \in S'_c \wedge attributes(X') = \{A'_1, \ldots, A'_{n_1}\} \wedge content(X') = (R'_1, \ldots, R'_{n_2}) \tag{10}$$
$$\wedge\ repr'(X') = \lambda \Rightarrow re = (re_1, \ldots, re_{n_1+n_2}) \wedge$$
$$(\forall 1 \leq i \leq n_1)(re_i\ corresponds\ to\ A'_i) \wedge$$
$$(\forall 1 \leq i \leq n_2)(re_{n_1+i}\ corresponds\ to\ R'_i)$$

$$X' \in S'_c \wedge attributes(X') = \{A'_1, \ldots, A'_{n_1}\} \wedge content(X') = (R'_1, \ldots, R'_{n_2}) \tag{11}$$
$$\wedge\ repr'(X') \neq \lambda \Rightarrow re = (re_0, re_1, \ldots, re_{n_1+n_2}) \wedge$$
$$(re_0\ corresponds\ to\ repr'(X') \vee$$
$$(re_0 \in N_c \wedge I_{psm}(re_0) = repr'(X'))) \wedge$$
$$(\forall 1 \leq i \leq n_1)(re_i\ corresponds\ to\ A'_i) \wedge$$
$$(\forall 1 \leq i \leq n_2)(re_{n_1+i}\ corresponds\ to\ R'_i)$$

$$X' \in S'_a \Rightarrow re = Z^{card(X')} \wedge I_{psm}(Z) = X' \tag{12}$$

$$X' \in S'_r \wedge child'(X') \in S'_c \wedge name'(X') \neq \lambda \tag{13}$$
$$\Rightarrow re = Z^{card(X')} \wedge I_{psm}(Z) = child'(X')$$

$$X' \in S'_r \wedge child'(X') \in S'_c \wedge name'(X') = \lambda \tag{14}$$
$$\Rightarrow (re = Z^{card(X')} \wedge I_{psm}(Z) = child'(X')) \vee$$
$$(re = (re_r)^{card(X')} \wedge re_r\ corresponds\ to\ child'(X'))$$

$$X' \in S'_r \wedge child'(X') \in S'_m \Rightarrow re = (re_r)^{card(X')} \wedge re_r\ corresponds\ to\ child'(X') \tag{15}$$

$$X' \in S'_m\ is\ a\ sequence\ (or\ choice\ or\ set)\ with\ content(X') = (R'_1, \ldots, R'_n) \tag{16}$$
$$\Rightarrow re = (re_1, \ldots, re_n)\ (or\ re = re_1|\ldots|re_n$$
$$or\ re = \{re_1, \ldots, re_n\},\ respectively) \wedge$$
$$(\forall 1 \leq i \leq n)(re_i\ corresponds\ to\ child'(R'_i))$$

Rules (10) and (11) describe when re corresponds to a class C'. When C' is not a structural representative, rule (10) requires that re is composed of subexpressions which correspond to attributes and content of C'.

Rule (12) describes when re corresponds to an attribute A'. It simply requires that re is an XML attribute declaration or XML element declaration with a simple content repeated according to the cardinality of A'.

Rules (13), (14) and (15) describe when re corresponds to an association R'. If the child of R' is a class then there are two cases. If R' has a name then re is an XML element declaration with a complex content repeated according to the cardinality of R' (rule (13)). If R' does not have a name then the previous case is possible or re is a regular expression which corresponds to the child class and is repeated according to the cardinality of R' (rule (14)). If the child of R' is a content model then re is a regular expression which corresponds to the child content model and is repeated according to the cardinality of R' (rule (15)).

Finally, rule (16) describes when re corresponds to a content model M' and is similar to rule (10).

Example 10. Let us demonstrate Def. 13 on our sample regular tree grammar and PSM schema. Let $re = Login, Nm, Cont$ (from the production rule of non-terminal $Cust$). According to the definition, re corresponds to class $Customer'$. To prove this, we show that rule (10) is satisfied. We can easily see that $Login$ and Nm correspond to the attributes $Customer'.login'$ and $Customer'.name'$, respectively, according to rule (12) ($I_{psm}(Login) = Customer'.login'$ and $I_{psm}(Nm) = Customer'.name'$). We can also see that $Cont$ corresponds to the association R' going from $Customer'$ to $Contact'$ according to rule (14) ($I_{psm}(Cont) = Contact'$).

As another example, let $re = Code2, (Tester|Price), Amount$ (from the production rule of non-terminal $Item$). We show that re corresponds to class $Item'$ according to rule (10). $Item'$ has no attributes and three associations R_1, R_2 and R_3, respectively. $Code2$ corresponds to R_1 according to rule (14) (we can show that $Code2$ corresponds to class $Product'$ according to rule (10)). Similarly, $Amount$ corresponds to R_3 according to rule (14). $(Tester|Price)$ corresponds to R_2 according to rule (15). This is because $Tester|Price$ corresponds to the choice content model $child'(R_2)$ according to rule (16).

Definition 14. *A PSM schema S' specifies an XML language denoted $L(S')$ which is defined as $L(S') = L(G)$ where G is a regular tree grammar with an interpretation against S'.*

The correctness of Definition 14 is ensured by Lemma 1. It shows that the language $L(S')$ is specified unambiguously.

Lemma 1. *Let $G_1 = (N_1, T_1, P_1, S_1)$ and $G_2 = (N_2, T_2, P_2, S_2)$ be two regular tree grammars and S' be a PSM schema. Let $I_{1,psm}$ and $I_{2,psm}$ be interpretations of G_1 and G_2 against $\mathcal{M}_{psm}$, respectively. Then $L(G_1) = L(G_2)$.* □

Therefore, Def. 14 delimits algorithms which translate a PSM schema to a regular tree grammar or vice versa. It requires that there is an interpretation of the grammar against the PSM schema. The interpretation must meet the conditions given by Def. 12 and Def. 13. Lemma 1 ensures that algorithms which ensure the conditions derive equivalent regular tree grammars from the same PSM schema or equivalent PSM schemas from the same regular tree grammar. Due to space limitations, we omit concrete algorithms. These are left for our future work or future work.

5   Conclusions

In this paper, we developed a formal model which unifies the conceptual and grammatical direction of designing XML schemas. The model allows for expressing the schemas on both conceptual and grammatical levels in a formal way. Moreover, it allows for mapping between both levels. The results were implemented in a tool XCase [11].

The model may be applied for different purposes. For example, there may be specified more different regular tree grammars each describing a particular XML format. If these XML formats represent data from the same problem domain, they may be formally integrated by a conceptual schema. The model formally ensures consistency between both levels in a case when regular tree grammars are derived from the conceptual schema (which we discuss in [10]) or, vice versa, when a conceptual schema is derived from the grammars (which we discuss in [6]). Moreover, new transformation algorithms developed on the base of this work will ensure the same conditions.

References

1. Al-Kamha, R., Embley, D.W., Liddle, S.W.: Augmenting Traditional Conceptual Models to Accommodate XML Structural Constructs. In: Parent, C., Schewe, K.-D., Storey, V.C., Thalheim, B. (eds.) ER 2007. LNCS, vol. 4801, pp. 518–533. Springer, Heidelberg (2007)
2. Clark, J., Makoto, M.: RELAX NG Specification. Oasis (December 2001), http://www.oasis-open.org/committees/relax-ng/spec-20011203.html
3. Dobbie, G., Xiaoying, W., Ling, T., Lee, M.: ORA-SS: An Object-Relationship-Attribute Model for Semi-Structured Data. Technical Report, Department of Computer Science, National University of Singapore, Singapore (December 2000)
4. Dominguez, E., Lloret, J., Perez, B., Rodriguez, A., Rubio, A.L., Zapata, M.A.: A Survey of UML Models to XML Transformations. In: Benatallah, B., Casati, F., Georgakopoulos, D., Bartolini, C., Sadiq, W., Godart, C. (eds.) WISE 2007. LNCS, vol. 4831, pp. 184–195. Springer, Heidelberg (2007)
5. Klímek, J., Nečaský, M.: Integration and Evolution of XML Data via Common Data Model. In: Proceedings of the 2010 EDBT/ICDT Workshops, Lausanne, Switzerland, March 22-26. ACM, New York (2010)
6. Klímek, J., Nečaský, M.: Semi-automatic Integration of Web Service Interfaces. To appear in Proceedings of 8th International Conference on Web Services (ICWS 2010), Miami, Florida, USA, July 5-10. IEEE, Los Alamitos (2010)
7. Mani, M.: Semantic Data Modeling Using XML Schemas. In: Kunii, H.S., Jajodia, S., Sølvberg, A. (eds.) ER 2001. LNCS, vol. 2224, pp. 149–163. Springer, Heidelberg (2001)

8. Mani, M.: Erex: A conceptual model for xml. In: Proceedings of the Second International XML Database Symposium, Toronto, Canada, pp. 128–142 (August 2004)
9. Murata, M., Lee, D., Mani, M., Kawaguchi, K.: Taxonomy of XML schema languages using formal language theory. ACM Trans. Internet Technol. 5(4), 660–704 (2005)
10. Nečaský, M.: Conceptual Modeling for XML. Dissertations in Database and Information Systems Series, vol. 99. IOS Press/AKA Verlag (January 2009)
11. Nečaský, M., Klímek, J., Kopenec, L., Kučerová, L., Malý, J., Opočenská, K.: XCase – A Tool for XML Data Modeling (2008)
12. Nečaský, M., Mlýnková, I.: On Different Perspectives of XML Schema Evolution. In: FlexD-BIST 2009, Linz, Austria, pp. 422–426. IEEE, Los Alamitos (2009)
13. Routledge, N., Bird, L., Goodchild, A.: UML and XML Schema. In: Proceedings of 13th Australasian Database Conference (ADC 2002), ACS (2002)
14. Thompson, H.S., Beech, D., Maloney, M., Mendelsohn, N.: XML Schema Part 1: Structures, W3C, 2nd edn. (October 2004), http://www.w3.org/TR/xmlschema-1/

Crowdsourced Web Augmentation: A Security Model

Cristóbal Arellano, Oscar Díaz, and Jon Iturrioz

ONEKIN Research Group, University of the Basque Country,
San Sebastián, Spain
{cristobal.arellano,oscar.diaz,jon.iturrioz}@ehu.es
http://www.onekin.org/

Abstract. Web augmentation alters the rendering of *existing* Web applications at the back of these applications. Changing the layout, adding/removing content or providing additional hyperlinks/*widgets* are examples of Web augmentation that account for a more personalized user experience. *Crowdsourced* Web augmentation considers end users not only the beneficiaries but also the contributors of augmentation scripts. The fundamental problem with so augmented Web applications is that code from numerous and possibly untrusted users are placed into the same security domain, hence, raising security and integrity concerns. Current solutions either coexist with the danger (e.g. *Greasemonkey*, where scripts work on the same security domain that the hosting application) or limit augmentation possibilities (e.g. *virtual iframes* in *Google's Caja*, where the widget is prevented from accessing the application space). This work introduces *Modding Interfaces*: application-specific interfaces that regulate inflow and outflow communication between the *hosting code* and the *user code*. The paper shows how the combined use of sandboxed *iframes* and "modding-interface" *HTML5 channels* ensures application integrity while permitting controlled augmentation on the hosting application.

Keywords: Augmentation, Sandbox, JavaScript, Crowdsourcing, Web2.0.

1 Introduction

The evolution of Web applications can be staged based on the degree of layman's involvement. Web 1.0 limits laymen activities to mainly reading and form filling. Next, Web 2.0 puts *content authoring* in the user's hand: blogging, tagging or wiki editing are nowadays common practices. This work is about the last frontier of layman participation: *application authoring*. The challenge is for end users to provide their own functionality on top of existing Web applications. Mashups [19] and Web augmentation [3] illustrate this approach. Mashup techniques are available for laymen to build *new* applications out of existing Web resources (e.g. APIs, RSS feeds). By contrast, Web augmentation does not create a new application. Rather, the rendering of the hosting application is augmented to change the user experience. An example is the *Skype* add-on, a plugin that turns any phone number found in a web page to a button that launches *Skype* to call that number. No new application is created. Rather, the hosting application is augmented with a *Skype* button.

L. Chen, P. Triantafillou, and T. Suel (Eds.): WISE 2010, LNCS 6488, pp. 294–307, 2010.

This paper focuses on *crowdsourced* Web augmentation whereby end users are not only the beneficiaries but also the contributors of augmentation scripts. *Greasemonkey* scripts illustrate this approach that permits end users change the content/rendering/layout of the current page on the fly [1]. With over 32,000,000 downloads, *Greasemonkey* evidences the success of this approach.

Crowdsourced augmentation implies user-provided code (i.e. the scripts) to co-exist with hosting code (e.g. the *HTML page*). This is risky. Placing different resources for numerous and possibly untrusted or malicious sources into the same security domain raises security and integrity concerns. Threats include: creation of/redirection to phishing pages, stealing history information (or sensitive data stored on either pages or cookies), or port scanning upon the user's local network [10]. Traditionally, the solution is to place content from multiple untrusted sources in different security domain. If scripts are not in the same domain, *JavaScript*'s Same Origin Policy prohibits them from communicating with each other. This ensures application integrity but at the expense of limiting the communication between the script and the hosting document. This is a severe limitation for augmentation where the added value comes from the amendment being seamlessly and interspersedly intermingled with the hosting code. **The challenge is then, balancing security *versus* augmentation expressiveness.**

In a previous work [6], we introduced the notion of *"Modding Interface" (MI)* as a means to isolate user scripts from changes in the hosting application. The aim was that changes in the rendering of the hosting application (a certainty in the always-evolving Web world) do not make scripts stop working. From the script perspective, *MI* provides a stable base on which to set the script. However, from the application viewpoint, interfaces also prevent scripts from peering at the hosting code. Hence, even if the application never changes, there is still a case for *MI* as a means to protect the hosting code.

This paper introduces an MI-based architecture for safe co-existence of hosting code and augmentation scripts. Set and managed by the hosting application, this interface regulates both the outflow (i.e. what "hosting data" can flow to augmentation scripts) and the inflow (i.e. what "hosting rendering aspects" can be subject to augmentation). Akin to the *JavaScript* programming model, this interface is event-based. Traditionally, scripts have open access to hosting rendering through *DOM events*. Now, scripts are "sandboxed" so that interaction can only be through *MI events*. Notice however, that no change is needed in how scripts are programmed. Rather than subscribing to *DOM events*, augmentation scripts subscribe to *MI events*. Through a running example, the paper introduces the notion of *MI*, the architecture, and discusses on the sought balance between security and augmentation expressiveness. Next section introduces the sample case.

2   Crowdsourced Augmentation: A Sample Case

Conferences addressing Web issues can tap on their attendees to enhance the conference site itself. The vision is to regard the conference site as a platform for attendees to enhance. As an example, consider the conference website for *ICWE'09*. Figure 1 (left side) depicts a screenshot for the page on accepted submissions, located at *http://icwe2009.webengineering.org/Accepted.aspx*. On deciding which presentations

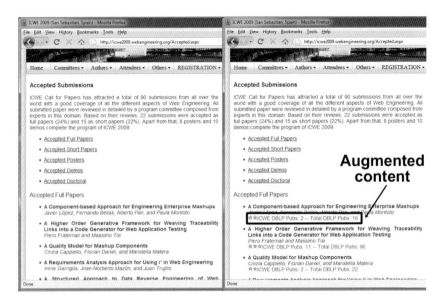

Fig. 1. Raw page (left side) vs. Augmented page (right side)

to attend, an attendee can augment this content with data obtained from Michael Ley's DBLP site[1] so that each accepted paper is augmented with data about previous publications from the paper's authors. To this end, the attendee writes the *dblpFigures* script[2] (see Figure 2, left side). The outcome (see Figure 1, right side) shows how *"host markup"* is intermingled with *"augmented markup"* produced by the script.

So far, the blend of host markup and augmented markup is conducted at the client through a *weaver*. The *weaver* is realized as a browser plugin[3]. The *weaver* places both markups in the same domain so that user scripts can react to events raised by the hosting application. The process is broadly illustrated for the script in Figure 2, left side:

- interacting with a page triggers UI events (e.g. *load*),
- the script *reacts* to this event by triggering a handle (e.g. *"init"*, line 19). An event is bound to a handler through the *addEventListener( )* function (line 38),
- a handler can access any node of the page (using *DOM functions* such as *"document.evaluate( )"* in lines 21-23 and 26-28). It can also create *HTML* markup (e.g. *dblpFiguresPanel* in lines 10-17),
- a handler can also *change* the *DOM* structure at wish by injecting *HTML* markups. In the example, a *dblpFiguresPanel* markup is injected at the point identified by an *XPath* expression on the *DOM* structure (i.e. the injection point). *DOM functions* are used for this purpose (e.g. *"papers[i].appendChild(dblpFiguresPanel)"* in line 34).

---

[1] http://dblp.uni-trier.de/db/index.html

[2] Available at *http://userscripts.org/scripts/source/76472.user.js.*

[3] Script *weavers* are available for Firefox (e.g. *Greasemonkey*), Internet Explorer (e.g. *IE7Pro* or *Turnabout*), Safari (e.g. *SIMBL+GreaseKit*). In Opera and Google Chrome, it is supported natively.

```
 1  // ==UserScript==                                      1  // ==UserScript==
 2  // @name          Simplified version of dblpFigu       2  // @name          Mod version of dblpFigures
 3  // @description    DBLP past publications for pap       3  // @description    DBLP past publications for pap
 4  // @include        http://icwe2009.webengineering       4  // @include        http://icwe2009.webengineering
 5  // ==/UserScript==                                      5  // ==/UserScript==
 6                                                          6
 7  var doc=window.document;                                7  var doc=window.document;
 8                                                          8
 9  //MOD-LOGIC                                             9  //MOD-LOGIC
10 ⊟function createDblpFiguresPanel(author){              10 ⊟function createDblpFiguresPanel(author){
11   var dblpFiguresPanel=doc.createElement("span");      11   var dblpFiguresPanel=doc.createElement("span");
12   //A panel with a past publications of selected       12   //A panel with a past publications of selected
13   //author from DBLP is created and assigned to        13   //author from DBLP is created and assigned to
14   //the dblpFigures variable                           14   //the dblpFigures variable
15   ...                                                  15   ...
16   return dblpFiguresPanel;                             16   return dblpFiguresPanel;
17  }                                                     17  }
18                                                        18
19 ⊟function init(){                                      19 ⊟function init(loadPaperOcc){
20   //HTML SCRAPING                                      20   //EVENT PARAMETER RETRIEVAL
21   var papers=document.evaluate(                        21   var paper=loadPaperOcc.currentTarget;
22   "//*[@class='paper']",document,null,                 22   var firstAuthor=
23  XPathResult.UNORDERED_NODE_SNAPSHOT_TYPE,null);       23   paper.getElementsByTagName("author").item(0);
24   for(var i=0;i<papers.snapshotLength;i++){            24   //MOD-LOGIC CALL
25    //HTML SCRAPING                                     25   var dblpFiguresPanel=
26    var firstAuthor=document.evaluate(                  26   createDblpFiguresPanel(firstAuthor);
27    ".//*[@class='author']",papers[i],null,             27   //EVENT DISPATCH FOR HTML INJECTION
28  XPathResult.UNORDERED_NODE_SNAPSHOT_TYPE,null).       28   var appendChildPaperOcc=
29  item(0);                                              29   doc.createEvent("ProcessingEvents");
30    //MOD-LOGIC CALL                                    30   appendChildPaperOcc.initProcessingEvent(
31    var dblpFiguresPanel=                               31     "appendChildPaper",paper,dblpFiguresPanel);
32    createDblpFiguresPanel(firstAuthor);                32   doc.dispatchEvent(appendChildPaperOcc);
33    //HTML INJECTION                                    33  }
34    papers[i].appendChild(dblpFiguresPanel);            34
35   }                                                    35
36  }                                                     36
37             DOM Event                                  37            Conceptual Event
38  doc.addEventListener( 'load' ,init,true);             38  doc.addEventListener( 'loadPaper' ,init,true);
```

Fig. 2. Two versions of the *dblpFigures* script: using *DOM Events* (left side) vs. using *Conceptual Events* (rigth side)

This process takes place at the client. The hosting application is completely unaware of this process: no responsibility is taken on certifying or disseminating augmentation scripts among its users. Script safety is not validated, hence, script users are exposed to malware.

This is certainly bad news for users but so is it for Web application owners. Although augmentation can threat the business models of some sites (e.g. by removing banners), in other cases, augmentation accounts for honest enhancements that serve a small set of users the application cannot afford to support their requirements, but leaves external users to fill the gap. After all, popular sites such as *Facebook*, encourage their users to build and share *Facebook applications* on the certitude that this increases the stickiness and usefulness of the site [9,12]. Customer loyalty, engagement and satisfaction are among the rewards. The vision is to create an open ecosystem between the hosting application and the augmentation contributors. However, this vision is undermined by the lack of an architecture where the hosting code can safely coexist with user-provided scripts. Next section looks at related work on this issue, mainly related with *widgets*.

3   Related Work

Certification Approach. *Widget* code is certified. External parties need to get validated before their widgets being made accessible to customers. This approach is exhibited by component-based development [16]. However, it imposes an important burden to third parties. This contrasts to the openness and freewill that characterise crowdsourcing. As a Web master, it is in your own benefit to reduce the hurdles for offering contributions, more to the point, if no financial incentives exist. Reducing the contribution hurdles calls for run-time, automated solutions that permit uploading user contribution while preserving the integrity of the hosting application. That is, *the application architecture itself should ensure this property.*

Iframe-jail Architecture. *Widget* code is isolated. The *widget* is loaded inside of its own *iframe* tag [8]. The *src* attribute of each iframe is a randomly generated subdomain of the hosting application. Being in different domains, *JavaScript*'s Same Origin Policy ensures that each code runs in complete isolation. Isolation brings security but also severely limits the interconnection of the *widget* with the hosting application.

Sanitization Architecture. *Widget* code is sanitized. *Widgets* run in the very same domain that the hosting code. This imposes *widgets* should first go sanitized. That is, *widgets* are compiled into "safe *widgets*" by removing and restricting some *JavaScript* functions. Because the code of the compiled *widget* is not pure *JavaScript*, an interpreter is needed to run it. Once sanitized, the host functionality and the *widget* are executed in the same domain. As an additional precaution, the *DOM API* is re-written (i.e. *tamed API*) to prevent *widget* output to expand over the assigned area (the "virtual iframe" in *Google*'s parlance). This approach is followed by *MS' Web Sandbox* [4] and *Google's Caja* [10].

Validation Architecture. *Widget* code is validated, i.e. code is checked for unsafe instructions. A list of unsafe instructions can be found at *Dojo*'s home page [7]. If valid, *widget* markup is constrained to a "virtual iframe" using a *tamed API* ("sandbox" in *Dojo*'s terminology).

Figure 3 depicts an architecture along the lines of the aforementioned approaches. *Widgets* are kept at the server. *Google/Microsoft* first sanitizes the *widget*'s code, and then, makes it available for the application to load at runtime. By contrast, *Dojo* moves the validation process at the client: the *widget* is loaded, and then, validated. If passed then, the *widget* is ready for enactment.

However, augmentation scripts pose more stringent demands than *widgets*. According to Wikipedia, a *widget* "is a portable chunk of code that can be installed and executed within any separate *HTML*-based web page by an end user without requiring additional compilation". *Widgets* are Web components (i.e. potentially reusable in different applications) and "single-markuped" (i.e. only one piece of markup is delivered). By contrast, the scope of augmentation scripts can be a single application, and scripts can deliver a set of markups. Additionally, *widgets* are thought to be rendered in a pre-set cell. This contrast with the dynamic binding that characterise the selection of the place where the script output is placed. This puts aside the *Caja* approach, where *widgets* can only manipulate a bounded portion (virtual iframe). Additionally,

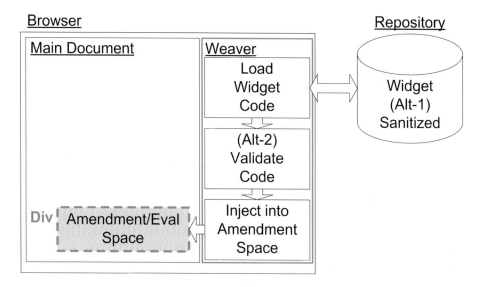

Fig. 3. An architecture for *widget* secure execution: sanitization approach ("Alt-1") and validation approach ("Alt-2")

Caja widgets are coded in *Valija* (a subset of *JavaScript*) but debugged in *Cajita* (the compiled code in which Valija is compiled to). Since bugs are noted in Cajita, the programmer is forced to move back and forth between *Valija* and *Cajita* to debug the code. Finally, *Valija* and its *tamed API* are built on top of *HTML* and *JavaScript* (*EcmaScript5*). Upgrades on these standards can require new amendments to *Valija* and its *tamed API*. Our approach departs from creating a new framework for safely executing third-party code but to use the *HTML5* security characteristics. This is the *MI Architecture*[4].

4   The Modding-Interface Architecture

Figure 4 outlines the runtime evolution of a document with embedded scripts. On loading, the document becomes a *DOM tree*. Initially, *DOM nodes* stand for the raw content of the page. Additionally, some nodes contain "cells" (denoted as doted-lined rectangles in Figure 4). A cell is realized as either an *HMTLDivElement* (i.e. *<div> HTML tag*) or an *HTMLIFrameElement* (i.e. *<iframe> HTML tag*) element that holds the script. Enacting the script can result on augmenting the *DOM tree* (denoted as a dot-filled circle in Figure 4). This figure illustrates the existence of two spaces: "the eval space" where the script is enacted (doted-lined rectangles), and "the amendment space" where the script markup is placed. In widget-oriented architectures both spaces coincides.

---

[4] The term "modding" is borrowed from the video-game industry where mods are introduced by vendors for players to tune the appearance, weapons, and even the strategy of the game [14]. Similar terminology is used for cars with a similar meaning: product personalization by the customers themselves.

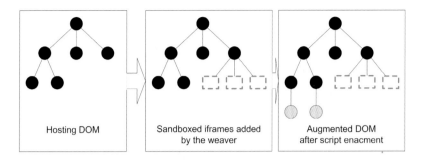

Fig. 4. Augmentation at run-time: *DOM tree* evolution

The *Modding-Interface Architecture (MI Architecture)* (see Figure 5) clearly distinguishes between the amendment space and the eval space. The amendment space is contained within the hosting document. The eval space is placed within an *"iframe jail"*. The eval space is sandboxed from the hosting document so that access is not permitted. The novelty comes from the communication model. A publish/subscribe communication model regulates the interaction between the eval space and the amendment space. A *Modding Interface* describes the messages permitted between these two spaces. Being event-driven, a *weaver* regulates publish/subscribe messages. However, and unlike *Greasemonkey*-like approaches, now the *weaver* is part of the hosting application itself. No browser plugin is required.

Therefore, engineering a Web application for augmentation requires (1) a *Modding Interface* as a means to preserve application integrity and, (2) a generic *weaver* that mediates between the amendment space and the eval space. Next sections address these topics.

5   Modding Interface Specification

This interface encapsulates the hosting code. It regulates both (1) what data can be obtained and (2), what amendments are permitted on the hosting code. Interfaces are commonly specified in terms of operations defined upon data types. However, *JavaScript* favours event-based programming, i.e. handlers are associated to UI events. Unlike operations, handlers are not explicitly called but *triggered* when the associated event occurs. Akin to the *JavaScript* approach, *Modding Interfaces* are to be described in terms of events rather than operations, but they will act upon *concepts* (e.g. *Paper*) rather than *DOM nodes*. In this way, scripts can subscribe to the event *loadPaper* (rather than the *DOM event, load*) and obtain *Paper* data as event payload rather than scraping the *DOM tree*. Scripts can also publish the event *appendChildPaper* to add an *HTML fragment* as a child of a *Paper* (rather than using an *XPath expression*).

The right side of Figure 2 shows the *dblpFigures* augmentation script but now using *Conceptual Events*. The augmentation logic is the same (lines 10-17). Differences rest on (1) *HTML* scraping being substituted by event parameter recovering (lines 21-23) and, (2) amendment spaces described by the point where *Conceptual Events* occur (lines 28-32) rather than *XPath expressions*.

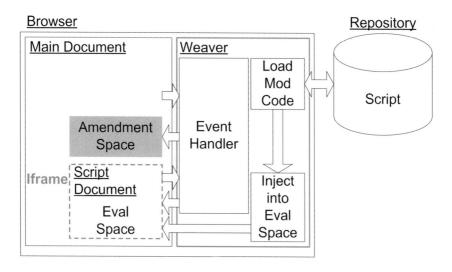

Fig. 5. The *Modding-Interface* Architecture

Therefore, a *Modding Interface* encapsulates a Web application in terms of its concepts, and provides a set of services to "read" and to "write" these concepts. The "read" part realizes the required interface as the set of events the interface just signals but leaves to the scripts the event processing (a.k.a. **Publishing Events**). As for the "write" part, it identifies the amendment space in terms of concept occurrences rather than through *DOM nodes*. Rather than using *XPath* on *DOM trees*, the amendment space is identified by the target of **Processing Events**. For instance, the event *appendChildPaper* indicates that *Paper* denotes an amendment point. Scripts can now raise *appendChildPaper* to inject its *HTML markup* into this amendment point. *Processing Events* then realize the provided interface[5].

The *Modding Interface* is described as an *OWL* document [15]. *OWL* permits to describe concepts, properties related to these concepts and associations between concepts. The aim of *OWL* is to provide a way to exchange information between applications with a specific semantic. Such aim aligns with our purposes. Next paragraphs describe the *"Concepts"*, *"PublishingEvents"* and *"ProcessingEvents"*[6] for our sample case (see Figure 6).

Concepts. The *ICWE* Website is seen as renderer of a set of concepts: *ConferenceEvent*, *Paper*, *Person*, etc. (lines 14-16). The *<Ontology>* element contains the description for these concepts. Concept description includes *<DataTypeProperty>* and *<ObjectProperty>* (i.e. title and author (lines 17-24)). It is possible to import the ontology. For conference description, an external ontology is available at [13].

---

[5] The terminology of "processing events" and "publishing events" is widely used for event-based components such as portlets [11].

[6] A schema for defining *Modding Interfaces* is available at *http://userscripts.org/scripts/source/61129.user.js.*

```
 1  <!DOCTYPE rdf:RDF [
 2  <!ENTITY xsd "http://www.w3.org/2001/XMLSchema#">
 3  <!ENTITY rdf "http://www.w3.org/1999/02/22-rdf-syntax-ns#">
 4  <!ENTITY rdfs "http://www.w3.org/2000/01/rdf-schema#">
 5  <!ENTITY owl "http://www.w3.org/2002/07/owl#">
 6  <!ENTITY swc "http://data.semanticweb.org/ns/swc/ontology#">
 7  <!ENTITY swrc "http://swrc.ontoware.org/ontology#">
 8  <!ENTITY mod "http://userscripts.org/scripts/source/61129.user.js#">
 9  <!ENTITY icwe "http://icwe2009.webengineering.org/">
10  ]>
11  <rdf:RDF xmlns:base="&icwe;" xmlns:owl="&owl;" xmlns:rdf="&rdf;"
12  xmlns:rdfs="&rdfs;" xmlns:mod="&mod;">
13  <owl:Ontology>
14    <owl:Class rdf:about="&swc;ConferenceEvent"/>
15    <owl:Class rdf:about="&swc;Paper"/>
16    <owl:Class rdf:about="&swrc;Person"/>
17    <owl:DatatypeProperty rdf:about="&swrc;title">
18      <rdfs:domain rdf:resource="&swc;Paper"/>
19      <rdfs:range rdf:resource="&xsd;string"/>
20    </owl:DatatypeProperty>
21    <owl:ObjectProperty rdf:about="&swrc;author"/>
22      <rdfs:domain rdf:resource="&swc;Paper"/>
23      <rdfs:range rdf:resource="&swrc;Person"/>
24    </owl:ObjectProperty>
25  </owl:Ontology>
26
27  <mod:PublishingEvent rdf:ID="loadPaper">
28    <mod:payloadType rdf:resource="&swc;Paper"/>
29    <mod:uiEventType rdf:resource="&mod;load"/>
30    <mod:cancelable>false</mod:cancelable>
31  </mod:PublishingEvent>
32
33  <mod:ProcessingEvent rdf:ID="appendChildPaper">
34    <mod:payloadType rdf:resource="&mod;HTMLSpanElement"/>
35    <mod:operationType rdf:resource="&mod;appendChild"/>
36    <mod:targetConcept rdf:resource="&swc;Paper"/>
37  </mod:ProcessingEvent>
38  </rdf:RDF>
```

Fig. 6. *ICWE Website Modding Interface*

Publishing Events. These events notify "concepts" being delivered by the Web application. In other words, the payload of a *Publishing Event* is a concept of the Web application at hand. But events are happenings of interest, i.e. they are instants of time. An event cannot be described just by its associated concept but needs to include what happens to this concept, e.g. loading, selecting, de-selecting the concept, etc. Therefore, *Publishing Events* are described by both the event payload (*"payloadType"* property), and the time when the event arises (*"uiEventType"* property). The values for *"uiEventType"* are taken from the *W3C's DOM Level 2 Events* specification [17]. Additionally, a *"cancelable"* property is added that mimics the namesake property available for *JavaScript* events whereby an event is liable to be called off by a handler so that the occurrence is no longer propagated to other handlers. As specified in Figure 6 (lines 27-31), *loadPaper* is introduced as a *Publishing Event* to occur every time a *Paper* is loaded.

Processing Events. A website determines *what* can be augmented but leaves to the scripter to decide *when* and *how* is to be augmented. The *what* refers to the concept that denotes the amendment space (*"targetConcept"* property). But being a layout issue, the concept alone is not enough. We need to indicate the position w.r.t. the concept (*"operationType"* property) through a reference to the *W3C's DOM Level 2 Core* operations [17]. Figure 6 shows an example where the concept *Paper* is used to pinpoint the amendment space. The *"operationType"* indicates that augmented content is to be rendered as children of the *Paper* at hand (i.e. *appendChildPaper* (lines 33-37)).

As for the *how*, traditional scripts can inject any *HTML fragment* on the premise that the disclosure of the page implementation makes them acknowledgeable about what would be the right fragment code. This approach may work for simple pages but is hardly scalable as pages become more complex. We cannot rely on end users peering on *HTML code* to ascertain what would be a wrong fragment to be injected. We resort to *HTML types* [17]. The augmentation markup should be compliant to an *HTML type* (*"payloadType"* property). This type restricts how rendering can be augmented. For instance, augmenting a *"Paper"* is set to be of type *HTMLSpanElement,* meaning that augmentation markup on *Papers* need to be compliant with this type. This introduces a type-like mechanism for regulating augmentation to existing Web application. The *weaver* can then check whether this *payloadType* is fulfilled, and if not so, ignores the script markup but still renders the rest of the page. This is akin to browser practices where wrong *HTML tags* do not prevent the browser from rendering the page.

6   A Weaver for Augmentation Scripts

The *weaver* mediates between the main document and the script document (see Figure 5). Specifically, the *weaver*'s duties include (1) loading the augmentation scripts for the current user, and (2), managing *Conceptual Events*. This section outlines the implementation of these functions. The code has been tested for *Google's Chrome,* using extensively *HTLM5* new features [18].

Loading Augmentation Scripts. Customers of the Web application have previously registered their interests in some augmentation scripts. These preferences are kept locally through a *localStorage* variable at the browser: *augmentationConfiguration.* Scripts are kept at the server. Figure 7 lists the *weaver*'s code that loads the scripts.

On loading the Web application, the *weaver*'s first duty is to load the script identifiers kept at *augmentationConfiguration* (lines 2-3). For each script, the *weaver* creates an *iframe* (line 7-11). An *iframe* holds a generic document (*src* attribute) that is parameterized with the identifier of the script at hand. This document has no rendering counterpart (i.e. *"display:none"*). *Iframes* are sandboxed. When the *iframe* is added to the page (line 12), the script is downloaded and evaluated. Being sandboxed, the script cannot access the hosting page (i.e. the script cannot subscribe to UI events from the main document). Interactions are restricted to occur through a channel (line 14-17). A message channel is an *HTML5* object that enables the direct communication of independent pieces of code (e.g. running in different browsing contexts). This interaction follows a publish/subscribe pattern based on *Conceptual Events*.

```
1  // Load configuration
2  var installedScripts=localStorage.getItem("augmentationConfiguration");
3  installedScripts=installedScripts?JSON.parse(installedScripts):[];
4  // Load scripts
5  for(script in installedScripts){
6    // Create of sandbox
7    var iframe=document.createElement("iframe");
8    iframe.src="http://icwe2009.webengineering.org/mod_document.html?"+
9              installedScripts[script].id;
10   iframe.style.setProperty("display","none","important");
11   iframe.sandbox="allow-scripts";
12   document.body.appendChild(iframe);
13   // Initialize communication
14   var channel = new MessageChannel();
15   iframe.addEventListener("load",function(){
16       iframe.contentWindow.postMessage("initChannel",[channel.port2],
17       "http://icwe2009.webengineering.org/"); },true);
18   ...
```

Fig. 7. *Weaver*'s code: loading augmentation scripts

Managing Conceptual Events. When the *iframe* space is initialized, the main document and the script document are ready for exchanging *Conceptual Events*. However, these *Conceptual Events* are to be produced/handled by the *weaver*. The *weaver* has two main duties: raising *Publishing Events* in the main document, and handling *Processing Events* as signalled by *script* documents.

The process goes as follows. The UI event (e.g. loading a page) is first notified to the *weaver*. The *weaver* constructs and raises the *Conceptual Event* (e.g. *loadPaper*) along the indications of the *Modding Interface*. *Conceptual Events* are captured by the *script* that recovers the event payload, constructs an *HTML fragment*, and dispatches the appropriate *Processing Event* (e.g. *appendChildPaper*). This *Processing Event* is then de-constructed in terms of UI operations by the *weaver* according to the indications of the *Modding Interface*. These UI operations causes the main document (the page you see) to be augmented.

7   Discussion

7.1   Impact on Security

The work's first motivation was safe coexistence of heterogeneous-sourced code. Both redirection to phishing pages or stealing sensitive data are avoided by running the script inside an *"iframe jail"*. On the other side, we can prevent port scanning and history sniffing by using the same approach as *Google Caja*: a *monkey patch* [2]. *Monkey patch* is a way to extent/modify runtime code in dynamic languages. This technique can be applied to dynamically replace/extend script functions liable to content malware with others that block such malware. Finally, browser blocking can be alleviated as in *MS' Web Sandbox*, i.e. using a QoS Layer [5]: a wrapper-like mechanism that imposes some limits on the consumption of shared resources. Exceeding these thresholds (e.g. CPU consumption) makes the script be blocked.

```
<head>
 <link rel="moddingInterface" href="/modding_interface.owl" />
 <link rel="transformation" href="/extractor.xsl" />
 <script type="text/javascript" src="/weaver.js"></script>
 ...
```

Fig. 8. Augmentation-enabled page: meta-data about the *Modding Interface*

7.2 Impact on Users

In a crowdsourcing setting, the viability of an approach heavily rests on causing minimal disturbance to the involved parties: Web application programmers and script programmers. As for the former, the *MI Architecture* imposes almost no disruption. Augmentation-enabled HTML pages differ from traditional pages in that they keep three links: two to the *MI* files, another to the *weaver* script (see Figure 8). Apart from that, these pages do not differ from "traditional pages".

From a script-programmer perspective, *MI* implies notification/publication to be based on *Conceptual Events* rather than *DOM events*. Otherwise, native *JavaScript* mechanisms are used to handle *Conceptual Events* with no variations w.r.t. traditional script development. From the start of this work, we have been very conscious about reducing the hurdles for offering contributions. Next paragraphs provide evidence that programming on top of a *Modding Interface,* causes minimal deviation from traditional practices.

Notification of Processing Events. *JavaScript* follows an event-based approach where handlers can be associated with *DOM-based events*. Operations are available for creation of event occurrences (e.g. *createEvent("MouseEvents")*), assigning the payload to an occurrence (e.g. *initMouseEvent("eventInstance", "eventParameters")*), or raising the event manually (e.g. *dispatchEvent(eventOccurrence)*). Raising of *Conceptual Events* uses these standard *JavaScript* operations. Back to our running example, a *dblpFiguresPanel* (i.e. an *HTML fragment*) is to be injected as a child of a *Paper*. Figure 2 (right side) show the code along the following pattern: *createEvent* (lines 28-29), obtain concept (line 21), *initProcessingEvents* (line 30-31), *dispatchEvent* on this concept (line 32). This is standard *JavaScript* code. The only difference with traditional scripting is that now the injection point is not a *DOM node* but the current concept. This current concept is to be obtained through a *Publishing Event*.

Subscription to Publishing Events. *JavaScript* achieves event subscription by registering a handler through the *addEventListener* method. Subscription to *Conceptual Events* is accomplished in the very same way: associating a handler. For instance, instruction (line 38 in Figure 2 (right side)) *"doc.addEventListener("loadPaper", init, true)"* adds a handler to the *loadPaper* event, i.e. occurrences of *loadPaper* will trigger the *init()* function. The difference rests on handlers being associated to the whole document (i.e. variable *doc*) rather than to *DOM nodes* (e.g. a *checkbox*). This highlights the fact of events being risen by acting on *Papers* rather than on *DOM nodes* (i.e. the circumstantial representation of these *Papers*).

7.3   Impact on Performance

All our measurements are realized in Windows 7 x64 running on Intel Core2 Duo 2.20 GHz CPU with 4GB of memory. The experiments have been realized with a domestic 6Mbps WIFI LAN bandwidth.

Loading time. The *Greasemonkey* architecture keeps scripts at the client. So, no loading penalty at the time the script is enacted. By contrast, our approach makes scripts a valuable asset of the Web application which becomes a partner on disseminating these resources among its user base. Therefore, the *MI Architecture* maintains scripts at the server as site resources. When application pages are loaded, so are the appropriate scripts (as any other page resource such as associated images). Compared with *Greasemonkey*, this certainly imposes an overhead. However, script files tend not to be very large, and its cost is similar to loading a *"jpg"* thumbnail file. Additionally, the *weaver* and *Modding Interface* file are loaded on accessing the first page. The size of the *weaver* file is 3.8kb (no obfuscated) which approximately accounts for a 100 millisecond delay (less if the *weaver* is cached by the browser). The size of the *Modding Interface* for a given page is similar to a script. On the upside, this approach frees users from installing any plugin (as it is the case for *Greasemonkey*).

Enactment time. Script enactment takes place at the client (no server impact). *Greasemonkey* scripts act upon *DOM events*. By contrast, interface-aware scripts rest on *Conceptual Events*. This imposes an indirection: *Conceptual Events* need first to be (de)constructed from *DOM events* and send over the channels that connect the script space with the hosting application space. A first experiment has been conducted for the *dblpFigures* sample realized as both a *Greasemonkey* script and an interface-aware script. The results show that the indirection and communication accounts for a delay of 30 and 2 milliseconds, respectively, when compared with the *Greasemonkey* alternative (i.e. acting directly upon *DOM events*).

8   Conclusions

Web augmentation requires hosting markup and user-provided markup to run together. This raises integrity issues where malware can cause important damages on both end users and the reputation of hosting applications. We tackled this issue by combining *"iframe jails"* and "modding-interface" *HTML5 channels*. The approach benefits Web applications that now can be safely augmented. So does for script contributors that now achieve greater visibility by having their scripts uploaded at the hosting application.

Safe coexistence of external code is just one of the issues raised by crowdsourcing augmentation. The construction of communities around Web applications implies promoting/rewarding contributions, disseminating contributions through the Web application, facilitating end users to suggest augmentations, and so on. In the same way that Web2.0 APIs open data silos to achieve application composition at the back-end, we envision *Modding Interfaces* "to open" application markup to crowdsourced, front-end composition.

Acknowledgments. This work is co-supported by the Spanish Ministry of Education, and the European Social Fund under contract TIN2008-06507-C02-01/TIN (MODELINE). Arellano has a doctoral grant from the Spanish Ministry of Science & Education.

References

1. Greasemonkey Homepage, http://www.greasespot.net/
2. Monkey patch, http://en.wikipedia.org/wiki/Monkey_patch
3. Bouvin, N.O.: Unifying Strategies for Web Augmentation. In: Proceedings of the 10th ACM Conference on Hypertext and Hypermedia, HYPERTEXT 1999 (1999)
4. Microsoft Corp. Microsoft Web Sandbox, http://websandbox.livelabs.com/
5. Microsoft Corp. Microsoft Web Sandbox: QoS Layer, http://websandbox.livelabs.com/documentation/vm_qos.aspx
6. Díaz, O., Arellano, C., Iturrioz, J.: Layman Tuning of Websites: Facing Change Resilience. In: The 17th International Conference on World Wide Web, WWW 2008 (2008)
7. Dojo. Secure Mashups with dojox.secure, http://www.sitepen.com/blog/2008/08/01/secure-mashups-with-dojoxsecure/
8. Hoffman, B., Sullivan, B.: Web Mashups and Aggregators. In: AJAX Security, ch. 11, pp. 295–327. Addison-Wesley, Reading (2007)
9. Facebook Inc. Apps on Facebook, http://developers.facebook.com/docs/guides/canvas/
10. Google Inc. Google Caja: A source-to-source translator for securing Javascript-based web content, http://code.google.com/p/google-caja/
11. JCP. JSR 168: Portlet Specification Version 1.0 (2003), http://www.jcp.org/en/jsr/detail?id=168
12. Maver, J., Popp, C.: Essential Facebook Development: Build Successful Applications for the Facebook Platform. Addison-Wesley, Reading (2009)
13. Möller, K., Bechhofer, S., Heath, T.: Semantic Web Conference Ontology (2007), http://data.semanticweb.org/ns/swc/ontology
14. Scacchi, W.: Computer Game Mods, Modders, Modding, and the Mod Scene. First Monday 15 (2010)
15. Smith, M.K., Welty, C., McGuinness, D.L.: OWL Web Ontology Language Guide. W3C Recommendation (2004), http://www.w3.org/TR/owl-guide/
16. Voas, J.: Certification: Reducing the Hidden Costs of Poor Quality. IEEE Software 16, 22–25 (1999)
17. W3CDOMWG. W3C DOM Level 2, http://www.w3.org/DOM/DOMTR#dom2
18. W3CHTML5WG. HTML5, http://www.w3.org/TR/html5/
19. Yu, J., Benatallah, B., Casati, F., Daniel, F.: Understanding Mashup Development. IEEE Internet Computing 12, 44–52 (2008)

Design of Negotiation Agents Based on Behavior Models

Kivanc Ozonat and Sharad Singhal

Hewlett-Packard Labs, Palo Alto, CA, USA
kivanc.ozonat@hp.com, sharad.singhal@hp.com

Abstract. Despite the widespread adoption of e-commerce and online purchasing by consumers over the last two decades, automated software agents that can negotiate the issues of an e-commerce transaction with consumers still do not exist. A major challenge in designing automated agents lies in the ability to predict the consumer's behavior adequately throughout the negotiation process. We employ switching linear dynamical systems (SLDS) within a minimum description length framework to predict the consumer's behavior. Based on the SLDS prediction model, we design software agents that negotiate e-commerce transactions with consumers on behalf of online merchants. We evaluate the agents through simulations of typical negotiation behavior models discussed in the negotiation literature and actual buyer behavior from an agent-human negotiation experiment.

1 Introduction

Negotiation is the process of reaching an agreement between two (or more) parties on the issues underlying a transaction. Negotiations are an integral part of business life, and are instrumental in reaching agreements among the parties.

Over the last two decades, electronic commerce has become widely adopted, providing consumers with the ability to purchase products and services from businesses online. Electronic commerce offers many advantages, including increased operational efficiency for the businesses, reduction of the inventory costs, and availability of the product and service 24 hours a day. Yet, after two decades, electronic commerce systems still lack the ability to enable the businesses to negotiate with consumers during the purchases. There is a need for automated, online software agents that can negotiate with consumers on behalf of businesses.

Designing automated agents for negotiation has been explored across multiple disciplines, in artificial intelligence [5, 6, 7], human psychology [16, 17, 18, 21] and statistical learning [3, 9]. One form of negotiation that has received much attention in these disciplines is the *bilateral sequential negotiation*. A bilateral sequential negotiation typically starts with one party (e.g., the *buyer*) making an offer on each of the negotiated issues. The issues can include, for instance, the price of a product or service, its delivery time and its quantity. The opposing party (e.g., the *seller*) has three options: (i) it accepts the offer, (ii) it rejects the offer and makes a counter offer on each issue, or (iii) it rejects the offer and

L. Chen, P. Triantafillou, and T. Suel (Eds.): WISE 2010, LNCS 6488, pp. 308–321, 2010.
© Springer-Verlag Berlin Heidelberg 2010

terminates the negotiation. The process continues in rounds, where, the buyer and the seller make and respond to counter offers.

The artificial intelligence community has focused on finding game-theoretic solutions to the bilateral sequential negotiation problem. Each of the two parties is assumed to have a *utility function* of the negotiated issues. The utility function measures how preferable an offer is to the party. In the game-theoretic model, each party is assumed to have the goal of maximizing its utility function, and the aim is to find an agreement that is *Pareto-optimal*, i.e., an agreement where the utility of one party cannot be increased any further without decreasing the utility of the other [6].

Many game-theoretic models of negotiation assume that each party knows the opposing party's utility function [6], while more realistic models employ statistical algorithms that *learn* the opposing party's utility function [3, 5, 9]. The learning can be based, for instance, on a training set of issue offers from the parties. In either case, a fundamental problem, when one of the parties is a human, is that people rarely construct mathematical functions, such as utility functions, to quantify their issue preferences in a negotiation. Moreover, humans are not "super-rational beings"; during a negotiation, it is unlikely that a person will always give the most rational decisions in terms of maximizing a mathematical function.

We take an alternative approach, and aim to predict the behavior of a human buyer in the issue space instead of estimating a utility function. In [5], the human buyer's negotiation behavior has been modeled directly in the issue space rather than in the utility space. Multiple sets of parameterized functions, which imitate buyers' behaviors in the issue space, are used to construct trajectories that represent the buyer's movement in the issue space from its initial offer to the final (or target) offer. A typical trajectory consists of a sequence of these parametrized functions such that the parameters of the functions, the positions of the functions in the sequence, and the time of switching from one function to another in the sequence are determined based on prior probabilities.

We devise and employ statistical learning-based prediction techniques based on the behavior models developed in [5]. Our goal is to predict the future trajectories of the buyer in the issue space based on the history of the buyer's offers. Linear dynamical systems have been used in literature to predict trajectories in fields as diverse as communications, aeronautics, finance, and navigation [11]. We employ an LDS-based prediction model to predict the buyer's future trajectory. The models in [5] consist of sequences of alternating and changing functions, suggesting a prediction model that accounts for switching behavior. We therefore approach the prediction problem through *switching* linear dynamical systems (SLDSs). A challenge is that one cannot assume a prior knowledge of the SLDS model size. To address this challenge, we extend the SLDS learning to a setting that allows the model size grow with new offers. We then design agents based on the SLDS prediction model, and evaluate them using two sets of experiments: The first set tests the agents against trajectories generated according to the models in [5], and the second set tests them against actual buyer behavior from an agent-human negotiation experiment.

2   Negotiation Behavior of Buyers

A (bilateral sequential) negotiation occurs between two parties, the buyer and the seller of some product or service. The two parties negotiate over D issues such as the price of the product, the delivery time and the product quantity. At each negotiation round n, $n \geq 1$, the buyer makes an offer, y_n, to the seller. The offer y_n is a D-dimensional real-valued vector with its dth element $y_{n,d}$ denoting the buyer's offer for the dth issue. The seller responds with one of three options: (i) continue the negotiation with a counter offer, which, analogous to y_n, is also a D-dimensional vector, (ii) terminate the negotiation by accepting y_n (i.e., a deal is reached), (iii) terminate the negotiation by rejecting y_n (i.e., no deal reached). We assume that the seller is a software agent, and the buyer is a human.

The buyer has a set of *targets* that it does not disclose to the seller. A target is an offer that the buyer aspires the seller would accept by the termination of the negotiation. Often, the buyer has multiple targets. For instance, in a scenario with 2 negotiated issues, price and delivery time of a product, the buyer may have targets of {240 dollars and 2 weeks}, {220 dollars and 3 weeks}, and {160 dollars and 4 weeks}.

The buyer's negotiation behavior has been modeled in [5]. According to the models discussed in [5], the buyer rarely offers one of its targets to the seller in the initial round, i.e., at $n = 1$. Instead, at $n = 1$, it makes an offer that is more advantageous to itself than its targets. For instance, a buyer with a target of {240 dollars and 2 weeks} may offer only {200 dollars and 1 week} at $n = 1$. The buyer then follows a trajectory in the issue space from this initial offer towards one of its targets. Often, the buyer's trajectory never reaches the target (i.e., the buyer never offers the target) with the intention of obtaining more favorable counter offers from the seller.

The *offer strategy* of a buyer specifies the movement in the issue space of the buyer from one round to the next. The negotiation may be viewed as consisting of multiple phases or states, $s = 1, 2, 3...$, where, at each state s, the buyer follows some offer strategy. Each time the buyer changes its strategy, a new state starts. The buyer can return to a previously-visited state. For instance, the buyer might make small concessions both in price and delivery time for $1 \leq n \leq 5$ (state $s = 1$), make large concessions in price while holding the delivery time constant for $6 \leq n \leq 8$ (state $s = 2$), and, make small concessions both in price and delivery time for $n > 8$ (back to state $s = 1$).

We design a seller's agent that negotiates with the buyer. At every round n, the agent makes a counter offer to the buyer based on a prediction model that utilizes the history of the buyer's offers up to and including round n. Thus, the counter offer is based on the future trajectory of the buyer as predicted by the agent. The agent estimates the likelihood of the buyer's future steps in the issue space, and selects a counter offer that is likely to be along the buyer's future trajectory. The prediction model and the agent are discussed in sections 3 through 5.

3   Buyer Prediction Model

Linear dynamical systems (LDSs) have been used to model dynamical phenomena in multiple disciplines as diverse as communications, aeronautics, finance, tracking and navigation [11]. While the trajectory of the buyer in the issue space in a given state s can be captured by an LDS, the fixed (Gaussian) model of the LDS is insufficient to explain the buyer's behavior across multiple states.

A switching linear dynamical system (SLDS) is a variation on the LDS, where the fixed Gaussian model is replaced with a conditional Gaussian model [11,12]. The underlying conditional Gaussian model provides the flexibility needed to describe dynamical phenomena that exhibit structural changes over time. We model the buyer's negotiation behavior through an SLDS with the transition model given by

$$x_n = A(s_n)x_{n-1} + v_n(s_n), \tag{1}$$
$$y_n = Cx_n + w_n, \tag{2}$$
$$p_{i,j} = P(s_n = i | s_{n-1} = j), \tag{3}$$

where $x_n \in R^D$ is the hidden LDS state at round n, $y_n \in R^D$ is the observation at round n, $w_n \in R^D$ is the Gaussian measurement noise, $v_n(s_n) \in R^D$ is the Gaussian state noise, $A(s_n) \in R^{D \times D}$ is the LDS state transition matrix, and $C \in R^{D \times D}$ is the LDS observation matrix. x_1 is Gaussian.

The joint distribution of X,Y and S is given by

$$P(y_1^n, x_1^n, s_1^n) = \prod P(y_n|x_n)P(x_1|s_1) \prod P(x_n|x_{n-1}, s_n)P(s_1) \prod P(s_n|s_{n-1}), \tag{4}$$

where y_1^n denotes the sequence $y_1, ..., y_n$, and similarly for x_1^n and s_1^n.

While the SLDS learning problem has been addressed in the literature [11,12], the focus has been on learning an SLDS with the number of switching states known a priori. In our setting, it is not possible to assume a priori knowledge of the model size. The model size can grow with new issue offers from buyers. We address this problem in the next section where we discuss the *infinite SLDS* within the minimum description length framework.

4   Infinite SLDS Modeling with MDL Principle

We focus on the principle of minimum description length (MDL), with its roots in the works Kolmogorov, Solomonoff and Shannon [10,13,14], to model and learn the parameters of an infinite SLDS. We use Shannon-optimal code lengths to compute the minimum code length required for coding the observations y_1^T. Any code requires a *codebook*, i.e., a lookup table that shows the mapping between the data to be coded and the code. In the SLDS setting, the codebook grows as the number of switching states increases. We find the SLDS model that minimizes the description length, which is the sum of the Shannon-optimal code lengths and the size of the codebook. Thus, the number of switching states of the SLDS does not have to be pre-defined; it is discovered through the MDL minimization.

4.1   Related Work on Infinite-Length HMMs

The standard learning procedures for hidden Markov models (HMMs) assume that the number of hidden states is finite and known a priori. Beal [2], Teh [15] and Fox [8] extended HMMs to have a countably infinite number of hidden states through the hierarchical Dirichlet process. The hierarchical Dirichlet process has three hyper-parameters, two of them controlling the tendency of the process to enter a new state, and one parameter controlling the tendency of a state to self-transition. The Dirichlet model allows the process to discover new states through these hyper-parameters, and further ensures that the process can exit from a newly discovered state with non-zero probability.

While the hierarchical Dirichlet model makes it possible to learn infinite HMMs, the inherent assumption is that the hidden state process is first-order Markov. The first-order Markov assumption is not valid for the hidden state process of the SLDS due to the presence of the switching state s.

4.2   MDL for the SLDS

The principle of minimum description length (MDL) stems from Shannon's coding theory [1, 13, 19]. Given a set of hypotheses indexed by h, the observed data y can be described with the hypothesis h using total code length $l(y|h)+l(h)$. Here, $l(h)$ is the minimum code length to describe the selected hypothesis h, and $l(y|h)$ is the minimum code length to describe y according to the selected hypothesis h. Each hypothesis h induces a probability distribution on the observed data. Based on the Kraft's inequality, $l(Y|h) = -\log P(y|h)$ and $l(h) = -\log P(h)$. The MDL principle selects the hypothesis $h = h^*$ if h^* minimizes $l(y|h) + l(h)$ among all h.

We let the switching state s play the role of the hypothesis h. This leads to the following code length for the SLDS given in (1)-(4) at time T,

$$\sum_{n=1}^{T} l_{\Theta_1}(y_n|y_1^{n-1}, s_1^n) + \sum_{n=1}^{T} l_{\Theta_2}(s_n|s_{n-1}), \tag{5}$$

where Θ_1 are the parameters $A(s)$, C, and the covariance matrices for $v(s)$ and w, Θ_2 is the pmf $p_{i,j}$, l_{Θ_1} maps $y_n|(y_1^{n-1}, s_1^n)$ to its minimum code length under Θ_1, and l_{Θ_2} maps $s_n|s_{n-1}$ to its minimum code length under Θ_2.

A *codebook* is a lookup table used for coding and decoding; it shows the mapping between the values of y and s and the code. In the infinite SLDS setting, the size of the codebook is not a constant. As new switching states are discovered, the codebook grows. Thus, the size of the codebook should be included in the description length. The description length therefore is

$$\sum_{n=1}^{T} l_{\Theta_1}(y_n|y_1^{n-1}, s_1^n) + \sum_{n=1}^{T} l_{\Theta_2}(s_n|s_{n-1}) + C_{\Theta_1} + C_{\Theta_2} \tag{6}$$

where C_{Θ_1} and C_{Θ_2} denote the size of the codebook for the mapping between $y_n|(y_1^n, s_1^n)$ and its code, and the size of the codebook for the mapping between $s_n|s_{n-1}$ and its code, respectively.

Learning

The parameters Θ_1 and Θ_2 are learned to minimize (6) through the generalized EM algorithm. The E-step is the inference, i.e., the estimation of the posterior probabilities of the the hidden states x_1^T and the switching states s_1^T given the observations y_1^T and the parameters Θ_1 and Θ_2. The M-step consists of the parameter updates.

We focus on the case with a fixed number of switching states in this section (i.e., the codebook sizes are constant), and extend it to the infinite setting in the next section. Suppose $n = T$ and there are K states. During the E-step, the switching states s_1^T are set, and the posterior probabilities of x_1^T are computed. We denote by $H_{T,k}$ the minimum value of (5), when the switching state sequence s_1^T ends with $s_T = k$. $H_{T,k}$ is computed recursively as

$$H_{n,k} = \min_j H_{n-1,j} + l_{\Theta_1}(y_n|y_1^{n-1}, s_1^{n-2,*}(j), s_{n-1} = j, s_n = k) + l_{\Theta_2}(s_n = k|s_{n-1} = j), \tag{7}$$

where $1 \le n \le T$, and $s_1^{n-2,*}(j)$ is the switching state sequence that minimizes $H_{n-1,j}$ among all state sequences up to time $n - 2$.

Based on the MDL, the second term on the right side of (7) is given by $-\log P(y_n|y_1^{n-1}, s_1^{n-2,*}(j), s_{n-1} = j, s_n = k)$. This probability is computed through the standard forward algorithm for the Gaussian LDS inference based on the Θ_1 parameters estimated in the M-step. The standard LDS inference can be used since the hidden states x_1^n of the SLDS are first-order Markov when conditioned on the switching state sequence s_1^n. The third term on the right side of (7), which is $-\log P(s_n = k|s_{n-1} = j)$, is computed through the Θ_2 estimate of the M-step.

Once all the costs H up to time $n = T$ are computed for every state, the index of the switching state that minimizes $H_{T,k}$ is selected, and the switching states are back-traced to find the best sequence of s_1^T.

The M-step updates the parameters, Θ_1 and Θ_2, based on the posterior probabilities (from the standard LDS inference of the E-step) and the switching states from the E-step.

New Switching States

According to the MDL, an observation is said to be unexplainable if, for every hypothesis in the set of hypotheses, the description length of the observation conditioned on the hypothesis is greater than the unconditional description length of the observation. Consequently, if

$$\arg\min_k l_{\Theta_1}(y_n|y_1^{n-1}, s_1^{n-1}, s_n = k) > -\log(P(y_n)), \tag{8}$$

for every state k, $1 \leq k \leq K$, the existing switching model is insufficient to explain the observation y_n. If (8) is satisfied, we add the time index n to the set P, the set of time steps with observations not explained by the switching model.

Suppose there exist K states prior to time T, and the codebooks are C_{Θ_1} and C_{Θ_2} for the existing K-state model. We seek to find a subset $p \in P$ and new codebooks (for the (K+1)-state model) $\tilde{C}_{\Theta_1}$ and $\tilde{C}_{\Theta_2}$ such that (6) decreases when the observations in p are assigned to the new state $K + 1$. In particular, we check if

$$\min_k H_{T,k} + C_{\Theta_1} + C_{\Theta_2} > \min_k \tilde{H}_{T,k} + \tilde{C}_{\Theta_1} + \tilde{C}_{\Theta_2}, \tag{9}$$

where $\tilde{H}_{T,k}$ has the same switching state sequence path as $H_{T,k}$ except that the members of the subset p are assigned to the new state $K + 1$. If (9) holds true, then at time T, a new state $K + 1$ is formed.

The best subset p is selected by trying multiple permutations of the elements, i.e., subsets, of P, and selecting the subset that minimizes the right-hand side of (9). Although the number of permutations to find a good subset p may be large (e.g., it takes $2^{|P|}$ permutations if one wants to try all possible subsets), one may always initialize the EM based on a subset selected from a small number of permutations. The approach provides a convenient way to initialize the EM algorithm for the new model with the $K+1$ switching states. The EM algorithm described in the previous section is then applied to estimate the parameters, the switching states, and the posterior probabilities.

Code Lengths

We use Shannon-optimal code lengths in computing the length functions l [4,13]. Thus,

$$l_{\Theta_2}(s_n = k|s_{n-1} = j) = -\log(P(s_n = k|s_{n-1} = j)). \tag{10}$$

As $y_n|(y_1^{n-1}, s_1^n)$ is real-valued, we need to quantize it before coding. Since $y_n|(y_1^{n-1}, s_1^n)$ is Gaussian, it can be shown that its quantized version can be coded with a Shannon optimal code length of approximately [4]

$$\frac{1}{2}\log(2\pi e)^D |\Sigma_{y_n|(y_1^{n-1}, s_1^n)}| - \log \Delta, \tag{11}$$

where $\Sigma_{y_n|(y_1^{n-1},s_1^n)}$ is the covariance matrix, and Δ is the volume of the quanti-zation bin. Here, Δ is assumed to be small. In the SLDS setting, the quantization bin volume Δ acts as a parameter that controls the tendency to discover new states. The function of the parameter Δ is similar to that of the hyper-parameters of the Dirichlet model of [2, 8, 15] that control the tendency of the process to transition into new states. It may appear as if Δ does not have an effect on the minimization in (6) since $\log \Delta$ is an additive constant. However, Δ influences the codebook size. As Δ is increased, the quantization gets coarser, and thus the expected code lengths for $y_n|(y_1^{n-1},s_1^n)$ decrease. This decreases the size of the codebook, leading to more switched states.

More importantly, however, the parameter Δ has an actual meaning in the negotiation setting. For instance, the quantization bin width should be set small enough to make sure that significant differences in issue space are preserved. If the seller of a laptop computer knows that a price difference of more than 50 dollars is very important to a buyer in the transaction, then Δ should be set at a value below 50.

5   Experiments with Simulated Trajectories

5.1   The Agent

The seller's agent is a software agent that negotiates the issues with buyers on behalf the seller. The agent is available to it a rank-ordered list of the issue vectors such that vectors that rank higher are more advantageous to the seller. The agent is also provided with some cut-off point in the list; vectors that rank higher than the cut-off are target vectors of the seller. The agent is provided with neither the rank-ordered list of the buyer nor the buyer's target vectors.

The goal of the agent is to make smart counter offers to the buyer, i.e., counter offers that are target vectors for the seller and are likely to be along the buyer's trajectory. Hence, the agent should be able to predict the future trajectory of the buyer, and make counter offers that are compatible with the predicted trajectory.

Given the history of the buyer's offers up to (and including) time T, the agent needs to predict the buyer's offers for $n > T$. Prediction models, such as the SLDS, provide conditional probabilities of the future trajectories of the buyer for $n > T$ given the history of the buyer's offers for $n \leq T$. The agent's goal is to make a counter offer that satisfies two conditions: First, the counter offer should be a target vector for the seller, and second, the counter offer should have a high likelihood of being along the future trajectory of the buyer.

Towards this end, at time T, for each target vector u of the seller, we compute the probability, $q_{u,t}$, that the buyer's trajectory will pass through target u at or before time $T + t$. This probability is computed based on a prediction model (e.g., LDS or SLDS) that predicts the buyer's future trajectory based on its offer history. We then set a probability threshold Q such that any target u with probability $q_{u,t} > Q$ is considered a candidate counter offer. Of the candidate counter offers, the agent selects the one with the lowest value of t. This ensures

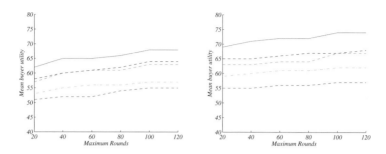

Fig. 1. Buyer utility values of the agent's counter offers for the SLDS and other algorithms. Figures are for the conceder behavior. R=10 (left) and R=30 (right). SLDS (red, solid), infinite LDS (black, dashed), finite LDS (magenta, dashed), auto-regressive (green, dash-dot), replication (blue, dashed).

that the counter offer has the property of having a high likelihood of being in the buyer's path. A smaller value of t implies that the selected offer is more favorable to the buyer's current state of mind.

5.2   Results

In [5], the buyers' negotiation behaviors have been modeled through a sequence of functions; each function imitates the buyer's offers in the issue space. The functions are monotonic in the issue values and belong to families of parametrized functions. The parameters determine whether the buyer's behavior is *boulware* or *conceder*. We use the families of paramterized functions in [5] to generate buyer trajectories in the multi-issue space.

We evaluate the performance of the seller's agent. The agent negotiates issues with buyers on behalf the seller, where the buyer's offers on the issues are generated through the simulations. The agent is available to it a rank-ordered list of the issue vectors with the higher-ranking vectors being more advantageous to the seller. The cut-off point (described in section 5.1) is set such that R percent of the issue vectors are seller's target vectors for some R. The buyer has a different rank-ordered list of the issue vectors, and its cut-off point is set such that 20 percent of the issues vectors are buyer's target vectors.

At each time $n = T$, the agent computes each $q_{u,t}$, as described in section 5.1, where t is set so that $t + T = t_{\max}$, the total number of rounds. The probability threshold Q is set such that the top 10 percent of the $q_{u,t}$ values are selected as candidate offers.

We compute the probability $q_{u,t}$ using five prediction models: Replication ($y_{T+t} = y_T$), auto-regressive modeling ($y_{T+t} = a_t * y_T + z_T$, for Gaussian noise Z), fixed-length LDS, infinite-length LDS and the SLDS. The fixed-length LDS learning was done through an EM framework with the E-step being a standard LDS forward inference. The infinite-length LDS learning was done through the Dirichlet formulation of [2, 8, 15]. For the Dirichlet formulation of the infinite

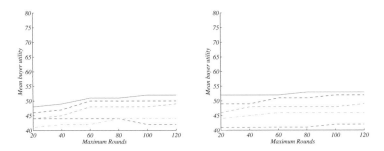

Fig. 2. Buyer utility values of the agent's counter offers for the SLDS and other algorithms. Figures are for the boulware behavior. R=10 (left) and R=30 (right). SLDS (red, solid), infinite LDS (black, dashed), finite LDS (magenta, dashed), auto-regressive (green, dash-dot), replication (blue, dashed).

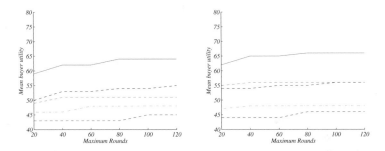

Fig. 3. Buyer utility values of the agent's counter offers for the SLDS and other algorithms. Figures are for the linear concession behavior. R=10 (left) and R=30 (right). SLDS (red, solid), infinite LDS (black, dashed), finite LDS (magenta, dashed), auto-regressive (green, dash-dot), replication (blue, dashed).

LDS and the SLDS, we selected the hyperparameters (for the LDS) and the Δ values (for the SLDS) to get the best outcomes.

In Fig. 1, we assigned a *utility score* (in the 0-100 range) to each issue point with higher-ranking points (in the buyer's ranked-ordered list) with higher utility scores, and computed the mean of the utility values of the agent's counter offers. In Fig. 1, the concession behavior of the buyer is that of a conceder. The SLDS performed best, followed by the LDS (fixed and infinite). The performance of the simple replication and auto-regressive models were poor compared to the SLDS. Thus, the SLDS was able to provide the best counter offers to the buyer, in terms of maximizing the buyer's utility while ensuring that the counter offers are always selected from the seller's target offers.

As the maximum number of rounds is increased, the performance of each prediction algorithm improves as there is more offer history upon which the agent can build the predictions. As the value of R is increased (from left to right), the agent allows more target vectors for the seller, and thus, the utility for the buyer increases.

In Fig. 2, the experiment is repeated for a buyer with a boulware behavior. Similar to the case with the conceder behavior, the SLDS performed best, followed by the LDS (fixed and infinite). The performance of the simple replication and auto-regressive models were poor compared to the SLDS. The mean utility values are lower than those for the conceder behavior; this is expected because it is more difficult for the agent to predict the behavior of a boulware buyer.

Finally, in Fig. 3, the experiment was repeated for a buyer with a linear concession behavior. In this case, the performance difference (in terms of mean buyer utility values) of the SLDS with the other algorithms is more significant. This implies that the SLDS performs best (when compared to the other algorithms) in a linear concession setting.

6   Experiments with Actual Buyer Behavior

We also validated the model against actual buyer behavior from an agent-human negotiation experiment [20]. For each case of the agent-human negotiation experiment, the buyer is assumed by a human subject, and was asked to maximize its own utility in the final agreement with the seller as high as possible, while meeting a minimum threshold at 44 utility points. There was no restriction of any specific offer and acceptance strategies that the buyer should use. The seller was assumed by a software agent coded with four negotiation strategies [20]. The seller agent has a same minimum threshold at 44 utility points. In all the four strategy manipulation conditions, the agent proposed a final offer to the buyer for decision. Therefore, each data point comprises a buyer's offer history at all negotiation rounds, up to a maximum of 9 negotiation rounds.

We compared the p-step prediction error of the infinite SLDS with other approaches for $p = 1$ and $p = 2$. The comparisons are included in Fig. 4. We varied the parameter Δ from 0.01 to 1, and, for comparison purposes, converted each issue range to a non-integer scale from 0 to 10. The prediction errors in Fig. 4 are the squared errors averaged over all rounds, 110 subjects and 4 issues. For the very small values of Δ, the infinite SLDS model size grew very slow with too few switching states s, leading to a poor prediction accuracy. As Δ was increased, the prediction accuracy improved until $.1 < \Delta < .2$. Then, for $\Delta > .2$, the cost of forming new states started to become too low and the model started to produce too many new states, lowering the prediction accuracy. At $\Delta = .2$, the parameter value that minimized the prediction error, we observed an unexplainable issue offer (implying a new state is necessary) at a rate of approximately once in every 30 offers. It would be interesting to explore if this rate has any significance in conjunction with the human psychology community.

The other four prediction models included in the comparisons in Fig. 4 were simple replication ($y_{t+p} = y_t$), auto-regressive modeling ($y_{t+p} = a_p * y_t + z_n$, for Gaussian noise Z), fixed-length LDS and infinite-length LDS. The simple replication and the auto-regressive model led to high prediction errors, implying the need for a state-space based model.

We quantified how well a "smart agent" would perform, where a smart agent is one that makes counter offers by predicting the buyer's trajectory. The selected

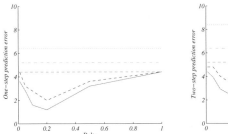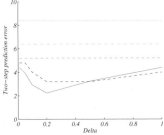

Fig. 4. Prediction comparison of SLDS with other algorithms. 1-step (left) and 2-step (right) prediction errors. SLDS (red, solid), infinite LDS (black, dashed), finite LDS (magenta, dashed), auto-regressive (green, dash-dot), replication (blue, dotted).

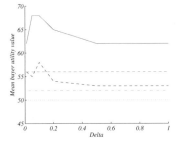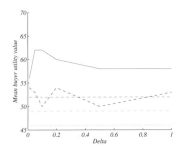

Fig. 5. Buyer's utility value for counter-offers. Seller's th_A low (left) and seller's th_A high (right). SLDS (red, solid), infinite LDS (black, dashed), finite LDS (magenta, dashed), auto-regressive (green, dash-dot), replication (blue, dotted).

counter offer is the one that is closest to the predicted trajectory (closest in Euclidean distance after converting each issue range to the 0 to 10 range) of the buyer, provided it is above the threshold th_A (defined in section 5.1) in the agent's list. 9 counter offers per buyer is made.

In Fig. 5, we assigned a *utility score* (in the 0-100 range) to each issue point with higher-ranking points (in the buyer's ranked-ordered list) with higher utility scores, and computed the mean utility value of agent's best counter offer (of the 9 counter offers). The SLDS performed best, followed by the LDS (fixed and infinite). The performance of the simple replication and auto-regressive models were poor compared to the SLDS. Thus, the SLDS was able to provide the best counter offers to the buyer, in terms of maximizing the buyer's utility while ensuring that the counter offers are above rank th_A.

7   Conclusions

Electronic commerce provides consumers with the ability to purchase products and services from businesses online. Despite their widespread adoption, electronic commerce systems still lack the ability to enable the businesses to negotiate with consumers during the purchases. We develop a seller's agent, a software agent that negotiates the issues with buyers on behalf the online merchant. The agent makes smart counter offers to the buyer, i.e., counter offers that are profitable for the seller and are likely to be accepted by the buyer. For this, the agent uses prediction models to predict the future trajectory of the buyer, and makes counter offers that are compatible with the predicted trajectory.

We predict the behavior of the human buyer in the issue space; the human negotiation behavior models in [5] are used to build the predictors. Our approach is more realistic than the more traditional game-theoretic approach of estimating utility functions since human buyers rarely construct mathematical functions prior to negotiation to quantify their issue preferences. Further, humans are not super-rational beings; it is unlikely that a human will always give the most rational decisions in terms of maximizing a mathematical function. To capture the switching behavior, we extend the LDS to the SLDS prediction. We allow the model size to grow with new offers through the MDL setting.

We are working towards a new experiment, where the actual human subjects will negotiate with the agent in section 5. We are also exploring how an agent's counter offers influence the buyer behavior. Yet another research thread for us is to utilize the findings of the buyer negotiation pshycology research in designing software agents.

Acknowledgements

We would like to thank Yinping Yang and Yunjie Xu for the data used in this paper from their negotiation experiments [20].

References

[1] Barron, A., Rissanen, J., Yu, B.: The minimum description length principle in coding and modeling. IEEE Transactions in Information Theory 44(6), 2743–2765 (1998)

[2] Beal, M., Ghahramani, Z., Rasmussen, C.E.: The infinite hidden Markov model. In: NIPS (2002)

[3] Coehoorn, R.M., Jennings, N.: Learning an opponent's preferences to make effective multi-issue negotiation trade-offs. In: ICEC (2004)

[4] Cover, T., Thomas, J.: Elements of Information Theory. Wiley, New York (1991)

[5] Faratin, P., Sierra, C., Jennings, N.R.: Negotiation decision functions for autonomous agents. Int. Journal of Robotics and Au-tonomous Systems 24(3-4), 159–182 (1998)

[6] Fatima, S., Wooldridge, M., Jennings, N.R.: Multi-issue negotiation under time constraints. In: AAMAS (2002)

[7] Fatima, S., Wooldridge, M., Jennings, N.R.: An agenda based framework for multi-issues negotiation. Artificial Intelligence Journal 152(1), 1–45 (2004)

[8] Fox, E.B., Sudderth, E.B., Jordan, M.I., Willsky, A.S.: An HDP-HMM for systems with state persistence. In: ICML (2008)

 [9] Ito, T., Hattori, H., Klein, M.: Multi-issue negotiation protocol for agents: Exploring nonlinear utility spaces. In: IJCAI (2007)
[10] Kolmogorov, A.N.: Three approaches to the quantitative definition of information. Probl. Inform. Transm. 1, 4–7 (1965)
[11] Pavlovic, V., Rehg, J.M., Cham, T.J., Murphy, K.P.: A dynamic bayesian network approach to figure tracking using learned dynamic models. In: Intl. Conf. Computer Vision (1999)
[12] Pavlovic, V., Rehg, J., MacCormick, J.: Learning switching linear models of human motion. In: NIPS (2001)
[13] Shannon, C.E., Weaver, W.: The mathematical theory of communication. University of Illinois Press, Urbana (1949)
[14] Solomonoff, R.J.: A formal theory of inductive inference I,II. Inform. and Control 7, 1–22, 224–254 (1964)
[15] Teh, Y.W., Jordan, M.I., Beal, M.J., Blei, D.M.: Hierarchical Dirichlet processes. Journal of American Statistics Association 101(476), 1566–1581 (2006)
[16] Van Kleef, G.A., De Dreu, C.K.W., Manstead, A.S.R.: The interpersonal effects of anger and happiness in negotiations. Journal of Personality and Social Psychology 86, 57–76 (2004)
[17] Van Kleef, G.A., De Dreu, C.K.W., Manstead, A.S.R.: The interpersonal effects of emotions in negotiations: A motivated information processing approach. Journal of Personality and Social Psychology 87, 510–528 (2004)
[18] Vetschera, R.: Preference structures and negotiator behavior in electronic negotiations. Decision Support Systems 44, 135–146 (2007)
[19] Wallace, C., Dowe, D.L.: Minimum mesaage length and Kolmogorov complexity. The Computer Journal 42(4), 270–283 (1999)
[20] Yang, Y., Singhal, Y.S., Xu, Y.: Offer with Choices and Accept with Delay: A Win-Win Strategy Model for Agent-Based Automated Negotiation. In: ICIS 2009 (2009)
[21] Yukl, G.A.: Effects of situational variables and opponent concessions on a bargainer's perception, aspirations, and concessions. Journal of Personality and Social Psychology 29, 227–236 (1974)

High Availability Data Model for P2P Storage Network

BangYu Wu[1], Chi-Hung Chi[1], Cong Liu[1],
ZhiHeng Xie[1], and Chen Ding[2]

[1] School of Software, Tsinghua University, Beijing, China
[2] Department of Computing Science, Ryerson University, Canada

Abstract. With the goal to provide high data availability, replicas of data will be distributed based on the idea of threshold, meaning that data service is guaranteed to be available so long as any k out of n peers are online. The key distribution algorithm of the model as well as its scalability, management, and other availability-related factors are presented and analyzed.

Keywords: P2P Network, Storage, Data Availability Model.

1 Introduction

Peer-to-Peer (P2P) has been proven to be one of the most effective and popular networks for large scale content sharing on Internet. Unlike traditional client-server environment where data is managed in a few highly reliable centralized servers, P2P network distributes the burden of data storage among its hundreds and thousands of peers. This results in the hard problem of data availability in the P2P network. Although ad hoc solutions based on redundancy are proposed, no guarantee of data availability is made.

In this paper, we would like to investigate and improve the data availability problem of P2P network. In particular, we focus on the following issues: (i) how should the replicas placement and distribution be managed so that it can cope consistently with the statistics of the peer failure, and (ii) despite the constant changes in the peer online membership, what should be the replica placement scheme so that files are highly available without excessively replicating them over lots of peers. To answer these questions, we propose a new high availability model for data storage in P2P network. The key idea behind is to distribute replicas based on threshold, an idea borrowed from secret sharing in security [7]. Analysis shows that our threshold model is very efficient in enhancing data availability in P2P network.

2 Related Work

To ensure data availability in P2P network, redundancy is often added to the original data. Reed Solomon erasure coding is proposed to distribute bulk of data to millions of users through multicast and broadcast. Its basic principle is to divide a file into m fragments and recode them into n fragments, where $m < n$. At the other end, the file can be reassembled back from any m fragments with an aggregated size equal to the original file size [5]. Various strategies for replication and placement have been

L. Chen, P. Triantafillou, and T. Suel (Eds.): WISE 2010, LNCS 6488, pp. 322–327, 2010.

proposed to address the data availability problem of P2P network. PAST [6] maintains a strict replication rule: for each data item in the network, there should be k available replicas stored in k NodeID contagious living peers. OceanStore [2] employs multiple hashing functions for replication. It hashes data item's ID with a few different seed numbers and assigns them to multiple target peers. PlanetP [1] and [3] generate redundant packets when a member needs to increase the availability of a file during periodic estimation of data availability. Path replication method [4] demands that the requested data is replicated in all the peers that are along the data transmission path between the peer requesting the data and the peer having the data.

While all these replication rules and strategies give good foundation in data availability for P2P networks, they are usually either too restrictive with the replica number or too complicated to be used in practical systems.

3 Threshold-Based Model for High Availability Data Storage

Under P2P networks with ad-hoc peer nature, we argue that there are at least four main design requirements for any threshold-based high availability data storage model:

- Differential Data Reliability and Availability Setting. Data availability of each file in a P2P network should be uniquely defined based on the degree of its significance.
- Reconstruction of Original Data File through Simple Re-assembling of Data Chunks, but Without Complicated Computational Overhead. Minimizing data storage overhead should be achieved without sacrificing retrieval performance.
- Robustness of Data Availability Support. Good threshold-based data storage model should be as robust as possible.
- Easy Management of Data Chunk Replicas. Good threshold-based data storage model should provide file update without complicated, dynamic replication and migration. All data chunks should be managed, maintained, and searched easily.

3.1 Assumptions and Notation

Before we go into the model, we should like to list down the assumptions made in our threshold-based data availability model. In the P2P network of our interest, every peer is delegated with equal responsibility. All peers are uniform in their storage capacity resources to permit our threshold model to deploy data chunk replicas freely. Furthermore, the P2P network is relatively stable in that peers eventually rejoin the community after they go offline. And the online time of peers is relatively much longer than their offline time. With the last assumption, it makes sense for us to talk about the average availability of a peer.

To simplify the rest of our discussion, we define the following notations used in the paper. Let a file with size l is available in a P2P network with n peers and average peer availability p. The P2P network should provide file service with availability a when any k peers are online (which we call this model (k, n)). To achieve this goal, the file will be divided into m data chunks with size s each. Among them, r data chunks should be located in each peer so that when any k peers are available, the original file can be reconstructed back by assembling data chunks in those k peers.

In general, since every peer in the network might have different probability for being online, say p_i for each peer i, it would be too expensive to compute the exact file availability. Instead, we use the following approximation that uses the average probability of being online:

$$p_i = \frac{\text{average online time}}{(\text{average online time} + \text{average offline time})}, \text{ and } p = \frac{1}{n}\sum_{i=1}^{n}p_i .$$

3.2 Threshold Constraints

The overall description of our threshold model is as follows. There are n peers in the P2P network. For each file stored in the network, it is divided into many chunks. The constraint of our model is that if any k or more peers are online, the original file is guaranteed to be reconstructed back, else the file reconstruction will fail. What we target is to achieve the minimum of average storage overhead in each peer. The challenge is: how to divide the file and how to distribute the divided file chunks among peers? About the threshold value k, it should be computed in terms of the required system availability. Given a, k, m, and n that constrain the distribution of file chunks, for our (k, n) model, the availability function $a(n,k,p)$ of a P2P network should be of one of the following forms:

$$a(n,k,p) = \sum_{i=k}^{n} C_n^i p^i (1-p)^{n-i} \tag{1}$$

$$= \sum_{i=0}^{n-k} C_n^i p^{n-i} (1-p)^i \tag{2}$$

$$= 1 - \sum_{i=0}^{k-1} C_n^i p^i (1-p)^{n-i} \tag{3}$$

Equation (1) explains the file availability probability from the viewpoint that at least k peers are online. Equation (2) explains the file availability probability from the opposite viewpoint that at most $n-k$ peers are offline. And Equation (3) gives the availability probability when the total number of online peers is more than k. These three equations are identical in semantic.

3.3 Model Prototype

Let $D = \{d_1, d_2, ..., d_m\}$ be the set of m data chunks that a given file is divided into. A P2P network peers set is denoted by $E = \{e_1, e_2, ..., e_n\}$. A data chunk distribution $n*m$ matrix $W = [w_{ij}]$ is obtained by multiplying a 0-1 matrix $G = [g_{ij}]$ with size $n*m$ to a diagonal matrix H with size $m*m$ formed by d_i. The rows of W act as the data chunk vectors distributed among peers. It is $W = G * H$, Therefore,

$$
\begin{bmatrix}
w_{1,1} & w_{1,2} & w_{1,3} & \cdots & w_{1,m} \\
w_{2,1} & w_{2,2} & w_{2,3} & \cdots & w_{2,m} \\
\vdots & \vdots & \vdots & \vdots & \vdots \\
w_{n,1} & w_{n,2} & w_{n,3} & \cdots & w_{n,m}
\end{bmatrix}
=
\begin{bmatrix}
g_{1,1} & g_{1,2} & g_{1,3} & \cdots & g_{1,m} \\
g_{2,1} & g_{2,2} & g_{2,3} & \cdots & g_{2,m} \\
\vdots & \vdots & \vdots & \vdots & \vdots \\
g_{n,1} & g_{n,2} & g_{n,3} & \cdots & g_{n,m}
\end{bmatrix}
*
\begin{bmatrix}
d_1 & 0 & 0 & \ldots & 0 \\
0 & d_2 & 0 & \ldots & 0 \\
\vdots & \vdots & \vdots & & \vdots \\
0 & 0 & 0 & \ldots & d_m
\end{bmatrix}
$$

where $w_{ij} \in D \cup \{0\}$. If $w_{ij} = d_j$ ($1 \le i \le n$, $1 \le j \le m$), then d_j will be distributed over peer i. Otherwise, if $w_{ij} = 0$, then d_j will not be distributed over peer i. $g_{ij} \in \{0,1\}$. Implementing a threshold model means that it needs to compute the distribution matrix W. Since matrix H is known, the construction of the 0-1 matrix G will be the key problem.

Theorem 1. Data chunk distribution matrix W that is consistent with the threshold model is accomplished if and only if three conditions about G are satisfied. They are,

(1) For any j, $\sum_{i=1}^{n}(1 - g_{i,j}) \le k - 1, 1 \le j \le m$.

(2) $m \ge C_n^{k-1}$.

(3) There must exist C_n^{k-1} columns in matrix G such that the number of columns with 0 is k-1, and that with 1 is n-(k-1). These columns together represent all the combinations with k-1 zeros distributed in n locations. Each column belongs to any one of these combinations, and all these columns are different from each other.

Proof
Data chunk distribution based on our threshold model must satisfy the above three conditions. Condition (1) guarantees that a file can be reassembled by not less than k peers. Condition (2) and Condition (3) guarantee that the file reconstruction cannot be implemented by less than k peers.

In terms of the (k, n) model, a file can be reconstructed when the number of online peers is not less than k. So, if we pick any k rows from G, at least one element 1 representing some data chunk will appear in each column of these k rows. If Condition (1) is not correct, there will exist a column j such that the number of zero elements is larger than k-1 and we can pick up at least k rows of W such that elements are zero in column j. In this case, the file cannot be reconstructed from these rows since at least one data chunk is missing even if the number of online peers are more than k-1. This result is paradoxical with the definition of the threshold definition, so Condition (1) is correct.

For Condition (2), the original file will lack at least one data chunk to be reconstructed from all data chunks located in any k-1 peers, and any missing data chunks from any two different subsets of k-1 peers are different. Suppose there are two different sets A and B with $|A| = |B| = k$-1. If the reconstruction of file from them both miss data chunk d_1, then the file will also lack of the same data chunk d_1 even if it is reconstructed from more than k-1 peers in $A \cup B$. This situation is contrary to the (k, n) model. So in the threshold model, a file should be divided into at least C_n^{k-1} data chunks.

For Condition (3), still using the reduction to absurdity, we assume that there does not exist such columns satisfying the combinations where k-1 zero elements are distributed over n different locations. Suppose one combination $X = (x_1, x_2, \ldots, x_n)$ is missing in G, let $y_1, y_2, \ldots, y_{k-1}$ be the subscripts when $x_{y_1}, x_{y_2} \ldots x_{y_{k-1}}$ in X are zero. So there does not exist such column in G that its elements are zero in row $y_1, y_2, \ldots, y_{k-1}$. We pick up the corresponding row $y_1, y_2, \ldots, y_{k-1}$ in W, and the original file can be constructed from these k-1 rows since each data chunk is located at least in one peer. This result is inconsistent with the threshold model, so Condition (3) is proved.

Hence, the proof is completed.

Theorem 2. There are $m!/n!$ alternatives of data chunk distribution methods for our threshold model.

Proof

There are m columns in the 0-1 matrix G, and G still satisfies the threshold requirement no matter how all these columns are replaced with each other. This is because each data chunk d_i is independent with each other, and the distribution method for d_i can also be used for other data chunks. So we have $m!$ placement methods for all columns.

On the other hand, there are n rows in the 0-1 matrix G, with each row representing the data chunks located in the corresponding peer. G still satisfies the threshold requirement no matter how all rows are replaced with each other. Since each peer e_i is independent of each other, the data chunks located in peer e_i can also be distributed to other peers. But exchanging rows do not change the distribution of data chunks, it only influences peers distribution. So, we have $n!$ placement methods for all rows.

Hence, there are $m!/n!$ alternatives of data chunk distribution methods for our threshold model, and the proof is completed.

3.4 Algorithm of Distributing Data Chunks

From above analysis, we solve the problem of how many data chunks a file should be divided into and how many data chunks should be located in each peer. We understand that all data chunks are not scattered casually to peers, and there are $(m!/n!)$ alternatives of distribution methods. So the last question is on how to distribute the divided data chunks among peers? The data chunks distribution matrix W can be computed when the 0-1 matrix G is solved. To compute matrix G, we construct C_n^{k-1} combinations. The algorithm first determines the distribution of all zero elements, then set 1 to all the remaining elements in G. The algorithm of producing G is given in Fig. 1, its complexity is $O(C_n^{k-1})$.

Step 1. Initialization: Suppose the first combination generates the first column. There are continuous $k-1$ zero elements in the first column, and their row numbers are $y_1 = 1$, $y_2 = 2$, ..., $y_{k-1}=k-1$ respectively. Let $y_1, y_2, ..., y_{k-1}$ be the current combination.

Step2. For all the remaining $C_n^{k-1}-1$ columns, compute the next combination of y_1, $y_2, ..., y_{k-1}$ according to the current combination as follows:

   S1. $i = \max\{j \mid y_j < n-(k-1)+j\}$

   S2. $y_i = y_i +1$

   S3. $y_j = y_{j-1} +1, j = i+1, i+2, ..., k-1$

   S4. $y_1, y_2, ..., y_{k-1}$ forms the current combination.

Step 3. According to each combination $y_1, y_2, ..., y_{k-1}$ above, construct columns in G in this way:

   Set 0 to row elements whose row numbers are $y_1 = 1$, $y_2 = 2$, ..., $y_{k-1} = k-1$, and set 1 to the remaining elements.

Fig 1. Algorithm to Compute Matrix G

Finally, the distribution matrix W is calculated by $W=G*H$.

4 Conclusion

This paper proposes a theoretical model to achieve high availability for P2P system by partitioning file into chunks and distributing these chunks over peers. The availability is implemented based on the threshold model such that file is available by reassembling blocks as long as any k peers are online. The algorithms of partitioning and distribution are carefully designed to guarantee that the file chunks located in any k peers can be used to reconstruct back the original file. By comparing the threshold model with simple replication mechanism, we can show that our model provides more flexible availability and download performance than simple replication mechanism which just throws k replicas to k selected peers in a dynamic P2P network.

Acknowledgement

This work is supported by the China National 863 project #2008AA01Z129.

References

1. Cuenca-Acuna, F.M., Peery, C., Martin, R.P., Nguyen, T.D.: PlanetP: Using Gossiping to Build Content Addressable Peer-to-Peer Information Sharing Communities. In: Proceedings of the IEEE International Symposium on High Performance Distributed Computing (2003)
2. Kubiatowicz, J., Bindel, D., Chen, Y., Eaton, P., Geels, D., Gummadi, R., Rhea, S., Weatherspoon, H., Weimer, W., Wells, C., Zhao, B.: Oceanstore: An Architecture for Global-Scale Persistent Storage. In: Proceedings of ACM ASPLOS Conference (2000)
3. Liu, X.Z., Yang, G.W., Wang, D.X.: Stationary and Adaptive Replication Approach to Data Availability in Structured Peer-to-Peer Overlay Networks. In: Proceedings of the 11th IEEE International Conference on Networks (2003)
4. Lv, Q., Cao, P., Cohen, E., Li, K., Shenker, S.: Search and Replication in Unstructured Peer-to-Peer Networks. In: Proceedings of the 16th ACM International Conference on Supercomputing, New York (June 2002)
5. Ray, S., Francis, P., Handey, M., Karp, R., Shenker, S.: A Scalable Content-Addressable Network. In: Proceedings of the ACM SIGCOMM Conference (2001)
6. Rowstran, A., Druschel, P.: Storage Management and Caching in PAST, a Large-Scale, Persistent Peer-to-Peer Storage Utility. In: Proceedings of ACM Symposium on Operating Systems Principles (October 2001)
7. Shamir, A.: How to Share a Secret. Communications of the ACM 22(11), 612–613 (1979)

Modeling Multiple Users' Purchase over a Single Account for Collaborative Filtering

Yutaka Kabutoya[1], Tomoharu Iwata[2], and Ko Fujimura[1]

[1] NTT Cyber Solutions Laboratories, NTT Corporation,
1-1 Hikari-no-oka, Yokosuka-shi, Kanagawa, 239-0847 Japan
[2] NTT Communication Science Laboratories, NTT Corporation,
2-4 Hikaridai, Seika-cho, Soraku-gun, Kyoto, 619-0237 Japan

Abstract. We propose a probabilistic topic model for enhancing recommender systems to handle multiple users that share a single account. In several web services, since multiple individuals may share one account (e.g. a family), individual preferences cannot be estimated from a simple perusal of the purchase history of the account, thus it is difficult to accurately recommend items to those who share an account. We tackle this problem by assuming latent users sharing an account and establish a model by extending Probabilistic Latent Semantic Analysis (PLSA). Experiments on real log datasets from online movie services and artificial datasets created from these real datasets by combining the purchase histories of two accounts demonstrate high prediction accuracy of users and higher recommendation accuracy than conventional methods.

1 Introduction

Many recent websites such as Amazon[1] emphasize the services provided by their recommender systems. They are very useful in increasing sales in e-commerce, click rates on websites, and visitor satisfaction in general. Existing solutions, however, suffer several problems.

In this article, we focus on one of the problems, their inability to handle accounts shared by multiple users. Where several individuals share one account, conventional recommender systems can only estimate the preference of the account, not the individuals, because individual preferences are corrupted by the purchase histories of other individuals. This problem was pointed by Koren [10] and Tösher [16], members of the winners of the Netflix Prize[2].

We now give an example of this problem as follows: We consider the case of recommending items to a family in a service offering video-on-demand (VOD). The family consists of three people, a father, a mother, and a son. They share a single account. The mother often views drama videos on a weekday afternoon. The son often views animation videos on a weekday evening. The father often views sports videos at night. In this case, the preference extracted from the account is a mixture of three independent preferences.

---

[1] http://www.amazon.com
[2] http://www.netflixprize.com

L. Chen, P. Triantafillou, and T. Suel (Eds.): WISE 2010, LNCS 6488, pp. 328–341, 2010.
© Springer-Verlag Berlin Heidelberg 2010

It is generally true that the purchase histories gathered by online shops do not differentiate the individuals sharing the account. Our target is, therefore, to recommend items that an account would purchase next more accurately by forming hypotheses about the individuals (the latent users) sharing the account.

To achieve this target, we propose a probabilistic topic model for analyzing and extracting the independent preferences of users sharing a single account. The proposed model is an extension of Probabilistic Latent Semantic Analysis (PLSA) [6]. Whereas latent topics (which represent preferences in traditional memory-based collaborative filtering [11]) are assumed to be generated by the account in PLSA, the topics are assumed to be generated by the latent users in the proposed model. In this paper, time-of-day of each purchase is assumed to be generated also by indicative of the latent user making the purchase, so the proposed model predicts latent users from the purchased items and time-of-day.

In our evaluation section, we apply our method to real datasets of web services offering movies and artificial datasets created from these real datasets by combining the purchase histories of two accounts. It is assumed that each account is generated by a single user. We conduct two evaluations: First, to verify if our method can properly predict latent users, we determine the prediction accuracy of users against the artificial data. This experiment shows that the proposed model offers high user prediction accuracy. Second, to confirm that the prediction of latent users improves recommendation accuracy, we compare the accuracy of recommender systems based on the proposed model with conventional methods found in the literature. Experiments show that a recommender system based on our model outperforms conventional methods on real and artificial datasets.

This article provides the following contributions:

- Describes the proposed model as an extension of PLSA in terms of the assumption of latent users (Section 2).
- Reports an evaluation of user prediction from artificial datasets created by combining the purchase histories of two accounts as taken from real datasets of online movie services (Section 3).
- Compares a recommender systems based on our model to systems based on conventional methods over both real datasets and artificial datasets (Section 3).
- Demonstrates examples of recommendation results yielded by the proposed model (Section 3).

We describe the relationship of the current study to previous work in Section 4.

2   Proposed Method

2.1   Notation

We start with a set of U accounts, where each account consists of pairs of item and time-of-day $(\boldsymbol{i}_u, \boldsymbol{t}_u)$, where $\boldsymbol{i}_u = \{i_{um}\}_{m=1}^{M_u}$ is the set of items that the account purchased, and $\boldsymbol{t}_u = \{t_{um}\}_{m=1}^{M_u}$ is the set of time-of-day when the account made the purchase. Our notation is summarized in Table 1.

Table 1. Notation

Symbol	Description
U	number of accounts
N	number of unique items
K	number of topics
V	number of users sharing an account
M_u	number of items purchased by the uth account
i_{um}	mth item purchased by the uth account, $i_{um} \in \{1, \ldots, N\}$
t_{um}	time-of-day of mth purchase by the uth account
z_{um}	topic of the mth purchase by the uth account, $z_{um} \in \{1, \ldots, K\}$
v_{um}	user of the mth purchase by the uth account, $v_{um} \in \{1, \ldots, V\}$

2.2   Model

Before introducing our model, we review PLSA, which forms the basis of our model. In PLSA, each account has topic proportions $\boldsymbol{\xi}_u$, which represent the preferences of the account. For each of the M_u purchases of the account, topic z_{um} is chosen from the topic proportions, and then item i_{um} is generated from a topic-specific multinomial distribution over items $\boldsymbol{\phi}_{z_{um}}$.

Our model assumes that multiple users share an account. We introduce multiple latent users for each account, and each latent user has its topic proportions $\boldsymbol{\theta}_{uv}$. For each of the M_u purchases of the account, latent user v_{um} is chosen from user proportions $\boldsymbol{\psi}_u$. The purchased item is generated using the same process with PLSA given the topic proportions. In particular, topic z_{um} is chosen from the topic proportions $\boldsymbol{\theta}_{uv_{um}}$, and then item i_{um} is generated from topic-specific multinomial distribution $\boldsymbol{\phi}_{z_{um}}$. In addition to purchased items, our model generates the time-of-day for each purchase. The time-of-day is generated from a user-specific Gaussian distribution with mean $\tau_{uv_{um}}$ and variance $\sigma^2_{uv_{um}}$.

In summary, the proposed model assumes the following generative process for a set of account purchases $\{(\boldsymbol{i}_u, \boldsymbol{t}_u)\}_{u=1}^U$,

1. For each topic $z = 1, \ldots, K$:
    (a) Draw item probability $\boldsymbol{\phi}_z \sim \mathrm{Dirichlet}(\beta)$
2. For each account $u = 1, \ldots, U$:
    (a) Draw user proportions $\boldsymbol{\psi}_u \sim \mathrm{Dirichlet}(\gamma)$
    (b) For each user $v = 1, \ldots, V$:
        i. Draw topic proportions $\boldsymbol{\theta}_{uv} \sim \mathrm{Dirichlet}(\alpha)$
    (c) For each item $m = 1, \ldots, M_u$:
        i. Draw user $v_{um} \sim \mathrm{Multinomial}(\boldsymbol{\psi}_u)$
        ii. Draw time-of-day $t_{um} \sim \mathrm{Normal}(\tau_{uv_{um}}, \sigma^2_{uv_{um}})$
        iii. Draw topic $z_{um} \sim \mathrm{Multinomial}(\boldsymbol{\theta}_{uv_{um}})$
        iv. Draw item $i_{um} \sim \mathrm{Multinomial}(\boldsymbol{\phi}_{z_{um}})$

here we assume that multinomial parameters $\boldsymbol{\phi}_z$, $\boldsymbol{\psi}_u$ and $\boldsymbol{\theta}_{uv}$ are generated from Dirichlet distributions because of the conjugacy. We used $\alpha = 0.001$, $\beta = 0.001$, and $\gamma = 0.001$ as the parameters of the Dirichlet distributions in all experiments

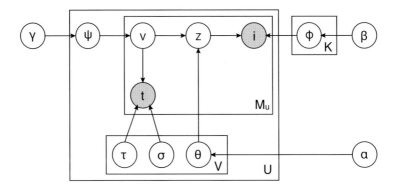

Fig. 1. Graphical representation of the proposed model

described in Section 3. Figure 1 shows a graphical representation of the proposed model, where shaded and unshaded nodes indicate observed and latent variables, respectively.

The probability of generating account purchase $(\boldsymbol{i}_u, \boldsymbol{t}_u)$ given a user proportion, a set of purchase time-of-day means, a set of purchase time-of-day standard deviations, a set of topic proportions, and a set of item probabilities is described as follows:

$$
P(\boldsymbol{i}_u, \boldsymbol{t}_u | \boldsymbol{\psi}_u, \boldsymbol{\tau}_u, \boldsymbol{\sigma}_u, \boldsymbol{\Theta}_u, \boldsymbol{\Phi})
$$
$$
= \prod_{m=1}^{M_u} \sum_{v=1}^{V} P(v|\boldsymbol{\psi}_u) p(t_{um}|v, \boldsymbol{\tau}_u, \boldsymbol{\sigma}_u) \sum_{z=1}^{K} P(z|v, \boldsymbol{\Theta}_u) P(i_{um}|z, \boldsymbol{\Phi}), \quad (1)
$$

where $\psi_{uv} = P(v|\boldsymbol{\psi}_u)$, $\boldsymbol{\tau}_u = \{\tau_{uv}\}_{v=1}^{V}$, $\boldsymbol{\sigma}_u = \{\sigma_{uv}\}_{v=1}^{V}$, $\boldsymbol{\Theta}_u = \{\boldsymbol{\theta}_{uv}\}_{v=1}^{V}$, $\theta_{uvz} = P(z|v, \boldsymbol{\Theta}_u)$, $\boldsymbol{\Phi} = \{\boldsymbol{\phi}_z\}_{z=1}^{K}$, $\phi_{zi} = P(i|z, \boldsymbol{\Phi})$, and the second term on the right hand side of (1) is given by the following Gaussian distribution:

$$
p(t_{um}|v, \boldsymbol{\tau}_u, \boldsymbol{\sigma}_u) = \frac{1}{\sqrt{2\pi}\sigma_{uv}} \exp\left(-\frac{|t_{um} - \tau_{uv}|^2}{2\sigma_{uv}^2}\right), \quad (2)
$$

where $|t_{um} - \tau_{uv}|$ is used to denote the distance between t_{um} and τ_{uv}, e.g. $|00{:}00{:}00 - 23{:}00{:}00| = 1{:}00{:}00$.

2.3   Parameter Estimation

The latent parameters $\boldsymbol{\Psi} = \{\boldsymbol{\psi}_u\}_{u=1}^{U}$, $\boldsymbol{\mathcal{T}} = \{\boldsymbol{\tau}_u\}_{u=1}^{U}$, $\boldsymbol{\Sigma} = \{\boldsymbol{\sigma}_u\}_{u=1}^{U}$, $\boldsymbol{\Theta} = \{\boldsymbol{\Theta}_u\}_{u=1}^{U}$, and $\boldsymbol{\Phi}$ can be estimated by maximizing a posteriori (MAP). Here, let us denote the set of all latent parameters by $\boldsymbol{\Lambda} = \{\boldsymbol{\Psi}, \boldsymbol{\mathcal{T}}, \boldsymbol{\Sigma}, \boldsymbol{\Theta}, \boldsymbol{\Phi}\}$. The log likelihood of the latent parameters $\boldsymbol{\Lambda}$ given the set of accounts $\boldsymbol{U}$ is described as follows:

$$
L(\boldsymbol{\Lambda}|\boldsymbol{U}) = \sum_{u=1}^{U} \sum_{m=1}^{M_u} \log \sum_{v=1}^{V} P(v|\boldsymbol{\psi}_u) p(t_{um}|v, \boldsymbol{\tau}_u, \boldsymbol{\sigma}_u) \sum_{z=1}^{K} P(z|v, \boldsymbol{\Theta}_u) P(i_{um}|z, \boldsymbol{\Phi}), \quad (3)
$$

where the number of users V and that of topics K are already known.

Following the generative process described in the previous subsection, we use a Dirichlet prior for item probability ϕ_z, user proportions ψ_u, and topic proportions θ_{uv} as:

$$p(\phi_z) = \frac{\Gamma((\beta+1)N)}{\Gamma(\beta+1)^N} \prod_{i=1}^{N} \phi_{zi}^{\beta}, \tag{4}$$

$$p(\psi_u) = \frac{\Gamma((\gamma+1)V)}{\Gamma(\gamma+1)^V} \prod_{v=1}^{V} \psi_{uv}^{\gamma}, \tag{5}$$

$$p(\theta_{uv}) = \frac{\Gamma((\alpha+1)K)}{\Gamma(\alpha+1)^K} \prod_{z=1}^{K} \theta_{uvz}^{\alpha}. \tag{6}$$

The estimation of latent parameters Λ can be computed more easily by maximizing the following Q function based on the EM algorithm [4] than by maximizing the posterior probability directly:

$$Q(\Lambda|\hat{\Lambda}) =$$
$$\sum_{u=1}^{U}\sum_{m=1}^{M_u}\sum_{v=1}^{V}\sum_{z=1}^{K} P(v,z|u,m;\hat{\Lambda}) \log P(v|\psi_u)p(t_{um}|v,\tau_u,\sigma_u)P(z|v,\Theta_u)P(i_{un}|z,\Phi)$$
$$+ \sum_{u=1}^{U} \log p(\psi_u) + \sum_{u=1}^{U}\sum_{v=1}^{V} \log p(\theta_{uv}) + \sum_{z=1}^{K} \log p(\phi_z), \tag{7}$$

where $P(v,z|u,m;\hat{\Lambda})$ represents the posterior probability of a user and a topic given the mth purchase of the uth account. In E-steps, the posterior probability of a user and a topic is calculated as:

$$P(v,z|u,m;\hat{\Lambda}) = \frac{P(v|\psi_u)p(t_{um}|v,\tau_u,\sigma_u)P(z|v,\Theta_u)P(i_{um}|z,\Phi)}{\sum_{v'=1}^{V} P(v'|\psi_u)p(t_{um}|v',\tau_u,\sigma_u)\sum_{z'=1}^{K} P(z'|v',\Theta_u)P(i_{um}|z',\Phi)}, \tag{8}$$

where $p(t_{um}|v,\tau_u,\sigma_u)$ can be calculated by (2). In M-steps, under the constraints of $\sum_{v=1}^{V}\psi_{uv} = 1$, $\sum_{z=1}^{K}\theta_{uvz} = 1$, and $\sum_{i=1}^{N}\phi_{zi} = 1$, the estimated values of latent parameters $\hat{\psi}_{uv}$, $\hat{\tau}_{uv}$, $\hat{\sigma}_{uv}$, $\hat{\theta}_{uvz}$, and $\hat{\phi}_{zi}$ can be computed by maximizing $Q(\Lambda|\hat{\Lambda})$ for ψ_{uv}, τ_{uv}, σ_{uv}, θ_{uvz}, and ϕ_{zi}, respectively, as follows:

$$\hat{\phi}_{zi} = \frac{\sum_{u=1}^{U}\sum_{m=1}^{M_u}\sum_{v=1}^{V} I(i_{um}=i)P(v,z|u,m;\hat{\Lambda}) + \beta}{\sum_{i'=1}^{N}\sum_{u=1}^{U}\sum_{m=1}^{M_u}\sum_{v=1}^{V} I(i_{um}=i')P(v,z|u,m;\hat{\Lambda}) + \beta N}, \tag{9}$$

$$\hat{\psi}_{uv} = \frac{\sum_{m=1}^{M_u}\sum_{z=1}^{K} P(v,z|u,m;\hat{\Lambda}) + \gamma}{\sum_{v'=1}^{V}\sum_{m=1}^{M_u}\sum_{z=1}^{K} P(v',z|u,m;\hat{\Lambda}) + \gamma V}, \tag{10}$$

$$\hat{\tau}_{uv} = \frac{\sum_{m=1}^{M_u}\sum_{z=1}^{K} P(v,z|u,m;\hat{\Lambda})t_{um}}{\sum_{m=1}^{M_u}\sum_{z=1}^{K} P(v,z|u,m;\hat{\Lambda})}, \tag{11}$$

$$\hat{\sigma}_{uv}^2 = \frac{\sum_{m=1}^{M_u} \sum_{z=1}^{K} P(v, z|u, m; \hat{\mathbf{\Lambda}}) |t_{um} - \hat{\tau}_{uv}|^2}{\sum_{m=1}^{M_u} \sum_{z=1}^{K} P(v, z|u, m; \hat{\mathbf{\Lambda}})}, \tag{12}$$

$$\hat{\theta}_{uvz} = \frac{\sum_{m=1}^{M_u} P(v, z|u, m; \hat{\mathbf{\Lambda}}) + \alpha}{\sum_{z'=1}^{K} \sum_{m=1}^{M_u} P(v, z'|u, m; \hat{\mathbf{\Lambda}}) + \alpha K}, \tag{13}$$

where $I(\cdot)$ is used to denote the indicator function, i.e., $I(A) = 1$ if A is true, and $I(A) = 0$ otherwise. By iterating the E-step and M-step alternately until the log likelihood represented in (3) converges, the latent values $\mathbf{\Lambda}$ are optimized locally.

3 Experiments

3.1 Datasets

The proposed model was evaluated experimentally using the following two datasets taken from real-world online movie services: (1) the *EachMovie* dataset provided by the Compaq Systems Research Center and (2) *MovieLens* dataset, provided by the MovieLens research project[3]. Though the datasets originally represent reputations of movies, we regarded each data as purchase information. Each dataset also contains information about reputation time (We regarded this as purchase time). We sampled items that were each purchased at least ten times and accounts that each purchased at least five times in each dataset.

Furthermore, we created artificial datasets from the real datasets above by combining the purchase histories of pairs of accounts. The artificial datasets are based on the assumption that each account in the real datasets was created by a single user. Where two accounts purchased the same item, we deleted the duplicate entry from one account selected at random.

We summarize the datasets in Table 2.

Table 2. Datasets

(i) Real datasets.

	# of accounts	# of items	# of purchases
EachMovie	14,155	1,249	625,424
MovieLens	943	1,152	97,953

(ii) Artificial datasets.

	# of accounts	# of items	# of purchases
EachMovie2	7,077	1,249	569,171
MovieLens2	471	1,152	88,826

---

[3] http://www.grouplens.org/node/73

3.2 Predicting Users

Evaluation Metric. To measure how accurately the proposed model predicts users sharing a single account, we conducted experiments based on the artificial datasets. By regarding the original accounts as users to predict, we can estimate the user-prediction quality of the proposed model.

Let $\bar{v}_{um}(\in \{1,2\})$ represent the user actually made the mth purchase of the uth account. We now evaluated the probability of this user given the mth purchase of the uth account by the following equation:

$$P(\bar{v}_{um}|u,m) = \sum_{z=1}^{K} P(\bar{v}_{um}, z|u, m; \hat{\mathbf{\Lambda}}), \qquad (14)$$

where the number of topics K is selected from ten candidates $\{10, 20, \ldots, 100\}$ and $P(v, z|u, m; \hat{\mathbf{\Lambda}})$ is computed using equation (8). Since each account is used by two users, the prediction of the user for the mth purchase of the uth account can be regarded correct if $P(\bar{v}_{um}|u, m) > 0.5$.

Results. Figure 2 shows the prediction rates of the proposed model for the (a) EachMovie and (b) MovieLens datasets. Baseline means the rate with random user assignment, thus the rate equals to average of the bigger value of the purchase rates of two users for each pair. The results demonstrate that our model achieved statistically significantly higher rates than baseline for all topics on both datasets ($p < .01$, sign test).

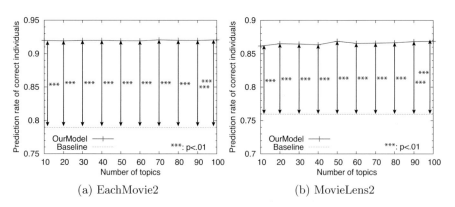

(a) EachMovie2   (b) MovieLens2

Fig. 2. Prediction rates of the proposed model. Baseline means random user assignment for each purchase of pairs.

3.3 Recommendation

Evaluation Metric. The target of this evaluation is to confirm whether predicting the latent users sharing an account improves recommendation accuracy. In this evaluation, we used real datasets as well as artificial datasets.

We used the top-n accuracy to determine the recommendation accuracy. The procedure used to calculate the top-n accuracy is as follows: First, the most recently purchased item in account u is regarded as test data $\bar{i}_u$ for the account. Items other than $\bar{i}_u$ are regarded as training data for u. Next, relevancy of account u against each item is calculated by each recommendation method. When $\bar{i}_u$ is included in the n most relevant items to u, the correct items are considered to have been recommended to u. The top-n accuracy is the fraction of accounts to which correct items were recommended.

Compared Methods. In each recommender, relevancy of an account against an item is given by the probability that the account purchases the item, $P(i|u)$. We now describe how to compute the probability with each recommender:

1. *OurModel.* The probability that an account purchases an item is calculated as follows:

$$P(i|u) \propto \sum_{v=1}^{V} P(v|\psi_u) p(\bar{t}_u|v, \tau_u, \sigma_u) \sum_{z=1}^{K} P(z|v, \Theta_u) P(i|z, \Phi), \qquad (15)$$

where $\bar{t}_u$ is the time-of-day of the account's last purchase, i.e., time-of-day of test data.

2. *Unigram.* Under the unigram model, the items that every account purchases are drawn independently from a single multinomial distribution as follows:

$$\begin{aligned} P(i|u) &\propto P(i) \\ &= \frac{\sum_{u=1}^{U} \sum_{m=1}^{M_u} I(i_{um} = i)}{\sum_{i'=1}^{N} \sum_{u=1}^{U} \sum_{m=1}^{M_u} I(i_{um} = i')}. \end{aligned} \qquad (16)$$

This model recommends the item purchased most frequently to every account.

3. *UserCF.* User-based collaborative filtering. This scheme, applied to the recommender systems "GroupLens," proposed by Resnick et al. [11], is the most traditional recommender approach. First, the similarity between one account, u, and another, u', is given by Pearson's product-moment coefficient represented by the following equation:

$$\text{UserSim}(u, u') = \frac{\sum_{i=1}^{N} (r_{ui} - \bar{r}_u)(r_{u'i} - \bar{r}_{u'})}{\sqrt{\sum_{i=1}^{N} (r_{ui} - \bar{r}_u)^2} \sqrt{\sum_{i=1}^{N} (r_{u'i} - \bar{r}_{u'})^2}}, \qquad (17)$$

where $r_{ui} = 1$ if the account purchased the item, $r_{ui} = 0$ otherwise, and $\bar{r}_u = M_u/N$ is the average of the account's purchase. The probability of purchasing an item is given by the following equation:

$$P(i|u) \propto \bar{r}_u + \frac{\sum_{u' \in U_{\setminus u}} \text{UserSim}(u, u')(r_{u'i} - \bar{r}_{u'})}{\sum_{u' \in U_{\setminus u}} |\text{UserSim}(u, u')|}, \qquad (18)$$

where $U_{\setminus u} = \{1, \dots, U\} - \{u\}$.

4. *ItemCF*. Item-based collaborative filtering. The recommendation scheme proposed by Sarwar et al. [12], offers reasonable computation cost. First, the similarity between one item i and another i' is given by the adjusted cosine similarity represented by the following equation:

$$\text{ItemSim}(i, i') = \frac{\sum_{u=1}^{U}(r_{ui} - \bar{r}_u)(r_{ui'} - \bar{r}_u)}{\sqrt{\sum_{u=1}^{U}(r_{ui} - \bar{r}_u)^2}\sqrt{\sum_{u=1}^{U}(r_{ui'} - \bar{r}_u)^2}}. \tag{19}$$

The probability of purchasing an item is proportional to the summation of the similarities of items in the purchase history of the account as follows:

$$P(i|u) \propto \frac{\sum_{i' \in \boldsymbol{I}_{\setminus i}} r_{ui'}\text{ItemSim}(i, i')}{\sum_{i' \in \boldsymbol{I}_{\setminus i}} |\text{ItemSim}(i, i')|}, \tag{20}$$

where $\boldsymbol{I}_{\setminus i} = \{1, \ldots, N\} - \{i\}$.

5. *PLSA*. Probabilistic latent semantic analysis [6]. Under PLSA, a set of topic proportions is directly estimated as $\boldsymbol{\Xi} = \{\boldsymbol{\xi}_u\}_{u=1}^{U}$, and $\xi_{uz} = P(z|\boldsymbol{\xi}_u)$ is the zth topic proportion of the uth account. The probability of $\boldsymbol{i}_u$ given $\boldsymbol{\xi}_u$ and $\boldsymbol{\Phi}$ is as follows:

$$P(\boldsymbol{i}_u|\boldsymbol{\xi}_u, \boldsymbol{\Phi}) = \prod_{m=1}^{M_u} \sum_{z=1}^{K} P(z|\boldsymbol{\xi}_u)P(i_{um}|z, \boldsymbol{\Phi}). \tag{21}$$

The unknown parameters in PLSA $\boldsymbol{\Upsilon}$ are the set of topic proportions $\boldsymbol{\Xi}$ and the set of item probabilities $\boldsymbol{\Phi}$. They can be estimated by maximizing the following likelihood with the EM algorithm:

$$L(\boldsymbol{\Upsilon}|\boldsymbol{U}) = \sum_{u=1}^{U} \sum_{m=1}^{M_u} \log \sum_{z=1}^{K} P(z|\boldsymbol{\xi}_u)P(i_{um}|z, \boldsymbol{\Phi}). \tag{22}$$

In E-step, the class posterior probability given the current estimate can be computed as follows:

$$P(z|u, m; \hat{\boldsymbol{\Upsilon}}) = \frac{P(z|\hat{\boldsymbol{\xi}}_u)P(i_{um}|z, \hat{\boldsymbol{\Phi}})}{\sum_{z'=1}^{K} P(z'|\hat{\boldsymbol{\xi}}_u)P(i_{um}|z', \hat{\boldsymbol{\Phi}})}. \tag{23}$$

In M-step, the next estimate of topic proportion $\hat{\xi}_{uz}$ and that of item probability $\hat{\phi}_{zi}$ are respectively given by:

$$\hat{\xi}_{uz} = \frac{\sum_{m=1}^{M_u} P(z|u, m; \hat{\boldsymbol{\Upsilon}})}{\sum_{z'=1}^{K} \sum_{m=1}^{M_u} P(z'|u, m; \hat{\boldsymbol{\Upsilon}})}, \tag{24}$$

$$\hat{\phi}_{zi} = \frac{\sum_{u=1}^{U} \sum_{m=1}^{M_u} I(i_{um} = i)P(z|u, m; \hat{\boldsymbol{\Upsilon}})}{\sum_{i'=1}^{N} \sum_{u=1}^{U} \sum_{m=1}^{M_u} I(i_{um} = i')P(z|u, m; \hat{\boldsymbol{\Upsilon}})}, \tag{25}$$

We used Dirichlet priors for the topic proportions and item probabilities as in the proposed model to evaluate the efficiency of latent users accurately.

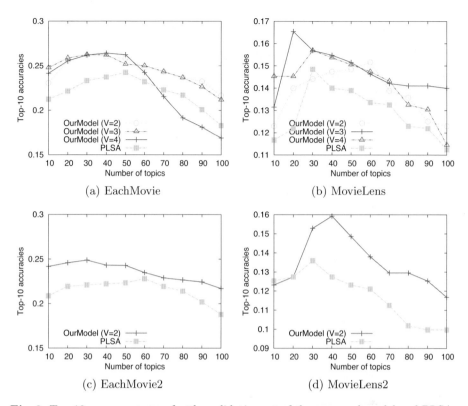

Fig. 3. Top-10 accuracy scores for the validation set of the proposed model and PLSA

The probability that an account purchases an item is described as follows:

$$P(i|u) \propto \sum_{z=1}^{K} P(z|\boldsymbol{\xi}_u)P(i|z, \boldsymbol{\Phi}). \qquad (26)$$

Results. We used a validation set to determine (V, K) in our model and K in PLSA. We used the second last purchased item in each account as the validation set. Figure 3 shows the top-ten accuracy scores with different numbers of topics for the validation set. From this result we chose our model where $(V, K) = (4, 40)$ and the PLSA model where $K = 50$ on EachMovie datasets, our model where $(V, K) = (3, 20)$ and the PLSA model where $K = 30$ on MovieLens datasets, our model where $K = 30$ and the PLSA model where $K = 60$ on EachMovie2 datasets, and our model where $K = 40$ and the PLSA model where $K = 30$ on MovieLens2 datasets.

We now show top-n accuracy scores for the test data of each method in Figure 4. Our model achieved higher prediction accuracy than the other four conventional schemes on EachMovie2 datasets ($n \geq 2$, $p < .01$, sign test) and MovieLens2 datasets (though the difference is not statistically significant since the

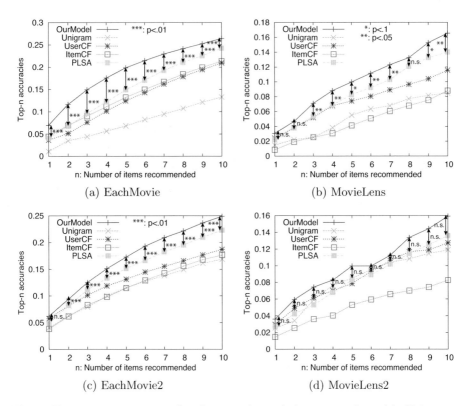

Fig. 4. Top-n accuracy scores for the test data of the proposed model, Unigram, UserCF, ItemCF, and PLSA

datasets were not large enough). These results indicate that the assumption of latent users is effective in improving recommendation accuracy.

The recommender systems based on our model outperform conventional methods on real datasets; EachMovie ($p < .01$, sign test) and MovieLens datasets ($n = 3, \ldots, 7, 9, 10$, $p < .1$, sign test). Since it is difficult to consider that each account in real-world reputation logs are accessed by multiple users, from this result our model is considered to well express varying preferences of a single user depending on time-of-day.

Next, we confirm what recommendation lists are provided to the latent users by the recommender systems based on our model. Table 3 shows the purchase histories of accounts in the leftmost column, and recommendation lists that are relevant to the users predicted from four time-of-day periods (0 a.m., 6 a.m., 0 p.m., and 6 p.m.) in the other columns using the proposed model. These results show that our model realizes recommender systems that consider latent users estimated from the time-of-day: To the first account in (a), action films and comedy films are recommended at 6 a.m. and at 0 p.m., and animation films such as "The Lion King" are also recommended at 6 p.m. Items corresponding to the latent user estimated from the time-of-day are also recommended to the second account in (a). To the first account in table (b), action films and SF films

Table 3. The purchase history of users (leftmost column), and the recommendation lists at 0 a.m., 6 a.m., 0 p.m., and 6 p.m. (other columns), provided by the proposed model using $(V, K) = (4, 40)$ on EachMovie datasets and $(V, K) = (3, 20)$ on MovieLens datasets, respectively. Each column represents one time

(a) EachMovie

History	At 0 a.m.	At 6 a.m.	At 0 p.m.	At 6 p.m.
Speed	Shawshank Rep.	Mission Imp.	Mission Imp.	Shawshank Rep.
Santa Clause	Die Hard	Muriel's Wedding	Muriel's Wedding	Die Hard
Batman	Smoke	Free Willy	Free Willy	Smoke
Pretty Woman	American Pres.	Phenomenon	Phenomenon	American Pres.
Toy Story	Lion King	Disclosure	Disclosure	Lion King
Bad Boys	Jumanji	Jumanji	Jumanji	Batman
Ghost	Tombstone	Tombstone	Tombstone	White Squall
Mission Imp.	Naked Gun	Naked Gun	Naked Gun	Space Jam
Dave	Serial Mom	Serial Mom	Serial Mom	Cable Guy
Celtic Pride	The Paper	The Paper	The Paper	The Crow

(b) MovieLens

Wizard of OZ	2001: A Space Od.	Home Alone	Manhattan	Manhattan
Mask	Godfather: Part II	Grease	Boot Das	Boot Das
Batman	Die Hard	Addams Family	Local Hero	Local Hero
Apollo 13	Cool Hand Luke	Ghost	Power Ranger	Power Ranger
Toy Story	Patton	Brady Bunch	Airheads	Airheads
Die Hard	Good Fellas	Die Hard 2	Die Hard 2	Die Hard 2
GoldenEye	Se7en	Batman Returns	Batman Returns	Batman Returns
E.T.	Psycho	Addams Family	Addams Family	Addams Family
Home Alone	Braveheart	Mask	Mask	Mask
Batman	Brazil	Santa Clause	Santa Clause	Santa Clause

are recommended at 0 a.m., comedy films at 6 p.m., and films for children are recommended in the evening. This account suggests a typical family. The second account in table (b) suggests two users. This indicates that our model is effective even with large numbers of users.

4   Related Work

Temporal information is often used to enhance the quality of recommender systems. Many works are based on the assumption that the most recent user interests are most effective in predicting the next purchase. This is a kind of concept drift, a gradual one. Adomavicius et al. [1] extended the traditional memory-based collaborative filtering approach to take the contextual information, which includes temporal information, into consideration. Sugiyama et al. [14] assumed that the user profile evolved over time in proposing a personalized web search engine. Ding et al. [5] suggested a time weighting scheme for a similarity-based collaborative filtering approach. Iwata [8] estimated how strongly the freshness of a purchase history influenced the next purchase based on maximum entropy models and Markov models. Koren [10], one of the winners of the Netflix Prize, also focused on the drift in users' preferences over time. Whereas gradual concept drift has been gathering more attention to improve recommender systems, there are still few works on recommendation improvement through the use of other types of concept drift, e.g. periodic concept drift like ours.

Generative topic models interest many researchers, because they can be used in a wide variety of applications. Probabilistic semantic analysis, proposed by Hofmann [6], is one of the most representative topic models. It is used for information retrieval and collaborative filtering [7]. Si and Jin [13] proposed a model for collaborative filtering by extending PLSA. Iwata et al. [9] used topic models to visualize discrete data. Blei et al. [2] analyzed and extracted annotations from noisy, annotated, discrete data sets such as web pages and images. This paper predicts users from discrete purchase histories of accounts.

5 Conclusion

We proposed a topic model for predicting the purchases of multiple individuals through a single account from discrete purchase history data. We have confirmed experimentally the efficiency of the proposed method as follows: First, the proposed model can predict the individual users of an account from its purchase history. Second, our model can predict the next purchase more accurately than conventional recommender techniques. Finally, recommender systems based on our model can provide recommendation lists that correspond to users predicted from time-of-day associations. The results strongly suggest that the proposed model is promising.

In future work, we will extend the current proposal in the two ways: First, we hope to achieve more robust estimation by using the Bayesian approach, instead of MAP estimation, as in Latent Dirichlet Allocation (LDA) [3]. Second, we will extend the proposal to a nonparametric Bayesian model such as the Dirichlet process mixture model [15] to determine the optimal number of users for each account (in the current model it is same for every account).

References

1. Adomavicius, G., Sankaranarayanan, R., Sen, S., Tuzhilin, A.: Incorporating contextual information in recommender systems using a multidimensional approach. ACM Transactions on Information Systems 23(1), 103–145 (2005)
2. Blei, D.M., Jordan, M.I.: Modeling annotated data. In: Proceedings of the 26th Annual International ACM SIGIR Conference on Research and Development in Informaion Retrieval, pp. 127–134. ACM, New York (2003)
3. Blei, D.M., Ng, A.Y., Jordan, M.I.: Latent Dirichlet allocation. Journal of Machine Learning Research 3, 993–1022 (2003)
4. Dempster, A., Laird, N., Rubin, D.: Maximum likelihood from incomplete data via the EM algorithm. Journal of the Royal Statistical Society, Series B 39(1), 1–38 (1977)

5. Ding, Y., Li, X.: Time weight collaborative filtering. In: Proceedings of the 14th ACM International Conference on Information and Knowledge Management, pp. 485–492. ACM, New York (2005)
6. Hofmann, T.: Probabilistic latent semantic indexing. In: Proceedings of the 22nd Annual International ACM SIGIR Conference on Research and Development in Information Retrieval, pp. 50–57. ACM, New York (1999)
7. Hofmann, T.: Collaborative filtering via Gaussian probabilistic latent semantic analysis. In: Proceedings of the 26th Annual International ACM SIGIR Conference on Research and Development in Information Retrieval, pp. 259–266. ACM, New York (2003)
8. Iwata, T.: Probabilistic user behavior models in online stores for recommender systems. Ph.D. thesis, Kyoto University (2008)
9. Iwata, T., Yamada, T., Ueda, N.: Probabilistic latent semantic visualization: topic model for visualizing documents. In: Proceedings of the 14th ACM SIGKDD International Conference on Knowledge Discovery and Data Mining, pp. 363–371. ACM, New York (2008)
10. Koren, Y.: Collaborative filtering with temporal dynamics. In: Proceedings of the 15th ACM SIGKDD International Conference on Knowledge Discovery and Data Mining, pp. 447–456. ACM, New York (2009)
11. Resnick, P., Iacovou, N., Suchak, M., Bergstrom, P., Riedl, J.: GroupLens: an open architecture for collaborative filtering of netnews. In: Proceedings of the 1994 ACM Conference on Computer Supported Cooperative Work, pp. 175–186. ACM, New York (1994)
12. Sarwar, B., Karypis, G., Konstan, J., Reidl, J.: Item-based collaborative filtering recommendation algorithms. In: Proceedings of the 10th International Conference on World Wide Web, pp. 285–295. ACM, New York (2001)
13. Si, L., Jin, R.: Flexible mixture model for collaborative filtering. In: Proceedings of the 20th International Conference on Machine Learning, pp. 704–711. AAAI Press, Menlo Park (2003)
14. Sugiyama, K., Hatano, K., Yoshikawa, M.: Adaptive web search based on user profile constructed without any effort from users. In: Proceedings of the 13th International Conference on World Wide Web, pp. 675–684. ACM, New York (2004)
15. Teh, Y.W., Jordan, M.I., Beal, M.J., Blei, D.M.: Hierarchical Dirichlet processes. Journal of the American Statistical Association 101(476), 1566–1581 (2006)
16. Tösher, A., Jahrer, M.: The BigChaos solution to the Netflix grand prize (2009), http://www.netflixprize.com/assets/GrandPrize2009_BPC_BigChaos.pdf

Interaction-Based Collaborative Filtering Methods for Recommendation in Online Dating

Alfred Krzywicki, Wayne Wobcke, Xiongcai Cai, Ashesh Mahidadia,
Michael Bain, Paul Compton, and Yang Sok Kim

School of Computer Science and Engineering
University of New South Wales
Sydney NSW 2052, Australia
{alfredk,wobcke,xcai,ashesh,mike,compton,yskim}@cse.unsw.edu.au

Abstract. We consider the problem of developing a recommender system for
suggesting suitable matches in an online dating web site. The main problem to be
solved is that matches must be highly personalized. Moreover, in contrast to typical product recommender systems, it is unhelpful to recommend popular items:
matches must be extremely specific to the tastes and interests of the user, but it
is difficult to generate such matches because of the two way nature of the interactions (user initiated contacts may be rejected by the recipient). In this paper,
we show that collaborative filtering based on interactions between users is a viable approach in this domain. We propose a number of new methods and metrics
to measure and predict potential improvement in user interaction success, which
may lead to increased user satisfaction with the dating site. We use these metrics
to rigorously evaluate the proposed methods on historical data collected from a
commercial online dating web site. The evaluation showed that, had users been
able to follow the top 20 recommendations of our best method, their success rate
would have improved by a factor of around 2.3.

1 Introduction

Online dating is a suitable application domain for recommender systems because it
usually has abundant information about users and their behaviour. The key to developing a recommender system is that matches must be personalized. Moreover, in contrast
to typical product recommender systems, it is unhelpful to recommend popular users;
matches must be extremely specific to the tastes and interests of the user. This is particularly important because, unlike product recommender systems, the receivers of messages may reject invitations from the user, whereas products are typically commodity
items that can be reproduced and sold any number of times.

The search and simple user profile matching technologies used in many dating sites
have serious limitations. The main limitations are that some users receive large amounts
of potentially unwanted communication (especially, perhaps, if they are the most popular users), whereas others may receive very low amounts of communication (e.g. if their
profiles are not easily found via common search keywords). Search interfaces typically
do not rank matches by suitability, requiring users to manually filter the results in order
to find a potential partner. These aspects reduce the usability of the site, in different
ways for different classes of users, and may lead to an increased attrition rate of users.

L. Chen, P. Triantafillou, and T. Suel (Eds.): WISE 2010, LNCS 6488, pp. 342–356, 2010.

A recommender system can improve user satisfaction by addressing some of the limitations of search and user profile matching. By ranking potential matches, some of the burden faced by users in searching and filtering through candidates can be reduced, ideally increasing interactions and communication with suitable matches, plus increasing satisfaction for those less popular users who are not prominent on the site. In addition, by recommending such less popular users (where appropriate), the large volumes of unwanted communication may be reduced, though not eliminated, as such a recommender system would not replace, but augment, the existing search interfaces.

This work concerns the type of online dating site where users can freely and easily browse large numbers of user profiles. In broad terms, the typical "user experience" with such an online dating site evolves from initial setting up of a profile, to sending messages (usually selected from a predefined set) to prospective partners, who may accept or reject these initial approaches, then, for those accepted, proceeding to full e-mail communication. At each stage, there is the potential for users to receive recommendations and as the quantity of information increases, the quality of the matches should also increase. Helping users to progress to the later stages in this process can achieve the aims of increasing user retention, maintaining the pool of candidate matches and improving user satisfaction. In this paper, we focus on the stage in this process where users initiate contact with potential partners, since we have reliable data concerning these interactions, and suitable metrics for evaluating the success of recommendations.

Content-based recommender systems make recommendations for a user based on a profile built up by analysing the content of items that users have chosen in the past. These content-based methods stem from information retrieval and machine learning research and employ many standard techniques for extracting and comparing user profiles and item content [6]. Initial experiments showed that it is difficult to use the user profiles for content-based recommendation because users do not always provide adequate information about themselves or provide such information in the form of free text, which is hard to analyse reliably. Another well-known significant limitation of content-based recommender systems is that they have no inherent method for generating recommendations of novel items of interest to the user, because only items matching the user's past preferences are ever recommended [8,2,1].

Collaborative filtering recommender systems address some of the problems with content-based approaches by utilizing aggregated data about customers' habits or preferences, recommending items to users based on the tastes of other "similar" users [8]. In this paper we show that collaborative filtering based on interactions between users is a viable approach to recommendation in online dating. This result is surprising because our collaborative filtering methods rely only on interactions between users and do not explicitly incorporate user profile information. We propose a number of new collaborative filtering methods and metrics specifically designed to measure user success on an online dating site, and show that a "voting" method used to rank candidates works efficiently and generates reasonably good recommendations. This paper is organized as follows. In the next section we provide a brief review of related research in this domain. Then we describe new collaborative filtering methods applicable in the online dating domain. In Section 4, we define a set of metrics and evaluate and compare all methods, then conclude with an outline of future research.

2   Related Work

We are not aware of any published work that uses pure collaboration filtering in the online dating domain. By pure collaboration filtering we mean user-to-user or item-to-item methods that are based only on user interactions. Brŏzovský and Petřiček [3] evaluate a number of collaborative filtering methods in an online dating service for estimating the "attractiveness" rating of a user for a given user, which is calculated from the ratings of similar users. Their results, however, are not directly comparable with our research since we focus on the problem of predicting *successful* interactions, which requires taking into account the interests of the receiver of a message in addition to the preferences of the sender, which is a two way interaction, in contrast to "attractiveness" ratings which only reflect the viewer's interests. In particular, whereas many people are likely to rate the same people highly, recommending these same people to many users is likely to result in a large number of unsuccessful interactions, since typically any person is limited in the number of successful responses they can give.

Generally, collaborative filtering methods differ in how the recommendations are derived and how they are ranked. The item-to-item collaborative filtering method of Amazon.com [5] uses a similarity table computed offline and the ranking is based on the similarity between items, which is related to the number of customers who purchased the two items together. For all of our methods, we compute both a similarity table and a recommendation table offline and rank the candidates not according to their similarity, but based on number of unique "votes" as described in the next section. Other methods of calculating similarity used in the literature include Pearson correlation coefficient and recency [9], cosine and conditional probability [4], and cosine, correlation and sum of user ratings weighted by similarity [7].

3   Methods

3.1   Basic Collaborative Filtering

The *Basic Collaborative Filtering* (Basic CF) item-to-item model is based on similar items: two items are similar if they have been purchased together in a shopping basket by a large number of customers, and (simplifying) the number of such purchases indicates the degree of similarity between the two items. By analogy, in the context of initiating contact in an online dating site there are similar senders and similar recipients, i.e. instead of purchasing items, the basic interaction is that users send messages to other users which may be successful (accepted) or unsuccessful (rejected). Two distinct users are *similar recipients* to the extent that they have received messages from the same senders; two distinct users are *similar senders* to the extent that they have sent messages to the same recipients. More formally, let $u \rightarrow v$ denote an interaction where u sends a message to v. The similar recipients $u \overset{r}{\sim} v$ and similar senders $u \overset{s}{\sim} v$ relations can be defined as follows:

$$u \overset{r}{\sim} v : \exists w \, (w \rightarrow u \wedge w \rightarrow v) \tag{1}$$

$$u \overset{s}{\sim} v : \exists w \, (u \rightarrow w \wedge v \rightarrow w) \tag{2}$$

Both types of similarity can be used to make recommendations, as shown in Figure 1. In these diagrams, rectangles (blue background) typically indicate senders, ellipses (pink background) indicate recipients. Unbroken (black) arrows indicate messages sent between users, and (green) lines indicate the similarity of users (as recipients and senders, respectively).

In the first diagram in Figure 1, all pairs of recipients except r1 and r4 are similar recipients, since for any pair, both users in the pair have received at least one message from the same sender; the number of such messages indicates the degree of similarity as recipients. So for example, r2 is a similar recipient to r3 to degree 2 since both have received contacts from s1 and s2. The same reasoning implies that s1 and s2 are similar senders; s1 and s2 are similar senders to degree 2 since they have both sent messages to r2 and r3.

Now consider another user u who has already sent messages to r3 and r4. Basic CF item-to-item recommendation says that recipients similar to either r3 or r4 should be recommended to u. In general, the method generates a set of candidate pairs $\langle u, c \rangle$, giving a set of candidates c for each of a set of users u. If $u \nrightarrow v$ means that u has not contacted v, the set of candidate pairs can be defined by the following expression:

$$C = \{\langle u, r \rangle : \exists r' \, (u \to r' \wedge r' \overset{r}{\sim} r \wedge u \nrightarrow r \wedge r \nrightarrow u)\} \tag{3}$$

where u is the user to whom we recommend another user r.

These recommendations can be ranked by various means, however, we choose to rank them in the following way: each candidate recipient r who is a similar recipient to either r3 or r4 has a number of "votes" being the number of recipients (out of r3 and r4) to which r is a similar recipient; the ranking of r is just the number of these votes. In the first diagram in Figure 1, since r1 and r2 are similar to either r3 or r4, both can be recommended to u; r1 has 1 vote since r1 is similar to r3 but not r4, and r2 has 2 votes since r2 is similar to both r3 and r4. Where $| \cdot |$ denotes set cardinality, the number of votes of a candidate r for a user u is defined as follows:

$$votes(u, r) = |\{r' : (r \overset{r}{\sim} r' \wedge u \to r')\}| \tag{4}$$

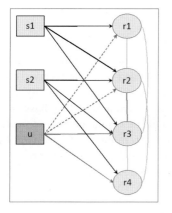 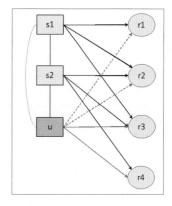

Fig. 1. Basic CF Recommendation with Similar Recipients and Similar Senders

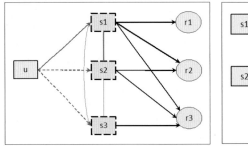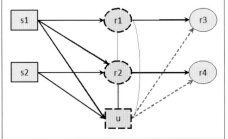

Fig. 2. Inverted CF Recommendation with Similar Senders and Similar Recipients

Recommendation using similar senders yields exactly the same set of candidates, but with a different ranking, where this ranking is determined by the votes of similar senders.

By symmetry, there are two other collaborative filtering recommendation methods, which we call "inverted" methods, for both similar senders and similar recipients. In Figure 2, thick broken line bordered shapes (green background) represent users who are both senders and recipients (for the purposes of the discussion, though of course senders also receive messages from other users and vice versa). In the first diagram in Figure 2, as in Basic CF recommendation using similar recipients, users are recommended to u if they are similar to u's contacts; however, in contrast to the standard approach, similarity here is based on similarity *as senders*. In the example, u has contacted s1, so s2 and s3 can be recommended to u, since they are both similar as senders to s1. The ranking is the number of u's contacts to whom they are similar (as senders), in this case both 1. One intuition underlying this inverted scheme is that it makes sense to recommend to u people with similar preferences (as determined by their sending patterns) to the contacts of u, as ideally, if those preferences indicate an interest in u, then those recommended users should also be highly inclined to reply positively to an invitation from u. Thus this method is a very strong filter on candidates compared to the other methods, in using the preferences of recipients as the basis of the recommendations. More formally, the candidate set and the vote calculation for this *Inverted CF Sender* model is as follows:

$$C = \{\langle u, s \rangle : \exists s' \, (u \to s' \wedge s' \overset{s}{\sim} s \wedge u \nrightarrow s \wedge s \nrightarrow u)\} \qquad (5)$$

$$votes(u, s) = |\{s' : (s' \overset{s}{\sim} s \wedge u \to s')\}| \qquad (6)$$

Another inverted method is illustrated in the second diagram in Figure 2. As in the Inverted Sender method, for a user u, users are recommended who have received contacts from similar users to u, but this time the similarity to u is based on similarity *as recipients*. In the example both r3 and r4 can be recommended to u as they have received contacts from users similar to u. The rank of a candidate r is the number of users who are similar as recipients to u and who have contacted r, in this case 1 for both r3 and r4. An intuition here is that u and the similar users are perceived as similar by prospective senders, hence (again inverting the Basic CF scheme) are likely to receive positive replies from similar people (for example, such senders). The candidate set and the vote calculation for this *Inverted CF Recipient* model is shown below:

$$C = \{\langle u, r \rangle : \exists r' \, (r' \overset{r}{\sim} u \wedge r' \to r \wedge u \nrightarrow r \wedge r \nrightarrow u)\} \qquad (7)$$

$$votes(u, r) = |\{r' : (r' \overset{r}{\sim} u \wedge r' \rightarrow r)\}| \tag{8}$$

An interesting point about this method is that users do not have to have sent any messages in order to receive recommendations; it is sufficient that they are similar as recipients to other users. Also note that for the inverted methods u and r may be both senders and recipients.

3.2   Extended Methods

The methods described in the previous section take all interactions (both successful and unsuccessful) into consideration in the calculation of "similarity" and in the ranking of candidates. In this section, we discuss variants of the basic methods that treat such positive and negative interactions separately for constructing the sets of candidates. We also consider combining candidate sets from different basic models. By a *positive* or *successful* interaction, we mean an interaction where user u sent a message to user v and received a reply message (again drawn from a predefined set) that was considered to encourage further communication. Other interactions are *negative* or *unsuccessful*.

We now consider variants of the basic and inverted collaborative filtering methods, denoted *Basic CF+*, *Inverted CF+ Sender* and *Inverted CF+ Recipient*, that consider only positive interactions for the generation of similarity and candidate sets. In addition, *Combined CF+* unifies the above three methods by adding the votes for candidate pairs that occur more than once. We also tried other methods for combining votes, such as taking the maximum and average of the votes of the methods, but found that simple addition gives the best results. Finally, *Best Two CF+* combines only Basic CF+ and Inverted CF+ Recipient in one candidate set, adding the votes for candidates generated by both methods. Another method called *Two Way CF+* is based on combining the two similarity relations by taking their intersection, so that pairs are similar as both senders and recipients. Recommendations for a user are generated from both senders similar to the user and recipients similar to the user's contacts. The idea was that this would create stronger candidates for recommendation. The formal definitions are as follows, where $u \overset{+}{\rightarrow} v$ means a positive contact from u to v.

$$u \overset{rs}{\sim} v \overset{\text{def}}{=} u \overset{r}{\sim} v \wedge u \overset{s}{\sim} v \tag{9}$$

$$C = C_r \cup C_s \tag{10}$$

where

$$C_r = \{\langle u, r \rangle : \exists r' \, (u \overset{+}{\rightarrow} r' \wedge r' \overset{rs}{\sim} r \wedge u \not\rightarrow r \wedge r \not\rightarrow u)\} \tag{11}$$

and

$$C_s = \{\langle u, r \rangle : \exists s \, (s \overset{+}{\rightarrow} r \wedge s \overset{rs}{\sim} u \wedge u \not\rightarrow r \wedge r \not\rightarrow u)\} \tag{12}$$

We also tested methods that use both positive and negative interactions. These methods construct two candidate sets: one based on positive votes using the Best Two CF+ method, and one using negative votes, also using the Best Two CF+ method but created using negative interactions. The two sets are then combined by subtracting the negative votes from the positive votes for a given candidate pair. Intuitively, the negative votes for a candidate to be recommended to a user count against the recommendation, in that the candidate has previously rejected other users who are similar to that user. We considered two ways of weighting the negative votes. In *Best Two CF+/-*, negative votes are weighted equally to positive votes. However, since there are many more negative interactions than positive interactions, this weighting is high. So we also considered a method *Best Two CF+/-0.2* in which negative votes were weighted 0.2 (there are roughly 5 times as many negative interactions as positive interactions, so the average number of votes for a moderate candidate is around 0).

4   Evaluation

In this section, we focus on evaluating the collaborative filtering methods described in Section 3 on historical data collected from a commercial dating web site. This dating site records user profiles and the interactions between users. User profiles contain basic information about users, such as age, location, education and occupation, information about user interests in the form of free text and information about a user's previous searches. The interaction information contains details of messages sent between users.

Note that users of the site did not have access to our recommendations, so the results are with respect to the users' actual behaviour on the site, with a view to comparing the various methods using several metrics, discussed below.

4.1   Evaluation Metrics

Any interaction instance that actually occurs has a correct classification (known from the historical data) as either successful or unsuccessful. However, in recommender systems, we are interested in the generation of sufficiently many *new* candidates with sufficiently high confidence of them being successful. In particular, in evaluating different methods, we are interested in three questions: (i) how well the method conforms to the reality of user interactions, (ii) how many recommendations can be made and with what distribution over the set of users, and (iii) how well the candidates are ranked. Note that since only historical data is used, it is not our objective to predict the interactions that actually occurred in the data (since our eventual aim is to recommend new candidates to users), but rather to use suitable metrics to compare different methods.

More formally, each method considers a set of users P in some training set, and generates a set of potential successful contacts between members of P, typically many more contacts than occurred. Let C be the set of candidate pairs generated by a given method. This implicitly gives a set of senders S and two sets of recipients, R (those potential recipients of contacts from S) and Q (those recipients who actually received contacts from S in a test period), where a contact from S means a contact from any member of S. Let $m(C)$ be the set of contacts in C that actually occurred in a given test set, $nm(C)$ be the size of $m(C)$, and $nm(C, +)$ be the number of contacts in $m(C)$ that were successful. Similarly, $n(S)$ is the size of S, $ns(S, R)$ is the number of contacts from a member of S to a member of R and $ns(S, R, +)$ is the number of successful contacts from S to R.

The first question is how well a method conforms to the user interactions occurring in the test period. Borrowing from information retrieval, the metrics are:

– *Precision:* The proportion of those predicted successful contacts C that occurred which were successful, i.e.

$$Precision = nm(C, +)/nm(C) \qquad (13)$$

– *Recall:* The proportion of successful contacts sent by S that are predicted by the method, i.e.

$$Recall = nm(C, +)/ns(S, Q, +) \qquad (14)$$

A related question is how many users can receive recommendations using the various collaborative filtering methods. This is a kind of coverage measure, but oriented towards users rather than contacts. This is measured with respect to the set of users who sent or received at least one message in the training period (i.e. those users to whom candidates could potentially be recommended) rather than with respect to the users in the test set. Since there is a high overlap between the senders and receivers during the training period (around half the receivers are also senders), there are around 128,000 such users. Let M be this set of users and let N be the subset of M of users who receive recommendations using a collaborative filtering method. The metric is:

– *Coverage:* The proportion of users from M who receive recommendations, i.e.

$$Cov = n(N)/n(M) \qquad (15)$$

The second question is how many potential interactions are generated by a given method (and in particular, how many candidates are recommended for each user). For this we simply count the number of candidates generated for the different users, and plot the distribution of these counts.

The third question is how well the candidates are ranked by the "voting" schemes used with the collaborative filtering methods. The metric we use here is Success Rate Improvement, defined as follows.

– *Baseline Success Rate:* The proportion of contacts from senders S to all recipients Q that were successful, i.e.

$$SR(S, Q) = ns(S, Q, +)/ns(S, Q) \qquad (16)$$

– *Success Rate:* The precision, the proportion of those predicted successful contacts C that occurred which were successful, i.e.

$$SR(C) = nm(C, +)/nm(C) \qquad (17)$$

– *Success Rate Improvement (SRI):* Success Rate/Baseline Success Rate, i.e.

$$SRI = SR(C)/SR(S, Q) \qquad (18)$$

The success rate improvement is designed specifically to measure how a particular method improves the chances of successful interactions between users as determined on the test set. To evaluate the ranking of candidates, the set of recommended candidates for each user is limited to various bounds, 20, 40, 60, 80 and 100, and SRI is calculated over only those candidates. That is, the top N success rate is calculated using only at most N candidates per user and not the set of all candidate pairs in C.

4.2   Evaluation Methodology

All collaborative filtering methods were evaluated by constructing a candidate set using some initial training set and then evaluating the candidates on a smaller test set, disjoint from the training set. The aim is to measure the effectiveness of the different recommendation methods with reference to the successful interactions that occurred.

The training set we used consisted of about 1.3 million interactions covering 4 weeks. The test set consisted of about 300,000 interactions over a period of six days immediately following the training set period. We also did preliminary evaluation using training periods of 3, 2 and 1 week before the test period, but found the 4 week period optimal in terms of precision and coverage.

A major requirement is that candidate generation be reasonably efficient. There is clearly a trade-off between the period of prior contact data used in the methods to calculate similarity, and the number of users who can be recommended and for whom recommendations can be provided. This trade-off is amplified due to the large numbers of interactions for a small minority of "popular" users. If all users are considered, a very large number of candidates will be the already popular users and will be provided for the already popular users, since they are similar (as both senders and recipients) to very many users. However, this is counter to the purpose of a recommender system, since these are the users least likely to want a recommendation or to be recommended to anyone else, as they are already the users with the highest number of communications. In the evaluations reported here, we therefore remove all popular users as both candidates for recommendation and for receiving recommendations, where *popular* means "has received more than 50 messages in the last 30 days".

The training dataset was prepared by selecting all contact events created in a four week period, selected as the training period, and excluding those made by such popular users or sent to such popular users (or both). The test dataset consisted of contact events occurring in the six day period following the training period, called the test period, for which both sender and recipient were active during the training period (i.e. ignoring interactions between users who started sending or receiving messages in the test period). This leaves around 700,000 interactions in the training set and around 120,000 in the test set. There were around 60,000 senders and 110,000 recipients in the training set, and 25,000 senders and 47,000 recipients in the test set (there is a high overlap between senders and recipients).

4.3   Results for Different Methods

Table 1 shows some basic statistics for the similarity tables computed using the different definitions of similarity, after removing popular users. Shown are the number of similar pairs of users and the average and maximum number of similar users per user, followed by the number of users with only 1 similar, 2 similar, 3 similar and between 3 and 50 users similar. Recipient+ and Sender+ refer to similarity as recipients and senders but taking into account only positive interactions, and R+ ∩ S+ is their intersection, used in the Two Way CF+ method. Since there are many more recipients than senders (a ratio of around 2 to 1), and many more contacts than successful contacts (a ratio of around 6 to 1), the relativities are not surprising. However, the large absolute numbers indicate that there is a high overlap in sending patterns (intuitively, senders contacting the

Table 1. Similarity of Users Under Various Schemes

	Recipient	Sender	Recipient+	Sender+	R+ ∩ S+
Similar pairs	57,859,748	7,789,252	2,240,656	616,568	9,680
Unique users	114,044	61,024	50,939	35,808	4,916
Average similar users	507	127	49	18	2
Maximum similar users	8,043	7,789	1,004	740	29
Users with 1 similar	957	980	3,059	3,218	2,931
Users with 2 similar	1,046	1,032	2,753	2,837	992
Users with 3 similar	1,081	971	2,550	2,513	404
Users with 3–50 similar	32,426	27,748	32,854	27,347	993

same groups of receivers). These numbers suggest that the main collaborative filtering methods can generate large numbers of candidates, since even users with only one other similar user can potentially receive a recommendation.

Table 2 shows a summary of results for the candidate generation methods if all generated candidates for each user are recommended (the "All" column) and if the top 20 are recommended (the "Top 20" column). Note that the candidates do not include those possible candidates whom the user has already contacted in the training period. Unsurprisingly, the Basic CF method generates the most candidates, while the Inverted CF methods generate perhaps more than expected. What is surprising is the recall of all the methods, which indicate that the contacting patterns of the users largely conform to the underlying assumptions of collaborative filtering (i.e. similar users really are contacting similar recipients). As expected, the methods based on positive only interactions generate fewer candidates, but with higher precision and SRI than the methods based on all contacts. Note that the Basic CF and Inverted CF Sender methods (and the corresponding methods based on positive contacts) can only generate recommendations for users who sent messages during the training period; the test set of users includes, in addition, a substantial number of users who, whilst not sending any messages during the training period, sent messages in the test period. Their messages can never count towards the recall for these methods. In contrast, the Inverted CF Recipient methods do generate candidates for a large number of such users.

Considering only individual basic and inverted collaborative filtering methods, it is clear that methods based on positive interactions (Basic CF+, Inverted CF+ Recipient and Inverted CF+ Sender) outperform those based on all contacts. Of these three methods, the method of choice is Basic CF+, which has a high SRI and a reasonable recall for those users for whom it provides recommendations (much higher, at 57%, when considering only the senders in training set). Since this has an SRI of 1.64, if those users followed the recommendations of Basic CF+, their success rate would improve by 64%. This degree of SRI is somewhat surprising, since it implies that users, once they have replied positively to some contacts, will continue to do so for contacts initiated by similar senders. The results for Inverted CF+ Recipient are more surprising, though the recall is much lower than for Basic CF+.

Since Inverted CF+ Recipient generates candidates for users who have sent no messages (but have replied positively to some messages), combinations of the above methods were analysed. The next two lines in Table 2 show the results for all three methods

Table 2. Summary of Results for Collaborative Filtering Methods

Method	Baseline SR	All				Top 20		
		SR	Recall	Coverage	SRI	SR	Recall	SRI
Basic CF	0.161	0.174	0.565	44.6%	1.08	0.184	0.0236	1.14
Inverted CF Recipient	0.179	0.22	0.269	82.6%	1.23	0.2	0.0098	1.12
Inverted CF Sender	0.161	0.216	0.188	38.6%	1.35	0.274	0.0098	1.7
Basic CF+	0.165	0.271	0.174	25.8%	1.64	0.359	0.016	2.18
Inverted CF+ Recipient	0.189	0.378	0.036	35.1%	2.00	0.45	0.0105	2.38
Inverted CF+ Sender	0.168	0.309	0.024	19.9%	1.83	0.308	0.0064	1.84
All Combined CF+	0.173	0.346	0.071	47.1%	2.00	0.399	0.0134	2.31
Best Two CF+	0.169	0.282	0.179	46.6%	1.67	0.394	0.0202	2.34
Best Two CF+/-	0.169	0.282	0.179	46.6%	1.67	0.386	0.0162	2.29
Best Two CF+/-0.2	0.169	0.282	0.179	46.6%	1.67	0.399	0.019	2.36
Two Way CF+	0.157	0.247	0.027	18.3%	1.58	0.273	0.0044	1.75

based on positive contacts combined and for the best two combined (Basic CF+ and Inverted CF+ Recipient). It can be seen that Basic CF+ and Inverted CF+ Recipient complement each other very well, with the coverage increasing significantly to around 47%, maintaining the precision of Basic CF+ with an increase in recall. However, the combination of all three methods performs similarly to the best two combined.

The next two lines are for methods that combine positive and negative votes to give different rankings, and are discussed further below. The results for Two Way CF+ (with ranking determined by adding the votes of senders similar to the user and recipients similar to the user's contacts) are shown in the last row of the table. Despite the fact that this method is based on a stronger notion of similarity, the SRI and recall are lower than for the methods based on one way similarity. This is probably because the sparsity of the data makes two way similarity too strict a criterion for generating candidates, but also there is an asymmetry between senders and recipients, in that some users will respond favourably to contacts received but will not initiate contacts. Two Way CF+ therefore fails to generate many potential successful interactions found by the other methods.

The second question of interest is the distribution of the counts of candidates generated for the users. We show here the results for the Basic CF+ method (perhaps the best single method, not considering the combined methods). The distribution of counts is shown in Figure 3 for bands of 1000. The distribution is as expected, with a small number of users having a large number of candidates, and the number of candidates dropping rapidly for a larger number of users. For example, around 35,000 users receive 1000 or fewer recommendations, and most users receive 2000 or fewer recommendations.

Figure 4 shows the distribution of candidate counts for bands of 10 up to 100. Over 4000 users receive 10 or fewer recommendations; 3000 between 10 and 20, etc. However, this still means that over half of the approximately 60,000 senders from the training set receive more than 100 recommendations, and around 90% of senders from the training set receive more than 20 recommendations. These results show that Basic CF+ is capable of generating large numbers of candidates, and therefore the same applies to the combined methods, which generate even more candidates.

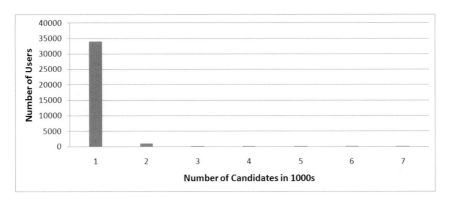

Fig. 3. Distribution of Candidate Counts for Basic CF+

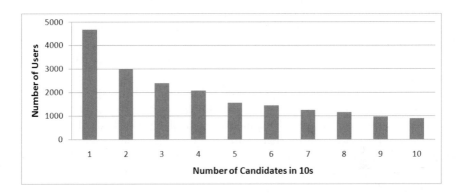

Fig. 4. Distribution of Candidate Counts within the 1-100 Range for Basic CF+

The final question is the ability of the "voting" methods used with collaborative filtering to rank candidates for recommendation. The question is whether the higher ranked candidates, being "closer" to the user's successful contacts in terms of interaction patterns, give a higher SRI than the recommended candidates in general. To answer this question, we calculated the SRI considering only the top N ranked candidates for each user. If the ranking is effective, the SRI should be higher for smaller values of N; while for large values of N, this SRI should be similar to the SRI when all candidates are considered. The baseline SRI of 1 represents the user's actual behaviour.

Figure 5 shows the SRI for the top N candidates for various values of N up to 100 for methods based on all contacts (for Basic CF, the ranking is from the votes of similar recipients): 100 is the likely maximum number of candidates presented to any user, 20 is the number of candidates that fit in a single e-mail message. Inverted CF Sender provides some improvement in SRI, probably because recipient interest in the senders is taken into account, however this method has a low coverage. The ranking functions for the other CF methods show little improvement over the baseline. This is because these methods recommend many candidates who replied negatively to similar users and so also reply negatively to the user receiving the recommendation.

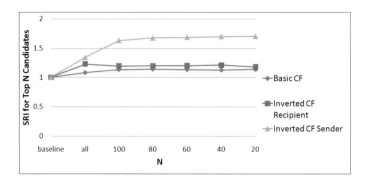

Fig. 5. SRI for Methods Based on All Contacts

Figure 6 shows the SRI for the top N candidates for the methods based on positive contacts (with the ranking for Basic CF+ derived from the votes of similar recipients). As expected, the corresponding SRI values are all higher than in Figure 5. Surprisingly, Basic CF+ gives a high SRI with a reasonable coverage and the ranking function provides even more improvement over the baseline. More surprisingly, Inverted CF+ Recipient gives a high SRI with the ranking providing additional improvement, and the coverage is reasonable. The combination of the best two methods (Basic CF+ and Inverted CF+ Recipient), shown in Figure 7 as Best Two CF+, gives even better results (with the ranking calculated by adding the votes from the two methods). For the top 20 candidates of Best Two CF+, the SRI is around 2.3, meaning that if users had followed these recommendations, their success rate would improve from around 17% to nearly 40%. Moreover, as observed above, the coverage of users is quite high, around 50% of all senders and receivers in the training set.

The combination of all three methods (Figure 6), again adding the votes together to give the ranking, whilst having a slightly higher coverage, has a consistently lower SRI

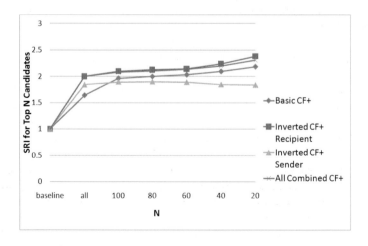

Fig. 6. SRI for Methods Based on Positive Contacts

than the best two methods combined. As shown in Figure 7, Best Two CF+/- and Best Two CF+/-0.2 have similar SRI to Best Two CF+ but the recall for the top 20 is lower than that for Best Two CF+. Hence using negative votes does not improve the resulting recommendations, indicating that the fact that a similar user has been rejected by a candidate does not affect whether the user will be rejected by the candidate. The reduction in recall (Table 2) is more pronounced when the negative votes are weighted the same as positive votes compared to when they are weighted 0.2 (where there is probably very little difference in the final ranking of candidates). Thus the method combining Basic CF+ and Inverted CF+ Recipient is the clear method of choice as the basis of a recommender system.

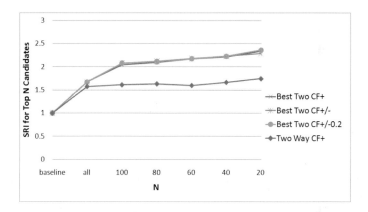

Fig. 7. SRI for Combined Methods

5   Conclusion and Future Work

We developed and evaluated a number of collaborative filtering methods designed for recommendation in online dating sites. We applied these methods to historical data from a dating web site and proposed a number of metrics to determine the effectiveness of our recommendation methods. Using these metrics we showed that collaborative filtering methods work very well when evaluated against the true (historical) behaviour of the users, not only for the generation of suitable candidates, but also to provide a substantial improvement in success rate over the baseline. This result is interesting because our collaborative filtering methods work only with user interactions and do not incorporate profile information. Our results imply that our methods (singly or combined) can be used as the basis of a recommender system for an online dating service. In particular, we found that combining the best two methods based on positive interactions provides an acceptable trade-off between success rate improvement, recall and coverage of users.

Future work includes applying machine learning methods to improve the ranking of recommended candidates, and hybrids of collaborative filtering and content-based methods to generate recommendations for users who do not have any successful interactions.

Acknowledgements

This work was funded by Smart Services Cooperative Research Centre. We would also like to thank our industry partners for providing the datasets.

References

1. Adomavicius, G., Kwon, Y.: New Recommendation Techniques for Multicriteria Rating Systems. IEEE Intelligent Systems 22(3), 48–55 (2007)
2. Balabanović, M., Shoham, Y.: Fab: Content-Based, Collaborative Recommendation. Communications of the ACM 40(3), 66–72 (1997)
3. Brŏzovský, L., Petřček, V.: Recommender System for Online Dating Service. In: Proceedings of Znalosti 2007 (2007)
4. Karypis, G.: Evaluation of Item-Based *Top-N* Recommendation Algorithms. In: Proceedings of the Tenth International Conference on Information and Knowledge Management, pp. 247–254 (2001)
5. Linden, G., Smith, B., York, J.: Amazon.com Recommendations: Item-to-Item Collaborative Filtering. IEEE Internet Computing 7(1), 76–80 (2003)
6. Pazzani, M., Billsus, D.: Content-Based Recommendation Systems. In: Brusilovsky, P., Kobsa, A., Nejdl, W. (eds.) Adaptive Web 2007. LNCS, vol. 4321, pp. 325–341. Springer, Heidelberg (2007)
7. Sarwar, B., Karypis, G., Konstan, J., Riedl, J.: Item-Based Collaborative Filtering Recommendation Algorithms. In: Proceedings of the 10th International World Wide Web Conference, pp. 285–295 (2001)
8. Shardanand, U., Maes, P.: Social Information Filtering: Algorithms for Automating "Word of Mouth". In: Proceedings of the ACM Conference on Human Factors in Computing Systems (CHI 1995), pp. 210–217 (1995)
9. Zhou, D.X., Resnick, P.: Assessment of Conversation Co-mentions as a Resource for Software Module Recommendation. In: Proceedings of the 2009 ACM Conference on Recommender Systems, pp. 133–140 (2009)

Developing Trust Networks Based on User Tagging Information for Recommendation Making

Touhid Bhuiyan[1], Yue Xu[1], Audun Jøsang[2],
Huizhi Liang[1], and Clive Cox[3]

[1] Faculty of Science and Technology
Queensland University of Technology, Australia
t.bhuiyan@qut.edu.au, yue.xu@qut.edu.au,
oklianghuizi@gmail.com
[2] UniK Graduate School, University of Oslo, Norway
josang@unik.no
[3] Rummble.com, Cambridge, England
clive.cox@rummble.com

Abstract. Recommender systems are one of the recent inventions to deal with ever growing information overload. Collaborative filtering seems to be the most popular technique in recommender systems. With sufficient background information of item ratings, its performance is promising enough. But research shows that it performs very poor in a cold start situation where previous rating data is sparse. As an alternative, trust can be used for neighbor formation to generate automated recommendation. User assigned explicit trust rating such as how much they trust each other is used for this purpose. However, reliable explicit trust data is not always available. In this paper we propose a new method of developing trust networks based on user's interest similarity in the absence of explicit trust data. To identify the interest similarity, we have used user's personalized tagging information. This trust network can be used to find the neighbors to make automated recommendations. Our experiment result shows that the proposed trust based method outperforms the traditional collaborative filtering approach which uses users rating data. Its performance improves even further when we utilize trust propagation techniques to broaden the range of neighborhood.

Keywords: Trust Networks, Interest Similarity, Recommender Systems, Social Networks and Tag.

1 Introduction

Recommender systems have been an active research area for more than a decade. There are many different techniques and systems have already been developed and implemented in different domains including online social networks. But most of the existing research on recommender systems focuses on developing techniques to better utilize the available information resources to achieve better recommendation quality. Because of the amount of available data and information remains insufficient, these

L. Chen, P. Triantafillou, and T. Suel (Eds.): WISE 2010, LNCS 6488, pp. 357–364, 2010.
© Springer-Verlag Berlin Heidelberg 2010

techniques have achieved only limited improvements to overall recommendation quality [1]. In recent years, incorporating trust models into recommender systems attracts attention of many researchers [2-6]. They emphasize on generating recommendation based upon trust peers opinion, instead of traditional most similar users opinion. Massa and Avesani [2] presented with evidence that, trust based recommender systems can be more effective than traditional collaborative filtering based systems. They argued that data sparsity causes the serious weakness in collaborative filtering system. However, they have assumed the trust network is already exists with the users explicit rating data. They did not consider a situation where the trust network is not available. Ziegler [3] has proposed frameworks for analyzing the correlation between interpersonal trust and interest similarity and suggested the positive interaction between the two. They have assumed that if two people have similar interests, they most likely trust each other. In a recent work, Bhuiyan [4] presents a survey on the relationship between trust and interest similarity in a social network; which results also strongly support Ziegler's hypothesis. Inspired by these findings, we have used users' interest similarity to form the trust network among the users irrespective of personal relationship; only based on utility. The existing trust based recommender works have assumed that the trust network is already exists. In this work, we have proposed to use trust as an alternative method in the absence of explicit rating data to find the neighbors and replace the first step of traditional collaborative filtering system where it finds the neighbors based on overlapped or common previous ratings data. Based on the results obtained from the experiment conducted in the work, it has been found that the proposed techniques have achieved promising improvement in the overall recommendation making in terms of correct recommendation.

The rest of the paper is organized in following ways. In section 2, we have discussed other related work in this field of study. Chapter 3 presented the proposed algorithm for trust estimation. Chapter 4 presented the results of the experiments and discussed about the findings and the paper is concluded in chapter 5.

2   Related Work

Collaborative filtering is the most popular techniques for recommender systems which collects opinions from customers' in the form of ratings on items, services or service providers'. But the collaborative filtering system performs poor when there is not sufficient previous common rating available between users [7]. To overcome this data sparse problem, trust based approach to recommendation has emerged. This approach assumes a trust network among users and makes recommendations based on the ratings of the users that are directly or indirectly trusted by the target user [5]. There are very few sites such as www.epinions.com, www.allconsuming.net, http://trust.mindswap.org/FilmTrust/, http://www.rummble.com/ etc. allow members to express which other agents they trust; by which the explicit trust value is collected. But most of the social networks do not collect explicit rating about trust among the user [8]. Though there are a good number of works are available in the field of recommender systems using collaborative filtering, very few researchers consider using tag information to make recommendation [9-13]. Tso-Sutter [9] used tag information as a supplementary source to extend the rating data. They did not use it as the replacement of the explicit rating information. Liang et.al [10] proposed to integrate social tags with item taxonomy to make personalized user recommendations. Other recent works include integrating tag information with content based recommender systems [13], extending the user-item matrix to user-item-tag matrix to collaborative filtering item recommendations [11], combining users' explicit ratings with the predicted users' preferences for item-based on their inferred preferences for tags [14] etc. However, using tagging information to estimate users' trust for generating user trust network has not drawn adequate attention.

3   Proposed Trust Estimation

To describe the proposed approach, we define some concepts used in this paper as below.

- **Users:** $U = \{u_1, u_2, \ldots, u_{|U|}\}$ contains all users in an online community who have used tags to label and organize items.
- **Items (i.e., Products, Resources):** $P = \{p_1, p_2, \ldots, p_{|P|}\}$ contains all items tagged by users in U. Items could be any type of online information resources or products in an online community such as web pages, videos, music tracks, photos, academic papers, documents and books etc.
- **Tags:** $T = \{t_1, t_2, \ldots, t_{|T|}\}$ contains all tags used by users in U. A tag is a piece of textural information given by users to label or collect items.

As mentioned before, we believe the trustworthiness between users is useful for making recommendations. However, the trust information is not always available, and even available, it may change over time. In this research, we propose to automatically construct the trustworthiness between users based on users' online information and online behaviour.

The current research on tags is mainly focusing on how to build better collaborative tagging systems, personalize search using tag information [9] and recommending items [14] to users etc. However,   tags are free-style vocabulary that users used to classify or label their items. Since there is no any restriction, boundary, or pre-specified vocabulary on selecting words for tagging items, the tags used by users lack in standardization and unification and also contain a lot of ambiguity.   Moreover, usually the tags are short containing only one or two words, which make it even harder to truly get the semantic meaning of the tags. To solve this problem, we propose an approach to extract the semantic meaning of a tag based on the description of the items in that tag. For each item, we assume that there is a set of keywords or topics which describe the content of the item. This assumption is usually true in reality. For most products, normally there is a product description along with the product. From the product description, by using text mining techniques such as *tf-idf* keywords extraction method, we can generate a set of keywords to represent the content of the item.

In a tag, a set of items are gathered together according to users' viewpoint. We believe that there must be some correlation between the user's tag and the content of the items in that tag. Otherwise the user may not classify the items into that tag. Thus, using text mining techniques, from the descriptions of the items in the tags, we can derive a set of keywords or topics to represent the semantic meaning of each tag.

For user $u_i \in U$, let $T_i = \{t_{i1},...,t_{il}\} \subseteq T$ be a set of tags that are used by u_i, for a tag $t_{ij} \in T_i$, by using text mining techniques such as *tf-idf* method, from the descriptions of the items in t_{ij}, we can generate a set of frequent keywords denoted as $W_{ij} = \{w_1,...,w_n\}$ to represent the semantic meaning of the tag. The frequency of the keywords, denoted as $v_{ij} =< f_1,....,f_n >$ where f_k is the frequency of the kth keyword, measures the strength of each keyword in tag t_{ij} to represent the meaning of the tag. Also the vector v_{ij} can be used to calculate the similarity of two tags in terms of their semantic meaning

$\forall u_i, u_j \in U$, let $T_i = \{t_{i1},...,t_{il}\}, T_j = \{t_{j1},...,t_{jl}\} \subseteq T$ be the set of tags which were used by user u_i and u_j, respectively. Corresponding to T_i and T_j, $W_i = \{w_{i1}, ..., w_{in}\}$ and $W_j = \{w_{j1},...w_{jm}\}$ are the collection of keyword sets for the tags in T_i and T_j, respectively, and $V_i = \{v_{i1}, ..., v_{in}\}$ and $V_j = \{v_{j1},...v_{jm}\}$ are the corresponding vectors of keyword frequency. For example, w_{il} is the set of keywords derived from the items descriptions in tag t_{il} and v_{il} is the vector of frequency of the keywords in w_{il}. Let $sim(v_{ip}, v_{iq})$ be the similarity between v_{ip} and v_{iq}, if $sim(v_{ip}, v_{iq})$ is larger than a pre-specified threshold, the two tags t_{ip} and t_{iq} are considered similar.

The aim is to build the conditional probability $p(u_i / u_j)$ estimating the likelihood that user u_i is similar to user u_j in terms of user u_j's information interests. The following equation is defined to calculate how similar user u_i is interested in keyword k given that user u_j is interested in the keyword k:

$$p_k(u_i / u_j) = \frac{n_{ij}^k}{n_j^k} \tag{1}$$

Where, n_j^k denotes the number of tags in W_j that contain keyword w_k, n_{ij}^k denotes the number of tags in W_i that contain keyword w_k and are similar to some tags in W_j that contain keyword w_k as well. After calculating this for every keyword, the average of the probability $p_k(u_i / u_j)$ is used to estimate the probability $p(u_i / u_j)$:

$$p(u_i / u_j) = \left(\sum_{k \in W} p_k(u_i / u_j)\right) / |W| \tag{2}$$

where, $W = \{w_1,..., w_r\}$ is the set of all keywords in W_i or W_j.

In this paper, we use the conditional probability $p(u_i / u_j)$ to measure the trust from user u_j to user u_i. Given u_j, the higher the $p(u_i / u_j)$, the higher user u_j trusts u_i since user u_i has similar interest as u_j.

4 Experiment

We have used the traditional collaborative filtering algorithm to make recommendations. The traditional collaborative filtering algorithm has two steps. First, it finds the similar neighbors based on the overlap of previous ratings and in the second steps, it calculates to predict an item to recommend to a target user. For all of the experiment data, we have used the same method for the second part of the algorithm. But, we have used our proposed trust network based algorithm to find the neighbors and make recommendations. Then compare those recommendation results with the traditional collaborative filtering method using Jaccard's coefficient to find the neighbors. For our experiment; we have used the book dataset downloaded from www.amazon.com. User tag data and book taxonomy data, both are obtained from Amazon site. The book data have some significant difference between other data about movies, games or videos. Every published book has a unique ISBN, which makes it easy to ensure interoperability and gather supplementary information from various other sources, e.g., taxonomy or category descriptors from Amazon for any given ISBN. The dataset consists of 3,872 unique users, 29,069 unique books and total 54,091 records. The tree structure Amazon book taxonomy contains 9,919 unique topics. Each book in Amazon dataset has several taxonomic descriptors, each of which is a set of categories in the book taxonomy. In this experiment, we extract keywords for each tag from the descriptors of the books in the tag.

4.1 Traditional Approach of Collaborative Filtering

Collaborative filtering recommender systems try to predict the utility of items for a particular user based on the items previously rated by other users [6]. More formally, the utility $u(c,s)$ of item s for user c is estimated based on the utilities $u(c_j,s)$ assigned to item s by those users $c_j \in C$ who are "similar" to user c. The value of the unknown rating $r_{c,s}$ for user c and item s is usually computed as an aggregate of the ratings of some other users for the same item s:

$$r_{c,s} = aggr_{c' \in C} r_{c's} \tag{3}$$

Where C denotes the set of N users that are the most similar to user c and who have rated item s (N can range anywhere from 1 to the number of all users). We have used the following aggregation function:

$$r_{c,s} = k \sum_{c' \in C} sim(c, c') \times r_{c',s} \tag{4}$$

Where multiplier k serves as a normalizing factor and is selected as

$$k = 1/\sum_{c' \in C} |sim(c, c')| \tag{5}$$

In the case of binary value (eg. Either an item is rated or not), Jaccard's coefficient is used to measure similarity. For some applications, the existence of S in Simple Matching makes no sense because it represents double absence. Jaccard's coefficient removes the S from simple matching coefficient to become Formula

$$S_{ij} = \frac{p}{p+q+r} \tag{6}$$

Where; p = number of variables that positive for both objects; q = number of variables that positive for the ith objects and negative for the jth object; r = number of variables that negative for the ith objects and positive for the jth object and s = number of variables that negative for both objects.

4.2 Evaluation Metrics

The "Precision and Recall" method is used to evaluate the recommendation performance. This evaluation method has been initially suggested by Cleverdon as evaluation metrics for information retrieval systems [15]. Due to the simplicity and the popular uses of these two metrics, they have been also adopted for recommender system evaluations [7]. The top-N items are recommended to the users. For comparison, we have used N= 5, 10, 15, 20 and 25. The training data contains users used books and tag information but the testing dataset only used user books information. Precision and Recall for an item list recommended to user u_i is computed based on the following equations:

$$\text{Precision} = \frac{|T_i \cap P_i|}{|P_i|} \qquad\qquad \text{Recall} = \frac{|T_i \cap P_i|}{|T_i|} \tag{7}$$

Where T_i is the set of all items preferred by user u_i and P_i is the set of all recommended items generated by the recommender system. Based on the equation 11 and 12, it can be observed that the values of precision and recall are sensitive to the size of the recommended item list. Since, precision and recall are inversely correlated and are dependent on the size of the recommended item list, they must be considered together to evaluate completely the performance of a recommender [16]. F1 Metric is one of the most popular techniques for combining precision and recall together in recommender system evaluation which can be computed by the formula 13 is used for our evaluation.

$$F1 = \frac{2 \times \text{Precision} \times \text{Recall}}{\text{Precision} + \text{Recall}} \tag{8}$$

4.3 Experiment Results and Discussion

We have used the same dataset for making recommendations. We let each of the four techniques to recommend a list of N items to each of these 3,872 users, and different values for N ranging from 5 to 25 are tested. Figure 3 shows the precision values of recommendation made among our proposed tag-based Similarity Trust approach (ST) and the Jaccard's coefficient based traditional Collaborative Filtering (CF) approach. We have also extend the neighbor range by propagating trust using Golbeck's Tidal Trust (TT) algorithm and our previously proposed DSPG using Subjective Logic (SL) propagation algorithm [17,18]. The results of the experiment are shown in Figure 1, Figure 2 and Figure 3. It can be observed from the figures that for all three evaluation metrics, the proposed propagated SL technique achieved the best result among all the four techniques compared. The two propagated method TT and SL performed closely but SL is slightly better among these two. Among the proposed ST and traditional CF recommender, our proposed ST performed significantly better than the traditional CF method. Both of these methods used the same recommendations techniques but the difference is in the finding neighbors' technique. The results clearly show that when we have used the traditional collaborative filtering approach for finding neighbors, it performed the worst. Our tag-based similarity trust approach performed better than the traditional approach.

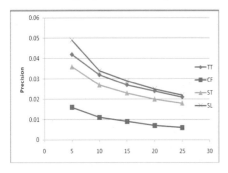

Fig. 1. Evaluation with precision metric

The results improved further when we have extended the boundary of the neighbors using trust propagation algorithms. We have compared our proposed DSPG algorithm with widely know Tidal Trust propagation algorithm. The results showed that our approach outperformed the Tidal Trust propagation techniques.

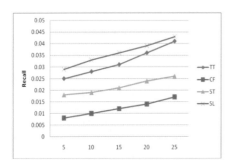
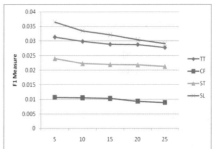

Fig. 2. Evaluation with recall metric   **Fig. 3.** Evaluation with F1 metric

Figure 2 shows the Recall values between the same four approaches. And finally in Figure 3, the F-Measure based on Precision and Recall of the four approaches is presented.

5 Conclusion

We have presented a new algorithm for generating trust networks based on user tagging information to make recommendation in the online environment. The experiment results showed that this tag-based similarity approach performs better while making recommendations than the traditional collaborative filtering based approach. This proposed technique will be very helpful to deal with data sparsity problem; even the explicit trust rating data is unavailable. The finding will contribute in the area of recommender system by improving the overall quality of automated recommendation making.

References

1. Park, S.T., Pennock, D., Good, N., Decoste, D.: Naïve Filterbots for Robust Cold-start Recommendations. In: Proceedings of the 12th International Conference on Knowledge Discovery and Data Mining, Philadelphia, PA, USA (2006)
2. Massa, P., Avesani, P.: Trust-aware Recommender Systems. In: ACM Recommender Systems Conference, Minneapolis, MN, USA (2007)
3. Ziegler, C.N., Golbeck, J.: Investigating interactions of trust and interest similarity. Decision Support Systems 43(2), 460–475 (2007)
4. Bhuiyan, T.: A Survey on the Relationship between Trust and Interest Similarity in Online Social Networks. To appear in the Journal of Emerging Technologies in Web Intelligence 3 (2010)
5. Golbeck, J.: Computing and Applying Trust in Web-based Social Networks. PhD thesis, University of Maryland College Park (2005)
6. Adomavicius, G., Tuzhilin, A.: Toward the Next Generation of Recommender Systems: A Survey of the State-of-the-Art and Possible Extentions. IEEE Transactions on Knowledge and Data Engineering 17(6), 734–749 (2005)
7. Sarwar, B.M., Karypis, G., Konstan, J.A., Reidl, J.: Analysis of Recommendation algorithms for e-commerce. In: The Second ACM Conference on Electronic Commerce, Minneapolis, Minnesota, USA, pp. 158–167 (2000)
8. Fu, B.: Trust Management in Online Social Networks. MSc Dissertation, Trinity College, University of Dublin, Ireland (2007)
9. Tso-Sutter, K.H.L., Marinho, L.B., Schmidt-Thieme, L.: Tag-awer Recommender Systems by Fusion of Collaborative Filtering Algorithms. In: Proceedings of the ACM Symposium on Applied Computing, USA, pp. 1995–1999 (2008)
10. Liang, H., Xu, Y., Li, Y., Nayak, R., Weng, L.T.: Personalized Recommender Systems Integrating Social Tags and Item Taxonomy. In: Proceedings of the Joint Conference on Web Intelligence and Intelligent Agent Technology, Italy, pp. 540–547 (2009)
11. Liang, H., Xu, Y., Li, Y., Nayak, R.: Collaborative Filtering Recommender Systems based on Popular Tags. In: Proceedings of the 14th Australasian Document Computing Symposium, Sydney (2009)
12. Heymann, P., Ramage, D., Garcia-Molina, H.: Social Tag Prediction. In: Proceedings of the 31st Annual International ACM SIGIR Conference on Research and Development in Informational Retrieval, USA, pp. 531–538 (2008)
13. Gemmis, M.D., Lops, P., Semeraro, G., Basile, P.: Integrating Tags in a Semantic Content-based Recommender. In: Proceedings of ACM Conference on Recommender Systems 2008, pp. 163–170 (2008)
14. Sen, S., Vig, J., Riedl, J.: Tagomenders: Connecting Users to Items through Tags. In: Proceedings of WWW 2009, pp. 671–680 (2009)
15. Cleverdon, C.W., Mills, J., Keen, M.: Factors Determining the Performance of Indexing Systems. ASLIB Cranfield Project, Cranfield (1966)
16. Herlocker, J.L., Konstan, J.A., Terveen, L.G., Riedl, J.T.: Evaluating Collaborative Filtering Recommender Systems. ACM Transactions on Information Systems 22, 5–53 (2004)
17. Jøsang, A., Hayward, R., Pope, S.: Trust Network Analysis with Subjective Logic. In: Proceedings of the 29th Australasian Computer Science Conference, CRPIT, Hobart, Australia, vol. 48 (2006)
18. Jøsang, A., Bhuiyan, T.: Optimal Trust Network Analysis with Subjective Logic. In: The Second International Conference on Emerging Security Information, Systems and Technologies, Cap Esterel, France (2008)

Towards Three-Stage Recommender Support for Online Consumers: Implications from a User Study

Li Chen

Department of Computer Science, Hong Kong Baptist University
Hong Kong, China
lichen@comp.hkbu.edu.hk

Abstract. In this paper, a three-stage recommender support was implied from a user study. The purpose of the user study was to understand how to best utilize different types of social information (e.g., product popularity, user reviews) for facilitating online consumers' decision-making process in the e-commerce environment. Through both of in-depth tracking users' objective behavior and qualitative interviewing their reflective thoughts, we have not only refined a traditional two-stage decision process into a more precise three-stage process, but also identified at each stage what information users are inclined to seek for. Based on the study's results, suggestions were made to related recommender systems about their practical roles in the three-stage framework and how they can more effectively support users' information needs.

Keywords: user study, e-commerce, recommender supports, Flickr camera finder, complex decision making, high-value products, users' information needs.

1 Introduction

Social content has always been recognized to play important role in a consumer's hybrid decision process in which the decision maker seeks for advices for the purpose of reducing the uncertainty and the amount of information that must be processed to make a decision [9]. As a matter of fact, in recent years, recommender systems have been broadly developed in order to suggest items that users may be interested in [2]. For high-value, infrequently experienced products (e.g. a digital camera, a computer, and a house), the typically applied recommendation in existing e-stores is "people viewed this product also view others …" (or "people bought this product also bought …"). These recommended products are normally computed by the item-based method [2], which identifies a set of items that are most similar to the user's currently viewed one, in combination with other users' clicking or purchase histories.

However, most of related works have just focused on providing recommendations from the perspective of algorithm improvement. No much work has studied on when these recommendations will be most relevant to the user's actual needs within her/his whole decision process. Indeed, according to Adaptive Decision Theory [8], human's decision making is in nature a constructive and adaptive process, which basically follows two stages: 1) to screen down the number of available alternatives to a

L. Chen, P. Triantafillou, and T. Suel (Eds.): WISE 2010, LNCS 6488, pp. 365–375, 2010.
© Springer-Verlag Berlin Heidelberg 2010

reduced consideration set, and 2) to in-depth examine the selected candidates and obtain the final choice. The questions are then that, at each stage, what kinds of social content (i.e., other users' generated product content) will be particularly interesting to the current user, and whether a recommender support that incorporates the required social resources could more effectively benefit the user.

With these questions, we have launched a user study with the goal of observing users' decision behavior when they are looking for a high-value product to buy. More concretely, two objectives of the study are: 1) tracing users' decision-making process to refine the basic two-stage model; and 2) understanding users' social information needs at each stage. Two kinds of social resources were especially investigated in the study. One is *product reviews*, which are what other users generate according to their post-purchase evaluation experiences with the product. The reviews are expressed in form of numerical ratings or natural languages such as in Yahoo shopping and Amazon. Another is *product popularity* info (e.g., "the top products"). In this regard, we have involved a newly released Flickr Camera Finder, since it primarily provides the popularity data based on the statistics of Flickr community members who have uploaded images or videos with a particular camera over a certain time. Its involvement is not only because the product domain is what we emphasize (i.e., the high-value product), but also its usage-driven popularity generation is different from traditional purchases or promotion driven method (as been used in most e-commerce sites). Through the empirical study, we can hence observe whether users could in reality perceive the difference between the two types of popularity data (i.e., one from social media site, and another from standard e-commerce site), and know whether the novel type could support users to make a more confident decision.

As a result of the user study, a three-stage recommender framework is established that suggests a set of directions regarding how to improve related decision systems and best assist users when they are at different decision stages. In the following, we will introduce our research questions and describe how our experiment was setup and the set of suggestions concluded from the user study.

2   Research Questions and Experiment Setup

2.2   Research Questions

The amount of cognitive effort applied to the decision process is in essence related to the level of importance that the consumers place on the specific product [5]. As for high-value products that are expensive and infrequently experienced (e.g., cameras, computers, cars), extensive decision-making effort is commonly spent by consumers in seeking information and deciding. Accordingly, as mentioned earlier, researchers from the psychology field describe a two-stage process in this condition, where the depth of cognitive load and information processing varies [8].

In the last decade, recommender systems have been widely developed, but most works were oriented to low-value, public taste products (e.g., music, movie, webpage, book) [6,7]. Little has exploited the effect of social information (i.e., user-generated content) on producing recommendations for high-value products. In this area, most systems have been still based on products' static attributes (e.g., the camera's optical zoom, megapixels) to model users' information needs. For instance, the critiquing-based recommenders such as Dynamic-Critiquing, FindMe and Example-Critiquing [4] can support the incremental refinement of user preferences via providing the critiquing facility (e.g., "I would like something cheaper", "with faster processor speed"), but the user's preference model is purely built on static attribute values. [10] has conducted a tentative experiment that discovered merits of product ratings and reviews as part of recommendations to influence users' searching strategies. However, it is still not clear how the social content can be integrated with static product attributes to construct more accurate user model.

Given the existing limitations, our objective was to identify the relative importance of social content (e.g., product popularity, product reviews, etc.) in users' complex decision process. Specifically, we have targeted to answer the following research questions by means of a user study:

1) How would consumers in practice follow the basic two-stage process when interacting with the e-stores? Would the two-stage be further refined, and at each stage, how would different kinds of social content act to assist the user? More deeply, we have examined the following three aspects:

a. How would users perceive the so called personalized recommendations that suggest items based on the user's current click (e.g., *"people viewed this product also viewed others"*)?

b. Would there be any differences of users' perception and usage of the *product popularity info* as from social media sites (e.g., Flickr), relative to from the standard e-commerce sites? And if there are differences, which type of popularity would be more effective?

c. As for *product reviews*, what would be their primary role in users' decision process and which kind of reviews (i.e., positive vs. negative reviews) would be more interesting to the buyer?

2) How would we develop more useful recommender supports applicable at each stage so as to maximize the benefits of social content that users will intentionally rely on?

2.2 Experiment Setup

In order to obtain answers to above questions, we have conducted an experiment that recorded users' decision behavior when they were assigned the task of looking for an item to buy. As mentioned before, we used Flickr Camera Finder as one experiment material because it provides the unique source of usage-based popularity info. Besides, a standard e-commerce site was also offered, which provides the complete set of static attribute values, the traditional popularity data (so as to be compared with Flickr's), product reviews and personalized recommendations. For this site, we have considered a number of options, including Amazon, Yahoo Shopping, shopping.com, etc, and finally selected Yahoo Shopping because it is not only based on the same product database that Flickr Camera Finder (CF) uses, but also can be representative of other e-commerce sites regarding information amount and information diversity.

The experiment was concretely designed in a free-choice scenario where users were allowed to freely select and examine any product info that can be obtained from the two sites: Flickr Camera Finder (www.flickr.com/cameras/, Flickr CF for short) and Yahoo Shopping (shopping.yahoo.com) for Cameras (Yahoo CF for short). Note that this is not a comparative study and we were not to identify which site is better. Our goal was to reveal what kind of social info that the two sites provide can be in reality adopted by users.

We have finally recruited twelve motivated volunteers (three females) because they were interested in buying a digital camera at the time of our experiment. They are Master or PhD students in our department with ages between 20 and 40. They have often visited e-stores (at least once every three months), which indicates that they can be representative of online consumers to some extent. It is also worth noting that all subjects were first-time encounters to the two websites (i.e., Flickr CF and Yahoo CF), so that their behavior would not be biased by any of previous usage experience.

In the experiment, an initial warm-up period (10 minutes) was first given to each participant for her/him to be familiar with the two sites' facilities as much as possible. Then s/he was assigned the task: "*Imagine you are prepared to buy a digital camera, please use the provided two sites to examine product information and find a product that best meets your needs.*" During the formal trial, all of the user's interaction events, including on-screen mouse moves, clicks and keyboard inputs, were automatically captured by a screen observer software (Morae). Then, after the participant finished the task, a semi-structured interview was conducted by the administrator in order to obtain the participant's reflective thoughts and opinions on various examined aspects.

3   Results Analysis and System Implications

3.1   Result 1: Three-Stage Decision Process

The analysis of all users' decision-making behavior surprisingly shows that they all exhibited a more precise, three-stage process: 1) to screen all alternatives and select one for in-depth evaluation; 2) to view the selected product's details and save it in wish list if near-satisfactory; 3) to compare candidates in the wish list and make the final choice. Moreover, the transition between these stages did not follow a sequential order, but was iterative in nature and the size of consideration set gradually decreased. Concretely, at the start, all users were with some initial preferences in mind, e.g., looking for a camera that is "*easy to use*", "*easy to carry*", "*with colorful images*", "*of high cost performance*", "*better for night scenes*", or "*better for long distance picture-taking*". Some users (6) had specific criteria on product's static attributes (e.g., on type, megapixels, screen size, battery, focal length). As for price, 5 users expressed the constraint (e.g., "cheap", "not expensive", "under $500").

Their interaction logs further show that they maximally considered four brands within the whole decision session. Their information-seeking process was typically brand-based. That is, they first looked into one preferred brand. If anyone product(s) interested them when they browsed the brand's product list (Stage 1, see Figure 1 left), they went to examine the product's details and saved it into the wish list if the details were near-satisfying (Stage 2). After reviewing one brand's products, they switched to another preferred brand and performed the similar browsing process. The iterative cycle between stages 1 and 2 continued until a set of candidates was determined (the average number of candidates is 4. See the analysis later). At this point, they entered into Stage 3 to compare candidates in their wish list and confirmed the final choice. Due to the common behavior exhibited by all participants, we came up with the three-stage decision process model (see Figure 1 left) with the processed input and output at each stage.

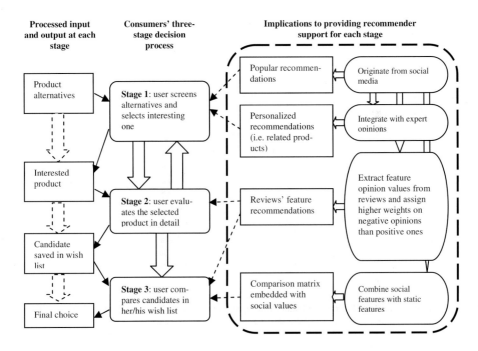

Fig. 1. Three-stage recommender framework, with the goal of facilitating an online buyer's information seeking need at each decision stage

3.2 Result 2: Implications to Providing Recommender Supports at Each Stage

It is then interesting to know what product information (i.e., social/static content) that users have processed at each stage and what kind of recommender supports could be relevant to their information need. Figure 1 (right) summarizes system implications. In the following, we will in detail explain how the implications were derived from the user study's results.

3.2.1 Stage 1: Screening Out Interesting Products

3.2.1.1 Users' Objective Behavior

At this stage, we measured how many products were selected (i.e. clicked by the user for details) and from where these products were located. It indicates that on average 9.67 (*St.d.* = 4.78) products were chosen to view details, among which 5.42 were located in Yahoo, and 4.25 were in Flickr CF. Figure 2 (left) concretely shows the distribution of these products' locations. In fact, basic static features provided by

370   L. Chen

Yahoo for the browsing/filtering got the highest chance that enabled the average user to obtain 39.79% interesting products. The second and third winners came to Flickr CF's popularity-based sorting list (27.51%) and brand popular products (12.18%) respectively. In comparison, Yahoo's popularity list got much less hits (5.28%). There were in fact only 2 participants who accessed "Top Digital Cameras" in Yahoo, against 9 in Flickr CF. As for the remaining selected products, they were either found through keyword-based search (for example, the user input a model's name) (6.53%), or through Flickr's photo-related products (4.83%), or Yahoo's recommendations (i.e., *"shopper who viewed this also viewed ..."*, 3.89%).

Thus, in total, above half of interested products (53.69%) were stemming from social sources (see Figure 2, right). Product popularity was shown particularly more active relative to the other social contents at this stage. As one user said, *"popularity is a suitable proxy to measure the product's quality when I am not familiar with a brand or uncertain about what I want"*.

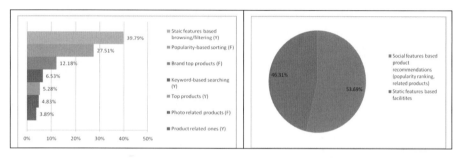

Fig. 2. (Stage 1) Distribution of product locations in Flickr CF (*F*) and Yahoo CF (*Y*), and their overall distribution as from social sources vs. from static sources

3.2.1.2 Users' Qualitative Comments
Users' qualitative comments further exposed their reflective thoughts on two aspects:

Credibility of product popularity. When being asked why they went to Flickr CF for accessing "product popularity", they replied that because it was perceived more trustworthy: *"I trust the information on the social forum." "I trust Flickr's popularity information because of its large amount of users." "Flickr is more neutral and credible." "Although this is my first-time using this website, the information sounds credible since it should be based on actual usages."* As for the product popularity on Yahoo CF, they commented *"the 'top products' in Yahoo may be only dependent on users' clicks or for companies' promotion purpose." "The popularity information in Yahoo may be faked. It looks more trustworthy and real in Flickr." "Flickr is more neutral because it is a consumer-operated website. The information on Yahoo may be not so real because it is more commercial-oriented."* It can be hence seen that users propend to trust the data from the social media site like Flickr, because it is more dependent on a large community's real usages and less of commercial interests.

Additionally, users suggested several ways to improve the generation of product popularity. One user suggested involving the geographical distribution of community members, because *"one camera model is suitable for European, but probably not for*

Chinese." "People from the same cultural background may have similar preferences." Another user proposed to take into account of the time dimension, given that *"it should be easier to compare the popularity values of different products if they were released at the same time."*

Expert opinions to be integrated into "Related Products". As for "Related Products" (e.g., "shoppers who viewed this product also viewed …"), users commented that if they could integrate with experts' professional opinions on the relevance of currently viewed product with others, they can be more meaningful than being purely dependent on other consumers' clicking behavior. As one user said, *"imagine the friends around you all use XX, you would be not familiar with YY. But if there is a comparison table from an expert explaining what their differences are, I will go to see YY's products."* Another user also noted *"because people sometimes just randomly clicked, the information from 'shopper viewed this product also viewed others' cannot be so credible. Experts' suggestions can be more useful to be regarded as important references."* Moreover, users suggested that the recommendation can be even better if it incorporates their hard constraints, like on price range and product type, because *"the 'related products' are useful, but I will not be interested in them if they are beyond my price expectation." "If the products are of the type that I prefer, I will more likely consider them."* Users' comments can hence explain why they did not select many items from "Related Products" (as shown above). It also infers if this type of recommendation could be well integrated with experts' opinions and be matched to the user's constraints if any, the adoption degree will be potentially highly increased.

3.2.1.3 System Implications for this Stage

Thus, the findings indicate two system implications for the first stage (see Figure 1).

1) It is suggested that "product popularity" should better originate from the social media site, because in such platform, it can reflect real usages from a community of like-minded users (like from Flickr), and can be hence perceived more credible than in standard e-commerce sites. Moreover, a popularity-based recommender tool should be particularly referential and helpful when users have not formed their clear targets at the beginning. To provide such support, the popularity can be customized to involve contextual factors like contributors' regional properties and products' releasing time, so as to dynamically match to the current user's context.

2) "Related Products" can be likely enhanced by means of involving expert opinions and users' stated attribute constraints, so as to augment users' adoption degree of such recommendation. Furthermore, we believe that the provision of a hybrid mechanism that well unifies benefits from popularity-based recommendation and "Related Products" (i.e., the personalized recommendation) should be capable of facilitating users to locate more interesting products at the first stage.

3.2.2 Stage 2: Evaluating Product in Detail

3.2.2.1 Users' Objective Behavior

At the 2nd stage, we were interested in knowing what detailed info that users have evaluated after selecting a product from the 1st stage. The analysis of their page visits shows that on average 42.86% of selected products were in detail evaluated on Yahoo (that provides the product's full specifications and consumer ratings/reviews), 30.44% on Flickr CF (that provides the product's usage trend and resulting photos), and 26.70% on both sites' product pages. Among the evaluated products, 45.82% were finally saved into the average user's wish list (i.e., mean = 4 products, *St.d.* = 1.95). The page evaluations respectively contributed 39.09% (1.50 products), 6.25% (0.25) and 91.67% (2.17), to establishing the wish list (the % means the percent of products saved as candidates after the corresponding page evaluations. See Figure 3). It hence infers that the evaluation of product details from both Flickr CF and Yahoo CF can most likely convince the user to take the product as a candidate. The correlation is indeed highly significant ($p < 0.001$) by Pearson Coefficient. Another fact is that 91.7% users' final choice was a product that underwent this combined evaluation.

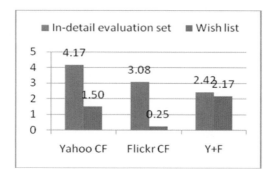

Fig. 3. (Stage 2) Correlation between evaluated products and saved ones to the wish list

3.2.2.2 Users' Qualitative Comments

Product reviews: negative versus positive reviews. Most of users noted that product ratings/reviews were very important when they examined a product, since *"they help me judge the product's true quality."* As for ratings, the extreme ratings are more helpful than neutral ones: *"the middle rate cannot say anything."* *"Human normally has vague consistency on the interpretation of middle scores, so extreme scales should be more useful."* They said that when the product rating was low, they read user reviews (i.e., the textual commens). They liked the separation of reviews into *pros* and *cons* categories because it facilitates the comparison: *"The motivation of buying a product is not because it is very perfect, but is whether you can stand its drawbacks."* *"Every product should have flaws, and what I want to get from user reviews is whether they can disclose these negative aspects."* All participants agreed that *"I will not buy a product only because it has positive ratings and reviews, but will certainly not buy it if it has negative reviews, especially on features that I am concerned about."* Moreover, the quantity of user reviews also takes a certain effect: *"less number of reviews will have lower credibility"*, but it is still better than zero because *"in the case that two products both have few numbers of user reviews, I will still read the reviews to get the feeling of which product should be better."*

3.2.2.3 System Implications for this Stage

Users' favor on negative comments than positive ones can be supported by a previous claim that "when information about an object or firm comes through the opinions of another person, negative information can be more credible and generalizable than positive information" [11]. Thus, we believe that the mining and exposure of negative opinion values and the indication of their relevance to users' interests can be likely to support more informed and accurate product evaluation at the stage 2 (see Figure 1). To achieve this goal, the sentimental analysis and opinion mining techniques [12] can be first applied to extract features (e.g., ease of use, image quality) from reviews and then associate opinion scores (i.e., positive, neutral and negative) to the features. Then, the recommendation can be returned in form of *{feature, opinion score}* sets, placing higher weights on negative opinions and further personalizing the sets with the user's specific feature concerns.

3.2.3 Stage 3: Comparing Candidates and Confirming the Final Choice

At the last stage when users nearly made the "purchase" decision, they compared candidates in their wish list and then identified the best one. In order to know which factors they mainly considered during the comparison, we recorded items they have viewed after their wish list was established. It shows that 66.7% (8 out of 12) users went to Flickr CF to compare candidate products' usage trends or photos (as taken by Flickr community), and 33.3% emphasized product specifications or reviews provided in Yahoo CF (see Figure 4 left).

From Figure 4 (right), we can see that totally 75% participants replied on social features (i.e., usage trends, product reviews, and community photos) to accomplish the choice confirmation. The social features are hence demonstrated more influential than static features (i.e., specifications) in convincing users about their "purchase" decision. Users' comments reflected that "*I would like to rely on the social content to identify which product should be better than others.*" "*The product's usage trend can help me form a correct judgment and decrease the uncertainty from purely evaluating its static specifications.*"

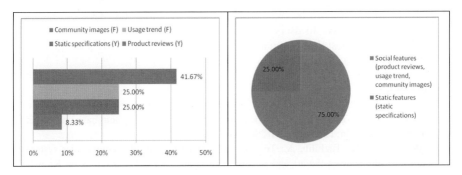

Fig. 4. (Stage 3) Major factors that users compared among candidates before they made the final choice, and these factors' overall distribution as from social sources vs. from static sources

3.2.3.1   System Implications for this Stage

For this stage, the implication is mainly about the comparison support (see Figure 1). Many existing e-commerce sites provide a comparison matrix, by which users can perform pair-wise comparison between multiple products in an *alternatives* (rows) x *attributes* (columns) matrix. Although this method has been demonstrated to enable higher decision quality than the condition without it [8], most of its applications are still limited to the simple display of products' static feature values. As implied from above findings, it will be likely more effective if the traditional comparison matrix can be improved to embed products' social values. For instance, the extracted *{feature, opinion}* pairs (implied in Section 3.2.2.3) can be also adopted here to complement standard *{feature, static value}* pairs. Besides, other types of social content (e.g., usage trend) can be also included in the matrix to facilitate users' comparison actions.

4   Conclusion

Thus, our research questions were well answered through the user study. All subjects' common behavior reflects how an online consumer normally interacts with e-stores and what social info s/he does require during a complex decision process (i.e., looking for a high-value product to buy). Based on the study's observations, we suggest a three-stage recommender framework (as illustrated in Figure 1): 1) at the 1st stage when users browse alternatives, show popular recommendations that are derived from usage-driven social media. Personalized recommendation is also fit for this stage, which can be potentially more likely adopted by users if being integrated with expert opinions and users' personal constraints; 2) at the 2nd stage when users evaluate a product in detail, provide *{feature, opinion}* recommendations that are extracted from product reviews and place higher weights on negative opinions; 3) at the 3rd stage when users make comparison among candidates that were saved in their wish list, include social feature values in a comparison matrix to facilitate them confirming the final choice.

We believe that these implications can be very useful to related works, including ones that have attempted to fuse product popularity and reviews into recommenders [1,3]. In the future, we will be engaged in building the three-stage recommender system so as to be adaptive to users' various information needs. We will also conduct more user studies to test the system and consolidate the current study's findings.

References

1. Aciar, S., Zhang, D., Simoff, S., Debenham, J.: Recommender system based on consumer product reviews. In: Proc. WI-IAT 2006, pp. 719–723 (2006)
2. Adomavicius, G., Tuzhilin, A.: Toward the next generation of recommender systems: a survey of the state-of-the-art and possible extensions. IEEE Transactions on Knowledge and Data Engineering 17(6), 734–749 (2005)
3. Amatriain, X., Lathia, N., Pujol, J.M., Kwak, H., Oliver, N.: The wisdom of the few: a collaborative filtering approach based on expert opinions from the web. In: Proc. SIGIR 2009, pp. 532–539 (2009)
4. Chen, L., Pu, P.: Evaluating critiquing-based recommender agents. In: Proc. AAAI 2006, pp. 157–162 (2006)

5. Engel, J.F., Blackwell, R.D., Miniard, P.W.: Consumer Behavior. Dryden Press, Orlando (1990)
6. Groh, G., Ehmig, C.: Recommendations in taste related domains: collaborative filtering vs. social filtering. In: Proc. GROUP 2007, pp. 127–136 (2007)
7. Guy, I., Chen, L., Zhou, M.X.: Workshop on social recommender systems. In: Proc. IUI 2010, pp. 433–434 (2010)
8. Häubl, G., Trifts, V.: Consumer decision making in online shopping environments: the effects of interactive decision aids. Marketing Science 19(1), 4–21 (2000)
9. Kim, Y.A., Srivastava, J.: Impact of social influence in e-commerce decision making. In: Proc. ICEC 2007, 293-302 (2007)
10. Leino, J., Räihä, K.: Case Amazon: ratings and reviews as part of recommendations. In: Proc. RecSys 2007, pp. 137–140. ACM Press, New York (2007)
11. Mizerski, R.: An attribution explanation of the disproportionate influence of unfavorable information. Journal of Consumer Research 9, 301–310 (1982)
12. Pang, B., Lee, L.: Opinion mining and sentiment analysis. Foundations and Trends in Information Retrieval l 2(1-2), 1–135 (2008)

Query Relaxation for Star Queries on RDF

Hai Huang and Chengfei Liu

Faculty of ICT, Swinburne University of Technology
VIC 3122, Australia
{hhuang,cliu}@swin.edu.au

Abstract. Query relaxation is an important problem for querying RDF data flexibly. The previous work mainly uses ontology information for relaxing user queries. The ranking models proposed, however, are either non-quantifiable or imprecise. Furthermore, the recommended relaxed queries may return no results. In this paper, we aim to solve these problems by proposing a new ranking model. The model ranks the relaxed queries according to their similarities to the original user query. The similarity of a relaxed query to the original query is measured based on the difference of their estimated results. To compute similarity values for star queries efficiently and precisely, Bayesian networks are employed to estimate the result numbers of relaxed queries. An algorithm is also proposed for answering top-k queries. At last experiments validate the effectiveness of our method.

Keywords: Query Relaxation, RDF query Processing.

1 Introduction

The Resource description Framework (RDF) is a data model used on the Semantic Web. Recently, more and more data is represented and stored in RDF format. RDF data is a set of triples and each triple called statement is of the form (*subject*, *property*, *object*). This data representation is general and flexible. Almost any kind of data can be represented in this format. To query RDF data, many query languages on RDF data have been proposed and implemented such as SPARQL. SPARQL is an RDF query language, which has a SQL-like style. A SPARQL query usually consists of several triple patterns.

With RDF data growing in size and complexity, several RDF repositories have also been developed in recent years such as Jena, Sesame. Generally, users do not know the whole RDF database and often post the *failure queries*, which do not return any answers. In this case, it is desirable to relax the original user query to obtain some approximate answers. Given the RDFs ontology, a failure query can be relaxed to more general queries.

For example, Fig.1 shows an RDFs ontology and a SPARQL query Q, which retrieves the academic staff members who are lecturers and also the reviewers of conference "WWW". Suppose Q returns no answers (or not enough answers) and we try to relax this query to achieve some approximate answers. Using the ontology information in Fig.1, query Q can be generalized to capture approximate

L. Chen, P. Triantafillou, and T. Suel (Eds.): WISE 2010, LNCS 6488, pp. 376–389, 2010.

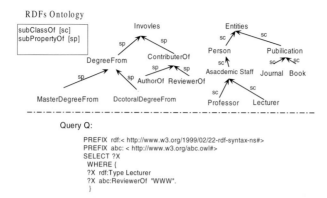

RDFs Ontology

Fig. 1. The Example of RDFs Ontology and a SPARQL Query Q

answers in many ways. Queries Q'(?X, type, AcademicStaff)(?X, ReviewerOf, "WWW") and Q''(?X, Type, Lecturer)(?X, ReviewerOf, ?Y) are both the relaxed queries of Q. The answers of each relaxed query are approximate answers of Q. Since the number of possible relaxations is huge, we hope to rank them.

To rank these relaxations, the authors in [1] introduced the ranking model on the relaxed queries that have the *subsume* relationship among them. Given two relaxations Q_i and Q_j of the original query Q, if Q_i is subsumed by Q_j, then Q_i is better. However, this ranking model is non-quantifiable; moreover, it fails to rank Q' and Q'' in the above example because they have no subsume relationship between them.

In [5], Huang and Liu proposed a method to rank the relaxed queries depending on the extent to which the concepts and properties in the relaxed queries are similar to the user query. First, this ranking model is a little coarse-grained because it is only based on the ontology information. Furthermore, the ranking model does not consider the real data distribution in the database, so it is possible that relaxed queries may still have no answers in the database, which would lead to unnecessary execution cost. To solve this problem, we propose a fine-grained ranking model and consider the selectivity of relaxed queries in our ranking model.

In this paper, we propose new methods to rank the relaxed queries and compute the top-k approximate answers. The contributions of this paper can be summarized as follows:

- We construct a ranking model which ranks the relaxed queries according to their similarities to the original user query. The similarity of a relaxed query to the original query is measured based on the difference of their estimated results.

- We propose a method to compute the similarity values of relaxed star queries efficiently using Bayesian networks.

- We develop an algorithm to compute the top-k approximate answers according to our ranking model based on best-first search.

The remainder of this paper is organized as follows. Section 2 introduces some preliminary knowledge. In Section 3, we construct the ranking model. Section 4 presents the method to compute the similarity value of relaxed queries efficiently using Bayesian Networks. In Section 5 we present the algorithm to compute the top-k approximate answers for queries. Section 6 describes an experimental evaluation of our approach. Some related work is discussed in Section 7. At last, in Section 8, we conclude our work in this paper and discuss the future work.

2   Preliminary

2.1   Basic Concepts and Problem Definition

A *triple* $(s, p, o) \in (I \cup B) \times (I \cup B) \times (I \cup B \cup L)$ is called an RDF triple, where I is a set of IRIs (Internationalized URIs), B a set of blank nodes and L a set of literals. In the triple, s is called subject, p the property (or predicate), and o the object or property value. An RDF *triple pattern* $(s, p, o) \in (I \cup V) \times (I \cup V) \times (I \cup V \cup L)$, where V is a set of variables disjoint from the sets I, B and L. An *RDF graph pattern* $G = (q_1, q_2, ..., q_n)$, $q_i \in T$, where T is a set of triple patterns.

In this paper, we focus on star SPARQL query patterns, which are common in SPARQL query patterns. The star query pattern retrieves instances of a class with some conditions. Fig.1 shows an example of star SPARQL query pattern.

Definition 1 (Star Query patterns): The star query pattern has the form of a number of triple patterns with different properties sharing the same subject.

2.2   Query Relaxation Model

Hurtado *et al.* in [1] propose two kinds of relaxation for triple pattern which exploit RDF entailment to relax the queries.

Simple relaxation on triple pattern: Given RDF graphs G_1, G_2, a map from G_1 to G_2 is a function u from the terms (IRIs ,blank nodes and literals) in G_1 to the terms in G_2, preserving IRIs and literals, such that for each triple $(s_1, p_1, o_1) \in G_1$, we have $(u(s_1), u(p_1), u(o_1)) \in G_2$. This simple relaxation exploits the RDF simple entailment. An RDF graph G_1 simply entails G_2, denoted by $G_1 \xrightarrow{\text{simple}} G_2$, if and only if there is a map u from G_2 to G_1: $G_2 \xrightarrow{u} G_1 : \frac{G_1}{G_2}$. We call triple pattern t_2 the simple relaxation of t_1, denoted by $t_1 \xrightarrow{\prec}_{\text{simple}} t_2$, if $t_1 \xrightarrow{\text{simple}} t_2$ via a map u that preserves variables in t_1.

For example, there is a map u from the terms of triple pattern (?X, type, ?Y) to (?X, type, Lecturer) that makes u("?X")= "?X", u("type")= "type" and u("?Y")= "Lecturer". So (?X, type, Lecturer) can be relaxed to (?X, type, ?Y) by replacing "Lecturer" with the variable "?Y". We have (?X, type, Lecturer) $\xrightarrow{\prec}_{\text{simple}}$ (?X, type, ?Y).

Ontology relaxation on triple pattern: This type of relaxation exploits RDFS entailment in the context of an ontology (denoted by onto). We call $G_1 \xrightarrow{\text{RDFs}} G_2$, if G_2 can be derived from G_1 by iteratively applying rules in groups (A), (B) (C)(sc, sp are rdfs:subclassOf and rdfs:subpropertyOf for short):

- Group A (Subproperty) (1) $\frac{\text{(a sp b)(b sp c)}}{\text{(a sp c)}}$; (2) $\frac{\text{(a sp b)(x a y)}}{\text{(x b y)}}$

- Group B (Subclass) (3) $\frac{\text{(a sc b)(b sc c)}}{\text{(a sc c)}}$; (4) $\frac{\text{(a sc b)(x type a)}}{\text{(x type b)}}$

- Group C (Typing) (5) $\frac{\text{(a dom c)(x a y)}}{\text{(x type c)}}$; (6) $\frac{\text{(a range d)(x a y)}}{\text{(y type d)}}$

Let t, t' be triple patterns, where $t \notin closure(onto)$, $t' \notin closure(onto)$. We call t' an ontology relaxation of t, denoted by $t \overset{\prec}{\text{onto}} t'$, if $t \cup onto \overset{\Longrightarrow}{\text{RDFs}} t'$. It includes relaxing type conditions and properties such as: (1) replacing a triple pattern (a, type, b) with (a, type, c), where (b, sc, c) $\in$ *closure(onto)*. For example, given the ontology in Fig.1, the triple pattern (?X, type, Lecturer) can be relaxed to (?X, type, AcademicStaff). (2) replacing a triple pattern (a, p1, b) with (a, p, b), where (p1, sp, p) $\in$ *closure(onto)*. For example, the triple pattern (?X, ReviewerOf, "WWW") can be relaxed to (?X, ContributerOf, "WWW"). (3) replacing a triple pattern (x, p, y) with (x, type, c), where (p, domain, c) $\in$ *closure(onto)*. For example, the triple pattern (?X, ReviewerOf, "WWW") can be relaxed to (?X, Type, Person).

Definition 2 (Relaxed Triple Pattern): Given a triple pattern t, t' is a relaxed pattern obtained from t, denoted by $t \prec t'$, through applying a sequence of zero or more of the two types of relaxations: *simple relaxation* and *ontology relaxation*.

Definition 3 (Relaxed Query Pattern): Given a user query $Q(q_1, q_2, \cdots, q_n)$, $Q'(q'_1, q'_2, \cdots, q'_n)$ is a relaxed query of Q, denoted by $Q \prec Q'$, if there exists $q_i = q'_i$ or $q_i \prec q'_i$ for each pair (q_i, q'_i).

Definition 4 (Approximate Answers): An approximate answer to query Q is defined as a match of a relaxed query of Q.

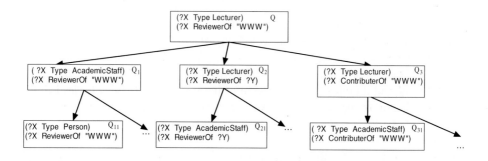

Fig. 2. Query Q and its relaxation graph

3   Ranking Model for Relaxations

User queries can be relaxed in many ways by ontology relaxation and simple relaxation. For instance, in Fig.2, query Q has relaxed queries Q_1=(?X, Type,

AcademicStaff)(?X, ReviewerOf, "WWW"), Q_2=(?X, Type, Lecturer)(?X, ReviewerOf, ?Y) and so on. Since the number of relaxations could be large, returning all answers of these relaxed queries is not desirable. So we need to rank the relaxed queries and execute the better one first. Given the user query Q and its relaxed queries Q_1, Q_2, if Q_1 is ranked prior to Q_2, then we return the results Q_1 prior to the results of Q_2. In this section, we discuss the ranking model for relaxed queries.

Intuitively, for a relaxation Q' of user query Q, the more similar Q' is to Q, the better Q' is. Thus, we hope to rank the relaxed queries based on their similarities to the original query. Several similarity measures for concepts have been proposed such as information content measure [4], distance based measure [6]. Each method assumes a particular domain. For example, distance based measure is used in a concept network and Dice coefficient measure is applicable when the concepts can be represented as numerical feature vectors. To our problem, the goal is to measure similarity between Q and its relaxations Q_i and our objects are queries not concepts. Thus, these similarity measures can not be directly applied in our problem. We need to find an appropriate method to measure the similarity between the relaxed queries and original query. Before giving the method to quantify the similarity value, we first clarify some helpful intuitions as follows:

Intuition 1: The similarity between Q and its relaxation Q_i is proportional to their commonality. From Fig. 3, we can see that the more answers Q and Q_i share in common, the more similar they are.

Intuition 2: The similarity between Q and its relaxation Q_i ranges from 0 to 1 and if Q and Q_i are identical (i.e., they have the same result set), the similarity value should be 1.

Intuition 3: Given Q and its relaxation Q_i and Q_j, if $Q_i \prec Q_j$, then Q_i should be more similar to Q. It indicates that the ranking model should be consistent.

We know that the information contained by a query Q is its answers in the database. We define $p(Q)$ to be the probability of encountering an answer of query Q in the database. According to information theory, the *information content* of query Q, $I(Q)$ can be quantified as the negative log likelihood, $I(Q) = -log\, p(Q)$. For query Q and its relaxation Q_i, the information shared by two queries is indicated by the information content of the query that subsume them.

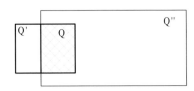

Fig. 3. Query Q and its relaxations Q' and Q''. The shaded area shows the answers they share in common.

Here, since $Q \prec Q_i$, we have:

$$Common(Q, Q_i) = I(Q_i) = -log \, Pr(Q_i)$$

Where $Pr(Q)$ stands for the probability of encountering an answer of query Q in the database. And we have:

$$Pr(Q) = \frac{|Ans(Q)|}{|Res|}$$

where $Ans(Q)$ denotes the answer set of query Q and Res denotes the resource set of the database.

Given a query Q and its relaxation Q_i, we use the ratio between $Common(Q, Q_i)$ and the information content of Q to define the similarity value $Sim(Q, Q_i)$ as follows:

$$Sim(Q, Q_i) = \frac{Common(Q, Q_i)}{I(Q)}$$

$$= \frac{I(Q_i)}{I(Q)} = \frac{log \, Pr(Q_i)}{log \, Pr(Q)} \tag{1}$$

Since for Q, $Pr(Q)$ is fixed, we also have:

$$Sim(Q, Q_i) \propto -log \, Pr(Q_i)$$

$$\propto \frac{1}{Pr(Q_i)}$$

$$= \frac{1}{\frac{|Ans(Q_i)|}{|Res|}} \tag{2}$$

Given two relaxations Q_i and Q_j of query Q if $Sim(Q, Q_i) \geqslant Sim(Q, Q_j)$, results of Q_i are returned prior to Q_j. Note that if query Q has no answers, then $Pr(Q) = 0$ and formula (1) would make no sense. In this case, for computing $Sim(Q, Q_i)$, we can use a constant $Pr(Q) = \frac{1}{|Res|}$ to replace 0 in formula (1).

When relaxed query Q_i has no answers in the database, i.e. $Pr(Q_i) = 0$, we have $Sim(Q, Q_i) = 0$ according to formula (1). This agrees with the fact that a relaxed query with no answers in the database is not what we want.

We also have propositions as follows:

Proposition 1.(Consistency) Given Q and its relaxation Q_i and Q_j, if $Q_i \prec Q_j$, then $Sim(Q, Q_i) \geqslant Sim(Q, Q_j)$.

Proof. According to formula (2), $Sim(Q, Q_i)$ and $Sim(Q, Q_j)$ are proportional to $\frac{1}{Pr(Q_i)}$ and $\frac{1}{Pr(Q_j)}$. Obviously, if $Q_i \prec Q_j$, $Pr(Q_i) \leqslant Pr(Q_j)$. Thus, we have $Sim(Q, Q_i) \geqslant Sim(Q, Q_j)$. □

Proposition 2. Given Q and its relaxation Q_i, if Q and Q_i are identical, then $Sim(Q, Q_i) = 1$.

From formula (2), we can see that the similarity between query Q and its relaxation Q_i is proportional to $\frac{1}{Pr(Q_i)}$. However, computing $Pr(Q_i)$ is not straightforward, we will discuss it in the next section.

4   Computing the Similarity Efficiently

In this section, we will present the method for computing the similarities between the original query and its relaxations efficiently.

We have given the formula to compute the similarity value of relaxed queries. From formula (1), we can observe that for computing $Sim(Q, Q_i)$, we have to know $\Pr(Q_i)$, which is equal to $\frac{|Ans(Q_i)|}{|Res|}$. In the RDF database, $|Res|$ is the number of resources, which is a constant and easy to obtain. $|Ans(Q_i)|$ is the number of answers of Q_i. The naive way to compute $|Ans(Q_i)|$ is executing query Q_i in the database. However, it would not be time efficient because the number of possible relaxed queries is usually huge. Thus, we resort to approximate methods to compute $|Ans(Q_i)|$ efficiently without executing Q_i.

4.1   Selectivity Estimation for Star Patterns Using Bayesian Networks

Selectivity estimation technology can be employed to estimate $|Ans(Q_i)|$. We use $sel(Q_i)$ to denote an estimation of $|Ans(Q_i)|$. Some work has been done on estimating the selectivity of RDF triple patterns. In [2,11] the join uniformity assumption is made when estimating the joined triple patterns with bound subjects or objects (i.e., the subjects or objects are concrete values). It assumes that each triple satisfying a triple pattern is equally likely to join with the triples satisfying the other triple pattern. But this assumption does not fit the reality and would lead to great errors because it never considers the correlations among properties in a star query pattern. Thus we propose a method to estimate the selectivity more precisely with considering the correlations among properties.

For a group of subjects, choose a set of single-valued properties that often occur together. And we can construct a table R called cluster-property table for these properties. For example, Fig.4 shows a cluster-property table with three attributes *ResearchTopic*, *TeacherOf*, *ReviewerOf*. Each row of the table corresponds to a subject with values for the three properties.

Given a star pattern Q with predicates $prop_1, prop_2, \cdots, prop_n$, the results of query Q should fall in the corresponding cluster-property table R. We denote the joint probability distribution over values of properties $prop_1, prop_2, \cdots, prop_n$ as $\Pr(prop_1 = o_1, prop_2 = o_2, \cdots, prop_n = o_n)$, where o_i is the value of property $prop_i$. We have:

$$sel(Q) = \Pr(prop_1 = o_1, prop_2 = o_2, ..., prop_n = o_n) \cdot |T_R| \qquad (3)$$

where $sel(Q)$ is the number of results of query Q and $|T_R|$ is the number of rows in the cluster-property table R.

Notice that the possible combinations of values of properties would be exponential and it is impossible to explicitly store $\Pr(prop_1 = o_1, prop_2 = o_2, \cdots, prop_n = o_n)$. Thus, we need an appropriate structure to *approximately* store the joint probability distribution information. The Bayesian network [12] can

Subject	ResearchTopic	ReviewerOf	TeacherOf
S_1	Web	WWW	course1
S_2	Web	WISE	course1
S_3	Web	WISE	course2
S_4	Web	WWW	course1
S_5	Software	ICSE	course2
S_6	Machine Learning	ICML	course3
...	....	....	....

Fig. 4. Cluster property table

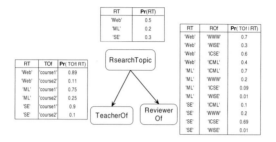

Fig. 5. Bayesian Networks

approximately represent the probability distribution over a set of variables using a little space. Bayesian networks make use of Bayes' Rule and conditional independence assumption to compactly represent the full joint probability distribution. Let X, Y, Z be three discrete valued random variables. We say that X is conditionally independent of Y given Z if the probability distribution of X is independent of the value of Y given a value for Z; that is:

$$\Pr(X = x_i | Y = y_j, Z = z_k) = \Pr(X = x_i | Z = z_k)$$

where x_i, y_j, z_k are values of variables X, Y, Z. The conditional independence assumptions associated with a Bayesian network and conditional probability tables (CPTs), determine a joint probability distribution. For example, in Fig.5, a Bayesian network is constructed on cluster property table in Fig.4. We can see that properties *TeacherOf* and *ReviewerOf* are conditional independent given condition *ResearchTopic*, which means if we already know the research topic of some person, knowing his teaching information does not make any difference to our beliefs about his review information.

For a star query Q with properties $prop_1 = o_1$, $prop_2 = o_2$, $\cdots$, $prop_n = o_n$ and a Bayesian network β, we have:

$$Pr_\beta(prop_1 = o_1, prop_2 = o_2, ..., prop_n = o_n)$$
$$= \prod_{i=1}^{n} \Pr(prop_i = o_i \mid Parents(prop_i) = \boldsymbol{o_k})$$

where $Parents(prop_i)$ denotes the set of immediate predecessors of $prop_i$ in the network β and $\boldsymbol{o_k}$ denotes the set of values of $Parents(prop_i)$. Note that for computing $\Pr(prop_i \mid parents(prop_i) = \mathbf{o_k})$, we only need to know the values of $prop_i$'s parent properties, which would save a lot of space in practice. So given the Bayesian network β, we can use $\Pr_\beta(prop_1 = o_1, prop_2 = o_2, ..., prop_n = o_n)$ to *approximately* represent $\Pr(prop_1 = o_1, prop_2 = o_2, ..., prop_n = o_n)$. We have:

$$
\begin{aligned}
sel(Q) &= \Pr(prop_1 = o_1, prop_2 = o_2, ..., prop_n = o_n) \cdot |T_R| \\
&\approx \Pr_\beta(prop_1 = o_1, prop_2 = o_2, ..., prop_n = o_n) \cdot |T_R| \\
&= \prod_{i=1}^{n} \Pr(prop_i = o_i \mid Parents(prop_i) = \mathbf{o_k}) \cdot |T_R| \quad (4)
\end{aligned}
$$

For example, given the star pattern Q(?x, ResearchTopic, 'Web')(?x, TeacherOf, 'course1') (?x, ReviewerOf, 'WWW') Bayesian network described in Fig.5, we compute the selectivity of Q as follows:

$$
\begin{aligned}
sel(Q) &= \Pr(ResearchTopic = \text{'Web'}, TeacherOf = \text{'course1'}, \\
&\qquad ReviewerOf = \text{'WWW'}) \cdot |T_R| \\
&= \Pr(TeacherOf = \text{'course1'} \mid ResearchTopic = \text{'Web'}) \\
&\quad \cdot \Pr(ReviewerOf = \text{'WWW'} \mid ResearchTopic = \text{'Web'}) \\
&\quad \cdot \Pr(ResearchTopic = \text{'Web'}) \cdot |T_R| \\
&= 0.89 \cdot 0.7 \cdot 0.5 \cdot |T_R| = 0.3115 \cdot |T_R|
\end{aligned}
$$

It should be noted that relaxed queries may contain some properties which subsume properties in a property cluster table. Suppose that we have a property table R containing properties $p_1, p_2, \cdots p_n$ and we construct a Bayesian network on this table. Given a relaxed query Q (?x p o_1)(?x p_3 o_3)$\cdots$(?x p_n o_n), if property p subsumes p_1 and p_2, we estimate the selectivity of Q as follows:

$$
\begin{aligned}
sel(Q) &= \Pr(p = o_1, p_3 = o_3, ..., p_n = o_n) \cdot |T_R| \\
&\approx \Pr_\beta(p_1 = o_1, p_3 = o_3, ..., p_n = o_n) \cdot |T_R| \\
&\quad + \Pr_\beta(p_2 = o_1, p_3 = o_3, ..., p_n = o_n) \cdot |T_R| \\
&= \sum_{p_i \prec p} \left(\prod_{j \neq 2, \text{if } i=1; j \neq 1, \text{if } i=2} \Pr(p_j = o_j \mid Parents(p_j) = \mathbf{o_k}) \right) \cdot |T_R|
\end{aligned}
$$

4.2   Learning Bayesian Networks

To approximately represent the joint probability distribution of property values for selectivity estimation, we need to construct Bayesian networks for each cluster-property table. Before building Bayesian networks, the domain values are first discretized and clustered into equi-width subsets.

Bayesian network learning includes learning the structure of the DAG in a Bayesian network and parameters (i.e., conditional probability tables). A *score* *criterion score$_\beta$* is a function that assigns a score to each possible DAG G' of the Bayesian network β under consideration based on the data.

$$score_\beta(d, G') = \prod_{i=1}^{n} \Pr(d, X_i, Parents(X_i)^{(G')})$$

where d is the data in the the cluster-property table and $Parents(X_i)^{(G')}$ is the set of parents of variable X_i in DAG G'. Bayesian network learning is to find a DAG that best fits the data (with the highest *score$_\beta$*), which is also called *model selection*. Much research has been conducted in Bayesian network learning [13,14]. We use K2 [15] algorithm to learn the structure of Bayesian networks. It adopts a greedy search that maximizes *score$_\beta$*(d, G') approximately. For each variable X_i, it locally finds a value of $Parents(X_i)$ that maximizes *score$_\beta$*$(d, X_i, Parents(X_i))$. This method needs to provide the temporal ordering of variables, which can significantly reduce the search space. In our implementation, we input the temporal ordering of variables by hand.

5   Top-K Query Processing

In this section, we discuss how to acquire top-k approximate answers through query relaxation. Fig.6 shows a relaxation graph of query Q. From the relaxation graph of query Q, we can see that there are 9 relaxed queries of Q and $Q_1 \prec Q_{11}$, $Q_1 \prec Q_{12} \prec Q_{121}$, $Q_2 \prec Q_{21}$, $Q_3 \prec Q_{31}$, $Q_3 \prec Q_{32}$. In the relaxation graph, there exists an edge from relaxed queries Q_i to Q_j if Q_i is directly subsumed by Q_j(i.e., $Q_i \prec Q_j$ and $\nexists Q_m$ s.t. $Q_i \prec Q_m \prec Q_j$). We also call Q_j is a child of Q_i in the relaxation graph.

Suppose query Q has no answers returned and we want to acquire top-k approximate answers of Q. The relaxation process for obtaining top-k approximate answers is implemented by the best-first search. We first add the children of Q to the *Candidates* set, which are promising to be the best relaxation of query Q. Then select the best candidate Q' (with the highest similarity value) in the *Candidates* set to execute and add the children of Q' to the *Candidates* set. Repeat this process until we get enough answers. This process is described in Algorithm 1-*topkAlgorithm*.

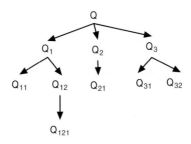

Fig. 6. The Relaxation Graph of Query Q

Algorithm 1. *topkAlgorithm*

Input: Query Q; k(the number of answers required);
Output: top-k approximate answers of query Q;

1: *Answers* $= \Omega$; *Candidates* $= \Omega$;
2: Add Q's child nodes into *Candidates*;
3: **while** $|Answers| < k$ and $|Candidates| <> \Omega$ **do**
4:    Select the Q' with the best similarity value from *Candidates*;
5:    Add child nodes of Q' to *Candidates*
6:    Add answers of Q' to *Answers*
7:    *Candidates* $=$ *Candidates* $\setminus Q'$
8: **return** *Answers*;

---

6   Experiments

In this section, we conduct experiments to verify our methods.

Experiment setup. The data is stored in and managed by Mysql 5.0. All algorithms are implemented using Jena (http://jena.sourceforge.net/). We run all algorithms on a windows XP professional system with P4 3G CPU and 2 GB RAM.

Data sets. We construct the dataset using the Lehigh University Benchmark LUBM [3]. LUBM consists of a university domain ontology containing 32 properties and 43 classes. The dataset used in the experiments contains 600k triples.

Query Load. We develop 5 star queries shown in Fig.7.

Bayesian Networks. We first construct Bayesian Networks from the dataset for estimating the selectivity of relaxed queries. Fig.8 shows one of Bayesian Networks Learned from the dataset for "Professor" entity. We can see that "Professor" entity has 6 properties. Properties "MastersDegreeFrom"and "DoctoralDegreeFrom"are independent with other properties. The values of property "ResearchInterest"depends on the values of properties "type" and "worksFor".

	Queries
Q1	Select ?y Where { ?y researchInterest 'Research17'. ?y worksFor Deaprt0.Univ0. ?y type VisitingProfessor ?y. ?y DoctoralDegreeFrom University503. }
Q2	Select ?y Where { ?y worksFor Deaprt0.Univ0. ?y doctoralDegreeFrom Univ476. ?y type Lecturer.}
Q3	Select ?y Where { ?y undergraduateDegreeFrom Univ476. ?y type GraduateStudent. ?y takesCourse GraduateCourse65}
Q4	select ?x where { ?x type TeachingAssitant. ?x teachingAssistantOf Department0.University0/Course2. ?x mastersDegreeFrom Department0.University0}
Q5	select ?x where { ?x worksFor Department0.University0. ?x takesCourse Department0.University0/GraduateCourse16. ?x advisor AssistantProfessor3}

Fig. 7. Queries Used In the Experiments

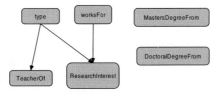

Fig. 8. One of Bayesian Networks learned from the dataset for "Professor" entity

According to this Bayesian Network, we can easily estimate the selectivity of star queries on "Professor" entity.

Performance. We conduct experiments to evaluate the performance of our relaxation algorithm. We first fix the number of approximate answers K=50. Fig.9 (a) and (b) show the running time and the number of relaxed queries executed for the relaxation algorithm. The similarity values of the approximate answers of Q_1 to Q_5 are $\geq$0.61, $\geq$0.67, $\geq$0.79, $\geq$0.71, $\geq$0.69, respectively. Our approach can avoid some unnecessary relaxations. For example, query Q_1 has four triple patterns. There are many ways to relax Q_1, but some of them have no answers. From the Bayesian network constructed (shown in Fig.8), we know the data distribution in the database, which indicates that there is no visiting professor who works for "Department0.University0" and is interested in "Research 17". Thus, there will be no answers by relaxing triple pattern(?x DoctoralDegreeFrom University503) to

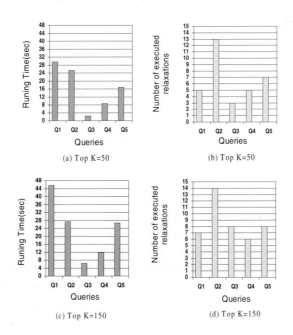

Fig. 9. The performance of our relaxation algorithm

(?x DoctoralDegreeFrom ?w). We relax Q_1 through replacing "VisitingProfessor" to "Professor"(super class of"VisitingProfessor"), which is close to the original query and avoids the empty answer set.

When K =150, from Fig.9 (c) and (d), we can see that the running time and the number of relaxed queries executed increase. On the contrary, the similarity values of the approximate answers of Q_1 to Q_5 are ≥ 0.57, ≥ 0.64, ≥ 0.69, ≥ 0.66, ≥ 0.63, respectively. Our relaxation algorithm returns approximate answers incrementally and users can stop the relaxation process at any time when they are satisfied with the answers generated.

7   Related Work

Hurtado *et al.* [1] proposed the RDF query relaxation method through RDF(s) entailment and relax the original user query using the ontology information. They also proposed the ranking model to rank the relaxed queries according to the subsume relationship among them. However, for some relaxed queries, if there does not exit subsume relationship among them, it fails to rank them.

Similarity measures for concepts have been developed in the previous decades. Generally, these methods can be categorized into three groups: edge counting-based methods [6], information theory-based methods [4]. For evaluating the similarity of queries, in [5], the authors proposed the ranking model using distance based measure on RDFs ontology. However, this measure is only based on ontology information without considering the real data distribution in the database, it is possible that relaxed queries ranked high still have no answers.

Some related work has been done, such as query relaxation on XML data [9,8,7]. Amer-Yahia *et al.* [10] presented a method for computing approximate answers for weighted patterns by encoding the relaxations in join evaluation plans. The techniques of approximate XML query matching are mainly based on structure relaxation and can not be applied to query relaxation on RDF data. Cooperative query answering [16] is designed to automatically relax user queries when the selection criteria is too restrictive to retrieve enough answers. Such relaxation is usually based on user preferences and values.

8   Conclusion

Query relaxation is important for querying RDF data flexibly. In this paper, we presented a ranking model which considers the real data distribution in the RDF database to rank the relaxed queries. The relaxations are ranked based on their similarity values to the original query. The similarity values of relaxed star queries are computed efficiently using Bayesian networks. We also presented the algorithm to answer top-k queries according to our ranking model.

Acknowledgments

This work was partially supported by the Australian Research Council Discovery Project under the grant number DP0878405.

References

1. Hurtado, C.A., Poulovassilis, A., Wood, P.T.: A Relaxed Approach to RDF Query-ing. In: Cruz, I., Decker, S., Allemang, D., Preist, C., Schwabe, D., Mika, P., Uschold, M., Aroyo, L.M. (eds.) ISWC 2006. LNCS, vol. 4273, pp. 314–328. Springer, Heidelberg (2006)
2. Stocker, M., Seaborne, A., Bernstein, A., Kiefer, C., Reynolds, D.: SPARQL basic graph pattern optimization using selectivity estimation. In: WWW, pp. 595–604 (2008)
3. Guo, Y., Pan, Z., Heflin, J.: An Evaluation of Knowledge Base Systems for Large OWL Datasets. In: McIlraith, S.A., Plexousakis, D., van Harmelen, F. (eds.) ISWC 2004. LNCS, vol. 3298, pp. 274–288. Springer, Heidelberg (2004)
4. Resnik, P.: Using Information Content to Evaluate Semantic Similarity in a Tax-onomy. In: IJCAI, pp. 448–453 (1995)
5. Huang, H., Liu, C., Zhou, X.: Computing Relaxed Answers on RDF Databases. In: Bailey, J., Maier, D., Schewe, K.-D., Thalheim, B., Wang, X.S. (eds.) WISE 2008. LNCS, vol. 5175, pp. 163–175. Springer, Heidelberg (2008)
6. Ho, L.J.: Information Retrieval Based on Conceptual Distance in Is-A Hierarchies. Journal of Documentation 49(2), 188–207 (1993)
7. Fuhr, N., Grobjohann, K.: XIRQL: A Query Language for Information Retrieval in XML Documents. In: SIGIR 2001, pp. 172–180 (2001)
8. Query Relaxation for XML Model. PhD thesis, Department of Computer Science, University of Califonia Los angeles (2002)
9. Kanza, Y., Sagiv, Y.: Flexible Queries Over Semistructured Data. In: PODS 2001 (2001)
10. Amer-Yahia, S., Cho, S., Srivastava, D.: Tree Pattern Relaxation. In: Jensen, C.S., Jeffery, K., Pokorný, J., Šaltenis, S., Hwang, J., Böhm, K., Jarke, M. (eds.) EDBT 2002. LNCS, vol. 2287, pp. 496–513. Springer, Heidelberg (2002)
11. Neumann, T., Weikum, G.: RDF-3X: a RISC-style engine for RDF. PVLDB 1(1), 647–659 (2008)
12. Pearl, J.: Probabilistic reasoning in intelligent systems: Networks of plausible in-ference. Morgan-Kaufmann, San Mateo (1988)
13. Wray, L.: Buntine: Operations for Learning with Graphical Models. J. Artif. Intell. Res. (JAIR) 2, 159–225 (1994)
14. Heckerman, D., Geiger, D., Maxwell Chickering, D.: Learning Bayesian Networks: The Combination of Knowledge and Statistical Data. Machine Learning 20(3), 197–243 (1995)
15. Cooper, G.F., Herskovits, E.: A Bayesian Method for the Induction of Probabilistic Networks from Data. Machine Learning 9, 309–347 (1992)
16. Kleinberg, J.M.: Authoritative Sources in a Hyperlinked Environment. J. ACM 46(5), 604–632 (1999)

Efficient and Adaptable Query Workload-Aware Management for RDF Data

Hooran MahmoudiNasab[1] and Sherif Sakr[2]

[1] Macquarie University, Australia
hooran@ics.mq.edu.au
[2] National ICT Australia (NICTA) and
University of New South Wales, Australia
ssakr@cse.unsw.edu.au

Abstract. This paper presents a *flexible* and *adaptable* approach for achieving efficient and scalable management of RDF using relational databases. The main motivation behind our approach is that several benchmarking studies have shown that each RDF dataset requires a tailored table schema in order to achieve efficient performance during query processing. We present a **two-phase** approach for designing efficient tailored but flexible storage solution for RDF data based on its query workload, namely: 1) a workload-aware **vertical partitioning** phase. 2) an automated **adjustment** phase that reacts to the changes in the characteristics of the continuous stream of query workloads. We perform comprehensive experiments on two real-world RDF data sets to demonstrate that our approach is superior to the state-of-the-art techniques in this domain.

1 Introduction

The *Semantic Web* term is coined by the W3C founder Tim Berners-Lee in a Scientific American article describing the future of the Web [4]. In general, the main goal of the Semantic Web vision is to provide a common framework for data-sharing across applications, enterprises, and communities. The core of the Semantic Web is built on the Resource Description Framework (RDF) data model [15]. RDF describes a particular resource using a set of RDF statements of the form (subject, predicate, object) triples. The *subject* is the resource, the *predicate* is the characteristic being described, and the *object* is the value for that characteristic. Efficient and scalable management of RDF data is a fundamental challenge at the core of the Semantic Web. Several research efforts have been proposed to address these challenges using the infrastructure of Relational Database Management Systems (RDBMSs) [3,13,18,22]. In practice, RDBMSs have repeatedly shown that they are very efficient and scalable in hosting different types of data such complex objects [21], spatio-temporal data [11] and XML data [12]. In practice, RDMBSs derive much of their performance from sophisticated optimizer components which makes use of physical properties that

L. Chen, P. Triantafillou, and T. Suel (Eds.): WISE 2010, LNCS 6488, pp. 390–399, 2010.

are specific to the relational model such as: sortedness and proper join ordering. Generally, the relational RDF stores can be classified to the following main categories:

- *Vertical (triple) table stores:* where each RDF triple is stored directly in a three-column table (subject, predicate, object) [18,22].
- *Property (n-ary) table stores:* where multiple RDF properties are modeled as n-ary table columns for the same subject [13].
- *Horizontal (binary) table stores:* where RDF triples are modeled as one horizontal table or into a set of vertically partitioned binary tables (one table for each RDF property) [3].

Recent benchmarking projects [5,14,20] have shown that there is no approach which is dominant for the different types of RDF queries and none of these approaches can compete with a purely relational model. Therefore, we believe that each RDF dataset requires a *tailored* but *flexible* table schema based on its query workload in order to achieve efficient and scalable query processing.

In this paper, we present a novel *adaptable* solution for designing efficient and scalable query performance for RDF queries. Our approach goes through two main phases: 1) A workload-aware **vertical partitioning** phase. 2) An automated **adjustment** phase that reacts to the changes in the characteristics of the continuous stream of query workloads. The aim of the vertical partitioning phase is to reduce the number of join operations in the query evaluation process. It uses a mathematical model to analyze the initial query workload in order to detect groups of frequently-queried and related properties that can be gathered into *n-ary* relations (property tables) while the rest of attributes can remain stored in a conventional relational triple store (a single table with *subject, predicate* and *object* fields). In our approach, we monitor and analyze the changes on the characteristics of the query workloads in order to decide on the changes that are required in the underlying schema of the property tables. Hence, the adjustment phase exploits the power of flexibility provided by the relational *pivot/unpivot* operators [10] in order to maintain the efficiency of the query performance by adapting the underlying schema to cope with the dynamic nature of the query workloads in an automated and efficient way.

The remainder of this paper is organized as follows. Section 2 discusses the state-of-the-art of the relational approaches for storing and querying RDF data. Section 3 describes our adaptive approach for designing a tailored relational schema for RDF dataset based on the characteristics of its query workloads. We evaluate our approach by conducting an extensive set of experiments which are described in Section 4 before we conclude the paper in Section 5.

2   Related Work

Several RDF storage solutions have relied on relational databases to achieve scalable and efficient query processing [19]. The naive way to store a set of RDF statements using RDBMS is to design a single table consists of three columns that store the *subject*, *predicate*, and *object* information. While simple, this schema quickly hits scalability limitations. Therefore, several approaches have been proposed to deal with this limitation by using extensive set of indexes or by using selectivity estimation information to optimize the join ordering. The RDF-3X query engine [18] tries to overcome the criticism that triples stores incurs too many expensive self-joins. It builds indexes over all 6 permutations of the three dimensions that constitute an RDF triple and follows RISC-style design philosophy [7] by relying mostly on merge joins over sorted index lists. The Hexastore engine [22] does not discriminate against any RDF element and treats subjects, properties and objects equally. Each RDF element type have its special index structures built around it. Each index structure in a Hexastore centers around one RDF element and defines a prioritization between the other two elements. Hence, in total six distinct indices which materialize all possible orders of precedence of the three RDF elements are used for indexing the RDF data. A clear disadvantage of these approaches is that they feature a worst-case five-fold storage increase in comparison to a conventional triples table.

SW-Store [3] is an RDF storage system which uses a fully decomposed storage model (DSM) [8]. In this approach, the triples table is rewritten into n two-column tables where n is the number of unique properties in the data. An advantage of this approach is that the algorithm for creating the binary tables is straightforward and need not change over time. However, the main limitations of this approach are: 1) High cost of inserting new tuples since multiple distinct locations on disk (for different tables) have to be updated for each tuple. 2) Increased tuple reconstruction costs as it requires joining multiple tables to re-gather the attributes of a single object.

In general, both of the *Vertical* and *Horizontal* approaches for RDF data management represent a generic solution that follows the "*One Size Fits All*" principle. Although the design mechanisms of both approaches are quite simple and straightforward, recent benchmarking studies [5,14,20] have shown that they do not achieve the best performance in terms of their query processing time. The query performance of both approaches have shown to be very sensitive to the characteristics of the input RDF datasets and the query workload. Therefore, the *property tables-based* RDF stores represent a balance between the two extremes of a single triple stores and binary tables. The main idea of this approach is to create separate *n-ary* tables (property tables) for subjects that tend to have common properties. For example, Levandoski and Mokbel [13] have presented a property table approach for storing RDF data without any assumption about the query workload statistics. It scans the RDF data to automatically discover groups of related properties and using a support threshold, each set of n properties which are grouped together in the same cluster. Our work belongs to the *property tables-based* category. However, our work distinguishes itself by building a tailored but flexible relational schema for each RDF dataset based on its query workload thus following "*One Size Does Not Fit All*" principle.

3   Adaptable Workload-Aware Property Tables

3.1   Vertical Partitioning Phase

Workload information has shown to be a crucial component for implementing effective query optimizers [6]. Essentially, it provides the means to understand user behavior and system usage. The *Vertical partitioning* process in the physical database design refers to the operation of subdividing a set of attributes into a number of groups (fragments) where each group is represented by a separate physical entity (table). An important quality of the vertical partitioning process is that the attributes of each fragment must closely match that of the workload requirements. The main goal of this quality is to optimize I/O performance by minimizing the number of fragments which are required to be accessed for evaluating the result of the seed query and avoid the need of accessing non-required attributes. Ideally, the accessed attributes by each query matches to the attributes of a single fragment. In principle, the vertical partitioning problem can be generally formulated as follows. Given a set of relations $R = \{R_1, R_2, ..., R_n\}$ and a set of queries $Q = \{Q_1, Q_2, ..., Q_m\}$, we need to identify a set of fragments $F = \{F_1, F_2, ..., F_x\}$ such that:

1. Every fragment $F_i \in F$ represents a subset of the attributes of a relation $R_j \in R$ in addition to the primary key column(s).
2. Each attribute is contained in exactly *one* fragment.
3. The workload cost when executed on top of the partitioned schema is less than the cost of executing the same workload on top of the original schema.

In our context, we have only a single input relation (a triple store with *subject*, *property* and *object* columns) where the attributes of the object and their values are represented as tuples in this triple store. Therefore, the main goal in the vertical partitioning phase is to utilize the query workload information to select the set of *best candidate properties* from the triple store in order to derive effective property tables that can efficiently improve the performance of the workload query processing. In this work, we apply a *modified* version of the vertical partitioning algorithm presented by Navathe et al. in [17]. The main idea of this approach is to exploit the *affinity* measure (how often a group of attributes are accessed together in the queries of a representative workload) within a set of attributes. Using a clustering algorithm, it produces a set of fragments that reasonably group related attributes to vertical fragments. In this section we describe the steps of our modified version of the algorithms to generate the useful vertical partitions based on the characteristics of the RDF dataset and its query workload as follows.

Attribute Usage Matrix. The first step in the vertical partitioning phase is to build the Attribute Usage Matrix (AUM). Figure 1(a) illustrates a sample of the AUM matrix where each row represents a query q in the access workload, each column represents a unique attribute of the RDF dataset and each 0/1 entry in the matrix reflects whether a given attribute is accessed by a given query or not.

$$AUM = \left\{ \begin{array}{c|cccc} & a_1 & a_2 & a_3 & a_4 & \cdots \\ \hline Q_1 & 1 & 1 & 0 & 0 & \cdots \\ Q_2 & 0 & 1 & 1 & 0 & \cdots \\ Q_3 & 0 & 0 & 1 & 1 & \cdots \\ Q_4 & 1 & 1 & 1 & 0 & \cdots \\ & \cdots & \cdots & \cdots & \cdots & \cdots \end{array} \right\} \qquad AAM = \left\{ \begin{array}{c|cccc} & a_1 & a_2 & a_3 & a_4 & \cdots \\ \hline a_1 & 30 & 10 & 0 & 5 & \cdots \\ a_2 & 10 & 40 & 18 & 4 & \cdots \\ a_3 & 0 & 18 & 32 & 7 & \cdots \\ a_4 & 5 & 4 & 7 & 42 & \cdots \\ & \cdots & \cdots & \cdots & \cdots & \cdots \end{array} \right\}$$

(a) Attribute Usage Matrix   (b) Attribute Affinity Matrix

Fig. 1. Examples of Attribute Usage Matrix and Attribute Affinity Matrix

Attribute Affinity Matrix. Using the attribute usage matrix (AUM), we derive the Attribute Affinity Matrix (AAM) which is a symmetric square matrix (each row/column represents a unique attribute of the RDF dataset) that records the *affinity* among each pair of attributes (a_i , a_j) as a single number $(aff_{(i,j)})$. This number represents the count of the workload queries that simultaneously accesses the pair of attributes (a_i , a_j). Figure 1(b) illustrates a sample attribute affinity matrix (AAM). Apparently, the higher the affinity value for any two attributes (a_i , a_j), the higher the indication that these two attributes should belong to the same fragment (This fact will be considered by the next *Clustering* and *Partitioning* steps). The diagonal of this matrix $(aff_{(i,i)})$ represents the total access number for each attribute a_i in the whole query workload. In general, the lower the total access number $(aff_{(i,i)})$ for any attribute a_i, the higher the indication that this attribute should remain in the general triple store and not to be moved to any property table. Therefore, we apply a *filtering* process where we reduce the number of attributes in the AAM. This process go through the following two main steps:

1. We select the top K attributes with the highest total access number (a_i , a_i) while the remaining attributes are removed from the matrix. The value of K can be defined either as a fixed user-defined value or a percentage variable of the total number of attributes in the RDF dataset (columns of the matrix).
2. We remove any attribute (from the selected top K attributes) with a total access number (a_i , a_i) that is less than a user-defined threshold N which is defined as a percentage variable of the total number of workload queries.

Hence, after applying the filtering process, the AAM contains the set of attributes $(\leq K)$ with the highest total access numbers which are guaranteed to be higher than a defined threshold. Therefore, we can ensure that this group of selected attributes are the *best candidates* for constructing effective property tables as we will show next.

Clustering. In this step, the attribute affinity matrix (AAM) is *clustered* by employing the Bond Energy Algorithm (BEA) [16] which is a general procedure for permuting rows and columns of a square matrix in order to obtain a semiblock diagonal form. Therefore, the algorithm cluster groups of attributes with high affinity in the same fragment, and keep attributes with low affinity in separate fragments. The objective function used by the BEA tends to surround large values of the matrix with large values, and the small ones with small values. It aims to maximize the following expression:

$$CAM = \left\{ \begin{array}{c} \text{...} \end{array} \right\}$$

	Upper							**Lower**		
	a_6	a_2	a_3	a_7	a_{10}	a_8	a_1	a_9	a_4	a_5
a_6	90	30	25	25	15	15	10	0	0	0
a_2	30	105	60	40	20	15	15	0	0	0
a_3	25	60	40	30	15	15	10	0	0	0
a_7	25	40	30	65	25	15	10	0	0	0
a_{10}	15	20	15	25	70	10	10	0	0	0
a_8	15	15	15	15	10	60	10	18	18	18
a_1	10	15	10	10	10	10	50	18	18	18
a_9	0	0	0	0	0	18	18	40	40	40
a_4	0	0	0	0	0	18	18	40	40	40
a_5	0	0	0	0	0	18	18	40	40	40

Fig. 2. Example of a Clustered Affinity Matrix with a Sample Splitting Point

$$\sum_{ij} aff_{(i,j)}(aff_{(i,j-1)} + aff_{(i,j+1)} + aff_{(i-1,j)} + aff_{(i+1,j)})$$

where $aff_{(i,0)} = aff_{(0,j)} = aff_{(i,n+1)} = aff_{(n+1,j)} = 0$ and n represent the number of attributes in (AAM). Figure 2 illustrates an example of a Clustered Affinity Matrix (CAM). In principle, the main goal of the BEA clustering algorithm is to surround large values of the matrix with large values, and the small ones with small values. The rationale behind this is to ease the job of the partitioning step which is described next.

Partitioning. The partitioning step uses the clustered affinity matrix (CAM) to produce two non-overlapping fragments. It selects one of the points along the main diagonal of the CAM. In principle, each point P in the main diagonal splits the CAM into two main blocks: an upper one (**U**) and a lower one (**L**) (see Figure 2) where each block specifies the set of attributes of a vertical fragment (property table). Therefore, given a clustered affinity matrix (CAM) with n attributes, the partitioning step considers the $(n-1)$ possible locations of the point P along the diagonal. The partitions point is then decided by selecting the point that maximizes the Z value which is computed using the following expression:

$$Z = (QU * QL) - QI^2$$

where QU represents the total number of queries that need to only access the attributes of the *upper* fragment, QL represents the total number of queries that need to only access the attributes of the *lower* fragment and QI represents the total number of queries which need to access attributes from both fragments. Having negative Z values for all of the splitting points gives the sign that the splitting process should not be done as it is not going to improve the query performance. The partitioning process is then recursively applied for each fragment until no further improvement can be achieved.

3.2 Pivoting/Unpivoting Phase

The main aim of the vertical partitioning phase is to utilize the query workload information in order to group frequently-queried and related properties and produce a set of property tables. This set of property tables should reduce the I/O cost and improve the performance of query processing. However, in dynamic environments, the characteristics of the query workloads are continuously changing. Hence, the current configuration of the underlying physical schema might become *suboptimal*. This calls for the need of applying *reactive* adjustments in the underlying schema structure in order to maintain the performance efficiency of RDF queries. Therefore, we periodically trigger an adjustment mechanism that re-analyze the workload information and re-apply the steps of the vertical partitioning phase in order to *recommend* the required changes (if any) to the structure of the property tables in the underlying schema. For instance, let us consider the situation with a previous query workload W_o where applying the steps of the vertical partitioning algorithm yield to a set of property tables P_o and a recent query workload W_n where re-applying the steps of the vertical partitioning algorithm yield to a set of property tables P_n. Examples of situations that can cause differences between the schemas of the two sets of property tables (P_o and P_n) are:

1. A group of attributes is frequently accessed in W_n in a related manner while they were not frequently accessed in W_o. As a consequence, P_n may contain a new property table which does not exist in P_n. Similarly, a frequently accessed group of attributes in W_o may become rarely accessed in W_n. Hence, their associated property table in P_o may not exist in P_n.
2. An attribute (a) has become frequently accessed in W_n while it was not frequently accessed in W_o. Moreover, the access behavior of (a) is correlated with a set of attributes that constituted a property table PT in P_o. Therefore, the associated table for PT in P_n may include (a) as an extra attribute.
3. According to W_o, an attribute (a) belongs to a property table P_{o1} where $P_{o1} \in P_o$. According to W_n, attribute (a) should belong to another property table P_{n1} where $P_{n1} \in P_n$.

A naive way to deal with such type of changes in the underlying schema is to rebuild the whole set of property tables from scratch. However, this solution is quite expensive especially in dynamic environments. *Pivot* and *Unpivot* are data manipulation operators that are used to exchange the role of rows and columns in relational tables [10]. The *Pivot* operator is mainly used to transform a series of rows into a series of fewer rows with additional columns. The *Unpivot* operator provides the inverse operation. It removes a number of columns and creates additional rows that capture the column names and values of the unpivotted columns. In practice, the main advantage of the *Pivot* and *Unpivot* operators is that they allow the a *priori* requirement for specifying the set of columns that constitute a relational table to be relaxed. In [9] Cunningham et al. have presented an approach for efficient implementation of both operators as first-class RDBMS operations. In our context, we use the *Pivot* and *Unpivot* operators for executing the required changes to *adjust* the underlying schema efficiently. They provide a very cheap solution in comparison to the complete rebuilding of the whole set of property tables. In particular, we use the pivot operators to move the data from the triple store to the property tables, the unpivot operator to move back the data from the property tables to the triple store and a combination of them to move an attribute from one property table to another.

4   Experiments

In this section, we provide the experimental evidence that our *tailored* approach outperforms the generic "*One Size Fits All*" approaches (Vertical Triple Stores - Horizontal Binary Table Stores). We conducted our experiments using the IBM DB2 DBMS running on a PC with 3.2 GHZ Intel Xeon processors, 4 GB of main memory storage and 250 GB of SCSI secondary storage. We used two real-world data sets in our experiments: 1) *SwetoDBLP* [1] which represents the RDF version of the famous database of bibliographic information of computer science journals and conference proceedings, DBLP. It contains approximately 14M triples where the subjects are described by 30 unique properties. 2) *Uniprot* [2] which represents a large-scale database of protein sequence and annotation data. It contains approximately 11M triples where the subjects are described by 86 unique properties.

For each dataset, four query sets are generated, each of which has 1000 queries. These 1000 queries are designed to randomly access m attributes of n subjects from the underlying RDF dataset. Each query set is denoted by the value of its parameters as $Q_{m,n}$. We compared the performance of our approach with two main approaches: 1) *Vertical Triple Stores (VS)* where we followed the Hexastore [22] indexing mechanism where we build B-tree indexes over all 6 permutations of the three dimensions that constitute the RDF triples. 2) *Horizontal Binary Table Stores (HS)* where B-tree indexes have been built over the 2 permutations of the columns of each binary table. For each dataset, we used the generated query workloads information to create *different* property-tables storage schemes (PS) based on different values for the K parameter (most frequently-queried and related properties) and a fixed value for the N parameter (least total access number) equals to 100 (10% of the total workload queries). Each created schema is denoted by the value of its K parameter as PS_K. For each n-*ary* property table in the created schemas, we build $(n\text{-}1)$ B-tree indexes where each index consists of the *subject* column and one of the *property* columns in the table.

Figure 3 illustrates the average query processing times of the 1000 queries of different query sets over the *SwetoDBLP* dataset (Figure 3(a)) and the *Uniprot* dataset (Figure 3(b)). For the *SwetoDBLP* dataset, we created two different schemas of *workload-aware property tables* (PS). For these two schemas, we specified the value of the K parameter as 8 and 16 respectively. Similarly, for the *Uniprot* datatset, we created two schemas where the value of the K parameter is 15 and 30. The results of our experiments confirm that our approach outperforms both of the two generic approaches: vertical triple stores (VS) and the horizontal binary table stores (HS). The main reason behind this is that our approach groups query-related properties into a single relation which reduces the disk access cost and also reduces the number of the required expensive join operations.

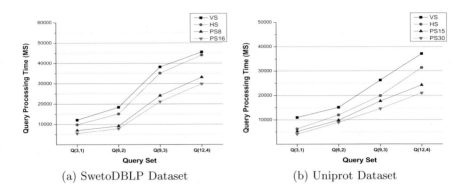

(a) SwetoDBLP Dataset   (b) Uniprot Dataset

Fig. 3. Query Processing Time

For example, let us assume a sample query which accesses n properties of the RDF dataset, the VS approach will involve joining n instances of the giant triple store and the HS approach will involve joining n binary tables. Clearly, the size of the binary tables is much smaller than the size of the single triple store which explains why the the HS approach can outperform the VS approach. In our approach, in the best case we may need to access only *one* property table (if the n attributes are query-related) while in the worst case, we may need to join n instances of a combination between the triple store and the property tables (if the n attributes are not related at all). The main advantage of our approach is that such situation can not re-occur for long if the same access pattern re-appears frequently because by analyzing the workload information we will be able to detect the frequency of such patterns and adjust the underlying schema to group their related properties and thus reduce their access cost to a minimum. On one hand, designing workload-aware property tables based on a larger value of the K parameter improves the query processing time. On the other hand, it increases the cost of the required disk storage space due to the associated increase on the sparseness of the property tables. For example, in the *SwetoDBLP* datatset, the sparseness of the PS_8 and PS_{16} is 34% and 48% respectively while in the *Uniprot* datatset, the sparseness of the PS_{15} and PS_{30} is 46% and 71% respectively.

5   Conclusion

In this paper, we presented an approach for designing a flexible and adaptable approach for achieving efficient and scalable performance for RDF queries using the infrastructures of RDBMSs. Our approach derives the relational schema in a tailored way for each RDF dataset based on its query workload. A vertical partitioning algorithm is applied to group the query related properties in the data set into *n-ary* tables while the remaining properties reside in a generic triple store. To maintain the efficiency of the designed partitions, we monitor the characteristics of the query workloads to recompute the vertical partitions and transparently adjust the underlying relational schema by applying the necessary changes. We evaluated the performance of our approach with extensive experiments using two real datasets. The results show that our approach outperforms the current state-of-the-art of RDF stores.

References

1. SwetoDBLP ontology, http://lsdis.cs.uga.edu/projects/semdis/swetodblp/
2. Uniprot RDF Data Set, http://dev.isb-sib.ch/projects/uniprot-rdf/
3. Abadi, D., Marcus, A., Madden, S., Hollenbach, K.: SW-Store: a vertically partitioned DBMS for Semantic Web data management. VLDB J. 18(2) (2009)
4. Berners-Lee, T., Hendler, J., Lassila, O.: The Semantic Web: A new form of Web content that is meaningful to computers will unleash a revolution of new possibilities. Scientific American (2001)
5. Bizer, C., Schultz, A.: Benchmarking the Performance of Storage Systems that expose SPARQL Endpoints. In: SSWS (2008)
6. Chaudhuri, S., Ganesan, P., Narasayya, V.: Primitives for Workload Summarization and Implications for SQL. In: VLDB (2003)
7. Chaudhuri, S., Weikum, G.: Rethinking Database System Architecture: Towards a Self-Tuning RISC-Style Database System. In: VLDB (2000)
8. Copeland, G., Khoshafian, S.: A Decomposition Storage Model. In: SIGMOD, pp. 268–279 (1985)
9. Cunningham, C., Graefe, G., Galindo-Legaria, C.: PIVOT and UNPIVOT: Optimization and Execution Strategies in an RDBMS. In: VLDB (2004)
10. Gray, J., et al.: Data Cube: A Relational Aggregation Operator Generalizing Group-by, Cross-Tab, and Sub Totals. Data Min. Knowl. Discov. 1(1) (1997)
11. Botea, V., et al.: PIST: An Efficient and Practical Indexing Technique for Historical Spatio-Temporal Point Data. GeoInformatica 12(2) (2008)
12. Grust, T., Sakr, S., Teubner, J.: XQuery on SQL Hosts. In: VLDB, pp. 252–263 (2004)
13. Levandoski, J., Mokbel, M.: RDF Data-Centric Storage. In: ICWS (2009)
14. MahmoudiNasab, H., Sakr, S.: An Experimental Evaluation of Relational RDF Storage and Querying Techniques. In: BenchmarX 2010 (2010)
15. Manola, F., Miller, E.: RDF Primer, W3C Recommendation (February 2004), http://www.w3.org/TR/REC-rdf-syntax/
16. Mccormick, W., Schweitzer, P., White, T.: Problem Decomposition and Data Reorganization by a Clustering Technique. Operations Research 20(5) (1972)
17. Navathe, S., Ceri, S., Wiederhold, G., Dou, J.: Vertical Partitioning Algorithms for Database Design. TODS 9(4) (1984)
18. Neumann, T., Weikum, G.: RDF-3X: a RISC-style engine for RDF. PVLDB 1(1) (2008)
19. Sakr, S., Al-Naymat, G.: Relational Processing of RDF Queries: A Survey. SIGMOD Record 38(4) (2009)
20. Schmidt, M., Hornung, T., Lausen, G., Pinkel, C.: SP2Bench: A SPARQL Performance Benchmark. In: ICDE (2009)
21. Türker, C., Gertz, M.: Semantic integrity support in SQL: 1999 and commercial (object-)relational database management systems. VLDB J. 10(4) (2001)
22. Weiss, C., Karras, P., Bernstein, A.: Hexastore: sextuple indexing for semantic web data management. PVLDB 1(1) (2008)

RaUL: RDFa User Interface Language – A Data Processing Model for Web Applications

Armin Haller[1], Jürgen Umbrich[2], and Michael Hausenblas[2]

[1] CSIRO ICT Centre, Canberra, Australia
armin.haller@csiro.au
[2] DERI Galway, Ireland
firstname.lastname@deri.org

Abstract. In this paper we introduce RaUL, the RDFa User Interface Language, a user interface markup ontology that is used to describe the structure of a web form as RDF statements. RaUL separates the markup of the control elements on a web form, the *form model*, from the *data model* that the form controls operate on. Form controls and the data model are connected via a data binding mechanism. The form elements include references to an RDF graph defining the data model. For the rendering of the instances of a RaUL model on the client-side we propose ActiveRaUL, a processor that generates XHTML+RDFa elements for displaying the model on the client.

1 Introduction

Traditional Web applications and in particular Web forms are the most common way to interact with a server on the Web for data processing. However, traditional web forms do not separate the purpose of the form from its presentation. Any backend application processing the input data needs to process untyped key/value pairs and needs to render HTML or XHTML in return. For XHTML, XForms [3] was introduced to separate the rendering from the purpose of a Web form. With the advent of RDFa [4], a language that allows the user to embed structured information in the format of RDF [12] triples within XHTML documents, the presentation layer and metadata layer are similarly interweaved. Currently, when annotating Web pages with semantic information, the user first needs to define the structure and presentation of the page in XHTML and then use RDFa to annotate parts of the document with semantic concepts. The data processing in the backend requires a method to bind the data input to its presentation. For example, an input field for a first name has to be bound to an RDF triple stating a `foaf:firstname` relation over this property. When returning data from the server, the back-end application also needs to deal with and create XHTML as well as RDFa code.

In this paper we introduce RaUL, the <u>RDFa</u> <u>U</u>ser Interface <u>L</u>anguage, a model and language that enables to build semantic Web applications by separating the markup for the presentation layer from the data model that this presentation

L. Chen, P. Triantafillou, and T. Suel (Eds.): WISE 2010, LNCS 6488, pp. 400–410, 2010.

layer operates on. The RaUL model (ontology) itself is described in RDF(s). For the rendering of the RaUL model instances on the client-side we propose ActiveRaUL, a processor that generates XHTML+RDFa elements for displaying the model on the client.

2 Related Work

Annotating forms with semantics is a relatively new research topic in the Web engineering realm. We are aware of some earlier attempts concerning form-based editing of RDF data [5] as well as mapping between RDF and forms [9].

In [16,6] we proposed a read/write-enabled Web of Data through utilizing RDForms [11], a way for a Web browser to communicate structured updates to a SPARQL endpoint. RDForms consists of an XHTML form, annotated with the RDForms vocabulary[1] in RDFa [4], and an RDForms processor that gleans the triples from the form to create a SPARQL Update[2] statement, which is then sent to a SPARQL endpoint. However, RDForms is bound to a domain-agnostic model—that is, it describes the fields as key/value pairs—requiring a mapping from the domain ontology (FOAF, DC, SIOC, etc.) and hence is not able to address all the use cases we have in mind.

Dietzold [10] propose a JavaScript library, which provides a more extensive way for viewing and editing RDFa semantic content independently from the remainder of the application. Further, they propose update and synchronization methods based on automatic client requests. Though their model is rich and addresses many use cases, it is restricted to a fixed environment (the Wiki), and hence not generally applicable.

Further, there are other, remotely related works, such as SWEET [13], which is about semantic annotations of Web APIs, as well as Fresnel [1], providing a vocabulary to customize the rendering of RDF data in specific browser (at time of writing, there are implementations for five browsers available). Eventually, we found backplanejs [7] appealing; this is a JavaScript library that provides cross-browser for XForms, RDFa, and SMIL as well as a Fresnel integration and jSPARQL, a JSON serialization of SPARQL.

3 Motivating Example

Interaction with forms is ubiquitous on the Web as we experience it every day. From ordering a book at Amazon over updating our Google calendar; from booking a flight via Dohop or filing bugs; from uploading and tagging pictures on Flickr to commenting on blog posts.

The example in Figure 1 is essentially a slightly simplified version of the common user registration forms of social networking sites, such as Facebook or MySpace. Figure 1 shows a Web form on the left side and its encoding in pure XHTML on the right side. We will use this example to exemplify the data binding in RaUL in Section 4.1.

---

[1] http://rdfs.org/ns/rdforms
[2] http://www.w3.org/TR/sparql11-update/

```
<html>
  <head> <title>Registration</title> </head>
  <body>
    <form method="post" action="" id="register" >
    <span>First Name:</span>
    <input type="text" class="inputtext" id="firstname" name="firstname" />
    <br /> <span>Last Name:</span>
    <input type="text" class="inputtext" id="lastname" name="lastname" />
    <br />Your Email:
    <input type="text" class="inputtext" id="reg_email" name="reg_email" />
    <br />Password:
    <input type="password" class="inputpassword" id="reg_pwd" name="reg_pwd" />
    <br />I am:
    <input type="radio" name="sex" value="male" /> <span>Male</span>
    <input type="radio" name="sex" value="male" /> <span>Female</span>
    <br />Birthday: <select id="birthday_day" name="birthday_day" >
      <option value="0">Day:</option> /* List of days */
    </select> <select id="birthday_month" name="birthday_month" >
      <option value="0">Month:</option> /* List of months */
    </select> <select name="birthday_year" id="birthday_year" >
      <option value="0">Year:</option> /* List of years */
    </select> <br /> <br />
    <input type="submit" name="submit" value="Submit" onClick="javascript:
      parseAndSend('body')" typeof="raul:Button" />
    </form>
  </body>
</html>
```

Fig. 1. Example Social Networking Registration Form

4   The RDFa User Interface Language (RaUL)

The RDFa User Interface Language provides a standard set of visual controls that are replacing the XHTML form controls. The RaUL form controls are directly usable to define a web page in the back-end with RDF statements. The automatically rendered webpage in XHTML+RDFa from RDF statements according to the RaUL vocabulary consists of two parts:

1. *Form model*: One or many form model instances which are rendered as an XHTML page including RDFa annotations describing the structure of the form according to the RaUL vocabulary and
2. a *data model* describing the structure of the exchanged data, also expressed as RDFa annotations which are referenced from the *form model* instance via a data binding mechanism.

The *form model* and *data model* parts make RaUL forms more tractable by referencing the values in the forms with the structure of the data defined by an ontology. It also eases reuse of forms, since the underlying essential part of a form is no longer irretrievably bound to the page it is used in. Further, the data model represented by the form can be accessed by external applications, ie. semantic Web crawlers or software agents supporting users in their daily tasks.

4.1   Form Model

RaUL defines a device-neutral, platform-independent set of form controls suitable for general-purpose use. We have implemented a mapping to XHTML form elements. However, similar to the design of XForms the form controls can be bound to other languages than XHTML forms as well. A user interface described in RDF triples according to the RaUL vocabulary is not hard-coded to,

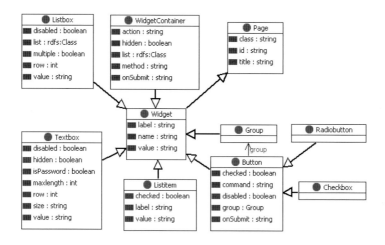

Fig. 2. RaUL form model

for example, display radio buttons as they are in XHTML, but it can be bound to any host language (e.g. proprietary UIs for mobile devices). As mentioned, our current implementation of the ActiveRaUL processor supports the rendering of XHTML+RDFa.

A form control in RaUL is an element that acts as a direct point of user interaction and provides write access to the triples in the *data model*. The controls are bound to the *data model* via a binding mechanism. Every form control element has a *value* property that is associated to a reified triple or named graph in the *data model* (see Section 4.2). Form controls, when rendered by the ActiveRaUL processor, display the underlying data values to which they are bound.

In the following we describe the RaUL form controls, including their attributes. They are declared using the RaUL form ontology. Figure 2 shows a class diagram like overview of the main classes in the ontology. The full ontology can be found at: `http://purl.org/NET/raul#`.

Page: The *Page* class acts as the main container for all other content in a RaUL document. In the mapping to XHTML a *Page* instance maps to a `<body>` element.

Widget: All form controls are a subclass of the *Widget* class inheriting its standard properties, a *label* and a *name*. The Widget class includes a *value* property that is used to associate triples in the *data model* to a control element.

WidgetContainer: A *WidgetContainer* groups *Widgets* together. It is similar to a form container in XHTML and in the mapping to XHTML every *WidgetContainer* is rendered as a `<form>` in XHTML. It defines a *method* and *action* property to define the form submission. The ordering of the form controls in a *WidgetContainer* is defined with an RDF collection. When rendering the XHTML+RDFa code, the ActiveRaUL processor determines the positioning of the control elements based on the ordering in the RDF collection.

Textbox: The *Textbox* form control enables free-form data entry. The constraints on the input type are obtained from the underlying triples associated with the *Textbox* control element. In XHTML a *Textbox* is rendered as an *input* box of type *text*. Properties of the Textbox are, *disabled*, *hidden*, *isPassword*, *maxlength*, *row* and *size*. These properties are straightforwardly mapped to their equivalents in the XHTML model.

Listbox: This form control allows the user to make one or multiple selections from a set of choices. Special properties include *list*, *multiple*, *row* and *disabled*. In case of the *row* and *multiple* properties the rendering in XHTML is straightforward. In both cases it is rendered as a *select* input box displayed as a multi-row selection box (*row*) and with the ability to select more than one value (*multiple*). The *list* property is used to associate the *Listbox* to a collection of *Listitems*. *Listitems* are required to define a *value* and *label* property and can be defined as *checked*.

Button: The *Button* form control is used for actions. Beyond the common attributes inherited from its superclasses it defines the following properties, *checked*, *command*, *disabled* and *group*. The mapping to XHTML creates either a normal push button (ie. an input field of type *button*) or a submit button in case the *command* property is set to "submit". A submit button is used to trigger the action defined in the form element. The *Radiobutton* and the *Checkbox* are subclasses of a *Button* in the RaUL vocabulary. They are mapped to their respective counterparts in XHTML, input controls of type *radio* or of type *checkbox*, respectively. To group buttons together and determine the selected values, the group property of the *Button* class can be used. For the data binding any *Group* of *Buttons*, ie. either *Radiobuttons* or *Checkboxes*, which values are not literals, must bind to an RDF collection with the same number of node elements as there are *Buttons* in the respective group. After the submission of the *Button* control element the JavaScript processor creates a *checked* relation for all selected *Checkboxes* or for the selected *Radiobutton* denoting the user selection of the respective control element.

4.2   Data Model

The purpose of XHTML forms is to collect data and submit it to the server. In contrast to the untyped data in XHTML forms, in RaUL this data is submitted in a structured way as RDF data according to some user defined schema. The data, defined in the backend as an RDF graph, is also encoded in the generated XHTML+RDFa document as statements within a *WidgetContainer*. This approach gives the user full flexibility in defining the structure of the model. Empty `rdf:object` fields serve as place-holders in the RDF statement describing the data for a control element (see the rdf:object property in Figure 3) and are filled at runtime by the JavaScript processor with the user input. Initial values can be provided in the `rdf:object` field which is used by the ActiveRaUL processor in the initial rendering to fill the value field of the XHTML form.

In our motivating example, the data structure of the *firstname* element is given by the FOAF ontology [8], a vocabulary to describe persons, their activities and their relations to each other. We use RDF reification to associate this triple in the RaUL *form model* instance in the *firstname textbox*. Reification in RDF describes the ability to treat a statement as a resource, and hence to make assertions about that statement. Listing 3 shows how the triple *<http://sw.deri.org/ haller/foaf.rdf#ah><foaf:name><"">* which defines the data structure of the #valuefirstname is described as a resource using RDF reification.

```
<span about="#valuefirstname" typeof="foaf:Person">
 <span rel="rdf:subject" resource="http://sw.deri.org/~haller/foaf.rdf#ah" />
 <span rel="rdf:predicate" resource="foaf:name" />
 <span rel="rdf:object" resource="" />
</span>
```

Fig. 3. RDFa reified triple for a foaf:firstname object

This new resource is then associated in the *form model* instance with the `raul:value` property as shown in Listing 4. Initial values for the instance data

```
<span about="#firstname">
 <span property="raul:label">First Name:</span>
 <span property="raul:value" content="#valuefirstname" />
 <span property="raul:class" content="inputtext" />
 <span property="raul:id" content="firstname" />
 <span property="raul:name" content="firstname" />
</span>
```

Fig. 4. Value association in the *firstname* textbox

may be provided or left empty. As such, the data model essentially holds a skeleton RDF document that gets updated as the user fills out the form. It gives the author full control on the structure of the submitted RDF data model, including the reference of external vocabulary. When the form is submitted, the instance data is serialized as RDF triples.

5   Architecture and Interaction

The general architecture as shown in Figure 5 follows the Model-View-Controller (MVC) pattern [14]. It uses a Java servlet container as the controller part, RDF as the general model for the domain logic and the ActiveRaUL Processor as the responsible for the view creation (GUI). A client interacts with RaUL by triggering a submit action of a particular XHTML form. Before the form submission to the server, the *parseAndSend(...)* function of the *JavaScript (JS)*

Processor handles the *RDFa parsing* on the client side by extracting the embedded RDF content and the HTTP request method from the DOM (Document Object Model) tree. The implementation of the parser and the data binding is a pure Ajax implementation. Once the request is submitted to the server, the *Controller Servlet* handles the RDF input and the related HTTP response message. The input content (the structure/form and instance information) will be validated and processed according to the domain-logic as implemented in the backend application by the service provider. The response is modeled again in RDF and forwarded to the *ActiveRaUL Processor*. It controls the presentation of the response at the server side by generating the XHTML+RDFa representation. Finally, the Controller Servlet streams the generated data back to the client. This allows not only browsers to interact with the server but also Web crawlers or automated agents to ingest the data since they receive the full XHTML+RDFa content.

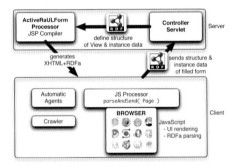

Fig. 5. Architecture diagram

6   Evaluation

The evaluation is split into two parts: 1) a comparison between existing solutions based on features and 2) benchmarks of the overhead added by the RDFa markup to the existing XHTML forms.

6.1   Feature Comparison

We compare our approach 1) RaUL with 2) the native XHTML form/POST mechanism, 3) RDForms, 4) XForms and 5) XUL.

The first category of features are the *complexity* of the development *on the client and server side*. The complexity on the client-side is only low for XHTML forms as it is supported by all browsers. XForms requires Ajax scripts or browser plug-ins (for Firefox) rendering the data. RaUL and RDForms provide third-party JavaScript libraries which parse and process RDF statements from the RDFa annotations and for the latter transforms the statements into an SPARQL update operation. Processing XUL on the client side requires a high complexity; it is a proprietary language that has to be rendered by client-side code (ie. in Firefox). Looking at the complexity for the server-side handling of the request, XHTML forms have a medium complexity. Although there exists an abundance of CGI scripts that handle the requests and submitted parameters, the data is submitted as untyped key/value pairs only which have to be handled in custom built code. RDForms require a medium complexity since the server has to understand SPARQL update requests. RaUL has a low complexity, since it provides a generic ActiveRaUL service that handles the RDF serialization. The web application developer only needs to deal with triples for which many data abstraction layers exist. The complexity for XForms is low as the data is encoded in XML and for XUL all logic resides on the client-side and thus the server side complexity is low.

The second feature category are general characteristics of the approaches. The first characteristic is that the solution should be *representation agnostic* with respect to the RDF input/output, meaning whatever RDF serialization is supplied (RDF/XML, Turtle, RDFa, microformats+GRDDL), the system should be able to handle it. This is not possible with any language but RDForms. The current implementation of ActiveRaUL only supports RDFa, but in future work we intend to support other formats as well. The second characteristic is the support for an explicit *data model*: XHTML does not define a data model, the data is send as key/value pairs. RaUL and RDForms explicitly define the data model in

Table 1. Comparison of different update interfaces

	RaUL	XHTML form	RDForms	XForms	XUL
1. Complexity					
1.1 Client	medium	low	medium	medium	high
1.2 Server	low	medium	medium	low	low
2. Characteristics					
2.1 Agnostic	no	no	yes	no	no
2.2 Data Model	yes	no	yes	yes	yes
2.3 Form Model	yes	limited	limited	limited	yes
2.4 Browser support	all	all	all	some	one

Table 2. Comparison of added overhead by RDFa markup. Min and max values in the XHTML and XHTML+RDF columns are the content size in Bytes.

Form Element	XHTML		XHTML+RDFa		# triples		Overhead in %	
	min	max	min	max	min	max	min	max
Page	39	89	121	279	1	4	210%	213%
WidgetContainer	72	115	251	376	3	6	248%	226%
Textbox	41	100	88	377	1	7	114%	277%
ListBox	49	125	228	459	4	8	365%	367%
Button	48	81	98	415	1	8	104%	412%

RDF. XUL and XForms define it in XML. The third characteristic is the *form model* - which is, if the client request contains the form model as structured data. Only RaUL and XUL encode the form model explicitly. XHTML forms do not define the form model as structured data, but offer DOM manipulation. RDForms do not define the form structure explicitly either, but the mapping logic between the RDF triples defining the form and the XHTML rendering is hard-coded in the transformation tool. The last characteristic is the *browser support* of the approaches. The first four solutions are browser independent approaches; RaUL, RDForms and XForms also require that the browser supports and enables JavaScript. XUL is only supported by Firefox.

6.2   Overhead Benchmark

From a performance point of view it is of interest how much overhead in the data size is added by the RDFa annotated forms. The processing and parsing speed of the page content depends on the file size wrt. to two factors. 1) The resulting download time (file size / available bandwidth) and 2) how efficient an RDFa parser can handle the content; the RDFa parser used in our implementation – as with most other RDFa parsers – needs to parse the whole DOM structure into memory to recreate the RDF structure. Thus, we measured the additional overhead in bytes for the form elements in Table 2.

Minimal (min) in the table denotes the minimal RaUL model to generate the respective form element (or page). The maximal (max) value denotes a model that uses all properties of the respective form element in its annotation. The number of triples column denotes how many RDF triples are required in the backend and are encoded in the resulting web page as annotations. Whenever instance identifiers are required in the RDFa annotations we assumed a two digit identifier (which allows a page to include at least 3844 identifiers if we consider case sensitive alphanumeric combinations).

Results: The *Page* element is considerably bigger than its pure XHTML counterpart (header and body) when rendered in RDFa because of the namespace definition (at least the RaUL namespace has to be defined). However, it is only required once for every page and as such the bytes in the table are the absolute overhead per page. The *WidgetContainer* element is also adding about 2 $^1/_2$ times the overhead to a pure form container in XHTML. Again, only one *WidgetContainer* for every form is required (typically one per page) and as such the added bytes in the table are in most cases only added once per page. An annotated *Textbox* takes about twice the size of the pure XHTML form and only requires one triple in the backend. Adding all properties (in total 7 RDF statements) to a *Textbox* causes an overhead of about 277%. The *Listbox* control rendered in RDFa adds more than 3 $^1/_2$ times the size of the pure XHTML form. This is due to the fact that there is at least one Listitem associated to a *Listbox*, which is modeled as a class in RaUL. As such, a Listbox needs at least four statements (RDF triples). Similar to the *Textbox* an annotated *Button* takes about twice the size of the pure XHTML control element and only requires one triple in the backend. Again, adding all properties to a *Button* adds a considerable amount of space (more than four times the pure XHTML element) to the page due to the *Group* class which can be used to associate multiple buttons together (see Section 4.1). However, it also includes 8 triples in the annotation.

Discussion: The overhead seems to be significant in size, especially if all properties of a form control element are used. Whereas a *Textbox* and a *Button* only about double the size of the pure XHTML form element in its minimal configuration, a *Listbox* and a *Button* with all properties defined add about four times the size. Similar to adding div containers and CSS styles to XHTML, adding RDFa increases the size of the file. Since the RDFa annotations are additions to the XHTML code the size of the encoding is influenced by the verbose syntax of the RDFa model. As there are potentially many form controls in a web form, the user has the trade-off between the depth of the annotations and the size they consume. The bigger size, though, does not necessarily cause more packages delivered over the wire. However, if the data transfer volume is pivotal, the rendering of the document can be achieved by a JavaScript DOM generation algorithm that operates on pure XML/RDF. Another option is to install a client-side code (similar to the Firefox extensions for XUL) that does the rendering of pure XML/RDF.

7   Conclusion

In this paper we introduced RaUL, the RDFa User Interface Language, which provides a standard set of visual controls that are replacing the XHTML form controls. The RaUL form controls are directly usable to define a web page in the backend with RDF statements according to the RaUL ontology. RaUL form controls separate the functional aspects of the underlying control (the *data model*) from the presentational aspects (the *form model*). The *data model*, which describes the structure of the exchanged data (expressed as RDFa annotations) is referenced from the *form model* via a data binding mechanism. For the rendering of the instances of a RaUL model on the client-side we propose ActiveRaUL, a processor that generates XHTML+RDFa elements for displaying the model on the client. Only when data is submitted to the server a JSP servlet creates the RDF triples. Summarized, the advantages of RaUL in comparison to standard XHTML forms are:

1. **Data typing:** Submitted data is typed through the ontological model.
2. **RDF data submission:** The received RDF instance document can be directly validated and processed by the application back-end.
3. **Explicit form structure:** The form elements are explicitly modeled as RDF statements. The backend can manipulate and create forms by editing and creating RDF statements only.
4. **External schema augmentation:** This enables the RaUL web application author to reuse existing schemas in the modeling of the input *data model*.

Acknowledgements. This work is part of the water information research and development alliance between CSIRO's Water for a Healthy Country Flagship and the Bureau of Meteorology.

References

1. Fresnel, Display Vocabulary for RDF (2005),
   http://www.w3.org/2005/04/fresnel-info/
2. Cross-Origin Resource Sharing (2010), http://www.w3.org/TR/cors/
3. XForms Working Groups (2010), http://www.w3.org/MarkUp/Forms/
4. Adida, B., Birbeck, M., McCarron, S., Pemberton, S.: RDFa in XHTML: Syntax and Processing. In: W3C Recommendation, W3C Semantic Web Deployment Working Group (October 14, 2008)
5. Baker, M.: RDF Forms (2003), http://www.markbaker.ca/2003/05/RDF-Forms/
6. Berners-Lee, T., Cyganiak, R., Hausenblas, M., Presbrey, J., Sneviratne, O., Ureche, O.-E.: On integration issues of site-specific apis into the web of data. Technical report (2009)
7. Birbeck, M.: backplanejs (2010), http://code.google.com/p/backplanejs/
8. Brickley, D., Miller, L.: FOAF Vocabulary Specification 0.91. Namespace document (November 2007)
9. de hOra, B.: Automated mapping between RDF and forms (2005),
   http://www.dehora.net/journal/2005/08/automated_mapping_between_rdf_and_forms_part_i.html
10. Dietzold, S., Hellmann, S., Peklo, M.: Using javascript rdfa widgets for model/view separation inside read/write websites. In: Proceedings of the 4th Workshop on Scripting for the Semantic Web (2008)
11. Hausenblas, M.: RDForms Vocabulary (2010),
    http://rdfs.org/ns/rdforms/html
12. Klyne, G., Carroll, J.J., McBride, B.: RDF/XML Syntax Specification (Revised). W3C Recommendation, RDF Core Working Group (2004)
13. Maleshkova, M., Pedrinaci, C., Domingue, J.: Semantic Annotation of Web APIs with SWEET. In: Proceedings of the 6th Workshop on Scripting and Development for the Semantic Web (2010)
14. Reenskaug, T.: The original MVC reports (February 2007)
15. RFC2818. HTTP Over TLS (2000), http://www.ietf.org/rfc/rfc2818.txt
16. Ureche, O., Iqbal, A., Cyganiak, R., Hausenblas, M.: On Integration Issues of Site-Specific APIs into the Web of Data. In: Semantics for the Rest of Us Workshop (SemRUs) at ISWC 2009, Washington DC, USA (2009)
17. XMLHttpRequest. W3C Working Draft (2009),
    http://www.w3.org/TR/XMLHttpRequest/

Synchronising Personal Data with Web 2.0 Data Sources

Stefania Leone, Michael Grossniklaus,
Alexandre de Spindler, and Moira C. Norrie

Institute for Information Systems, ETH Zurich
CH-8092 Zurich, Switzerland
{leone,grossniklaus,despindler,norrie}@inf.ethz.ch

Abstract. Web 2.0 users may publish a rich variety of personal data to a number of sites by uploading personal desktop data or actually creating it on the Web 2.0 site. We present a framework and tools that address the resulting problems of information fragmentation and fragility by providing users with fine grain control over the processes of publishing and importing Web 2.0 data.

1   Introduction

An increasing amount of personal data is being published on Web 2.0 sites such as Facebook and Flickr as well as various forms of discussion forums and blogs. These sites provide easy-to-use solutions for sharing data with friends and user communities, and consumers of that data can contribute to the body of information through actions such as tagging, linking and writing comments. Sometimes, Web 2.0 sites also take over the primary role of managing a user's personal data by providing simple tools for creating, storing, organising and retrieving data, even when users are on the move.

Desktop applications for multimedia processing such as Adobe Photoshop Lightroom provide tools to support the publishing process. There is also an increasing number of tools to support cross publishing, meaning that information posted on one site is published automatically to another. For example, WordPress allows information published on Twitter to be automatically published as a blog article and vice versa. Although such tools facilitate the publishing process, they are only available for certain types of data and applications. Further, since it is now typical for active users of Web 2.0 sites to publish data to many different sites, they need to learn to use a variety of publishing tools, some of which require manual activation while others are fully automated.

Another problem is the *fragility* of information managed solely by Web 2.0 sites since many of these sites are considered to have an uncertain future. For example, the expenses incurred by YouTube far exceed their advertisement revenues which has led to speculation about its future [1]. It may therefore be desirable to *import* data from Web 2.0 sites into desktop applications to ensure offline and/or long-term access.

L. Chen, P. Triantafillou, and T. Suel (Eds.): WISE 2010, LNCS 6488, pp. 411–418, 2010.

To address these issues, there is a need for a single, general tool to support the processes of publishing and importing Web 2.0 data across applications. It should be easy to specify new synchronisation processes and data mappings through a graphical user interface. In addition, it should be possible to integrate existing tools for publishing and cross publishing, while allowing users to control the actual synchronisation processes. As a first step in this direction, we have developed a synchronisation framework which allows users to synchronise personal data with one or more Web 2.0 sites based in a plug-and-play style of selecting and configuring synchronisation modules.

Section 2 discusses the background to this work, while Sect. 3 provides an overview of our approach. Section 4 examines support for the required data mappings. Details of the synchronisation engine are given in Sect. 5. Concluding remarks are given in Sect. 6.

2   Background

Nowadays, it is common for users to manage a wide variety of personal information using a mix of desktop applications and Web 2.0 services. Data is often replicated with a user keeping a copy on the desktop as well as publishing it on one or more Web 2.0 sites. For example, a version of a photo may be published on Facebook and Flickr, and a friend on Facebook may also be connected to on LinkedIn. In other cases, the data is created and managed using the Web 2.0 service, leaving the user entirely dependent on that service for access to the data.

While Web 2.0 services offer many advantages over traditional desktop applications in terms of data sharing and ubiquitous access, there are also drawbacks resulting from the fragmented management of personal data across desktop applications and Web 2.0 services. To address this problem, researchers have proposed different ways of providing file system services over Web 2.0 data sources. One approach is to propose that Web 2.0 applications should use independent file system services [2] rather than managing their own data. A less radical approach is to propose a software layer that supports an integrated file system view of a user's data based on a common interface to Web 2.0 services [3]. In both cases, naming and security are major issues. There are also efforts in the industry sector to promote open protocols and standards such as DataPortability.org. OpenLink Data Spaces[1] (ODS) implements the recommendation from DataPortability.org and offers several components for personal data management such as an address book and picture gallery as well as integration with applications such as Facebook.

Other researchers have addressed the problem of extracting and aggregating data from Web 2.0 services, for example [4,5,6]. While these approaches mostly use data extraction techniques for gathering data, many Web 2.0 platforms nowadays provide an API which provides access to platform data. We note that a number of commercial data aggregator services are available for news, blogs and

---

[1] http://virtuoso.openlinksw.com/dataspace/dav/wiki/Main/Ods

social networking data, e.g. FriendFeed[2]. Content aggregator services such as GNIP[3] act as an intermediate data aggregator between Web 2.0 services and applications by offering developers a uniform API and polling services.

While these projects provide partial solutions, they do not deal with issues of data replication and volatility. Also, they pay little attention to the publishing process or the need to import Web 2.0 data in order to ensure both offline and long-term access. Further, in cases where data synchronisation is supported, it tends to be hard-wired in the sense that the developer predefines data mappings and means of handling conflict resolution.

Web 2.0 platforms provide opportunities to address data synchronisation issues in a novel way by passing greater responsibilities, and hence more control, to the user. Just as users have become actively involved in, not only the creation and sharing of content, but also the development and integration of applications, they should be empowered to take decisions on what, when and how data should be synchronised in order to meet their own information needs. Therefore, rather than providing a fully automated and transparent solution, we believe that it is important to provide a flexible and configurable framework that gives users the freedom to either select synchronisation modules developed/configured by others or to develop/configure their own modules.

3   Synchronisation Framework

Our synchronisation framework allows users to control all synchronisation of personal data with Web 2.0 data in a single place. In Fig. 1, we present an overview of the architecture of the framework. It consists of a synchronisation engine, a number of synchronisation modules and some utilities, and exposes an API that offers the synchronisation functionality. A synchronisation module defines a particular synchronisation process in terms of internal and external synchronisation endpoints, a data mapping and a synchronisation algorithm.

The *synchronisation endpoints* are adapters to data sources and encapsulate the access to these sources. An external synchronisation endpoint is the conceptual representation of a Web 2.0 data source that includes information about the data source's data model as well as other synchronisation specific information such as access and authentication characteristics. We make use of the APIs offered by Web 2.0 applications such as Facebook and Flickr to implement synchronisation endpoints and access their data. These platforms usually expose their data and functionality through REST-based APIs. An internal synchronisation endpoint is the conceptual representation of a desktop application. The framework is extensible in that synchronisation with additional external and internal data sources can easily be implemented by adding an associated synchronisation endpoint.

The *synchronisation engine* is responsible for the actual synchronisation process. It first configures the synchronisation process based on the specification

---

[2] http://friendfeed.com

[3] http://www.gnip.com/

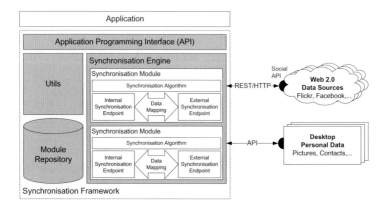

Fig. 1. Architecture of the synchronisation framework

provided by the synchronisation module and then executes it. The synchronisation module defines a mapping between the data models of the two data sources and also a synchronisation algorithm. Additionally, it specifies other configuration information such as when the synchronisation process should be executed and a conflict resolution strategy. Default mappings and specifications of the synchronisation processes are provided, but users can choose and configure a synchronisation module for a pair of endpoints or even develop their own.

Note that the distinction between internal and external endpoints is somewhat artificial, since the framework can be used to synchronise any two data sources. Thus, it could be used to support cross-publishing between two different Web 2.0 data sources such as Twitter and Facebook as well as between two desktop applications. However, since our main goal was to investigate support for publishing and importing Web 2.0 data, we have focussed so far on synchronising desktop data with social networking sites such as Facebook and Flickr.

4   Data Access and Data Mappings

The synchronisation endpoints provide access to Web 2.0 data sources through their APIs. While the Facebook API[4], Microsoft Live Services[5] and Flickr Services[6] are all platform-specific, the Google OpenSocial API[7] is a common API which can be implemented by any Web 2.0 service. Although several well-known social networking platforms, e.g. LinkedIn, implement the OpenSocial API, they usually implement only part of the API since many features are optional.

Some APIs only provide access to a platform's data while others allow applications or external data to be integrated. For example, Facebook allows users to publish their own applications developed using either the Facebook Markup

---

[4] http://developers.facebook.com/

[5] http://dev.live.com

[6] http://www.flickr.com/services/api/

[7] http://code.google.com/apis/opensocial/

Language (FBML) and Facebook Query Language (FQL) or a general programming language such as Java, PHP or Visual Basic. Similarly, services such as Orkut that implement the OpenSocial API can be extended by building gadgets using Google Gadget technology[8].

While Facebook and LinkedIn target different audiences and therefore offer different kinds of data and services, the data accessible through the APIs is actually quite similar in terms of the concepts and attributes. The main difference lies in "additional" attributes that can be associated with friends in the case of Facebook and business contacts for LinkedIn. Application developers are highly dependent on what the APIs offer in terms of functionality and data access. In the case of Facebook, a developer has read access to user profiles as well as to their immediate network of friends. However, the OpenSocial API offered by LinkedIn can only be used upon approval.

The data models for many Web 2.0 sites tend to be quite similar and also relatively simple and modular. Consequently, the data mappings between synchronisation endpoints tend to be much simpler than typical data integration scenarios tackled by the information systems community. This is something that we have been able to exploit in the data mapping tools provided for developers of synchronisation modules. We now describe how a data mapping between Outlook contacts and Facebook could be specified using our mapping tool.

Figure 2 shows parts of the two data models represented as ER models. The Outlook contact model shown on the left has a `Contact` entity which defines attributes such as `Full Name`, `Birth Date`, `Home Address` and `Business Address`. The central entity of the Facebook model shown on the right is `User` which defines attributes such as `Birthday`, `Name` and `HomeTown`. A user can be a member of one or more `Groups` and has a work history, which is a sequence of `WorkInfo` entities.

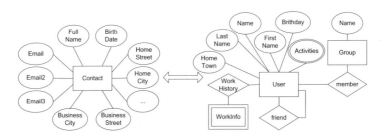

Fig. 2. Simplified Outlook and Facebook data models

The actual synchronisation of data is performed over data collections. To create a mapping between the data models, the collections of data to be synchronised must first be specified. Assume that we want to synchronise the collection of all contacts in Outlook with the collection of friends in Facebook. Using our mapping tool, we could define a mapping between the corresponding entity types

---

[8] http://code.google.com/apis/gadgets/

contact and user as shown in the screenshot in Fig. 3. The two data models are represented as tree structures, with the Outlook data model in the left window and that of Facebook on the right. Types and attributes can be mapped by simply dragging and dropping the tree nodes from one model over those of the other model. When creating such a mapping, the system checks whether the types of the two nodes are compatible. Basic transformations are done automatically by the system. For more complex transformations, the user has to select one of the provided data transformers.

The highlighted mapping defines a mapping between the attribute `fullName` of the local `contact` type to the attribute `name` of the Facebook `User` type. In the lower part of the screenshot, the navigation paths of that specific mapping are displayed. Note that the Facebook `User` type defines both attributes `firstName` and `lastName` as well as a composed `name` attribute. Since the local `contact` type only offers a composed `fullName` attribute, we chose the latter for the mapping. Attributes and entities which are mapped are tagged with a small lock label.

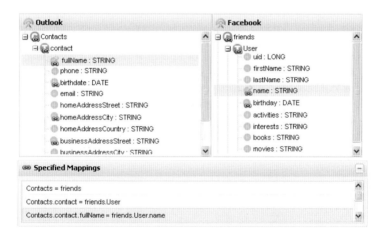

Fig. 3. Mapping tool screenshot

Structural mappings, such as mapping the first entity of Facebook's `Work-History` list to the business information in Outlook represented as a set of attributes defining an organisation's address, can also be performed.

5   Implementation

Figure 4 shows a simplified UML diagram of the system architecture. The synchronisation engine `SyncEngine` is the heart of the system and handles the synchronisation process. Parts of its functionality are extracted into configuration and synchronisation strategy objects allowing the engine to be adapted to the needs of an application and simplifying the implementation of extensions. A synchronisation strategy implements the interface `SyncStrategy` and provides the

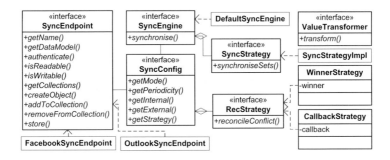

Fig. 4. Synchronisation framework UML diagram

actual synchronisation algorithm. We provide a naive synchronisation algorithm `SyncStrategyImpl` as default strategy, but more sophisticated strategies can be added by developing a class that implements the `SyncStrategy` interface.

`SyncEngine` uses `SyncConfig` to configure the synchronisation process. A `SyncConfig` object is associated with two `SyncEndpoint` objects `internal` and a reconciliation strategy object `RecStrategy` as well as configuration parameters such as `periodicity` and `mode`. Classes that implement the `RecStrategy` interface offer a strategy to handle conflicts that arise when synchronising. Currently, we provide two strategies. The `WinnerStrategy` can be configured to give priority to one of the endpoints, while the `CallbackStrategy` propagates the conflict resolution task to the user. Other conflict resolution strategies can be added by developing classes implementing the `RecStrategy` interface.

Synchronisation endpoints can be added by implementing the `SyncEndpoint` interface. An endpoint provides access to the data source's objects via one or more collections that group these objects. In our example, the Facebook endpoint offers access to a single collection `friends`, but a developer could use two collections, one for close friends and one for acquaintances.

Objects can be created as well as added to and removed from the actual collection to be synchronised. In order to synchronise data in an efficient way, synchronisation endpoints represent their data in a intermediate and uniform format based on object graphs following the JavaBeans conventions. For the data mappings specified through the user interface, we use OGNL (Object-Graph Navigation Language)[9]. Mapping expressions specify the mapping between the two models, both represented as object graphs with the collections to be synchronised as the root objects of the graphs. Mappings can either be created and manipulated using the visual mapping tool or through XML files. A simplified excerpt of a mapping for our example is shown below.

```
<cMapping iCollName="contacts" expr="friends"/>
<tMapping iName="contact" expr="User" key="fullName">
   <attrMapping iName="fullName" expr="name"/>
   <attrMapping iName="birthdate" expr="birthday"
      valueTransformerClass="..Birthdate2FBBirthday"/>
</tMapping>
```

---

[9] http://www.opensymphony.com/ognl/

cMapping maps the internal collection `contacts` to an external concept matching the OGNL expression `friends`. tMapping maps the internal type `contact` to a concept matching the expression `User` and defines its key to be `fullName`. There are two attribute mappings which map the local attributes `fullName` and `birthdate` to the expressions `name` and `birthday`, respectively. Both external expressions are evaluated by OGNL in the context of the the external `User` type. OGNL performs automatic type conversions between numeric types, booleans and strings. To support special conversions, we introduced value transformers that offer methods to convert values from one format to another. For example, the class `Birthdate2FBBirthday` transforms a birthdate value in the Outlook date format to the Facebook birthday format. Note that it is also possible to use transformer classes to convert from flat to nested structures and vice versa. Our framework provides several standard value transformer classes and additional transformers can easily be implemented.

6   Conclusion

We have presented a framework for synchronising personal data with Web 2.0 data sources that allows users to control where, when and how data is published to, and imported from, the Web. Tools are provided that enable users to configure the synchronisation process as well as to define new synchronisation processes along with the required data mappings. Some of the key distinguishing features of our approach stem from the fact that, instead of trying to develop new standards or architectures, our aim is to work with existing APIs and publishing services as well as taking into account the nature of data replication, data volatility and data inconsistencies evident in current Web 2.0 settings.

References

1. Wayne, B.: YouTube Is Doomed. The Business Insider (April 2009), http://www.businessinsider.com/is-youtube-doomed-2009-4
2. Hsu, F., Chen, H.: Secure File System Services for Web 2.0 Applications. In: Proc. ACM Cloud Computing Security Workshop, CCSW 2009 (2009)
3. Geambasu, R., Cheung, C., Moshchuk, A., Gribble, S.D., Levy, H.M.: Organizing and Sharing Distributed Personal Web-Service Data. In: Proc. Intl. World Wide Web Conf., WWW 2008 (2008)
4. Matsuo, Y., Hamasaki, M., Nakamura, Y., Nishimura, T., Hasida, K., Takeda, H., Mori, J., Bollegala, D., Ishizuka, M.: Spinning Multiple Social Networks for Semantic Web. In: Proc. Natl. Conf. on Artificial Intelligence, AAAI 2006 (2006)
5. Guy, I., Jacovi, M., Shahar, E., Meshulam, N., Soroka, V., Farrell, S.: Harvesting with SONAR: the Value of Aggregating Social Network Information. In: Proc. ACM Intl. Conf. on Human-Computer Interaction (CHI 2008), pp. 1017–1026 (2008)
6. Matsuo, Y., Mori, J., Hamasaki, M., Nishimura, T., Takeda, H., Hasida, K., Ishizuka, M.: POLYPHONET: an Advanced Social Network Extraction System from the Web. Web Semant. 5(4), 262–278 (2007)

An Artifact-Centric Approach to Generating Web-Based Business Process Driven User Interfaces

Sira Yongchareon, Chengfei Liu, Xiaohui Zhao, and Jiajie Xu

Faculty of Information and Communication Technologies
Swinburne University of Technology
Melbourne, Victoria, Australia
{syongchareon,cliu,xzhao,jxu}@swin.edu.au

Abstract. Workflow-based web applications are important in workflow management systems as they interact with users of business processes. With the Model-driven approach, user interfaces (UIs) of these applications can be partially generated based on functional and data requirements obtained from underlying process models. In traditional activity-centric modelling approaches, data models and relationships between tasks and data are not clearly defined in the process model; thus, it is left to UI modellers to manually identify data requirement in generated UIs. We observed that artifact-centric approaches can be applied to address the above problems. However, it brings in challenges to automatically generate UIs due to the declarative manner of describing the processes. In this paper, we propose a model-based automatic UI generation framework with related algorithms for deriving UIs from process models.

1 Introduction

Over the past several years, the use of workflow management system in organizations has been considered as a promising automation approach to enable them to design, execute, monitor and control their business processes. It is conceived that current workflow technologies support organization's own developed web applications for users to efficiently interact with the processes they involve. The interaction of users and these workflow-based applications are through user interfaces (UIs) that are designed and developed when workflows are modelled. This can bring an issue of coupled alignment between business processes and UIs, i.e., changes of the processes that impact on UIs are to be managed in an ad-hoc manner [9]. Several works [9-12] have been proposed the adoption of *Model Driven Approach* (MDA) that specify the association between models that support the propagating changes and control the alignment of business processes and UIs of underlying applications.

Traditionally, business processes are modelled by identifying units of work to be done and how these works can be carried out to achieve a particular business goal. This approach, so called activity-centric business process modelling, has been recognized as a traditional way of process modelling and it has also been used in many MDA approaches, e.g., in OOWS-navigational model [12], for the semi-automatic generation of UIs by deriving task-based models from business process models. These

L. Chen, P. Triantafillou, and T. Suel (Eds.): WISE 2010, LNCS 6488, pp. 419–427, 2010.
© Springer-Verlag Berlin Heidelberg 2010

approaches require UI modellers to know information that is needed to be inputted from users and then to manually assign it to corresponding UIs. Thus, the changes of the data requirements of any task are still not able to reflect to the UIs if the process changes, so a better approach is required.

We observed the traditional approaches of business process modelling and found that they have some drawbacks and are limited to support only partially automatic UI derivation. Especially, the data model and task model are defined independently and their relation may not be coherently captured in the activity-centric model. In addition, as the limitation of the model, the current derivation approaches can provide only one-to-one transformation, i.e., one task to one page of UI. A mechanism to combine multiple tasks to fit within a single page without losing control or breaking the integrity of the process, e.g. transaction, is not supported. To this end, we consider a new paradigm of process modelling called *artifact-centric approach* [1-4].

By using the artifact-centric approach, UIs can be automatically generated from business processes by deriving both *behavioural aspect* (navigational control flow relations between UIs) and *informational aspect* (related/required data in each UI) from the underlying processes. In this paper, we propose a web-based business process driven user interface framework. It comprises two models, *Artifact-Centric Business Process (ACP) model* and *User Interface Flow (UIF) model*, and a mechanism to derive UIF model from ACP model. The UIF model describes the constitution of UIs and their navigational control flows which can be derived from the underlying ACP model. In summary, our work makes the following contributions to the research in business process modelling and web engineering areas:

• Analyze the relations between artifact-centric web-based processes and UIs
• Facilitate the UIs derivation for processes with UIF models and algorithms

The remainder of this paper is organized as follows. Section 2 provides the formal model for artifact-centric business processes. Section 3 presents the approach for UIF model generation. Section 4 reviews the related works. Finally, the concluding remarks are given in Section 5 together with our future work.

2   Artifact-Centric Business Process Models

The concept of modelling artifact-centric processes has been established under the framework proposed in [4] with the formal model [5]. Our artifact-centric business process model (ACP model) extends their work. The model consists of three core constructs: *artifacts*, *services*, and *business rules*. An *artifact* is a business entity or an object involved in business process(es). A *service* is a task that requires input data from artifact(s) or users, and produces an output by performing an update on artifact(s). A *business rule* is used to associate service(s) with artifact(s) alike in a *Condition-Action-Role* style. To explain the model, we use a retailer business scenario consisting of two business processes: product ordering and shipping. The ordering process starts when a customer places an order to the retailer for a particular product and ends when the customer pays the invoice. The shipping process starts when the retailer creates a shipment and ends when the item arrives to the customer.

Figure 1 illustrates artifact lifecycle diagram of each artifact used in our business processes. We denote $l.v[m.s]$ for the transition to be triggered by a performer with role l invokes service v if artifact m is in state s.

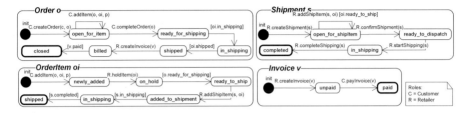

Fig. 1. Lifecycle diagram of each artifact used within our business scenario

2.1 Syntax and Components of ACP Model

Definition 1: (Artifact class). An *artifact class* abstracts a group of artifacts with their data attributes and states. An artifact class C is a tuple (A, S), where,

- $A = \{a_1, a_2, ..., a_x\}$, $a_i \in A(1 \le i \le x)$ is an attribute of a scalar-typed value (string and real number) or undefined value
- $S = \{s_1, s_2, ..., s_y\} \cup \{s^{init}\}$ is a finite set of states, where s^{init} denotes *initial state*

Definition 2: (Artifact schema). An *artifact schema* $\mathbb{Z}$ contains a set of artifact classes, i.e., $\mathbb{Z} = \{C_1, C_2, ..., C_n\}$ where $C_i \in \mathbb{Z}(1 \le i \le n)$ is an *artifact class*.

From our business scenario, we define a primary set of artifact classes as below.

- $Order = (\{orderID, customerID, grandTotal\}, \{open\_for\_item,$
  $ready\_for\_shipping, in\_shipping, shipped, billed, closed\})$
- $Shipment = (\{shipID, customerID, shipDate, shipCost\}, \{open\_for\_shipitem,$
  $ready\_to\_dispatch, in\_shipping, completed\})$
- $OrderItem = (\{orderID, productID, shipID, qty, price\}, \{newly\_added, on\_hold,$
  $ready\_to\_ship, added\_to\_shipment, in\_shipping, shipped\})$
- $Invoice = (\{invoiceID, ordereID, invoiceDate, amountPaid\}, \{unpaid, paid\})$

We also define two predicates over schema $\mathbb{Z}$ (1) *defined*(C, a) if the value of attribute $a \in C.A$ in artifact of class C is defined and (2) *instate*(C, s) if the current state of artifact of class C is s, where $s \in C.S$

Definition 3: (Service). A service or task provides a particular function. A service may involve with several artifacts of classes $C_1, C_2, ..., C_y$, where $C_i \in \mathbb{Z}(1 \le i \le y)$.

Definition 4: (Business Rule). A business rule regulates which service can be performed by whom, under what condition, and how artifacts' states change accordingly. Rule r can be defined as tuple (c, v, Γ), where,

- c is a conditional statement defined by a quantifier-free first-order logic formula (only AND connective ($\wedge$) and variables are allowed).
- $v \in V$ is a service to be invoked, and v can be *nil* if no service is required
- Γ is a set of *transition functions* where $\Gamma = \{\tau_1, \tau_2, ..., \tau_y\}$, each $\tau_i \in \Gamma (1 \leq i \leq y)$ denotes a function *chstate(C, s)* to assign the state $s \in C.S$ to the current state of an artifact of class C

Table 1 lists some business rules in our business scenario.

Table 1. Examples of business rules

r1 : *Customer c* requests to make an *Order o*	
Condition	*instate(o, init)* $\wedge$ ¬*defined(o.orderID)* $\wedge$ ¬*defined(o.customerID)* $\wedge$ *defined(c.customerID)*
Action	*createOrder(c, o)*, *chstate(o, open_for_item)*
r2: Add *OrderItem oi* of *Product p* with a quantity *qty* to *Order o*	
Condition	*instate(o, open_for_item)* $\wedge$ *instate(oi, init)* $\wedge$ *defined(p.productID)* $\wedge$ *defined(oi.productID)* $\wedge$ ¬*defined(oi.orderID)* $\wedge$ *defined(oi.qty)* $\wedge$ ¬*defined(oi.price)*
Action	*addItem(o, oi, p)*, *chstate(o, open_for_item)*, *chstate(oi, newly_added)*
r3: Complete *Order o*	
Condition	*instate(o, open_for_item)* $\wedge$ *o.grandTotal > 0*
Action	*completeOrder(o)*, *chstate(o, ready_for_shipping)*
r4: Pay *Invoice v* for *Order o*	
Condition	*instate(o, billed)* $\wedge$ *instate(v, unpaid)* $\wedge$ *defined(v.orderID)* $\wedge$ *o.orderID = v.orderID* $\wedge$ *o.grandTotal = v.amountPaid*
Action	*payInvoice(v, o)*, *chstate(v, paid)*

2.2 Artifact System for Artifact-Centric Processes

In this section we define *artifact system* as the operational model for capturing the *behavior* of artifact-centric processes. The artifact system is modeled by adopting the concept of state machine for describing behaviors of objects in a system [6].

Definition 5: (Artifact Machine). An *artifact machine* defines state transitions of an artifact class. An artifact machine m for an artifact of class C can be defined as tuple (S, s^{init}, T), where S is a set of states of C, $s^{init} \in S$ is the *initial state*, and $T \subseteq S \times V \times G \times S$ is a 4-ary relation of a set of states S, services V, and guards G. A transition $t = (s_s, v, g, s_t) \in T$ means that the state of the artifact will change from s_s to s_t if service v is invoked and condition g holds.

Definition 6: (Artifact System). An *artifact system* Λ is a tuple $(\mathbb{Z}, V, R, M)$ where $\mathbb{Z}$ is an artifact schema, V and R are sets of services and business rules over $\mathbb{Z}$, respectively, and M is a set of artifact machines, each for a class in $\mathbb{Z}$.

For an artifact class $C \in \mathbb{Z}$, given service set V and rule set R, its artifact machine $m \in M$ can be generated by deriving from corresponding business rules that are used to induce state transitions of C.

3 User Interface Flow Model Generation

In this section, we formally describe the constructs in *User Interface Flow Model* (UIF model), and propose an approach to derive UIF models from underlying artifact-centric process models. The model comprises (1) a set of web pages and (2) relations between these pages. A page may contain a single or multiple input forms. Each form contains input fields that user must fill in data to make a form completed. There are two abstract aspects of the UIF models: *behavioural aspect* (navigational control flow relations between UIs) and *informational aspect* (related/required data for each UI).

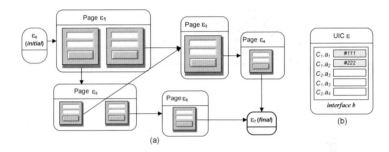

Fig. 2. (a) UIF Model, (b) UIC with an interface and its required attributes of artifacts

Figure 2 shows the components and structure of the *UIF model*. The round-rectangle depicts a *User Interface Container (UIC)*, which represents a single web page of UIs. An *Interface* represents a form comprising a set of required attributes of corresponding artifact that is used in the form. A single UIC may contain either empty interface (for the *final* or *initial* UIC), or a single or multiple interfaces (for *normal* UIC). The *Navigational Control Flow (NCF)* is used to indicate that once the interface with all required data has been submitted, then the action, e.g., service, corresponding to such interface is performed and the following UIC then becomes active. The UIF starts at the *initial* UIC and terminates when it reaches the *final* UIC.

3.1 Syntax of UIF Model

Definition 7: (Interface). An *interface* represents a form of web page. It contains a required set of attributes of artifact, as well as a role of users and a corresponding service that will be invoked if users complete the form. Let b denote an *interface* and it is defined as tuple (O, ∂, Δ, v), where

- $O \subseteq \mathbb{Z}$, is a finite set of artifact classes used in the interface
- $\partial \subseteq \bigcup_i o_i.A$ is a required attribute set, which can be inputted/edited by users

– Δ defines a set of current states of each artifact of class in O when they are in the interface b, i.e., $\forall s_j \in \Delta, \exists o_i \in O, \exists s_j \in o_i.S$, such that $instate(o_i, s_j)$. We use $s_{o_i}^j$ to denote the s_j state of artifact of class o_i.

– $v \in V$ is a corresponding service which can be performed after attributes in ∂ are all completed by users

Note that an interface may contain nothing, called *empty interface*, where O, ∂, $\Delta = \varnothing$ and $v = nil$. It is only used in the *initial* and the *final* UICs.

Definition 8: (User interface Flow Model or UIF Model). The *UIF model*, denoted as θ, represents UI components and their relations, and it is tuple (Σ, Ω, B, F) where,

– $\Sigma = \{\varepsilon_1, \varepsilon_2, ..., \varepsilon_x\}$, $\varepsilon_i \in \Sigma$ $(1 \leq i \leq x)$ is a UIC
– $B = \{b_1, b_2, ..., b_z\} \cup \{b^{nil}\}$, $b_i \in B$ $(1 \leq i \leq z)$ is an *interface*, where b^{nil} denotes an *empty interface*
– $\Omega \subseteq \Sigma \times B$ defines the relation between UICs and interfaces
– $F \subseteq \Omega \times \Sigma$ is a finite set of *Navigational Control Flow* (NCF) relations. A flow f $= ((\varepsilon_s, b_x), \varepsilon_t) \in F$ corresponds to a NCF relation between the source UIC ε_s and the target UIC ε_t, such that when ε_s is active and every attribute in ∂ of interface b_x is completed then ε_t is enabled (activated) and ε_s is disabled (deactivated).

According to two aspects of the UIF model, the *behavioral aspect* is represented by its UIs components and their NCF relations, while the *informational aspect* is represented by internal information of artifacts required for each interface. Once we defined ACP and UIF models, then the next step is to derive UIF models from underlying ACP models. Two main steps are required: (1) generating the interfaces and their NCF relations for constructing the behavior of the model and (2) mapping the required artifacts and their attributes for constructing the information for each interface. These steps are described in Section 3.2 and 3.3, respectively.

3.2 Constructing the Behavior of UIF Models

Every machine in the system is required to be composed into a single machine as to generate the entire behavior of the system, i.e., *behavioural aspect* of UIF models according to the control logic of underlying business processes. In this section we define *artifact machine system* for the completed composition of all machines in the artifact system by adapting the compositional technique presented in [6].

Definition 9: (Artifact system machine). Let $m_i \oplus m_j$ denote the result machine generated by combining artifact machine m_i and machine m_j. For an artifact system with machine set M for its artifacts, the combined artifact machine, i.e., $\oplus_i m_i$, is called *artifact system machine*. Combined machine $m_c = m_i \oplus m_j = (S_c, s_c^{init}, T_c)$, where set of states $S_c \subseteq m_i.S \times m_j.S$, initial state $s_c^{init} = (m_i.s^{init}, m_j.s^{init})$, transition relation $T_c \subseteq S_c \times V \times G_c \times S_c$, and G_c is guards, such that G_c contains no references to states in m_i and m_j.

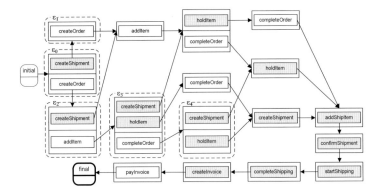

Fig. 3. The behavioral aspect of UIF model

After completing the composition for *artifact system machine*, we need to process a mapping from such machine to the UIF model. The mapping contains two steps: (1) states to UICs mapping and transitions to interfaces mapping. (2) NCF relation generation. There can be multiple interfaces in a single UIC if such state has multiple exit transitions. The result of mapping shows the *behavioral aspect* of the model. Figure 3 shows the result of applying this mapping to our business processes.

3.3 Mapping Information of Artifacts to Interfaces

Once we completed *behavioral aspect* mapping of UIF model, then we need to generate its *informational aspect* by assigning artifacts onto interfaces. In this step, we need to find all the corresponding artifacts required for each interface. We can classify the information needs for a particular interface into: (1) a set of artifacts to be read or updated, and (2) a set of required values to be assigned to attributes of such artifacts. We can simply find both sets by extracting them from every condition of business rules that corresponds to a service to be invoked of such interface. Note that the required attribute set and artifacts for each interface are minimal and sufficient. They can be extended if users would like to incorporate other related artifacts by adding them into the interface; however, these additional artifacts need to be validated as to ensure that the behaviour consistency between UIF and ACP models is preserved. Here, we can say that our proposed information mapping explicitly overcomes the drawbacks of current approaches in which activities, data and their relation are treated separately.

4   Related Work and Discussion

In the context of business process modelling, Küster Ryndina, & Gall [7] established a notion of business process model compliance with an object life cycle. They also proposed a technique for generating a compliant business process model from a set of given reference object life cycles in forms of state machines. Redding et al. [8] conducted a similar work, where they proposed the transformation from objects behavior model to process model by using the heuristic net for capturing the casual relations in the object model. Compared with our work, their transformations use an object behavior model as input, while our work uses the artifact process models. In addition, these approaches are different from ours in such way that they do not consider state dependency between artifacts but we do.

In the area of web engineering in user interfaces, both Sousa et al. [9] and Sukaviriya et al. [10] presented a model-driven approach to link and manage software requirements with business processes and UI models. With their approaches, a process model is mapped to a UI model, thus change propagation can be managed more efficiently. Guerrero et al. [11] and Torres et al. [12] applied the similar concept for developing UIs corresponding to workflow models. All these approaches considered traditional activity-centric process models and proposed approaches to define the internal components and functionalities of the UIs at different levels, e.g., task-base model, abstract UI, and concrete UI. In comparison with these approaches, we considered the artifact-centric model to capture data requirements and their relation with tasks, and propose an automatic generation framework to provide a highly-cohesive bridge between the operational back-end system of business processes and the front-end UI system. The generated UIs can be further customized by UI modelers without a concern of the integrity of business logic. Moreover, changes of data requirement that specified in the model can be reflected on UIs.

5   Conclusion and Future Work

This paper has proposed a model-based automatic UI generation framework for web-based business processes based on artifact-centric process modeling approach. In the framework, the ACP model and the UIF model are defined with a mechanism to derive the UIF model from the ACP model. The UIF models reflect the logic of business processes and intuitively represent what information is required during the processes. In the future, we plan to improve the model for supporting wider user interface requirements e.g., optional data elements, role-based configuration.

References

1. Nigam, A., Caswell, N.S.: Business artifacts: An approach to operational specification. IBM Syst. J. 42(3), 428–445 (2003)
2. Liu, R., Bhattacharya, K., Wu, F.: Modeling Business Contexture and Behavior Using Business Artifacts. In: Krogstie, J., Opdahl, A.L., Sindre, G. (eds.) CAiSE 2007 and WES 2007. LNCS, vol. 4495, pp. 324–339. Springer, Heidelberg (2007)
3. Bhattacharya, K., et al.: Artifact-centered operational modeling: Lessons from customer engagements. IBM Systems Journal, 703–721 (2007)
4. Hull, R.: Artifact-Centric Business Process Models: Brief Survey of Research Results and Challenges. In: Meersman, R., Tari, Z. (eds.) OTM 2008, Part I. LNCS, vol. 5331, pp. 1152–1163. Springer, Heidelberg (2008)
5. Bhattacharya, K., et al.: Towards Formal Analysis of Artifact-Centric Business Process Models. In: Alonso, G., Dadam, P., Rosemann, M. (eds.) BPM 2007. LNCS, vol. 4714, pp. 288–304. Springer, Heidelberg (2007)

6. Lind-Nielsen, J., et al.: Verification of Large State/Event Systems Using Compositionality and Dependency Analysis. Formal Methods in System Design 18(1), 5–23 (2001)
7. Küster, J., Ryndina, K., Gall, H.: Generation of Business Process Models for Object Life Cycle Compliance. In: Alonso, G., Dadam, P., Rosemann, M. (eds.) BPM 2007. LNCS, vol. 4714, pp. 165–181. Springer, Heidelberg (2007)
8. Redding, G., et al.: Generating business process models from object behavior models. Information Systems Management 25(4), 319–331 (2008)
9. Sousa, K., et al.: User interface derivation from business processes: A model-driven approach for organizational engineering. In: ACM SAC 2008, pp. 553–560 (2008)
10. Sukaviriya, N., et al.: Model-driven approach for managing human interface design life cycle. In: Engels, G., Opdyke, B., Schmidt, D.C., Weil, F. (eds.) MODELS 2007. LNCS, vol. 4735, pp. 226–240. Springer, Heidelberg (2007)
11. Guerrero, J., et al.: Modeling User Interfaces to Workflow Information Systems. In: ICAS 2008 (2008)
12. Torres, V., Pelechano, V.: Building Business Process Driven Web Applications. In: Dustdar, S., Fiadeiro, J.L., Sheth, A.P. (eds.) BPM 2006. LNCS, vol. 4102, pp. 322–337. Springer, Heidelberg (2006)

A Pattern-Based Temporal XML Query Language *

Xuhui Li[1,2], Mengchi Liu[1], Arif Ghafoor[2] and Philip C-Y. Sheu[3]

[1] State Key Lab of Software Engineering, Wuhan Univ.,
Wuhan, China
lixuhui@whu.edu.cn, mengchi@sklse.org
[2] School of Electronic and Computer Engineering, Purdue Univ.
West Lafayette, USA
{li472, ghafoor}@purdue.edu
[3] Department of Electr. Engi. and Comp. Sci., California Univ., Irvine
Irvine, USA
psheu@uci.edu

Abstract. The need to store large amount of temporal data in XML documents makes temporal XML document query an interesting and practical challenge. Researchers have proposed various temporal XML query languages with specific data models, however, these languages just extend XPath or XQuery with simple temporal operations, thus lacking both declarativeness and consistency in terms of usability and reasonability. In this paper we introduce TempXTQ, a pattern-based temporal XML query language, with a Set-based Temporal XML (STX) data model which uses hierarchically-grouped data sets to uniformly represent both temporal information and common XML data. TempXTQ deploys various patterns equipped with certain pattern restructuring mechanism to present requests on extracting and constructing temporal XML data. These patterns are hierarchically composed with certain operators like logic connectives, which enables TempXTQ to specify temporal queries consistently with the STX model and declaratively present various kinds of data manipulation requests. We further demonstrate that TempXTQ can present complicated temporal XML queries clearly and efficiently.

1 Introduction

With the wide spread of XML in various areas, more and more temporal information is presented with XML documents. How to efficiently process the temporal XML data has become an interesting and practical problem. For example, in experiment or multimedia data processing, it is often required to fetch snapshots at certain time instants; in historical information management, temporal information on various subjects is often compared and analyzed.

* This research is partially supported by the Fundamental Research Funds for the Central Universities of China under contract No.6082010, the Wuhan ChenGuang Youth Sci.&Tech. Project under contract No.200850731369, the 111 Project under contract No.B07037, and the NSF of China under contract No.60688201.

L. Chen, P. Triantafillou, and T. Suel (Eds.): WISE 2010, LNCS 6488, pp. 428–441, 2010.

Temporal data model and temporal query language are often regarded as fundamental aspects of temporal data processing. The former specifies a reasonable data structure associating temporal attributes to common data elements, and imposes integrity constraints on temporal data. The latter specifies how to present requests on extracting, comparing, composing and constructing temporal information based on data model. As the interface between users and database, temporal query language plays a central role in temporal data utilization.

Generally, an ideal query language is required to be both declarative and consistent. In terms of usability, it should be declarative so that users can focus on presenting their query purpose without considering the procedures to fulfill it. In terms of reasonability, it should also be consistent with the underlying data model, so that query programs can be easily resolved to data manipulations supported by data model and efficiently processed in database. The consistency of a query language lies in two aspects: a) Syntactically, query requests involving various kinds of data should be presented in a uniform style; b) Semantically, data manipulations represented by query program should correspond to the canonical data operations supported by data model. A typical example of such language is SQL, which originates from algebraic operations on the relational data model. However, for complicated data structure such as temporal XML documents, these two features cannot be easily achieved.

Several studies have proposed certain temporal models and some of them also proposed temporal XML query languages. To meet the standards on XML query, some literatures prefer to specify the temporal XML queries in standard XQuery based on regulating temporal information as common XML tags or attributes, the others propose new query languages by extending XPath or XQuery with certain temporal operations. However, existing temporal XML query languages are inadequate in either declarativeness or consistency, which would be elaborated in the next section. This lack motivates us to find an approach to declaratively and consistently present temporal XML queries.

In this paper, we will introduce a pattern-based temporal query language named TempXTQ. TempXTQ is a temporal extension of our previous study [13] which proposed XTQ, a declarative XML query language, based on hierarchical patterns composed with conjunctive and disjunctive operators. The major contributions of TempXTQ are: a) TempXTQ deploys a hierarchically organized data model named Set-based Temporal XML (STX) which consistently represents both temporal data and common XML data as sets. b) TempXTQ inherits hierarchical patterns from XTQ and extends them with temporal ingredients to declaratively specify temporal XML data query. c) TempXTQ seamlessly integrates the features of the STX model, such as temporal domains and temporal integrity constraints, with common XML data manipulation, which makes the language consistent with the data model.

The remainder of the paper is organized as follows. In Section 2 the related studies on temporal XML documents are briefly discussed. Section 3 describes the Set-based Temporal Context model. Section 4 introduces the major features of TempXTQ with certain examples. Section 5 concludes the paper.

2   Related Work

Temporal data processing is an extensively studied topic in conventional data management field. Recently temporal XML document was also concerned and some studies have tried various approaches to modeling and querying it.

Several temporal data models[1] have been proposed to organize and manage XML documents with temporal information. Amagasa[2] extends the XPath data model by attaching valid time labels to edges between element nodes introducing consistency of temporal information as integrity constraints. Dyreson[3] introduces the transaction time in XML document and proposes a snapshot-based temporal XML data model focused on linear history of web document versioning. Marian[4] deals with the versioning problem in building XML data warehouse. It adopts an incremental model to figure out the evolution of snapshots and uses a diff algorithm to calculate the changes of periodically updated data. Buneman[5] proposes an archival model to gather versional data based on identification keys. Wang[6] proposes using common XML document order to indicate versions. Rizzoio[7] proposed a graph-based temporal XML data model using duration-like time label to specify the temporal relationships between element nodes. Some studies also consider the issue of multi-dimensional temporal domains. Wang [8] introduces the bitemporal model common in conventional temporal database into XML modeling and uses specific attributes to denote transaction time and valid time. Grandi[9] considers a four-domain scenario in normative XML documents management. Gergatsoulis[10] even uses multiple dimensions to include any context parameters besides temporal domain.

Some of the above studies also consider how to present temporal XML query. Some of them [6,8,9] prefer to use XPath or XQuery with certain functions for temporal XML query by formulating temporal information as standard XML data, and the others try to extend XPath or XQuery with considerate temporal operations and thus form new temporal XML query languages. The language TTXPath[3] extends XPath with several new axes, node tests and temporal constructors to extract and generate temporal information. Another temporal extension of XPath named TXPath[7] deploys specific temporal attributes and operations to present basic temporal queries. τXQuery, a temporal extension of XQuery supporting valid time[11], uses the *cur* and *seq* modifiers to extend common XQuery program for temporal information extraction. τXQuery program is translated to conventional XQuery program for execution. Besides the extensions of XPath or XQuery, the studies in query temporal semi-structured data can also be used to present temporal XML query[12].

As previously mentioned, these temporal query languages lack either declarativeness or consistency. These lacks make it hard to clearly and efficiently present complicated query requests which are comprehensive in semantic queries on multimedia and scientific data. The lack of declarativeness, as elaborated in [13], stems from the navigational data extraction adopted by XPath. XPath expression navigates document tree and extracts a flat set of homogeneous data. When heterogenous data in hierarchical XML documents are to be extracted, compared and combined, e.g., to present temporal relationships between different elements, certain control flows in XQuery like *for* loops and *let* assignments, nested queries and sometimes conditional statements are necessary. To iterate the elements from XPath expressions and further fetch the data deeply inside, multiple *for* and *let* statements have to be involved in complex variable binding, which segments the original tree structure and reduces program readability. Therefore, users usually have to write complicated query program in a lengthy and nesting style, and data extraction and construction are often intertwined to describe the procedures of extracting data and constructing results. The essential lack in XPath makes the temporal extension of XPath or XQuery not "declarative" enough either for presenting complicated temporal queries.

The lack of consistency stems from both data model and language design. Many temporal XML data models extend the tree model (the XDM model) by attaching temporal information, e.g., duration, to document tree nodes or edges. Thus the temporal query languages deploy XPath or XQuery expression to find certain document nodes and use some syntax sugar-alike extensions to extract the temporal information attached. Syntactically, these temporal operations are not consistent with the XPath expression and the FLWOR expression. Semantically, temporal information and common XML document nodes are represented and manipulated differently, which makes it difficult to design a consistent set of operations to interpret the temporal data manipulations. An exception is *TT*XPath which uses a linear XDM model and introduces a set of axes specific to navigate temporal data. Although *TT*XPath is a consistent language, it is often too redundant to represent temporal data and present complicated query. The lack of consistency restricts the extent of temporal query to simple temporal data extraction. Those temporal operations can hardly be used in comprehensive temporal computation where temporal data is to be compared, merged, joined and even inter-transformed with common XML data.

3   Set-Based Temporal XML Data Model

Here we introduce a new temporal XML data model named Set-based Temporal XML (STX) model. The STX model is composed of Set-based Time Domain and Set-based Temporal XML Domain.

Definition 1. A **Set-based Time Domain**(STD) is a pair $<T, O>$ where
- T is a linear order set indicating the domain of time values;
- $O = \{$*before, equal, isAdjacent, getCurrent, isDuration, getDuration, from, to, durations*$\}$, is a set of basic temporal operations on T where
    ◇ *before, equal, isAdjacent*: $T \times T \to$ Bool. *before* judges whether the first time argument is before the second one; *equal* judges whether the two arguments represent the same time; *isAdjacent* judges whether the two arguments represents the adjacent time instants if the time domain is not dense.
    ◇ *getCurrent*: $\to T$ returns the current instant, i.e., *now*, in the domain.
    ◇ *isDuration*: $2^T \to$ Bool judges whether a time set is successive to form a duration in the domain.

⋄ *getDuration*: T × T → 2^T gets a duration whose end instants are specified by the two arguments.

⋄ *from, to* : 2^T → T. *from* returns the earliest time instant in a time set, and *to* returns the latest one.

⋄ *durations*: 2^T → 2^{2^T} analyzes a time set and returns a collection of time sets which are disjunct durations (a single time instant can also be treated as a duration).

⋄ *snapshots*: 2^T → 2^{2^T} analyzes a time set and returns a collections of single-element time sets; for discrete time domain the results are the time instants contained in the argument, for dense time domain the results are the endpoint time instants of the durations.

As Definition 1 shows, we allow any linear order set to be the time domain. A time domain needn't be dense (a dense order is the one in which for any two different element t_1 and t_2 there is an element t_3 between them). For example, a concrete time domain can be Greenwich mean time, or be a document order which implicitly indicates the temporal information. It is only required that the domain be associated with a set of basic temporal operations. That is, a concrete STD should implement an interface containing basic operations based on which various temporal functions can be defined. For example, the function *isIn(t,s)* judging whether a time instant t is in the durations represented by a time set s can be defined as ∃ d ∈ *durations(s)*. *(before(from(d), t)* ∨ *equal(from(d),t)*) ∧ *(before(t,to(d))* ∨ *equal(t, to(d))*.

Definition 2. A **Set-based Temporal XML** (STX) Domain is a tuple DS = <×TD_i, TS> where

• TD_i = <T_i, O_i > (i=1..n) is a list of STDs.
• TS is the set of document elements inductively defined as:

⋄ For a common XML content value (such as CDATA string) v and a time set tuple $(×ts_i)$ ∈ $(×2^{T_i})$, $(v, (×ts_i))$ ∈ TS. $(v, (×ts_i))$ is denoted as $v[(×ts_i)]$.

⋄ For an ordered multiset s of elements in TS, an XML tag t (attributes are treated as special tags and denoted with a prefix "@"), and a time set tuple $(×ts_i)$ ∈ $(×2^{T_i})$, $(t, (×ts_i), s)$ ∈ TS, if for each element e of s, the time set tuple $(×ts'_i)$ of e satisfies the *temporal integrity constraint* that ∀i∈(1..n). ts'_i ⊆ ts_i. $(t, (×ts_i), s)$ is named temporal context and denoted as $t[(×ts_i)]$⇒s.

STX domain is a hierarchically set-based model for temporal XML documents. As a tree model essentially, it adopts the set-based form instead of the tree-like one to denote element content, which treats both temporal data and common XML data as sets. It deploys multi-dimensional time domains ×TD_i, and the domains are distinguished by domain names. The associated time set of a tag or value indicates its available time in its parent context. A temporal document element can be presented as "$t[(tn_1)ts_1, ..., (tn_n)ts_n]$⇒s" and a document fragment is represented by a set of such elements. For convenience here we only consider the one-dimensional case and the domain name is often omitted by default. The multi-dimensional case is just a trivial extension.

Fig. 1 illustrates a scenario of a university architecture developing history as an example throughout this paper. As the figure shows, the university has a hierarchical architecture organized as several levels of units. Department of Computer Science was founded in 1972 as a second-level unit under School of Mathematics; in 1978 it became a first-level unit under university, and was renamed as School of Computer Science in 1996; in 1998 it was renamed back and combined as an second-level unit under School of Mathematics again; in 2001 it was renamed as Computer School and became a first-level unit again.

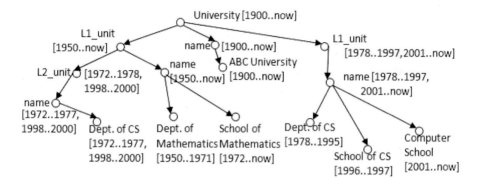

Fig. 1. A Scenario of Set-based Temporal XML Document

STX has several features distinct from existing temporal XML models. a) STX uses sets to consistently represent both common XML documents and temporal data, which lays the foundation for the consistency of the query language based on it. Furthermore, the set-based model is more appropriate to data management where data are often stored and manipulated in a set-oriented way. b) STX deploys a hierarchical structure rather than a graph structure where one node can exist in different context with different temporal information[7]. The hierarchical model is simple to present, to implement and to check integrity, and it is consistent with the hierarchical pattern we used in the TempXTQ language, which efficiently facilitates manipulating temporal information. c) STX is actually a temporal logical view of documents rather than a concrete specification of temporal XML documents. It imposes no restriction on the representation of temporal information in concrete XML documents. The temporal information can be represented as tags(or attributes) or just implicitly indicated by document order, depending on the implementation of the time domains in the model. This relative independency enables the flexibility of implementation and facilitates deployment in various situations, e.g., the distributed and heterogenous situations such as temporal information integration. d) STX allows temporal information be presented separately, that is, a time set of an STX element is not assumed to be successive, i.e., a duration, which enhances the flexibility of the representation and the storage of temporal elements. Temporal documents can be stored in a free way, e.g., snapshots and duration sketch can coexist, which

reduces the cost of maintaining consistency of temporal information and leaves
more room for database to optimize storage and query processing.

4 TempXTQ Language

4.1 Patterns in TempXTQ

TempXTQ is a pattern-based query language built on the STX model. It deploys various patterns to present request on data extraction and construction.
Temporal Logic Tree (TLT) pattern is the fundamental pattern in TempXTQ.
It extends LXT pattern in XTQ [13] with temporal patterns, inheriting the
declarativeness of specifying data extraction requests.

Definition 3. A Temporal Logic Tree (TLT) pattern is inductively defined as:
- An element pattern "*prefix tag_p temp_p* ⇒*content_p temp_p*" is a TLT
pattern, where
  ⋄ *prefix* can be "/", "//" or "@", which respectively indicates that the
element is a child, a descendant, or an attribute under current context;
  ⋄ *tag_p* is a valid tag name, a variable or a wildcard "*" indicating *anything*;
  ⋄ *temp_p* is a temporal pattern in the form of *[(tdname)time_p,..., (tdname)time_p]* where: *tdname* is a STD domain name, often omitted on default if
only one STD domain is involved; the time pattern *time_p* is a time set constant,
a variable or a wildcard; the suffix *temp_p* in the element pattern is an optional
one specific to a leaf node in STX document.
  ⋄ *content_p* is a content value, a variable or a wildcard.
- For an element pattern *rp* and a TLT pattern *bp*, the tree pattern *rp bp* is
a TLT pattern where *rp* is root pattern and *bp* is branch pattern.
- For the TLT patterns p_1, ..., p_n, the conjunctive pattern $(p_1, ..., p_n)$ is a
TLT pattern, the disjunctive pattern $(p_1 || ... || p_n)$ is a TLT pattern.
- For a TLT pattern p containing no variable, *not(p)* is a TLT pattern.

TLT is named after the composition types of element patterns. As Definition 3
indicates, TLT patterns can be composed conjunctively, disjunctively and negatively, which behaves like common logical operations. Especially, the tree-like
pattern composition is introduced to enable the pattern to directly sketch the
hierarchical STX document structure.

```
P1./l1_unit[(valid)$t]=>$u   P2./l1_unit[*]=>$u P3./l1_unit=>$u
P4.//name[1998y..now]=>$n[$t1] P5./$n[1998y]=>"Dept. of CS"
P6./l1_unit[$t]=>$u1/l2_unit=>$u2 P7.(/l2_unit[$t]=>$u,/name=>$n[$nt])
P8.(/l2_unit[$t]=>$u||/name=>$n[$nt])  P9.not(/name=>"Dept. of CS")
P10.//l1_unit[$t](/l2_unit[$t1]=>$u, /name=>$n[nt])
```

TLT pattern is to be matched with certain STX document fragments, which
would bind the values in the document to the variables in the pattern as the
extracted data. For the sample patterns listed above: P1 extracts the time sets
associated with the temporal domain "*valid*" and the content of each *l1_unit*

element in certain context; P2 only extracts the content of each *11_unit* by specifying temporal pattern as wildcard; P3 is a syntax sugar of P2 by omitting the wildcard part; P4 extracts the values with time sets of all the *name* descendants since 1998 in certain context; P5 extracts the name of an element whose content is the value *"Dept. of CS"* in 1998; P6 is a simple tree pattern, and for each *11_unit* P6 extracts its time set and content and then extract the content of its *12_unit* children; P7 and P8 both extract time set and content from *12_unit* and *name* elements, and the difference is that P7 composes the matching results from the two patterns conjunctively meanwhile P8 composes them disjunctively; P9 simply denotes a negation of a proposition that there is not a name element with content *"Dept. of CS"*; P10 is a tree pattern whose branch is a conjunctive pattern; for each *11_unit* descendent element it extracts the time set and then extract the time set and content of the subordinate *12_unit* and *name* elements.

As shown above, TempXTQ composes various values bound to variables in the pattern and forms the whole result of data extraction. To precisely figure out the composite structure of the extracted data, TempXTQ adopts **variable pattern** which was proposed in XTQ[13].

Definition 4. A variable pattern is an expression defined inductively as follows:
1. Any variable $x is an atomic variable pattern.
2. For a variable pattern p, a group (pattern) denoted by $\{p\}$ is a variable pattern.
3. For variable patterns $p_1, p_2, \ldots, p_n$, the tuple (pattern) denoted by $(p_1, p_2, \ldots, p_n)$, the option (pattern) denoted by $p_1 \| p_2 \| \ldots \| p_n$ and the enumeration (pattern) denoted by $<p_1 \| p_2 \| \ldots \| p_n >$ are variable patterns.

Variable pattern represents the hierarchically-nested groups which result from matching TLT pattern with STX documents. It inherits the hierarchical structure from both TLT pattern and STX documents and thus is consistent with the STX model. The raw variable pattern of a TLT pattern can be easily derived. For example, P1 has the variable pattern "$\{(\{\$t\},\$u)\}$", indicating that matching P1 with a document fragment yields a group in which each *11_unit* element yield a tuple containing a time set denoted by the group $\{\$t\}$ and a content denoted by u; P4 has the variable pattern $\{\{(\$n,\{\$t1\})\}\}$ indicating a group in which each *name* node yields a subordinate group of values with the associated time set; P6 has the variable pattern $\{(\{\$t\},\$u1,\{\$u2\})\}$, indicating that for each *11_unit* the time set (denoted by $\{\$t\}$) and the content (denoted by $\$u1$) is extracted and further the content of the subordinate *12_unit* elements are extracted and gathered as a group $\{\$u2\}$; P7 has the variable pattern $(\{(\{\$t\}, \$u)\}, \{\{(\$n, \{\$nt\})\}\})$, composing the two sub-patterns in a tuple; P8 has the variable pattern $<\{(\{\$t\}, \$u)\} \| \{\{(\$n, \{\$nt\})\}\}>$, composing two sub-patterns in an enumeration; P10 has the variable pattern $\{(\{\$t\},\{(\{\$t1\},\$u)\}, \{\{(\$n, \{\$nt\})\}\})\}$, composing the root pattern and the branch patterns in a tuple.

Variable pattern is distinct from other kinds of data structure description, e.g., data types, in that it can be flexibly restructured to indicate the request of transforming extracted data for data construction. The restructuring mechanism of variable patterns, a decidable rewriting system constituted by a set of restructuring rules, is specified in XTQ[13] and totally inherited by TempXTQ. A restructured pattern indicates the request of data construction, and the restructuring route can be resolved by the query parser, which enables users to declaratively present their query request. Certain policies are specified for users to simplify presenting common restructured patterns and leave the deduction of restructuring route to parser[13,14].

These restructuring rules include tuple stretching and distributing, group folding and flattening, group merging and splitting, etc., specifying how to transform hierarchical data structure meanwhile relatively maintain the logical relationships between the data. For example, the pattern "$(\{\$t\},\$u)$" can be restructured to "$(\{\$t\},\$u, \$u)$" or "$\{\$t\}$" by duplicating or eliminating $\$u$, which specifies the request of duplicating or hiding data in construction; the pattern "$(\{(\{\$t\}, \$u)\}, \{\{(\$n, \{\$nt\})\}\})$" can be restructured to "$\{(((\{\$t\}, \$u, \{\{(\$n, \{\$nt\})\}\})\}$ groupby $(\{\$t\},\$u)$" by distributing the elements in the group "$\{(\{\$t\}, \$u)\}$" to the remainder of the tuple and thus transform a tuple into a group, the groupby suffix indicating the provenance of distribution which is often be omitted if can be resolved by the parser; the pattern "$\{\{(\$n,\{\$t1\})\}\}$" can be restructured to "$\{\ ^\wedge\{(\$n,\{\$t1\})\}\}$" indicating a flattened group which eliminates the inner group layer and thus releases the elements to the outer group; the pattern "$\{(\$n,\{\$t1\})\}$" can be restructured to "$\{\{(\$n,\{\$t1\})\}$ groupby $\$n\%\}$" or "$\{(\{(\$n,\{\$t1\})\}, \$n\%\}$" indicating that the group can be folded by classifying elements into new groups by the distinct values of $\$n$; the pattern "$<\{(\{\$t\}, \$u)\} \mid\mid \{\{(\$n, \{\$nt\})\}\}>$" can be restructured to "$\{(\{\$t\}, \$u) \mid\mid \{(\$n, \{\$nt\})\}\}$" indicating that the two groups composed disjunctively can be merged to a large group, and reversely the latter can also be split to the former.

TempXTQ extends the **construct pattern** in XTQ with temporal ingredients to generate temporal document fragments from the extracted data. Here a construct pattern is in the form of "$tag[(tdname)timeset]\Rightarrow content$", as a function on variable pattern by inserting certain values and notations into it. For example, "$\{l1\_unit[\{\$t\}]\Rightarrow \{l2\_unit[\{\$t1\}]\Rightarrow\$u\}\}$" from P10 would generate a document fragment maintaining the l2_unit elements with original hierarchy. The temporal part of a construct pattern should be able to transformed into a flat time set by the underlying STX model. That means: a) The pattern not representing a time set, e.g., a single variable $\$t$, cannot be the temporal part; b) Any variable pattern, whether a time group pattern like $\{\$t\}$ or not, can be used as temporal part if it can be transformed into a time set.

TempXTQ strictly follows the temporal integrity constraint of the STX model. In a tree TLT pattern, the restrictions on the time pattern of the root would be automatically imposed on the subordinate patterns of the branches, and thus the extracted data would be adjusted to follow the integrity constraint. In data construction from a construct pattern like "$tag[timeset]\Rightarrow content$", the intersection operation would be automatically deployed to adjust the time sets of the content so as to make them be subsets of timeset. Especially, a construct pattern "$[ts]p$", e.g., "$[\{\$t\}]\$u$", is introduced specific to deploy temporal constraint to get a projection of the values of the pattern *p* on the time set *ts*. This pattern can be directly used to generate a snapshot like "*[2010y]$u*".

The pattern-based mechanisms listed above show that TempXTQ is a consistent language in that: a) The hierarchical patterns in TempXTQ accurately reflects the structural features of the STX model and uniformly manipulates the temporal data and common XML data. b) The pattern restructuring rules and the temporal operations of the STX model seamlessly work together to form a set of canonical data operations on the semantic model based on the STX model.

4.2   Synopsis of TempXTQ Lanugage

TempXTQ is a functional language built on the patterns previously described. Due to the space limit, we use some typical queries to briefly illustrate the syntax, usage and features of TempXTQ in the rest of the section. The full syntax and semantics of TempXTQ can refer to [14].

Example 1. 1. Find the snapshot of the *l1_unit* elements in university architecture of 2010. 2. Find the information of *l1_unit* elements during 2000-2010.

```
Q1. query doc("arch.xml")/university[2010y]/l1_unit=>$u
    construct notemp({university=>{l1_unit=>$u}})
Q2. query doc("arch.xml")/university[$t]/l1_unit=>$u
    where $t in 2000y..2010y
    construct {university[{$t}]=>{l1_unit=>$u}}
```

TempXTQ adopts *Query-Where-Construct* expression to present query. In a QWC expression, the *query* clause specifies data extraction through matching TLT patterns with data sources, the optional *where* clause uses condition to filter extracted data, and the *construct* clause uses construct pattern to specify data construction based on filtered data. QWC expression separately presents data extraction and data construction, enabling users to explicitly describe the purpose of query with a global sketch of data extraction and construction. TempXTQ allows function call on constants or bounded variables in all clauses.

In Q1, the query clause binds the TLT pattern to the data source specified by a common function call *doc()*, which fetches the snapshots of *l1_unit* elements whose time set is the single element set *2010y*. The construct clause uses these snapshots contained in {*$u*} to constitute the required snapshot of the university, and deploys a *notemp()* function to eliminate the temporal information and return a pure XML document fragment.

In Q2, a time variable *$t* is used to get the time values of the university in query clause. The where clause imposes a restriction on *$t* by specifying the condition "*$t in 2000y..2010y*". A condition is actually a boolean-value construct pattern to filter off the extracted data which makes the pattern *false*. Therefore, only the time values in the duration "*2000y..2010y*" would be maintained and be used in data construction. However, the condition "{*$t*}= *2000y..2010y*" is not fit for the request because it actually filters the time sets {*$t*} rather than the time values *$t*, that is, it would filter off all the time sets other than "*2000y..2010y*".

Example 2. For the *ll_unit* of Mathematics, 1. Find its sub-units which didn't appear **until** their parent changed its name to "*School of Mathematics*". 2. Find the sub-units which have never changed their names **since** their parent unit changed its name.

```
Q1. query doc("arch.xml")//ll_unit(/l2_unit[$t]=>$u,
                             /name=>"School of Mathematics"[$t1])
    where from({$t1}) <= from({$t})
    construct notemp({{name=>($u/name)}})
Q2. query doc("arch.xml")//ll_unit(/l2_unit/name=>$n[$t],
                             /name=>"School of Mathematics"[$t1])
    where from({$t1}) <= $t with isDuration({$t})
    construct notemp({name=>$n})
```

Queries on temporal relationships such as "since" and "until" are common in temporal data management. These queries often involves comparing temporal information, which can be easily presented with conjunctive TLT pattern with proper conditions, as the example shows. However, other temporal XML query languages can seldom present them easily due to the lack of declarativeness and consistency. To present this query in those languages, temporal information can only be fetched and compared using many nested loops, since the temporal operations of the languages can only extract temporal information and have to resort to FLWOR expression to carry out complicated temporal data manipulation.

This example also illustrates the subtle features of the conditions. In Q1, the time sets $\{\$t\}$ and $\{\$t1\}$ are compared and filtered. In Q2, in the first condition the time value $\$t$ is compared with $\{\$t1\}$, and those time values satisfying the condition would be maintained. These maintained time values form the group $\{\$t\}$ to be tested in the second condition whether it is a duration. In Q2 a condition connective "*with*" is deployed to combine the two conditions, signifying that the two conditions are to be processed in sequential, that is, the second condition should be tested based on the filtered result of the first one.

Further, the construct clause in Q1 uses the variable pattern $\{\{\$u\}\}$, whereas in Q2 the restructured variable pattern $\{\,\hat{}\,\{\$n\}\}$ is simplified as the pattern $\{\$n\}$ and the restructuring process would be resolved by the parser.

Example 3. Assume that the *ll_unit* elements have the "dean" or the "director" sub-elements whose content are temporal strings indicating the names of the unit chiefs and their terms. 1. Since a chief of a unit might be assigned to be the chief of another unit, for each person who has been a chief construct a temporal fragments of his chief records. 2. For the *ll_units* list the recent two deans' names, use an empty "et-al" element to indicate the case of more than two deans.

```
Q1. query doc("arch.xml")//ll_unit((/dean=>$dn[$t]||/director=>$dn[$t]),
                             /name=>$n[$t1])
    where overlap({$t},{$t1})
    construct chiefs=>{person[{{$t}}]=>(name=>$dn%,
                             records=>{unit[{$t}]=>{$n[{$t1}]}})}
Q2. query doc(arch.xml)//ll_unit((/dean=>$n1[$t1]||/dean=>$n2[$t2]),
                             /name=>$n[$t])
    where count({$n1}) = 1 || count({$n2})>=2
```

```
construct {deans=>(unit_name=>({$n[{$t}]} orderby(to({$t}))desc).(0),
            <dean_name=>$n1[{$t1}] ||
            (dean_names=>{$n2[{$t2}]} orderby (to({$t2}))desc).(0..1),
            et-al=>~)>)}
```

Due to the semi-structured feature of XML documents, handling heterogeneity becomes an important reqirement in processing XML documents in that heterognous data are often to be handled homogeneously and homogeneous data are to be handled heterogeneously.

This example shows how to deploy disjunctive TLT patterns to declaratively handle heterogeneity. In Q1, the patterns extracting the heterogeneous data, i.e., dean and director elements, are combined disjunctively, indicating that the extracted data are to be treated homogenously. On the other hand, in Q2, the disjunctive pattern extract the homogeneous data, i.e., dean elements, and yield two disjoint groups $\{\$n1\}$ and $\{\$n2\}$. The two groups would be treated heterogeneously with different restrictions, as the where clause indicates. Existing XML query languages, either the navigational ones like XQuery or the pattern-based ones like Xcerpt, have to handle heterogeneity in a procedural way because they lack a proper mechanism as the disjunctive pattern to separating or merging data flexibly. For example, XQuery often uses the typical procedural statement "if..then..else" to handle heterogeneity.

Another feature this example shows is the strong expressiveness and flexibility of pattern restructuring. In the data construction of Q1, the raw variable pattern $\{(<\{(\$dn,\{\$t\})\}||\{(\$dn,\{\$t\})\}>, \{(\$n,\{\$t1\})\})\}$ is automatically merged to $\{(\{(\$dn,\{\$t\})\}, \{(\$n,\{\$t1\})\})\}$ and then restructured to $\{\,\hat{}\{(\$dn, \{\$t\}, \{(\$n,\{\$t1\})\})\}\}$. The latter is flattened to $\{(\$dn, \{\$t\},\{(\$n,\{\$t1\})\})\}$, further folded to $\{(\$dn\%, \{(\{\$t\},\{(\$n,\{\$t1\})\})\})\}$ and duplicated to $\{(\$dn\%, \{(\{\$t\}\}, \{(\{\$t\},\{(\$n,\{\$t1\})\})\})\})\}$ which is used in the construct clause. This restructuring route is resolved by the parser according to the construct pattern. In the construct pattern, the temporal pattern $\{\{\$t\}\}$ is a folded group of time sets indicating the terms of a person in various 11_units; the tuple $(\{\$t\},\{(\$n,\{\$t1\})\})$ indicates one of his term and the names with the temporal information of the 11_unit during the term. Notice that in data construction the pattern $\{\$t1\}$ inside the scope of $\{\$t\}$ would be automatically adjusted to obey the temporal integrity constraint.

The construct clause of Q2 uses the *orderby* suffix of a group to manually order the elements in the group by the values of specified construct pattern. Thus the pattern $(\{\$n[\{\$t\}]\}$ *orderby(to(\{\$t\}))desc)* is to order the values of $\$n[\{\$t\}]$ by the descending order of the endpoint of the time set $\{\$t\}$.

Example 4. Show the history of university by listing the names of the 11_units at the beginning of each decade and the names of those 11_units newly-created.

```
declare getDecadePoints($ts) as (
    query type()/int^/item=>$i
    where $i>=0 and interval(from($ts),to($ts))<=$i*10y.
    construct {item=>single(postpone(from(ts),$i*10y))})
query doc("arch.xml")//l1_unit[$t](/name=>$n||/name=>$n1),
```

```
        funcall()/getDecadePoints(^/input=>$i, ^/output/item=>$ts)
where {$t}=$i and isnull([prevdecade($ts)]$n1) and
      notnull([$ts]$n1)
construct {history[{$ts}]=>{record[$ts]=> <arch=>{unitname=>{$n}}||
                                    newunits=>{name=>{$n1}}>}}
```

TempXTQ allows user define a function as a QWC expression, which makes the language self-fulfilling. Functions can be defined recursively, thus TempXTQ is Turing-completed. Function calls on variable patterns can be treated as special construct patterns and freely used in where and construct clauses. However, in query clause there is no bound variable available, TempXTQ deploys a special way to make function call in query clause by introducing a special data source *funcall()* which is a virtual documents containing function images in the form of "*(input, output)*" pairs. Another data source *type()* is also a virtual document containing the values of basic data types such as "*int*", "*bool*" and "*char*", which is used to provide constant values in query clause.

In the example, the function *getDecadePoints* is declared as a QWC expression which extracts the *type()* data source to get a subset of integers, i.e., {*$i*}, which can enumerate the decades over the range of argument time set *$ts*. The function returns a set of single value time sets which enumerate the beginning points of each decade. In the main QWC expression, the *getDecadePoints* function is called by combining the original data source and the *funcall()* data source conjunctively and specifying the join condition "{*$t*}=*$i*" in the where clause. The variable pattern "{(*$i*,{*$ts*})}" thus represents the result of the function call *getDecadePoints({$t})*. Notice that in the query clauses, some element patterns have a flattened prefix ^ which indicates that the group is unnecessary and would be flattened before data construction. For example, the pattern "{({*$i*},{{*$ts*}})}" of the function call is automatically flattened as "{(*$i*,{*$ts*})}" since for each (input, output) pair only one argument and one result would be involved.

The example also illustrates the consistency of the manipulation of temporal data and common XML data. In the declaration of the function, the argument time set is analyzed and a group of single-value time sets are generated. Further, in the construct clause, this group {*$ts*} is used in group distribution to generate a set of snapshots indicated by the *record* element.

As the above examples show, TempXTQ can declaratively present temporal queries with the expressive TLT patterns and the flexible restructuring mechanism. Its syntax and semantics is consistent with the underlying STX data model, which makes it fit for presenting various kinds of temporal queries.

5   Conclusion

Existing studies on temporal XML query extend XPath or XQuery with temporal features, however, they often lack a proper declarativeness and consistency in presenting complicated queries. In this paper, we introduced TempXTQ, a pattern-based temporal XML query language which can declaratively and consistently present temporal XML query based on a set-based temporal data model named STX. TempXTQ uses TLT pattern, a tree-like pattern expression with conjunctive and disjunctive operators, to present composite data extraction and handle heterogeneity. It deploys variable pattern to present the hierarchical structure of extracted data and a flexible restructuring mechanism to transform the data according to the request of data construction. These features enable TempXTQ to expressively present complicated temporal query which is quite difficult and trivial for other languages.

We have implemented core features of TempXTQ, based on the implementation of XTQ language, on a simple prototype of STX model which embeds in normal XML document a specific time element indicating temporal information. The prototype is still being improved and would be migrated to concrete XML database where performance of the queries can be analyzed and compared.

References

1. Ali, K.A., Pokorny, J.: A Comparison of XML-Based Temporal Models. In: Damiani, E., Yetongnon, K., Chbeir, R., Dipanda, A. (eds.) SITIS 2006. LNCS, vol. 4879, pp. 339–350. Springer, Heidelberg (2009)
2. Amagasa, T., Yoshikawa, M., Uemura, S.: A Data Model for Temporal XML Documents. In: Ibrahim, M., Küng, J., Revell, N. (eds.) DEXA 2000. LNCS, vol. 1873, pp. 334–344. Springer, Heidelberg (2000)
3. Dyreson, C.E.: Observing Transaction-time Semantics with TTXPath. In: Proc. of WISE 2001, pp. 193–202 (2001)
4. Marian, A., Abiteboul, S., Coben, G., Mignet, L.: Change-centric Management of Versions in an XML Warehouse. In: Proc. of VLDB 2001, pp. 81–590 (2001)
5. Buneman, P., Khanna, S., Tajima, K., Tan, W.: Archiving Scientific Data. In: Proc. of SIGMOD 2002, pp. 1–12 (2002)
6. Wang, F., Zaniolo, C.: Temporal queries and version management in XML-based document archives. Data and Knowledge Engineering 65(2), 304–324 (2008)
7. Rizzoio, F., Vaisman, A.A.: Temporal XML: Modeling, Indexing and Query Processing. The VLDB Journal 17, 1179–1212 (2008)
8. Wang, F., Zaniolo, C.: XBiT: an XML-based Bitemporal Data Model. In: Atzeni, P., Chu, W., Lu, H., Zhou, S., Ling, T.-W. (eds.) ER 2004. LNCS, vol. 3288, pp. 810–824. Springer, Heidelberg (2004)
9. Grandi, F., Mandreoli, F.: Temporal Modelling and Management of Normative Documents in XML Format. Data and Knowledge Engineering 54, 327–354 (2005)
10. Gergatsoulis, M., Stavrakas, Y.: Representing Changes in XML Documents Using Dimensions. In: Bellahsène, Z., Chaudhri, A.B., Rahm, E., Rys, M., Unland, R. (eds.) XSym 2003. LNCS, vol. 2824, pp. 208–222. Springer, Heidelberg (2003)
11. Gao, C., Snodgrass, R.: Syntax, Semantics and Query Evaluation in the τXQuery Temporal XML Query Language. Time Center Technical Report TR-72 (2003)
12. Combi, C., Lavarini, N., Oliboni, B.: Querying Semistructured Temporal Data. In: Grust, T., Höpfner, H., Illarramendi, A., Jablonski, S., Fischer, F., Müller, S., Patranjan, P.-L., Sattler, K.-U., Spiliopoulou, M., Wijsen, J. (eds.) EDBT 2006. LNCS, vol. 4254, pp. 625–636. Springer, Heidelberg (2006)
13. Li, X., Liu, M., Zhang, Y.: Towards a "More Declarative" XML Query Language. In: Bringas, P.G., Hameurlain, A., Quirchmayr, G. (eds.) DEXA 2010. LNCS, vol. 6262, pp. 375–390. Springer, Heidelberg (2010)
14. Li, X.: Syntax, Semantics and Examples of TempXTQ. Technical Report (2010), http://www.sklse.org:8080/tempxtq

A Data Mining Approach to XML Dissemination

Xiaoling Wang[1], Martin Ester[2], Weining Qian[1], and Aoying Zhou[1]

[1] Software Engineering Institute, East China Normal University, China
[2] School of Computing Science, Simon Fraser University, Burnaby, Canada
{xlwang,wnqian,ayzhou}@sei.ecnu.edu.cn
{ester}@cs.sfu.ca

Abstract. Currently user's interests are expressed by XPath or XQuery queries in XML dissemination applications. These queries require a good knowledge of the structure and contents of the documents that will arrive; As well as knowledge of XQuery which few consumers will have. In some cases, where the distinction of relevant and irrelevant documents requires the consideration of a large number of features, the query may be impossible. This paper introduces a data mining approach to XML dissemination that uses a given document collection of the user to automatically learn a classifier modelling of his/her information needs. Also discussed are the corresponding optimization methods that allow a dissemination server to execute a massive number of classifiers simultaneously. The experimental evaluation of several real XML document sets demonstrates the accuracy and efficiency of the proposed approach.

Keywords: XML Classification, pattern matching, XML dissemination, frequent structural pattern, feature vector.

1 Introduction

XML dissemination has become an important instance in the content-based publish/subscribe application[1], where the streams of XML documents are arriving at a rapid rate and the server is responsible for managing these documents and disseminating them to a pool of clients. CNN's RSS(Really Simple Syndication) news, new CS research papers (DBLP) and new movies all are examples. Consumers such as computer scientists want to be alerted to relevant news documents without being overwhelmed by "spam" documents. Current approaches require the clients to specify their interests as XPath or XQuery queries. Existing work [2,3,4] on XML dissemination considers only a subset of XPath queries. These may not allow accurate discrimination of relevant and irrelevant documents. Some cases where this distinction requires the consideration of a large number of features.

In order to provide a simple and easy way to express user's information needs we present a data mining approach to obtain user's information requirements. Our approach has more expressive capacity than single XPath/XQuery. Classification methods for XML documents have recently received some attention in

L. Chen, P. Triantafillou, and T. Suel (Eds.): WISE 2010, LNCS 6488, pp. 442–455, 2010.

the database and data mining communities [5,6], however these information dissemination scenarios create unique challenges for XML classification that have not yet been addressed in the literature.

1. Compared to the classical scenario of XML classification, the number of training examples is small, because the personal client has only a few hundred interested documents/news. This makes classifier construction much more difficult. Particulary in combination with the fact that the feature space is extremely high-dimensional.
2. In XML dissemination context, precision is more important than recall, this differs from traditional classification methods. In this kind of system there are many documents/news relevant to a user, and special attention is needed to avoid sending to many irrelevant documents to users.
3. Compared to a single classifier, this system applies a very large number(up to one million) different classifiers to each incoming document. Traditionally, the training efficiency of classifiers have been emphasized, while the runtime for applying a learned classifier was not a major concern. In information dissemination the efficiency of classifier application becomes of crucial importance.

To deal with small training sets and high-dimensional feature spaces, we explored XML classification methods for XML document dissemination. The main contributions of this paper are as follows:

- We introduce a data mining approach to XML dissemination. The corresponding client-server system architecture is also given.
- We present a novel XML classification method taking into account both the structure and contents of XML documents. The context for information dissemination includes a small number of training examples, very high efficiency at the test phrase, and the precision is also much more important than recall.
- Also discussed are optimization techniques on the server side. In order to efficiently execute a massive number of classifiers we explore methods to optimize the simultaneous execution classifiers to an incoming document.
- Our experimental evaluation on real XML document sets demonstrates the accuracy and efficiency of the proposed methods.

The rest of this paper is organized as follows. Related work is discussed in Section 2. The system architecture is given in Section 3. Section 4 introduces our novel XML classification method for XML documents. Particularly the training of these classifiers on the client side. Section 5 proposes our approach of executing large sets of such classifiers on the server side. The results of our experimental evaluation are presented in Section 6, followed by our conclusion in Section 7.

2   Related Work

There are two kinds of the related work to our proposed method. The first is XML dissemination techniques; and the second is XML classification methods.

XML Dissemination. XML filtering techniques are key for XML dissemination application. Most XML filtering systems [7,2,3] support a limited query language. Usually a fragment of XPath/XQuery is used to express user's information needs. In order to publish matched documents to corresponding users the incoming documents are processed through query structures such as prefix tree or NFA. YFilter[2] employs a Non-Deterministic Finite Automaton (NFA) to represent all path queries by sharing the common prefixes of the paths. BloomFilter[3] adapts BloomFiter to represent the simple XPath expression after treating each path as a string.

There are two main differences between XPath/XQuery-based information dissemination and our classification-based approach.

- Our classifier consists of a collection of rules rather than a single XPath/XQuery expression. These rules have the capacity to define more-complex decision surfaces. Even more our approach avoids forcing users to write complex XQuery expressions and can provide an alternative way to achieve the goal of document dissemination.
- Our method is an approximate method this varies from a XQuery-based document dissemination where only exact results are returned. In our approach the rules specify the distance of actual(test) feature vectors from prototype vectors. A similarly based approach is applied to find documents which are not supported by XQuery-based ones.

XML classification. Mining frequent sub-trees is one main method in XML classification. Former work [8,6] utilized a rules-based approach for frequent sub-tree mining. XRules [6] extend TreeMiner to find all frequent trees, and it is cost-sensitive and uses Bayesian rule based class decision making.

The efficiency of these methods is not very good. This is especially true for deep or complex XML document trees/graph. The number of candidate subtrees are huge; This causes the search space to be to large and time consuming when training or applying a learned classifier for new document. We verify this point by experimentation in Section 6.

The SVM approach [5] is another kind of method. SVM suffer from small training dataset. In the scenario of XML dissemination, there are not many examples for each client to train the classifier.

As discussed above current XML filtering systems lack good utility for users to express complex information needs. Traditional XML classification methods can't provide efficient solution to deal with challenges such as small training data set and high efficiency requirements in an information dissemination scenario.

3   System Architecture

The architecture of our proposed system is shown in Fig. 1. The procedure of this system is as follows.

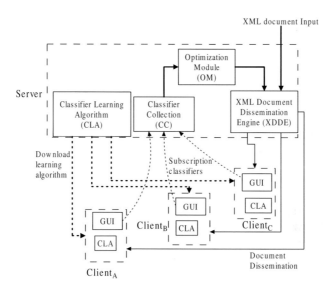

Fig. 1. System Architecture

1. A training collection of "positive" (relevant) XML documents on each client describes the client's (consumer's) interest. Using a sample of other documents as "negative" training documents, a two-class classifier (relevant / irrelevant) is trained on the client by Classifier Learning Algorithm(CLA) downloading from the server. GUI(Graphic User Interface) is a tool to help clients to download CLA from the server and subscribe learned classifiers to the server.
2. A client subscribes by sending his classifier to the server. Each classifier consists of some rules and each rule is a combination of frequent structural pattern, an aggregated term frequency vectors. The classification method is described in Section 4.
3. By Classifier Collection (CC) Module the server collects all learned classifiers from clients. The server applies these classifiers and forwards a document to all clients whose classifier predicts the document as relevant. The dissemination method in XDDE module are introduced in Section 5.
4. The client (consumer) receives a potentially relevant XML document, they then check the document for actual relevance and records relevance feedback (is/is not relevant).
5. The client maintains and updates the classifier. According to the client relevance feedback, the client re-trains the classifier on the current document collection. They then send a new classifier to the server.

In this framework the server doesn't record the documents of users/consumers. It only owns the rules which include structural pattern and frequent term vector. The revealed information from rules is similar to XPath/XQuery expressions and our approach doesn't violate the privacy concerns.

4   XML Classification Using Structure and Contents

In this section, followed by the discussion of XRules method, we present our XML classification method.

4.1   XRules Revisited

We first introduce XML document tree and revisit the XRules method [6], which is the state-of-the-art rule-based XML classifier.

An XML document can be modelled as a labelled, ordered and rooted tree $T = (V, E, root(T))$, where V is a set of nodes, $E \subset V \times V$ is a set of edges, and $root(T) \in V$ is the root node of T. A node $n_i \in V$ represents an element, an attribute, or a value.

XRules [6] employs the concept of structural rules and frequent sub-tree mining. Each rule or sub-tree is of the form $T \rightarrow c, (\pi, \delta)$, in which T there is a tree structure, c the class label, π the support of the rule, and δ the strength of the rule. To train the classifier XRules first mines all frequent structural rules with respect to a specific class c whose support and strength is larger than the pre-defined parameters for the class π_c^{min} and δ_c^{min}, using XMiner. Then all such rules are ordered according to a precedence relation. Thus the classifier is obtained.

We argue that XRules is not sufficient for XML document classification. The main reason is that XRules does not consider the content of XML documents. Only structural information is used to generate the rules. This is useless for XML documents that conform to a specific schema.

In order to handle both structure and content of XML document a direct approach is to extend XRules [6] by adding content leaf nodes into XML structure tree. This extension method of XRules calls XRules+. XRules can then be applied to mining frequent sub-trees from extended XML tree. Fig. 2 is an example of an extended XML tree for XRules+, where circle nodes are structure nodes and rectangle nodes are content nodes. Then, XRules can be applied to mining frequent sub-trees from extended XML tree.

For most XML documents the number of terms in content is much greater than the number of structure nodes, so there are many nodes in extended XML tree. XRules+ is equivalent to considering all possible combinations of a term and a structural pattern. This can lead to too many patterns may be generated,

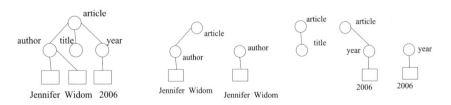

Fig. 2. XML for XRules+   **Fig. 3.** PV rules

only few of them will be frequent. With the increasing of nodes in XML trees the number of candidate sub-trees becomes larger, the efficiency of XRules+ is weakened.

4.2   PV: xPath-Vector Rule Approach

Aiming at the shortcomings of XRules+, a new method using feature vectors to handle content of XML documents is designed.

The PV rule format is $(P, V) \rightarrow c, (\pi, \epsilon)$, where P is the XPath pattern, V is a feature vector corresponding to P, c is the class label, π is a generalized support of the PV rule, and ϵ is a parameter to represent the boundary for the positive example. The corresponding PV rules of XML tree in Fig. 2 is shown in Fig. 3.

1. π is a generalized support. A path in XML document is *covered* by a raw pattern. If the path satisfying the structural pattern and the similarity between vector of the path, the raw pattern is no more than a predefined threshold. The number of covered paths of a specific raw pattern is the *generalized support* of the pattern.
2. ϵ is a parameter to represent the border for the positive examples. Since clients are only interested in finding positive test documents, only rules that have the positive class as the majority class, i.e. predict a matching test document as positive are needed. Instead of having two feature vectors(one for the positive and another one for the negative class), only one positive (prototype / mean) feature vector is needed. The role of the negative training examples is to limit the border of the class. How to obtain the border becomes important. Parameter ϵ is automatically learned as the minimum distance between the positive prototype feature vector and the closest negative feature vector among all documents matching the given structural pattern. The detailed algorithm is shown in Algorithm 1.

During the testing vector phase, when the distance between a test vector and a prototype is less than ϵ, we assign this test document as positive; otherwise, the test vector is considered negative. There are some exceptions, such as $v1$ in Fig. 4. $v1$ is nearer to the positive prototype, but is not located in the sphere

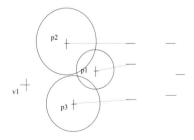

Fig. 4. Prototype and ϵ for Testing

Algorithm 1. Learning parameter ϵ to determine the border

Input: training examples D;
a set of prototype ps, which is obtained by k-means.
Output: ϵ

```
1  procedure LearnBorder(D, ps)  begin
2      for each prototype p in ps do
3          let minDis = MAXVALUE;
4          for each vector v of negative examples do
5              Calculate the distance dis of p and v;
6              if minDis < dis then
7                  minDis = dis;

8          ε = minDis/2;
9          Assign ε is the radius for prototype p;
10         Output p, ε;

11 end
```

---

of that prototype. This causes it to be regarded as negative. We think this case is reasonable in the XML dissemination context where precision is more important than recall. We need to avoid false positives more than false negatives (assuming that there are many relevant documents, the user will not be upset if he/she does not receive all of them). The pseudo-code of PV is shown in Algorithm 2.

5   Applying Many Classifiers for Document Dissemination at Server

According the format of rules above in last section, the server maintains the classifiers in two parts: structural filter and content filter. The structural filter is a general purpose path filtering engine. We use Bloom-filter-based engine [3] in current implementation. When a new XML document is received by the server, the document is first filtered by the structural filtering engine. Because structural filtering approaches have been studied in many former works [3], we don't discuss it further.

There may be hundreds or more content patterns for each structural pattern. Each vector may have more than a thousand dimensions. There were over 3000 dimensions of each vector in our experiment. This is the major problems to be addressed. In order to reduce the search space, we adopt clustering methods to get small number of representatives and use them to filter no-relevant input documents. We are then able to calculate the distances of examples in these candidate clusters in the refine phase. This two-phase method reduces the search space.

The pseudo-code is shown in Algorithm 3.

Algorithm 2. PV

    Input: training data set $Train$ and testing document d
    Output: PV rule set rs and Class label L
1 **procedure** PV($Train, d$) **begin**
2 | PathMiner($Train$);
3 | Test(d);
4 **end**
5 **procedure** PathMiner($Train$) **begin**
6 | Get all regular path pattern PathSet; Obtain the TF*IDF of terms and Sort them descending;
7 | Get the first 3000 terms to be dimensions of vector;
8 | **for** *each path p in the PathSet* **do**
9 | | Generate term vectors for p;
10 | | Determine positive prototypes by k-means;
11 | | Learn ϵ by Algorithm 2 by using negative examples to determine the radius for each prototype;
12 | | Prune rules by generalized support;
13 **end**
14 **procedure** Test(d) **begin**
15 | Let FreqRuleSet of the positive class = prs;
16 | **for** *each path p in prs* **do**
17 | | Find the set ps of pattern matching p;
18 | | TextSet = NULL;
19 | | **for** *each node n in ps* **do**
20 | | | Build TextVector v for n;
21 | | | **for** *each vector r corresponding to p* **do**
22 | | | | Calculate the distance dis between r to v;
23 | | | | **if** *dis < ϵ and generalized support > π* **then**
24 | | | | | Assign L as positive class and output L;
25 | | | | | /* if one rule match, then exit; */
26 | | | | | exit;
27 | **if** *no p matches d* **then**
28 | | Let L as negative class and output L;
29 **end**

Algorithm 3 first builds some clusters in the server at the initialization phase. Vectors of the same structural pattern are clustered by K-Means. We use *prototypevector* to represent the vector of each cluster. If tested vector belongs to this cluster only those vectors in this cluster are candidates. Only these candidates will further calculate the distance of tested vector to vectors in this cluster. These candidate vectors need to access from disk and to computer the distance with the incoming document. Using this filter-and-refine method we reduce the search space and I/O. Experiments in next section also show the efficiency of this approach.

Algorithm 3. Documents Dissemination at the server

    Input: rule set rs; testing document d;

    Output: Class Labels

1 **procedure** Test(d) **begin**

2   Extract all paths and corresponding vectors as pattern $p = <path, vector>$ from d;

3   **for** *each pattern $p = <path, vector>$* **do**

4   /*structural filtering */

5   Computer the containment relationship between *path* and structural bitstring bs;

6   **if** *path is contained by bs* **then**

7   Record d match the structural requirement of client Ci, whose BloomFilter value is bs;

8   /*content filtering */

9   Let *ClusterSet* be the clusters of the structure *path* of all learned classifiers;

10   Let $<prototypevector, radius>$ is the prototype and range of each cluster in *ClusterSet*;

11   **for** *each cluster in ClusterSet* **do**

12   Let $distance =$ EuclideanDistance($vector$,$prototypevector$);

13   **if** *distance < radius* **then**

14   /* checking each vector in the satisfied cluster and calculate the real distance to obtain the result classifier */

15   **for** *each vector in cluster* **do**

16   Get all patterns $(P, V) \rightarrow Ci, (\pi, \epsilon)$ from server storage;

17                     $distance =$ EuclideanDistance($vector$,V);

18   **if** *distance < ϵ* **then**

19   Assign d passes the content filtering of class Ci;

20   **if** *path is also matched Ci according to structural filtering* **then**

21   Assign the class label c to document d and output c;

22 **end**

---

6 Experiments

In this section, we report the results of our experimental evaluation using several real XML datasets. All the experiments are conducted on a Pentium IV 3.2 G CPU platform with 512MB of RAM.

6.1 Experimental Setting

Classification tasks

    The real data comes from DBLP XML Records[1]. We build two classification tasks:

---

[1] http://www.informatik.uni-trier.de/ ley/db/

1. Task1: SIGMOD-KDD classifier. We extract some documents from DBLP to build dataset I. For any given document, even for an expert, it is not easy to judge whether it comes from the SIGMOD proceedings or the KDD proceedings. This is because both conferences have papers focusing on data mining or data management. There are some hints, data mining people often publish papers in SIGMOD proceedings. We want to verify that classification method with structural and content can find such patterns.
2. Task2: DB-AI classifier. We extract documents from DBLP according to the ACM categories and build dataset II for task 2. DB and AI are two relatively distinct topics, this makes text mining a good method for classification. Integrated methods considering both structure and content are expected to increase the precision. We want to test this point by our experiments.

Data sets

– Dataset I comes from DBLP, the depth of the XML document tree is about 4 to 5. We extract sub-trees of "article" and "inproceedings" from "dblp" tree to build three datasets I(1), I(2) and I(3).

Table 1. Dataset I

Dataset I	#training	
	positive documents	negative documents
I(1)	1000 SIGMOD documents with the author and title elements	1000 KDD documents with the author and title elements
I(2)	200 SIGMOD documents with the author and title elements	200 KDD documents with the author and title elements
I(3)	200 SIGMOD documents with the author, title and abstract elements	200 KDD documents with the author, title and abstract elements

– Dataset II also comes from DBLP. We extracted articles for two ACM categories, "DB" and "AI". For these articles we added abstract information into each XML document. By changing the number of positive and negative examples we were able to obtain Dataset II(1), II(2), and II(3). We wanted to test the influence of the negative examples for the classification methods.

Table 2. Dataset II

Dataset II	#documents	
	positive documents	negative documents
II(1)	train: 80 DB articles test: 20 DB articles	train: 80 AI articles test: 20 AI articles
II(2)	train: 120 DB articles test: 20 DB articles	train: 60 AI articles test: 20 AI articles
II(3)	train: 120 DB articles test: 20 DB articles	train: 30 AI articles test: 20 AI articles

Classification Methods

Four methods for XML classification are compared.

1. XRules. This method considers only the structure of XML documents. The XRules implementation was downloaded from the website of its author [6].
2. NN. This method is to implement traditional nearest neighbor based text classification method, which only considers the content (term frequency vectors) of a document.
3. XRules+. As we discussed in Section 4.1, this method extends XRules and considers both structure and content. Each term in the content is treated as a separate leaf node.
4. PV. PV is proposed in Section 4.2. It considers path patterns $(T, V) \rightarrow c, (\pi, \delta)$, generalized support π, positive prototypes, and distance parameter δ.

Experiments were conducted concerning the accuracy and the efficiency of the compared methods. The evaluation measures include the precision, runtime, and the number of rules.

6.2   Performance of Classification Methods

All experiments were repeated 5 times and the results were averaged. Classification precision is the most important factor for XML document dissemination. First, we studied the precision of various methods. The precisions of XRules, PV, and NN method over Dataset I are shown in Fig. 5, Fig. 6 and Fig. 7. Dataset II is shown in Fig. 8, Fig. 9 and Fig. 10. These figures are always represented with respect to different minsupport values π. Table 3 shows the runtimes of all methods on all Datasets. Fig. 5 shows that when the number of training examples is large each method can obtain relatively good precision. However, with a decreasing number of training examples the precision of XRules, and NN suffers significantly. Results from Fig. 6 and Fig. 7 demonstrate this. In contrast, PV had quite robust a performance and was clearly best among the three methods.

In conclusion we observe that

– XRules is a good method for structural data classification. However, if the structure of examples are identical or similar XRules is unable to work. XRules is also sensitive to the number of the training examples. The smaller of the training examples, the lower the effectiveness and precision of the XRules methods.
– XRules+ considers both structure and content. This is an inprovment, but the number of candidate sub-trees becomes very large. This implies that its training phase is often very time-consuming and cannot terminate in a few days even for hundreds of documents.
– NN is a pure text classification method. Though the training and testing procedure is fast, the precision of NN is not acceptable. This is especially true for small data sets, such as dataset I(2), I(3), II(2), and II(3).

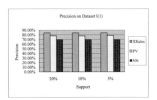

Fig. 5. Precision I(1)

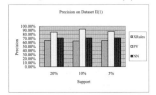

Fig. 6. Precision I(2)

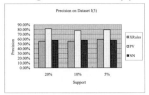

Fig. 7. Precision I(3)

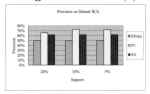

Fig. 8. Precision II(1)

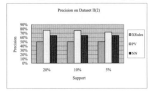

Fig. 9. Precision II(2)

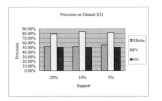

Fig. 10. Precision II(3)

- The PV approach is effective and efficient for XML document classification. The consideration of structure and contents achieves better precision than methods that take into account only structure, such as XRules, or only content, such as NN approach.

6.3 Performance of Dissemination Method

We also evaluated the performance of the dissemination at the server side. As discussed in Section 1, previous work has investigated structural filtering [2,3]. In this section we focused on content filtering and testing the efficiency of our method in Section 5.

We implement Algorithm 4. The scan method is as the baseline method, where it scans the collection of all feature vectors and computes the distance from a given test document vector v_0 to each client interest vector v on the disk. If the distance is less than ϵ, v_0 belongs to the positive class of v.

Table 4 shows the initialization time for all documents on the server. For the Scan method, the initialization time includes the times for parsing the XML documents and building the vectors. In our method, initialization includes parsing documents, building vectors, constructing the digital signature, and clustering.

Compared to the Scan Method, our method demands more initialization time because we need to build digital signatures and clusters. However, the benefit

Table 3. Runtime(seconds) of all methods on all Datasets

Dataset	SUPPORT	Runtime(s) (training/testing)			
		XRules	XRules+	PV	NN
I(1)	20%	0.23/0.02	0.49/0.04	2.06/0.45	1.36/0.33
I(1)	10%	0.45/0.02	0.83/0.07	2.12/0.46	1.36/0.33
I(1)	5%	1.71/0.03	2.72/0.18	2.12/0.46	1.36/0.33
I(2)	20%	0.12/0.01	/	0.26/0.09	0.19/0.09
I(2)	10%	0.57/0.01	/	0.26/0.09	0.19/0.09
I(2)	5%	1.50/0.01	/	0.27/0.10	0.19/0.09
I(3)	20%	0.37/0.01	/	0.74/0.26	0.78/0.26
I(3)	10%	1.93/0.01	/	0.76/0.26	0.78/0.26
I(3)	5%	4.75/0.01	/	0.74/0.26	0.78/0.26
II(1)	20%	0.08/0.01	46/0.67	0.39/0.16	0.45/0.13
II(1)	10%	0.16/0.01	32856.19/9.28	0.38/0.17	0.45/0.13
II(1)	5%	0.63/0.01	1566.98/462.41	0.40/0.17	0.45/0.13
II(2)	20%	0.29/0.01	94.61/3.84	0.46/0.19	0.33/0.09
II(2)	10%	1.59/0.01	/	0.45/0.18	0.33/0.09
II(2)	5%	38.3/0.02	/	0.47/0.19	0.33/0.09
II(3)	20%	0. 23/0.01	272.69/3.03	0.41/0.18	0.41/0.12
II(3)	10%	0.95/0.01	/	0.41/0.18	0.41/0.12
II(3)	5%	14.31/0.01	/	0.42/0.19	0.41/0.12

Table 4. Content Filtering Performance at Server

Dataset	#documents	Initialization (seconds)		Content Filtering Time /per document(seconds)	
		Scan	Our Approach	Scan	Our Approach
dblp	100	3.156	3.532	0.1688	0.0047
dblp	1000	30.296	30.766	1.3188	0.0031
dblp	10000	294.875	299.04	11.3328	0.0031
dblp	100000	3018.047	3046.75	126.914	0.0047

of our approach for document content filtering becomes obvious.When increasing the number of classifiers/documents our proposed method achieves great efficiency gains when compared to the Scan Method. For example, for 10^5 documents, our method obtains a speed-up factor of 20000. There are two main reasons: (1) our method uses filter-and-refine strategy, which greatly reduces the number of distance computations. (2) our method applies a hash-based method, where bit operations are much more efficient than string matching.

7   Conclusion

This paper introduced a data mining approach to XML dissemination that uses a given document collection of the user to automatically learn a classifier modelling the information needs on the client side. To deal with small training sets and high-dimensional feature spaces we introduced a novel classification method that represents both the structure and the contents of XML documents. In order to efficiently execute a massive number of classifiers at the dissemination server we explored methods that optimize the simultaneous execution of massive sets of classifiers. Our experimental evaluation on real XML document sets demonstrated the accuracy and efficiency of the proposed XML classification approach.

Acknowledgments. This work is supported by NSFC grants (No. 60773075), National Hi-Tech 863 program under grant (No. 2009AA01Z149), 973 program (No. 2010CB328106), Shanghai Education Project (No. 10ZZ33), Shanghai Leading Academic Discipline Project (No. B412) and Project(No. KF2009006) supported by Key Laboratory of Data Engineering and Knowledge Engineering (Renmin University of China), Ministry of Education.

References

1. Jacobsen, H.-A.: Content-based publish/subscribe. In: Liu, L., Tamer Ozsu, M. (eds.) Encyclopedia of Database Systems, pp. 464–466. Springer, Heidelberg (2009)
2. Diao, Y., Rizvi, S., Franklin, M.J.: Towards an internet-scale XML dissemination service. In: Nascimento, M.A., Özsu, M.T., Kossmann, D., Miller, R.J., Blakeley, J.A., Bernhard Schiefer, K. (eds.) VLDB, pp. 612–623. Morgan Kaufmann, San Francisco (2004)
3. Gong, X., Yan, Y., Qian, W., Zhou, A.: Bloom filter-based XML packets filtering for millions of path queries. In: ICDE, pp. 890–901. IEEE Computer Society, Los Alamitos (2005)
4. Kwon, J., Rao, P., Moon, B., Lee, S.: Fast xml document filtering by sequencing twig patterns. ACM Trans. Internet Techn. 9(4) (2009)
5. Theobald, M., Schenkel, R., Weikum, G.: Exploiting structure, annotation, and ontological knowledge for automatic classification of xml data. In: WebDB, pp. 1–6 (2003)
6. Zaki, M.J., Aggarwal, C.C.: Xrules: An effective algorithm for structural classification of xml data. Machine Learning 62(1-2), 137–170 (2006)
7. Hong, M., Demers, A.J., Gehrke, J., Koch, C., Riedewald, M., White, W.M.: Massively multi-query join processing in publish/subscribe systems. In: Chan, C.Y., Ooi, B.C., Zhou, A. (eds.) SIGMOD Conference, pp. 761–772. ACM, New York (2007)
8. Zaki, M.J.: Efficiently mining frequent trees in a forest: Algorithms and applications. IEEE Trans. Knowl. Data Eng. 17(8), 1021–1035 (2005)

Semantic Transformation Approach with Schema Constraints for XPath Query Axes

Dung Xuan Thi Le[1], Stephane Bressan[2], Eric Pardede[1],
Wenny Rahayu[1], and David Taniar[3]

[1] La Trobe University, Australia
{dx1le@students.,w.rahayu@,e.pardede@}latrobe.edu.au
[2] National University of Singapore
steph@nus.edu.sg
[3] Monash University, Australia
David.Taniar@infotech.monash.edu.au

Abstract. XPath queries are essentially composed of a succession of axes defining the navigation from a current context node. Among the XPath query axes family, **child**, **descendant, parent** can be optionally specified using the path notations {/,//,..} which have been commonly used. Axes such as **following-sibling and preceding-sibling** have unique functionalities which provide different required information that cannot be achieved by others. However, XPath query optimization using schema constraints does not yet consider these axes family.

The performance of queries denoting the same result by means of different axes may significantly differ. The difference in performance can be affected by some axes, but this can be avoided. In this paper, we propose a semantic transformation typology and algorithms that transform XPath queries using axes, with no optional path operators, into semantic equivalent XPath queries in the presence of an XML schema. The goal of the transformation is to replace whenever possible the axes that unnecessarily impact upon performance. We show how, by using our semantic transformation, the accessing of the database using such queries can be avoided in order to boost performance. We implement the proposed algorithms and empirically evaluate their efficiency and effectiveness as semantic query optimization devices.

Keywords: XML, XPath, Query Processing, Semantic XML Query Optimization.

1 Introduction

As the popularity of XML increases substantially, the importance of XML schemas for describing the structure and semantics of XML documents also increases. Semantic transformation of XPath queries based on XML schemas has become an important research area.

One of the existing problems with XML databases is the inefficient performance of the XPath queries including those that are specified with axes [1, 5]. Previous researches [3, 6, 7, 8, 9, 16] improved the performance of XPath queries involving

L. Chen, P. Triantafillou, and T. Suel (Eds.): WISE 2010, LNCS 6488, pp. 456–470, 2010.

XPath fragment [/,//,*,...,.]. Among the existing optimization techniques is semantic query transformation where schema constraints are used for query transformation. Work on XPath query axes that do not have optional operators is non-existent.

To increase semantic transformations of XPath query axes by utilizing XML schema information [2], we investigate a new semantic transformation typology to transform *preceding-* and/or *following-sibling* axes in XPath query.

Our **motivation** is to leverage constraints in the XML Schema to propose a semantic transformation typology for the rewriting of XPath axes queries. To the best of our knowledge, no work has yet to address the XPath axes transformation by purely using schema constraints [2,16]. One of the benefits of this semantic transformation is improved performance. In addition, we are motivated by the fact that not all the existing XML-Enabled databases can handle the execution of XPath axes queries efficiently. By performing such transformation, XPath axes queries can be alternatively executed to reduce resource utilization.

Assuming a self-explanatory XML data with a clear intended meaning of the element names and structure, let us consider:

EXAMPLE 1. *"list all the following-sibling of last name node of permanent employee"*

//perm/name/lastname/following-sibling::node()

The XPath query involves the axis *following-sibling* in order to determine the siblings of element 'lastname'. This XPath query can be transformed, under the knowledge of different constraints [2] such as *structure, cardinality, and sequence* in the XML schema. Semantic query transformation precisely takes advantage of such structural and explicit constraints defined in the XML schema to transform the query.

One of the benefits of semantic transformation is the detection and removal of any redundancy in the query that may impact upon performance. Assume that the constraints in the schema allow us to eliminate the location step following-sibling::node() from the XPath query, which now is simplified to *"//perm/name/firstname"*.

In example 1 above, the following-sibling axis and node () have been completely removed and 'lastname' is replaced with 'firstname'. Referring to Figure 1, under the permanent name, 'firstname' comes after 'lastname'. In XML query semantic transformation, the structure of the data must be taken into account: the semantic transformation involves decisions about both the tree pattern structure and the other explicit constraints of the data.

While there are 13 axes in XPath query [1, 5], existing works [4, 6, 7, 14, 16] in semantic optimization for XML query targeted only the *parent, child and descendant, descendant-or-self* axes. These addressed axes allow information to be navigated vertically. We learn that *following-* or/and *preceding-sibling* are also important axes as information can be navigated horizontally. Our proposed technique shows that these XPath query axes can also be transformed. In some cases, performance opportunities also exist.

Our **approach** to semantic transformation has two directions: (1) semantic transformation which transforms the location step that contains *following-* or *preceding-sibling* due the incompleteness of constraints used in an XPath query. In this case, the transformation can also determine if the XPath query returns an empty result set. This is to avoid the unnecessary accessing of the database that overloads the resources. (2)

Semantic transformation transforms the location step that contains *following-* or *preceding-sibling* if, and only if, the constraints used by the XPath query are fully compliant. Semantic transformation has the ability to remove the unnecessary components used in the XPath query to reduce the resource utilization.

During the transformation, we also detect any conflicts of constraints in the XPath axes query that may cause a query to be unsatisfied [10, 11, 12], which definitely will return an empty result. Our constraint conflict detection is based purely on the constraints defined in the XML schema, which are used to determine the validity of each location step.

EXAMPLE 2. *"list all the preceding-siblings of last name node of permanent employee"*

//perm/name/lastname/preceding-sibling::node()

For XPath query in example 2, under permanent employee name (refer to Figure 1), no member precedes the 'lastname'. Clearly, this XPath query returns an empty result set. Hence, the XPath query cannot be transformed and can be rejected prior to accessing the database. This proves that by using semantic transformation, not only does it increase the performance improvement opportunities, it also avoids an unnecessary utilization of resources.

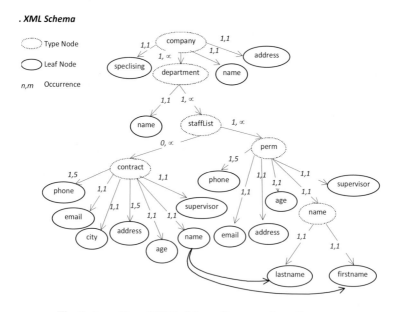

Fig. 1. An outline of XML Schema Fragment for a ***Company***

2 Preliminary and Framework

An XPath query is a hierarchical query consisting of multiple location steps, which are separated by a '/' or '//' hierarchical relationship. A location step comprises *axis*, *node test* and *predicate*.

Axes [1, 5] are the operators used for traverse the XML documents. A node test (denoted as nodetest) is a valid element name or '*', which must be true for each node (element) selected by the location step. The predicate [13] enclosed within square brackets [ ], is a condition to filter the unwanted sequence. A location step is specified in the form of *Axis::nodetest* [predicate$_1$][predicate$_2$]..[predicate$_n$].

EXAMPLE 3. *"list all following-sibling of first name node from the staff list such that the staff list must has one valid permanent employee"*

/descendant::staffList[child::perm[position()= 1]]/child::contract/ child::*/following-sibling::firstname

Figure 1 shows how some axes are specified in individual location steps. There are four location steps of XPath query in example 3, which are shown in Table 1. The first location step uses a descendant and child axis, the node tests are 'staffList' and 'perm' respectively. While 'staffList' has a predicate that contains a 'perm' node to restrict the returned information, the 'perm' node has a predicate that contains a node set function to affirm that 'perm' is the first node. The second and third locations share the same axis 'child' where their node tests are 'contract' and '*'. The fourth location step has a following-sibling axis whose node test is 'firstname'.

Table 1. Location Steps & Their Members

Location Step	Axis	Node Test
1	Descendant, child	staffList, perm
2,3	Child	*, contract
4	Following-sibling	firstname

Structural and order structural constraints are used in this semantic transformation typology. For this paper, we provide definition for these constraints as follows.

Definition 1. (Structural Constraint): *structural constraint is a path expression between a pair of elements having a parent-child or ancestor-descendant relationship in an XML schema.*

The common structural constraint in the schema is the path defined by the selector key, as shown by the following.

```
<xs:element name="company">
   <xs:key name="permKey">
     <xs:selector xpath=
        "./department/staffList/perm"/>
     <xs:field xpath="@id"/>
   </xs:key>
</xs:element>
```

Selector path for permKey is "./department/staffList/perm". As path that leads to @id is defined under "company", '.' can also be replaced by "company". The full path becomes "company/department/staffList/perm"

Definition 2. (Structural Order Constraint): *structural order constraint is a constraint that places an order on the children using a sequence constraint value on the parent.*

Referring to Figure 1, we show a fragment of XML schema that defines a 'contract' element type, using a sequence value to control the order of its children.

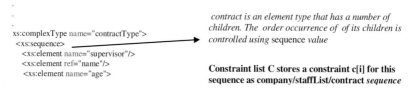

contract is an element type that has a number of children. The order occurrence of of its children is controlled using sequence value

```
xs:complexType name="contractType">
 <xs:sequence>
  <xs:element name="supervisor"/>
  <xs:element ref="name"/>
  <xs:element name="age">
```

Constraint list C stores a constraint c[i] for this sequence as company/staffList/contract *sequence*

Structural constraints are not an essential requirement in the XML schema. In some cases, they can be omitted. However, structural constraints are the central requirement for semantic transformation. Hence, we derive these constraints by using a unique path derivation technique [6].

3   Semantic Transformation Proposed Typology

Semantic transformation consists of two main tasks: parsing the location steps of XPath query and semantic transformation algorithm for *preceding-* or *following-sibling* axes.

3.1   Parsing the Location Steps of XPath Query

Referring to our preliminary section, it states that XPath query contains a sequence of location steps. In this parsing section, we propose a function ε that parses each location step of the input XPath query P using predefined information of O and $Axes$. O contains a set of operators {'/', '//',':', '*','..','.'}; and Axes contains a set of axis names {child, descendant, descendant-or-self, ancestor, ancestor-or-self, parent, following, following-sibling, preceding, preceding-sibling}. The parsing location steps are finally stored in xP as any array data structure.

$$\text{(O,Axes)}$$
$$\downarrow$$
$$\varepsilon(P) \longrightarrow xP$$

Due to page constraints, we are unable to provide a complete algorithm; however, we provide an example to demonstrate the expected xP for an input XPath query P.

EXAMPLE 4. Let us consider an XPath query P is

//contract/name/child::firstname[1]/following-sibling::lastname

After applying function ε, expected xP is as xP(0)= descendant-or-self::contract,   xP(1)= child::name,   xP(2)= child::firstname,   xP(3)=

following-sibling::lastname. In this example, *Parent* element is name, *sibling* element is firstname and *following-sibling* element is lastname.

3.2 Proposed Semantic Transformation and Algorithms

For each set of location steps, the semantic transformation *preceding-* or *following-sibling* location step involves a processing of three location steps containing a *parent*, a *sibling* and a *preceding-* or *following-sibling* element.

The transformation removes a *preceding-* or *following-sibling* axis according to structural constraint and order constraint are applied and their criteria are satisfied.

Let us consider a generic XPath query as below

$$child::b/child::c/\beta::*$$

child::b/child::c is the semantic path after it has passed a set of **rules** below. Let β be the preceding-sibling or following-sibling axis that is specified with element represented by '*'. c is the sibling and '*' represents all *following-* or *preceding-sibling* elements of c. The elimination of child::c and/or β such as following-sibling or preceding-sibling must have the following essential requirements.

1. c and elements represented by '*' share the same parent b;
2. b must have a sequence order constraint on c and elements represented by '*';
3. elements represented by '*' must occur after element c if β is a following-sibling or before element c if β is a preceding-sibling;
4. remove ` β::*' and replace with next sibling element if c has a minimal and miximal occurrence is 1
5. Add [postion() >1] or [position()<last] if c has a maximal occurrence is greater 1

EXAMPLE 5. "List all the preceding siblings of permanent employee first name in each staffList"

$$//perm/name/firstname/preceding-sibling::*$$

In this example, * represents a preceding-sibling element, firstname is the sibling element, and permanent name is the parent element. Refer to Figure 1, Unique Paths[1] Q and element constraint[2] list C have been derived to assist with seeking the order that exists between the sibling elements and the values of constraints other than the structural constraint of elements.

Q={company, company/specializing, company/department/, company/department/name, company department/staffList,...,**company/department/staffList/perm/name/lastname, company/department/staffList/ perm/name/firstname**,...}

---

[1] A unique path is a full path that traverses from the root of the XML schema tree to a target element. The path elements are separated by operator "/".
[2] Element constraint is the constraint that has values to restrict an element in XML schema. Namely a few, for example, are *inclusive, occurrence, enumeration, sequence*.

C={company occurrence 1 1, ..., company/staffList/perm/name sequence, company/department/staffList/perm/name/lastname occurrence 1 1,
company/department/staffList/perm/name/firstname occurrence 1 1,..}

List Q indicates that 'firstname' of each permanent employee comes after the 'lastname'. There exists a parent-child relationship between permanent 'name' and 'firstname' and between permanent 'name' and 'lastname'. In order to remove the sibling location step 'firstname' and *preceding-sibling* axis that contains '*', it requires a structural order constraint on the parent permanent 'name'.

Schema constraint list C indicates that the permanent name has a required structural order constraint 'company/staffList/perm/name sequence'. In List Q, it also shows that by sharing the same parent permanent 'name', only 'lastname' precedes 'firstname'. Therefore, //perm/name/firstname/preceding-sibling::* is transformed to //perm/name/lastname. Both XPath queries produce the same result set.

We propose Algorithm 1 for semantic transformation typology above. The semantic transformation only applies whenever there is an input of XPath query. The algorithm takes in the input list of location steps xP (Line 01) where xP is derived by function ε in section 3.1. Lists Q and C (Line 01) are derived by algorithm *preprocessingSchema* proposed in [6, 7, 8, 9]. While Q contains a list of unique paths, C contains a list of elements together with their constraint name and values. This *preprocessingSchema* only needs to run once until the schema is updated or changed; the algorithm needs to be re-run so that schema information in Q and C can be updated. The list semanticPath is the input information that contains semantic path queries. Every time a location step is transformed, the semanticPath is updated. The output of algorithm 1 is a semantic XPath query semanticPath (Line 02).

Algorithm 1 is intuitively designed with the objective of producing optimized performance. The function *verifiedConstraints* (Line 05) is called first to verify the names of the elements and structural expression of the Xpath query. If the conflict is found, the XPath query can be immediately rejected and unnecessary database access can be avoided. Otherwise, if no conflict is found (Line 07), the transformation begins with the last location step in list xP (Line 08) to the first location step. A targetNode parameter is set to false to start off (Line 09). When the *following-* or *preceding-sibling* appears in the last location step, its element is the target node[3] and the targetNode is set to true (Line 11). As mentioned earlier, our semantic transformation involves processing three location steps that contain *parent, sibling and following-sibling* or *preceding-sibling* elements which are carefully prepared (Line 12-19).

The transformation may also be terminated during the preparation of information (Line 17 and Line 19). This occurs only if it cannot find the respective parent and sibling. To find the correct parent and sibling, the element in the location step before the location step of *following-* or *preceding-sibling* element should be indicated with a child::. Refer to example 5 for more details. In the case where the parent location step and sibling location step are specified with other axes, e.g. descendant, they will be transformed first [6, 8].

---

[3] Target node is the element node of the last location step in an XPath query.

Algorithm 1. *transformationPrecedingFollowingSibling*

```
01: INPUT: xP, Q, C, semanticPath
02: OUTPUT: semanticPath
03: BEGIN
04: Let xp be a location step in xP, c be constraint row in C, axe be an array where axe ={ preceding-sibling::,
       following-sibling::}, op={*, node()} be an array type, parent be element of following and sibling, ,
       precedingSibling be element attached with axe[3], followingSibling be element attached with axe[4]
05: Boolean semanticConflict = verifiedConstraints(xP);
06: For (j is an array index start from 0; j still smaller than number of xp in xP; increase j)
       xP=transformPredicate(xP, Q, C);
07:    IF (!semanticConflict)
08:       FOR ( i is an array index start from last location step xp; i still greater than 0; decrease i )
09:          Boolean targetNode:= false
10:          IF xp contain axes[3] or axes[4]
11:           IF xp is the last location step in xP THEN targetNode:= true
12:           IF (xp contains axe[3] and contain op[0] OR op[2] THEN precedingSibling = op[0] OR
13:                xp contains axe[4] and contain op[0] OR op[2] THEN followingSibling = op[0]
14:           ELSE IF xp contains axe[3] AND not contain any op THEN  precedingSibling =  xp includes axe[3]
15:           ELSE IF xp contains axe[3] AND not contain any op THEN  followingSibling =  xp includes axe[4]
16:           IF (i-2) >=0 THEN  Parent = xp[i-2] excludes axe
17:           ELSE semanticPath set empty Msg :="cannot find parent"  EXIT
18:           IF (i-1) > 0 THEN sibling = xp[i-1] excludes axe
19:           ELSE semanticPath set empty Msg :="cannot find sibling" EXIT
20:       J=j-2
21:       FOR each constraint c of parent in C
22:          IF found parent has sequence order THEN
23:             SemanticPath = semanticTransformationFPSibling(parent, followingSibling, precedingSibling,
                                                        sibling, semanticPath, targetNode)
24:             EXIT
25:       IF (sequence order not found on parent && targetNode ) semanticPath = xP
26:       ELSE IF sequence order not found on parent && !targetNode semanticPath =semanticPath
27: END
```

A confirmation of structural order constraint on the parent parameter (Line 21) must be accomplished before the actual transformation is triggered by calling function *semanticTransformationFPSibling*, presented in the following pages, to produce new semanticPath (Line 22-24). The parent does not always have order sequence constraint; it may have all or choice constraint. In case this sequence constraint is not found and the precedingSibling or followingSibling is a target node, the semantic XPath query is the xP as no transformation is needed (Line 25). On the other hand, if the sequence constraint is not found on the parent and the precedingSibling or followingSibling is not a target node, semanticPath list is taken to be the semantic XPath query (Line 26) as the target node could be a child or a descendant and has been transformed earlier.

The input information for the function *semanticTransformationFPSibling* includes parent, followingSibling, precedingSibling, semanticPath and targetNode. The returned value is semanticPath list (Line F20 & F39). Notice that we use n as a set parameter to hold '*' and node(). The transformation of followingSibling is handled from (Line F03-F19), which separates into two tasks.

The first task (Line F03-F10) is to transform the location step that contains the fol-lowingSibling when its element is a targetNode. It starts from the top of unique path Q list to seek parent and sibling in a unique path q, (Line F04-F05), and set the cursor index to the next q (Line F06). From this point onwards (Line F07), the semantic path is found in one of two circumstances. The first circumstance is that the unique path q can become a semantic path if q contains parent and a followingSibling belongs to n. The second circumstance is that a unique path q can become a semantic path if q has a parent and the last node in q is the child of parent, then q is the semanticPath (Line F08-F09). When the search reaches the last q, the cursor index needs to move to the last q to avoid repeating the q by outer loop (Line F10).

Function *semanticTransformationFPSibling* (String parent, String followingSibling, String precedingSibling, String sibling, String [] semanticPath, Boolean targetNode)

```
F01:    BEGIN
F02:        Let n = {'*', node()} ; q be a unique path belong to Q;   ps be a semantic path
               belong to  semanticPath
F03:    IF  (targetNode AND followingSibling not empty) THEN
F04:        FOR (cursor from first q in Q; cursor not yet reach last q in Q; next q)
F05:          IF (parent found in q AND sibling found in q)
F06:            Move cursor to read next q
F07:            FOR (cursor from current q; cursor not yet reach last q in Q ; next q)
F08:              IF ((parent found in q AND followingSibling not in n AND followingSibling is last node in q ) ||
                    (parent  found in q  AND followingSibling  in n AND last node in q must be child of parent))
F09:                 insert q to end of semanticPath  list;
F10:            Cursor points to last q in Q
F11:      ELSE IF  (!targetNode AND followingSibling not empty) THEN
F12:      Let tempList be an empty array
F13:        FOR (cursor from first ps in semanticPath; cursor not yet reach last ps in semanticPath; next ps)
F14:          IF (parent found in ps AND sibling found in ps)
F15:            Move cursor to read next ps
F16:          FOR (cursor from current ps; cursor not yet reach last ps in semanticPath; next ps)
F17:            IF ((parent found in ps AND followingSibling not in n AND followingSibling is last node in ps) ||
                  parent found in ps AND followingSibling in n AND last node in ps must be child of parent))
F18:              Insert  ps to end of tempList
F19:            Cursor points to last ps in semanticPath
F20:      semanticPath:=tempList
F21:    IF  (targetNode AND preceding not empty) THEN
F22:        FOR (cursor from last q in Q; cursor not yet reach first q in Q;  move back earlier q)
F23:          IF (parent found in q AND sibling found in q)
F24:            Move cursor to read previous q
F25:            FOR (cursor from current q; cursor not yet reach first q in Q ; move to an earlier q)
F26:              IF ((parent found in q AND precedingSibling not in n AND precedingSibling is last node in q ) ||
                    (parent  found in q  AND precedingSibling  in n AND last node in q must be child of parent))
F27:                 insert q to top of semanticPath list;
F28:            Cursor points to last q in Q
F29:  ELSE IF  (!targetNode AND precedingSibling not empty) THEN
F30:    Let tempList be an empty array
F31:    FOR (cursor at last ps in semanticPath;cursor not yet reach first ps in semanticPath;move to ealier ps)
F32:      IF (parent found in ps AND sibling found in ps)
F33:        Move cursor to read ps before current ps
F34:        FOR (cursor at current ps; cursor not yet reach first ps in semanticPath; move to an earlier ps)
F35:          IF ((parent found in ps AND precedingSibling not in n AND precedingSibling is last node in ps) ||
                (parent found in ps AND precedingSibling in n AND last node in ps must be child of parent))
F36:              Insert  ps to top of tempList
F37:        Cursor points to first ps in semanticPath
F38:      semanticPath:=tempList
F39:  RETURN semanticPath
```

The second task (Line F11-F20) is to transform the location step that holds the fol-lowingSibling and it is not a targetNode. A tempList list is used to play the role of a tem-porary list (Line F12). It starts from the top of semanticPath list to find any semantic path ps that contains parent and sibling, the cursor is immediately moved to the next ps (Line F14-F15). When a followingSilbing is not specified with '*' or node() in a location step of original XPath query, semantic path is ps that contains parent and child other than its sibling *or* any semantic path ps that contains parent and its child must be followingSilbing as it is specified in the location step of the original xpath query (Line F17). Insert the identified semantic path ps into list tempList (Line F18) and cursor needs to move to the last semantic path ps to avoid the repeating path ps by outer loop (Line F20). tempList becomes the new semanticPath list.

The transformation of the location step containing precedingSibling is handled from (Line F21-F38). It is done in the same direction with two tasks in transforming the location step that contains the followingSibling. The only differences are: (1) instead of seeking the parent and sibling from the top of Q and semanticPath lists, it seeks from the end of the Q and semanticPath lists (Line F22 & F31), and (2) instead of seeking the followingSibling, it is now seeking the precedingSibling (Line F21-F38).

4 Implementation Overview and Empirical Evaluation

This section presents the implementation and result analysis using the proposed se-mantic transformation. We describe the implementation in terms of hardware and software: (1) the *hardware* includes a machine that has a configuration of Pentium 4 CPU 3.2GHz 4GB RAM and (2) the *software* includes Windows XP Professional OS and Java VM 1.6. We select a leading commercial XED[4] and use their provided data-base connection driver to connect to our algorithm modules. We use five synthetic datasets (compliant with the schema shown in Figure 1) of varying sizes: 20, 40, 60, 80 and 100 megabytes.

We divide the test queries into two groups. The first group contains the XPath que-ries where each has a valid semantic transformed query produced through our seman-tic transformation implementation. The second group contains another set of XPath queries where no semantic transformed queries have been produced due to various types of conflicts. For each query in the first group, we compare the performance of the original query and the transformed query.

Original Query 1: company/department/staffList/contract/name/following-sibling::*
Transformed Query 1: company/department/staffList/contract/supervisor
(The name/following-sibling:: axes have been replaced by /supervisor in regard to the structural and order constraints indicated in the schema. Namely, the following sibling of 'name' under 'contract' is always supervisor.)*

Original Query 2: company/descendant-or-self::perm/name/firstname/preceding-sibling::*
Transformed Query 2: company/descendant-or-self::perm/name/lastname
(The firstname/preceding-sibling:: axes have been replaced by /lastname in regard to the structural and order constraints indicated in the schema. Namely, the preceding sibling of 'firstname' under 'perm' is always 'lastname'.)*

---

[4] RDMS-XML Enabled Database Management System

Original Query 3: company/department/staffList/contract/following-sibling::perm[supervisor]/name

Transformed Query 3: company/department/staffList/perm/name

(*The* contract/following-sibling::perm[supervisor]/name *axes have been replaced by* /perm/name *in regard to the structural and order constraints indicated in the schema. Namely, the following sibling of 'perm' is always 'contract' and every employee, including permanent employee, must have a supervisor.*)

Original Query 4: company/*//department/staffList[1]/preceding-sibling::name

Transformed Query 4: company/*//department/name

(*The* /staffList[1]/preceding-sibling::name *axes have been replaced by* /name *in regard to the structural and order constraints and occurrence constraint indicated in the schema. Namely, the preceding-sibling of first 'staffList' is always 'name' under the 'department' and every department must have at least 1 staffList.*)

In **Query 1**, we demonstrate the use of the following-sibling axis with '*' specified in the location step. With reference to our Rule Set 1, it first validates the relationship between the contract 'name' and '*' if they share the same parent. In the case where contract 'name' and '*' share the same parent, the following-sibling relationship of '*' to contact 'name' is that '*' must come after contact 'name'. Based on Figure 1, 'supervisor' is the only sibling of contract 'name', so 'supervisor' now can be mapped to '*'. In addition, the minimal occurrence of 'supervisor' is 1; so for every contract name, there must be 1 supervisor following it. As these schema constraints are satisfied, the location elimination proceeds to XPath query 1.

The transformed query shows a sound performance compared with the original XPath query. It indicates a good improvement on the first dataset (20MB). However, there is a slight drop on the second dataset (40MB) and then picks up a higher improvement of between 15% and 28% for the last 3 datasets (60MB, 80MB, 10MB). We believe that, as the data-size increases, this transformation contributes to a better improvement overall.

In **Query 2**, we demonstrate the use of the preceding-sibling axis with * specified in the location step. Based on Rule Set 1 and Figure 1, 'firstname' has only one preceding sibling 'lastname'. In addition, the minimal and maximal occurrence constraint of 'firstname' is 1. This allows location elimination according to Rule Set 1.

The transformed XPath query 2 shows a fair improvement between 15% and 25% and as the values increase, the data size increases. Although there is a very tiny improvement in performance when the data size is 40MB, the rest showed a promising improvement. This could be caused by resources dependency but can also further be investigated using different database systems.

In **Query 3,** we demonstrate a preceding-sibling axis with a valid context node as well as a predicate that supports an element for a comparison. If we recall, our Rule Set 1 also supports predicate transformation. The predicate is transformed based on an element type specified. It is supported by rule 3 (a) in Rule Set 1. Although based on the structure in Figure 1, 'supervisor' is the child of a permanent employee, rule 3 (a) still needs the occurrence constraint on 'supervisor' and it must have a minimal occurrence value of 1. In this case, the information in the schema for 'supervisor' satisfies the constraint and criteria. Therefore, the predicate is removed. Performance- wise, we found a significant improvement between 79% and 85% for this particular XPath query structure. This indicates that when the XPath query is specified in this structure type, our semantic transformation makes a significant contribution to improved performance.

Similar to query 3, in **Query 4** we demonstrate a preceding-sibling axis with a valid node test 'name' as well as a predicate that supports an index ordered value 1 of the context node 'staffList'. The predicate is transformed based on the central focus, which is an ordered index of an element supported by rule 4 (a) in Rule Set 1. Based on the schema in Figure 1, element 'name' of department precedes the staffList and the minimal occurrence constraint of both department name and 'staffList' is 1. It is eligible to remove the staffList[1] and preceding-sibling axes supported by Rule set 1. Performance wise, we also found a significant improvement between 30% and 90% for the first 2 sets of data and a constant improvement of ~30% for the last 3 sets of data. Although we will further investigate this type of XPath query structure in the future, overall, the positive outcome for the time being is the good performance.

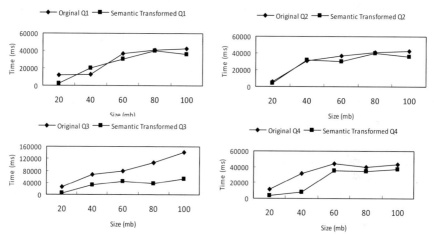

Fig. 2. Query Performance (With Transformed XPath Queries)

Original Query 5: //contract//supervisor/following-sibling::name/descendant::firstName
(*The transformation produces no semantic XPath query due to the conflict in regards to structural constraint indicated in the schema. Namely, 'firstname' is a leaf node that has no descendants.*)

Original Query 6: following-sibling::*//department
(*The transformation produces no semantic XPath query due to the conflict in regards to structural constraint indicated in the schema. Namely, following sibling elements must have a sibling.*)

Original Query 7: company/department/staffList/perm[3]/following-sibling::contract/*
(*The transformation produces no semantic XPath query due to the conflict in regards to structural constraint indicated in the schema. Namely, 'contract' employee must always be a preceding sibling of 'perm' employee.*)

Original Query 8: company/department/staffList/contract[supervisor]/preceding-sibling::perm
(*The transformation produces no semantic XPath query due to a conflict in regards to structural constraint indicated in the schema. Namely, 'perm' employee must always be a following sibling of 'contract' employee.*)

From **Query 5** to **Query 8**, we demonstrate various cases where transformation detects conflicts of constraints used in each query; hence, no transformation XPath query is produced. Our semantic transformation detects the invalid XPath queries, and removes the necessity for database access.

Referring to Figure 3, the graph shows the time taken by the original XPath query to access the database. As we can see, each XPath query requires an excessive amount of time before it can determine the answer to the query. This is in contrast to the amount of time that our semantic transformation uses to determine the conflict of constraints in the XPath query which is between 400 and 1008 milliseconds (we do not graph this value on each graph as the time is considerably small and uniform regardless of the sizes of the data).

Observation of Result Patterns: we observe that result patterns in Q1 and Q2 fluctuate. While the result pattern in the semantic transformed Q2 does not show the loss of performance for some data sizes, the result pattern in semantic transformed Q1 shows the loss of performance in data sizes between 20 and 60. The difference between the two queries is that we use more than one axes in Q2 (descendant-or-self and preceding-sibling), whereas in Q1 only one axis is adopted.

On the same observation for result patterns in Q3 and Q4, not only do we demonstrate the axes of following- and preceding-sibling axes, but we also demonstrate the predicate where pattern matching and value matching needs to be achieved for the transformation to be successful. There is a significant gain on the performance. Major benefit is gained from the semantic transformed Q3 and Q4 when predicates and following- preceding-siblings are removed from the queries.

Currently, we are obtaining results from similar pattern queries on the real XML (DBLP) datasets to advance our evaluation and result analysis. In future, we will also test our proposed semantic transformation on the second commercial XML-enabled DBMS for a comparison of performance versus the easting XML-enabled DBMS studied performance.

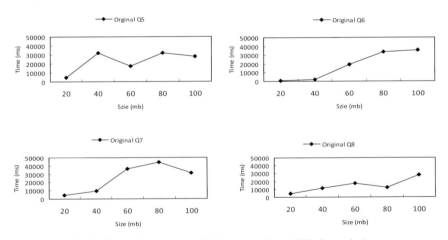

Fig. 3. Query Performance (With no transformed XPath queries)

5   Conclusion

In this paper, we have proposed a semantic query transformation for location steps that uses a *following-* or *preceding-sibling* axis to navigate the information. The work context in this paper is to transform such location steps by eliminating the location and replacing it with equivalent location step based on the axis used. We also show that, in the case where an XPath query cannot be transformed, it can be guaranteed to return a valid result data set. Moreover, an XPath query that returns an empty result-set can also be detected earlier. We show empirically that the proposed transformations provide promising opportunities to improve performance.

Our objective is to ultimately offer a comprehensive semantic transformation to rewrite 13 axes. Our future work is to continue to extend our current semantic transformation typology to address the remaining axes.

References

1. Berglund, A., Boag, S., Chamberlin, D., Fernándezark, M., Kay, M., Robie, J., Siméon, J.: Axes in XML Path Language, XPath 2.0 (2007),
   http://www.w3.org/TR/xpath20/#axes
2. Biron, P., Permanente, K., Malhotra, A.: XML Schema Part 2: Datatypes, 2nd edn. (2004),
   http://www.w3.org/TR/xmlschema-0/
3. Chan, Y., Fan, W., Zeng, Y.: Taming XPath Queries by Minimizing Wildcard Steps. In: Proceedings of the Thirtieth International Conference on Very Large Data Bases, pp. 156–167 (2004)
4. Che, D., Aberer, K., Özsu, M.T.: Query optimization in XML structured-document databases. The VLDB Journal The International Journal on Very Large Data Bases 15(3), 263–289 (2006)
5. Clark, J., DeRose, S.: XML Path Language, XPath 1.0 (1999),
   http://www.w3.org/TR/xpath20/#axes
6. Le, D., Bressan, S., Taniar, D., Rahayu, W.: Semantic XPath Query Transformation: Opportunities and Performance. In: Kotagiri, R., Radha Krishna, P., Mohania, M., Nantajeewarawat, E. (eds.) DASFAA 2007. LNCS, vol. 4443, pp. 994–1000. Springer, Heidelberg (2007)
7. Le, D., Bressan, S., Taniar, D., Rahayu, W.: A Utilization of Schema Constraints to Transform Predicates in XPath Query. In: Bringas, P.G., Hameurlain, A., Quirchmayr, G. (eds.) DEXA 2010. LNCS, vol. 6261, pp. 331–339. Springer, Heidelberg (2010) (in press)
8. Le, D.X.T., Pardede, E.: Towards Performance Efficiency in Safe XML Update. In: Benatallah, B., Casati, F., Georgakopoulos, D., Bartolini, C., Sadiq, W., Godart, C. (eds.) WISE 2007. LNCS, vol. 4831, pp. 563–572. Springer, Heidelberg (2007)
9. Le, D., Pardede, E.: On Using Semantic Transformation Algorithms for XML Safe Update. In: 8 th International Conference on Information Systems Technology and its Applications, ISTA 2009, pp. 367–378 (2009)
10. Groppe, S., Groppe, J.: A Prototype of a Schema-Based XPath Satisfiability Tester. In: Bressan, S., Küng, J., Wagner, R. (eds.) DEXA 2006. LNCS, vol. 4080, pp. 93–103. Springer, Heidelberg (2006)

11. Groppe, J., Groppe, S.: Satisfiability-Test, Rewriting and Refinement of Users' XPath Queries According to XML Schema Definitions. In: Manolopoulos, Y., Pokorný, J., Sellis, T.K. (eds.) ADBIS 2006. LNCS, vol. 4152, pp. 22–38. Springer, Heidelberg (2006)
12. Groppe, J., Groppe, S.: Filtering Unsatisfiable XPATH Queries. In: Proc. of the 8th Intl. Conf. on Enterprise Information Systems: Databases & Information Syst Integration, ICEIS 2006, vol. (2), pp. 157–162 (2006)
13. Gupta, K.A., Suciu, D.: Stream Processing of XPath Queries with Predicates. In: Proceedings of the 2003 ACM SIGMOD International Conference on Management of Data, pp. 419–430 (2003)
14. Su, H., Murali, M., Rundensteiner, E.: Semantic Query Optimization in an Automata Algebra Combined XQuery Engine over XML Streams. In: Proceedings of the 30th Very Large Data Bases (VLDB) Conference, Toronto, Canada, pp. 1293–1296 (2004)
15. Su, H., Rundensteiner, E., Mani, M.: Semantic Query Optimization for XQuery over XML Streams. In: Proceedings of the 31st Intl Conference on Very Large Data Bases (VLDB), pp. 277–282 (2005)
16. Wang, G., Liu, M., Yu, J.: Effective Schema-Based XML Query Optimization Techniques. In: Proceedings of the 7th Intl Database Engineering and Application Symposium (IDEAS), pp. 1–6 (2003)

Domain-Specific Language for Context-Aware Web Applications

Michael Nebeling, Michael Grossniklaus,
Stefania Leone, and Moira C. Norrie

Institute of Information Systems, ETH Zurich
CH-8092 Zurich, Switzerland
{nebeling,grossniklaus,leone,norrie}@inf.ethz.ch

Abstract. Context-awareness is a requirement in many modern web
applications. While most model-driven web engineering approaches have
been extended with support for adaptivity, state-of-the-art development
platforms generally provide only limited means for the specification of
adaptation and often completely lack a notion of context. We propose a
domain-specific language for context-aware web applications that builds
on a simple context model and powerful context matching expressions.

1 Introduction

Context-awareness has been recognised as an important requirement in several
application domains ranging from mobile and pervasive computing to web engineering. Particularly in web engineering, it has been proposed to address person-
alisation, internationalisation and multi-channel content delivery. With clients
ranging from desktop computers, laptops and mobile devices to digital TVs and
interactive tabletops or wall displays, web applications today have to support
many different device contexts and input modalities.

Researchers have mainly addressed the need for adaptivity in web appliations
on the conceptual level by adding support for specifying context-sensitive behaviour at design-time into existing methodologies. Most of these model-driven
approaches also feature code generators capable of deploying adaptive web applications, but often lack comprehensive run-time support for managing context-
aware sites. Practitioners, on the other hand, tend to build on development
platforms such as Silverlight, JavaFX and OpenLaszlo to cater for multiple runtime platforms; however, the methods and languages underlying these frameworks often follow a less systematic approach and, in particular, lack support
for context-aware constructs and mechanisms.

In this paper, we propose a domain-specific language for context-aware web
applications that extends and consolidates existing approaches based on context-
aware concepts in markup languages. We present a possible execution environment that builds on a database-driven system developed in previous work [1].
We begin in Sect. 2 by discussing related work. Our approach is described in
Sect. 3. Section 4 presents an implementation of the proposed language followed
by concluding remarks in Sect. 5.

L. Chen, P. Triantafillou, and T. Suel (Eds.): WISE 2010, LNCS 6488, pp. 471–479, 2010.

2   Background

Various model-based approaches have addressed the need for adaptivity in web appliations. For example, WebML [2] has introduced new primitives that allow the behaviour of context-aware and adaptive sites to be modelled at design-time [3]. In UWE [4], the Munich Reference Model [5] for adaptive hypermedia applications provides support for rule-based adaptation expressed in terms of the Object Constraint Language (OCL). The specification of adaptive behaviour is an integral part of the Hera methodology [6], where all design artefacts can be annotated with appearance conditions. These annotations will then be used to configure a run-time environment such as AMACONT [7] which supports run-time adaptation based on a layered component-based XML document format [8]. Finally, the Bellerofonte framework [9] is a more recent approach using the Event-Condition-Action (ECA) paradigm. The implementation separates context-aware rules from application execution by means of a decoupled rule engine with the goal of supporting adaptivity at both design and run-time.

In a second stream, research on user interface description languages has spawned a number of different models and methods with the most prominent example being the CAMELEON framework [10] which separates out different levels of user interface abstraction to be able to adapt to different user, platform and environment contexts. A diversity of languages and tools have been proposed (UsiXML [11], MARIA [12], etc.), and W3C is forming a working group on future standards of model-based user interfaces.[1]

State-of-the-art development practices show a significant gap between the methodical approaches at the conceptual level and existing implementation platforms at the technological level. For example, while researchers promote the importance of design processes such as task modelling, in practice, this step is often skipped and developers instead use visual tools such as Photoshop and Flash for rapid prototyping and to directly produce the concrete user interfaces. Moreover, even though platforms such as Silverlight, JavaFX and OpenLaszlo have been specifically designed to support multiple run-time environments, the underlying languages (Microsoft XAML, JavaFX Script, OpenLaszlo LZX, etc.) use many different approaches and lack common concepts and vocabulary, let alone a unified method, for the specification of context-aware adaptation.

Early research on Intensional HTML (IHTML) [13] demonstrated how context-aware concepts can be integrated into HTML with the goal of declaring versions of web content inside the same document. The concepts used in IHTML were later generalised to form the basis for Multi-dimensional XML (MXML), which in turn provided the foundation for Multi-dimensional Semi-structured Data (MSSD) [14]. However, these languages still suffer from two major design issues. First, the multi-dimensional concepts are not tightly integrated with the markup and need to be specified in separate blocks of proprietary syntax. This renders existing HTML/XML validators and related developer tools void as any attempt to parse IHTML/MXML documents will fail. Second, it becomes

---

[1] http://www.w3.org/2010/02/mbui/program.html

extremely difficult to manage such versioned documents, at least for a considerable number of contextual states, if the possible states are not clearly defined in some sort of context model.

3 Approach

To alleviate the aforementioned problems with IHTML and MXML, we designed XCML not to support versioning of single web documents, but instead to augment entire web sites with context-dependent behaviour based on the following three core principles.

- **XML-based markup language.** We base XCML on XML since many web development frameworks build on XML-based markup languages. XCML introduces a set of proprietary, but well-defined, tags that integrate context-aware concepts into the markup while still producing well-formed XML.
- **Application-specific context model.** Context-aware web sites based on XCML must first define the context dimensions and possible states specific to the respective application. To facilitate the management of larger version spaces, we enforce a clear separation between context state definitions in the header and context matching expressions in the body of XCML documents.
- **Context algebra and expression language.** A distinguishing feature of our approach is that context is a refining concept used to augment rather than completely specify content delivery. XCML is hence based on context *matching* expressions as opposed to statements formulated in an *if-then-else* manner. Instead of relying on rather strict ECA rules as used in [9,5], XCML requests trigger a matching process in which the best-matching variants are determined based on scoring and weighting functions as defined in [1].

To keep the specification of context models simple, we use a very general context representation based on the notion of context dimensions and states. A *context dimension* represents a set of characteristics or semantically grouped factors to be taken into account when compiling the context-aware web site, while a *context state* describes a valid allocation of such a dimension. The following example defines a simple context model to distinguish different browser contexts.

```
<xcml:context>
 <xcml:context-dimension name="Browser">
  <xcml:context-key name="agent" />
  <xcml:context-key name="version" />
 </xcml:context-dimension>
 <xcml:context-state name="old_IE" dimension="Browser">
  <xcml:context-property key="agent" value="IE" />
  <xcml:context-property key="version" value="3..6" />
 </xcml:context-state>
 <xcml:context-state name="new_IE" dimension="Browser">
  <xcml:context-property key="agent" value="IE" />
  <xcml:context-property key="version" value="7..8" />
```

```
</xcml:context-state>
<xcml:context-state name="Firefox" dimension="Browser">
 <xcml:context-property key="agent" value="Firefox" />
 <xcml:context-property key="version" value="1..3.8" />
</xcml:context-state>
[..] <-- other dimensions and states -->
</xcml:context>
```

The extract shows the definition of the Browser dimension in terms of agent and version. With the states defined thereafter, an application can distinguish between previous and current versions of Internet Explorer and Firefox. The potential advantages of our approach are clearer when looking at context models with multiple dimensions. For example, in a three-dimensional context space, an XCML body could be defined to distinguish not only browsers but also languages English and German as well as desktop and mobile platforms as follows.

```
<xcml:layout name="websiteLayout">
 <xcml:layout-variant match="Desktop">
  <!DOCTYPE html PUBLIC "-//W3C//DTD XHTML [..]//EN" [..]>
  <html>
   <head>
    <title> <xcml:attribute-value select="header/title" /> </title>
    <xcml:context match="Firefox">
     <link href="firefox.css" rel="stylesheet" type="text/css" />
    </xcml:context>
    <xcml:context match="old_IE or new_IE">[..]</xcml:context>
   </head>
   <body> <xcml:component select="*" />[..]</body>
  </html>
 </xcml:layout-variant>
 <xcml:layout-variant match="Mobile"> [..] </xcml:layout-variant>
</xcml:layout>
<xcml:component name="website">
 <xcml:component-variant layout="websiteLayout" match="Desktop">
  <xcml:child-components>
   <xcml:component name="header">
    <xcml:component-variant type="headerType"  match="English">
     <xcml:attribute-value name="title" value="XCML Site" /> [..]
    </xcml:component-variant>
    <xcml:component-variant match="German">[..]</xcml:component-variant>
   </xcml:component>
   [..] <!-- other components -->
  </xcml:child-components>
 </xcml:component-variant>
 <xcml:component-variant match="Mobile"> [..] </xcml:layout-variant>
</xcml:component>
```

As with many existing approaches, we separate presentation concepts from content and navigation. In XCML, we achieve this by means of a simple component-based model that allows for versioning on each layer in terms of so-called variants. The element websiteLayout in the example above defines variants to place the web site's title in the best-matching language, link stylesheets optimised for Internet Explorer or Firefox and recursively include all children of the associated components. The website component associated with the layout defines content variants such as the title in English and German and structure variants for the desktop and mobile platforms.

XCML supports versioning at the data, structure and layout levels for maximum flexibility. The essence of XCML is however the underlying context algebra that allows for the specification of context-dependent behaviour along simple but powerful context matching expressions. The match clause "Firefox" used in the example directly translates to $\{(agent, Firefox), (version, 1..3.8)\}$ and, in this case, constitutes a best-match for all Firefox browsers versions 1 to 3.8. XCML also supports more complex context expressions such as "old_IE or new_IE" and "not Firefox" or even those that span multiple dimensions as in "Desktop and (old_IE or new_IE)". The evaluation of such expressions involves two steps. First, each context state will be substituted with the respective context properties. This would mean that the latter expression evaluates to "{(type,desktop)} and ({(agent,IE), (version,7)} or {(agent,IE), (version,8)})". A dimensional analysis would then find that the left part of the given expression is of the device dimension (which defines the key type) while the right part translates to the browser dimension (which defines the two keys agent and version) for both the left and right operand. In the second step, all propositional connectives will be resolved to give {(type,desktop)} and {(agent,IE), (version,7..8)} and finally to {(type,desktop), (agent,IE), (version,7..8)}.

After evaluating each context expression, it is necessary to compare the resulting context states against the context for which the web site is compiled. Only if equal or at least a partial match will a variant associated with a context state become part of the whole version. In our example, this would mean that the variant associated with {(type,desktop), (agent,IE), (version,7..8)} would only be displayed if the web site is accessed from the desktop using either Internet Explorer 7 or 8 (independent of the selected language).

4   Implementation

We have implemented an execution environment for XCML based on a generalisation of the XCM system that we developed in previous work [1]. By mapping the XCML language concepts to the previously established concepts for context-aware data management, we can use the database as a cache for the context-dependent variants as shown in Fig. 1. We have defined several components for the processing of XCML, which now form the new *design-time* and complement the existing *run-time* of the XCM system.

The resulting platform supports two processes: (1) the compilation process of XCML to create context-dependent variants in the database and (2) the linking process triggered in response to client requests and the context in which these take place. The roles of the individual system components are as follows.

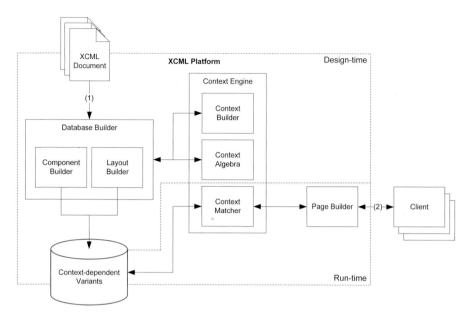

Fig. 1. Integration of the new design-time components with XCM run-time

- *Database Builder.* As the Database Builder parses the XCML documents, it employs the Context Builder to build the context space and creates the corresponding context-dependent variants in the database.
- *Context Engine.* The Context Engine has been extended with the Context Builder as part of the design-time to provide several methods to parse context information declared in XCML documents. The Context Algebra is represented by a utility class that implements the two-step evaluation process of context expressions described in Sect. 3. The role of the Context Matcher is to score all variants of a specific object and determine the best-matching version for a given context at run-time as described in detail in [1].
- *Page Builder.* In response to client requests, the Page Builder uses the Context Matcher to retrieve the best-matching variant for all context-aware elements stored in the database and assembles them into the requested version of the web site. Again, details are given [1].

We now explain how XCML documents are processed and compiled into the database (process (1) in Fig. 1). As shown in Fig. 2, presentation instructions within xcml:layout tags are handled differently from XCML component variants.

Components. For each xcml:component-variant, the Component Builder creates a version with the respective attribute value sets. Each version is appended to its component object and stored in the database together with the specific context that was evaluated from the associated match clause. This step repeats recursively for all child components contained in each variant.

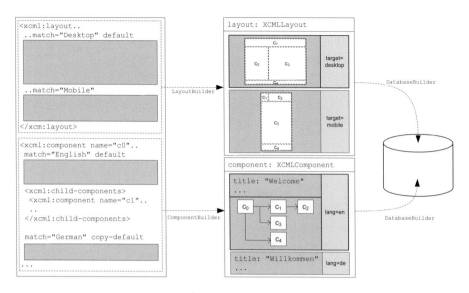

Fig. 2. Processing of XCML layouts and components

The processing of components is rather straightforward as we specify each variant separately using structured data in XCML attribute values. We have, however, defined an inheritance mechanism for component variants to inherit from the default variant using the **copy-default** attribute and only override the context-sensitive values. This kind of inheritance is reasonable as default variants are usually rich in detail, and some fields such as a person's first and last name in staff member pages are constant across all versions. This is in contrast to layouts where the underlying templates are typically semi-structured and patterns for reuse are somewhat different. For that reason, we allow the developer to define layout variants separately similar to components but without inheritance or to use nested context matching expressions within the same layout variant. If the differences between individual templates are not substantial, then nested variants are more practical in order not to duplicate large parts between versions that are essentially the same.

Layouts. For each xcml:layout defined in an XCML document, the Layout Builder preprocesses all specified variants to find, combine, and evaluate the individual context expressions that may have been nested inside the template. For each context evaluated from the expressions, it then runs a first XSL transformation on the template to generate a new XSL template that

(a) copies only the XML child nodes nested in matching xcml:context blocks and
(b) replaces each xcml:attribute-value/xcml:component select with a corresponding xsl:value-of select statement.

Step (a) therefore reduces the multi-dimensional XCML template to an XSL template specific to the context for which it was evaluated. Step (b) is required to resolve the XCML namespace and prepare the resulting XSL template for the run-time where all placeholders for nested attribute values and components will be substituted with the best-matching results evaluated for the specific context. Finally, each of the reduced XSL templates will be appended to the corresponding layout object and stored in the database together with the context.

5   Conclusion

We have presented a domain-specific language that promotes a simple context model and powerful context matching expressions. Our work was motivated by the lack of context-aware concepts in state-of-the-art web development platforms and languages. In future work, we want to improve the integration with existing frameworks and also design adequate tool support for the development and debugging of adaptive web sites based on our language.

References

1. Grossniklaus, M., Norrie, M.C.: An Object-Oriented Version Model for Context-Aware Data Management. In: Benatallah, B., Casati, F., Georgakopoulos, D., Bartolini, C., Sadiq, W., Godart, C. (eds.) WISE 2007. LNCS, vol. 4831, pp. 398–409. Springer, Heidelberg (2007)
2. Ceri, S., Fraternali, P., Bongio, A., Brambilla, M., Comai, S., Matera, M.: Designing Data-Intensive Web Applications. Morgan Kaufmann Publishers Inc., San Francisco (2002)
3. Ceri, S., Daniel, F., Matera, M., Facca, F.M.: Model-driven Development of Context-Aware Web Applications. TOIT 7(1), Article 2 (2007)
4. Hennicker, R., Koch, N.: A UML-Based Methodology for Hypermedia Design. In: Evans, A., Caskurlu, B., Selic, B. (eds.) UML 2000. LNCS, vol. 1939, pp. 410–424. Springer, Heidelberg (2000)
5. Koch, N., Wirsing, M.: The Munich Reference Model for Adaptive Hypermedia Applications. In: De Bra, P., Brusilovsky, P., Conejo, R. (eds.) AH 2002. LNCS, vol. 2347, pp. 213–222. Springer, Heidelberg (2002)
6. Houben, G.J., Barna, P., Frăsincar, F., Vdovják, R.: Hera: Development of Semantic Web Information Systems. In: Cueva Lovelle, J.M., Rodríguez, B.M.G., Gayo, J.E.L., Ruiz, M.d.P.P., Aguilar, L.J. (eds.) ICWE 2003. LNCS, vol. 2722, pp. 529–538. Springer, Heidelberg (2003)
7. Fiala, Z., Hinz, M., Houben, G.J., Frăsincar, F.: Design and Implementation of Component-based Adaptive Web Presentations. In: Proc. SAC, pp. 1698–1704 (2004)
8. Fiala, Z., Hinz, M., Meissner, K., Wehner, F.: A Component-based Approach for Adaptive, Dynamic Web Documents. JWE 2(1-2), 58–73 (2003)
9. Daniel, F., Matera, M., Pozzi, G.: Managing Runtime Adaptivity through Active Rules: the Bellerofonte Framework. JWE 7(3), 179–199 (2008)

10. Calvary, G., Coutaz, J., Thevenin, D., Limbourg, Q., Bouillon, L., Vanderdonckt, J.: A Unifying Reference Framework for Multi- Target User Interfaces. Interacting with Computers 15(3), 289–308 (2003)
11. Limbourg, Q., Vanderdonckt, J., Michotte, B., Bouillon, L., López-Jaquero, V.: UsiXML: UsiXML: a Language Supporting Multi-Path Development of User Interfaces. In: Proc. EHCI/DS-VIS, pp. 200–220 (2005)
12. Paternò, F., Santoro, C., Spano, L.: MARIA: A Universal Language for Service-Oriented Applications in Ubiquitous Environment. ACM Trans. on Computer-Human Interaction 16(4), Article 19 (2009)
13. Wadge, W.W., Brown, G., Schraefel, M.C., Yildirim, T.: Intensional HTML. In: Munson, E.V., Nicholas, C., Wood, D. (eds.) PODDP 1998 and PODP 1998. LNCS, vol. 1481, pp. 128–139. Springer, Heidelberg (1998)
14. Stavrakas, Y., Gergatsoulis, M.: Multidimensional Semistructured Data: Representing Context-Dependent Information on the Web. In: Pidduck, A.B., Mylopoulos, J., Woo, C.C., Ozsu, M.T. (eds.) CAiSE 2002. LNCS, vol. 2348, pp. 183–199. Springer, Heidelberg (2002)

Enishi: Searching Knowledge about Relations by Complementarily Utilizing Wikipedia and the Web

Xinpeng Zhang, Yasuhito Asano, and Masatoshi Yoshikawa

Kyoto University, Kyoto, Japan 606-8501
{xinpeng.zhang@db.soc.,asano@,yoshikawa@}i.kyoto-u.ac.jp

Abstract. *How global warming and agriculture mutually influence each other?* It is possible to answer the question by searching knowledge about the relation between global warming and agriculture. As exemplified by this question, strong demands exist for searching relations between objects. However, methods or systems for searching relations are not well studied. In this paper, we propose a relation search system named "Enishi." Enishi supplies a wealth of diverse multimedia information for deep understanding of relations between two objects by complementarily utilizing knowledge from Wikipedia and the Web. Enishi first mines elucidatory objects constituting relations between two objects from Wikipedia. We then propose new approaches for Enishi to search more multimedia information about relations on the Web using elucidatory objects. Finally, we confirm through experiments that our new methods can search useful information from the Web for deep understanding of relations.

Keywords: knowledge retrieval, Wikipedia mining, relation.

1 Introduction

What is the relation between petroleum and Japan? Why has the bankruptcy of Lehman Brothers Holdings Inc. so strongly influenced Japan? How global warming and agriculture mutually affect each other? Answering these questions demands knowledge about relations between objects, such as countries, products, people, and events. Our real life includes strong needs for querying knowledge about relations. For example, understanding the relation between two countries is useful to study history or politics. Discovering the relation between a country and a product can presumably help in making an investment decision. Surveying the connection between oneself and an unknown person to find paths to contact the unknown person through some mutual acquaintances.

The Web has become the most important source for seeking information. However, the knowledge existing on the Web is not well organized, in general (We consider Wikipedia separately from the Web in this paper). A web page might contain knowledge related to multiple objects; multiple different objects might be represented by the same word. Therefore, it is difficult to search knowledge about relations between objects on the Web. Recently, several semantic knowledge bases [1] are used for searching semantic relations between two objects. However, the semantic relations defined in these knowledge bases, such as "isCalled," "type," and "subClassOf," are far from covering relations existing in the real world. For example, questions described at the beginning of

L. Chen, P. Triantafillou, and T. Suel (Eds.): WISE 2010, LNCS 6488, pp. 480–495, 2010.

this section are intended to obtain diverse knowledge about relations, rather than simple semantic relations. Because of need for and lack of methods for searching relations, we aim to establish a means to search knowledge about relations in this paper.

In Wikipedia, the knowledge associated with an object is well organized on a single page. Wikipedia contains knowledge of objects in numerous categories such as people, science, geography, politic, and history. The link structure between Wikipedia pages represents the relations between objects. Therefore, Wikipedia is efficient for searching knowledge about objects and relations between objects. However, the knowledge contained in Wikipedia is still much less than that existing on the Web. Especially, multimedia information such as images is lacking in Wikipedia. It is desired to search more knowledge, including images on the Web, to complement the knowledge that is available from Wikipedia.

In this paper, we propose a relation search system named "Enishi[1]." Enishi uses the knowledge in Wikipedia and that of the Web complementarily, to search knowledge about relations. Generally, relations existing in the real world are complicated and diverse. To help users to deeply understand relations between pairs of objects, we aim to supply a wealth of multimedia information about relations. Given two objects, Enishi (a) mines knowledge about the relation between the two objects from Wikipedia; and (b) searches for knowledge, including images, explaining the relation on the Web. We use the method proposed by Zhang et al.[2,3] to mine relations in Wikipedia. To achieve the goal of (b), as a contribution of this paper, we propose approaches to search knowledge including images explaining relations on the Web, using knowledge mined from Wikipedia. Finally, we were able to complement the knowledge that is lacking in Wikipedia using the knowledge retrieved from the Web for deeply understanding a relation. We confirm through experiments described in Section 5 that our approaches were useful to search knowledge that is applicable for understanding relations. To the best of our knowledge, Enishi is the first system that is used for searching diverse knowledge including images about relations on the web. Although there are services on the Web for searching people related to a certain people [4,5].

The rest of this paper is organized as follows. Section 2 discusses the system in detail. Section 3 introduces the method proposed by the authors for mining relations in Wikipedia. Section 4 presents approaches for searching knowledge about relations on the Web. Section 5 reports experiments used for evaluating the method described in Section 4. Section 6 reviews related work. Section 7 concludes this paper.

2   Enishi Search System

The Enishi (relation) Search System offers means to search knowledge about a relation between two objects for deep understanding of the relation. Enishi is implemented based on Wikipedia. Consequently, it only searches the relations between objects existing in Wikipedia. Fig. 1 portrays the interface of Enishi, in which we search the relation between "Global warming" and "Agriculture." Enishi searches knowledge of the following three types about the relation between two given objects.

---

[1] Enishi is a Japanese word meaning relation.
The system is accessible at http://www.db.soc.i.kyoto-u.ac.jp/enishi/enishi.html.

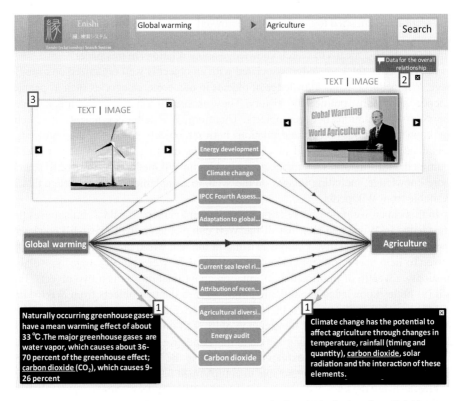

Fig. 1. Searching relation between "Global warming" and "Agriculture" on Enishi

(1) Disjoint paths formed by links in Wikipedia that are important for explaining the relation.
(2) Images with surrounding text retrieved on the web, which include knowledge about the overall relation.
(3) Images with surrounding text retrieved on the web, which include knowledge about each path for the relation of type (1).

We introduce knowledge of the three types in detail below.

2.1   Disjoint Paths Mined from Wikipedia

Two kinds of relations exist: "explicit relations" and "implicit relations." In Wikipedia, an explicit relation is represented as a link. For example, an explicit relation between global warming and carbon dioxide might be represented by a link from the page "Global warming" to the page "Carbon dioxide." A user could understand its meaning by reading the text "greenhouse gases have a mean warming effect...The major greenhouse gases are...carbon dioxide..." surrounding the anchor text "carbon dioxide" on page "Global warming." In Wikipedia, multiple links and pages represent an implicit relation. For example, global warming affects agriculture through changes in carbon

U.S. President George W. Bush Meets With Japanese Prime Minister Junichiro Koizumi

CRAWFORD, TX - MAY 23: U.S. President George W. Bush (R) and Japanese Prime Minister Junichiro Koizumi walk together after driving up in Bush's pickup truck before a media conference May 23, 2003 on Bush's ranch near Crawford, Texas. President Bush and Prime Minister Koizumi met to discuss international issues such as North Korea and Iraq.

Fig. 2. George W. Bush with Junichiro Koizumi

dioxide. This fact might be an implicit relation represented by two links in Wikipedia: one between "Global warming" and "Carbon dioxide" and the other one between "Agriculture" and "Carbon dioxide."

We define a *Wikipedia information network*, whose vertices are pages of Wikipedia and whose edges are links between pages. We then mine disjoint paths between two objects on a Wikipedia information network for explaining the relation between the two objects using the method proposed by Zhang et al. [3]. Enishi displays the top-10 important paths explaining a relation. For example, Fig. 1 shows the top-10 paths for the relation between "Global warming" and "Agriculture." A user can understand the meaning of a path easily by tracing the links in the path from left to right. Tracing each link can be done by understanding the meaning of an explicit relation represented by the link. For each link (u, v), Enishi extracts a snippet surrounding the anchor text of link v on page u for explaining the link. In Enishi, a balloon displaying the snippet for a link is popped up by a mouse rollover on the link. For example, in Fig. 1, if users read the snippets indicated by the number 1, then they can understand the bottom path containing "Carbon dioxide."

Users prefer not to read the same objects repeatedly in the mined paths, and they might desire to obtain knowledge of various kinds by understanding a few paths. Therefore, Zhang et al. [3] mine disjoint paths to avoid outputting redundant objects in the mined paths. In Wikipedia, a path constituted by edges of different directions might be important for a relation, such as the the bottom path in Fig. 1. Therefore, Zhang et al. [3] mine both paths constituted by edges of the same or opposite directions. Zhang et al. [3] confirmed through experiments that the method can mine many paths important for understanding relations.

2.2   Images with Surrounding Text for the Overall Relation

It is desired to search more knowledge about a relation on the web, especially images lacking in Wikipedia. Enishi searches sets of "image with surrounding text" (hereafter abbreviated as $IwST$) on the Web containing knowledge about a relation, using methods which will be discussed in Section 4. An image itself might make it difficult to infer a relation. However, by reading the surrounding text of the image, users were able to understand the relation. For example, Fig. 2 portrays an image in which both George W. Bush and Junichiro Koizumi appear. It is difficult to understand the relation between the two by watching the image. We were able to understand that they had discussed international issues by reading the surrounding text explaining the image shown on the right side. Nevertheless, the image could be an evidence of the fact described in the

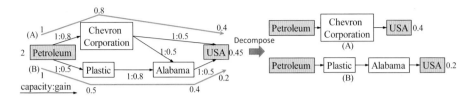

Fig. 3. A generalized max-flow and its decomposition

surrounding text, and is helpful for understanding the text. Therefore, we search images with surrounding text by observing which users could understand a relation.

In Enishi, a window is popped up by pressing the button "Data for the overall relation," on which users were able to browse images associated with a relation. Users are able to click an image to link to the Web page containing the image. By reading the surrounding text of the image on the page, users could understand knowledge about a relation. For example, in Fig. 1, the window indicated by the number 2 displays an image about the relation between "Global warming" and "Agriculture." Users could click the triangles on the window to browse other images associated with the relation.

2.3   Images with Surrounding Text for Each Path

We extract snippets from Wikipedia to explain every link in a path mined from Wikipedia. Searching more knowledge about a path is useful for deepening understanding of the path. Therefore, Enishi also searches images on the Web with surrounding text for explaining a path. Precisely, for a path of a relation between two objects s and t, Enishi searches $IwST$s containing knowledge about the relations between objects in each edge of the path, or knowledge about how important the intermediate objects in the path are to the relation between s and t. In Enishi, a window is popped up by clicking a path. Thereby, users were able to brows images associated with the path. Users could click an image to view its surrounding text. For example, in Fig. 1, the window indicated by number 3 displays an image for the top path containing "Energy development."

3   Methods for Mining Knowledge in Wikipedia

In this section, we introduce a method for mining disjoint paths explaining a relation in Wikipedia, as proposed by Zhang et al. [3]. Before that, we first explain its basis: the generalized max-flow model proposed by Zhang et al. [2] for measuring relations.

3.1   Generalized Max-Flow Model for Measuring Relations

Zhang et al. [2] model a relation between two objects in a Wikipedia information network using a generalized max-flow. The generalized max-flow problem [6] is identical to the classical max-flow problem except that every edge e has a gain $\gamma(e) > 0$; the value of a flow sent along edge e is multiplied by $\gamma(e)$. Let $f(e) \geq 0$ be the amount of flow f on edge e, and $\mu(e) \geq 0$ be the capacity of edge e. The capacity constraint

$f(e) \leq \mu(e)$ must hold for every edge e. The goal of the problem is to send a flow emanating from the source into the destination to the greatest extent possible, subject to the capacity constraints. Let *generalized network* $G = (V, E, s, t, \mu, \gamma)$ be information network (V, E) with the source $s \in V$, the destination $t \in V$, the capacity function μ, and the gain function γ. Fig. 3 depicts an example of a generalized max-flow. 0.4 units and 0.2 units of the flow arrive at "USA" along path (A) and path (B), respectively.

To measure the strength of a relation from object s to object t, Zhang et al. [2] use the value of a generalized maximum flow emanating from s as the source into t as the destination; a larger value signifies a stronger relation. Such a flow seldom uses an edge whose direction is opposite that from the source to the destination. As discussed in Section 2, a path constituted by edges of different directions could be important for a relation in Wikipedia. To use such paths, Zhang et al. construct a doubled network [2] by adding to every original edge a reversed edge whose direction is opposite to the original one. It is desired to assign larger gains to edges that are important for the relation in the model. To realize such a gain assignment, Zhang et al. [2] proposed an edge gain function using the category structure of Wikipedia. We omit details of the gain function here because of space limitations. Zhang et al. [2] ascertained the model can measure the strength of relations more correctly than previous methods [7,8] can.

3.2   Generalized Flow Based Method for Mining Disjoint Paths

Based on the generalized max-flow model, Zhang et al. proposed a method to mine disjoint paths important for a relation from object s to object t in Wikipedia [3]. Zhang et al. also proposed a new technique of setting vertex capacities to force the flow to be sent along disjoint paths [3].

We introduce the method presented below. (1) Construct a generalized network $G = (V, E, s, t, \mu, \gamma)$ using pages and links within, at most, m hop links from s or t in Wikipedia. (2) Construct the doubled network G' for G, determine the edge gain γ, and set vertex capacities. (3) Compute a generalized max-flow f emanating from s into t on G'. (4) Decompose the flow f into flows on a set P of paths. Let $df(p_i)$ denote the value of flow on a path p_i, for $i = 1, 2, ..., |P|$. For example, the flow on the network depicted in Fig. 3 is decomposed into flows on two paths (A) and (B). The value of the decomposed flow on path (A) is 0.4; that on path (B) is 0.2. (5) Output the top-k paths in decreasing order of $df(p_i)$. (6) For each edge (u, v) in the top-k paths, extract an explanatory snippet, i.e., text surrounding the anchor text of link v on page u.

4   Searching Knowledge Explaining Relations on the Web

We now present techniques for searching images on the Web with surrounding text about a relation, using knowledge mined from Wikipedia. We first propose a evidence-based method for searching $IwSTs$ associated with a relation. We then propose a Top-Down approach and a Bottom-Up approach using the method to search $IwSTs$ for the overall of a relation and $IwSTs$ for each path of a relation.

4.1 Evidence-Based Method

We aim to search $IwST$s containing knowledge about a relation between two objects. Keyword image search engines such as Google image search engine[2] offer an easy way for searching images related to two objects. For example, we could search images related to "petroleum" and "Japan" using a query "petroleum Japan," or "petroleum and Japan." Some images in the search results generated using these queries are useful for understanding the relation between "petroleum" and "Japan" by reading the images and the surrounding text of the images. However, some images and their surrounding text contain no knowledge about the relation, such as images of products using oil from shopping sites, images of books talking about petroleum, or images even almost unrelated to "petroleum" or "Japan." The engine returns images whose pages containing words in the queries, but it does not search images associated with a relation. Therefore, we propose an evidence-based method for searching $IwST$s associated with a relation. The method regards the image as relevant to the relation if the surrounding text of an image contains knowledge about a relation.

The method introduced in Section 3.2 mines the top-k paths that are important for explaining a relation in Wikipedia. We define *elucidatory objects* of a relation as objects in the top-k paths, except the source and destination. Every elucidatory object o in a path p is assigned a weight $0 < w(o) < 1$, which equals to the value of the decomposed flow on path p. A high weight signifies that the elucidatory object plays an important role in the relation. If the surrounding text of an image contains two objects s and t, and many elucidatory objects of the relation between s and t, we then infer that the text contains knowledge about the relation. That is, elucidatory objects are evidence that is useful for judging whether a text contains knowledge about a relation.

We present the method **EBM**(s, t, k, m, n) for searching $IwST$s associated with the relation between two objects s and t as follows.

Input: objects s and t, integer parameters k, m, and n.
(1) Mine the top-k paths that are important for the relation between s and t using the method discussed in Section 3.2.
(2) Obtain a set O of elucidatory objects in the top-k paths.
(3) Search the top-m images, say $m = 300$, using a keyword image search engine with query "$s\ t$."
(4) Extract the surrounding text of each image. Let I be the set of the top-m images with surrounding text.
(5) Remove $IwST$s whose surrounding text contains no s or t from I.
(6) Compute a score $s(i)$ for every $i \in I$ to

$$s(i) = \sum_{o \in s, t, O'} w(o) \times log_e(e + f(o)), \qquad (1)$$

---

[2] http://images.google.com/

where $O' \subseteq O$ is the set of elucidatory objects appearing in the surrounding text of i, and $f(o)$ is the appearance frequency of o in i. The weight $w(s)$ and $w(t)$ is set to the maximum weight of all objects in O.

(7) **Output:** the top-n $IwST$s $i \in I$ having high scores.

An object might be represented by multiple alternative words on the Web. For example, "USA," "U.S.A.," "U.S.," "America," and "United States of America," all represent the "United States." We count the appearance frequency of an object including the instances of appearance of its alternative words at step (6). We obtain alternative words representing an object using redirect pages in Wikipedia. For example, searching "USA" will redirect you to the page "United States" in Wikipedia.

A text in which many elucidatory objects appear a few times tends to contain more knowledge about a relation, than a text in which few elucidatory objects appear many times. Therefore, we take the logarithm of the appearance frequency in Equation (1) to alleviate the influence of the high appearance frequency of a single elucidatory object.

4.2   Top-Down Approach

As discussed in Section 2, the Enishi system searches both $IwST$s explaining the overall of a relation and those explaining each path that is important for a relation mined from Wikipedia. In this section, using the evidence-based method, we propose a Top-Down approach that first searches $IwST$s for the overall relation, then it constructs $IwST$s for each path of the relation. We present the Top-Down approach as follows.

Input: objects s and t, integer parameters k, m, n, and l.
(1) Search the top-n $IwST$s associated with the relation r between s and t using the evidence-based method **EBM**(s, t, k, m, n). Let TI be the set of the top-n $IwST$s.
(2) **Output:** TI as the $IwST$s for explaining the overall of relation r.
(3) For every path p that is important for relation r, do step (4) and (5).
(4) Compute a score $s(i)$ for every $i \in TI$ toward path p using the following equations. Let $O'_p \subseteq O$ be the set of elucidatory objects which exist in path p and appear in the surrounding text of i; then

$$s'(i) = \sum_{o \in O'_p} w(o) \times log_e(e + f(o)).$$

$$if\ s'(i) = 0,\ then\ s(i) = 0;$$

$$else\ s(i) = s'(i) + \sum_{o \in s,t} w(o) \times log_e(e + f(o)).$$

(5) **Output:** the top-l $IwST$s $i \in TI$ having high scores as the $IwST$s explaining p.

The Top-Down approach first searches a set TI of $IwST$s for the overall relation using the evidence-based method. The approach then designates an $IwST$ in TI for a path if the surrounding text of the $IwST$ contains elucidatory objects in the path. The $IwST$s for every path constitute a subset of the $IwST$s for the overall relation. If the surrounding text of any $IwST$ in TI contains no elucidatory object in a path, then the Top-Down method can not search $IwST$s for the path.

4.3   Bottom-Up Method

We now propose the Bottom-Up approach; it first searches $IwST$s for each path, then constructs $IwST$s for the overall relation. We present the approach as follows.

Input: objects s and t, integer parameters k, m, n, and l.
(1) For every top-k path $p = (o_0 = s, o_1, ..., o_n = t)$ that is important for the relation r between s and t, do step (2) and (3).
(2) For every edge (o_i, o_{i+1}) in p, search the top-n $IwST$s associated with the relation between o_i and o_{i+1} using the evidence-based method **EBM**(o_i, o_{i+1}, k, m, n).
(3) **Output:** the top-l $IwST$s having high scores among the $IwST$s for all edges in p, as the $IwST$s explaining path p.
(4) **Output:** the top-n $IwST$s having high scores among the $IwST$s for all top-k paths, as the $IwST$s explaining the overall of relation r.

To search $IwST$s for the overall of relation r, the Bottom-Up approach first have to search $IwST$s for every relation represented by each edge in the top-k paths that are important for relation r. In contrast to the Top-Down method, the $IwST$s for the overall relation constitute a subset of the $IwST$s for all the paths. We compare the Top-Down method and the Bottom-Up method in detail through experiments in Section 5.

5   Experiments

In this section, we describe experiments conducted to evaluate the methods for searching $IwST$s associated with a relation on the Web, as discussed in Section 4.

5.1   Dataset

We select 40 relations between two objects of the following four types for evaluation: (1) global warming and an industry, (2) petroleum and a country, (3) a politician and a country, and (4) two politicians. From a English Wikipedia dataset (2010/03/12 snapshot) we mined the disjoint paths important for every relation. To search $IwST$s for the overall and each path of the 40 relations, the evidence based method totally gathered 400,000 images and their Web pages.

5.2   Evaluation of Searching Knowledge on the Web

We consider the following questions to evaluate the Top-Down and the Bottom-Up approaches. (Q1) Are the $IwST$s searched by the two methods useful for understanding a relation or a path of a relation? (Q2) Which method is better for searching $IwST$s associated with the overall relation or for each path of a relation? (Q3) Do the $IwST$s searched on the Web contain new knowledge that does not exist in Wikipedia?

Table 1. Selected relations for human subjects

ID	Relation	ID	Relation
1	Global warming - Agriculture	5	Vladimir Putin - Ukraine
2	Global warming - Energy industry	6	Hu Jintao - North Korea
3	Petroleum - Saudi Arabia	7	Jacques Chirac - Wen Jiabao
4	Petroleum - Japan	8	George W. Bush - Junichiro Koizumi

Human Subjects. We first answer questions (Q1) and (Q2) using evaluations by human subjects. We randomly select 8 from the 40 relations (two from each of the four types) explained above, which are listed in Table 1. For each relation, we output the top-10 $IwST$s for the overall relation and the top-5 $IwST$s for each of the top-5 paths of the relation, respectively using Top-Down approach and Bottom-Up approach. We set $k = 100$ for both approaches, that is using the objects in the top-100 paths as elucidatory objects. The evidence-based method outputs $IwST$s associated with a relation between s and t from the top-300 images searched using Google image search engine using a query "s t." We regard the top-10 images searched using Google for a relation as the $IwST$s explaining the overall relation outputted by Google; we compare the $IwST$s with those searched using our approaches. Totally, we obtained 417 $IwST$s for the eight relations (different approaches output some identical $IwST$s).

We then ask 10 testers to evaluate every $IwST$s. To each $IwST$ for the overall relation, every tester assigns an integer score of 0, 1, or 2 representing whether the $IwST$ is useful for understanding the relation. To each $IwST$ for a path of a relation r, every tester assigns an integer score of 0, 1, or 2 representing whether the $IwST$ is useful for understanding any relation represented by each edge in the paths, or understanding the importance of the elucidatory objects in the path to the relation. A higher score represents utility for understanding the overall or a path of a relation better or more deeply. Score 0 means that an $IwST$ is useless. Each tester assigns scores independently of the others by reading every image with its surrounding text. We then compute the average of the scores given by the 10 testers of every $IwST$ for each relation.

Fig. 4 presents the average scores of the $IwST$s for each overall relation obtained using each approach. A bar is combined by the lower dark-colored part and the upper light-colored part. Let $S(i)$ be the sum of scores $i \in 0, 1, 2$ assigned by the testers to the $IwST$s outputted by a approach for a relation. The dark-colored part and the light-colored part indicate the proportion $p(2) = \frac{S(2)}{S(2)+S(1)}$ and $p(1) = \frac{S(1)}{S(2)+S(1)}$, respectively. The Top-Down approach produced the highest average for all relations. As shown by the dark-colored parts of the bars, the $IwST$s outputted by the Top-Down approach also received more scores 2 (on average for the 8 relations $p(2) = 0.69$) than other approaches. By contrast, 75.5% of the scores assigned to the $IwST$s for all the 8 relations searched using Google are 0 or 1. For example, many top-10 images searched using Google for the relation between George W. Bush and Junichiro Koizumi are assigned score 0. The surrounding text of these images contain almost no description about the relation, although the two politicians appear in the images. As another example, for the relation between petroleum and Japan, the surrounding text of many top-10 images searched using Google contain knowledge related only to petroleum or Japan, or no meaningful knowledge about the relation at all. However, the Top-Down approach

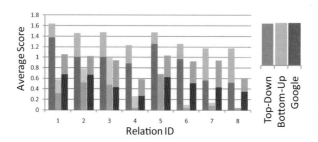

Fig. 4. Average scores of $IwST$s for each overall relation

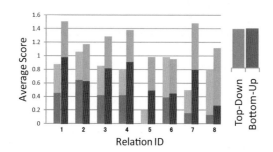

Fig. 5. Average scores of $IwST$s for paths of each relation

were able to ensure that the surrounding text of result $IwST$s contain knowledge about a relation using elucidatory objects. The performance of the Bottom-Up approach was inferior. The Bottom-up approach searches $IwST$s for an overall relation from $IwST$s explaining each path of the relation. Testers consider that most $IwST$s explaining a path are too minute to understanding the overall relation generally. For example, the $IwST$ of ID 2 presented in Fig.6 explains the path containing "Current sea level rise" of the relation between "Global warming" and "Agriculture." However, the $IwST$ is not helpful for understanding the overall relation without knowing the path. Therefore, we conclude that the Top-Down approach is sufficient for searching $IwST$s for the overall relation. Furthermore, the top-10 images searched using Google for a relation between s and t using an query "s t" are inappropriate for understanding the relation.

Fig. 5 presents average scores of $IwST$s for each of the top-5 path of the eight relations. Similarly to Fig. 4, the dark-colored part indicates the proportion $\frac{S(2)}{S(2)+S(1)}$. The Bottom-Up approach produced much higher average than the Top-Down approach did for all relations except the relation of ID 6 between Hu Jintao and North Korea. The Bottom-Up approach also gained more scores 2 than the Top-Down approach. The Top-Down approach searches $IwST$s for a path of a relation among $IwST$s explaining the overall relation. However, experiment results revealed that many of these $IwST$s contains little or no knowledge strongly related to a single path. On the other hand, the Bottom-Up approach searches for $IwST$s for a path that contain knowledge about any relation represented by each edge in the path. Such $IwST$s are useful for understanding a path in detail. For example, Fig.6 presents two $IwST$s for a path containing "Current

ID	Image	URL of Web page	Surrounding text extracted	Score
1		http://globalwar mingforkids.com /what-is-global- warming	...Greenhouse gasses trap the rays inside the earth, making us hotter...The warmer oceans lead to ice melt. The ice melt leads to rising sea levels worldwide. The rising sea levels affect clean drinking water supplies, people's homes, agriculture...	1.2
2		http://wattsupwi ththat.com/2009 /03/06/basic- geology-part-3- sea-level-rises- during-inter glacial-periods/	One of the most cited "proofs" of global warming is that sea level is rising...sea level may rise two meters this century, which would be a rate of 22mm/year, nearly seven times faster than current rates...very little change in sea level rise rates over the last 100 years...	2

Fig. 6. $IwST$s for the path $(Globalwarming, Currentsealevelrise, Agriculture)$

sea level rise" of the relation between "Global warming" and "Agriculture." The $IwST$ of ID 1 is searched using the Top-Down approach, which contains a little knowledge that global warming affects agriculture through rising sea levels. The $IwST$ of ID 2 is searched using the Bottom-Up approach, which contains detailed knowledge of the relation between "Global warming" and "Current sea level rise." Testers consider the $IwST$ of ID 2 is more useful for understanding the path, and assign a score 2 to it. Therefore, we conclude that the Bottom-Up approach is more appropriate to search $IwST$s for a path of a relation.

We determine to implement Enishi using the Top-Down and the Bottom-Up approach to search $IwST$s for the overall relation and each path of a relation, respectively.

Novelty of Knowledge Retrieved from the Web. Next, we answer question (Q3) presented at the first of Section 5.1 about the novelty of knowledge retrieved from the Web compared to that existing in Wikipedia. For a relation, we extract a set N_{Wiki} of nouns from the Wikipedia pages of the objects in the top-100 paths mined for the relation; and a set N_{Web} of nouns from the surrounding text of the $IwST$s searched for the overall relation using the Top-Down approach. We use a part-of-speech tagger[3] to extract nouns from text. We consider that the knowledge about the relation searched on the Web is not contained in Wikipedia if few nouns in N_{Web} are contained in N_{Wiki}. For every relation, we obtain two sets of N_{Web}, denoted by $N_{Web}(T10)$ and $N_{Web}(> 0)$, respectively from the top-10 $IwST$s and all the $IwST$s having positive scores. On average, for all 40 relations, 28.9% nouns in $N_{Web}(T10)$, and 38.5% nouns in $N_{Web}(> 0)$, do not belong to N_{Wiki}. The average number of nouns in $N_{Web}(T10)$ and $N_{Web}(> 0)$ are 3080 and 4798, respectively. Therefore, there is a possibility that the $IwST$s retrieved from the Web contain new knowledge which is unavailable in Wikipedia.

We also actually confirmed that the surrounding text contain knowledge which is not contained in Wikipedia. Fig.8 presents the top-5 $IwST$s for the overall relation between "Global warming" and "Agriculture," searched using the Top-Down approach. The surrounding text of the second image consists of 77 sentences, some of which are

---

[3] TreeTagger: http://www.ims.uni-stuttgart.de/projekte/corplex/TreeTagger/

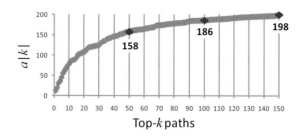

Fig. 7. Average number of $IwST$s containing any elucidatory object in the top-k paths

presented in the "Surrounding text extracted" column. We checked every sentence carefully for whether the knowledge represented by the sentence is contained in Wikipedia. We found that 69 of the 77 sentences contain knowledge unavailable from Wikipedia. Especially, the text contains many precise data which are not described in Wikipedia. Such as, "Global warming is exposing an additional 69 million to 91 million people to food shortages by 2085;" "Due to development and agriculture, the forested area of the world is expected to fall 25 percent to 30 percent by 2050 and the area of coastal wetlands are expected to decline 40 percent by 2085;" and "Between now and 2085, global warming could increase forested areas by 5 percent; but it could reduce the area of coastal wetlands another 13 percent. " Similarly, we also find many other surrounding text segments retrieved from the Web contain knowledge unavailable from Wikipedia.

As a future challenge, we plan to extract knowledge which is not contained in Wikipedia, from the surrounding text segments retrieved from the Web. We then could add new information into Wikipedia automatically using the extracted knowledge; or help users to edit Wikipedia by presenting the extracted knowledge.

Observation of Elucidatory Objects. We examine the effect of elucidatory objects on the evidence-based method. For a relation, we obtain the top-300 $IwST$s using Google; we then count the number of $IwST$s whose text segment contains at least one elucidatory object in the top-k paths, where $1 \leq k \leq 150$. Fig. 7 plots the average number $a(k)$ of the 40 relations for each k. The number $a(k)$ is monotonically increasing, although the increase becomes slow when k becomes larger. For example, the difference 28 between $a(100) = 186$ and $a(50) = 158$ is fairly large, while $a(150) - a(100)$ is only 12. On the other hand, the evidence-based method costs much time when the number k of used paths is large. Furthermore, in preliminary experiments, we also observed that $IwST$s searched using the top-150 paths, are almost identical to those searched using the top-100 paths. Therefore, we set $k = 100$ for the evidence-based method.

We also observed that about 40.5% of the elucidatory objects on average for the 40 relations appear in at least one $IwST$s. Therefore, a fairly large number of elucidatory objects are actually used in the evidence-based method.

5.3   Case Study: Relation between Global Warming and Agriculture

Fig.8 presents the top-5 $IwST$s for the overall relation between "Global warming" and "Agriculture," searched using the Top-Down approach. The "Surrounding text

Ranking Ours	Ranking Google	Image	URL of Web page	Surrounding text extracted	Score
1	160		http://www.di changeagro.or g/about-this-conference/ab stract/prof-m-c-varshneya/	..Almost 20% of total greenhouse-gas emissions were from deforestation in 2007 ... based on the opportunity costs of the land use that would no longer be available for agriculture if deforestation were avoided, emission savings from avoided deforestation could potentially reduce CO_2 emissions for under $5/t$CO_2$...World energy usage will rise in the next few decades...	1.2
2	159		http://www.nc pa.org/pub/st 278?pg=3	..Development and agriculture could cause a greater reduction in global forests and wetlands than global Warming for the foreseeable future...Enhancing agricultural productivity would reduce the costs associated with carbon sequestration... Reducing habitat loss and fragmentation would allow ecosystems to adapt naturally to climate change.	1.8
3	9		http://www.e df.org/page.cf m?tagid=1092	..agriculture practices can offset or avoid greenhouse-gas emissions...Some promising examples include: Soil Carbon Sequestration, Methane Captured from Manure , Direct greenhouse-gas Emission Reductions, and Reforestation...	1.8
4	157		http://koopym an-lobalwarmi ng.blogspot.co m/2007/10/gl obal-warming-and-climatic-concerns.html	..Rapid climate change will impact on a range of human activities including agriculture... Agriculture would not be spared, since climate change has been blamed by many environmentalists and scientists for the severity of violent storms such as hurricanes that have repeatedly ravaged our region...	1.6
5	18		http://www.sc iencedaily.com /releases/200 8/04/0804101 53658.htm	..Fifteen hundred years ago, tribes people mixed their soil with charcoal derived from animal bone and tree bark...Now this ancient, remarkably simple farming technique seems far ahead of the curve, holding promise as a carbon-negative strategy to rein in world hunger as well as greenhouse gases...	1.0

Fig. 8. $IwST$s explaining the overall relation between "Global warming" and "Agriculture"

extracted" column presents some sentences in the surrounding text of each image extracted using our program. The numbers in the "score" column are the average scores obtained by a human subject. As discussed in Section 2, an image itself is insufficient for understanding a relation. However, reading the images with their surrounding text, users could understand the relation. By reading the summaries of surrounding text, we ascertained that all the five $IwST$s contain knowledge about the relation between "Global warming" and "Agriculture." Especially, we were able to understand the relation deeply by reading $IwST$s ranked 2nd, 3rd, and 4th. We list some of the 17 elucidatory objects appearing in the $IwST$s ranked 2nd here: Habitat conservation, Biodiversity, Population growth, Emissions trading, Kyoto Protocol, Intensive farming, Climate change.

The image of the $IwST$ ranked 1st does not relate to the relation directly, although its surrounding text contains knowledge about the relation. The $IwST$ ranked 5th presents a farming technique that is intended to solve the greenhouse gas problem. The two $IwST$s ranked 1st and 5th contains fewer knowledge about how global warming and agriculture affect each other than the other three Imgs do. Therefore, testers assigned lower scores for the two $IwST$s. The "Google" column presents rankings generated by the Google image search engine using a query "Global warming Agriculture." The images ranked 2nd and 4th using our approach are ranked very low by Google, although the two contain a wealth of knowledge about the relation.

6   Related Work

To the best of our knowledge, no method was proposed for searching knowledge including images about relations between objects on the Web. On the other hand, many methods extract knowledge from Wikipedia. YAGO [1,9] is a semantic knowledge base extracted from Wikipedia and WordNet. YAGO contains about two million objects and about 20 million binary relations about them, such as "Politician" is a subclass of "person", and "Max Planck" has won the "Nobel Prize". Kasneci et al. [10] then proposed a search engine to search semantic relations between two objects using YAGO.

Zhu et al. [4] extract semantic relations between pairs of people from the Web. Bollegala et al. [5] proposed a method to extract semantic relations between any objects from the Web, such as **X** was born in **Y**, **X** is a CEO of **Y**, and **X** buys **Y**. As discussed in Section 1, seeking semantic relations is insufficient for understanding complicated relations in the real world.

Recently, Taneva et al. [11] proposed methods for searching photos of objects, such as people, building, or mountains. They search photos of an object using a traditional image search engine by queries extended with objects related to the object in YAGO. For example, searching photos of a mountain uses queries constituted by its name and its location or height. Similar to the methods, we search knowledge on the Web using knowledge extracted from Wikipedia. However, we search knowledge including images associated with relations between objects rather than that about single objects. As another related work, KORU [12] is a search engine, which extends queries using Wikipedia to help users to search what they did not know how to search.

7   Conclusion

We proposed a relation search system named Enishi, which supplies a wealth of diverse multimedia information retrieved both from Wikipedia and the Web for deep understanding of relations. Enishi first mines disjoint paths on a Wikipedia information network that are important for a relation, using the method proposed by Zhang et al. [3]. We then proposed approaches for Enishi to search images on the Web with surrounding text for a relation using paths mined from Wikipedia for the relation. Enishi searches both images with surrounding text containing knowledge associated with the overall of a relation or that related to each path explaining a relation mined from Wikipedia.

Our experiments revealed that our approaches can search useful images with surrounding text for understanding a relation. Furthermore, the information retrieved from the Web contain new knowledge which is not contained in Wikipedia. Therefore, Enishi would enable users to understand relations deeply.

Acknowledgment

This work was supported in part by the National Institute of Information and Communications Technology, Japan.

References

1. Suchanek, F.M., Kasneci, G., Weikum, G.: Yago: a core of semantic knowledge. In: Proc. of 16th WWW, pp. 697–706 (2007)
2. Zhang, X., Asano, Y., Yoshikawa, M.: Analysis of implicit relations on Wikipedia: Measuring strength through mining elucidatory objects. In: Kitagawa, H., Ishikawa, Y., Li, Q., Watanabe, C. (eds.) DASFAA 2010. LNCS, vol. 5981, pp. 460–475. Springer, Heidelberg (2010)
3. Zhang, X., Asano, Y., Yoshikawa, M.: Mining and explaining relationships in Wikipedia. In: Bringas, P.G., Hameurlain, A., Quirchmayr, G. (eds.) DEXA 2010, Part II. LNCS, vol. 6262, pp. 1–16. Springer, Heidelberg (2010)
4. Zhu, J., Nie, Z., Liu, X., Zhang, B., Wen, J.R.: Statsnowball: a statistical approach to extracting entity relationships. In: Proc. of 18th WWW, pp. 101–110 (2009)
5. Bollegala, D., Matsuo, Y., Ishizuka, M.: Relational duality: unsupervised extraction of semantic relations between entities on the web. In: Proc. of 19th WWW, pp. 151–160 (2010)
6. Wayne, K.D.: Generalized Maximum Flow Algorithm. PhD thesis, Cornell University, New York, U.S (January 1999)
7. Nakayama, K., Hara, T., Nishio, S.: Wikipedia mining for an association web thesaurus construction. In: Benatallah, B., Casati, F., Georgakopoulos, D., Bartolini, C., Sadiq, W., Godart, C. (eds.) WISE 2007. LNCS, vol. 4831, pp. 322–334. Springer, Heidelberg (2007)
8. Koren, Y., North, S.C., Volinsky, C.: Measuring and extracting proximity in networks. In: Proc. of 12th ACM SIGKDD Conference (2006)
9. Weikum, G., Kasneci, G., Ramanath, M., Suchanek, F.: Database and information-retrieval methods for knowledge discovery. Commun. ACM 52(4), 56–64 (2009)
10. Kasneci, G., Suchanek, F.M., Ifrim, G., Ramanath, M., Weikum, G.: Naga: Searching and ranking knowledge. In: Proc. of 24th ICDE, pp. 953–962 (2008)
11. Taneva, B., Kacimi, M., Weikum, G.: Gathering and ranking photos of named entities with high precision, high recall, and diversity. In: Proceedings of 3rd WSDM, pp. 431–440. ACM, New York (2010)
12. Milne, D.N., Witten, I.H., Nichols, D.M.: A knowledge-based search engine powered by Wikipedia. In: Proc. of 16th CIKM, pp. 445–454. ACM, New York (2007)

Potential Role Based Entity Matching for Dataspaces Search[*]

Yue Kou, Derong Shen, Tiezheng Nie, and Ge Yu

School of Information Science & Engineering, Northeastern University,
Shenyang 110004, China
{kouyue,shenderong,nietiezheng,yuge}@ise.neu.edu.cn

Abstract. Explosion of the amount of personal information has made the technique of dataspaces search become a hot topic. However current search engines for dataspaces are becoming increasingly inadequate for the query tasks with users' diverse preferences. In this paper, we present a potential role based entity matching model called POEM. We respectively propose the strategies of homologous entity matching and heterogeneous entity matching, which can better utilize both entity features and entity relationships to match and organize entities. By combining homologous entity matching and heterogeneous entity matching, the query result can be adaptable to the diverse needs from different users. The experiments demonstrate the feasibility and effectiveness of the key techniques of POEM.

Keywords: Dataspaces Search, Potential Role, Entity Matching.

1 Introduction

Explosion of the amount of personal information has made the technique of dataspaces search become a hot topic. Here dataspaces are large collections of heterogeneous and related data resources. In this paper, we consider the various entities (e.g. paper PDF, author homepage or conference website) as search units. An important goal of dataspaces search is to provide users with relevant entities, that is, users can not only acquire a complete result set but also drill-down to one specific part to meet their needs.

1.1 Motivating Scenarios and Challenges

Let us consider the following motivating scenarios (as shown in Figure 1).

Scenario 1. Someone wants to find some papers about 'dataspaces'. Due to the limited ways of entity matching, current search engines for dataspaces may return the paper entities according to the keyword 'dataspaces'. In the query result, the papers with different feature (such as *affiliation* or *conference*) values are mixed together.

[*] Supported by the National Natural Science Foundation of China (No. 60973021, 61003059), the National High Technology Development 863 Program of China (No. 2009AA01Z131) and the Fundamental Research Funds for the Central Universities (No. 90304005).

L. Chen, P. Triantafillou, and T. Suel (Eds.): WISE 2010, LNCS 6488, pp. 496–509, 2010.

Therefore, if these homologous entities (with the same entity type) can be matched and organized according to their features, users will quickly drill-down to one specific part meeting their personal needs.

Scenario 2. Someone wants to search for all the relevant entities about a NSFC project. However many relevant entities such as relevant papers, author homepages or conference websites may be missed by traditional approaches. Therefore, if the relationships among the heterogeneous entities (with the different entity types) can be considered during entity matching, the result set will be more complete.

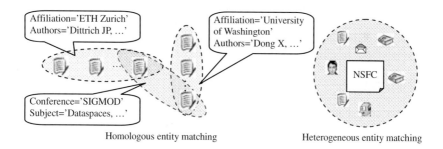

Homologous entity matching   Heterogeneous entity matching

Fig. 1. Demonstration of motivating scenarios about potential role based entity matching

From the above scenarios, we can observe two significant properties for entity matching.

- Firstly, the entities owning similar characteristics in function can represent a potential role. For example, a group of papers written by the same organization (owning similar *affiliation* and *authors*) often represent a certain potential role which can satisfy the user's need for tracing one institution's research result. Each potential role can be represented through two aspects: entity feature and entity relationship. The entities with the same potential role often own similar features and tight relationships. If the entities can be matched and clustered according to their potential roles, users can easily drill-down to one specific part to meet their needs. Therefore compared with the traditional approaches, potential role based entity matching can be more appealing to users.

- Secondly, different entity features or entity relationships have different effects for entities' potential roles. For example, the features *affiliation* and *authors* are more important for the potential role of tracing one institution's research result. While the features *conference* and *subjects* are more important for the potential role of determining one conference's subjects or evaluating one conference's authority. Therefore entity features and entity relationships should be selected or quantified for entity matching based on different potential roles.

The high importance of entity matching for dataspaces search has triggered a huge amount of research. Related prototype systems or techniques have been developed in recent years. But for entity matching, most current work is based on either keywords or simple semantics, which may not reflect entities' potential roles in the result set.

Besides the overview of entire result set, the user would like to acquire its several subsets categorized by entities' potential roles further. On the other hand, most previous work only considers the local characteristics of entities, while their semantic relationships have not been utilized better during matching. Although some techniques about social networks utilize entity relationships for mining community, most of them are limited to analyze homologous entities. However, if not only homologous entities but also heterogeneous ones can be matched and organized, the result set will be more complete. So in this paper, we present a potential role based entity matching model called POEM, which can efficiently match and organize relevant entities to satisfy users' diverse preferences.

However, because of the large-scale data resources in dataspaces, there are some challenges for entity matching: (1) How to formalize the entities' potential roles? (2) How to select entity features to represent different potential roles? (3) How to evaluate the importance of different entity relationships for different potential roles?

1.2 Contributions

The primary contributions of this paper are as follows.

- A potential role based entity matching model called POEM is presented. Unlike traditional approaches, POEM better utilizes both entity features and entity relationships to match and organize entities.
- The strategies of homologous entity matching and heterogeneous entity matching are proposed respectively. By combining homologous entity matching and heterogeneous entity matching, the query result can be adaptable to the diverse needs from different users.
- An experimental study is proposed to determine the feasibility and effectiveness of the key techniques of POEM.

The rest of this paper is organized as follows. Section 2 discusses related work. Section 3 describes the overview of POEM. In Section 4-5, we introduce the strategies of homologous entity matching and heterogeneous entity matching respectively. Section 6 shows the experimental results and Section 7 concludes.

2 Related Work

The section overviews the related research efforts including dataspaces search, potential role and social networks. We then conclude by stating how our work distinguishes itself from these efforts.

Some dataspaces prototype systems such as iMeMex [1-4], SEMEX [5-8], MyLifeBit [9], HayStack [10], OrientSpace [11] have been developed. iMeMex was the first implementation of a personal dataspaces management system. In order to implement entity matching, iMeMex defined a new query language (iQL) to express query request and presented a quality-driven query processing strategy. However, it acquires the semantic information only through parsing entities' local physical characteristics such as format or path. Therefore its ability to support semantic query was limited. SEMEX was another platform for personal information management. It supported three ways of entity matching: keywords based matching, advanced form based matching and association based matching. For advanced form based matching, user must express his query by choosing a domain and entering specific values of some attributes. For association based matching, user can express the associations among entities in his query request. The ways of entity matching provided by SE-MEX were abundant and flexible, but these techniques mainly relied on the query requests in specific patterns, which were possibly at the cost of complicating users' operations to some extent. MyLifeBit was a database of resources and links, which was used to store all of one's digital data. It presented the integration technique for text and multimedia data. HayStack was a project to develop several applications around personal information management and the semantic Web. It emphasized to annotate personal data by analyzing users' preference.

Potential role based entity matching and clustering has been studied to address the problem of result organization (e.g. [12-14]). In [12], a hybrid subspace clustering algorithm was proposed to match the entities returned from Web search engines. It considered Web characteristics (such as the co-occurrence of entity pairs) to analyze entities' potential roles. In [13], temporal information embedded in Web pages was used to analyze Web documents' potential roles. But these techniques only focused on the Web environment, which were not adaptable to dataspaces.

Social networks analysis has attracted much attention in recent years. Some techniques of community mining on social networks have been proposed to identify the communities (e.g. [15-17]). In [15], a method for mining hidden community in heterogeneous social networks was presented. The idea was to learn an optimal linear combination of relations which can best meet user's expectation. In [16], a model was proposed to represent the topic-level social influence on social networks, so that entities with the same topic could be mined. But most of these techniques are limited to analyze the relationships among homologous entities in social networks.

Our work is to match and organize entities based on entities' potential role for dataspaces search. The differences from existing work are as follows. Firstly, during entity matching we use not only entities' local characteristics but also semantic relationships to determine their potential roles. Secondly, we consider entity features and entity relationships as influence factors which are more adaptable to dataspaces. Thirdly, we use some existing social networks techniques to quantify the relationships, but the considered relationships are not limited within homologous entities.

3   Overview of POEM

In this section, we briefly describe the overview of POEM. Firstly this section states preliminaries on modeling POEM. Then a sandglass model of POEM is presented.

3.1   Preliminaries

We consider the whole dataspaces as a repository of entities E (e.g. paper PDF, author homepage or conference website), where each entity (defined in Definition 1) is a concise search unit in dataspaces.

Definition 1 (Entity). An entity $e \in E$ is represented both as an entity type e_T and a feature value vector $FV=(fv_1,\ldots, fv_n)$ on a domain-specific feature set $F=\{f_1,\ldots,f_n\}$.

We use an entity type and a set of specific features to describe an entity. For example, a paper entity can be described as a FV ('Indexing dataspaces', 'Dong X, Halevy A', 'University of Washington', 'Sigmod', 2007, 'Dataspace') on F {*title, authors, affiliation, conference, year, subject*}. The process of entity extraction can be done by using the existing extraction techniques (e.g. [18, 19]). This paper focuses on the model and strategies of entity matching.

The feature values can describe the local characteristics of an entity. According to the feature values, entities can be matched and organized into several clusters, each of which is composed of the entities with some similar feature values. In a sense, each cluster represents a potential role with respect to a certain need as viewed from entity features. Such a similar feature set is called Support Feature Subset (*SFS*, defined in Definition 2) and is a certain guideline for clustering. For example, the papers written by the same research organization constitute a cluster. The cluster will be supported by the *SFS*:{*authors, affiliation*}, because the entities in the cluster own the similar feature values on the two features. If the cluster is composed of the papers published by a certain conference subject, the *SFS* of the cluster will be {*conference, subject*}. The more features a *SFS* includes and the more similar the feature values on *SFS* are, the more relevant the entities in a cluster will be.

Definition 2 (Support Feature Subset, SFS). A support feature subset $SFS=\{f_1,\ldots, f_k\}$ is composed of k features ($k \in [0,n]$) of an entity type, which acts as a guideline for clustering and represents a certain potential role of homologous entities.

In a specific domain, there always exist various relationships among entities, such as written-by, published-by, and attended-by. Besides the local characteristics, entity relationships are also important to capture entity's potential role. Given a set of entities and their relationships which can be represented by a graph G(V, E) called entity relationship graph (ERG), V is the set of entity nodes and E is the set of relationship edges. The adjacent edges between two entities constitute a relationship link. A *SRL* is such a relationship link satisfying Definition 3. As shown in Figure 2, an ERG includes different types of entity nodes (painted as different colors) and four kinds of *SRLs* are illustrated. Each *SRL*'s endpoints are two homologous entities.

Definition 3 (Support Relationship Link, SRL). A support relationship link *SRL* is represented both as a weight w ($w \in [0,1]$) and a relationship vector $L=(r_1,\ldots, r_m)$ composed of m relationships, where the relationships in L are adjacent in ERG and the endpoints connected by L are two homologous entity nodes.

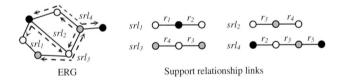

ERG   Support relationship links

Fig. 2. Demonstration of support relationship links

3.2 Sandglass Model of POEM

As can be seen in Figure 3, a sandglass model of POEM is presented, which consists of three layers: initial entity matching layer, homologous entity matching layer and heterogeneous entity matching layer.

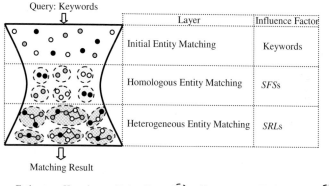

Fig. 3. Sandglass model of POEM

Firstly, the entities in dataspaces are matched initially according to the query keywords. The initial entity matching layer contains these matched entities. And the influence factor of this layer is the inputted keywords.

Secondly, all the homologous entities are classified together. And for each entity type, these homologous entities are matched and clustered further according to their features. The homologous entity matching layer contains all such homologous entity clusters. The influence factor of this layer is a series of *SFS*s.

Thirdly, the clustered homologous entities are matched and extended further according to their relationships. For each homologous entity cluster, the heterogeneous entities which have tighter relationships with the entities in the cluster will be added to generate a heterogeneous entity cluster. The heterogeneous entity matching layer contains all such heterogeneous entity clusters. The influence factor of this layer is a series of *SRL*s.

4 Homologous Entity Matching

Based on the initial entity matching, we can acquire a set of candidate entities containing the query keywords. These candidate entities can be easily classified according to their types e_T. Each class is composed of homologous entities representing different potential roles. The goal of homologous entity matching is to further match these homologous entities for each class according to their potential roles (represented by the selected *SFS*s). The process of homologous entity matching includes offline *SFS* selection and online *SFS* based homologous entity matching.

4.1 *SFS* Selection

As the basis of homologous entity matching, *SFS* is a set of features (e.g. {*authors*, *affiliation*} or {*conference*, *subject*}) which can represent a certain potential role. How to select *SFS*s to represent different potential roles? Firstly, we consider arbitrary combinations of entities' features as candidate *SFS*s. And then we compute their support degrees. The candidates with higher support degrees will be selected as the final *SFS*s. We now discuss how to quantify the support degree of a candidate *SFS*.

Before *SFS* selection, we consider some homologous entities as the training data and cluster them according to our intuition. Then a series of standard clusters are generated, each of which represents an actual potential role. The goal of *SFS* selection is to determine which feature subsets can better support these standard clusters. In other words, the *SFS*s which are most consistent with user's perception will be selected. Suppose one of current candidate *SFS*s is $sfs = \{f_1, \ldots, f_k\}$, its support degree for a standard cluster (denoted as sc) is calculated as follows.

Step 1. For each feature f_i and each entity pair (e.g. e and e') in sc, calculate the similarity (denoted as $S(e, e', f_i)$) of their feature values on f_i ($i \in [1, k]$). Depending on application semantics, the function S can be calculated by calling some suitable functions (such as text similarity functions, numeric similarity functions and date similarity functions etc).

Step 2. For each feature f_i, calculate its support degree for sc. The support degree of f_i for sc is calculated by accumulating all the similarities of entity pairs in sc (described in Formula 1).

$$Support(sc, f_i) = \frac{\sum_{e_x \in sc, e_y \in sc, x \neq y} S(e_x, e_y, f_i) \times 2}{|sc| \times (|sc| - 1)} \tag{1}$$

Step 3. Calculate the support of sfs for sc. The support of sfs for sc is calculated by accumulating the support degree of each feature in sfs (described in Formula 2).

$$Support(sc, sfs) = \frac{\sum_{i \in [1,k]} Support(sc, f_i)}{k} \tag{2}$$

Step 4. Determine whether sfs can be selected as one of the final *SFS*s. Intuitively, the higher its support degree is, the more relevant the entities in a cluster are. In addition, the more features it has, the more relevant the entities in a cluster are. Therefore if sfs can be selected, it should satisfy two constraints: On the one hand, its support should exceed a threshold δ_{min}, that is, $Support(sc, sfs) \geq \delta_{min}$. On the other hand, it includes at least k_{min} features, that is, $k \geq k_{min}$. The setting of these two parameters will be discussed in the section of experiments.

4.2 *SFS* Based Homologous Entity Matching

Based on the selected *SFS*s, the candidate homologous entities can be matched and clustered online. The result of homologous entity matching is composed of a series of homologous clusters, each of which can represent a specific user need. In this paper, we adopt bottom-up hierarchical clustering algorithm to generate the matching result (as shown in Figure 4). The basic idea of our algorithm is: each entity starts in its own cluster, and pairs of clusters can be merged as one to move up the hierarchy.

Algorithm: HomoMatch(*E*, *sfs*, δ_{min}.)

Input:

  A set of homologous entities $E=\{e_1,...,e_n\}$ with query q; A selected *sfs*; A similarity threshold δ_{min}.

Output:

  A set of homologous clusters C.

1. $C \leftarrow \{\}$;
2. *queue* $\leftarrow \{\}$;//Sorted queue to store cluster pairs $<c_i, c_j>$ in the descending order of Support($c_i \cup c_j$, *sfs*).
3. for $i=1$ to n do
4. $c_i=\{e_i\}$; //Each entity starts in its own cluster
5. C.add(c_i);
6. end for
7. for $i=1$ to n-1 do   //Pairs of clusters will be determined whether they can be merged.
8.   for $j=i+1$ to n do
9.   *queue*.add($<c_i, c_j>$); //Add a cluster pair to *queue*.
10.   end for
11. end for
12. *queue*.sort(); //Sort *queue* in the descending order of support.
13. while Support(*queue*.hasNext(), *sfs*)$\geq\delta_{min}$ do //Move the cluster pair $<c, c'>$ with the highest support from *queue*. Determine whether $c \cup c'$ satisfies the constraints of *sfs*.
14.   Merge $<c, c'>$ and update C;
15.   Update *queue* and sort it again;
16. end while
17. return C;

Fig. 4. *SFS* based homologous entity matching algorithm

Firstly, the algorithm treats each candidate entity as a singleton cluster. And then pairs of clusters are merged successively if they can satisfy the constraints of a certain *SFS*. In other words, the merged cluster should insure that its *SFS* will not violate the conditions of *SFS* selection. The process of merging will be continued until there are not pairs of clusters satisfying the merging condition. Finally, for each *SFS*, a series of homologous clusters are generated.

Now we provide some instances to illustrate the process of homologous entity matching. Suppose we want to query the entities relevant to "personal information system". Table 1 shows the relevant data set. Suppose one of the selected *SFS*s is {*authors*, *affiliation*}, the result of *SFS* based homologous entity matching will be $c_1=\{e_1, e_2\}$, $c_2=\{e_3\}$, $c_3=\{e_4\}$ and $c_4=\{e_5, e_6\}$. Each cluster is composed of entities with similar feature values on the *SFS*: {*authors*, *affiliation*}. In fact the *SFS* has a potential role of tracing one institution's research.

Table 1. Some instances to illustrate the process of entity matching

Id	Title	Authors	Affiliation	...
e_1	A platform for personal information management and integration	Dong X, Halevy A	University of Washington	...
e_2	Answering structured queries on unstructured data	Liu J, Dong X, Halevy A	University of Washington	...
e_3	Haystack: A customizable general-purpose information management tool for end users of semi-structured data	Karger DR, Bakshi K, Huynh D	MIT	...
e_4	Bootstrapping pay-as-you-go data integration systems	Sarma A, Dong X, Halevy A	Stanford University	...
e_5	A dataspace odyssey: the iMeMex personal dataspace management system	Lukas B, Dittrich J, Girard O	ETH Zurich	...
e_6	Managing personal information using iTrails	SallesMAV, Dittrich J, Lukas B,	ETH Zurich	...

5 Heterogeneous Entity Matching

The goal of heterogeneous entity matching is to match and extend the generated homologous clusters further according to entity relationships. The process of heterogeneous entity matching includes offline *SRL* quantification and online *SRL* based heterogeneous entity matching.

5.1 *SRL* Quantification

In ERG, the adjacent edges between two homologous entity nodes constitute a *SRL*. Different *SRLs* have different roles with respect to a certain need. For each *SRL*, we should quantify how much it can support a potential role having been represented by a *SFS*. Notice if a *SRL* includes other *SRLs*, we only consider the included ones.

Our approach is based on the idea for community mining presented in [15]. In that paper, different social relationships were modeled by different matrixes. And a linear regression based algorithm was presented to find a linear combination of the existing matrixes to optimally approximate the target social relationships. Similarly, in our setting, different *SRLs* correspond to different social relationships, and the result of *SFS* based homologous entity matching is considered as the target social relationship. Our goal is to find a combined *SRLs* which can make the relationship between the intra-cluster entities as tight as possible and at the same time the relationship between the inter-cluster entities as loose as possible. Here the cluster means a homologous entity cluster corresponding to a certain *SFS*.

Such consideration can be shown in the following example (Figure 5). As for the entities in Table 1, for the *SFS*:{*authors, affiliation*}, some homologous clusters (e.g. $c_1 \sim c_5$) have been generated. Then the process of *SRL* quantification is as follows.

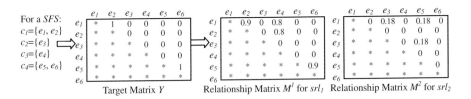

Fig. 5. Demonstration of *SRL* quantification for a *SFS*

Step 1. Construct the target matrix Y, where Y_{ij} indicates the relationship between e_i and e_j ($e_i, e_j \in c_1 \cup c_2 \cup c_3 \cup c_4 \cup c_5$) (Formula 3).

$$Y_{ij} = \begin{cases} 1, & e_i \text{ and } e_j \text{ are in the same cluster;} \\ 0, & \text{otherwise.} \end{cases} \tag{3}$$

Step 2. Find all possible *SRLs* which can connect two entities in $c_1 \cup c_2 \cup c_3 \cup c_4 \cup c_5$ (e.g. srl_1 and srl_2). And for each *SRL* (denoted as srl_k), construct a relationship matrix M^k, where M_{ij}^k indicates the connection strength between e_i and e_j connected by srl_k (denoted as cs_{ij}^k). Here e_i and e_j are two endpoints connected by srl_k. In [20], we have proposed some algorithms to calculate the connection strength between two entities in ERG. So we can use the result to construct each M^k (Formula 4).

$$M_{ij}^k = \begin{cases} cs_{ij}^k, & e_i \text{ and } e_j \text{ are two endpoints connected by } srl_k; \\ 0, & \text{otherwise.} \end{cases} \qquad (4)$$

Step 3. Find a linear combination of all such M^k (e.g. M^1 and M^2) to optimally approximate Y. Also a constraint $\sum w_k \leq 1$ should be satisfied, where w_k is the weight assigned for M^k. The combination coefficients are the weights of *SRL*s (Formula 5).

$$W = \arg\min_w \text{distance}(Y, \sum_k M^k w_k) \qquad (5)$$

Notice that the above process is only one way of *SRL*'s weight assignment for a certain *SFS*. As for different *SFS*s, a *SRL* may have different weight assignments for different needs.

5.2 *SRL* Based Heterogeneous Entity Matching

The basic idea of heterogeneous entity matching is to extend more entities for each homologous cluster. If two entities in a homologous cluster have tighter relationships, the entities along their path will be added to the cluster. The extended cluster will become a heterogeneous cluster. Here the entity relationships can be acquired from ERG. So the problem is transformed as: given a set of homologous entities, how to find the relevant nodes in ERG between the arbitrary homologous entity pair. This section discusses two approaches to matching heterogeneous entities:

Connectivity based Approach. Suppose e_i and e_j are two entities in a cluster c through homologous entity matching. If they are connected in ERG, all the entities along their paths will be added to c.

Cohesion based Approach (*SRL* based Approach). There may be a series of *SRL*s (denoted as $srl_1, \ldots, srl_n$) corresponding to a *SFS* (denoted as *sfs*), each of which has a weight w_i. Suppose e_i and e_j are two entities in a homologous cluster c generated by *sfs*. And srl_k is one of their relationship links, then its cohesion co_{ij}^k is defined by considering both connection strength between entities and the weight for srl_k (Formula 6). If e_i and e_j have higher cohesion, entities along their path will be added to c.

$$co_{ij}^k = cs_{ij}^k \times w_k \qquad (6)$$

The first approach only considers whether the entities are connected in ERG. If there are multiple *SRL*s between two entities, all the entities appearing in these *SRL*s will be added equally. So it does not differentiate the potential roles of *SRL*s. On the contrary, the second approach considers entity relationships' different roles for different needs. The *SRL*s which are unrelated to the current cluster should be filtered out. Only the entities along the *SRL*s with higher cohesions will be added. The result can most satisfy users' diverse preferences. We also call it *SRL* based approach.

For example, as for the entities in Table 1, in ERG there are a co-author relationship among $\{e_1, e_2, e_4\}$, $\{e_5, e_6\}$, and a co-conference relationship between $\{e_1, e_3, e_5\}$. As for the potential role of tracing one institution's research result, via *SRL* quantification, we can see that the co-author relationship is more important than co-conference relationship. The result of Connectivity-approach and *SRL*-approach is $c_1=\{e_1, e_2, e_3, e_4, e_5, e_6\}$ and $c_1=\{e_1, e_2, e_4\}$, $c_2=\{e_3\}$, $c_3=\{e_5, e_6\}$ respectively. The latter approach is more precise because e_3, e_5, e_6 have different potential roles from others, and should not be clustered into c_1.

5.3 Adaptive Tuning

As already stated in the previous sections, through homologous entity matching and heterogeneous entity matching the entities can be matched and clustered according to their potential roles. But the process of heterogeneous entity matching depends on the result of homologous entity matching. That is, both *SRL* quantification and cluster extending are on the promise that the generated homologous clusters are complete enough. However we only consider the similarity of feature values to generate homologous clusters. Besides the similar entities, the relevant entities which are homologous with them should also be added to the current homologous cluster.

For example, as for the entities in Table 1, the initial homologous clusters are $c_1=\{e_1, e_2\}$, $c_2=\{e_3\}$, $c_3=\{e_4\}$ and $c_4=\{e_5, e_6\}$. Subsequently some *SRL*s are quantified based on them. After heterogeneous entity matching, besides heterogeneous entities, some new homologous entities may possibly be added too (e.g. e_4 is added to c_1). Although e_4 is not similar to e_1 or e_2, they appear in certain *SRL*s which make e_1 and e_2 have higher cohesions. Now it is indicated that the initial homologous cluster c_1 is not complete enough. We should replace c_1 by c_1': $\{e_1, e_2, e_4\}$ and tune the weights of *SRL*s again. The above process will be iterated adaptively until c' is close to c. Because the number of homologous entities relevant to c is limited in ERG, the above iteration is a convergent process. The algorithm of adaptive tuning is as Figure 6.

Algorithm: AdaptiveTuning(*c, Y, SRL*s,)

Input:
    A homologous cluster c; The target matrix Y; $SRLs=\{srl_1,\ldots,srl_k,\ldots,srl_n\}$ each of which is relevant to c.
Output:
    The tuned weights $W=\{w_1,\ldots,w_k,\ldots,w_n\}$ for *SRL*s.
1. $c' \leftarrow c$;
2.   for $k=1$ to n do
3.   Construct a relationship matrix M^k;
4.   end for
5. $W=\mathrm{argmin}_W \mathrm{distance}(Y, \sum_{k=1}^{n} M^k w_k)$; // Find a linear combination of M^k to optimally approximate Y.
6.   for $k=1$ to n do
7.   for each pair(e_i, e_j) in c do
8.   if (e_i and e_j have higher cohesion)
9.   Add to c' the homologous entities from e_i and e_j along srl_k; // Extend the current cluster.
10.   end if
11.   end for
12. end for
13. if(c' is not close to c)
14.   Update the current *SRL*s and Y based on c';
15. $W=$AdaptiveTuning(c', Y, *SRL*s); // Replace c by c' and quantify *SRL*s again.
16. end if
17. return W;

Fig. 6. Adaptive tuning algorithm

6 Experiments

In this section, we present various experiments to evaluate the efficiency and effectiveness of the key techniques of POEM.

6.1 Datasets and Setup

We focus on matching the entities of paper type in digital libraries domain. Via the entity matching methods adopted in POEM, the paper entities will be matched and organized according to their potential roles. Via entity extraction in advance, the data set containing about 500 papers, 750 authors and 100 conferences is obtained from a personal computer. The whole data set is divided into two parts: analysis set (50%) and matching set (50%). The former is used to select *SFS*s and quantify *SRL*s offline, while the latter is used to perform online entity matching.

Based on the dataset, we can estimate the performance of entity matching methods presented in this paper. Here we consider two aspects of performance: role entropy $H(R)$ and cluster entropy $H(C)$ (defined as Formula 7 and 8 respectively).

$$H(R) = \overline{H(cr)} = \sum_{i=1}^{m} \frac{|c_i \cap cr|}{|cr|} \log_2 \overline{|c_i \cap cr|/|c_i \cap cr|} \tag{7}$$

$$H(C) = \overline{H(c)} = \sum_{i=1}^{m} \frac{|cr_i \cap c|}{|c|} \log_2 \overline{|cr_i \cap c|/|cr_i \cap c|} \tag{8}$$

$H(R)$ is used to measure the dispersion of entity matching. Assume cr is a standard cluster composed of the entities owning the same potential role r. Suppose c_i ($i \leq m$) is a cluster generated via entity matching which has intersection with cr. The ideal value of $H(R)$ is 0 when all the entities of cr are actually clustered into a certain c_i. Similar to $H(R)$, we define cluster entropy $H(C)$ to measure the diversity of entities assigned to a cluster c, i.e., it is probable that a cluster is composed of entities with different potential roles. The ideal value of $H(C)$ is 0 when all the entities assigned to c refer to the same role. The lower the value of $H(R)$ (or $H(C)$), the higher the quality of entity matching will be. We use *entropy* $((H(R) + H(C))/2)$ to measure the trade-off between $H(R)$ and $H(C)$.

We conduct the experiments on a 3.16GHz Pentium 4 machine with 4GB RAM and 500GB of disk.

6.2 Parameter Setting

During *SFS* selection, if k ($k \geq k_{min}$) features of two entities in a standard cluster are similar enough (exceeding δ_{min}), the k features will constitute a *SFS*. Therefore each *SFS* is selected according to the parameters k_{min} and δ_{min}. If they are set too low (or high), the constraint for *SFS* selection will be too loose (or tight). Consequently the selected *SFS*s will influence the result of homologous entity matching. Our goal is to determine the two parameters by finding the best trade-off between $H(R)$ and $H(C)$.

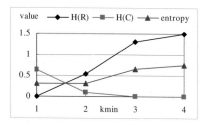 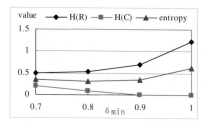

Fig. 7. Parameter setting about k_{min} **Fig. 8.** Parameter setting about δ_{min}

As shown in Figure 7 (or 8), the increase of k_{min} (or δ_{min}) leads to creating little clusters each of which tends to include entities representing different potential roles. So $H(R)$ is better, but $H(C)$ becomes worse. To get the trade-off between $H(R)$ and $H(C)$, parameters should be set based on the lowest *entropy*. Finally the best k_{min} and δ_{min} are set as 2 and 0.8 respectively.

6.3 Performance Comparisons

We compare the performance of three entity matching approaches:

- *SFS*-approach: *SFS* based homologous entity matching.
- Connectivity-approach: *SFS*-approach + Connectivity based heterogeneous entity matching.
- *SRL*-approach (our approach): *SFS*-approach + *SRL* based heterogeneous entity matching.

Here the latter two approaches are performed by combining homologous entity matching and heterogeneous entity matching. As shown in Figure 9, for $H(R)$, *SFS*-approach is worse than others because only performing homologous entity matching may easily result in incomplete clusters. While the latter two approaches are almost equal because neither of them partitions the current clusters further. For $H(C)$, Connectivity-approach is worse than *SRL*-approach, because Connectivity-approach considers all the entity relationships as equal, while *SRL*-approach distinguishes them according to potential roles. Finally our approach acquires better *entropy* by comparing $H(R)$ and $H(C)$ synthetically.

We also compare the performance of the methods without adaptive tuning (NAT) and with adaptive tuning (AT). Figure 10 illustrates their $H(R)$, $H(C)$ and entropy respectively. Owing to performing AT method to adaptively tuning the weights of *SRL*s, its quality of entity matching is superior to NAT method.

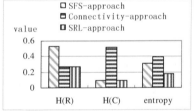
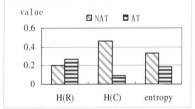

Fig. 9. Performance comparisons among different entity matching approaches

Fig. 10. Performance comparisons between NAT and AT methods

7   Conclusion

In this paper, we present a potential role based entity matching model called POEM for dataspaces search. We respectively propose the strategies of homologous entity matching and heterogeneous entity matching, which can better utilize both entity features and entity relationships to match and organize entities. By combining homologous entity matching and heterogeneous entity matching, the query result can be adaptable to the diverse needs from different users. The experiments have demonstrated the feasibility and effectiveness of the key techniques of POEM. Next we will consider more influencing factors from user (e.g. user's query intent from his request) for entity matching. Also we will make a further research on quantifying the matching degree between user's query intent and entity's potential role.

References

1. Dittrich, J.P., Blunschi, L., Girard, O.: From personal desktops to personal dataspaces: a report on building the iMeMex personal dataspace management system. In: BTW 2007, pp. 292–308. LNI Press, Germany (2007)
2. Salles, M.A.V., Jens, D., Lukas, B.: Adding structure to Web search with iTrails. In: II-MAS 2008, Mexico (2008)
3. Salles, M.A.V., Dittrich, J.P., Karakashian, S.K.: iTrails: Pay-as-you-go information integration in dataspaces. In: VLDB 2007, pp. 663–674. ACM Press, New York (2007)
4. Salles, M.A.V., Dittrich, J.P., Blunschi, L.: Intensional associations in dataspaces. In: ICDE 2010, pp. 984–987. IEEE Press, New York (2010)
5. Liu, J., Dong, X., Halevy, A.: Answering structured queries on unstructured data. In: WebDB 2006, pp. 20–25. ACM Press, New York (2006)
6. Dong, X., Halevy, A.: Malleable schemas: a preliminary report. In: WebDB 2005, pp. 139–144. ACM Press, New York (2005)
7. Jeffery, S., Franklin, M., Halevy, A.: Pay-as-you-go user feedback in dataspace systems. In: SIGMOD 2008, pp. 847–860. ACM Press, New York (2008)
8. Sarma, A., Dong, X., Halevy, A.: Bootstrapping pay-as-you-go data integration systems. In: SIGMOD 2008, pp. 861–874. ACM Press, New York (2008)
9. Gemmell, J., Bell, G., Lueder, R.: MyLifeBits: Fulfilling the memex vision. In: Proceedings of ACM Multimedia 2002, pp. 235–238. ACM Press, New York (2002)
10. Karger, D.R., Bakshi, K., Huynh, D.: Haystack: A customizable general-purpose information management tool for end users of semistructured data. In: CIDR 2005, pp. 13–26. ACM Press, New York (2005)
11. Li, Y.K., Meng, X.F.: Research on personal dataspace management. In: SIGMOD PhD Workshop 2008, pp. 7–12. ACM Press, New York (2008)
12. Lee, J., Hwang, S., Nie, Z.: Query result clustering for object-level search. In: KDD 2009, pp. 1205–1214. ACM Press, New York (2009)
13. Alonso, O., Gertz, M., Baezayates, R.: Clustering and exploring search results using timeline constructions. In: CIKM 2009, pp. 97–106. ACM Press, New York (2009)
14. Zamir, O., Etzioni, O.: Web document clustering: A feasibility demonstration. In: SIGIR 1998, pp. 46–54. ACM Press, New York (1998)
15. Cai, D., Shao, Z., He, X.: Mining hidden community in heterogeneous social networks. In: KDD 2005, pp. 58–65. ACM Press, New York (2005)
16. Tang, J., Sun, J., Wang, C.: Social influence analysis in large-scale networks. In: KDD 2009, pp. 807–816. ACM Press, New York (2009)
17. Yin, X., Han, J., Philip, Y.: CrossClus: user-guided multi-relational clustering. In: DMKD 2007, vol. 15(3), pp. 321–348. ACM Press, New York (2007)
18. Zhao, H.K., Meng, W.Y., Yu, C.: Automatic extraction of dynamic record sections from search engine result pages. In: VLDB 2006, pp. 989–1000. ACM Press, New York (2006)
19. Minkov, E., Cohen, W.W., Ng, A.Y.: Contextual search and name disambiguation in email using graphs. In: SIGIR 2006, pp. 27–34. ACM Press, New York (2006)
20. Kou, Y.: Improving the accuracy of entity identification through refinement. In: EDBT Ph.D. Workshop 2008, pp. 39–48. ACM Press, New York (2008)

Personalized Resource Search by Tag-Based User Profile and Resource Profile

Yi Cai, Qing Li, Haoran Xie and Lijuan Yu

Department of Computer Science
City University of Hong Kong
Hong Kong, China
{yicai3,itqli}@cityu.edu.hk

Abstract. With the increase of media-sharing web sites such as YouTube [1] and Flickr [2], there are more and more shared multimedia resources on the Web. Multimedia search becomes more important and challenging, as users demand higher retrieval quality. To achieve this goal, multimedia search needs to take users' personalized information into consideration. Collaborative tagging systems allow users to annotate resources with their own tags, which provide a simple but powerful way for organizing, retrieving and sharing different types of social media. The user profiles obtained from collaborative tagging systems should be very useful for resource retrieval. In this paper, we propose a new method to model user profiles and resource profiles from wider perspectives and apply them to personalized resource search in a collaborative tagging environment. We implement a prototype system named as FMRS. Experiments in FMRS show that our proposed method outperforms baseline methods.

1 Introduction

The advent of media-sharing web sites like Flickr and YouTube allows users to share multimedia resources with others and has brought huge amount of multimedia resources to the web. Multimedia search becomes more important and challenging, as users demand higher retrieval quality. Current multimedia search methods mainly depend on the relevant match of the query and resource descriptions. Although different users input the same query, they may have different information need corresponding to their own preferences. Thus, it is necessary to implement personalized multimedia retrieval based on users' profiles so as to obtain more effective and useful search results.

Currently, collaborative tagging systems become more and more popular and many social media sites support tagging mechanism. For example, bookmarks on Del.icio.us [3] may be tagged in terms of interest topics to the user. The resources and the tags posted by Web users to these systems are supposed to be highly dependent on their interests, and the tags given by users provide rich information for building more accurate and specific user profiles. Currently, only a few studies in the literature try to construct user

---

[1] http://www.youtube.com
[2] http://www.flickr.com
[3] http://delicious.com

L. Chen, P. Triantafillou, and T. Suel (Eds.): WISE 2010, LNCS 6488, pp. 510–523, 2010.

profiles from data in collaborative tagging systems [5] [8], and usually only a set of popular tags given by a user are used to represent the user interests. Such a simple user profile may not reflect a user's preference properties adequately. We observe that not only the tags explicitly given by users can indicate the users' interests but also the resources tagged by them. If a user annotates a resource, it implicitly indicates that the user may be interested in some features of the resource. Besides, while current methods mainly use only frequency of tags to model user profile, we consider that the position of a tag and the TF/IRF (tag frequency/inverse resource frequency) of a tag could indicate the user preference degree on a feature (tag) of the resource better.

In this paper, we present a new user profile model from wider perspectives and apply it to assist personalized multimedia search. We narrow down our focus on a particular domain, namely, Recipes. Currently, a recipe can be introduced to people through many multimedia resources such as texts of recipe introduction, videos which demonstrate the cooking procedure of the recipe, or photos about dishes of the recipe. There are many multimedia resources associated to a recipe.

In multimedia resource search, how to model the user profiles and the resources is the key to improve the personalized resource search. In our opinion, we propose a new method to model user profiles from wider perspectives, not only the TF/IRF and positions of collaborative tags explicitly given by users but also tags extracted from resources can indicate users' implicit preference. Similarly, we model a resource by collaborative tags given by users and features extracted from its description. Based on our user profiles and resource profiles, we propose a personalized resource search method. We also implement a prototype personalized resource retrieval system named as Folksonomy-based Multimedia Retrieval system (FMRS). Experiments in FMRS show that the proposed method using user profiles and resource (i.e., recipe) profiles outperforms baseline methods in personalized resource search. In addition, experiments show that a user profile is more sensitive than a resource profile on personalized search results, and plays an important role in improving the personalized search.

The structure of the paper is as follows: Section 2 introduces the background and related work. We propose a new method to model user and resource profiles from the collaborative tagging environment in section 3. In section 4 we introduce how to achieve personalized resource search using the user profiles and resource profiles. We run experiments to evaluate our proposed method and introduce a prototype recipe system in section 5. Section 6 concludes the paper and introduces potential future works.

2   Background and Related Works

2.1   Collaborative Tagging and Tag-Based User Profile

Collaborative tagging, also called as folksonomy, has gained popularity on the Web 2.0 services in recent years. Existing research on collaborative tagging can be divided into two aspects: The first type focuses on investigating and analyzing the features of tags. Golder and Huberman [7] analyzed the tag usage patterns, user activities and annotated resources in collaborative tagging systems. Bischoff et al. [2] did a deeper survey to systems like Del.icio.us, and Last.fm[4], to discover useful tags for information access.

---

[4] http://www.last.fm/

The second type mainly tried to explore social annotations and link structures in folksonomy for various applications. Social annotations are usually good summarizing of tagged resources, thus can be used for information search. Bao et al. [1] proposed two algorithms, named as SocialSimRank and SocialPageRank respectively, to explore the latent semantics of tags to optimize web search. A recent work done by Markines et al. [11] did evaluations of various similarity measures for semantics of social tags. The tag based profile can be represented by naive approach, co-occurrence approach, or adaptive approach, etc., as proposed in [12]. However, there are drawbacks of this profiling method, either due to the lack of user input which makes discovering co-occurred tags difficult, or the difference between personal preferred vocabulary which leads to problems like polysemous and synonymy tags, and thus influences the precision of learned profiles.

2.2   Personalized Search

The motivation of proposing personalized search is to return users with customized retrieval results according to their various information needs. The strategies of personalized search fall into two categories:

- One is query expansion such as [3], which refers to modify the original query either by expanding it with other terms, or assigning different weights to the terms in the query.
- Another category is result processing, primarily re-ranking, which adapts the search results to a particular user's preference. Most re-ranking strategies attempted to construct user profile from a user's historical behavior, and used the profile to filter out resources that unmatched with his/her interest. Pretschner and Gauch [15] structured user profiles with an ontology consisting of 4,400 nodes. Chirita et al. [4] modeled both user profiles and resources as topic vectors from ODP[5] hierarchy, thus the matching between user interest and content can be measured by their vector distance. A personalized PageRank algorithm was proposed in [14], which was a modification to the global PageRank, and the search results were personalized based on the hyperlink structure of web. Besides learning user profile based on only his/her own browsing history, Sugiyama et al. [17] also explored social information to refine search results with the help of like-minded neighbors. Dou et al. [6] did comparisons between various personalization approaches, like clicked based, profile based, long-term based, short-term based, etc., and also proposed an evaluation framework for the strategies. In this paper, we also adopt the profile based approach for personalized search.

2.3   Personalized Search in Collaborative Tagging Systems

With the recent development of collaborative tagging systems, some works are proposed on personalized search in the collaborative tagging environment. Noll and Meinel [13] propose an approach to explore user's and resource's related tags based on term frequency, and re-rank the non-personalized search results based on these related tags.

---

[5] http://www.dmoz.org/

Their work is rather simple while it is effective. Xu et al., [18] propose a topic-based personalized search in folksonomy, in which the personalized search is conducted by ranking the resources based on not only term similarity matching but also topic similarity matching. In their work, instead of using term frequency, term frequency/inverse document frequency (TF-IDF) or BM25 are used to construct user and resource profiles. However, they do not consider the positions of tags and the implicit tags of users and resources.

3   User- and Resource- Profiling from Collaborative Tagging

3.1   Modeling User

We consider that there are two kinds of tags that a user may be interested in. The first one kind of tags are those tags that tagged by the user. If a user has used a tag to describe a resource, it explicitly indicates that the user may consider the information of the tag is interesting (or attractive) for him. Especially in recipe resources, according to the users' feedback in our system, most of the users would like to use those tags that can indicate the preference features of a resource for tagging. We name the set of tags that used by a user to annotate resources as *explicit interested tag vector* for the user.

Definition 1. An **explicit interested tag vector** for a user i denoted by $\overrightarrow{T_i'}$ is a vector of tag:value pairs as follows:

$$\overrightarrow{T_i'} = (t_{i,1}' : v_{i,1}', t_{i,2}' : v_{i,2}', \cdots, t_{i,n}' : v_{i,n}')$$

where $t_{i,x}'$ is a tag that used by a user i to tag resources, n is the number of tags that user i had given for resources and $v_{i,x}'$ is a value obtained by the following equation:

$$v_{i,x} = I_1(i, x) \times I_2(i, x) \tag{1}$$

where

$$I_1(i, x) = N_{i,x} \times ln\frac{|U|}{N_x} \tag{2}$$

and

$$I_2(i, x) = \frac{1}{N_{i,x}} \times \sum_{j=1}^{N_{i,x}} \frac{len(R_j) - pos(t_{i,x}', R_j) + 1}{len(R_j)} \tag{3}$$

In equations 1-3, $N_{i,x}$ is the number of times of user i using tag x for description, $|U|$ is the number of users, N_x is the number of users who use tag x to describe resources, $len(R_j)$ is the size of annotated tag set of user i for the resource j and $pos(t_{i,x}', R_j)$ is the position of tag x in the resource j. The intuition of equation 2 is that if a user has used a tag frequently to tag resources, then the tag is interesting for him. The intuition of equation 3 is that if a user use a tag to describe a resource prior to others, then the tag will be more important than others for the user.

As mentioned above, most users would like to use those tags that can indicate the preference features of them to tag resources. Thus, we name $v_{i,x}'$ as *explicit preference degree* for user i on the tag $t'_{i,x}$ because users explicitly give tags to describe resources and those tags can indicate users' preference.

The other kind of tags that a user may be interested in are those tags extracted from resources for which a user has given tags. The intuition is that if a user chooses a resource to give tags, it can implicitly reflect that the user is interested in the resource, more specifically, he is interested in some features of the resource. Thus, we consider it would be helpful to model a user's preference in his profile by taking the tagged resources (i.e., the features of the resources) into consideration. For a resource, it can be represented by a set of features. The features could be extracted from the description of a resource. For example, in the recipe domain, we extract features such as ingredients and tastes of a recipe from its text description. We name the set of tags that extracted from resources tagged by a user as *implicit interested tag vector* for the user.

Definition 2. An **implicit interested tag vector** of a user i denoted by $\overrightarrow{T_i''}$ is a vector of tag:value pairs as follows:

$$\overrightarrow{T_i''} = (t''_{i,1} : v''_{i,1}, t''_{i,2} : v''_{i,2}, \cdots, t''_{i,n} : v''_{i,n})$$

where $t''_{i,x}$ is a tag that extracted from resources tagged by a user i, n is the number of tags that extracted from resources tagged by user i, and $v''_{i,x}$ is a value which indicates the importance degree of tag $t''_{i,x}$. In our method, we consider that all tags extracted from resources have the same importance degree. Thus, the values of all $v''_{i,x}$ are given as 1.

We consider that the tags in a user profile are the conjunction of the *explicit interested tag vector* for the user and *implicit interested tag vector* for the user. A user is represented by a user profile defined below:

Definition 3. A **user profile** of a user i denoted by $\overrightarrow{U_i}$ is a vector of tag:value pairs as follows:

$$\overrightarrow{U_i} = (t_{i,1} : v_{i,1}, t_{i,2} : v_{i,2}, \cdots, t_{i,n} : v_{i,n}), \forall x, t_{i,x} \in \overrightarrow{T_i'} \cup \overrightarrow{T_i''}$$

where $t_{i,x}$ is a tag that may be interesting for user i, $v_{i,x}$ is the preference degree for user i on the tag $t_{i,x}$ and n is the number of tags that user i may be interested in.

For a user profile, an important point is to obtain the preference degree of a user on tags so as to model the user's preference. To what a degree a user being interested in a tag (a tag can indicate a user's preference to what an extent) depends on the aggregation of explicit preference degrees in *explicit interested tag set* and implicit preference degrees in *implicit interested tag set* for the user.

The preference degree for user i on the tag $t_{i,x}$ is obtained by an aggregation function as follows:

$$v_{i,x} = (\alpha) \cdot v'_{i,x} + (1 - \alpha) \cdot v''_{i,x} \tag{4}$$

where α is a parameter in the range of [0,1], which is used to adjust the weight of tags in explicit interested tag set and that in implicit interested tag set.

3.2   Modeling Resources

For a resource, we can extract some features (can be considered as some tags) to represent it and name the set of extracted features of a resource as its **extracted tag vector**. Besides, in the collaborative tagging environment, we also can represent a resource by the set of tags which are given by users and name the set of tags given by users as its **collaborative tag vector**. Their formal definitions are as follows:

Definition 4. The **extracted tag vector** of a resource p denoted by $\overrightarrow{R_p^e}$ is a vector of tag:value pairs:

$$\overrightarrow{R_p^e} = (t_{p,1}^e : w_{p,1}^e, t_{p,2}^e : w_{p,2}^e, \cdots, t_{p,n}^e : w_{p,n}^e)$$

where $t_{p,x}^e$ is a tag that extracted from the description of resource p, n is the number of tags that extracted from the description of resource p, and $w_{p,x}^e$ is a value which indicate the importance degree of tag $t_{p,x}^e$ for resource p. In our method, we consider that all tags extracted from resources have the same importance degree. Thus, the values of all $w_{p,x}^e$ are given as 1.

Definition 5. The **collaborative tag vector** of a resource p denoted by $\overrightarrow{R_p^c}$ is a vector of tag:value pairs:

$$\overrightarrow{R_p^c} = (t_{p,1}^c : w_{p,1}^c, t_{p,2}^c : w_{p,2}^c, \cdots, t_{p,n}^c : w_{p,n}^c)$$

where $t_{p,x}^c$ is a tag used by any user to tag the resource p, n is the number of tags that used by any user to tag the resource p and $w_{p,x}^c$ is a value obtained by the following equation:

$$w_{p,x}^c = M_{p,x} \times ln\frac{|R|}{M_x} \tag{5}$$

where $M_{p,x}$ is the times of users use tag x to describe resource p, $|R|$ is the number of resources, M_x is the number of resources which are annotated by tag x.

In our method, we model a resource from two perspectives. We consider that a resource profile is a tag vector which is a combination of its **extracted tag vector** and **collaborative tag vector**.

Definition 6. A **resource profile** denoted by $\overrightarrow{R_p}$ is a vector of tag:value pairs:

$$\overrightarrow{R_p} = (t_{p,1} : w_{p,1}, t_{p,2} : w_{p,2}, \cdots, t_{p,n} : w_{p,n}), \forall x, t_{p,x} \in \overrightarrow{R_p^e} \cup \overrightarrow{R_p^e}$$

where $t_{p,x}$ is a tag used to describe a resource p, $w_{p,x}$ is the value to which resource p possesses the tag (feature) $t_{p,x}$ and n is the number of tags used to describe the resource p.

The degree of resource p possesses the tag $t_{p,x}$ is obtained by an aggregation function as follows:

$$w_{p,x} = (\beta) \cdot w_{p,x}^c + (1 - \beta) \cdot w_{p,x}^e \tag{6}$$

where β is a parameter in the range of [0,1,] which is used to adjust the weight of tags in extracted tag set and that in collaborative tag set.

4   Personalized Resource Search Using Tag-Based User Profiles

In our personalization framework, we first compute the content relevance between the query and resources, then measure how the resources match with a user's interest, and finally aggregate the two relevance scores to get the final rankings of personalized search results.

4.1   Content Relevance Measurement

In our method, users' queries are usually in the format of a term or term vector.

Definition 7. A query given by a user i, denoted by $\overrightarrow{q_i}$, is a vector of terms as follows:

$$\overrightarrow{q_i} = (t_{i,1}^q, t_{i,2}^q, \cdots, t_{i,m}^q)$$

where $t_{i,x}^q$ is a term in the query given by user i, and m is the total number of terms in the query. For example, a user may issue a query like "chicken" if he wants to find a chicken dish, or "spicy fish" to search for dishes made of fish with spicy flavor and so on.

The basic objective of search is to get relevant resources to a given query. We do this by matching the query with resource profiles. Since both the query and resource profiles in our system are modeled as term vectors, the *content relevance* between them can be directly calculated based on their cosine similarity:

$$Sim(\overrightarrow{q_i}, \overrightarrow{R_p}) = \frac{\overrightarrow{q_i} \cdot \overrightarrow{R_p}}{|\overrightarrow{q_i}||\overrightarrow{R_p}|}$$

where $\overrightarrow{q_i}$ is a query, $\overrightarrow{R_p}$ is a resource profile.

A larger value of $Sim(\overrightarrow{q_i}, \overrightarrow{R_p})$ means a higher content relevance between the resource $\overrightarrow{R_p}$ and query $\overrightarrow{q_i}$.

4.2   User Interest Relevance Measurement

Resources with high content relevance to a query may not always be accepted by the user who issues the query, for they may deviate from his/her taste. The objective of interest matching is to identify how a user would be interested on a resource, and thus uses this information to filter out resources that would be disliked by the user.

As analyzed above, user profiles in our system are also represented by tag vectors indicating his/her preference of resources, thus the interest relevance between a user and a resource can be similarly computed through the cosine similarity between their profiles:

$$Sim(\overrightarrow{U_i}, \overrightarrow{R_p}) = \frac{\overrightarrow{U_i} \cdot \overrightarrow{R_p}}{|\overrightarrow{U_i}||\overrightarrow{R_p}|}$$

where $\overrightarrow{U_i}$ is a user profile, $\overrightarrow{R_p}$ is a resource profile.

4.3 Personalized Ranking

The final goal of personalized search is to get resources that match with both the query and user's interest. To do this, we choose to aggregate both the content relevance and interest relevance between a query and resources to generate final rankings of the result set for a particular user. The final personalized relevance between a resource and a query for user i is combined as follows:

$$RScore(\overrightarrow{q_i}, \overrightarrow{U_i}, \overrightarrow{R_p}) = Sim(\overrightarrow{q_i}, \overrightarrow{R_p}) \cdot Sim(\overrightarrow{U_i}, \overrightarrow{R_p}) \tag{7}$$

where $\overrightarrow{q_i}$ is a query, $\overrightarrow{R_p}$ is a resource profile, $\overrightarrow{U_i}$ is a user profile, all of which are in the format of term vectors.

5 Experiment

Since there are no existing recipe systems using collaborative tagging techniques, as part of our work, we implement a prototype system FRMS. In this section, we conduct experiments in our prototype system FRMS to compare the proposed method with two baseline methods.

5.1 Data Set

In FRMS, there is a database of 500 recipes and 203 users. Besides, there are 7889 user-resource-tag tuples and each user on average annotated 12 recipes. These tags mainly describe recipes from the aspects of ingredient, taste, nutrition and so on. As all the other collaborative tagging systems, there are lots of polysemy and synonym words in our collected tags, and a simple preprocessing was conducted to reduce ambiguity.

5.2 Evaluation Metrics

We use three metrics here for evaluation. The first one is imp presented in [16], which is a common evaluation metric adopted in recommender systems to measure how the personalization strategy improves the ranking of the target resources for a user in his/her result list when comparing to basic search methods. It is defined as:

$$imp(q_i) = \frac{1}{r_p} - \frac{1}{r_b}$$

where q_i is an issued query, r_b is the rank of the target resource in the baseline search approach, and r_p is the rank of the same resource returned by our personalized search. The overall *ranking improvement* is calculated as average query imp for all queries in the test data as follows:

$$imp = \frac{\sum_{i=1}^{m} imp(q_i)}{m}$$

where m is the number of queries. The larger value of imp indicates a greater improvement of the ranking for the target resource by our personalization approach.

The second metric we use is *hitrate* (HR) [8], which is used to measure how often a user's interested resources are in the recommendation or personalized search result list. It is defined as:

$$HR(u_i) = \frac{|T_{u_i} \cap X_{u_i}|}{|T_{u_i}|}$$

where T_{u_i} is the resources relevant to user i in test set, and X_{u_i} is the result set of top-N returned resources. The overall *hitrate* of the top-N results is computed as average personal $HR(u_i)$ for all users in the test data set as following:

$$HR = \frac{\sum_{i=1}^{n} HR(u_i)}{n}$$

where n is the number of users. The larger the average *hitrate* is, the more precise the personalized search model is.

The third metric is *Mean reciprocal rank (MRR)*, which is a statistic for evaluating a ranking to a query. The reciprocal rank of a query result is the multiplicative inverse of the rank of the first correct answer. The mean reciprocal rank is the average of the reciprocal ranks of results for a query. It is defined as follows:

$$MRR = \frac{1}{m} \sum_{i=1}^{m} \frac{1}{rank_i}$$

where m is the number of queries, $rank_i$ is the position of the correct answer (relevant resource) in the result ranking for the query i. The larger the average *MRR* is, the faster and easier for the user to find out the resources he or she wants.

5.3 Baseline Methods

We compare our approach with three baseline methods. The first one (Baseline-1) handles the search by only considering the relevance between a query and resource tag-based description. It tries to match terms in the query to user-generated tags assigned to a resource. Since no user profiles are used in the basic search, given a same query, all users would receive a same result set.

The second baseline (Baseline-2) is a personalized search using the method in [5] to construct users' profiles. A user profile is constructed from his/her historical used tags. The resource profile is constructed by a collection of all tags given by users on the resource. The relevance of a resource to a query for a user depends is measured by equation 7.

The third one (Baseline-3) is the method presented in [18], with the weights of tags in user profiles and resource profiles being based on TF-IDF values. It is a state-of-the-art method in personalized search using tag-based user and resource profiles in collaborative tagging systems.

5.4 Experiment Results

Performance Comparison. By comparing our method with baseline methods, we can find some interesting results. Figure 1 shows the comparison of our method and baseline

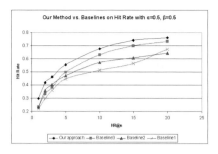

Fig. 1. Comparison of our method and baseline methods with different $HR@n$ on *hitrate*

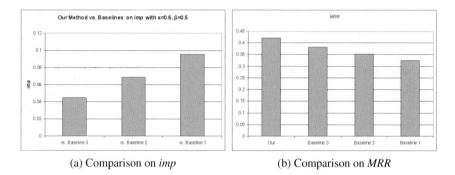

(a) Comparison on *imp*   (b) Comparison on *MRR*

Fig. 2. Comparison of our method and baseline methods on *imp* and *MRR*

methods with different n (i.e., size of result sets) on *hitrate* metric. $hitrate@n$ means that the size of the returned result set is n. We can find that our method outperforms all baseline methods for all n. With the increase of n, the *hitrate* of our method increases quickly and outperforms Baseline-1 at most 17.58%, and Baseline-2 at most 13.19% (while $n = 15$) and Baseline-3 at most 8.55% (while $n = 3$). The higher of n, the higher value of *hitrate* of all methods. We can find that when $n = 20$ (i.e., we only return 20 resources in the returned list), our approach can achieve a *hitrate* at 0.7619, which means most (76.19%) of users can find the resources they want exactly. Even when $n = 1$, our method still can obtain a *hitrate* at 0.2967.

Figure 2(a) shows the comparison of our method and baseline methods on *imp* metric. We can find that our method outperforms Baseline-1 by 9.54%, and Baseline-2 by 6.88% and Baseline-3 by 4.47% on *imp*, which means our method can push the user favorite resources to a more front position in the result ranking list so that user can find what they want earlier and less time-consuming in browsing the search results.

Figure 2(b) shows the comparison of our method and the other methods on *MRR*. According to figure 2(b), our method obtains the highest MRR value at 0.4218 while the other compared methods are less than 0.382. Our method outperforms the compared methods by at least 12.25% (the best baseline method is Baseline-3 whose MRR is 0.382). According to these results, we can conclude that for FRMS data set, our method can push the user favorite resources to a more front position in the result ranking list enabling users to find what they want easier and faster.

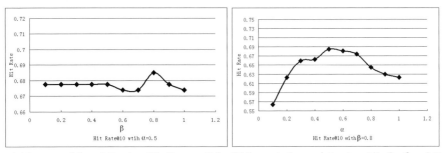

(a) Impact of β on $hitrate$ while fix $\alpha = 0.5$ (b) Impact of α on $hitrate$ while fix $\beta = 0.8$

Fig. 3. Impact of different values of α and β on $hitrate$

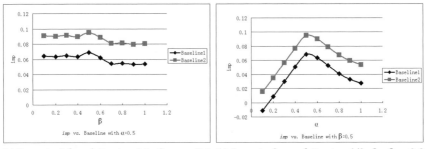

(a) Impact of β on $hitrate$ while fix $\alpha = 0.5$ (b) Impact of α on $hitrate$ while fix $\beta = 0.8$

Fig. 4. Impact of different values of α and β on imp

According to figures 1 and 2, our method performs the best among all three methods, while Baseline-3 obtains the second best performance, Baseline-3 performs worse and Baseline-1 is the worst one. The reason is that Baseline-1 only compares the query with the resource profile and does not take the user profile into consideration, and Baseline-2 not only measures how relevant a resource for a user query but also takes a simple user profile consisting of a set of tags into consideration. For Baseline-3, it constructs user profiles based on TF-IDF and does not consider the positions of tags and the implicit features of users and resources. Our method constructs users' profiles and resources' profiles from wider perspectives, thus we can obtain more personalization information about users' interests on features of resources so as to achieve the best performance among the three.

Influence of Parameters. In our method, there are two parameters which can affect the construction of user profile and resource profile so as to affect the search result. By testing all combinations of different parameters, we find that the best parameter combinations in our method for $hitrate$ metric are $\alpha = 0.5$ and $\beta = 0.8$, and for imp metric are $\alpha = 0.5$ and $\beta = 0.5$. We evaluate how a parameter affects the $hitrate$ and imp in our method by fixing another one. Figure 3(a) and 4(a) show the influence of the parameter β (we fix $\alpha = 0.5$) and the parameter α (we fix $\beta = 0.8$) on $hitrate$ correspondingly. Figures 3(b) and 4(b) show the influence of the parameter α (we fix $\beta = 0.5$) and the parameter β (we fix $\alpha = 0.5$) on imp correspondingly. We can find

(a) The snapshot of FMRS

(b) Multimedia and tags of a particular recipe

Fig. 5. Interface of FMRS

that there exist best parameter values for α and β to obtain the best performance on both *hitrate* and *imp*.

According to Figure 3(a), we can find that the effect of β on *hitrate* is not so obvious, i.e., with the change of β, the change of the value of *hitrate* is only slight. However, according to Figure 3(b), the effect of α on *hitrate* is more obvious than that of β, i.e., with the change of α, the change of *hitrate* value is more notable. In other words, α is more sensitive than β on the *hitrate* metric. Similarly, according to Figure 4(a) and 4(b), we can find that α is also more sensitive than β on the *imp* metric. The reason may be that α is used to adjust the weight of implicit tag vector and explicit tag vector of a user, and β is used to adjust the weight of extracted tag vector and collaborative tag vector of a resource. As the effect of user profile is more sensitive than that of resource profiles on both *hitrate* and *imp*, it is more important to select the value of parameter α. We observe that $\alpha = 0.5$ can obtain a stable and good performance.

5.5 Prototype System

As part of our research, a folksonomy-based multimedia retrieval prototype system (FMRS) has been implemented. This snapshot of FMRS in Figure 5(a) displays the main user interface integrating functions of browsing and retrieval. By giving a set of terms or keywords from a user, the retrieval result list will be displayed to the user in this page. One prominent characteristic of FMRS is that it allows users to tag and collect their preferred recipes collaboratively, which is similar to emerging Web 2.0 communities such as Delicious and Flickr. This is mainly due to the processes of user-profiling and item-profiling derivation, through which we extract features from collaborative tags in order to facilitate personalized multimedia recipe retrieval. Another important feature of FMRS is that diverse modalities of recipe media objects such as text, images and videos about a particular recipe are supported by FMRS. The Figure 5(b) illustrates how these kinds of media objects are organized in FMRS: the collaborative tags used more frequently are highlighted by bigger font size and text, image and video about a particular recipe are included in the same page. Figure 6 further demonstrates how above features are supported by the layered system architecture modified and enhanced from our previous work in [10]. The main functions of these three layers are summarized below:

– User Interaction - for accepting the request and delivering the result to user. This layer can be considered as an interface layer equivalently integrating the functions of recipe browsing, retrieval and user-profile editing.

– Retrieval and Profile Management - for processing user queries as discussed above and capturing the user tagging history and behaviors during user interactions.

– Data Collection and Management - for collecting recipe data from various recipe web-sites by a recipe crawler in [9] and converting them to recipe data records based on the templates of webpage automatically. A multimedia recipe database is then constructed to support the diverse modalities of media objects.

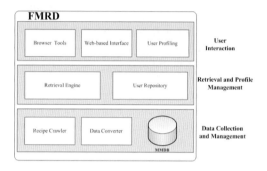

Fig. 6. The System Architecture of FMRS

6   Conclusion and Future Work

In this paper, we focus on exploring personalized resource search by tag-based user profiles and resource profiles in the recipe domain. We propose a new method to model user profiles and resource profiles from wider perspectives in the collaborative tagging environment. Based on our users' profiles and resources' profiles, we propose a personalized resource search method for multimedia resource search. A prototype personalized resource retrieval system named as Folksonomy-based Multimedia Retrieval system (FMRS) in a collaborative tagging environment is implemented. Experiments in FMRS show that the proposed method using user profiles and resource profiles outperforms baseline methods and can improve the personalized resources search effectively.

There are several potential future extensions of our current work. In our current work, we use a simple aggregation function to combine the query relevance and user interest relevance, and we will explore more on the aggregation function to study the effect of aggregation on the search result. Besides, in collaborative tagging systems, users can form different communities (e.g., based on their favorite resources). We believe it can be useful to take the community information into consideration in modeling user and resource profiles for enhancing the personalized search.

Acknowledgement

The research described in this paper has been supported primarily by the Research Grants Council of Hong Kong SAR through a general research grant (Project No. CityU 117608).

References

1. Bao, S., Xue, G., Wu, X., Yu, Y., Fei, B., Su, Z.: Optimizing web search using social annotations. In: WWW 2007, pp. 501–510. ACM, New York (2007)
2. Bischoff, K., Firan, C.S., Nejdl, W., Paiu, R.: Can all tags be used for search? In: CIKM 2008, pp. 193–202. ACM, New York (2008)
3. Chirita, P.A., Firan, C.S., Nejdl, W.: Personalized query expansion for the web. In: SIGIR 2007, pp. 7–14. ACM, New York (2007)
4. Chirita, P.A., Nejdl, W., Paiu, R., Kohlschütter, C.: Using odp metadata to personalize search. In: SIGIR 2005, pp. 178–185. ACM, New York (2005)
5. Diederich, J., Iofciu, T.: Finding communities of practice from user profiles based on folksonomies. In: Proceedings of TEL-CoPs 2006 (2006)
6. Dou, Z., Song, R., Wen, J.-R.: A large-scale evaluation and analysis of personalized search strategies. In: WWW 2007, pp. 581–590. ACM, New York (2007)
7. Golder, S.A., Huberman, B.A.: Usage patterns of collaborative tagging systems. J. Inf. Sci. 32(2), 198–208 (2006)
8. Ha, I.A., Jung, J.G., Jo, G.S., Kim, H.N.: User preference modeling from positive contents for personalized recommendation. In: Corruble, V., Takeda, M., Suzuki, E. (eds.) DS 2007. LNCS (LNAI), vol. 4755, pp. 116–126. Springer, Heidelberg (2007)
9. Li, Y., Meng, X., Wang, L., Li, Q.: RecipeCrawler: collecting recipe data from www incrementally. In: Yu, J.X., Kitsuregawa, M., Leong, H.-V. (eds.) WAIM 2006. LNCS, vol. 4016, pp. 263–274. Springer, Heidelberg (2006)
10. Wang, L.: CookingRecipe: towards a versatile and fully-fledged recipe analysis and learning system. In: Thesis of City University of Hong Kong (2008)
11. Markines, B., Cattuto, C., Menczer, F., Benz, D., Hotho, A., Stumme, G.: Evaluating similarity measures for emergent semantics of social tagging. In: WWW 2009, pp. 641–650. ACM, New York (2009)
12. Michlmayr, E.: Learning user profiles from tagging data and leveraging them for personal(ized) information access. In: Proceedings of the Workshop on Tagging and Metadata for Social Information Organization, WWW 2007 (2007)
13. Noll, M.G., Meinel, C.: Web search personalization via social bookmarking and tagging. In: Aberer, K., Choi, K.-S., Noy, N., Allemang, D., Lee, K.-I., Nixon, L.J.B., Golbeck, J., Mika, P., Maynard, D., Mizoguchi, R., Schreiber, G., Cudré-Mauroux, P. (eds.) ASWC 2007 and ISWC 2007. LNCS, vol. 4825, pp. 367–380. Springer, Heidelberg (2007)
14. Page, L., Brin, S., Motwani, R., Winograd, T.: The pagerank citation ranking: Bringing order to the web (1999)
15. Pretschner, A., Gauch, S.: Ontology based personalized search. pp. 391–398 (1999)
16. Shepitsen, A., Gemmell, J., Mobasher, B., Burke, R.: Personalized recommendation in social tagging systems using hierarchical clustering. In: RecSys 2008, pp. 259–266. ACM, New York (2008)
17. Sugiyama, K., Hatano, K., Yoshikawa, M.: Adaptive web search based on user profile constructed without any effort from users. In: WWW 2004, pp. 675–684. ACM, New York (2004)
18. Xu, S., Bao, S., Fei, B., Su, Z., Yu, Y.: Exploring folksonomy for personalized search. In: SIGIR 2008, pp. 155–162. ACM, New York (2008)

Incremental Structured Web Database Crawling via History Versions

Wei Liu[1] and Jianguo Xiao[2]

[1] Institute of Scientific and Technical Information of China, China 100038
[2] Institute of Computer Science & Technology, Peking University, China 100871
gue1976@gmail.com, xjg@icst.pku.edu.cn

Abstract. Web database crawling is one of the major kinds of design choices solution for Deep Web data integration. To the best of our knowledge, the existing works only focused on how to crawl all records in a web database at one time. Due to the high dynamic of web databases, it is not practical to always crawl the whole database in order to harvest a small proportion of new records. To this end, this paper studies the problem of incremental web database crawling, which targets at crawling the new records from a web database as many as possible while minimizing the communication costs. In our approach, a new graph model, an incremental crawling task is transformed into a graph traversal process. Based on this graph, appropriate queries are generated for crawling by analyzing the history versions of the web database. Extensive experimental evaluations over real Web databases validate the effectiveness of our techniques and provide insights for future efforts in this direction.

Keywords: Web database, Deep Web data integration, Web database crawling.

1 Introduction

The Deep Web refers to the data residing in web databases, and most of its content is in form of structured data records[1]. The Deep Web is believed to be the largest source of structured data on the Web and hence Deep Web data integration has been a long standing challenge in the field of Web data management. A popular solution for Deep Web data integration is web database crawling[2]. The crawling-based solution targets at gathering all the structured records from web databases to make users search and mine the Deep Web in a centralized manner. The rapid development of computer hardware and Internet makes this solution more practical than before.

To the best of our knowledge, all the previous efforts[3][4][5][6] focus on crawling the whole web database at one time with the goal of maximizing the coverage of the web database. We call this approach "full crawling". As it is widely known, most web databases are highly dynamic, e.g. new records are being inserted endlessly. To assure the local data in the Deep Web data integration system is consistent to that in the web databases, the maintenance operation has to be performed periodically. However, it is not affordable to always apply the full-crawling approach to only harvest a small quantity of new records(compared to the whole web database), which could result in

L. Chen, P. Triantafillou, and T. Suel (Eds.): WISE 2010, LNCS 6488, pp. 524–533, 2010.
© Springer-Verlag Berlin Heidelberg 2010

the heavy burdens for web databases, the network and the Deep Web data integration system. Therefore, a more reasonable solution is, the whole web database is crawled at beginning by employing the full-crawling approach, and then the incremental approach is applied to crawl the new records at regular intervals. For example, suppose the version of a Web database at time t_i has been crawled, the version of this Web database at time t_{i+1} can be crawled only by harvesting the new records which have been inserted during the period from t_i to t_{i+1}. In this way, the crawling cost including the issued queries and the returned records can be reduced greatly. In this paper, we study a crucial but largely unresolved problem in web database crawling: suppose the data in a web database has been obtained with the full-crawling approach, how to obtain the new records as many as possible while minimizing the issued queries and the old records returned?

To this end, we propose an efficient incremental-crawling approach. The basic idea of this approach is to crawl new records according to the history versions of the Web database. That is, we use the versions at time t_1, t_2,..., t_k, to infer the new records in the version at time t_{k+1}. The intuition behind our approach is the new records in a version are not completely random. Instead, most of these new records are close to those of the previous versions. For example, "cloud computing" is one of recent hot topics in research domain, so it is not surprising the "cloud computing" related papers in a Web database on research domain (e.g. DBLP) will be more and more. Therefore, more new records could be harvested if issuing the query "title= cloud computing" or "keyword= cloud computing".

As the initial effort to address the incremental web database crawling problem, the contribution of the paper is summarized as follows. First, we identify this novel problem of incremental web database crawling. Contrary to the full-crawling works, we demonstrate that a central issue of efficient web database crawling lies in the consistency between the local database and the integrated web databases. Second, we propose smart, efficient methods for the key problems which aims at generating promising queries to harvest the new records as many as possible while minimize the communication costs.

The rest of this paper is organized as follows: Section 2 presents the preliminaries.. Section 3 introduces the query-related graph model. The query selection method based on the query-related graph model is proposed in Section 4. We discuss our experimental findings in Section 5. Section 6 reviews the related works. Section 7 concludes this paper.

2 Preliminaries

2.1 Versions of WDB

As we have mentioned, the Deep Web is highly dynamic. We use $WDB(t_i)$ to denote the version of WDB at time t_i, i.e. all the records in WDB at time t_i. In this paper, we only consider the scenario of record insertion which is common in many domains, such as research domain and job domain. Suppose $t_i < t_j$, we use $WDB(t_j)-WDB(t_i)$ to denote the new records being inserted during the period from t_i to t_j. To formally describe the problem of this paper, we use $WDB(t_1)$, $WDB(t_2)$, ..., $WDB(t_k)$ to denote the k history versions of *WDB* and use $WDB(t_{k+1})$ to denote the current version of *WDB*. Any history version of one web database can be easily obtained from the local database because the crawled records were usually assigned a time stamp to indicate when they had been crawled. So given a time t_i, $WDB(t_i)$ is obtained by issuing SQL "select * from *WDB* where time_stamp$\leq t_i$" to the local database. As a result, any history version of *WDB* can be regarded as a virtual view of the local database.

2.2   Performance of Incremental Crawler

The goal of incremental web database crawling is to crawl as many new records as possible while minimizing the communication costs. To evaluate the crawling performance, two factors have to be considered: coverage rate and crawling cost. The former is used to measure the effectiveness, while the latter is used to measure the efficiency.

Coverage Rate

In the full-crawling works, coverage rate refers to the ratio between the number of crawled unique records and all the records of the web database, and it is denoted as *TCR* (**T**otal **C**overage **R**ate) in this paper. For the incremental web database crawling, the ratio between the number of crawled unique new records and the number of all new records is also considered, and it is denoted as *ICR*(**I**ncremental **C**overage **R**ate). Formally, the definitions of them are given below:

$$TCR = \frac{|\text{Crawled New Records}| + |WDB(t_i)|}{|WDB(t_{i+1})|} \tag{1}$$

$$ICR = \frac{|\text{Crawled New Records}|}{|WDB(t_{i+1})| - |WDB_i|} \tag{2}$$

Where $WDB(t_i)$ and $WDB(t_j)$ are two different versions of *WDB*, and $t_i < t_j$.

Crawling Cost

The crawling cost in this paper refers to the proportion of the new records in all the of returned records during the crawling process, and it is denoted as *PNR*(**P**roportion of **N**ew **R**ecords). The formal definition is given below:

$$PNR = \frac{|\text{Crawled New Records}|}{|\text{All Crawled Records}|} \tag{3}$$

Obviously, given *CCR* and *ICR*, larger *PNR* is, more efficient the crawler is. For instance, if 100 new records have been crawled and total 1000 records (include the duplicates returned by different queries) were returned in the incremental process, then $PNR=0.1$. Therefore, we must elaborately select the queries which can harvest more new records and maximize the ratio between the new records and all returned records.

3   Query-Related Graph Model

We use the query-related graph model[8] to represent a web database as an undirected graph, and the incremental crawling task is thus transformed into a graph traversal process. A query-related graph (QRG), $G(V,E)$, for WDB is an undirected graph that can be constructed as follows: for each record r_i, there exists a unique vertex $v_i \in V$. An undirected edge $(v_i, v_j) \in E$ i.f.f. r_i and r_j satisfy at least one query q can be represented in the query interface, i.e. both r_i and r_j are in the query result of q.

According to the query-related graph model, the graph complexity is determined by two factors: the number of records and the query capability of the query interface. Here the query capability refers to the attributes in the query interface and the value domains of the attributes. More attributes and more large size of the value domains of the attributes will lead to more edges in the graph. An interesting property of QRG is: if a vertex v is selected for query generation, the returned records must be its neighbors in the graph. Intuitively, if a vertex has many neighbors, more records are possible to be returned. We will use this character for vertex selection by approximating the number of the returned records.

By characterizing structured web databases using QRG, a incremental crawling task is transformed into a graph traversal process in which the crawler starts with the graph of the $WDB(t_i)$ and at each step a vertex v is selected for crawling(query formulation), and the new records which are the neighbors of v will be harvested and stored for future crawling. Obviously, different choices result in different coverage rate and crawling cost. In the next section, we will present our approach to crawl the new records based on QRG by making good choices.

4   Approach of Incremental Web Database Crawling

4.1   Approach Overview

In this section, we will introduce our approach for incremental web database crawling based on the query-related graph model. The basic idea of our approach is to crawl the new records from $WDB(t_{k+1})$ by analyzing the history versions. We assume the $k(\geq 2)$history versions of WDB have been obtained from the local database. Formally, our problem can be depicted as follows: given $WDB(t_1), WDB(t_2),\ldots, WDB(t_k)$, how to get the records$\in WDB(t_{k+1})-WDB(t_k)$?

Incremental Web Database Crawling Algorithm
Input: $WDB(t_1)$, $WDB(t_2)$,..., $WDB(t_k)$ //history k versions of WDB
Output: NRS // the set to store the crawled new records
Begin
1   Produce the query-related graphs $G(t_1)$,..., $G(t_k)$ for $WDB(t_1)$,..., $WDB(t_k)$;
2   While the terminate criteria is not met
3   r=RecordSelection($G(t_1)$,$G(t_2)$,..., $G(t_k)$);
4   q=QueryGeneration(r);
5   RS=RecordExtraction(q);
6   NRS.Add(RS);
7   Return NRS;
End

Fig. 2. Overview of the proposed approach

Fig.2 shows the overview of our approach, which is a loop process. The functions of the three components in each round are given below.

- RecordSelection: select a record from the local database based on the query-related graphs of k history versions of WDB.
- QueryGeneration: generate an appropriate query using the selected record.
- RecordExtraction: extract the records from the result pages.

This paper focuses on the key components record selection and query generation. Record extraction belongs to the research field of web data extraction, which has been widely studied, so no more discussion is for it. The stop criteria of the crawling process is no new records were crawled with recent 10 queries. In the rest part of this section, we will introduce our techniques for record selection and query generation.

4.2 Record Selection

When the query-related graphs $G(t_k)$ of $WDB(t_k)$ has been produced, the record selection problem is how to select a vertex from $G(t_k)$ whose neighbors are new records as many as possible. The intuition behind the method for this problem is that, if the neighbors of a vertex are very stable among the history versions of WDB, this vertex will have little chance to have new neighbors in the current version, and we call such vertexes "lazy vertex", otherwise we call them "active vertex". Obviously, "active vertexes" are more promising to harvest new-inserted records if they are selected to generate queries. We use the following formula to measure the activity degree of a vertex v during the period from t_1 to t_k:

$$\text{ACT}(v) = \sum_{i=2}^{k} \frac{1}{e^{k-i}} \cdot \frac{D_i(v) - D_{i-1}(v)}{D_i(v) + \varepsilon} \tag{4}$$

where $D_i(v)$ is the degree(i.e. the number of the neighbors) of v in $G(t_i)$, ε is a small positive number to avoid the denominator to be 0. According to the equation, the activity of v is the sum of the growth of the neighbors between any two adjacent versions. There are two parts in the equation. $\frac{1}{e^{k-i}}$ is the weight which means the activity of v between recent versions is more important than that between two old versions. $\frac{D_i(v)-D_{i-1}(v)}{D_i(v)+\varepsilon}$ is the activity of v between $WDB(t_{i-1})$ and $WDB(t_i)$: $D_i(v) - D_{i-1}(v)$ is the number of new neighbors of v in $WDB(t_i)$. $D_i(v)$ in the denominator is to make the number is between 0 and 1. For ε, an example is used to explain it. Suppose two vertexes v_1 and v_2, and $D_i(v_1) = 4, D_{i-1}(v_1) = 2, D_i(v_2) = 8, D_{i-1}(v_2) = 4$. Obviously, we prefer to select v_2 rather than v_1 because more new-inserted records would be harvested with one query. Based on such consideration, ε is used to make ACT(v_2)> ACT(v_1). If a vertex $v \notin G(t_i)$, $D_i(v) = 0$(i.e. the corresponding record has not been crawled at time t_i).

Record Selection Algorithm
Input: $G(t_1)$, $G(t_2)$,..., $G(t_k)$ //the query-related graphs of history k versions of WDB
Output: v // a vertex in $WDB(t_k)$
Begin
1   For each vertex v in $G(t_k)$ do
2   Compute ACT(v);
3   ACT_Sum=$\sum$ ACT(v_i) $v_i \in G(t_k)$
4   For each vertex v in $G(t_k)$ do
5   Prob(v)=$\frac{ACT(v)}{ACT\_Sum}$; //the probability of selecting v
6   Randomly select a vertex v from $G(t_k)$ according to Prob(v);
7   Return v;
8
End

Fig. 3. Record Selection Algorithm

In the algorithm, at the beginning, the query-related graphs are produced for $WDB(t_1)$,..., $WDB(t_k)$ respectively(line 1). In the implementation, $G(t_1)$ is produced first. Since $G(t_1)$ is a sub-graph of $G(t_2)$, $G(t_2)$ is produced by adding the vertexes whose responding records are in $WDB(t_2)$-$WDB(t_1)$ to $G(t_1)$ one by one. Next, the activities of the vertexes in $G(t_k)$ are computed(lines 2-3). A naïve method is selecting the vertex with max activity, but the selected records will be always in $WDB(t_k)$-$WDB(t_{k-1})$. To avoid such situation, we assign a probability(lines 4-6) to every vertex and select the vertex according to its activity.

4.3 Query Generation

After a vertex(record) has been selected, we will generate an appropriate query using the vertex. Each query can be modeled as a set of equality predicates, such as the query "position=Software Engineer and company=IBM" for searching jobs. Same to the existing works on web database crawling, we still focus on the simplest selection queries with only one equality predicate. Given a record, multiple queries can be generated. For example, three queries "A=a1", "B=b1", and "B=b2" can be generated using r_1 in Fig. 1.

Considering the crawling cost measure *ICR*, the goal of query generation is to select the most effective one from all possible queries. We define the effectiveness of a query below:

$$\text{eff}(q)= \frac{\text{count}(WDB(t_{k+1}),q)-\text{count}(WDB(t_k),q)}{\text{count}(WDB(t_{k+1}),q)} \tag{5}$$

where count(*WDB*, q) refers to the number of returned records when issuing q to *WDB*. count(*WDB*(t_{k+1}), q) can be got by extracting the hit number from the result page, such as 3,850 in "1 - 10 of **3,850**, (0.093) seconds". According to our statistics to a large number of web databases[16], more than 98.4% web databases present the hit numbers in their result pages to indicate the number of returned records for each query. And a method has been proposed by [15] to extract the hit numbers. count(*WDB*($_k$), q) can be got by issuing "select count(*) from *WDB*(t_k) where q" to the local database. As a result, count(*WDB*(t_{k+1}), q)-count(*WDB*(t_k), q) is the number of new records we will obtain.

Based on Eq. (5), a simple three-step method for query generation is described as follows. First, generate all one-equality-predicate queries using the selected record. Second, compute the effectiveness of each query. At last, the query with the max effectiveness is used for crawling.

5   Experimental Evaluation

In this section, we first present the dataset used in our experiment, then a set of experiments are conducted to evaluate the performance of the crawler implemented based on the proposed approach.

5.1   Dataset

To evaluate our incremental web database crawler precisely and objectively, our dataset is the history data of four real web databases in job and research domains which are provided by JobTong(www.jobtong.cn) and C-DBLP(www.cdblp.cn). The four web databases are Zhaopin.com(ZP), 51Job.com(5J), Journal of software(JOS), and Chinese Journal of Computers(CJC), which are depicted in Table 1. In Table 1, the second column presents their homepages, the third column shows the attributes that can be queried in their query interfaces. For each web database, we uniformly choose twelve time points, and obtain the history versions at these time points. For ZP and 5J, the time interval is one month. For JOS and CJC, the time interval is one month is half a year.

Table 1. The web databases used in our experiment

Web database	Home page	Queriable Attributes
Zhaopin.com(ZP)	www.zhaopin.com	Position, Company,
51Job.com(5J)	www.51job.com	Location
Journal of software(JOS)	www.jos.org.cn	
Chinese Journal of Computers(CJC)	cjc.ict.ac.cn	Title, Author, Key Words

5.2 Experiment Analysis

We evaluate the crawler on all the three measures(i.e. coverage rate(*CCR* and *ICR*) and crawling cost(*PNR*)) over the four web databases. A family of experiments are conducted, and Table 3 shows the experimental results. We use 4 consecutive versions to crawl the next version. For example, the head of the 3th column of Table 3(a) means we use versions 1, 2, 3 and 4 to crawl the new records in version 5. As it can be seen from Table 3, our approach can achieve very high performance on all three measures. For *CCR*, all ones are larger than 96.5% and some even close to 100%. This means almost all the records in current version of *WDB* can be crawled by us. For *ICR*, the ones(>80%) on job domain are better than those(>65%) on research domain. We think the reason is the sizes of job web databases are much larger than those of research web databases, which makes more edges exist among the records in their query related graphs. For *PNR*, all ones are in a stable range(job domain: 26.4%-50.8%; research domain: 18.9%-33.5%). This indicates that: (1)the crawler is very robust, which means its performances will not fluctuate significantly over different web databases; (2) the crawling cost is low enough for real applications.

Table 3. Experimental results coverage rate and crawling cost

WDB	Measure	History versions								AVG
		5	6	7	8	9	10	11	12	
ZP	CCR	97.6%	98.3%	99.1%	98.6%	99.4%	97.7%	98.5%	97.2%	98.3%
	ICR	88.2%	87.1%	84.6%	82.3%	90.9%	83.8%	81.5%	85.0%	85.4%
	PNR	30.4%	31.5%	26.4%	29.0%	28.5%	33.7%	46.5%	35.4%	32.7%
5J	CCR	97.6%	98.2%	98.7%	98.6%	97.4%	98.0%	98.4%	98.6%	98.2%
	ICR	91.5%	88.7%	89.3%	84.6%	81.5%	86.3%	94.1%	88.5%	88.1%
	PNR	50.8%	46.6%	45.9%	47.3%	42.4%	44.9%	47.7%	45.8%	46.4%

(a)   Job domain

WDB	Measure	History versions								AVG
		5	6	7	8	9	10	11	12	
JOS	CCR	97.5%	97.2%	98.3%	97.4%	96.7%	97.7%	98.0%	97.2%	97.5%
	ICR	75.4%	78.5%	82.5%	76.3%	67.5%	80.2%	78.7%	75.5%	76.8%
	PNR	21.5%	20.8%	23.6%	24.1%	18.9%	22.6%	21.2%	20.8%	21.7%
CJC	CCR	98.1%	97.9%	98.7%	98.6%	98.2%	98.5%	99.1%	98.3%	98.4%
	ICR	73.8%	78.0%	80.1%	77.2%	69.3%	75.4%	83.4%	78.4%	77.0%
	PNR	32.2%	33.5%	29.3%	27.6%	24.7%	29.3%	26.6%	24.5%	28.5%

(b) Research domain

Compared with the full-crawling approaches, our incremental-crawling approach is significantly superior to them on the measures *CCR* and *PNR*. To the best of our knowledge, [3] is the state-of-art work on full crawling. The best performance on *CCR* reported in its experimental result is only about 90%, which is far less than our average *CCR*(98.3% on job domain and 97.5% on research domain). According to the crawling cost reported in its experimental result, the returned records is 1.2-1.5 times of the crawled unique records. If the growth speed of *WDB* is 10%, its *ICR* is (1.1-1)/(1.1*1.2)<10%.

6 Related Work

[4] is the first work on web database crawling, which is presented to automatically extract and analyze the interface elements and submits queries through these query interfaces. [7] studies the construction of keyword queries to obtain documents from large web text collections. [5] reduced the problem of selecting an optimal set of queries from a sample of the data source into the well-known set-covering problem. [6] models web database crawling as a set covering problem and developed a new set covering algorithm to address this problem. All the works above focus on how to crawl the whole web database, which is not practical for highly dynamic web databases. Different with them, we study the incremental crawling problem. Though being a follow-up work to previous works, it is indispensable in the real deep web data integration systems.

Another important related area is web data extraction. Lots of solutions have been proposed for this issue. [9] and [10] study the problem of fully automatic data extraction by exploring the repeated patterns from multiple template generated result pages. [11] utilizes the information on the "detailed" record pages pointed by the current result page to identify and extract data records. Note that Web data extraction is orthogonal to the query selection problem investigated in this paper. There has been an active research interest in understanding the semantics of the query interfaces of the structured Web databases. [12] introduces WISE-integrator that employs comprehensive meta-data, such as element labels and default value of the elements to automatically identify matching attributes.

7 Conclusion

The high dynamic of web databases makes the full-crawling approach impractical to maintain the data consistency between the local database and the integrated web databases. In this paper, we studied the incremental web database crawling problem, and proposed an efficient and effective method to address this problem. The extensive experiment on four real Web databases shows our method can significantly reduce the crawling cost without the loss of coverage rate.

References

[1] He, B., Patel, M., Zhang, Z.: Accessing the Deep Web: A survey. Communications of the ACM 50(5) (2007)
[2] Madhavan, J., Afanasiev, L., Antova, L., Halevy, A.Y.: Harnessing the Deep Web: Present and Future. In: CIDR 2009 (2009)
[3] Wu, P., Wen, J.-R., Liu, H., Ma, W.-Y.: Query Selection Techniques for Efficient Crawling of Structured Web Sources. In: ICDE 2006 (2006)
[4] Raghavan, S., Garcia-Molina, H.: Crawling the Hidden Web. In: VLDB 2001, pp. 129–138 (2001)
[5] Lu, J., Wang, Y., Liang, J., Chen, J., Liu, J.: An Approach to Deep Web Crawling by Sampling. In: Web Intelligence 2008 (2008)

[6] Wang, Y., Lu, J., Chen, J.: Crawling Deep Web Using a New Set Covering Algorithm. In: Huang, R., Yang, Q., Pei, J., Gama, J., Meng, X., Li, X. (eds.) ADMA 2009. LNCS, vol. 5678, pp. 326–337. Springer, Heidelberg (2009)

[7] Barbosa, L., Freire, J.: Siphoning Hidden-Web Data through Keyword-Based Interfaces. In: SBBD 2004 (2004)

[8] Liu, W., Meng, X., Ling, Y.: Graph-based approach for Web database sampling. Journal of Software(Chinese) 19(2), 179–193 (2008)

[9] Zhao, H., Meng, W., Wu, Z., Raghavan, V.: Fully automatic wrapper generation for search engines. In: WWW 2005 (2005)

[10] Zhai, Y., Liu, B.: Web data extraction based on partial tree alignment. In: WWW 2005 (2005)

[11] Lerman, K., Getoor, L., Minton, S., Knoblock, C.: Using the structure of Web sites for automatic segmentation of tables. In: SIGMOD (2004)

[12] He, H., Meng, W., Yu, C., Wu, Z.: WISE-Integrator: an automatic integrator of Web search interfaces for E-commerce. In: VLDB (2003)

An Architectural Style for Process-Intensive Web Information Systems

Xiwei Xu[1,3], Liming Zhu[1,3], Udo Kannengiesser[1,3], and Yan Liu[3]

[1] NICTA*, Australian Technology Park, Eveleigh, Australia
[2] Pacific Northwest National Laboratory, USA
[3] School of Computer Science and Engineering, University of New South Wales, Australia
{xiwei.xu,liming.zhu,udo.kannengiesser}@nicta.com.au,
yan.liu@pnl.gov

Abstract. REpresentational State Transfer (REST) is the architecture style behind the World Wide Web (WWW), allowing for many desirable quality attributes such as adaptability and interoperability. However, as many process-intensive Web information systems do not make use of REST, they often do not achieve these qualities. This paper addresses this issue by proposing RESTful Business Processes (RESTfulBP), an architectural style that adapts REST principles to Web-based business processes. RESTfulBP views processes and activities as transferrable resources by representing them as process fragments associated with a set of standard operations. Distributed process fragments interoperate by adhering to these operations and exchanging process information. The process information contains basic workflow patterns that are used for dynamic process coordination at runtime. We validate our approach through an industry case study.

Keywords: REST, Resource-Oriented, Process-Intensive, Business Process.

1 Introduction

Web information systems are becoming increasingly complex as they need to support many sophisticated business processes. Such systems go beyond Create/Read/Update/Delete (CRUD) operations on information models. They need to connect services and processes, often in a distributed manner using the Web infrastructure. We call these systems process-intensive Web information systems. As the underlying Web infrastructure was originally designed for data manipulation and transfer, its application to processes has often been established based on principles of Service-Oriented Architecture (SOA) and workflow engines.

SOA provides a set of technical standards for implementing processes by connecting exposed services (operations) within and across enterprises. In traditional operation-centric implementations, SOA has a layered architecture (Figure 1). Each layer

---

* NICTA is funded by the Australian Government as represented by the Department of Broadband, Communications and the Digital Economy and the Australian Research Council through the ICT Centre of Excellence program.

L. Chen, P. Triantafillou, and T. Suel (Eds.): WISE 2010, LNCS 6488, pp. 534–547, 2010.
© Springer-Verlag Berlin Heidelberg 2010

has XML-based standards such as SOAP and WSDL, which are used to achieve protocol interoperability and hide the underlying infrastructures (Web or otherwise) from the upper layers. The principles and constraints behind SOA are based on Remote Procedure Call (RPC) with fine-grained operations. In software architecture, the term "architectural style" is used to refer to the collective set of design principles and constraints that restrict architectural elements [9]. The architectural style of the Web infrastructure is embodied in the Web protocol (HTTP). However, most process-intensive Web information systems use HTTP merely as a tunneling protocol. This has introduced architectural mismatches between the underlying data-centric Web style and the upper-layer operation-centric RPC style [28].

To further connect distributed operations/services in a systematic way, many work-flow specification languages such as BPEL and WS-CDL were introduced to define and execute business processes composed of individual services. However, practitio-ners often find it difficult to use these standards to support flexible and dynamic proc-esses [29]. The process coordination logic and decision making is often centralized and located in a workflow engine rather than distributed, exposed and transferred around that would allow local decisions. This reduces process interoperability and adaptability and violates many constraints imposed by the underlying architectural style of the Web. Recently, more styles have been introduced to the upper layer to address these limitations. For example, the MEST [20] style restricts the definition of business processes to message exchange patterns at service endpoints. But it often causes complex patterns behind each endpoint as it advocates using a single, coarse-grained "ProcessMessage" operation to handle all messages related to a process. This restricts the upper layer without exploiting the underlying Web style.

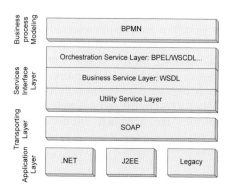

Fig. 1. SOA layered Architecture (adapted from [1])

The style that allows adaptability, interoperability and scalability of the WWW is often referred to as REpresentational State Transfer (REST) [9]. While REST was originally designed for data transfer, it has also been used to build process-intensive Web information systems. Most approaches extend SOA standards with RESTful interfaces [19, 21] or impose selected REST constraints on business process implementation [12, 16, 22]. However, they have a number of limitations:

- Methods for introducing additional constraints focus only on two constraints: uniform interfaces, and "hypermedia as the engine of application state" [24]. They ignore other useful mechanisms of the web infrastructure, such as content negotiation, rich metadata and transfer of process fragments to enable truly distributed and localized process execution.
- There is a lack of full support of workflow patterns.
- Confusion arises when different and sometimes conflicting additional constraints are proposed without a clear definition of the new style and associated methodology for implementing it.

This paper proposes RESTfulBP, an architectural style that includes additional constraints and tailors them to both the process endpoints and the orchestration layer. It has several distinguishing features:

- Unlike traditional SOA that promotes fine-grained domain-specific operations, RESTfulBP introduces the constraint of using a limited set of standard operations on "nounified" domain-specific actions for both the endpoint and the orchestration. These operations are based on the semantics of HTTP verbs.
- RESTfulBP includes mappings between web-based business process concepts and HTTP to provide guidance for implementation.
- RESTfulBP describes processes in a declarative rather than imperative way and exposes the coordination of tasks at design time using the notion of process fragments. Process fragments are primitive, reusable workflow patterns, representing possible next steps or sub-processes. They are exchanged at runtime among participants to dynamically coordinate their actions.

This paper extends our earlier work [28, 29] by 1) defining a new architectural style, 2) providing support for workflow patterns represented in a Microformat, 3) proposing a mapping between business processes and HTTP (verbs and status codes) and 4) supplying methodological guidance for the design of process-intensive web information systems. We demonstrate the benefits of the proposed architectural style through a case study. This paper is organized as follows: Section 2 discusses related work. Section 3 provides details of RESTfulBP and illustrates it using a property valuation process. Section 4 evaluates the approach against traditional SOA. Section 5 concludes the paper.

2 Related Work

Although REST [9] and Resource-Oriented Architecture [24] principles are well established, their potential use in developing web-based business process systems has not been well understood. Research focuses on different layers of SOA (Figure 1) to fill the gap between SOA and REST from different perspectives. At the transport layer, CREST [8] extends REST by modeling computation as a resource. The representation communicated is mobile code rather than higher-level abstractions such as process fragments. Some CREST principles, including data and computing bundling, could be applied to RESTfulBP.

At the orchestration layer, many traditional process languages and runtime environments evolved towards leveraging traditional Web Service and RESTful Web Service coordination. For example, many extensions of BPEL are proposed to support the RESTful Web Service invocation [10, 19, 21]. All of these BPEL-based solutions are separated from the services themselves and promote a centralized process coordination strategy. They are not able to communicate process information at runtime, which significantly reduces flexibility and visibility of the service.

At the process modeling layer, there are also efforts to align REST with web-based business processes. An example is the information-centric process modeling approach proposed in [13]. It views business processes as the evolution of values of a collection of business entities through state transitions. The behaviors of the business entities are modeled using a state machine. This kind of modeling can be implemented straightforwardly in a RESTful architecture.

3   The RESTfulBP Architectural Style

This section discusses a set of architectural constraints that aim to establish communication and coordination mechanisms among participants in a business process from a peer-to-peer and distributed point of view (rather than using centralized coordination supported by traditional workflow engines). Small pieces of workflow/process logic – process fragments – live within each endpoint and can be transferred to other endpoints.

We use a property valuation process in the Australian lending industry as an example to illustrate our approach. This process requires interactions among several parties, forming a complex process with a mixture of manual and automated tasks. Figure 2 shows only a high-level model of the process in BPMN. Here, a valuation request (received from a lender) is processed through several stages until the valuation is completed. Throughout the process, the lender and the valuation firm communicate with each other to deal with status updates, cancellation requests, fee renegotiation, and so forth. Each stage of the process has a status code, such as "Instructed", "Assigned", "Inspected" and "Valuation Completed", according to standard definitions provided by the LIXI (Lending Industry XML Initiative) consortium [15]. Figure 2 adds the status codes as annotations to the process model.

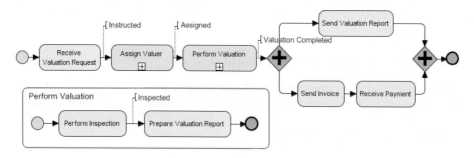

Fig. 2. LIXI property valuation process

3.1   Identify Process Elements by URI

The business process can be defined by a business manager using an existing modeling notation such as BPMN [7]. The purpose of modeling is to identify all the tasks at different levels of abstraction, parties (roles) and control dependencies (i.e., sequencing, parallel) between these tasks. The process model may also be a tree structure with loosely connected process fragments. Explicit conditional routing (e.g., using gateways in BPMN) may be defined in the model; however, it is also possible to attach such information to a task as process fragment information or rules to be exchanged among participants. A task may have different routing information attached to support a variety of situations as needed. Choosing an appropriate degree of granularity is a common issue in process modeling that requires considering a number of factors, such as desired flexibility [11] and integration effort [5]. Since our focus is on process realization rather than process modeling, addressing this issue in more detail is beyond the scope of this paper.

RESTfulBP realizes processes and tasks as a set of resources identified by declarative URIs [24]. In our property valuation process, the URI of the overall process can be represented as a resource identified by the URI www.restfulbp.com/lixi/valuationprocess/. The URI of a case entity (process instance) is created by putting the case ID behind the process resource URI, e.g. www.restfulbp.com/lixi/valuationprocess/1234.

The hierarchical task structure can be directly mapped to the hierarchical URI path syntax. Take the decomposed "Perform Valuation" task in Figure 2 as an example. The relative URI of this task is /valuationprocess/valuationperform, and the URI of the sub-task "Perform Inspection" is /valuationprocess/valuationperform/inspection/. The reserved slash ("/") character in a URI [6] denotes a separation between two hierarchical levels. Exposure of states is useful for visibility of running cases. We model each state of the process model as an abstract entity. For example, the information represented by the relative URI /valuationprocess/instructed/ includes the entire list of requests that can be accepted by the valuation firm.

The four kinds of resources discussed above (i.e., process, case, task and state) are required for process execution by the runtime engine. However, adding other resources is allowed to get better process management and process visibility. For example, an ActiveRequests entity can be designed and mapped to a resource identified by the URI /valuationprocess/activerequests, holding information of all the requests being processed.

3.2   Manipulate Entities through Uniform Methods

RESTfulBP represents manipulations of entities through a set of uniform methods. Our preliminary set of methods resemble the HTTP standard methods but with important differences. We outline our adaptation of these HTTP verbs in Table 1. The idempotent GET exposes relevant information in a consistent way. Every resource provides a GET method by default to supply different information according to authentication. The PUT method is used to implement tasks. Its conditional support by a case resource means that this resource is allowed to be updated only under certain conditions. The POST method creates a new case from a process resource. The DELETE method allows the client to cancel the generated case. These operations could be further extended; for example, the task resource can be designed to support a DELETE method to allow an administrator to dynamically remove tasks. But these extensions should be limited, to avoid including specific RPC-styled operations as used in traditional approaches.

Table 1. Manipulations of entities represented using HTTP methods

	Resource$_{process}$	Resource$_{case}$	Resource$_{task}$	Resource$_{state}$
GET	Provide the specification of the process	Get the status of the process instance	Provide specification of the task	Get the entire cases at that state
PUT	×	Conditional	Implement the task	×
POST	Create a new case	×	×	×
DELETE	×	Cancel the case or Delete the artifacts uploaded by client	×	×

3.3   Represent Connections between Entities through Microformat

In the WWW, Microformat is a popular and lightweight approach where information intended for end-users is embedded in existing XTHML tags with additional semantics. RESTfulBP uses Microformat-based messages to communicate routing information (next steps, task connections, and sequence/parallel semantics) at run time. Although the URI of each task is recommended to be as descriptive as possible, a Microformat is needed to supply additional semantics for tasks. For example, when a server needs to generate a page to stop repeated access to a URI to ensure idempotence, the URI is usually opaque to the client, and the semantics provided by the Microformat become useful. Generally, the communication format is based on the link element of XHTML with specific annotations on the attributes (e.g. class, and rel) and extra constructs that represent various workflow patterns [27].

Rather than describing the complete control flow, RESTfulBP describes only the relationship of two related steps. Basic constructs are used to communicate the information of five basic workflow patterns shown in Figure 3. In the Exclusive Choice pattern, the runtime engine removes the remaining task from the case representation after the client chooses either task2 or task3. In the Parallel pattern, the runtime engine keeps the remaining task in the case representation after the client chooses one. The Microformat can support all 20 basic workflow patterns proposed by [27], although some of the patterns need workarounds. Designers can add more constructs to our basic Microformat to support more advanced patterns if needed.

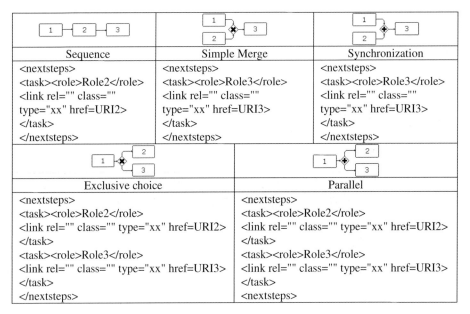

Sequence	Simple Merge	Synchronization
<nextsteps> <task><role>Role2</role> <link rel="" class="" type="xx" href=URI2> </task> </nextsteps>	<nextsteps> <task><role>Role3</role> <link rel="" class="" type="xx" href=URI3> </task> </nextsteps>	<nextsteps> <task><role>Role3</role> <link rel="" class="" type="xx" href=URI3> </task> </nextsteps>

Exclusive choice	Parallel
<nextsteps> <task><role>Role2</role> <link rel="" class="" type="xx" href=URI2> </task> <task><role>Role3</role> <link rel="" class="" type="xx" href=URI3> </task> </nextsteps>	<nextsteps> <task><role>Role2</role> <link rel="" class="" type="xx" href=URI2> </task> <task><role>Role3</role> <link rel="" class="" type="xx" href=URI3> </task> <nextsteps>

Fig. 3. Workflow patterns support example in the case representation

3.4 Communicate Context Information by Content Negotiation Mechanism

When a method on an entity is called, information about its status is returned. The information may contain meta-information or status information about an entity (task/process/case/state) and process coordination information such as possible next steps and connections among relevant tasks. The output of a method on an entity can be different according to different roles. For example, a case entity may have two kinds of outputs. One represents the current status of a case, which is often returned by the GET method to a process participant. This ensures the visibility of a running process instance to all parties involved for more informed local decision making. The other kind of output represents all possible next steps for a participant who is permitted to advance the case. This information can only be returned to the relevant counterparties with appropriate authentication so that they can advance the case by requesting the PUT method on the URI provided in each possible next step.

RESTfulBP uses content negotiation and authentication mechanism provided by the Web infrastructure to choose an appropriate representation of a resource for the client. Firstly, the server can estimate the desired representation format by the ACCEPT request header, which indicates the preferred media type. The existing media types are limited since they only describe the file format; however, two representations of one resource can be in the same format but with different routing information for different roles. We extend the two-level media type with a third level to represent the role of the requesting participant (e.g. Application/XML/lender). In our example, there are three kinds of roles: A "Lender" issues the valuation request, a "Valuation Firm" processes the valuation request, and a "Valuer" is the person performing the valuation. Based on the given format and role, the server can authenticate the participant, and return the corresponding information. For example, consider that the participant in the "Valuer" role attempts to get a list of those cases that are ready to be evaluated, by sending the following request but without any authentication:

```
GET /valuationprocess/inspected/ HTTP/1.1
Host: http://www.restfulbp.com
Accept: application/xml/valuer
```

The server will return the following response:

```
HTTP/1.1 401 Unauthorized
WWW-Authenticate: Digest
realm="valuer@www.restfulbp.com",
nonce="dcd98b7102dd2f0e8b11d0f600bfb0c093"
```

The "realm" value denotes which username and password to use and the "nonce" value is a server-specified data string which is uniquely generated. Once the participant provides the correct authentication, the server will return the representation with required information for the valuer to advance the case.

3.5 Communicate Exception by Message Header

Exception handling in business process execution is a key to ensure reliability of application systems [25]. In normal web applications, when an exception occurs at runtime inside the server, it will return to the client a response header with a "500 Internal Server Error" status code without any meta-data for the exception. Sometimes the description of the exception is also included in the response body. RESTfulBP communicates the exception information by a response header. Following the HTTP extension framework [17], we extend the HTTP protocol with an optional new header field "Exception" to denote the type of exception that can occur. This extension is only used in task resource realization. A client application that understands the extension can send requests as below to check the status of a case:

```
GET /recruitment/1234
HOST: www.restfulbp.com
UPGRADE: HTTP/1.2
IF-NOT-MATCH: 1234
IF-MODIFIED-SINCE: May 6 2008 13:32:08 GMT
ACCEPT: application/xml/lender
```

If an exception occurs during task completion, the response may look like this:

```
HTTP/1.2 500 Internal server error
Date: May 6 2008 14:32:08 GMT
Opt: http://www.restbp.org/exception; ns=57
Location: http://www.restfulbp.com/recruitment/1234
ETAG: 1236
LAST-MODIFIED: May 6 2008 14:14:08 GMT
Content-length: 123
Content-Type: application/xml/lender
57-exception: "resource unavailable"
```

Since the extension is optional, a browser that gets this response can simply ignore the extension. It is up to the designer to decide the values of the new header field

"Exception". In our example, the data model provided by LIXI identifies 10 delay reasons for the "Delayed" stage, such as "Incorrect Request Data", "Access Denied" and "Awaiting Fee Change Response". These delay reasons are modeled as values of the optional header field. Thus, the client can get useful information about the reason for a delay from the message header.

4   Case Study

The goal of the LIXI case study is to conduct a comparative evaluation of the proposed architectural style against the traditional SOA architectural style with respect to adaptability and interoperability. LIXI provides a simple BPEL-based reference implementation of the valuation process and its associated web information system. We implemented the process in a Property Valuation System (PVS) of a company.

Case Study Procedure. The industry project has two versions—PVS2.0 and PVS3.0. Both versions have been implemented using traditional SOA and RESTfulBP, respectively. In the traditional SOA implementation, the Apache Axis 2.0 [2] Eclipse plug-in was used to design the service interface, the BPEL designer [4] from an Eclipse incubation project was used to visualize the process models, and Apache ODE [3] was used to execute the process. In the RESTfulBP implementation, our Eclipse-based BPMN modeler extension (an Eclipse plug-in of the BPMN modeler to annotate BPMN diagrams with REST information) was used to model the process diagram provided by LIXI and to generate the basic code. Our application framework (extensions of the RESTlet framework [23]) was used to support and execute the generated code. Some additional information was added after code generation, such as registering exceptions and exposing selected monitoring information.

The two approaches are compared with respect to interoperability and adaptability. For interoperability, we followed several measurement approaches [26] on syntactic, semantic and protocol levels. For adaptability, a scenario-based architecture evaluation approach was followed. Here, we used process change scenarios that are realistic in the lending industry. The number of places affected and the complexity of each change are measured and compared.

Interoperability. Interoperability is the ability of a collection of communicating entities to share specific information and operate on it according to an agreed-upon operational semantics [18]. In a web-based business process implementation using SOA, the communicating entities are web services that are atomic or composite (process).

Table 2 compares three kinds of information at three different levels shared between two services (task information) and two processes (process information and meta-info of process entities). These kinds of information are essential to improve interoperability. Using HTTP as an application protocol standardizes many aspects of business process realization and execution. Firstly, it supports uniform methods rather than fine-grained operations defined in WSDL/BPEL, which can reduce the design burden. However, the semantics of the data are still an issue for both uniform and non-uniform operations, Secondly, by sharing the case itself as an accessible resource and mapping the HTTP headers, RESTfulBP can provide a unified business process management facility. Thirdly, the task connectedness facilitates local decision making according to the process context. The alignment between infrastructure and business process makes it possible to reuse existing HTTP mechanisms such as authentication and content negotiation. Due to space limitations, we will discuss only some of the issues shown in Table 2.

Table 2. Sharing information

		Syntax-Level	Semantics-Level	Protocol-Level
Meta-Info of Process entities	RESTful BP	Mapping to HTTP header	Mapping to HTTP headers	• URI templates as instance correlation • HTTP content negotiation • Status as accessible resource
	WSDL/ BPEL	Ad-hoc	Ad-hoc	• Correlation set as instance correlation • Ad-hoc content negotiation • Ad-hoc management facilities
Task (Operation)	RESTful BP	Resource Representation	Mapping to HTTP limited set of verbs	Process fragments
	WSDL/ BPEL	WSDL operation definition	Fine-grained Ad-hoc (e.g. using other standards like WSDL-S)	Proprietary and application specific
Process	RESTful BP	Light weight Microformat	Microformat	Transferable process fragment
	WSDL/ BPEL	BPEL process definition	Ad-hoc BPEL semantics	N/A

• *Monitoring the process instance status*

Currently, some industrial BPEL engines provide ad-hoc management functionality, but the API of the management functionality is not yet standardized by the BPEL specification [14]. The Apache ODE used in our case study provides a process instance management API, for example, to list all process instances, to suspend a process instance, and to delete a completed process instance. However, these functions are too general to be used by the participants who take part into a specific business process. Some workarounds were used in the traditional SOA implementation for the specific requirements around process instance monitoring. A separate operation was added to the process Web service interface to provide the monitoring function. An event handler was added to the BPEL definition to handle monitoring requests. RESTfulBP, however, allows the participants (with proper authentication) to check the status of any process instance by querying the corresponding resource and its meta-information through the standard HTTP protocol. This improves process visibility compared to the limited and specialized management functionality provided by Apache ODE.

- *Exchange process fragments at runtime*

Neither the BPEL specification nor any vendor-specific runtime engine provides a mechanism to share control-flow information at runtime. In traditional SOA, all allowed conversations have to be pre-defined and shared (through documentation and abstract BPEL) at design time. Each participant models the corresponding process according to these protocols for collaboration. How a valuation can proceed is fixed either through a very complex pre-defined process or a typical process throwing a lot of exceptions with undefined handling processes. RESTfulBP, however, can communicate the process coordination information at runtime. This facilitates more flexible decision making based on local, runtime information, and establish interoperability between process participants.

Adaptability. Adaptability is the ease with which a system may be changed to fit changed requirements [18]. SOA can improve this attribute by hiding implementation details behind interfaces driven by standard WS protocols. Although the protocols are standardized, the semantics and protocols of each fine-grained operation must be defined in ad-hoc ways. Changes in these fine-grained operations often break existing systems and incur costs in learning the new semantics/protocols of the new operation. RESTfulBP improves the situation by further standardizing the semantics/protocols by a limited set of operations, which only leaves the business data to be evolved. These benefits will be evaluated through a change-scenario.

In the LIXI case study, version 2.0 introduced a set of process-related new requirements, including task allocation automation, request auto-creation through PDF, and fee renegotiation. We compared the adaptability of the traditional SOA implementation and the RESTfulBP implementation by modifying them according to the new requirement. This paper describes this for the scenario of fee renegotiation.

In the traditional SOA implementation, fee renegotiation is handled as a type of delay. During the "Perform Inspection" task, a valuation firm may send a fee renegotiation request to the lender. The valuation process is then delayed, and might be active again until the two parties reach an agreement on the new fee, or is canceled if both parties cannot reach an agreement. In reality, only if the property is much more complex than expected (thus a much higher new fee), the valuation firm will send a fee renegotiation request to the lender *before* inspection. If the expected fee change is minimal, the valuation firm may send the fee renegotiation request after completing the inspection, or even after completing the valuation report. This is a typical scenario of the interleaved parallel routing workflow pattern [27]. This pattern can be implemented by the event handler and serializable scope in BPEL.

The RESTfulBP implementation of the system is shown in Figure 4. Every task resource may have several possible process fragments defined in a separate file. The initial fee estimation is included in the inspection task resource. First, the lender advances the case by PUTting it to the inspection resource. During the process, the valuation firm may find that the required fee is more than the lender provided, thus responding to the lender with an Exception header to indicate the delay reason (Awaiting Fee Change Response). The status of the corresponding case is updated to Delayed. The lender can then send another request to the inspection resource with its decision on the fee renegotiation. The inspection resource updates the case resource according to the lender's decision. To implement the new requirement, the fee negotiation service provided by Lender is exposed as a resource. Its link is added to the next steps of both "Perform Inspection" and "Prepare Valuation Report" tasks. The representation of the case resource is updated accordingly. Therefore, after the lender PUTs the case representation to the inspection resource, the valuer can send a GET request to the case resource to get the link of the fee renegotiation resource and communicate with the lender directly to renegotiate the fee without suspending the valuation process.

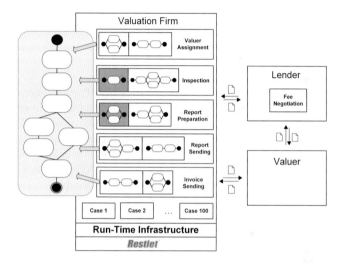

Fig. 4. RESTfulBP implementation of new requirement (the blue color is the updated part)

The change scenario focuses on the modification of control flow. In the traditional SOA implementation, the change leads to the addition of a fee renegotiation task and the modification of the two tasks "Perform Inspection" and "Prepare Valuation Report" that may cause the fee renegotiation. In addition to the three main modifications of the BPEL file, there are some modifications of the endpoint service code inducing the fee renegotiation service and the two tasks. This is due to the new workflows that were not expected in the original BPEL model and the ripple effects across the BPEL-based process model. In RESTfulBP, the change affects the fee renegotiation task only at the process level. By exposing the fee renegotiation service as a resource, the valuer can communicate with the lender directly at runtime. The key to achieving adaptability at both design time and runtime is the dynamic exchange of next-step information. Another example concerns the monitoring facilities mentioned earlier. The traditional BPEL/SOA implementation has to pre-define where and how such information can be accessed. In RESTfulBP, we simply provide the relevant monitoring URI in the response representation on a case-by-case basis. A lender can check the process instance status without having to know the control-flow information in advance.

The change scenario is a typical kind of change that can be handled more easily using RESTfulBP than using WSDL/BPEL. A change in a small part of a BPEL-based, centralized approach may affect the whole process. In RESTfulBP, however, the whole process is generated at runtime by distributed process fragments. Thus, the modification of one process fragment does not affect other fragments.

5 Conclusion

In this paper, we proposed an architectural style for realizing process-intensive web information systems. Our approach is based on REST principles by turning process constituents into resources and attaching a limited set of standard operations to them. Distributed processes interoperate by adhering to these standard operations and exchanging process information, which contains not only the typical meta-information about a process but also process fragments that indicate possible next steps and interconnectedness for process coordination purposes.

We implemented a framework to support RESTfulBP execution, based on an open-source REST framework called RESTlet. It has a tool that allows designers to annotate an existing BPMN model with REST-related information. The generated RESTfulBP systems use "Web" principles for process-intensive systems. RESTfulBP differs significantly from traditional service-oriented paradigms for process-intensive systems and other REST-based business processes. RESTfulBP uses standardized HTTP verbs rather than ad-hoc verbs, and represents all concepts (including tasks) as resources with unique identifiers. It uses a Microformat to dynamically communicate process information during execution, rather than relying on centralized process models and pre-determined control flows. RESTfulBP exposes all information as resources and promotes context-free, stateless and high-visibility communications, in contrast to existing business process engines that maintain internal state and expose only limited process information. Unlike existing REST-based business process modeling which introduces only the uniform interface architectural constraint, and "hypermedia as the engine of application state" into web-based business process, RESTfulBP fully utilizes other useful mechanisms of the web infrastructure, such as fee negotiation and authentication. Our evaluation has shown that this approach can enhance interoperability and adaptability of process-intensive Web information systems.

References

1. Arsanjani, A.: Service-Oriented Modeling and Architecture (November 09, 2004), IBM developerWorks (accessed: October 05, 2009)
2. Apache Axis2/Java – Next Generation Web Services, http://ws.apache.org/axis2/ (accessed: March 17, 2010)
3. Apache ODE, http://ode.apache.org/ (accessed: March 17, 2010)
4. BPEL Project, http://www.eclipse.org/bpel/ (accessed: March 17, 2010)
5. Brereton, P., Budgen, D.: Component-based Systems: a Classification of Issues. Computer 33(11), 54–62 (2000)
6. Burners-lee, T., Masinter, L., McCahill, M.: Uniform Resource Locators (URL). RF1738 (1994)
7. Business Process Model and Notation (BPMN), http://www.omg.org/spec/BPMN/ (accessed: March 17, 2010)

8. Erenkrantz, J.R., Gorlick, M., Suryanarayana, G., Taylor, R.N.: From Representation to Computations: the Evolution of Web Architectures. In: 6th Joint Meeting of the European Software Eng. Conference and the ACM SIGSOFT Symposium on The Foundations of Software Eng., pp. 255–264 (2007)
9. Fielding, R.: Architectural Styles and the Design of Network-based Software Architectures. Doctoral Dissertation, Univ. of California, Irvine (2000)
10. Hitzsche, J., Lessen, T.V., Karastoyanova, D., Leymann, F.: BPELlight. In: Alonso, G., Dadam, P., Rosemann, M. (eds.) BPM 2007. LNCS, vol. 4714, pp. 214–229. Springer, Heidelberg (2007)
11. Kannengiesser, U.: Process Flexibility: A Design View and Specification Schema. In: Enterprise Modeling and Information Systems Architectures 2009, University of Ulm, Germany, pp. 111–124 (2009)
12. Webber, J., Parastatidis, S., Robinson, I.: How to GET a Cup of Coffee (October 02, 2008), http://www.infoq.com/articles/webber-rest-workflow (accessed: October 05, 2009)
13. Kumaran, S.: A RESTful Architecture for Service-Oriented Business Process Execution. In: 4th Int'l Conf. e-Business Engineering, pp. 197–204 (2008)
14. Kuleshow, I., Rogovich, V.: Interoperability Challenges for WS-BPEL Standard. In: 4th IEEE Int'l Workshop on Intelligent Data Acquisition and Advanced Computing Systems: Technology and Applications, pp. 503–505 (2007)
15. Lending Industry XML Initiative (LIXI), http://www.lixi.org.au (accessed: March 17, 2010)
16. Muehlen, M., Nickerson, J.V., Swenson, K.D.: Developing web services choreography standards: the case of REST vs. SOAP. Decision Support Systems 40(1), 9–29 (2005)
17. Nielsen, H.F., Leach, P., Lawrence, S.: An HTTP Extension Framework, RFC 2774 (1999)
18. O'Brien, W., Bass, L., Merson, P.: Quality Attributes and Service-Oriented Architectures. Tech. Rep. CMU/SEI-2005-TN-014 (2005)
19. Overdick, H.: Towards resource-oriented BPEL. In: 2nd ECOWS Workshop on Emerging Web Services Technology, pp. 129–140 (2007)
20. Parastatidis, S., Webber, J., Woodman, S., Kuo, D., Greenfield, P.: An Introduction to the SOAP Service Description Language. Tech. Rep. CSTR-898, School of Computing Science, University of Newcastle (2005)
21. Pautasso, C.: BPEL for REST. In: Dumas, M., Reichert, M., Shan, M.-C. (eds.) BPM 2008. LNCS, vol. 5240, pp. 278–293. Springer, Heidelberg (2008)
22. Rest-client, http://github.com/caelum/rest-client (accessed: June 06, 2010)
23. RESTlet, http://www.restlet.org/ (accessed: March 17, 2010)
24. Richardson, L., Ruby, S.: RESTful Web Services. O'Reilly Media, USA (2007)
25. Shang, Z., Cui, L., Wang, H.: A Collaborative Framework for Exception Handling in Business Process Execution. In: 11th Int'l Conf. computer Supported Cooperative work in Design, pp. 914–919 (2007)
26. Vallecillo, A., Hernández, J., Troya, J.M.: Component Interoperability. Tech. Rep. ITI-2000-37, Dept.de Lenguajes y Ciencias de la Computacion, Univ. of Aalaga (2000)
27. van der Aalst, W.M.P., et al.: Workflow Patterns. Distributed and Parallel Databases 14(3), 5–51 (2003)
28. Xu, X., Zhu, L., Liu, Y., Staples, M.: Resource-Oriented Architecture for Business Process. In: 15th Asia-Pacific Conf. Software Engineering, pp. 395–402 (2008)
29. Xu, X., Zhu, L., Staples, M., Liu, Y.: An Architecting Method for Distributed Process-Intensive Systems. In: Joint Working IEEE/IFIP Conf. Software Architecture and European Conf. Software Architecture, pp. 277–280 (2009)

Model-Driven Development of Adaptive Service-Based Systems with Aspects and Rules

Jian Yu, Quan Z. Sheng, and Joshua K.Y. Swee

School of Computer Science
The University of Adelaide, SA 5005, Australia
{jian.yu01,qsheng,kheng.swee}@adelaide.edu.au

Abstract. Service-oriented computing (SOC) has become a dominant paradigm in developing distributed Web-based software systems. Besides the benefits such as interoperability and flexibility brought by SOC, modern service-based software systems are frequently required to be highly adaptable in order to cope with rapid changes and evolution of business goals, requirements, as well as physical context in a dynamic business environment. Unfortunately, adaptive systems are still difficult to build due to its high complexity. In this paper, we propose a novel approach called MoDAR to support the development of dynamically adaptive service-based systems (DASS). Especially in this approach, we first model the functionality of a system by two constituent parts: i) a stable part called the *base model* described using business processes, and ii) a volatile part called the *variable model* described using business rules. This model reflects the fact that business processes and rules are two significant and complementary aspects of business requirements, and business rules are usually much more volatile than business processes. We then use an aspect-oriented approach to weave the base model and variable model together so that they can evolve independently without interfering with each other. A model-driven platform has been implemented to support the development lifecycle of a DASS from specification, design, to deployment and execution. Systems developed with the MoDAR platform are running on top of a BPEL process engine and a Drools rule engine. Experimentation shows that our approach brings high adaptability and maintainability to service-based systems with reasonable performance overhead.

Keywords: Adaptive systems, Web service, aspect-oriented methodology, model-driven development, business processes, business rules.

1 Introduction

The service-oriented computing (SOC) paradigm, which promotes the idea of assembling autonomous and platform-independent services in a loosely coupled manner to create flexible business processes and applications that span organizational boundaries, presently has a dominant position in developing distributed, especially Web-based, software systems [10,16]. Web services are the

L. Chen, P. Triantafillou, and T. Suel (Eds.): WISE 2010, LNCS 6488, pp. 548–563, 2010.

most promising technology for implementing SOC, and Web Services Business Process Execution Language (WS-BPEL, or BPEL in short) [19] has become a de facto industry standard to create composite service processes and applications.

One of the key research challenges in SOC is *dynamically adaptive processes*. As stated in [16], "*services and processes should equip themselves with adaptive service capabilities so that they can continually morph themselves to respond to environmental demands and changes without compromising operational and financial efficiencies*". By *dynamically adaptive*, we mean a process is able to change its behavior at runtime in accordance with the changes in the external environment. To be specific, environmental changes could be classified into *execution* environment changes and *requirements* environment changes. Execution environment changes often demand an adaptive service-based system (DASS) to reconfigure itself in order to keep up with the *prescribed* quality-of-service (QoS) requirements. For example, a DASS may dynamically increase the number of service instances if its current response time is slower than the required criterion. On the other hand, requirements environment changes raise *new* requirements to a DASS. The requirements changes could be either *non-functional*, which also lead to system reconfiguration, or *functional*, which requires a DASS to include/exclude/replace component services or change its internal structure to expose a new functional behavior. For example, a business promotion campaign may require a travel booking process to include free car rental services or give discount to specific itineraries.

In this paper, we propose a novel approach called MoDAR (Model-Driven Development of DASS with Aspects and Rules) to facilitate the development of DASS that are able to dynamically adapt their behaviors in line with functional requirement changes. Based on the three-layer abstraction of service-based systems [15], at the top business process management (BPM) layer, we use business processes to model the relatively stable part of a DASS, which is called the *base model*, and use business rules, which are much easier to change than business processes [18], to model the volatile part, which is called the *variable model*. Such separation of concerns is crucial to manage the complexity of a DASS. An aspect-oriented approach is used to integrate business processes and rules so that they can keep their modularity and evolve independently. At the middle service composition and coordination layer, we use BPEL to represent business processes and encapsulate business rules as Web services. The invocations from a BPEL process to rule services are established based on the aspect model defined at the BPM layer. At the bottom service infrastructure layer, a BPEL engine and a Drools[1] rule engine work together to execute the DASS. A model-driven approach is adopted to automate the model transformation between layers. A service-based system created using MoDAR is dynamically adaptive in the sense that any changes made to the variable model of the system can be reflected dynamically in the running system.

Except for bringing dynamic adaptivity to service-based systems, another innovative feature of MoDAR is to use a model-driven approach to facilitate the

---

[1] www.jboss.org/drools/

modeling of DASS at different layers and the (semi-)automatic translation from higher-level models to executable systems, which eases both development and maintenance of DASS. A graphical model-driven platform has been developed to support the modeling, transformation, and deployment of DASS, which significantly reduces the system development time and cost.

The rest of the paper is organized as follows. Section 2 briefly overviews some basic concepts used in the paper and gives a motivating scenario that will be referred throughout the paper to illustrate our approach. Section 3 is dedicated to the MoDAR methodology. We first overview the whole approach and then explain in detail the three major layers of MoDAR and the model transformation between layers. Section 4 introduces the MoDAR platform and its performance evaluation. Finally, we discuss related work in Section 5 and conclude the paper in Section 6.

2   Background

In this section, we first briefly overview some basic concepts, namely *business rules*, *aspect-oriented methodology*, and *model-driven development*. We then present a motivating scenario to highlight the challenges in developing DASS.

2.1   Business Rules

Business rules are statements that define or constrain some aspects of a business [12]. They make the knowledge about *regulations*, *policies*, and *decisions* explicit and traceable [19]. Business rules can be classified into four groups:

- *constraint rule*: A statement that expresses an unconditional circumstance that must be true or false.
- *action enabler rule*: A statement that checks conditions and initiates some actions upon finding the conditions true.
- *computation rule*: A statement that checks a condition and when the result is true, provides an algorithm to calculate the value of a term.
- *inference rule*: A statement that tests conditions and establishes the truth of a new fact upon finding the conditions true.

2.2   Aspect-Oriented Methodology

Aspect-oriented methodology (AOM) is a strand of software development paradigm that models scattered crosscutting system concerns as first-class elements, and then weaves them together into a coherent model or program. Concerns can be high-level notations like security and quality of service, low-level notations like caching and buffering, functional elements such as features or business rules, or non-functional elements such as synchronization and transaction management [8]. AOM can bring better modularity: it allows developers to

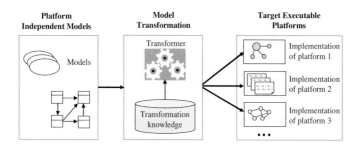

Fig. 1. Model-driven development

encapsulate system behavior that does not fit cleanly into the particular programming model in use [3]. Referring to Aspect4J [11], an aspect-oriented program usually consists of base modules and a number of aspects that modularizes crosscutting concerns. Basic concepts in Aspect4J include *aspect*, *pointcut*, and *advice*. An aspect wraps up pointcuts and advices. A pointcut picks out certain join points, which are well-defined points (e.g., method calls) in the program flow. An advice is a piece of code that is executed when a join point is reached. There are three types of advice:

- A *before advice* runs just before the join points picked out by the pointcut.
- An *after advice* runs just after each join point picked out by the pointcut.
- An *around advice* runs instead of the picked join point.

2.3   Model-Driven Development

Model-driven development (MDD) [9] is an approach that supports system development by employing a model-centric and generative development process. The basic idea of MDD is illustrated in Figure 1. Adopting a higher-level of abstraction, software systems can be specified in platform independent models (PIMs), which are then (semi)automatically transformed into platform specific models (PSMs) of target executable platforms using some transformation tools. The same PIM can be transformed into different executable platforms (i.e., multiple PSMs), thus considerably simplifying software development.

2.4   A Motivating Scenario

In this section, we present a context-aware travel planning scenario in the mobile commerce domain to motivate the challenges in developing DASS.

Suppose that a travel agency StarTravel wants to provide a mobile application called SmartTravel to its customers, as depicted in Figure 2. With SmartTravel, a customer can send travel requests to StarTravel via her mobile phone. When receiving a request, SmartTravel first validates the request, e.g., to see if the request contains valid departure and destination cities, and if the validation fails, the customer will receive an SMS explaining why the request is failed. Otherwise,

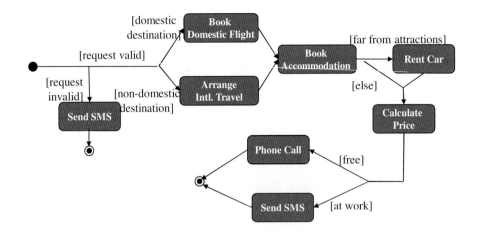

Fig. 2. The SmartTravel business process

based on the destination of the travel request, SmartTravel will either book a domestic flight or arrange an international travel. SmartTravel will also book the accommodation and rent a car for the customer based on her preferences. Finally, the customer could get a 10% discount if she is a member of StarTravel and the detailed travel plan will be sent back to the customer based on her presence: if she is free, a staff will give her a call; and if she is at work, an SMS will be sent instead.

The constant change in business environment, organization policies, and also user preferences and context demand the business logic of SmartTravel to evolve dynamically to be in line with the new requirements. For example, following requirements may crop up at anytime in the future:

- Booking domestic flight and arranging international travel are merged to a single service because of company policy adjustment.
- The customer wants to rent a car only if the location of her accommodation is far from tour attractions.
- If there is a promotion, the cost needs to be re-calculated using the promotion rule.
- The customer wants to receive the travel plan via email when she is in a meeting.

A major challenge is how to make SmartTravel dynamically adaptive so that when a new requirement emerges it is able to promptly adjust its behavior with minimal effort. If we follow the common SOC approach to implement Smart-Travel as an executable BPEL process, the process structure may need to change in many new situations, which results in *re-compilation* and *re-deployment* of the process. For example, if the business policy has changed and domestic and international travels need to be merged into one service, we have to review the whole process logic, and then create a new service to accommodate the functions of

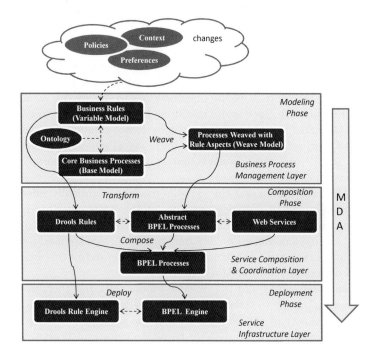

Fig. 3. MoDAR Overview

both services, and then adjust the process structure. Another challenge is that if we encode all parts of the requirements into business processes, the business rules in the requirements are hard-coded into the processes and their modularity and traceability are lost: there is no explicit association between a business rule and the process segment that implements this business rules, and if we change a business rule, the whole process needs to be changed.

3   The MoDAR Approach

3.1   Overview

In this section, we briefly introduce the MoDAR approach and discuss several key principles used in the design of this approach.

As illustrated in Figure 3, the MoDAR approach consists of three main phases, namely the *modeling phase*, the *composition phase*, and the *deployment phase*, based on the three-layer abstraction of service-based systems [15], and a model-driven approach is used to transform models between different phases.

Since DASS are generally difficult to specify, build, and validate due to its high complexity and variability [21,14], in the modeling phase, we adopt the basic principle of *separation of concerns* to manage complexity and variability: a system model is divided into a base model and a variable model. The base model represents the stable and constant part of a system and the variable model represents the volatile part. Specifically, business processes are used to specify the base model and business rules are used to specify the variable model. Business processes and business rules complement each other in a way that business processes capture the *flow* logic while business rules capture the *decision* aspect of a requirement. The reason that we use business rules to specify the variable model is that business rules are much easier to change than business processes [18,19,5]. As stated in [18], *"the most significant changes do not come from re-engineering workflow, but from rethinking rules"*. Furthermore, a business process even becomes more stable and reusable if we manage to abstract its complex branch structures into rules [19].

To make the base model and the variable model semantically interoperable, we use a minimum set of ontology concepts as the basic elements in defining activity parameters in processes and also in defining rule entities. To minimize the effort of building a domain ontology, only named classes are used to construct classes and any other class constructors such as *intersectionOf*, *unionOf*, and *someValueFrom* are not used in MoDAR.

To make the base model and the variable model work together, we adopt an aspect-oriented approach to integrate them into a weave model. This approach ensures the modularity of the base model and the variable model so that they can evolve independently. If we directly translate rules into process structures and insert them into a process, both modularity and traceability of the rules are lost. In MoDAR, we can specify cutting points on a process where rules can be woven in. The three models will be described in details in Section 3.2.

In the composition phase, the variable model is translated into Drools rules and the weave model is transformed into an abstract BPEL process. In this abstract process, at every join point, the invocation to a rule aspect is translated to a Web service invocation. The process is abstract in a way that all the service invocations translated from the activities in the based model are still abstract. After all the abstract services in the process are associated with concrete Web services to implement their functionalities, the process then can be transformed into an executable BPEL process.

Finally in the deployment phase, the BPEL process and the Drools rules are deployed to their corresponding engines. Dynamic adaptivity is achieved in a way that as long as the interface between a rule and the process does not change, we can freely change the rule in the modeling phase and then translate and redeploy the rule without even stop the execution of the process.

3.2   The MoDAR Models

In this section, we discuss in detail how to create the base model and the variable model, and how to weave these two models using the running example in Section 2.

The Base Model. The base model represents the stable part of business requirements. In MoDAR, a base model is a simplified business process that can be defined by the tuple $< \mathcal{T}, t_s, t_e, Seq : \mathcal{T} \times \mathcal{T}, Par : \mathcal{T} \times \mathcal{T} >$, where $\mathcal{T}$ is the set of business activities, $t_s \in \mathcal{T}$ is the start event, $t_e \in \mathcal{T}$ is the end event, Seq is the sequential flow between two activities, and Par is the parallel flow between two activities. So we remove all the complex process control constructs that are related to conditional decisions, which are usually the volatile part of a process.

We define a business activity as a trinary tuple of *name*, *inputs*, and *outputs*: $t =< name, I : Name \times \mathcal{C}, \mathcal{O} : Name \times \mathcal{C} >$, where $Name$ is a finite set of names, $name \in Name$, and every input or output of a business activity has a name and a type/class, and the type must be a concept in an ontology: $\mathcal{C} \in Concept_{onto}$. The purpose that we associate an I/O parameter with an ontology concept is twofold: first, the ontology serves as the common ground between the base model and the variable model and thus makes these two models semantically interoperable; second, the semantics attached to business activities later can be used to semantically discover services that implement business activities.

To create a base model, we can start from scratch to describe the general steps of a business process and then connect these steps using sequential or parallel flow. For example, the general steps of the SmartTravel process can be illustrated by Figure 4.

Fig. 4. The SmartTravel Base Model

Alternatively, we can also re-engineering legacy business processes to a base model using the concept of *variation point* ($\mathcal{VP}$). A $\mathcal{VP}$ includes any conditional structure (which is called *conditional $\mathcal{VP}$*) or activities that tend to change (which is called *activity $\mathcal{VP}$*) in a process. For example, in Figure 5, all the conditional structures have been identified as conditional $\mathcal{VP}$ and the `CalculatePrice` activity has been identified as an activity $\mathcal{VP}$.

For conditional $\mathcal{VP}$, we have two strategies to convert them to sequential flow:

- If this $\mathcal{VP}$ is served as a precondition/constraint for further steps of the process (which means the other branch terminates the process), we replace this $\mathcal{VP}$ with a sequence flow. Later on, a rule will be created based on this $\mathcal{VP}$ and weaved to the process. For example, in Figure 5, we can apply this strategy on $\mathcal{VP}_1$, which is used to check the validity of requests.
- If the $\mathcal{VP}$ is served as a branch for alternative activities, which means different branches will eventually merge, we use an abstract activity to replace this branch structure. Later on, a rule will be created to do the actual activity based on the $\mathcal{VP}$ conditions. For example, in Figure 5, we can apply this strategy to $\mathcal{VP}_2$, $\mathcal{VP}_3$, and $\mathcal{VP}_5$.

For an activity $\mathcal{VP}$, we keep it intact but later on we will weave rules onto it to adjust its behavior. By identifying $\mathcal{VP}$ and applying the above strategies, we can convert the SmartTravel process (Figure 2) to a MoDAR base model as shown in Figure 4.

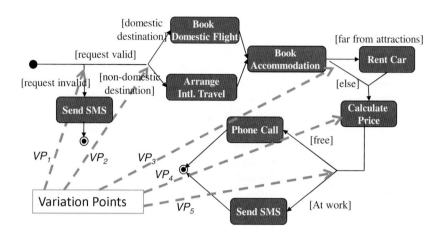

Fig. 5. Variation Points in the SmartTravel Business Process

The Variable Model. A variable model consists of a set of rules where each rule is defined by a binary tuple: $r = <condition, action>$. The *action* is enabled only if the *condition* is true. For any variable v in a condition or action expression, its type must be a concept in the same ontology used by the base model. If the type of a variable can uniquely identify a variable, we do not explicitly define this variable but use its type to represent it. An action expression can be any numeric or string expressions. For example, $Price * 10\%$, $FirstName + '.' + LastName$. To facilitate the definition of process related actions, we also define some standard actions with prescribed semantics:

- [SkipActivity]: skip an activity;
- [SkipActivityThenAction]: skip an activity then do an action. The action could be calling a service;
- [Abort]: abort the running of the process;
- [ActionThenAbort]: do an action before abort the running of the process.

The following are two rule examples:

> *If the departure airport or arrival airport of a travel request is empty, stop processing the request:*
> *Condition:* DepartureAirport equals to "" or ArrivalAirport equals to ""
> *Action:* [ActionThenAbort(SendSMS)]

> *If a customer is a frequent flyer, give him a 5% discount:*
> *Condition:* Customer has FrequentFlyer equals to true
> *Action:* Price equals to Price * 0.95

The first example is a constraint rule. The ontology concepts DepartureAirport and ArrivalAirport implicitly represents two variables, and the action is

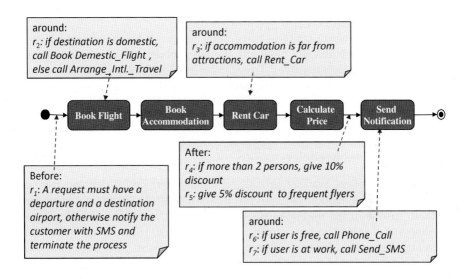

Fig. 6. The SmartTravel Weave Model

to send an SMS then abort the execution. The second example is a computation rule. In its condition, `Customer` is a complex class, and one of its property is `FrequentFlyer`.

The Weave Model. In MoDAR, an aspect weaves a rule to a process. It can be defined by: $a \in \{Before, Around, After\} \times \mathcal{T} \times \mathcal{R}$, where $\mathcal{R}$ is the set of rules. Similar to Aspect4J, we also identify three types of aspect: *before aspects*, *around aspects*, and *after aspects*. An aspect is always associated with a business activity, and it is activated either before, after the execution of the activity or replace the execution of the activity. From another perspective, $event \in \{Before, Around, After\} \times \mathcal{T}$ becomes the triggering event of a rule.

It is worth noting that the interoperability between an activity and its associated rules is established through the predefined ontology. For example, the input parameters of the `Book Flight` activity must contain two parameters having semantic annotation `DepartureAirport` and `ArrivalAirport`, so that the associated rule (see the first rule in the variable model) can use these two concepts in defining itself.

Figure 6 shows a weave model of the SmartTravel process, where the base model is weaved with rules from the variable model. As we can find in this weave model, all the requirement changes specified in Section 2.4 have been introduced by rules inside the aspects.

4   Implementation and Performance Evaluation

In this section, we discuss the implementation of the MoDAR platform, including how to transform models defined in the modeling phase (i.e., the base model, the variable model, and the weave model) to executable code. We also report some preliminary results on performance study.

The MoDAR platform contains three main components: the *Process Modeler* (the Modeler) for graphically modeling the base model, the rules, and also the weave model; the *Association and Transformation Tool* (AT Tool) for associating Web services with activities in the base model and actions in rules, and for generating and deploying executable code; and the *Business Domain Explorer* (Explorer) for exploring domain ontologies.

4.1   The Process Modeler

The Business Process Modeler provides a visual interface for defining the base model, the variable model, and also the weave model. As shown in Figure 7(a), the main canvas in the middle of the graphical interface of the Modeler is used to define and display the base model, and essential BPMN constructs including `Activity`, `Parallel Gateway`, `Start Event`, `End Event`, and `Sequence Flow` are displayed as a list of toolbar buttons on top of the canvas for visually creating the base model. If we click a specific activity in the base model, we can define its I/O semantics in the bottom pane. If we select a concept from the drop-down menu that contains all the concepts in the domain ontology as the type for a parameter, a variable is automatically created to represent this parameter. As shown in the snapshot, the `BookFlight` activity has two input parameters with type `DepartureAirport` and `DestinationAirport`, and one output parameter with type `FightNo`.

The middle canvas is also used to define the weave model. As shown in the snapshot, the left pane shows the list of business rules that can be drag-n-drop to an activity and becomes its *before/around/after rule*. It is worth noting that if a rule is associated with an activity, all the parameter variables of this activity are also visible to the rule.

A new business rule can be created in the left rule repository pane. As shown in Figure 7(b), the rule editor use the concepts in the domain ontology to define the *condition* and *action* components of a rule. To facilitate domain/application experts to create a correct business rule, the editor has a very nice auto-complete[2] feature: at every step in composing a condition or action, all the possible inputs are listed in a drop-down menu for the user to select. For example as shown in Figure 7(b), to write the constraint rule (the first one) as we discussed in Section 3.2, first we select the `DepartureAirport` concept from the drop-down menu, and then all the operators that are applicable to this concept are displayed in the follow-on menu, and so on. It is worth noting that all the I/O parameter variables in the based model that are visible to a rule will be automatically bind to the corresponding concepts in the rule, and if a rule needs to use variables other than the activity parameters it attached to, we can use the *variable visibility* tab in the bottom pane to make additional variable visible to this rule.

---

[2] http://en.wikipedia.org/wiki/Autocomplete

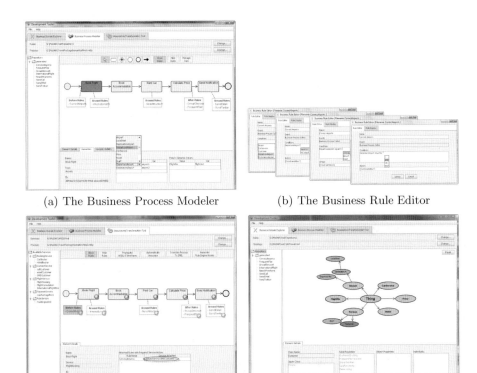

(a) The Business Process Modeler

(b) The Business Rule Editor

(c) The Association and Transformation Tool

(d) The Business Domain Explorer

Fig. 7. Main Components of the MoDAR Platform

4.2 The Association and Transformation Tool

The Association and Transformation Tool has two main functionalities. First, it provides a visual interface for associating Web services to process activities and rule actions. Second, it is used to automatically generate executable code and deployment scripts from the models defined in the modeling phase. It is worth noting that BPEL is selected as the targeting executable process language and Drools is selected as the targeting executable rule language.

Figure 7(c) shows the graphical interface of the AT Tool. The left pane shows a directory of Web services. The user can drag-n-drop a Web service to an activity or rule in the middle canvas. Since semantic service discovery and match is not the focus of our work, currently we only support a basic manual association method: a Web service can be associated with an activity/action if and only if they have the same number of inputs and outputs.

Transforming a weaved process model to a BPEL process is straightforward: BPMN constructs has a nice match with BPEL constructs, and invocation to an aspect is wrapped in a Web service call, either before, replace, or after calling the activity that the aspect attaches to. Figure 8 shows a BPEL code snippet

```
<!-- Book Flight Before Rules Service Execution-->
<bpws:invoke inputVariable ="rulerequest"
 name="RequestRuleService0"  operation="RuleEngineWS"
 outputVariable="ruleresponse" partnerLink="rule"
 portType="rls:RuleService"/>
<!-- Book Flight Before Rules Variable Propagation -->
<bpws:assign name="Propagate_BookFlightBeforeRules" validate="no">
...</bpws:assign>
<!-- Book Flight Before Rules Logic Copy -->
...
<!--  Book Flight Before Rules Logic Test - Abort Action -->
<bpws:if><bpws:condition>bpel:contains($boolTest, "true")
   </bpws:condition><bpws:sequence>
      <!-- Alternate Service Executed on Rule Engine -->
      <bpws:throw name="cancelProcess" faultName="cancelation"/>
</bpws:sequence></bpws:if>
```

Fig. 8. SmartTravel BPEL Code Snippet

that is generated from the weave model. In the code snippet, first the before rule service of the Book Flight service is invoked, and if the rule service returns true, which means abort the process, the execution of the process will be aborted by throwing a cancelProcess exception. It is worth noting that if the action part of a rule needs to invoke a Web service, the invocation is included in the wrapping Web service of the rule so that Drools engine is self-contained and does not need to have the capability to invoke Web services.

All the business rules are automatically translated to Drools rules. Finally, an ant^3 script is generated for deploying the BPEL code and the Drools rules code to the corresponding engines.

4.3   The Business Domain Explorer

Figure 7(d) is a snapshot of the Explore graphical interface. It displays the domain ontology that the SmartTravel application is based on. This ontology only contains the minimum concepts that are used to define the semantics for the I/O parameters of business activities and the variables in business rules. From this interface, the process designer can get familiar with the concepts in the domain. OWL [13] is used as the ontology language.

4.4   Performance Evaluation

This section reports some initial performance evaluation of the MoDAR platform. A PC with Intel Core i7 860 CPU and 4GB RAM is used as the test-bed server. JBPM-BPEL-1.1.1[4] running on top of JBOSS-Application-Server-4.2.2 is used as the BPEL engine and Drools-5.0[5] is used as the rule engine.

---

[3] http://ant.apache.org/
[4] http://www.jboss.org/jbpm
[5] http://www.jboss.org/drools

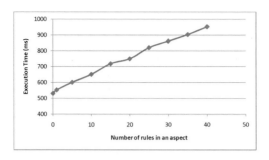

Fig. 9. Execution time of an aspect

The overhead of a weaved BPEL process mainly lies in its aspects, which are implemented as Web service invocations. The first experiment we have conducted is to test the performance impact of the number of rules in a single aspect. As illustrated in Figure 9, the execution time of an aspect increases steadily from around 552ms for one rule to 952ms for 40 rules.

Next we compare the execution time of two implementations of the running scenario: one is a manually composed BPEL process and the other is the weaved process generated by the MoDAR platform, with all the activity Web services deployed locally. Since the scenario includes some complex services such as `Book Flight` and `Book Accommodation` that need to communicate with external systems, on average, the pure BPEL process takes about 18.125s to execute and the weaved process takes about 18.715s. We can see that the overhead of the weaved process compared to the pure BPEL process is 590ms, or 3.3%.

From the above experiments, we can conclude that the MoDAR platform is suitable for dealing with processes that involve complex business logic. It dramatically simplifies the development process and makes the executable processes dynamically adaptable with a very small performance overhead.

5   Related Work

Dynamic and adaptive processes are one of the key research challenges in service-oriented computing [16]. Existing approaches can be classified into two main categories: adaptation to non-functional changes and adaptation to functional changes. While the former investigates the problem of how to adapt service configurations to satisfy multiple QoS requirements [20,2], the latter focuses on making the service-based systems adaptable to any business and environmental changes that require the system to change its functional behavior. Clearly, our work falls in the latter case.

In [5], the authors propose AO4BPEL, where BPEL is extended with aspect-oriented features: pointcuts can be defined on a process to weave aspects, and BPEL constructs such as <SWITCH> and <ASSIGN> are used to compose rule aspects. In [17], the authors propose a runtime environment where both a BPEL engine and a rule engine are connected to a service bus and BPEL invocation activities are intercepted so that rules can be applied either before or after a service invocation. In [6], the authors propose SCENE, a service composition execution environment that supports the execution of dynamic processes using rules to guide the binding of operations. While all these approaches focus on providing executable languages and environments to improve the adaptability of service-based systems, MoDAR is a model-driven approach that supports the development of service-based systems from high-level modeling to low-level code generation and execution. Furthermore, AO4BPEL does not support dynamic adaptation since all the rules are part of the process and thus cannot be managed and changed independently.

In the general research area of software architecture, model-driven development is also combined with aspect-oriented methodology to manage the vast number of possible runtime system configurations of dynamic adaptive systems. In [14], the authors put system configurations in aspects and weave them at a model level to reduce the number of artifacts needed to realize dynamic variability. In [7], the authors propose an aspect-oriented approach for implementing self-adaptive system on the Fractal platform [4]. They separate adaptation policies from the business logic and use low-level reconfiguration scripts to do the actual adaptation actions. While existing approaches in this area usually need the support of an adaptive middleware such as Fractal [4] and OSGI [1], our targeting platform is BPEL, the de facto industry standard language for service composition.

6   Conclusion

In this paper, we have presented MoDAR: a model-driven approach for developing dynamic and adaptive service-based systems using aspects and rules. We have introduced the methodology of how to separate the variable part from a stable process and model the variable part as business rules, and how to weave these rules into a base process. We have also introduced a platform that supports the model-driven development of rule-weaved BPEL processes. The targeting system is dynamically adaptive in the sense that rules as the variable part of a system can change at runtime without affecting the base process. Experimental results show that systems generated with MoDAR have a reasonable performance overhead while significantly simplifies the development process.

In the future, we plan to develop real-life service-based applications to further validate effectiveness of MoDAR and test its overall usability. Performance optimization is another focus of our future research work.

References

1. The OSGi Alliance. OSGi Service Platform Core Specification Release 4.1 (2007), http://www.osgi.org/Specifications/
2. Ardagna, D., Pernici, B.: Adaptive Service Composition in Flexible Processes. IEEE Transactions on Software Engineering 33(6), 369–384 (2007)

3. Baniassad, E.L.A., Clarke, S.: Theme: An Approach for Aspect-Oriented Analysis and Design. In: Proc. of the 26th International Conference on Software Engineering (ICSE 2004), pp. 158–167 (2004)
4. Bruneton, E., Coupaye, T., Leclercq, M., Quema, V., Stefani, J.: The FRACTAL Component Model and its Support in Java. Software Practice and Experience 36, 1257–1284 (2006)
5. Charfi, A., Mezini, M.: Hybrid Web Service Composition: Business Processes Meet Business Rules. In: Proc. of the 2nd International Conference on Service Oriented Computing (ICSOC 2004), vol. 30–38 (2004)
6. Colombo, M., Di Nitto, E., Mauri, M.: SCENE: A Service Composition Execution Environment Supporting Dynamic Changes Disciplined Through Rules. In: Dan, A., Lamersdorf, W. (eds.) ICSOC 2006. LNCS, vol. 4294, pp. 191–202. Springer, Heidelberg (2006)
7. David, P., Ledoux, T.: An Aspect-Oriented Approach for Developing Self-Adaptive Fractal Components. In: Löwe, W., Südholt, M. (eds.) SC 2006. LNCS, vol. 4089, pp. 82–97. Springer, Heidelberg (2006)
8. Elrad, T., Filman, R.E., Bader, A.: Aspect-Oriented Programming: Introduction. Commun. ACM 44(10), 29–32 (2001)
9. Frankel, D.S.: Model Driven Architecture^{TM}: Applying MDA^{TM} to Enterprise Computing. John Wiley & Sons, Chichester (2003)
10. Georgakopoulos, D., Papazoglou, M.P.: Service-Oriented Computing. The MIT Press, Cambridge (2008)
11. Gradecki, J.D., Lesiecki, N.: Mastering AspectJ: Aspect-Oriented Programming in Java. Wiley, Chichester (2003)
12. The Business Rules Group. Defining Business Rules, What Are They Really? (2001), http://www.businessrulesgroup.org
13. McKinley, P.K., Sadjadi, S.M., Kasten, E.P., Chen, B.H.C.: Composing Adaptive Software. IEEE Computer 37(7), 56–74 (2004)
14. Morin, B., Barais, O., Nain, G., Jezequel, J.M.: Taming Dynamically Adaptive Systems using Models and Aspects. In: Proc. of the 31st International Conference on Software Engineering (ICSE 2009), pp. 122–132 (2009)
15. Di Nitto, E., Ghezzi, C., Metzger, A., Papazoglou, M.P., Pohl, K.: A Journey to Highly Dynamic, Self-Adaptive Service-Based Applications. Automated Software Engineering 15(3-4), 313–341 (2008)
16. Papazoglou, M.P., Traverso, P., Dustdar, S., Leymann, F.: Service-Oriented Computing: State of the Art and Research Challenges. Computer 40(11), 38–45 (2007)
17. Rosenberg, F., Dustdar, S.: Usiness Rules Integration in BPEL - a Service-Oriented Approach. In: Proc. of the 7th IEEE International Conference on E-Commerce Technology, pp. 476–479 (2005)
18. Ross, R.G.: Principles of the Business Rules Approach. Addison-Wesley, Reading (2003)
19. Vanthienen, J., Goedertier, S.: How Business Rules Define Business Processes. Business Rules Journal 8(3) (2007)
20. Yau, S.S., Ye, N., Sarjoughian, H.S., Huang, D., Roontiva, A., Baydogan, M.G., Muqsith, M.A.: Toward Development of Adaptive Service-Based Software Systems. IEEE Transactions on Services Computing 2(3), 247–260 (2009)
21. Zhang, J., Cheng, B.H.C.: Model-Based Development of Dynamically Adaptive Software. In: Proc. of the 28th International Conference on Software Engineering (ICSE 2006), pp. 371–380 (2006)

An Incremental Approach for Building Accessible and Usable Web Applications

Nuria Medina Medina[1], Juan Burella[2,4], Gustavo Rossi[3,4],
Julián Grigera[3], and Esteban Robles Luna[3]

[1] Departamento de Lenguajes y Sistemas Informáticos, Universidad de Granada, España
nmedina@ugr.es
[2] Departamento de Computación, Universidad de Buenos Aires, Argentina
jburella@dc.uba.ar
[3] LIFIA, Facultad de Informática, UNLP, La Plata, Argentina
{gustavo,julian.grigera,esteban.robles}@lifia.info.unlp.edu.ar
[4] Also at CONICET

Abstract. Building accessible Web applications is difficult, moreover considering the fact that they are constantly evolving. To make matters more critical, an application which conforms to the well-known W3C accessibility standards is not necessarily usable for handicapped persons. In fact, the user experience, when accessing a complex Web application, using for example screen readers, tends to be far from friendly. In this paper we present an approach to safely transform Web applications into usable and accessible ones. The approach is based on an adaptation of the well-known software refactoring technique. We show how to apply accessibility refactorings to improve usability in accessible applications, and how to make the process of obtaining this "new" application cost-effective, by adapting an agile development process.

Keywords: Accessibility Visually Impaired, Web engineering, TDD, Web requirements.

1 Introduction

Building usable Web applications is difficult, particularly if they are meant for users with physical, visual, auditory, or cognitive disabilities. For these disadvantaged users, usability often seems an overly ambitious quality attribute, and efforts in the scientific community have been generally limited to ensure accessibility. We think accessibility is a good first step, but not the end of the road. Usability and accessibility should go hand in hand, so disabled users can access information in a usable way, since it is not fair to pursue usability for regular users and settle with accessibility for disabled users. Thus we consider that the term Web accessibility falls short and should be replaced by the term "usable web accessibility" or "universal usability", whose definition can be obtained from the combination of the two quality attributes. As an example, let us suppose a blind person accessing a (simplified) Web application

L. Chen, P. Triantafillou, and T. Suel (Eds.): WISE 2010, LNCS 6488, pp. 564–577, 2010.

like the one shown in Figure 1, using a screen reader [1]. To enforce our statement, we assume that this application fulfils the maximum level of accessibility, AAA, according to the Web Content Accessibility Guidelines (WCAG) [2]. This means that the HTML source of the application satisfies all the verification points, which check that all the information is accessible despite any user's disabilities. However, using the screen reader, it will be difficult for the blind user to go directly to the central area of the page where the books' information is placed. On the contrary, he will be forced to listen (or jump) one by one all the links before that information can be listened (even when he does not want to use them).

Fig. 1. Accessible but not usable page

The main problem, that we will elaborate later, is that this page has been designed to be usable by sighted users and "only" accessible by blind users. In this paper we present an approach to systematically and safely transform an accessible web application into a usable and accessible (UA) one. The approach consists in applying a set of atomic transformations, which we call accessibility refactorings, to the navigational and interface structure of the accessible application. With these transformations we obtain a new application that can be accessed in a more friendly way when using, for example, a screen reader. We show, in the context of an agile approach, how this strategy can be made cost feasible, particularly when the application evolves, e.g. when there are new requirements. Our ideas are presented in the context of the WebTDD development approach [3], but they can be applied either in model-driven or coding-based approaches without much changes. Also, while the accessibility refactorings we describe are focused on people with sight problems, the approach can be used to improve usability for any kind of disability.

The main contributions of the paper are the following: a) we introduce the concept of usable accessibility b) we show a way to obtain a UA web application by applying small, behaviour-preserving, transformations to its navigation and interface structures; c) we demonstrate the feasibility of the approach by showing not only how to generate the UA application but also how to reduce efforts when the application evolves. The rest of the paper is structured as follows: Section 2 discusses some related work in building accessible applications; section 3 presents the concept of accessibility refactoring and briefly outlines some refactorings from our catalogue. In section 4 we present the core of the approach using the example of Figure 1, and finally section 5 concludes the paper and discusses some further work we are pursuing.

2 Related Work

Web usability for visually impaired users is a problem that is far from being solved. The starting point of the proposed solutions leans on two basic supports: the WCAG Guidelines [2] and the screen readers [1] used together with traditional or "talking" browsers [4]. Then, the methods and proposed tools to achieve accessibility and usability in the Web diverge in two directions [5]: assessment or transformation. In the first group, the automated evaluation tools, such as Bobby (Watchfire) [6], analyze the HTML code to ensure that it conforms to accessibility or usability guidelines. In the second group, automated transformation tools help end users, rather than Web application developers. These tools dynamically modify web pages to better meet accessibility guidelines or the specific needs of the users.

Automated transformation tools are usually supported by some middleware, and they act as an intermediary between the Web page stored in the server, and the Web page shown in the client. Thus, in the middleware, diverse transcodings are performed. An example is the middleware presented in [7], which is able to adapt the Web content on-the-fly, applying a transcoding to expand the context of a link (the context is inferred from the text surrounding the link), and other transcoding to expand the preview of the link (processing the destination of the link). Other example is the proposal in [8], in which semantic information is automatically determined from the HTML structure. Using these semantics, the tool is able to identify blocks and reorganize the page (grouping similar blocks, i.e. all the menus, all the content areas, etc). This will create sections within the page that allow users to know the structure of the page and move easily between sections (ignoring non essential information for him). However, none of these automatic transcodings are enough to properly reduce the overhead of textual and graphic elements, as well as links, which clutter most pages (making their reading through a screen reader very noisy). This is because discerning meaningful from accessory content is a task that must be manually performed. A basic example of "manual" transcoding is the accessible method proposed in [9], which uses stylesheets to hide text (marked with a special label) from the page prepared for sighted users. Another interesting example is Dante [10], a semi-automated tool capable of analyzing Web pages to extract objects which are meaningful for the handicapped person during navigation, discover their roles, annotate them and transform pages based on the annotations. In [11] meanwhile, Dante annotations are automatically generated in the design process. In this case, the intervention of the designer is performed in the phase of modeling, but still needed.

We believe that the problem of usability for impaired people must be attacked from the early stages of applications design. Furthermore, all stakeholders (customers, designers and users) must be involved in the process. Hence, instead of proposing an automatic transcoding tool, we provide a catalogue of refactorings that the designer can apply during the development process, and later during the evolution of the Web application. The catalogue is independent of the underlying methodology and development environment, so refactorings can be integrated into traditional life cycle models or agile methodologies. However, to emphasize our point we show how a wise combination of agile and model-driven approaches can improve the process and allow the generation of two different applications, one for "normal" users and another, which provides usable accessibility for impaired users.

3 Making Accessible Web Applications More Usable

Achieving universal usability is a gradual and interdisciplinary process in which we should involve all application's stakeholders. In addition, we think that it is a user-centred process that must be considered in early phases of the design of Web applications. For the sake of conciseness, however, we will stress out the techniques we use, more than the process issues, which will be briefly commented in Section 4.

The key concept in our approach is refactoring for accessibility. Refactoring [12] was originally conceived as a technique to improve the design of object-oriented programs and models by applying small, behaviour-preserving, transformations to the code base, to obtain a more modular program. In [3] we extended the idea for Web applications with some slight differences with respect to the original approach: the transformations are applied to the navigational or presentation structures, and with the aim of improving usability rather than modularity. In this context we defined an initial catalogue of refactorings, which must be applied when a bad usability smell [13] is detected. More recently, in [14] we extended the catalogue incorporating a new intent: usable accessibility. As said before, we will concentrate on those refactorings targeted to sight disabled persons. Subsequently, section 3.1 briefly describes the specific catalogue of refactorings to achieve UA for sight impaired users.

3.1 The Refactoring Catalogue

Each refactoring in our catalogue to improve UA, specifies a concrete and practical solution to improve the usability of a Web application, that will be accessed by a visually impaired user. Each UA refactoring is uniformly specified with a standard template, so it can be an effective means of communication between designer and developers. The basic points included in the template are three: purpose, bad smells and mechanics. The purpose, defined in terms of objectives and goals, establishes the property of usability to be achieved with the application of the refactoring. The bad smells are sample scenarios in which it is appropriate to apply the refactoring, that is, elements or features of the Web site which generate a usability problem. Finally, each mechanics explains, step by step, the transformation process needed to apply the refactoring and thus solve the existing usability problem.

The refactorings included in the catalogue are divided in two groups: Navigation Refactorings and Presentation Refactorings. Navigation refactorings try to solve usability problems related to the navigational structure of the Web application. Therefore, the changes proposed by this first type of refactorings modify the nodes and links of the application. Presentation refactorings meanwhile propose solutions to usability problems whose origin are the pages' interfaces. Therefore this second type of refactorings implies changes in the appearance of the Web pages.

Concretely, the navigation refactorings included in the UA catalogue allow: to split a complex node, to join two small nodes whose contents are deeply related, to make easier the access between nodes creating new links between them, to remove an unnecessary link, information or functionality with the aim to simplify the node without losing significant content or, conversely, to repeat a link, functionality or information contained on a node in another node where it is also necessary and its inclusion does not overload the resulting node, etc.

Presentation refactorings included in the UA catalogue determine when and how: divide a complex and heterogeneous page in a structure of simpler pages, combine two atomic pages in a cohesive page, add needed anchors, remove superfluous anchors, add contextual information such as size indicators in dynamic list and tables, distribute or duplicate the options of a general menu for each one of the items valid for the menu, replace pictures and graphics for an equivalent specific text or remove the figure if it is purely aesthetic, reorder the information and functionality on the page in a coherent order to read and use, reorganize panels and sections to be read from top to bottom and from left to right, fix the floating elements, transform nested menus into linear tables more easier to read, etc.

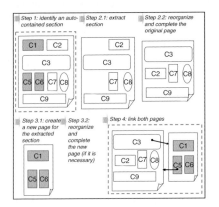
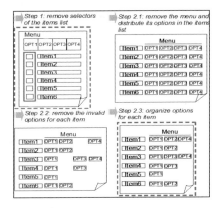

Fig. 2a. "Split Page"   **Fig. 2b.** "Distribute General Menu"

Figure 2a shows the steps needed to put into operation the "Split Page" presentation refactoring. As shown in the figure, the application of this refactoring involves simplifying an existing page, identifying and extracting self-contained blocks of information and functionality (steps 1 and 2.1), and then, creating one o more pages with the information / functionality extracted from the original page (step 3.1). Both, the original page (step 2.2) as the news pages (step 3.2) must be structured (that is, to organize the information for their appropriate reading and viewing), and can be supplemented or not with other additional information. Finally, the original page and the new pages must be linked together (step 4). Most refactorings allow several alternatives for certain steps in their mechanics. For the sake of conciseness Figure 2a shows the "normal" course of the "Split Page" refactoring. In Section 4 we illustrate the use of this refactoring in a concrete example.

Figure 2b shows the mechanism of the refactoring "Distribute General Menu", which proposes to remove the general menu affecting a list of elements by adding the menu actions to each element. The selectors of elements (e.g. checkboxes) are also removed in the container as the operations are now locally applied to each element.

In most cases, when solving a usability problem we need to update both navigation and presentation levels. Thus, many navigation refactorings have associated an automatic mechanism for changes propagation, which implies the execution of one or more presentation refactorings. More details can be read at [14]. We next show how we use the ideas behind accessibility refactorings in an agile development process.

4   Our Approach in a Nutshell

Along this section, we will show how we use the catalogue of accessibility refactorings to make the development of UA Web Applications easier. In a coarse grained description of our approach, we can say that it has roughly the same steps that any refactoring-based development process has (e.g. see [12]), namely: (a) capture application requirements, (b) develop the application according to the WACG accessibility guidelines, (c) detect bad smells (in this case UA bad smells), and (d) refactor the application to obtain an application that does not smell that way, i.e. which is more usable, besides being accessible.

Notice that step *b* (application development) may be performed in a model-based way, i.e. creating models and deriving the application, or in a code-based fashion, therefore developing the application by "just" programming. Step *c* (detecting bad smells) may be done "manually", either by inspecting the application, by performing usability tests with users, or by using automated tools. Finally, step *d*, when refactorings are applied, may be manually performed following the corresponding mechanics (See Figure 2), or automatically performed by means of transformations upon the models or the programming modules. A relevant difference with regard to the general process proposed in [12] is that in our approach step *d* is only applied when the application is in a stable step (e.g. a new release is going to be published) and not each time we add a new requirement. Anyway, for each accessibility refactoring we perform a short cycle, to improve the application incrementally.

One important concern that might arise regarding this process is that it might be costly, particularly during evolution. Therefore, we have developed an agile and flexible development process, and a set of associated tools which guarantee that we can handle evolution in a cost-effective way [15]. For the sake of conciseness, we will focus only on the features related to accessibility rather than evolution issues which are outside the scope of the paper. We discuss them as part of our further work.

Concretely, we use WebTDD [3], an agile method that puts much emphasis in the continuous involvement of customers, and comprises short development cycles in which stakeholders agree on the current application state. WebTDD uses specific artefacts to represent navigation and interaction requirements, which we consider to be essential for accessible applications. Similarly to Test-Driven Development (TDD) [16], WebTDD uses tests created before the application is developed to "drive" the development process. These tests are used later to verify that requirements have been corrected fulfilled. Different from "conventional" TDD, we complement unit tests with interaction and navigation tests using tools like Selenium (http://seleniumhq.org/). Figure 3 shows a simplified sketch of the development process. In the first step (1) we "pick" a requirement (e.g. represented with use cases or user stories) and in (2) we agree on the look and feel of the application using mockups. We capture navigation and interaction requirements, and represent them using WebSpec [17], a domain-specific language (DSL) which allows automatic test generation and tracking of requirement changes. At this point we can exercise mockups and simulate the application, either using a browser or a screen reader; therefore we can check accessibility guidelines and have early information on the need to refactor to improve usable accessibility in the step 7. Next (3), we derive the interaction tests from the WebSpec diagrams, and run them (4); it's likely that these tests will fail, indicating the starting point to begin the development to make tests pass. As said before, step 5 might imply dealing with models (generating code automatically), coding or a combination of both. In step 6, we run the tests again and iterate the process until all tests pass. Once we have the current version of the application ready we repeat the cycle with a new requirement (steps 1 to 6).

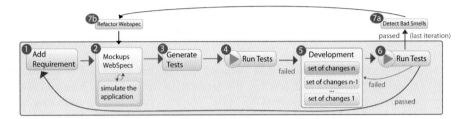

Fig. 3. WebTDD process for UA applications

After we reach a stable state of the application and we want to publish its current version, we look for bad accessibility smells and identify the need for UA refactorings (7.a). Next, we perform short cycles by applying each refactoring to the WebSpec diagrams containing the detected bad smells (7.b). The altered WebSpecs will generate new tests that check the new accessibility/usability features, and propagate the changes to the code (or model) to obtain the new UA application (steps 2 to 6).

In the next subsections we explain some of these aspects in a more detailed way focusing on UA development. To illustrate our approach, we will use the development of a simplified online book store (as the one shown in Figure 1), and when possible, we ignore the activities related with tests since they are outside the scope of the paper.

4.1   Gathering Navigation and Interface Requirements

Navigation and interface requirements are captured early in the development cycle through mockups and WebSpecs (step 2 in Figure 3). User interface Mockups help to establish the look and feel of the applications, along with other broad interaction aspects. They can be elaborated using plain HTML or commercial tools such as Balsamiq (http://www.balsamiq.com). Mockups can be easily adjusted to comply with accessibility guidelines. Figure 1 showed a mockup for our example's homepage.

WebSpecs are simple state machines that represent interactions as states and navigations as transitions, adding the formal power of preconditions and invariants to assert properties in states. An "interaction" represents a point where the user consumes information (expressed with interface widgets), and interacts with the application by using some of its widgets. Some actions (clicking a button, adding some text in a text field, etc) might produce "navigation" from one "interaction" to another and, as a consequence, the user moves through the application's navigation space.

Figure 4 shows a simplified WebSpec diagram that specifies the navigations paths from the BookList interaction and is related with the mockup of Figure 1. In the *BookList* "interaction" the user can authenticate, add books to the cart or to the wish list and search books. This diagram is the starting point for developing our simplified book store application, as it has key information to specify (at least partial) navigational models (as shown in [3]). Additionally, WebSpecs allow the automatic generation of navigation tests for the piece of functionality it represents, and with the aid of a tool suite, it records requirements changes, to trace and simplify implementation changes. In this sense, every changes made in a WebSpec (even the "initial" WebSpec as a whole) are recorded as "first class" change objects (as shown in Figure 4); these objects are later related with the corresponding model (or implementation) artefacts to improve traceability and automatic change management, using effect managers as explained in [15]. Specifically, each feature of the WebSpec in Figure 4 is traced to the corresponding modelling elements; in this way, when some of these features change, there is a way to automatically (or with minor designer intervention) change the corresponding models or programs. Therefore, step 5 in Figure 3 is viewed as the incremental application of these changes to the current implementation.

Fig. 4. *BookList* interaction in the *Books Store* WebSpec

4.2   Deriving an Accessible Application

In our approach we do not prescribe any particular development style, though we have experienced with model-driven (specifically, WebML [18]) and code-based (with Seaside- www.seaside.st/) approaches. Once we run the tests for a requirement and noticed that they fail, we build the corresponding models to satisfy such tests and derive the application (steps 3 to 6, in Figure 3). The construction of these models is an incremental task, managing the effects of each recorded change (step 5 in Figure 3). In addition to the changes log, we record the relationship between the WebSpec elements and its counterpart in the model, necessary for the automation of future changes on these elements. This recording is done at this stage, when change effects are managed and the model is built. For example, when the 'add to cart' addition link is managed, a counterpart element is added to the model and the relationship between both elements is recorded. Then, if we need to manage a change to configure any property of this element (e.g. its value), this can be automated since the change management tool knows its representation in the model.

   In this stage, accessibility can be addressed using any of the approaches cited in Section 2, for example by incorporating Dante annotations in the corresponding model-driven approach (See [10, 11]). Alternatively, we can "manually" work on the resulting application by improving the HTML pages to make them fully accessible. In both cases, given the nature of the WebTDD approach, the improvement is incremental; in each cycle we produce an accessible version of the application.

In traditional Web application development, accessibility is tackled as a monolithic requirement that must be satisfied by the application which is checked by running accessibility tests such as TAW (http://www.tawdis.net/). What is different in our approach is that we do not try to make the application accessible in one step; instead, we can decide which tests must be run on each development iteration and specific page. Therefore, in the first iteration we may want to make the "BookList" page accessible and satisfying the accessibility test "Page Titled" (Web pages must have titles that describe topic or purpose) and in the second iteration we may want to do the same with page "Best sellers". Our approach follows the very nature of agile development trying to incrementally improve the accessibility of an existing application. Stakeholders' involvement obviously helps in this process. To achieve this goal, we can specify which tests must be run on a specific interaction for each WebSpec diagram. For instance, in the diagram of Figure 4, we can initially run the "Page Titled" test during the first iteration, and the "Headings and Labels" (headings and labels describe topic or purpose) test during the second iteration. This approach helps to improve times during development and allows focusing on a specific accessibility requirement, though we can still execute all accessibility tests for every "interaction" like in traditional Web application development if necessary. From an implementation point of view, this "selective" testing is performed using a Javascript version of the WGAC accessibility tests and executing them depending on the tests selected on the WebSpec diagrams.

4.3 Detecting Bad Accessibility Smells

By following the WebTDD cycle (steps 1 to 6 in Figure 3), we will obtain an accessible application, but not necessarily a UA application. For example, if we analyze the page shown in Figure 1 (accessible according the WCAG), we can see that it presents several bad smells contemplated in the UA catalogue that have been outlined in Section 3.1.

First, the page mixes concepts and functions that are not closely related, such as: shopping cart, wish list, information on books, access to other products and user registration. A sighted user quickly disregards the information in which he is not interested (e.g. the registration if he just wants to take a look) and goes quickly to the area that contains what he wants (e.g. the central area where the available books are listed). However, a blind user does not have the ability to look through; when accessing the page using a screen reader which sequentially reads the page content, he will be forced to listen to a lot of information and functionality in what he may be not interested before reaching the desired content. In order to eliminate this bad smell, the refactoring "Split Page" can be applied. Besides, the actions provided to operate with the products listed in the central area of the page (books in this moment) refer to the selected books in the list; this implies that before applying an action in this menu (for example, add a book to the cart), the book or books must be selected by using checkboxes. This task is trivial for a sighted user, but it is considerably more complicated for a user who is accessing through a screen reader, as the reader reads the actions first and then the book list. Even though it is possible to scroll through the links on the page with the use of navigation buttons (provided by most readers), moving back (e.g. to look for the option once you have marked the products), can cause confusion and be tedious if the list is long. In order to eliminate this bad smell, the refactoring "Distribute General Menu" can be applied.

Therefore, we conclude that we need to apply some refactorings to obtain a better application. This could be done manually on the final application but it might be difficult to check that we didn't break any application behaviour. Next, we show how to make this process safer and compatible with the underlying WebTDD process and at the same time settle the basis to simplify evolution.

4.4 Applying Refactorings to the WebSpec Diagrams

As a solution to safely produce UA Web applications from existing ones, we propose to apply accessibility refactorings to the navigation and interaction requirements specifications (step 7 in Figure 3). Since WebSpec is a DSL formally defined in a metamodel [17], these refactorings are essentially model transformations of WebSpec's concepts. Each transformation comprises a sequence of changes on a WebSpec diagram, which are aimed to eliminate a specific bad smell. Moreover, as shown in [15,17] and explained before, these changes are also recorded in change objects that can be used to semi automatically upgrade the application as we will show in Section 4.5.

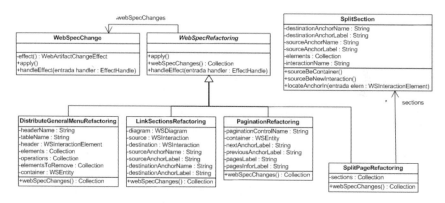

Fig. 5. Refactoring's metamodel

Usability refactorings are also conceptualized in a metamodel, part of which is shown in Figure 5. Refactorings classes provide an extension to the WebSpec metamodel; this allows grouping a set of changes with a coherent meaning. An interesting point to remark is the fact that these refactorings transform WebSpec diagrams instead of models (or code). This has several advantages; for example, with these diagrams and the corresponding new mockups, we can simulate the application. The new mockups could be automatically generated if they are also "imported" from a metamodel like we do in [19]. Also we automatically generate the new navigation tests to assess if the implementation changes were implemented correctly.

As an example, let us consider the application of the "Distribute General Menu" refactoring to the WebSpec of Figure 4. This refactoring takes as input an item container, elements and menu options to be distributed into this container, and elements to be eliminated for each item. In our example, we configure this refactoring with the "book" element as container, the elements "title" and "description", the menu options "Add to cart" and "Add to Wish List" and the checkbox to be removed. In Figure 6 we show the result of applying the refactoring, where the "Add to cart" and "Add to Wish List" options are added in each book item (in order to simplify the diagram we only shows the BookList interaction).

Fig. 6. "Distribute Menu" refactoring example

4.5   Deriving the UA Web Application

Once we applied a refactoring to the needed WebSpec specification, we proceed with the cycle (Figure 3). From now on, we work as we did with "normal" requirements (steps 2 to 6). Once we agreed the new look and feel of the refactored application with the customer (step 2), we generate and run the navigation tests to drive the implementation of these changes (steps 3 and 4). We run these tests; they obviously fail and the process continues with the refactoring effect management (step 5). As we previously explained, refactorings introduce changes in the corresponding WebSpecs, and their explicit representation as first-class objects helps us manage the changes to be applied to the application. As a refactoring is a group of WebSpec changes, we "visit" each of these changes to manage its effects. We start delegating these changes to a Change Management tool, which can automatically (or with some programmer intervention) alter either the models that will in turn generate the new application, or the code in a code-based approach.

In our tool suite we deal with these refactorings in the same way that we manage the changes generated by any new requirement. For example as shown in Figure 7, the "Distribute General Menu" refactoring involves "Move operations" changes; when we manage these changes the "Add to cart" and "Add to Wish List" links are moved into each book item on the application.

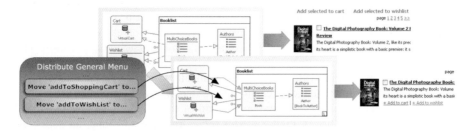

Fig. 7. Handle "Move operations" effects

We use these changes in the WebSpec specification to improve the development stage, with the aim of reducing the cost of their effects on the application, automating these effects in many cases. Additionally, we are able to determine which tests are affected by each change, to trim the set of required tests that must be performed (see the details of this change management process in [15,17]). Finally, a UA requirement is completed when all tests pass (step 6).

In our example, one of the bad smells detected in 4.2 is the way the interaction with the items on the book list is performed: a checklist with general operations to apply on the selected books. In this situation, the refactoring "Distribute General Menu" can be applied in order to improve the usability.

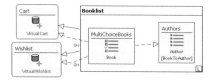
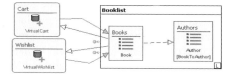

Fig. 8a. General Menu with checklists   **Fig. 8b.** Distributed Menu

Figure 8a shows a WebML diagram for the page that lists all books and lets the user add books to a Shopping Cart or a Whish list. Since the book list is presented as a checkbox set (using a specific WebML unit called "Multi Choice Index Unit"), the user has the ability to check different books and select an action to perform on the selected group, as seen in the units "Cart" and "Whishlist". The application of this refactoring generates automatically the diagram in Figure 8b, where the book list becomes a simple list (replacing the "Multi Choice Index Unit" unit with a plain "Index Unit"); the actions "Cart" and "Whishlist" are now directly linked from the list, and therefore every item on the Index Unit called "Books" gets individual links to each action. From this new navigational model and the corresponding interface template (derived from the mockup), we are able to derive a UA version of the home page shown in Figure 1. Figure 9a shows the result of the process.

Fig. 9a. Distributed Menu in book list   **Fig. 9b.** A new, usable and accessible home

Another bad smell detected is the mixed up contents on the bookstore's homepage. To overcome this problem, the "Split Page" refactoring is applied. For the sake of conciseness we only show the final result of applying the refactoring in the final application. Figure 9b shows the result of the new iteration, where the initial page has been cleaned, extracting in three new pages the information and functionality needed to: list products (BookList page), manage the wish list (WishList page) and manage the shopping cart (ShoppingCart page). After finishing this process we end with two Web applications: the "normal" one and the UA application. From now on, evolution can be tackled in two different ways: by treating the two applications separately or by working on the WebSpecs of the original one, following the WebTDD cycle and then re-applying the "old" refactorings to the modified specifications when needed.

5   Concluding Remarks and Further Work

In this paper we faced the problem of improving the usability of accessible Web applications. We consider that an application that has been developed to be usable for regular users is generally not usable (even if accessible) for handicapped users, and vice versa. In order to provide a solution for such important problem, we have presented an approach supported by there pillars: a) a catalogue of refactoring special-ized in UA problems for blind and visually impaired users; b) a test-driven develop-ment process, which uses mockups and Webspecs to simulate the application and to generate the set of tests to assure that all the requirements are satisfied (included the accessibility requirements) and c) a metamodel capable of internally representing the elements of the application and the changes upon these elements (included changes resulting from refactoring) in the same way; this makes easier the evolution of both, the normal application and the UA application. As further work, we are considering how to define catalogues of refactorings for other types of disabilities, for example: hearing impairments, physical disabilities, speech disabilities and cognitive and neu-rological disabilities. In turn, we are working in order to specialize each catalogue according to the particular type of disability. In addition, the catalogues may be also specialized according to the type of web application: communication applications (facebook, twitter, etc.), electronic commerce (amazon, e-bay, etc.), e-learning, etc. On the other hand, we are considering how to gather and represent usable and acces-sibility requirements. In this way, the UA refactorings could be applied at any itera-tion of the development cycle. For this to be feasible (and not too costly), we need to improve the change effect management, to automate the propagation of most changes from the original application to the UA one. Finally, we are improving out tool support to simplify evolution when new requirements affect those pages which were refactored during the usability improvement process. In this sense we need to have a smart composition strategy to be able to compose the "new" change objects with those which appeared in the refactoring stage.

Acknowledgements. This research is supported by the Spanish MCYT R+D project TIN2008-06596-C02-02 and by the Andalusian Government R+D project P08-TIC-03717. It has been also funded by Argentinian Mincyt Project PICT 2187.

References

1. Barnicle, K.: Usability Testing with Screen Reading Technology in a Windows Environ-ment. In: Conf. on Universal Usability, pp. 102–109. ACM Press, New York (2000)
2. W3C.: Web Content Accessibility Guidelines 2.0 (December 2008),
   http://www.w3.org/TR/WCAG20/

3. Robles Luna, E., Grigera, J., Rossi, G.: Bridging Test and Model-Driven Approaches in Web Engineering. In: Gaedke, M., Grossniklaus, M. (eds.) Web Engineering. LNCS, vol. 5648, pp. 136–150. Springer, Heidelberg (2009)

4. Zajicek, M., Venetsanopoulos, I., Morrissey, W.: Web Access for Visually Impaired People using Active Accessibility. In: Int. Ergonomics Association 2000/HFES, San Diego, pp. 445–448 (2000)

5. Ivory, M., Mankoff, J., Le, A.: Using Automated Tools to Improve Web Site Usage by Users with Diverse Abilities. Journal IT & Society 1, 195–236 (2003)

6. IBM: Watchfire's Bobby, http://www.watchfire.com

7. Harper, S., Goble, C., Steven, R., Yesilada, Y.: Middleware to Expand Context and Preview in Hypertext. In: ASSETS 2004, pp. 63–70. ACM Press, New York (2004)

8. Fernandes, A., Carvalho, A., Almeida, J., Simoes, A.: Transcoding for Web Accessibility for the Blind: Semantics from Structure. In: ELPUB 2006 Conference on Electronic Publishing, Bansko, pp. 123–133 (2006)

9. Bohman, P.R., Anderson, S.: An Accessible Method of Hiding Html Content. In: The International Cross-Disciplinary Workshop on Web Accessibility (W4A), pp. 39–43. ACM Press, New York (2004)

10. Yesilada, Y., Stevens, R., Harper, S., Goble, C.: Evaluating DANTE: Semantic Transcoding for Visually Disabled Users. ACM Transactions on Computer-Human Interaction (TOCHI) 14, 14-es (2007)

11. Plessers, P., Casteleyn, S., Yesilada, Y., De Troyer, O., Stevens, R., Harper, S., Goble, C.: Accessibility: A Web Engineering Approach. In: 14th International World Wide Web Conference (WWW 2005), pp. 353–362. ACM, New York (2005)

12. Fowler, M., Beck, K.: Refactoring: Improving the Design of Existing Code. Addison-Wesley Professional, Reading (1999)

13. Garrido, A., Rossi, G., Distante, D.: Model Refactoring in Web Applications. In: 9th IEEE Int. Workshop on Web Site Evolution, pp. 89–96. IEEE CS Press, Washington (2007)

14. Medina-Medina, N., Rossi, G., Garrido, A., Grigera, J.: Refactoring for Accessibility in Web Applications. In: Proceedings of the XI Congreso Internacional de Interacción Persona-Ordenador (INTERACCIÓN 2010), Valencia, Spain, pp. 427–430 (2010)

15. Burella, J., Rossi, G., Robles Luna, E., Grigera, J.: Dealing with Navigation and Interaction Requirement Changes in a TDD-Based Web Engineering Approach. In: Sillitti, A. (ed.) XP 2010. LNCS, vol. 48, pp. 220–225. Springer, Heidelberg (2010)

16. Beck, K.: Test-driven development: by example. Addison-Wesley, Boston (2003)

17. Robles Luna, E., Garrigós, I., Grigera, J., Winckler, M.: Capture and Evolution of Web requirements using WebSpec. In: Benatallah, B., Casati, F., Kappel, G., Rossi, G. (eds.) ICWE 2010. LNCS, vol. 6189, pp. 173–188. Springer, Heidelberg (2010)

18. Acerbis, R., Bongio, A., Brambilla, M., Butti, S., Ceri, S., Fraternali, P.: Web Applications Design and Development with WebML and WebRatio 5.0. In: Objects, Components, Models and Patterns. LNCS, vol. 11, pp. 392–411. Springer, Heidelberg (2008)

19. Rivero, J., Rossi, G., Grigera, J., Burella, J., Robles Luna, E., Gordillo, S.: From Mockups to User Interface Models: An extensible Model Driven Approach. To be published in Proceedings of the 6th Workshop on MDWE. LNCS. Springer, Heidelberg (2010)

CPH-VoD: A Novel CDN–P2P-Hybrid Architecture Based VoD Scheme

Zhihui Lu[1], Jie Wu[1], Lijiang Chen[2], Sijia Huang[1], and Yi Huang[1]

[1] School of Computer Science, Fudan University, 200433 Shanghai, China
[2] Department of Computer Science, Peking University, 100871 Beijing, China
{lzh,jwu}@fudan.edu.cn, clj@pku.edu.cn,
{082024070,082024093}@fudan.edu.cn

Abstract. Taking advantages of both CDN and P2P networks has been considered as a feasible solution for large-scale video stream delivering systems. Recent researches have shown great interested in CDN-P2P-hybrid architecture
and ISP-friendly P2P content delivery. In this paper, we propose a novel VoD
scheme based on CDN-P2P-hybrid architecture. First, we design a multi-layer
CDN-P2P-hybrid overlay network architecture. Second, in order to provide a
seamless connection for different layers, we propose a super-node mechanism,
serving as connecting points. Third, as part of experiment works we discuss
CPH-VoD implementation scheme based on peer node local RTSP server mechanism, including peer download/upload module, and their working processes.
The experimental results show that VoD based on CDN-P2P-hybrid is superior
to either pure CDN approach or pure P2P approach.

Keywords: P2P, CDN, VoD, CDN-P2P-Hybrid, Streaming Content Delivery.

1 Introduction

The rapid development of broadband technology has driven a lot of streaming media
content consumption, such as live streaming and VoD streaming. Nowadays, it is
well-known that up to 60-70% of Internet traffic is contributed by P2P content delivery applications. Most P2P applications are not adaptable and friendly to ISP and
network provider. P2P approach is more scalable and needs less investment, as it
makes full use of clients' resource (bandwidth, storage, CPU, etc). Each client is not
only a service consumer, but also a service provider. The challenge is system management and QoS under peer churn. Peers usually perform selfishly. They only care
theirs own benefit and ignore the global back-bone consumption. As CDN (Content
Delivery Networks) saves a lot of backbone bandwidth through pushing the content to
the edge in advance, for the benefits of telecom carries and ISPs, it is desirable that
combining the features of P2P and CDN system, can best protect the previous investment and improve the service capability at the same time.

Therefore, in order to make full use of the stable backbone-to-edge transmission
capability of CDN and scalable last-mile transmission capability of P2P, at the same
time avoiding ISP-unfriendly and unlimited usage of P2P delivery, some researches

L. Chen, P. Triantafillou, and T. Suel (Eds.): WISE 2010, LNCS 6488, pp. 578–586, 2010.

have begun to focus on CDN-P2P-hybrid architecture and ISP-friendly P2P content delivery [5-12] in recent years.

In this paper, we propose CPH-VoD: a novel VoD scheme based on CDN–P2P-hybrid architecture. The main contributions of this paper are described as follows:

(1) We design CDN–P2P-hybrid overlay network, and Super Node acts as the connecting point between CDN framework layer and Peer node layer. This architecture can reduce the pressure from the central and cache server by using CDN, at the same time, constructing a topology-aware and ISP-friendly P2P overlay network.

(2) We focus on designing the dedicated Super Node's main components, including Socket Manager, Buffer Manager, Regional Peer Manager.

(3) Specifically we use RTSP (Real Time Streaming Protocol) to design Download /Upload Module of Local RTSP Server, our scheme can make CDN and P2P coupled seamlessly and closely through RTSP.

The rest of this paper is organized as follows. In Section 2, we make a brief survey on current related works of CDN-P2P-hybrid delivery technology. Section 3 presents CDN-P2P-hybrid overlay network architecture design. In Section 4, we focus on design of Super Node. And then we describe how to implement VoD based on CDN-P2P-hybrid Architecture Scheme using RTSP in Section 5. The experimental results show that CDN-P2P-hybrid VoD scheme has better performances than the pure CDN or P2P approach. In Section 6, we present our conclusion and future work.

2   Related Works

One of the earliest CDN-P2P-hybrid architectures has been presented by Dongyan Xu [5]. They proposed and analyzed a novel hybrid architecture that integrates both CDN- and P2P-based streaming media distribution. Our previous work [6] proposed a novel hybrid architecture -- PeerCDN to combine the two approaches seamlessly with their inherited excellent features. PeerCDN is a two-layer streaming architecture. Upper layer is a server layer which is composed of CDN servers including origin servers and replica servers. Lower layer consists of groups of clients who request the streaming services, each client is considered as a client peer in the group. We also proposed a novel Web Services-based content delivery service peering scheme [13] and Web Services standard-supported CDN-P2P loosely-coupled hybrid and management architecture [14]. Cheng Huang et al. [3] considered the impact of peer-assisted VoD on the cross-traffic among ISPs. Although this traffic is significant, if care is taken to localize the P2P traffic within the ISPs, they can eliminate the ISP cross traffic while still achieving important reductions in server bandwidth. David R. Choffnes et al. [7] presented Ono scheme, which recycles network views gathered at low cost from content distribution networks (CDN) to drive biased P2P neighbor selection without any path monitoring or probing.

To guarantee QoS and facilitate management in large scale high-performance media streaming, Hao Yin et al.[10] presented the design and deployment experiences with LiveSky, a commercially deployed hybrid CDN-P2P live streaming system. LiveSky inherits the best of both worlds: the quality control and reliability of a CDN and the inherent scalability of a P2P system. Cheng Huang et al. [11] quantified the potential gains of hybrid CDN-P2P for two of the leading CDN companies, Akamai and Limelight. They found that hybrid CDN-P2P can significantly reduce the cost of content distribution, even when peer sharing is localized within ISPs and further localized within regions of ISPs. Hai Jiang et al. [12] presented a hybrid content distribution network (HCDN) integrating complementary advantages of CDN and P2P, which is used to improve efficiency of large scale content distribution.

3   CDN-P2P-Hybrid Overlay Network Architecture Design

Based on the above analysis, we propose a novel CDN-P2P-hybrid architecture to support VoD scheme: CPH-VoD. Just as the Figure 1 described, the overlay network of our CPH-VoD architecture has two layers: the first layer is the CDN network layer from the center to edge of the network, and the second layer is the P2P overlay network layer. The connecting points between CDN layer and P2P are some CDN cache servers or dedicated servers, we call them Super Node. It's the Super Node's duty to coordinate the CDN and P2P service for each request. Through Super Nodes' guide, the P2P overlay network is divided into multiple ISP-friendly regions. Every region contains quite a few peer nodes, which have closer neighbor relationship. At the same time, topology-adjacent regions can be formed neighbor relationship. The peer nodes between two neighbor regions also have looser neighbor relationship.

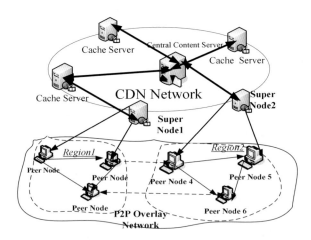

Fig. 1. The CDN-P2P-hybrid Overlay Network Architecture

As Figure 1 shows, CDN cache server distributes contents to Super Node servers located close to users, guaranteeing the fast response speed in the VoD startup stage. However, CDN service provider has to spend huge capital expenditure to deploy and maintain hardware device, Super Node servers' service capacity is limited by high cost and low expansibility.

Unlike most P2P schemes that construct arbitrary overlay networks, in CDN-P2P-hybrid VoD Architecture, the overlay network is constructed geographically to become topology-aware and ISP-friendly overlay network. Through the Super Node servers acting as leader of multiple peer nodes in each ISP region, we realize a P2P VoD model with controllable regional autonomy capability. Client peer node cache block of the VoD media content during they get VoD service, they may contribute their limited storage and up-link bandwidth to other peer nodes in the same and neighbor region through the leading and coordination of Super Node. This architecture increase delivery performance remarkably when a number of peers request the same VoD data at a short time, as peers can get VoD data directly from same region or neighbor region peers without much relying on Super Nodes. Therefore, it could greatly reduce loads of CDN cache server and backbone network, and make VoD services more scalable.

The topology-aware method reduces the consuming the valuable backbone bandwidth. At the same time, while startup or emergency, the clients can get content service from the nearest Super Node servers to ensure the reliability and robustness. Such a large-scale CDN-P2P-hybrid delivery architecture can considerably increase the total service capacity, at same time avoiding ISP-unfriendly and P2P abuse.

4   CDN and P2P Connecting Point: Super Node Design

4.1   Super Node Total Architecture

We can see from above Figure1, the Super Node servers act as the connecting point between CDN layer and P2P layer. Some CDN nodes can be used as Super Nodes, but in most cases we need to design a dedicated server as Super Node. Here we focus on designing the dedicated Super Node. Figure2 describes its main components, including Socket Manager, Buffer Manager, Regional Peer Manager, and so on.

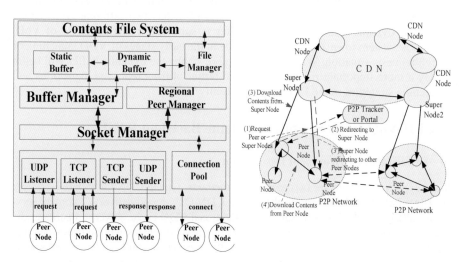

Fig. 2. The architecture of Super Node   **Fig. 3.** Super Node Content Serving Process

4.2   Peer Node Content Requesting and Super Node Content Serving Process

As a stable node, the Super Nodes firstly join P2P overlay network. When peer nodes enter the overlay network, they can firstly find these Super Nodes through P2P tracker or portal. Super Node can accept TCP or UDP Socket connection from peer node. In order to avoid the overhead of real-time creating TCP connection, the Super Node's TCP connection pool is a group of pre-created TCP connection. Super Nodes receive incoming peer nodes' requests through the Socket Manager. The Socket Manager forwards this request to Regional Peer Manager. Regional Peer Manager can make decision through scheduling algorithm, if they are the earliest-arriving peer nodes or some starved peer nodes in an emergency, it can directly provide content to them; if they are some other nodes, and it can guide them to the other earlier peer nodes of the same or neighbor region. Through this approach, incoming peer nodes gradually formed into different regions, at the same time everyone belongs to the entire overlay network.

Figure 3 illustrates peer node content request process:

(1)Peer request Contents from P2P portal or tracker.
(2) P2P portal or tracker redirecting it to Super Node.
(3) Download contents from Super node.
(3')Optionally, Super Node redirecting to other Peer Nodes.
(4')Download Contents from selected Peer Nodes.

If Socket Manager decides to directly send content to peer node, Socket Manager will send request to Buffer Manager, and then Buffer Manager can make interaction with media file system management module (File System Manager), finally load user-requested file blocks from media file system to the buffer, from where the media player can pull the media file blocks for VoD playing. Because the service capacity of Super Nodes is much larger than other ordinary nodes, and its buffer mechanism is more complex than other peer nodes, so we need to design its buffer mechanism and make the buffer have good performance. The buffer has two different types: static buffer and dynamic buffer. The static buffer is suitable for startup phase of VoD, making the system to ensure real-time service in the early stages of VOD launching or in high requests situation. When service is more stable, the dynamic buffer will begin to play a more important role. The dynamic buffer utilizes replacing algorithms and prefetching strategies of the LRU (Least Recently Used).

5   CPH-VoD Implementation Scheme Based on RTSP

5.1   Experiment Environment Description

In our experiment, there are two contrastive methods: typical CDN-Only-based delivery and CDN-P2P-hybrid delivery. In the typical CDN-Only-based delivery, 2 content servers, installed with Real Server, can act as traditional CDN cache server to distribute the cached content to client requesters, most VoD requests of clients will be responded and served by cache server. Only a small part of the contents depend on the central server. Nevertheless, in the CDN-P2P-hybrid delivery, 2 content servers, armed with our developed delivery software, can act as Super Node to distribute the cached content to client requesters. In the VoD startup phase, or in emergency situations, peer nodes can request contents from the Super Node. At the most other time, clients can act as peer node, and distribute the cached content to other peer nodes. The central server, two CDN cache servers, and two super nodes all have 10 media files with 800MByte size and 740kbps playing speed for VoD, requested by 400 client nodes.

5.2 Peer Node Local Server Download and Upload Module Design Based on RTSP

According to the Figure4, there are two main modules in our peer node to realize local RTSP delivery mechanism, download module and upload module. These two modules are all based on RTSP (Real Time Streaming Protocol) protocol [15]. Our scheme can make CDN and P2P coupled seamlessly and closely through RTSP.

1) Peer Node Download Module Design and Work Process Analysis

The process of downloading media data in peer node is listed below (No.(1)-(6) in the Figure4) :

(1) RealPlayer of one peer node will give a RTSP request to the listening thread of its local RTSP server, this listening thread will only receive request from local ip.

(2) The local RTSP server will parse the request. Sometimes it will directly give a response back without communicate with other peers or server.

(3) Check the cache, if there is no requested VoD data; forward the request to download module.

(4) The download module tries to find neighbors which own the VoD data; If yes, download from local RTSP servers of neighbor peers, if not, download from Super Node server.

(5) Write the VoD block data to cache.

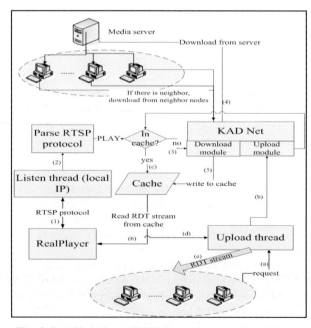

(6) The local RTSP server reads the data from cache and sends them to RealPlayer with RDT format

2) Peer Node Local RTSP Server Upload Module Design and Work Process Analysis

Upload module is an N-to-N upload, which means a peer can act as a local RTSP server and upload VoD media data to N peers, at the same time, it can download media data from N other local RTSP servers. The process of uploading media data is also listed below (No.(a)-(e) in the Figure4) :

Fig. 4. Peer Node Local RTSP Server Download and Upload Module and Work Process

(a) A peer node in CDN-P2P-hybrid network which need VoD data will send a request to the service port of several local RTSP servers of its neighbor peer nodes.

(b) The neighbor peer nodes' Local RTSP servers will parse the request and send it to upload module.

(c) The upload module checks the cache to find whether there are the requested VoD contents. If no, upload thread will return a fail response to the require peer. If yes, upload thread will go to the next step.

(d) Read the VoD media data from cache.

(e) Send these VoD data to requested peer node.

5.3 Contrastive Experiment Results Analysis

Figure 5 and 6 describe the contrastive experiment results respectively. In these figures, N is the number or peer nodes. P is proportion of the waiting time (T_{wait}) versus total playing time (T_{total}). From Figure 5 of CDN-Only-based VoD Architecture, we can observe: when 10-150 clients request from CDN Cache Server, the system works well. When the number of the clients arrives at 180 more, P is gradually used up.

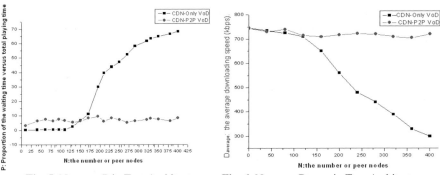

Fig. 5. N versus P in Two Architectures   **Fig. 6.** N versus $D_{average}$ in Two Architectures

When the number of the clients arrives at 200 more, the two Cache servers are overloaded, so the P is increasing very quickly. In the CDN-P2P-hybrid scenarios, we can see: when the number of the peer nodes is increasing from 1 to 400, P value is maintained at about 10%, as Super Node servers and peer node servers can evenly participate different delivery tasks. Peer nodes are not always stick to Super Node, as Super Node just provides startup blocks and emergency blocks for peer.

From Figure 6, we can observe: when 10-150 clients request from CDN Cache Server, the system works well in CDN-Only-based VoD Architecture. When the number of the clients arrives at 170 more, the average downloading speed-$D_{average}$ start to decrease and waiting time becomes slower. When the number of the clients arrives at more than 200, the total performance of the CDN Cache Server is deteriorating seriously. In the CDN-P2P-hybrid scenarios, we can observe: when the number of peer nodes is increasing from 1 to 400, the average downloading speed-$D_{average}$ has small change, as Super Node servers and peer node servers can take different distribution tasks when the clients number arriving at 200 more. The experiments show that CDN-P2P-hybrid VoD delivery scheme can dramatically increase content delivery performance while efficiently utilizing edge network and client access bandwidth.

6 Conclusion and Future Works

In this paper, we firstly make an analysis on emerging CDN-P2P-hybrid technologies. Based on the novel CDN-P2P-hybrid architecture, we propose a new VoD streaming scheme. We emphasize on designing of CDN-P2P-hybrid overlay network structure, Super Node system design and peer node local RTSP delivery mechanism. After that we carry out the prototype system experiment. Through the contrastive experiments between CDN-Only-based VoD and CDN-P2P-hybrid VoD, we have verified the CDN-P2P-hybrid mode can support better large-scale VoD content delivery performance.

Although P2P technology gives novel opportunities to define an efficient multimedia streaming application but meanwhile, it brings a set of technical challenges and issues due to its dynamic and heterogeneous nature. At the same time, although CDN is an effective means for content delivery, there are some barriers to making CDN a more common service, especially large extension cost. We vision that hybrid CDN-P2P content distribution can economically satisfy the exponential growth of Internet streaming content without placing an unacceptable burden on regional ISPs and previous CDN. But we also need to solve many problems, for instance, one important challenge is how to integrate existing CDN and P2P streaming systems, and how to make current CDN to accommodate all kinds of heterogonous P2P system, rather than building a fully-new hybrid system.

Acknowledgements

This work is supported by 2009 National Science Foundation of China (170167): Research on Model and Algorithm of New-Generation Controllable, Trustworthy, Network-friendly CDN-P2P-Hybrid Content Delivery.

References

1. Huang, Y., Fu, T.Z.J., Chiu, D.-M.: Challenges, Design and Analysis of a Large-scale P2P-VoD System. In: SIGCOMM 2008, Seattle, Washington, USA (2008)
2. Wang, J., Huang, C., Li, J.: On ISP-friendly rate allocation for peer-assisted VoD. In: ACM Multimedia 2008, pp. 279–288 (2008)
3. Huang, C., Li, J., Ross, K.W.: Can internet video-on-demand be profitable? In: Proceedings of SIGCOMM 2007, pp. 133–144 (2007)
4. Xie, H., Yang, Y.R., Krishnamurthy, A., et al.: P4P: Provider Portal for Applications. In: SIGCOMM 2008, Seattle, Washington, USA (2008)
5. Xu, D., Kulkarni, S.S., Rosenberg, C., Chai, H.-K.: Analysis of a CDN–P2P hybrid architecture for cost-effective streaming media distribution. Multimedia Systems 11(4), 383–399 (2006)

6. Wu, J., lu, Z., Liu, B., Zhang, S.: PeerCDN: A Novel P2P Network assisted Streaming Content Delivery Network Scheme. In: CIT 2008 (2008)
7. Choffnes, D.R., Bustamante, F.E.: Taming the Torrent A Practical Approach to Reducing Cross-ISP Traffic in Peer-to-Peer Systems. In: SIGCOMM 2008, Seattle, Washington, USA (2008)
8. Chen, Z., Yin, H., Lin, C., Liu, X., Chen, Y.: Towards a Trustworthy and Controllable Peer-Server-Peer Media Streaming: An Analytical Study and An Industrial Perspective. In: GLOBECOM 2007 (2007)
9. Liu, X., Yin, H., Lin, C., Liu, Y., Chen, Z., Xiao, X.: Performance Analysis and Industrial Practice of Peer-Assisted Content Distribution Network for Large-Scale Live Video Streaming. In: AINA 2008, pp. 568–574 (2008)
10. Yin, H., Liu, X., et al.: Design and Deployment of a Hybrid CDN-P2P System for Live Video Streaming: Experiences with LiveSky. In: ACM Multimedia 2009 (2009)
11. Huang, C., Wang, A.: Understanding Hybrid CDN-P2P:Why Limelight Needs its Own Red Swoosh. In: NOSSDAV 2008, Braunschweig, Germany (2008)
12. Jiang, H., Li, J., Li, Z., et al.: Efficient Large-scale Content Distribution with Combination of CDN and P2P Networks. International Journal of Hybrid Information Technology 2(2), 4 (2009)
13. Lu, Z., Wu, J., Xiao, C., Fu, W., Zhong, Y.: WS-CDSP: A Novel Web Services-based Content Delivery Service Peering Scheme. In: IEEE SCC 2009, pp. 348–355. IEEE Computer Society, India (2009)
14. Lu, Z., Wu, J., Fu, W.: Towards A Novel Web Services Standard-supported CDN-P2P Loosely-coupled Hybrid and Management Model. In: Proceeding of 2010 IEEE International Conference on Services Computing (SCC 2010). IEEE Computer Society, Miami (2010)
15. RFC 2326: Real Time Streaming Protocol (RTSP), http://www.ietf.org/rfc/rfc2326.txt

A Combined Semi-pipelined Query Processing Architecture for Distributed Full-Text Retrieval

Simon Jonassen and Svein Erik Bratsberg

Department of Computer and Information Science,
Norwegian University of Science and Technology,
Sem Sælands vei 7-9, NO-7491 Trondheim, Norway
{simonj,sveinbra}@idi.ntnu.no

Abstract. Term-partitioning is an efficient way to distribute a large inverted index. Two fundamentally different query processing approaches are pipelined and non-pipelined. While the pipelined approach provides higher query throughput, the non-pipelined approach provides shorter query latency. In this work we propose a third alternative, combining non-pipelined inverted index access, heuristic decision between pipelined and non-pipelined query execution and an improved query routing strategy. From our results, the method combines the advantages of both approaches and provides high throughput and short query latency. Our method increases the throughput by up to 26% compared to the non-pipelined approach and reduces the latency by up to 32% compared to the pipelined.

1 Introduction

Index organization and query processing are two of the most central and challenging areas within distributed information retrieval. Two fundamental distributed index organization methods for full-text indexing and retrieval are term- and document-based partitioning. With document-based partitioning each node stores a local inverted index for its own subset of documents. With term-partitioning each node stores a subset of a global inverted index. A large number of studies [1,2,7,11,12,13,15] have tried to compare these two organizations to each other. Each method has its advantages and disadvantages, and according to the published results, both methods may provide superior performance under different circumstances.

The main problems of document-based partitioning include a large number of query messages and inverted list accesses (as disk-seeks and lexicon lookups) for each query. Term-partitioning on the other hand reduces the number of query messages, improves inverted index accesses and provides more potential for concurrent queries, but it faces two difficult challenges: a high network load and a risk for a bottleneck at the *query processing node*.

Both challenges arise from the fact that the posting lists corresponding to the terms occurring in the same query may be assigned to two or more different nodes. In this case a simple solution is to transfer each list to the query processing

L. Chen, P. Triantafillou, and T. Suel (Eds.): WISE 2010, LNCS 6488, pp. 587–601, 2010.
© Springer-Verlag Berlin Heidelberg 2010

node and then process the posting lists. We refer to this solution as *a non-pipelined approach*. However, for a large document collection a single compressed posting list can occupy hundreds of megabytes and this processing model incurs high network load. Further, as a single node is chosen to process the posting lists the performance of this node is critical. Both challenges become more prominent as the document collection grows.

In order to solve these challenges, Moffat et al. [13] have proposed a pipelined architecture for distributed text-query processing. The main idea behind the method is to process a query gradually, by one node at a time. The performance improvement is achieved from work distribution and a dynamic space-limited pruning method [5].

Although the pipelined architecture reduces the network and processing load, it does not necessarily reduce the average query latency. In our opinion, considering query latency is important under evaluation of the system throughput, and with the pipelined approach the average query latency will be significantly longer. However, neither of the two papers [12,13] describing the pipelined approach look at the query latency, but only at the normalized query throughput and system load.

The contribution of this work is as follows. We look at query latency aspects of the pipelined and non-pipelined query processing approaches. We propose a novel query processing approach, which combines non-pipelined disk accesses, a heuristic method to choose between pipelined and non-pipelined posting-list processing, and an improved query routing strategy. We evaluate the improvement achieved with the proposed method with the GOV2 document collection and a large TREC query set. We show that our method outperforms both the pipelined and the non-pipelined approaches in the measured throughput and latency.

This paper is organized as follows. Section 2 gives a short overview of related work and limits the scope of this paper. Section 3 describes the pruning method and state of the art distributed query processing approaches. Section 4 presents our method. The experimental framework and results are given in Section 5. The final conclusions and directions for the further work follow in Section 6.

2   Related Work

A performance-oriented comparison between term- and document-partitioned distributed inverted indexes have been previously done by a large number of publications [1,2,7,11,12,13,15]. Most of these refer to term-based partitioning as a method with good potential, but practically complicated due to high network load and CPU/disk load imbalance.

The non-pipelined approach is an ad-hoc method considered by most of the books and publications within distributed IR. The pipelined approach was first presented by Moffat et al. [13]. The method is based on an efficient pruning algorithm for space-limited query evaluation originally presented by Lester et al. [5] and succeeds to provide a higher query throughput than the non-pipelined approach. However, due to load imbalance the method alone is unable to out-perform a corresponding query processing method for a document-partitioned index. The load balancing issues of the pipelined approach have been further considered by Moffat et al. [12] and Lucchese et al. [6].

While Moffat et al. looked only at maximizing query throughput, Lucchese et al. have combined both throughput and latency in a single optimization metric. In our work we minimize query latency and maximize query throughput from the algorithmic perspective alone, with no consideration to posting list assignment.

As an alternative to these methods, a large number of recent publications from Yahoo [8,9,10,11] study processing of conjunctive queries with a term-partitioned inverted index. However, from the description provided by Marin et al. [9] the methods considered in their work are based on impact-ordered inverted lists and bulk-synchronous programming model. Therefore, we do not consider this type of queries nor posting ordering in our work.

3   Preliminaries

In this paper we look at processing of disjunctive queries with a distributed term-partitioned disk-stored document-ordered inverted index [17]. Each posting list entry represents a document ID and *term frequency* $f_{d,t}$ denoting the number of occurrences. Consecutive document IDs are gap-coded and compressed with Gamma coding, term frequencies are encoded with unary coding. The vocabulary is also stored on disk. For each term it stores a posting list pointer and the corresponding *collection frequency* F_t and *document frequency* f_t. Document lengths and other global index statistics are stored in main memory.

Query processing is generally done by stemming and stop-word processing the query, looking-up the vocabulary, fetching and processing the corresponding posting list, followed by a final extraction, sorting and post-processing of the K best results.

Space-Limited Adaptive Pruning. In order to speed-up query processing an efficient pruning technique has been presented by Lester et al. [5]. The main idea is to process posting lists *term-at-a-time* according to an increasing F_t. For each posting the algorithm may either create a new or update an existing *accumulator* (a partial document score). Some of the existing accumulators might also be removed if their partial scores are too low. Further, the algorithm uses two threshold variables, *frequency threshold value* h_t and *score threshold value* v_t in order to restrict the maximum number of maintained accumulators by a *target value L*.

With a distributed term-partitioned inverted index each node stores a number of complete posting lists. All queries submitted to the system are forwarded to one of the nodes chosen in a round robin manner, which performs stemming and stop-word removal. To speed-up processing, each node stores a complete vocabulary. The set of stems corresponding to the query is further partitioned into a number of *sub-queries*, where each sub-query contains terms assigned to the same node.

Non-Pipelined Approach. Each sub-query can then be transferred to the node maintaining the corresponding posting lists. Each of these nodes will fetch compressed posting lists and send these back to the node responsible for the processing of the query, *the query processing node*. The query processing node will further receive all of the posting lists, decompress and process these according to the method described earlier. In order to reduce the network load the query processing node is chosen to be the node maintaining the longest inverted list.

Pipelined Approach. The non-pipelined method has two main problems: high network load and heavy workload on the query processing node. Alternatively to this approach, query processing can be done by routing a *query bundle* through all of the nodes storing the query terms. The query bundle is created by the node receiving the query and includes the original query, a number of sub-queries, routing information and an accumulator data structure.

Because of the pruning method the route is chosen according to increasing minimum F_t in each sub-query. When a node receives the query bundle it fetches and decompresses the corresponding posting lists and combines these with the existing accumulators according to the pruning method described earlier. Further, it stores the updated accumulator set into the bundle and forwards it to the next node. The last node in the route has to extract and post-process the top-results and return these to the query broker.

In addition to reduced processing cost due to pruned postings and a limited number of accumulators, the combination of the pruning method with the pipelined distributed query processing gives the following advantages:

1. Processing of a single query is distributed across a number of nodes. Any node receiving a bundle has to combine its posting lists with an existing result set, which is limited to L. The last node in the route has additionally to perform extraction of top-results, where the number of candidates is also limited to L.
2. The upper bound for the size of a query bundle transferred between any two nodes is proportional to L, which can be used to reduce the network load.

4   A Combined Semi-pipelined Approach

The pipelined approach has clearly a number of advantages compared to the non-pipelined approach. These include a more even distribution of work over all of the nodes involved in a query and reduced network and CPU loads. However, the method has a number of issues we want to address in this paper:

1. Non-parallel disk-accesses and overheads associated with query bundle processing result in long query latencies.

2. An accumulator set has less compression potential than an inverted list.
3. The pipelined approach does not always provide benefits compared to the non-pipelined.
4. The routing strategy does not minimize the number of transferred accumulators.

Now we are going to explain these problems in more detail and present three methods to deal with one or several of these at a time. The first method reduces the average query latency by performing fetching and decompression of posting lists in parallel. For the second method that deals with second and third issue, we use a decision heuristic to choose whether a query should be executed as a semi-pipelined or a non-pipelined query. As the third method, we apply an alternative routing strategy to the original $\min(F_t)$ to deal with the last issue. In the next section we show that a combination of all three methods provides the advantages of both the pipelined and non-pipelined approaches and significantly improves the resulting system performance.

4.1 Semi-pipelined Query Processing

With our first technique we are trying to combine latency hiding advantages of the non-pipelined approach with load reduction and distribution of the pipelined method. Using an idea similar to Lucchese et al. [6], we can express the query latency associated with the non-pipelined approach as:

$$T_{\text{non-pl}}(q) = T_{\substack{\text{query}\\\text{overhead}}} + \max_{q_i \in q} \sum_{t \in q_i} (T_{\text{disk}}(f_t\sigma) + T_{\text{trans}}(f_t\sigma)) + T_{\text{process}}(q) \qquad (1)$$

$$T_{\text{process}}(q) = \sum_{t \in q} (T_{\text{decomp}}(f_t) + T_{\text{compute}}(f_t, L)) + T_{\text{top}}(K, L) \qquad (2)$$

Where q is a set of sub-queries $\{q_1, .., q_l\}$. T_{disk} is the time to fetch an inverted list of f_t postings with an average compressed size of σ bytes per posting. T_{decomp} is the time to decompress f_t postings. T_{compute} is the time to merge (join) f_t postings into an accumulator set of L postings, and T_{top} is the time to extract, sort and post-process the K best results out of L candidates. T_{trans} is the time to transfer a data structure, in this case a posting list of size $f_t\sigma$. We include the transfer time into Eq. 1, since due to the Zipf's Law [16] it is reasonable to assume that transfer times of shorter sub-queries can be hidden within disk access times of the longest one. In the case when the assumption does not hold or with expensive network receive calls, the transfer time must be included into Eq. 2 instead.

Finally, we use $T_{\substack{\text{query}\\\text{overhead}}}$ to denote the time spent on transfer of the query and sub-queries itself, lexicon look-up, and result output. In future discussion we also neglect the time spent on transfer of an additional network message per sub-query, as this time is insignificantly small compared to the time spent on posting list fetching, transfer and processing.

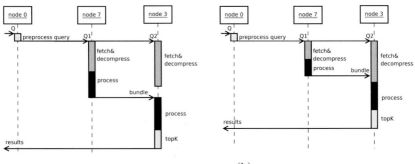

(a) Disk-Access/Decompression Latency Hiding   (b) Processing Latency Hiding

Fig. 1. Latency hiding with the semi-pipelined approach

Further, the query latency associated with the pipelined approach can be expressed as:

$$T_{\text{pl}}(q) = T_{\substack{\text{query} \\ \text{overhead}}} + \sum_{q_i \in q} \left(\sum_{t \in q_i} T_{\text{disk}}(f_t \sigma) + T_{\substack{\text{bundle} \\ \text{overhead}}}{}^{i,l}(L) \right) + T_{\text{process}}(q) \qquad (3)$$

Where T_{process} is defined by Eg. 2 and the bundle overhead can be expressed as the sum of compression, decompression and transfer of the accumulator set with a compression size of τ bytes per element:

$$T_{\substack{\text{bundle} \\ \text{overhead}}}{}^{i,l}(L) = \begin{cases} T_{\text{comp}}(L) + T_{\text{trans}}(L\tau) & \text{if } i = 1 \\ T_{\text{comp}}(L) + T_{\text{trans}}(L\tau) + T_{\text{decomp}}(L) & \text{if } i \neq 1, l \\ T_{\text{decomp}}(L) & \text{if } i = l \end{cases} \qquad (4)$$

Our idea is to use a different execution approach, where all of the posting-list fetches and posting list decompressions can be done in parallel for different sub-queries. Further, we can start a pipelined result accumulation with a query bundle as soon as all the data required for the first (shortest) sub-query is fetched and decompressed.

For a query consisting of only two sub-queries we have two possible scenarios. In the first case, as we illustrate in Figure 1(a), the node receiving the bundle has already fetched and decompressed the data, it can proceed straight to merging the posting data with the accumulated results. In this case, any latency associated with disk-access and decompression at the second node is hidden within posting list processing and query bundle overhead at the first node. Otherwise, as we illustrate in Figure 1(b), the node receiving the bundle still needs to wait for the data. In this case, the query processing latency at the first node is hidden within the disk-access and decompression latency at the second.

Algorithm 1. processQuery(q)

---

1 preprocess q;
2 partition q into $\{q_1, .., q_l\}$ and sort these by increasing $\min_{t \in q_i} F_t$ order;
3 create a new query bundle b;
4 mark b as *real*;
5 **for** *sub-query* q_i **do**
6   set b's counter idx to i;
7   **if** $i \neq 1$ *and* b *is marked as real* **then** mark b as *fake*;
8   send b to the corresponding node;

Algorithm 2. processBundle(b)

---

1 **if** b *is marked as fake* **then**
2   **for** $t \in q_{idx}$ **do** fetch and decompress I_t;
3   **if** *the corresponding accumulator set A is received* **then**
4   combine the posting data with A;
5   **if** $idx = l$ **then**
6   extract, sort and post-process the K best results; send the results to the query broker;
7   **else**
8   put A into b; increment idx; send b to the next node in the route;

9 **else if** b *is marked as real* **then**
10   **if** $idx = 1$ **then for** $t \in q_{idx}$ **do** fetch and decompress I_t;
11   extract A from the b;
12   **if** *all the required I_t are fetched and decompressed* **then**
13   similar to lines 4-8;

14

According to this logic, for an arbitrary number of sub-queries the latency can be expressed as:

$$T_{\text{semi-pl},i}(q) = \max\left(\sum_{t \in q_i}(T_{\text{disk}}(f_t \sigma) + T_{\text{decomp}}(f_t)), T_{\text{semi-pl},i-1}(q)\right)$$

$$+ \sum_{t \in q_i} T_{\text{compute}}(f_t, L) + T_{\text{bundle}_{i,l} \atop \text{overhead}}(L)$$

$$T_{\text{semi-pl},1}(q) = T_{\text{query} \atop \text{overhead}} + \sum_{t \in q_1}(T_{\text{disk}}(f_t \sigma) + T_{\text{decomp}}(f_t) + T_{\text{compute}}(f_t, L))$$

$$+ T_{\text{bundle}_{1,l} \atop \text{overhead}}(L)$$

$$T_{\text{semi-pl}}(q) = T_{\text{semi-pl},l}(q) + T_{\text{top}}(K, L) \tag{5}$$

Eq. 5 is hard to generalize without looking at the exact data distribution, compression ratios and processing, transfer and disk-access costs. However, from the equation we see that the method provides a significantly shorter query latency compared to the pipelined approach. The method is also comparable to the non-pipelined approach under disk-expensive settings and provides a significant improvement under network- and CPU-expensive settings. In addition to the model presented so far, which does not account for concurrent query execution, our approach provides the same load reduction and distribution as the pipelined method.

Algorithm 3. Changes to processQuery(q) line 4

---

1    $\hat{V}_{\text{semi}-\text{pl}} \leftarrow 0; \hat{V}_{\text{non}-\text{pl}} \leftarrow 0; tmp \leftarrow 0;$
2   **foreach** $q_i \in q_1, .., q_{l-1}$ **do**
3   **foreach** $t_j \in q_i$ **do**
4         $tmp \leftarrow tmp + f_{t_j};$
5         $\hat{V}_{\text{non}-\text{pl}} \leftarrow \hat{V}_{\text{non}-\text{pipelined}} + f_{t_j};$
6   **if** $tmp > L$ **then**
7         $tmp \leftarrow L;$
8       $\hat{V}_{\text{semi}-\text{pl}} \leftarrow \hat{V}_{\text{semi}-\text{pl}} + tmp;$
9   **if** $\hat{V}_{\text{semi}-\text{pl}} > \alpha \hat{V}_{\text{non}-\text{pl}}$ **then**
10   set the node with $argmax_i(\sum_{t \in q_i} f_t)$ as the query processing node;
11   mark b as *nobundle*;
12   **else**
13   mark b as *real*;

Algorithm 4. Append to processBundle(b) line 14

---

1   **else if** b *is marked as nobundle* **then**
2   **for** $t \in q_{idx}$ **do** fetch I_t;
3   **if** $l > 1$ **then**
4   put these into the b; mark the b as *nobundle_res* and send to the query processing node;
5   **else**
6   decompress the posting lists;
7   process the posting data; extract, sort and post-process the top-results; return these to the query broker;
8   **else if** b *is marked as nobundle_res* **then**
9   extract and decompress the posting lists from b;
10   **if** *all the other posting lists are received and decompressed* **then**
11   similar to line 7 of this algorithm;

We present our approach in Alg. 1-2. The node assembling the query bundle sends $l - 1$ replicas of the bundle marked as a *fake* to each of the sub-query processing nodes, except from the first node which receives the query bundle itself. The first node receiving the original query bundle fetches the posting lists and starts processing as normal. For any other node there are two different scenarios similar to the ones we have described earlier. With either of these the node will fetch and decompress the corresponding posting lists, extract the accumulator set and perform further query processing. Finally, the query bundle is either forwarded to the next node or the final results are extracted, post-processed and returned to the query broker.

4.2 Combination Heuristic

Besides the latency part, we can look at the next two observations on our list. First, document frequency values stored in postings tend to be small integers which can be compressed with unary coding or similar. The document IDs in postings are gap-coded and then compressed with Gamma coding. Thus long, dense posting lists achieve a high compression ratio. Document IDs in a pruned accumulator set tend to be sparse and irregular. Partial similarity scores stored in an accumulator data structure are single or double precision floats which cannot be compressed as good as small integer values. Therefore, even if it is possible to reduce the number of postings/accumulators transferred between two nodes, the data volume might not be reduced. Additional costs associated with compression/decompression must also be kept in mind.

Second, the pipe-lined or semi-pipelined approach itself will not necessarily reduce the number of scored postings or maintained accumulators. If none of the posting lists are longer than L, the algorithm will not give any savings for network or CPU loads. The network load will actually increase if the same data is transferred between every pair of nodes in the query route.

From these observations we suggest that, in order to reduce the query latency and network load, each query might be executed in either a non-pipelined or a semi-pipelined way. To choose between the two approaches we look at the amount of data transferred between the nodes. The upper bound for the amount of data can be approximated by Eq. 6.

$$\hat{V}_{\text{semi-pl}} = \tau \sum_{j=1}^{j=l-1} \min(L, \sum_{\substack{t \in \\ \{q_1,..,q_j\}}} f_t) \qquad \hat{V}_{\text{non-pl}} = \sigma \sum_{\substack{t \in \\ \{q_1,..,q_{l-1}\}}} f_t \qquad (6)$$

In this case, the non-pipelined approach is more advantageous whenever $\hat{V}_{\text{semi-pl}} > \theta \hat{V}_{\text{non-pl}}$, where θ is a system-dependent performance tuning parameter. Further we reduce the number of system-dependent parameters by introducing $\alpha = \frac{\sigma}{\tau}\theta$ and modify Alg. 1 with Alg. 3 and Alg. 2 with Alg. 4. This modification allows a query to be executed in either a semi-pipelined or a non-pipelined way, depending on the data distribution and the choice of α.

4.3   Alternative Routing Strategy

According to our last observation, the query route chosen by increasing $\min_{t \in q_i} F_t$ order is not always the optimal one. As each node is visited only once, the algorithm processes all the terms stored on this node before processing any other term. Assume four query terms t_1, t_2, t_3 and t_4 with collection frequencies $F_{t_1} < F_{t_2} < F_{t_3} < F_{t_4}$ and three sub-queries $sq_1 = \{t_1, t_4\}$, $sq_2 = \{t_2\}$ and $sq_3 = \{t_3\}$. In this case the algorithm will first process I_{t_1} and I_{t_4}, then process I_{t_2} and finally I_{t_3}. Now, if f_{t_4} is larger than L it will fill up the rest of the accumulator data structure. The number of accumulators to be transferred from this node is expected to be close to L. If both I_{t_2} and I_{t_3} are shorter than L, none of these will trigger a re-adjustment of h_t and v_t values used by the pruning method (see Sec. 3), and therefore at least L elements will be transferred also between the last two nodes.

There are two solutions to this problem. The first one is to modify the pruning method to be able to delete accumulators even if the current list is shorter than L. However, this will lead to loss of already created accumulators that still might be present in the result set. Instead, we can look at an alternative routing strategy.

In order to minimize the network load we propose to route the query not according to the increasing smallest term frequency, but according to *increasing longest posting list length* by replacing $\min_{t \in q_i} F_t$ with $\max_{t \in q_i} f_t$ in Alg. 1. In the next section we will show that this routing strategy incurs no significant reduction in the result quality, but does reduce the number of transferred accumulators and thus improves the average query latency and throughput.

5   Evaluation

In order to evaluate our ideas and compare the final method to the state-of-the-art methods we have extended the Terrier Search Engine [14] to a distributed search engine. We use all the original index data structures with minor modifications to provide support for distributed and concurrent query processing. Further we index the 426GB TREC GOV2 corpus and distribute the index across a cluster of 8 nodes. Each of the nodes has two 2.0GHz Quad-Core CPUs, 9GB memory, one 16GB SATA disk and runs Linux kernel v2.6.33.2, and Java SE v1.6.0_16. The nodes are interconnected with a 1Gb network.

For the document collection, we apply both stemming and stop-word removal. We distribute posting lists according to $hash(t)$ mod number of nodes. For the performance experiments we use the first 20000 queries from the Terabyte Track 05 Efficiency Topics [4], where the first 10000 are used as a warm-up and the next 10000 as an evaluation set. We make sure to drop OS disk caches before each experiment and all the experiment tests are performed twice, where the best of the two runs is considered as the result.

Each of the nodes runs an instance of our search engine, consisting of a modified version of Terrier, a communication manager, and a pool of query processing tasks. Processing on each node is done by 8 concurrent threads. The number is chosen according to the number of processor cores. One of the nodes works also as the query broker. For each experiment we set a fixed maximum number of queries to be executed concurrently, we refer to this value as *concurrency level*. The concurrency level is varied between 1 and 64, with steps at 8, 16, 24, 32, etc. The upper bound is chosen to be 64 since the maximum number of queries executing concurrently on each of the 8 nodes is 8. This means that for a concurrency level larger than 64, queries will be queued on at least one of the nodes, even with perfect load balancing.

Finally, as the reference methods we use implementations of the non-pipelined and the pipeline approaches based on the pruning method by Lester et al.. For all of the experiments we use $L = 400000$. As the document weighting model we use the variant of Okapi BM25 provided by Terrier.

The results are organized as follows. In Figure 2 we present a comparison of each of the proposed techniques, as well as the final combination, and compare these to the pipelined and non-pipelined approach. Each graph plots average query throughput against query latency at different concurrency levels. For every plot the lowest concurrency level (1) corresponds to the point with the shortest latency, and the highest concurrency level (64) corresponds to the

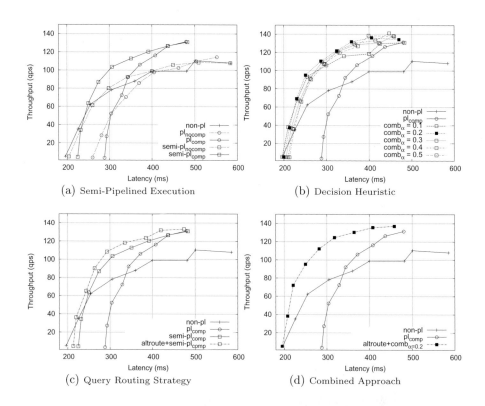

Fig. 2. Query latency and throughput with varied concurrency

point with the longest latency. Further details on throughput and latency for some of these methods are given in Table 2. For each of the methods the table presents the shortest latency, shortest latency with throughput over 100 QPS, highest throughput and the best throughput/latency ratio. Finally, in Table 1 we compare our routing strategy to the original one in terms of results quality and the number of transferred accumulators.

Semi-pipelined Execution. The plots presented in Figure 2(a) compare the performance of the semi-pipelined approach (*semi–pl*) to the pipelined (*pl*) and non-pipelined (*non–pl*). For the first two methods we also present the results with accumulator set compression applied (subscript *comp*) and without (*nocomp*). From the results the non-pipelined approach provides short query latency at low concurrency levels, but fails to achieve high throughput at higher concurrency levels as the nodes saturate on transfer, disk-access and posting list processing.

The pipelined approach with no compression has longer latencies to begin with, and achieves only a slightly better maximum throughput as the network and disk access become overloaded. With compression applied the method incurs even longer latencies at low concurrency levels, but succeeds to achieve a

Table 1. Effects of the routing method on the quality of results and average number of the final and transferred accumulators for 06topics.801-850 and qrels.tb06.top50, $L = 400000$

Method	mean average precision	average recall	final number of accs.	number of transferred accs.
Full (1 Node)	0.357	0.699	3130129	N/A
Lester (1 Node)	0.356	0.701	348268	N/A
TP MinTF 8 Nodes	0.356	0.701	347220	340419
TP MinMaxDF 8 Nodes	0.357	0.697	349430	**318603**
TP MinTF 4 Nodes	0.356	0.701	346782	272032
TP MinMaxDF 4 Nodes	0.356	0.700	350598	**247620**
TP MinTF 2 Nodes	0.356	0.701	354656	177166
TP MinMaxDF 2 Nodes	0.358	0.703	341732	**165953**

significantly higher throughput and shorter latency at higher concurrency levels. The improvement comes from a reduced network load.

The semi-pipelined approach with no compression performs quite similar to the non-pipelined approach, with a short query latency at lower concurrency levels, but not as good throughput at higher concurrency levels. However, with compression applied the method succeeds to achieve as high throughput as the pipelined approach, while it has a significantly shorter latency at lower concurrency levels. The method has also the best throughput/latency trade-off compared to the other methods, it succeeds to provide a throughput above 100 QPS with a query latency around 300 ms.

Decision Heuristic. In Figure 2(b) we plot the performance measurements of the decision heuristic ($comb_\alpha$) with different values of α. The method combines the semi-pipelined approach with compression applied together with the non-pipelined approach. The values of α we try are between 0.1 and 0.7 with a step at 0.1. We re-plot the pipelined and non-pipelined approaches to provide a reference point.

The figure shows that for the values $\alpha = 0.4$–0.6 the method provides the highest throughput (around or above 140 QPS), while $\alpha = 0.1$ minimizes latency at lower concurrency levels. The best overall performance, in our opinion, is obtained by $\alpha = 0.2$, with over 110 QPS in throughput at less than 300 ms in query latency.

Alternative Routing Strategy. In Figure 2(c) we show the performance improvement of the semi-pipelined approach with the alternative routing strategy ($altroute + semi$–pl). From the results, the method improves both latency and throughput by reducing the number of transferred accumulators.

In Table 1 we provide the results on the quality of results and the number of transferred accumulators. The results are obtained from the experiments with the TREC GOV2 corpus, the TREC Adhoc Retrieval Topics and Relevance Judgements 801-850 [3]. As the table shows, the alternative routing strategy does not reduce the quality of results, but reduces the number of transferred accumulators by 6.4 to 9.0%.

Table 2. The most important query latency and throughput measurements

Method	shortest latency	shortest latency 100QPS	highest throughput	best tp./lat.
$non-pl$	**5.1/196**	110.5/501	110.5/501	99.0/400
pl_{nocomp}	3.8/260	102.6/462	114.5/552	97.7/405
pl_{comp}	3.5/288	**106.1/372**	**131.4/480**	**126.4/437**
$semi-pl_{nocomp}$	4.9/204	106.0/449	108.6/509	92.4/343
$semi-pl_{comp}$	4.4/225	103.5/306	131.0/483	103.5/306
$comb_{\alpha=0.2}$	5.0/200	110.5/287	136.4/406	110.5/287
$altroute + semi-pl_{comp}$	4.7/213	108.4/293	133.3/474	108.4/293
$altroute + comb_{\alpha=0.2}$	**5.1/195**	**112.2/282**	**137.0/458**	**112.2/282**

However, it is not the average but the worst case performance we are trying to improve. As an example, Topic 807 - `'Sugar tariff-rate quotas'`, scheduled by smallest collection frequency will process {'quota' $F_t = 186108$, $f_t = 44395$, 'rate' $F_t = 10568900$, $f_t = 1641852$} as the first sub-query, {'sugar' $F_t = 281569$, $f_t = 109253$} as the second, and {'tariff' $F_t = 513121$, $f_t = 80017$} as the third. The total number of transferred accumulators will be 693244. Using our routing strategy the sub-query ordering will be {'tariff'}{'sugar'}{'quota rate'}. As the result, the number of transferred accumulators will be 265396, 61.7% less than with the original method, while the MAP and recall will be the same.

Combination of the Methods. In Figure 2(d) we compare a final combination ($altroute + comb$) of the semi-pipelined approach, decision heuristic with $\alpha = 0.2$ and the alternative routing strategy against the state-of-the-art approaches. More details for this method as well as the main methods discussed so far are given in Table 2. As the results show, our method outperforms both of the state-of-the-art method as it provides both a short query latency at lower concurrency levels and a high throughput at higher concurrency levels. Compared to the non-pipelined approach our method achieves up to 26% higher throughput and 22% shorter latency. Compared to the pipelined approach the method provides up to 47% higher throughput and up to 32% shorter latency. Our final method achieves also the best throughput/latency ratio, 112.2 QPS with 282ms in average response time per query. Finally, the best throughput value presented in Table 2 at 137.0 QPS was achieved by the final combination with a corresponding latency 458 ms. From the Figure 2(d), the method achieves 135.7 QSP with a corresponding latency at 408 ms, which can be considered as a quite good trade-off, 10.9% shorter latency for 0.9% lower throughput.

6 Conclusions and Further Work

In this paper we have presented an efficient alternative to the pipelined approach by Moffat et al. [13] and the ad-hoc non-pipelined approach. Our method combines non-pipelined disk-accesses, a heuristic method to choose between pipelined and non-pipelined posting list processing, and an efficient query routing strategy. According to the experimental result, our method provides a higher throughput than the pipelined approach, a shorter latency than the non-pipelined approach, and significantly improves the overall throughput/latency ratio.

Further improvements can be done by using a decision heuristic to choose wherever posting list fetch and decompression should be done beforehand, in a pipelined way, or just partially delayed. The decision depends on the number of queries executing concurrently, size of the posting list, size of the main memory or index access latency and decompression speed. The decision heuristic we use can be replaced by a more advanced one, based on an analytical performance model similar to the one presented by Lucchese et al. [6], which shall give a further improvement. The routing strategy we have presented is quite general and can be substituted by a more advanced approach that will account CPU and network load on different nodes or similar. Finally, from our results, compression is one of the main keys to achieve a high throughput and short query latencies at higher concurrency levels. However, there is a trade-off between the amount of work spent on compressing/decompressing the data and the actual benefit of it. Also not all queries/accumulator sets benefit from compression and, for example, at lower concurrency levels there might be no need to compress the data. These observations should be considered in the future.

References

1. Badue, C., Baeza-Yates, R., Ribeiro-Neto, B., Ziviani, N.: Distributed query processing using partitioned inverted files. In: Proc. SPIRE (2001)
2. Badue, C., Barbosa, R., Golgher, P., Ribeiro-Neto, B., Ziviani, N.: Basic issues on the processing of web queries. In: Proc. SIGIR. ACM Press, New York (2005)
3. Büttcher, S., Clarke, C., Soboroff, I.: The trec 2006 terabyte track. In: Proc. TREC (2006)
4. Clarke, C., Soboroff, I.: The trec 2005 terabyte track. In: Proc. TREC (2005)
5. Lester, N., Moffat, A., Webber, W., Zobel, J.: Space-limited ranked query evaluation using adaptive pruning. In: Ngu, A.H.H., Kitsuregawa, M., Neuhold, E.J., Chung, J.-Y., Sheng, Q.Z. (eds.) WISE 2005. LNCS, vol. 3806, pp. 470–477. Springer, Heidelberg (2005)
6. Lucchese, C., Orlando, S., Perego, R., Silvestri, F.: Mining query logs to optimize index partitioning in parallel web search engines. In: Proc. InfoScale (2007)
7. MacFarlane, A., McCann, J., Robertson, S.: Parallel search using partitioned inverted files. In: Proc. SPIRE. IEEE Computer Society, Los Alamitos (2000)
8. Marin, M., Gil-Costa, V.: High-performance distributed inverted files. In: Proc. CIKM. ACM, New York (2007)
9. Marin, M., Gil-Costa, V., Bonacic, C., Baeza-Yates, R., Scherson, I.: Sync/Async parallel search for the efficient design and construction of web search engines. Parallel Computing 36(4) (2010)
10. Marin, M., Gomez, C.: Load balancing distributed inverted files. In: Proc. WIDM. ACM, New York (2007)

11. Marin, M., Gomez-Pantoja, C., Gonzalez, S., Gil-Costa, V.: Scheduling intersection queries in term partitioned inverted files. In: Luque, E., Margalef, T., Benítez, D. (eds.) Euro-Par 2008. LNCS, vol. 5168, pp. 434–443. Springer, Heidelberg (2008)
12. Moffat, A., Webber, W., Zobel, J.: Load balancing for term-distributed parallel retrieval. In: Proc. SIGIR. ACM Press, New York (2006)
13. Moffat, A., Webber, W., Zobel, J., Baeza-Yates, R.: A pipelined architecture for distributed text query evaluation. Inf. Retr. 10(3) (2007)
14. Ounis, I., Amati, G., Plachouras, V., He, B., Macdonald, C., Lioma, C.: Terrier: A High Performance and Scalable Information Retrieval Platform. In: OSIR Workshop, SIGIR (2006)
15. Ribeiro-Neto, B., Barbosa, R.: Query performance for tightly coupled distributed digital libraries. In: Proc. DL. ACM Press, New York (1998)
16. Zipf, G.: Human behavior and the principle of least-effort. Journal of the American Society for Information Science and Technology (1949)
17. Zobel, J., Moffat, A.: Inverted files for text search engines. ACM Comput. Surv. 38(2) (2006)

Towards Flexible Mashup of Web Applications Based on Information Extraction and Transfer

Junxia Guo[1,2], Hao Han[3], and Takehiro Tokuda[1]

[1] Department of Computer Science, Tokyo Institute of Technology
Ookayama 2-12-1-W8-71, Meguro, Tokyo 152-8552, Japan
`{guo,tokuda}@tt.cs.titech.ac.jp`
[2] College of Information Science and Technology, Beijing University of Chemical Technology, 15 beisanhuan east road, chaoyang district, Beijing 100029, China
`gjxia@mail.buct.edu.cn`
[3] Digital Content and Media Sciences Research Division, National Institute of Informatics, 2-1-2 Hitotsubashi, Chiyoda, Tokyo 101-8430, Japan
`han@nii.ac.jp`

Abstract. Mashup combines information or functionality from two or more existing Web sources to create a new Web page or application. The Web sources that are used to build mashup applications mainly include Web applications and Web services. The traditional way of building mashup applications is using Web services by writing a script or a program to invoke those Web services. To help the users without programming experience to build flexible mashup applications, we propose a mashup approach of Web applications in this paper. Our approach allows users to build mashup applications with existing Web applications without programming. In addition, with our approach users can transfer information between Web applications to implement consecutive query mashup applications. This approach is based on the information extraction, information transfer and functionality emulation methods. Our implementation shows that general Web applications can also be used to build mashup applications easily, without programming.

Keywords: Web application, flexible mashup, information extraction, information transfer.

1   Introduction

In recent years, mashup applications have brought new creativity and functionality to Web applications by combining data or functionality from two or more existing Web sources. The Web sources that are used to build mashup applications mainly include Web applications and Web services. Although Web services are most commonly used Web sources of mashup, many existing Web sites do not provide Web services, such as the BBC Country Profiles [9]. Thus, it is not easy to build mashup applications with existing Web applications, especially for the users without programming experience.

L. Chen, P. Triantafillou, and T. Suel (Eds.): WISE 2010, LNCS 6488, pp. 602–615, 2010.

Most of the existing personal-use-oriented mashup tools for users without programming experience cannot address two problems. One problem is that they cannot use dynamic Web content of Web pages as mashup components. Here we use the term static Web content to represent the content shown on the Web page that can be found directly in the HTML source file, for example text. And we use the term dynamic Web content to represent the content that is created by client-side scripts dynamically, which is shown on the Web page but cannot be found directly in the HTML source file. The other problem is that they do not support the information transfer between the mashup components. Here we mean transfer the information extracted from mashup component A to mashup component B as the input keyword. And in this paper, the mashup component specifies the component that is generated from existing Web applications.

In this paper, we present an approach that allows users to build mashup applications with any parts of existing Web applications and to transfer information between mashup components without programming. To implement our approach, we need to address two main problems. One is how to wrap part(s) of a Web application into a mashup component. The other one is how to transfer information between the mashup components. The technique that we use to wrap part(s) of a Web application into a mashup component is based on information extraction and functionality emulation methods. We record the information of mashup components into a description file, which we call Mashup Component Description Language file (MCDL file) to configure the locations and scopes of target parts of Web applications. To transfer information between mashup components, we also record the information about the data that will be delivered to other components in MCDL file. Then, we obtain the data using information extraction method and transfer the data using the program hidden in the client-side mashup application.

The rest of this paper is structured as follows. In Section 2, we present the motivation of our research and an overview of the related work. In Section 3, we explain the details of our approach. In Section 4, we introduce an example which is implemented as a Web application. In Section 5, we give an evaluation of our approach. Finally, we conclude our approach and discuss our future work.

2   Motivation and Related Work

The research work on methods to help users without programming experience to build mashups easily and efficiently has been improved rapidly recent years. Most of the existing mashup tools build mashups only with Web services and Web feeds and do not support the using of Web applications, for example Yahoo Pipes [22] and so on.

Some mashup tools can integrate parts of Web applications without Web services, but they have certain limitations. C3W [2] can reuse text and functional parts of Web pages by clipping, connecting and cloning to build mashups. However, C3W is inaccessible to many users because of a special browser to clip Web content from Web pages and a special container named DerivationPad for the clipped Web content.

Dapp Factory [10] can use information extracted from Web pages and RSS to build mashups, but it cannot extract dynamic Web content from Web pages, for example the clock part of Localtimes.info [15].

Florian Daniel et al. proposed an approach in [1] that allows users to compose mashup applications using UI components. But it has a premise that component developers (IT specialists) need to componentize UI components from the Web applications and publish them on the Web.

Based on the above analysis, we can conclude that most of the existing personal-use-oriented mashup approaches towards the users without programming experience do not support the information transfer between mashup components. Additionally, building mashup applications with general Web applications always has some kind of restriction. Compared with the existing work, our approach has the following benefits.

- Our approach can easily implement information transfer between mashup components extracted from Web applications without programming.
- Our approach can extract not only static Web content, but also dynamic Web content from Web pages and use the extracted results to build mashups.
- Our approach uses MCDL files to describe the information of target parts of Web applications which is used to build mashups and the information about the data which will transfer to other components. Users can use any parts of any applications to build mashups and transfer appropriate information to other mashup components by editing the MCDL file.

3   Architecture of Description-Based Mashup

In this section we present our approach based on the generating process of a mashup application example. As shown in Figure 1 , the whole process inclueds the following steps.

Step 1: Users create a MCDL file which includes the information of mashup components that will be used to build mashup application and the information that will be transferred between the components. We provide an application to ease the creation of the MCDL file.

Step 2: Users send requests from the "Client-side Mashup Application" to the "Extractor".

Step 3: The "Extractor" reads the MCDL file, then according to the given keyword searches and extracts the target parts which are specified in the MCDL file.

Step 4: We integrate the extracted components and arrange the layouts as the response result.

We build an example mashup application which integrates four components that are extracted from four Web applications: Localtimes.info[15], BBC Country Profiles[9], Weatherbonk.com[17] and abc News[8].

After giving a country name, for example "Japan", and clicking the "Search" button, we can get a resulting Web page as shown in Figure 2, where following components can be found.

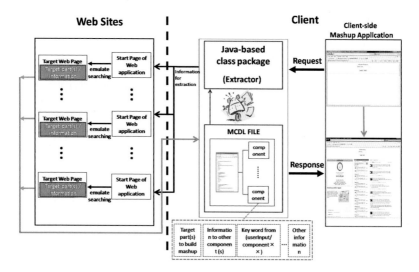

Fig. 1. Overview of Our Approach

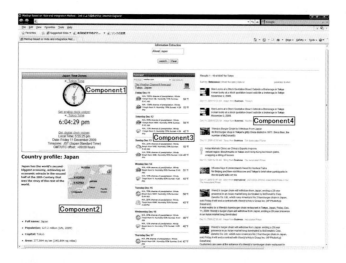

Fig. 2. Example of Mashup Web Application

Component 1. The local clock of given country's capital city from the Local-times.info Web site. The local clock is created by client-side script which changes every second.

Component 2. Given a country's name, introduction, location map and basic information from BBC Country Profiles. The capital city's name extracted here will deliver to other components as searching keyword.

Component 3. The capital city's weather information from Weatherbonk.com. The capital city's name is from "Component 2".

Component 4. The latest news articles from abc News about the capital city. The capital city's name is also from "Component 2".

3.1   Mashup Components Description Language (MCDL)

The components that we use to build mashup application are extracted from general Web applications. To describe the necessary information for the extraction, emulation and information transfer, we propose Mashup Components Description Language (MCDL). It is XML-based file, where following items are defined for each mashup component.

- id: We specify the unique name of each mashup component.
- type: We specify the type of the mashup component. It should be one of {Application, Feed, SOAP and REST}. In this paper, we focus on the "Application" type.
- startPageURL: We specify the URL of a Web page where we can submit the searching request to get the target Web page.
- inputType: We specify the type of searching method on the start page. It should be one of {LinkList (anchor list), OptionList (drop-down option list in selectbox) or InputBox (text input field) and None}. We just support these four basic input types by now. We would like do experiments about other input types, for example radio button, as future work.
- keywordFrom: We specify how to get the searching keyword of emulation. It could be the input value from user interface, or be the value transferred from other component. When the searching keyword is from other component, we use the "id" of the component to describe it.
- inputArea: We specify the path of the input Web content on the start page. If there is other content with the same "inputType" in the start page, we need to specify which one is going to be used. Users can give a value of "null" instead of the path, which will be treated as the path of *body* node by default. The path that we use here and other items are XPath-like expressions. The value of path can be gotten easily by using the tool we supply.
- targetContent: We specify the Web content to be extracted by path. Here we support multiple Web contents extraction.
- targetContentType: We specify the data type of the target Web content. It could be dynamic content (mashup, flash or function part) or static content (body, bullet, bullet list, definition list bullet, definition list, div, form, heading, input, label, text, image, link, object, select, paragraph, span, table or text area).
- newStyleforContent: We specify the new style information of target Web content in the mashup resulting page. However, the style can only be changed for static Web contents. According to the type of target Web content, different style attributes can be specified. We process this based on the attribute of different tags defined in HTML. The "newStyleforContent" information

is optional information, and when it is "null", extracted Web content will be shown in its original style. For dynamic Web contents, the extracted parts are shown in their original styles and the "newStyleforContent" value is "null".
- valueToOtherComponent: We specify the information that will be transferred to other component by path.

The above ten items describe how to get a component. A MCDL file describes a series of components from different Web sources, like a batch file. Each "component" defined in the MCDL file will be shown in one container. The data information of the MCDL file used in our mashup Web application example shown in Figure 2 is listed in Section 4.

In the above ten items, "type", "inputType", "keywordFrom", "targetContentType" and "newStyleforContent" are selected from the predefined list. "startPageURL" is copied from browser. "inputArea", "targetContent" and "valueToOtherComponent" can be copied from our Path-Reader Tool, as shown in Figure 3, which can generate path strings of target Web content by GUI. Users can get the paths easily by clicking the target parts and then *COPY* the path into MCDL file. Users do not need to read the HTML source code manually.

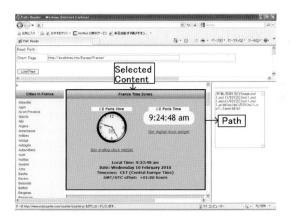

Fig. 3. Example of Path Reader

3.2 Generating Mashup Components

According to the description in the MCDL file, we search for the target Web pages to get the target Web contents from the Web applications. For each component, the process is as follows. First, we go to the start Web page of the Web application that is specified in the MCDL file. Then according to the input type and searching keyword, we emulate the searching function of the application to get the target Web page. Next, we search for the target Web contents by paths specified in "targetContent" of the MCDL file. Finally, we get the value that is delivered to other components if the "valueToOtherComponent" item is not "null".

3.2.1 Obtaining Target Web Page

Usually, in order to search for target information, Web applications provide request-submit functions for users. For example, search engine applications provide a text input field on the Web page for users to input the searching keywords. Three basic types of methods are widely used. One is entering a keyword into the input field by keyboard and clicking the submit button by mouse to get the target Web page. Another one is clicking an option in drop-down list in a Web page by mouse to view the target Web page. The third one is clicking a link in link list in a Web page by mouse to go to the target Web page. To submit a request, there are POST method and GET method, also encrypted codes or randomly generated codes are used by some Web sites. In order to get the target Web pages from various kinds of Web sites, we specify four kinds of input type and use HtmlUnit [11] to emulate the submitting operation.

When a user sends a request, according to the values of "keywordFrom" specified in the MCDL file, we can obtain the searching keyword from either the input box of the user interface or a certain component specified by component id. Then using the values of "inputType" item and "inputArea", we emulate the searching function to get the target Web page. The details of the emulation are explained in [3].

3.2.2 Contents Extraction and Visibility Control

The generation of mashup components we proposed here is based on the information extraction method. Hence, the accuracy of extraction is a very important factor of the mashup generating process. The nested structure of HTML document tags in a Web page naturally forms a tree structure. Each path represents a root node of subtree and each subtree represents a part of the Web page. We use this path to describe the target parts of a Web page. Usually, the response Web pages have the same or similar layouts if the requests are sent to the same request-submit function. Based on this fact, we can extract the specified part with different giving keywords.

The Xpath [19] of HTML document is widely used to describe the HTML nodes. The id attribute of tag in HTML document usually is thought as an unique value. Hence, the path format that begin with the last ancestor node which has id attribute is widely used. This kind of path is short, but the experiments we have done shows that the id attribute of tag in HTML document is not always unique, for example, the Web pages in Infoplease [14] Web site.

Through the experiments, we found that although the class attribute of tag in HTML document usually is not unique, the class attribute is widely used. According to the experiments that we have done, to ensure the extracting accuracy, we use HTML document's Xpath and the id, name, class attributes of tag in the following format to record the paths of "inputArea", "targetContent" and "valueToOtherComponent".

/ HTML/ BODY/ node1[index1] (id1, name1, class1)/ node2[index2] (id2, name2, class2)/ .../ node(N-1)[index(N-1)] (id(N-1), name(N-1), class(N-1))/ nodeN[index] (idN, nameN, classN)

Here, *nodeN* is the tag name of the *N-th* node, *[indexN]* is the order of the *N-th* node among the sibling nodes that have the same tag name as *nodeN*, *idN* is the ID attribute's value of the *N-th* node, *nameN* is the NAME attribute's value of the *N-th* node, *classN* is the CLASS attribute's value of the *N-th* node, and the *node(N-1)* is the parent node of *nodeN*.

The experimental results of information extraction are listed in Table 1, Section 5.

For the dynamic Web content, we use hide-and-integration method instead of using the extracting method of static Web content to keep the functionality of dynamic Web content. This is because the scripts usually use DOM operation to control the dynamic parts of Web pages and sometimes access the elements outside the target parts such as the hidden values. Therefore, if we just extract the target parts, the original executing environment of scripts may be broken. The hide-and-integration method we proposed works as following. First, we extract the whole HTML document of the target Web page. Then, we hide all HTML nodes in the extracted HTML document of the target Web page. Finally, we show the target Web contents that specified as "targetContent" in the MCDL file and do the necessary settings. The details of the hide-and-integration method can be found in [3].

For static Web content, we supply two kinds of extracting methods. One method is to extract the target part in text format excluding the tags of the HTML document. For example, the extracted information of a picture is the value of attribute "src" of node <img>, and the extracted information of a link is the value of attribute "href" of node <a>. The details of the extracting algorithm can be found in [4]. Another method is to use the same method as dynamic Web content, if the users want to keep the same layout of target parts with the original Web page. Users can choose the second method by setting the value of "valueToOtherComponent" into *dynamic*.

3.2.3 Information Transfer

The process of information transfer is shown in Figure 4. The whole process includes the following steps.

Step 1: We use the same algorithm as searching the target Web content to find the HTML node.

Step 2: We extract the text excluding the HTML tags of the node.

Step 3: We store the data into a storage space indexed by the component id.

Step 4: If certain component's value of "keyWordFrom" is the id of a component, we use the data stored in the storage space to do the emulation and obtain the target Web page.

Step 5: We transfer the extracted target part(s) to the client-side mashup application as response result through the "Extractor".

We classify the information transfer between mashup components into following two models.

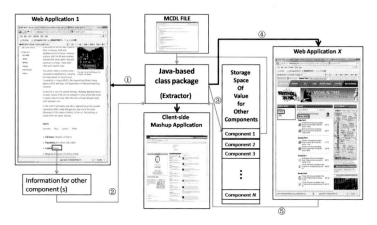

Fig. 4. Process of Information Transfer

- Parallel Model. If the information transfer between the mashup components in a mashup application is from one component to other all components, we call it Parallel Model. For example, information is transferred from component A to component B, C, ..., and N.
- Sequential Model. If the information transfer between the mashup components in a mashup application is from one to one sequentially, we call it Sequential Model. For example, information is transferred from component A to component B, then from component B to component C, and so on.

We create the mashup components one by one according to the recorded order in the MCDL file. As a result, the components can only use the information extracted from the component which is recorded before itself as the searching keyword. We support the parallel model and the sequential model information transfer with the above restriction. We also support the mix of parallel model and sequential model information transfer in one mashup application.

3.3   Integration and Layout Arrangement

Finally, we integrate the components extracted from different Web applications as a resulting Web page. We use IFrame [12] as the container of the mashup components. IFrame (Inline Frame) is an HTML element which makes it possible to embed an HTML document inside another HTML document. While regular frames are typically used to logically subdivide the Web content of one Web page, IFrames are more commonly used to insert Web content from another website into the current Web page. Also, IFrames are supported by all major browsers.

According to the MCDL file, we create as many IFrames as the number of "component" tags when the resulting page is loading. By default, we show two IFrames per row. When getting the returned HTML document processed by the visibility control method, we show it on the resulting Web page in an IFrame.

After the resulting page is shown completely, users can move iframes by dragging and dropping to adjust the location as shown in Figure 2.

4   Implementation

We build several client-side Java Applet mashup applications to implement our approach using Java and JavaScript. One of them is shown in Figure 2. The client-side mashup application has two parts. One part is an HTML file which is used as the user interface. It specifies the name of the MCDL file and manages the storage of the information which would transfer to other components. It also invokes the program interface (Extractor) we provide. The other part is the Java-based class package we provide which we call "Extractor". It includes the program interface which actually excutes extraction.

It works well with Internet Explorer7/8 and JDK 1.6 under Windows XP SP3 and Windows Vista SP1. It can work well on platforms that fully support Java and JavaScript.

Part of the MCDL file's information which is used to build the example shown in Figure 2 is listed as follows.

5   Evaluation

This section evaluates our approach. Our work aims to allow users with no or little programming experience to build mashup applications easily and efficiently. We built several mashup applications with various parts from different kinds of Web applications, and proved that our approach is applicable to general Web applications. We evaluate our approach from two aspects. First, we evaluate the extracting accuracy, which is the base of generating mashup components. Then we evaluate the efficiency of our approach by comparing with other existing research works.

We chose several Web applications to generate components and build mashup applications. The experimental results about extracting accuracy are given in Table 1.

As the experimental results show, the average extracting accuracy is more than 90%. We checked the extracting results manually and counted the parts that had been extracted correctly. Our approach is based on the assumption that the searching result pages of the same request function are the same or similar within a Web application. But some of the searching result pages may be very different from other pages. As a result, among the all 201 countries and regions that we have checked in the BBC Country Profiles, we got correct results from 194 web pages. At present we support one path to describe the location of "targetContent". We would like to support multiple paths to describe the location of "targetContent" and allow the users to decide how many paths they would like to use according to the accuracy they want.

We delivered the name of the capital city that extracted from BBC Country Profiles to the abc News and Weatherbonk. We got 194 (100%) correct results

```
<?xml version="1.0" encoding="ISO-8859-1"?>
<channel>
  <component>
    <id> Component1 </id>
    ...
  </component>
  <component>
    <id> Component2 </id>
    <type> Application </type>
    <strartPageURL>
      http://news.bbc.co.uk/1/hi/country_profiles/default.stm
    </strartPageURL>
    <inputType> OptionList </inputType>
    <keywordFrom> userInput </keywordFrom>
    <inputArea> </inputArea>
    <targetContent>
      <content>
        /HTML/BODY/DIV(null,null,centerbody)/DIV[5](null,null,mainwrapper)/
        TABLE(null,null,main)/TR(null,null,null)/TD[2](null,null,contentwr-
        apper)/TABLE[2](null,null,storycontent)/TR(null,null,null)/TD(null,
        null,null)/DIV(null,null,mxb)/H1(null,null,null)/SPAN(null,null,null)
      </content>
      <content>
        /HTML/BODY/DIV(null,null,centerbody)/DIV[5](null,null,mainwrapper)/
        TABLE(null,null,main)/TR(null,null,null)/TD[2](null,null,contentwr-
        apper)/TABLE[2](null,null,storycontent)/TR[2](null,null,null)/TD(n-
        ull,null,storybody)/DIV[3](quickguide,null,guidecontainer)/DIV(null,
        null,guidemain)/DIV(content,null,content)/UL(null,null,bulletList)
      </content>
    </targetContent>
    <targetContentType>
      <contentType> dynamic content </contentType>
      ...
      <contentType> dynamic content </contentType>
    </targetContentType>
    <newStyleforContent>
      <newStyle > </newStyle >
    </newStyleforContent>
    <valueToOtherComponent> </valueToOtherComponent>
  </component>
  <component>
    <id> Component3 </id>
    <type> Application </type>
    <strartPageURL> http://abcnews.go.com/ </strartPageURL>
    <inputType> InputBox </inputType>
    <keywordFrom> Component2 </keywordFrom>
    <inputArea>
      /HTML/BODY/DIV[3](null,null,window)/DIV(null,null,showbg)/DIV(null,
      null,headerbg)/DIV(null,null,header)/DIV(null,null,upperheader)/
      DIV[2](shownav,null,null)/DIV[2](search,null,null)/
      FORM(searchForm,null,null)
    </inputArea>
    <targetContent>
      <content>
        /HTML/BODY/DIV(null,null,window)/DIV(null,null,showbg)/DIV[3](null,
        null,bodycontainer)/DIV(null,null,mainsection)/DIV(null,null,main-
        Widgets)/DIV(null,null,widgetsColumnAB)/DIV[3](null,null,null)/
        DIV[2](searchResultsPane,null,null)
      </content>
    </targetContent>
    <targetContentType>
      <contentType> dynamic content </contentType>
    </targetContentType>
    <newStyleforContent>
      <newStyle > </newStyle >
    </newStyleforContent>
    <valueToOtherComponent> </valueToOtherComponent>
  </component>
  <component>
    <id> Component4 </id>
    ...
  </component>
</channel>
```

Fig. 5. An Example of MCDL File Information

Table 1. Experimental Results of Extracting Accuracy

Web Application	Input Type for Search	Input Parameter	Num. of Checked Web Pages	Num. of Extracting Parts	Num. of Correct Pages	Percent of Correct Pages
abc News	Input Box	city name (Beijing, Tokyo ...)	194	1 (text, link)	194	100%
BBC Country Profiles	Option List	country name (Japan, Italy ...)	201	4 (text, image, bullet list)	194	96%
Localtimes	Link List	country name (Canada, Russia ...)	72	1 (dynamic content)	72	100%
Weatherbonk	Input Box	city name (Beijing, Tokyo ...)	194	1 (dynamic content)	189	97%
Yahoo Finance [20]	Input Box	stock name (AAPL, GOOG ...)	36	1 (dynamic content)	36	100%
Infoplease [14]	Link List	country name (Canada, Russia ...)	204	5 (image, text)	186	91%
Yahoo News [21]	Input Box	city name (Beijing, Tokyo ...)	189	1 (text, link)	189	100%
Total	—	—	1090	—	1060	97%

from the abc News, and 189 (97%) correct results from the Weatherbonk. The cause of error is that some country/region name are the same as their capital city names. But in the Weatherbonk Web application, when we input the city's name, it was treated as a country name and we get a different page.

There are mainly two kinds of personal-use-oriented mashup approaches. The first kind of approach is to integrate the Web sources in predefined library using GUI interface as Yahoo Pipes, iGoogle [13], Mixup [7] and so on. The users do not need programming, but they cannot add new Web sources into the library. The second kind of approach is to use static contents extracted from Web applications, as Dapp Factory [10], C3W [2] and so on. Users can use information extracted from Web applications and other Web sources to do lightweight integration, but they cannot extract dynamic contents.

The MCDL we proposed is a short and simple description format. It is applicable to the description of general Web applications and easier to read, write, reuse and update any part of mashup application than user programming methods. Although some simple manual work is needed in the generation of MCDL file, by using the PathReader we supplied, users can generate the MCDL file easily without analyzing the HTML document manually or having the knowledge about the HTML tag event definition. By user study, the average time that an user (first time and unfamiliar with our system) builds a MCDL file including three mashup components is about 7 minutes. Our implementation and tests show that our approach is applicable to almost all kinds of Web applications. It allows users with no or little programming experience to implement the integration of components extracted from various Web applications without Web service APIs. Additionally, it allows the information transfer between mashup components. The range of mashup components are extended from Web services to the general Web applications.

However, we have the following tasks to address.

- The information transfer that we do is mainly towards the public Web applications without strong consideration of security problems. For example, if the users want to build mashup applications including the functionality of credit card payment, our approach cannot ensure the security of the information now.
- We support two kinds of information transfer models between mashup components now, parallel model and sequential model. They are not enough. To make users feel flexible, we need to support other kinds of models.
- We generate the mashup components one by one to make sure the information transfer. Hence, the processing time of the whole mashup application may be long if it has many components.
- We use HtmlUnit to emulate the submitting request and get the response automatically without manual analysis of the URLs. But the emulating process of HtmlUnit is slow for some Web sites if the Web sites use lots of external scripts in the submitting process.
- The input types we support now are the basic types in Web application. If the searching method in the Web application is generated by interacting with a Flash Plug-in object, our approach cannot process it currently. Also, if the target Web content is on a page that shows after a login operation, our method cannot currently process it.

6   Conclusion

In this paper, we have presented a novel approach that allows users to integrate any parts from any Web applications without programming and allows the value delivery between mashup components to realize the continual queries for personal use. Our approach uses a description language (MCDL) to describe the information about mashup components which is selected to be extracted form Web applications and the information about information transfer between mashup components.

In future work, we would like to explore more flexible ways of integrating Web applications and Web services. For example, explore an approach to help the users without programming experience to build mashup applications with SOAP type Web services easily and efficiently without restriction. We also would like to supply a tool to generate the MCDL file without the manual work.

References

1. Daniel, F., Matera, M.: Turning Web Applications into Mashup Components: Issues, Models, and Solutions. In: Proceeding of the 9th International Conference on Web Engineering, ICWE 2009 (2009)
2. Fujima, J., Lunzer, A., Hornbak, K., Tanaka, Y.: C3W: Clipping, Connecting and Cloning for the Web. In: Proceeding of the 13th International Conference on World Wide Web (2004)
3. Guo, J., Han, H., Tokuda, T.: A New Partial Information Extraction Method for Personal Mashup Construction. In: Proceeding of the 19th European-Japanese Conference on Information Modelling and Knowledge Bases (2009)
4. Han, H., Tokuda, T.: WIKE: A Web Information/Knowledge Extraction System for Web Service Generation. In: Proceeding of the 8th International Conference on Web Engineering (2008)
5. Kongdenfha, W., Benatallah, B., Vayssiere, J., Saint-Paul, R., Casati, F.: Rapid Development of Spreadsheet-based Web Mashups. In: Proceeding of the 18th International Conference on World Wide Web (2009)
6. Tatemura, J., Sawires, A., Po, O., Chen, S., Candan, K.S., Agrawal, D., Goveas, M.: Mashup Feeds: Continuous Queries over Web Services. In: Proceeding of the 2007 ACM SIGMOD International Coference on Management of Data (2007)
7. Yu, J., Benatallah, B., Saint-Paul, R., Casati, F., Daniel, F.: A Framework for Rapid Integration of Presentation Components. In: Proceeding of the 16th International Conference on World Wide Web (2007)
8. abc News, http://abcnews.go.com/
9. BBC Country Profiles, http://news.bbc.co.uk/2/hi/country_profiles/default.stm
10. Dapp Factory, http://www.dapper.net/
11. HtmlUnit, http://htmlunit.sourceforge.net
12. IFrame, http://en.wikipedia.org/wiki/iframe
13. iGoogle, http://www.google.com/ig/
14. Infoplease, http://www.infoplease.com/countries.html
15. Localtimes.info, http://localtimes.info/
16. Mouseover DOM Inspector, http://slayeroffice.com/content/tools/modi.html
17. Weatherbonk.com, http://www.weatherbonk.com/
18. Wikipedia, http://en.wikipedia.org/wiki/Main_Page
19. XPath, http://www.w3.org/TR/xpath20/
20. Yahoo Finance, http://finance.yahoo.com
21. Yahoo News, http://news.yahoo.com/
22. Yahoo Pipes, http://pipes.yahoo.com/pipes/

On Maximal Contained Rewriting of Tree Pattern Queries Using Views

Junhu Wang[1] and Jeffrey Xu Yu[2]

[1] School of Information and Communication Technology
Griffith University, Gold Coast, Australia
J.Wang@griffith.edu.au
[2] Department of Systems Engineering and Engineering Management
The Chinese University of Hong Kong, China
yu@se.cuhk.edu.hk

Abstract. The problem of rewriting tree pattern queries using views has attracted much attention in recent years. Previous works have proposed algorithms for finding the maximal contained rewriting using views when the query and the view are limited to some special cases, e.g., tree patterns not having the wildcard *. In the general case, i.e, when both //-edges and * are present, the previous methods may fail to find the maximal contained rewriting. In this paper, we propose a method to find the maximal contained rewriting for the general case, as well as an extension of the previous method to more special cases.

1 Introduction

Answering queries using views has been well studied for relational databases and found applications in fields such as query optimization and data integration and query answering when the original data is inaccessible or very hard to access, e.g., in a network environment [8]. Motivated by similar potential applications, the problem of answering tree pattern queries using views has recently attracted attention from many researchers.

A main approach for answering tree pattern queries using views exploits the technique of query rewriting, which transforms a query Q into a set of new queries, and then evaluates these new queries over the materialized answers to the view [18,10,9,1,6]. These new queries are called rewritings. In the literature, there are two types of rewritings, equivalent rewritings and contained rewritings. An equivalent rewriting will produce all answers to the original query, and a contained rewriting may produce only part of the answers. Most previous works concentrated on equivalent rewritings. However, contained rewritings are important especially when equivalent rewritings do not exist. For contained rewritings of tree pattern queries using views, [9] gave a method to find the maximal contained rewriting (MCR) (i.e., the union of all contained rewritings) when both the view and query are in $P^{\{/,//,[]\}}$ (i.e, they do not have nodes labeled with the wildcard *), [15] showed how to extend the approach of [9] to views in $P^{\{/,//,*,[]\}}$ and queries that (1) do not have *-nodes incident on //-edges and (2) there are

L. Chen, P. Triantafillou, and T. Suel (Eds.): WISE 2010, LNCS 6488, pp. 616–629, 2010.

no leaf *-nodes. More recently [12] gave a method for finding the MCR for views in $P^{\{/,//,*,[]\}}$ and queries that satisfy condition (1) above and the condition (3) if a leaf node is labeled * then its parent is not.

In this paper, we study the problem of finding the MCR of tree pattern query Q using a view V again, and make the following contributions:

1. We show that the approach of [9] can be extended to the following cases: (a) $V \in P^{\{/,//,*,[]\}}$ and Q has no /-edges, or (b) $Q \in P^{\{/,//,*,[]\}}$ and V has no //-edges.
2. For the more general case where both Q and V can be *any* pattern in $P^{\{/,//,*,[]\}}$, we present a method to find the MCR of Q using V, and discuss optimization methods which enable us to find the MCR more quickly.

The rest of the paper is organized as follows. Section 2 gives the preliminaries. The main contributions of this paper are presented in Section 3. Section 3.1 discusses about the special cases. Section 3.2 studies the general case. Sections 3.3 and 3.4 discuss strategies to speedup the computation of the maximal contained rewriting in the general case. Section 4 compares our work with more related works. We conclude the paper in Section 5.

2   Preliminaries

2.1   XML Data Tree and Tree Patterns

Let Σ be an infinite set of tags which does not contain the symbol *. An XML *data tree*, or simply a *data tree*, is a tree in which every node is labeled with a tag in Σ. A *tree pattern*, or simply a *pattern*, in $P^{\{/,//,*,[]\}}$ is a tree in which every node is labeled with a symbol in $\Sigma \cup \{*\}$, every edge is labeled with either / or //, and there is a unique *selection node*. The path from the root to the selection node is called the *selection path*. Fig. 1 shows some example patterns, where single and double lines are used to represent /-edges and //-edges respectively, and a circle is used to indicate the selection node. A pattern corresponds to an XPath expression. The patterns in Fig. 1 (a) and (b) correspond to the XPath expressions $a[c]//b[//d]$ and $a[c]//b[x]/y$ respectively.

Let P be a pattern. We will use $\mathrm{sn}(P)$, and $\mathrm{sp}(P)$ to denote the selection node and the selection path of P respectively. For any tree T, we will use $N(T)$, $E(T)$ and $\mathrm{rt}(T)$ to denote the node set, the edge set, and the root of T respectively. We will also use $label(v)$ to denote the label of node v, and call a node labeled x an *x-node*. In addition, if (u, v) is a /-edge (resp. //-edge), we say v is a /-child (resp. //-child) of u.

A *matching* of pattern P in a data tree t is a mapping δ from $N(P)$ to $N(t)$ that is

(1) *root-preserving:* $\delta(\mathrm{rt}(P)) = \mathrm{rt}(t)$,
(2) *label-preserving:* $\forall v \in N(P)$, $label(v) = label(\delta(v))$ or $label(v) = *$, and
(3) *structure-preserving:* for every /-edge (x, y) in P, $\delta(y)$ is a child of $\delta(x)$; for every //-edge (x, y), $\delta(y)$ is a descendant of $\delta(x)$.

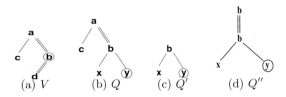

Fig. 1. View V, Query Q, and two contained rewritings Q' and Q''

Each matching δ produces a subtree of t rooted at $\delta(\mathsf{sn}(P))$, which is known as an *answer* to the pattern. We use $P(t)$ to denote the set of all answers of P over t. For a set S of data trees, we define $P(S) = \cup_{t \in S} P(t)$.

Let $P_1, \ldots, P_n$ be patterns. We use $P_1 \cup \cdots \cup P_n$ and $P_1 \cap \cdots \cap P_n$ to represent the *union* and *intersection* of $P_1, \ldots, P_n$ respectively. The meaning of union and intersection is as usual: for any set of data trees S,
$(P_1 \cup \cdots \cup P_n)(S) = P_1(S) \cup \cdots \cup P_n(S)$, and
$(P_1 \cap \cdots \cap P_n)(S) = P_1(S) \cap \cdots \cap P_n(S)$.

2.2 Tree Pattern Containment and Containment Mapping

Let P and Q be patterns. P is said to be *contained* in Q, denoted $P \subseteq Q$, if for every data tree t, $P(t) \subseteq Q(t)$. A *containment mapping* from Q to P is a mapping δ from $N(Q)$ to $N(P)$ that is label-preserving, root-preserving as discussed above, structure-preserving (which now means for any $/$-edge (x, y) in Q, $(\delta(x), \delta(y))$ is a $/$-edge in P, and for any $//$-edge (x, y), there is a path from $\delta(x)$ to $\delta(y)$ in P) and is *output-preserving*, which means $\delta(\mathsf{sn}(Q)) = \mathsf{sn}(P)$.

Several subsets of $P^{\{/,//,*,[]\}}$ are of special interest to us: $P^{\{/,//,[]\}}$ is the set of patterns that do not have *-nodes; $\widehat{P}^{\{/,//,[\ ],*\}}$ is the set of patterns in which all *-nodes are incident only on $/$-edges and are non-leaf; $P^{\{/,[\ ],*\}}$ is the set of patterns that do not have $//$-edges, and $P^{\{//,[\ ],*\}}$ is the set of patterns that do not have $/$-edges. Let P, Q be patterns in $P^{\{/,//,*,[]\}}$. It is well known that, in general, $P \subseteq Q$ if but not only if there is a containment mapping from Q to P. However, if $P \in P^{\{/,[\ ],*\}}$ or $Q \in \widehat{P}^{\{/,//,[\ ],*\}}$, then $P \subseteq Q$ iff there is a containment mapping from Q to P [11,16].

2.3 Contained Rewritings of Tree Pattern Queries Using Views

A *view* is an existing pattern. Let V be a view and Q be a pattern. In this paper we will implicitly assume that $label(\mathsf{rt}(Q)) = label(\mathsf{rt}(V)) \neq *$. A *contained rewriting* of Q using V is a pattern Q' such that

(1) for any data tree t, $Q'(V(t)) \subseteq Q(t)$;
(2) there exists a data tree t such that $Q'(V(t)) \neq \emptyset$;

Fig. 1 shows a view V, a pattern Q, and two contained rewritings, Q' and Q'', of Q using V. The *maximal contained rewriting (MCR)* of Q using V, denoted $\mathsf{MCR}(Q, V)$, is defined to be the union of all contained rewritings of Q using V.

Let Q' be a contained rewriting of Q using V. It is clear that $label(\mathrm{rt}(Q')) = label(\mathrm{sn}(V))$ or at least one of $label(\mathrm{rt}(Q'))$ and $label(\mathrm{sn}(V))$ will be *. We use $Q' \circ V$ to represent the pattern obtained by merging $\mathrm{rt}(Q')$ and $\mathrm{sn}(V)$. The merged node will be labeled with $label(\mathrm{rt}(Q'))$ if $label(\mathrm{rt}(Q')) \neq *$, otherwise it will be labeled with $label(\mathrm{sn}(V))$. The selection node of $Q' \circ V$ is that of Q'. Note that condition (1) in the definition of contained rewriting is equivalent to $Q' \circ V \subseteq Q$.

Given pattern Q and view V in $P^{\{/,//,[]\}}$, [9] shows that the existence of a contained rewriting of Q using V can be characterized by the existence of a *useful embedding* from Q to V. In brief, a useful embedding from Q to V is a partial mapping f from $N(Q)$ to $N(V)$ that is root-preserving, label-preserving, structure-preserving as defined in a containment mapping (except that they are required only for the nodes of Q on which the function f is defined), and that satisfies the following conditions.

(a) If $f(x)$ is defined, then f is defined on $parent(x)$ (the parent of x).
(b) For every node x on the selection path $\mathrm{sp}(Q)$, if $f(x)$ is defined, then $f(x) \in \mathrm{sp}(V)$, and if $f(\mathrm{sn}(Q))$ is defined, then $f(\mathrm{sn}(Q)) = \mathrm{sn}(V)$.
(c) For every path $p \in Q$, *either* p is fully embedded, i.e., f is defined on every node in p, *or* if x is the last node in p such that $f(x)$ is defined and $f(x) \in \mathrm{sp}(P)$, and y the child of x on p (call y the *anchor node*), then either $f(x) = \mathrm{sn}(V)$, or the edge (x, y) is a //-edge.

Given a useful embedding f from Q to V, a *Clip-Away Tree (CAT)*, denoted CAT_f, can be constructed as follows. (i) Construct the root of CAT_f and label it with $label(\mathrm{sn}(V))$. (ii) For each path $p \in Q$ that is not fully embedded, find the anchor node y and attach the subtree of Q rooted at y (denoted Q_y) under $\mathrm{rt}(\mathrm{CAT}_f)$ by connecting $\mathrm{rt}(Q_y)$ and $\mathrm{rt}(\mathrm{CAT}_f)$ with the same type of edge as that between y and its parent. The selection node of CAT_f is the node corresponding to $\mathrm{sn}(Q)$. For example, in Fig. 1 the query Q has two useful embeddings to V, both of them map the a-node and c-node in Q to the a-node and c-node in V, and one also maps the b-node in Q to the b-node in V, but the other does not. The corresponding CATs are shown in Fig. 1 (c) and (d) respectively. Each CAT is a contained rewriting of Q using V, and every contained rewriting of Q using V is contained in one of these CATs. Thus if $h_1, \ldots, h_n$ are all of the useful embeddings from Q to V, then $\mathrm{MCR}(Q, V) = \bigcup_{i \in [1,n]} \mathrm{CAT}_{h_i}$.

3   MCR for Tree Patterns in $P^{\{/,//,*,[]\}}$

In this section, we consider rewriting tree pattern queries using views when both the view V and the query Q are in $P^{\{/,//,*,[]\}}$. We make a slight modification to the definition of contained rewriting: in addition to the conditions (1) and (2) in Section 2.3, we add another condition as follows. (3) If $label(\mathrm{sn}(V)) \neq *$, then $label(\mathrm{rt}(Q')) \neq *$. This modification does not change the essence of maximal contained rewriting because, if Q' is a contained rewriting of Q using V and $\mathrm{rt}(Q')$ is labeled *, then we can change the label of $\mathrm{rt}(Q')$ to $label(\mathrm{sn}(V))$ to obtain another rewriting Q'', and $Q' \circ V = Q'' \circ V$. In other words, $Q'(V(t)) = Q''(V(t))$. The only purpose of this modification is to make our presentation easier.

We start with some special cases for which a simple extension to the approach of [9] is sufficient, then we propose a method for the more general case. We also discuss methods to make the computation of MCRs more efficient. Finally we discuss optimization techniques which enable us to find the MCR more quickly.

3.1 Some Special Cases

First we note that in the special case that $V \in P^{\{/,//,*,[]\}}$ and $Q \in \widehat{P}^{\{/,//,[\,],*\}}$, to find the $\mathrm{MCR}(Q, V)$ we can find all useful embeddings from Q to V and union the corresponding CATs [15], where the useful embeddings are slightly modified from the original definition in Section 2.3. The modification is to relax the requirement of label-preservation to *weak label-preservation* as defined in Definition 1.

Definition 1. *[15] A mapping h from $N(Q)$ to $N(V)$ is* weak label-preserving *if (1) for every $v \in N(Q)$, either $label(v) = *$, or $label(v) = label(h(v))$, or $h(v) = \mathrm{sn}(V)$ and $label(\mathrm{sn}(V)) = *$, (2) all non-* nodes that are mapped to $\mathrm{sn}(V)$ have the identical label.*

Correspondingly, if $label(\mathrm{sn}(V)) = *$, and f is a useful embedding from Q to V which maps some node u in Q to $\mathrm{sn}(V)$, and $label(u) \neq *$, then the root of CAT_f will be labeled with $label(u)$.

The purpose of the above modification is to handle the case where the selection node of V is labeled *. Its necessity will be clear if one considers the example where $V = a//*$, and $Q = a//b//*$. In this example, $b//*$ is a contained rewriting, and it can only be found using the modified definition of useful embeddings and CATs.

In the following we will consider two more special cases, and we will use the modified definition of useful embeddings and CATs.

Special Case 1. We first consider the special case where Q has no /-edges. We can prove the following result.

Proposition 1. *Let P, Q be patterns in $P^{\{/,//,*,[]\}}$. If Q has no /-edges, then $P \subseteq Q$ iff there is a containment mapping from Q to P.*

Proof. The 'if' direction is obvious. The 'only if' direction can be proved as follows. Consider the data tree t obtained from P by changing every //-edge to a /-edge and relabeling every *-node with a symbol z that is not present in Q. Clearly P has a matching in t. Since $P \subseteq Q$, Q also has a matching in t, which maps $\mathrm{sn}(Q)$ to $\mathrm{sn}(P)$. Let h be a matching of Q in t. Since the label z is not present in Q, only *-node in Q can be mapped to z-nodes in t. Since Q has no /-edge, and there is a 1:1 correspondence between the z-nodes in t and *-nodes in P, this matching h forms a containment mapping from Q to P.

Therefore, in the above special case, if Q' is a contained rewriting of Q using V, then there is a containment mapping from Q to $Q' \circ V$. This containment mapping implies a useful embedding from Q to V, and the corresponding CAT contains Q'. Therefore, the maximal contained rewriting of Q using V can be found by finding all useful embeddings from Q to V and unioning the corresponding CATs.

Special Case 2. We now consider the special case where $V \in P^{\{/,[\,],*\}}$ and $Q \in P^{\{/,//,*,[\,]\}}$. Although in this case $V \subseteq Q$ iff there is a containment mapping from Q to V, generally there may not exist a containment mapping from Q to $Q' \circ V$ for a contained rewriting Q' of Q using V because Q' may be a pattern in $P^{\{/,//,*,[\,]\}}$. Thus it is not straightforward to see whether the approach of useful embeddings can be used in this case. Next we will show that the approach of useful embeddings can indeed be used to find the MCR in this case. To prove it, we need some terms and notation first.

Let Q be a tree pattern, and u, v be any two nodes in Q. A *-*chain* between u and v, denoted $u \rightsquigarrow v$, is a path from u to v that has only /-edges and only *-nodes between u and v. Note that u and v may be labeled with * or any tag in Σ. The *length* of the *-chain, denoted $|u \rightsquigarrow v|$, is the number of *-nodes between u and v (not including u and v), which is equivalent to the number of edges between u and v minus 1. Note that a /-edge from u to v is a special *-chain of length 0. The length of the longest *-chain in Q is called the *-*length* of Q, denoted L_Q^*:

$$L_Q^* = max\{|u \rightsquigarrow v| \mid u \rightsquigarrow v \text{ is a *-chain in } Q\}$$

For example, the *-length of the pattern Q in Fig. 2 (b) is 0. It is easy to see that if there are no *-chains in Q, then there are no /-edges in Q, and the rewriting problem has been solved in the special cases discussed in Special Case 1.

It was shown in [11] that the containment of tree patterns in $P^{\{/,//,*,[\,]\}}$ can be reduced to the containment of *Boolean patterns* which are tree patterns that do not have selection nodes, and which will return either TRUE or FALSE when evaluated over a data tree t, depending on whether there is a matching of the pattern in t. For Boolean patterns P and Q, $P \subseteq Q$ means $\forall t, P(t) \Rightarrow Q(t)$, and $(P \cup Q)(t) \equiv P(t) \vee Q(t)$. Suppose P is a Boolean pattern and there are K //-edges in P. We choose a sequence of K non-negative integers $\overline{w} = (w_1, \ldots, w_K)$ and construct a pattern from P, called an *extension* of P and denoted $P(\overline{w})$, by replacing the i-th //-edge with a *-chain of length w_i (for $i = 1, \ldots, K$). It is shown in [11] that the following lemma holds.

Lemma 1. *[11] Let e be a matching of P in t. There exists a unique extension $P(\overline{w})$ and a unique matching e' of $P(\overline{w})$ in t such that $\forall v \in N(P), e(v) = e'(v)$.*

Let P be a Boolean pattern. We will use $mod(P)$ to denote the set of data trees in which P has a matching, i.e., $mod(P) = \{t | P(t) = true\}$. Let z be a tag in Σ. We will use $mod_m^z(P)$ to denote the set of data trees obtained from P by first replacing the i-th //-edge (u, v) with a *-chain between u and v of length l_i, where $0 \le l_i \le m$, and then relabeling all *-nodes with z. For example, for the pattern P in Fig. 2 (a), if $m = 1$, then $mod_m^z(P)$ contains 8 data trees, each of them is obtained from P by replacing the three $//$-edges with a *-chain of either 0 or 1. Clearly $mod_m^z(P)$ is a **finite** subset of $mod(P)$.

The following result is also proven in [11].

Lemma 2. *[11] Let P, Q be Boolean patterns, z be a label not present in Q, and $w = L_Q^*$, i.e., w be the *-length of Q. If $mod_{w+1}^z(P) \subseteq mod(Q)$, then $P \subseteq Q$.*

The above lemma indicates that to test whether $P \subseteq Q$, we only need to verify that Q has a matching in each of the data trees in $mod_{w+1}^z(P)$.

With similar reasoning, we can prove the following extension to Lemma 2.

Lemma 3. *Let P, $P_1, \ldots, P_n$ be Boolean patterns in $P^{\{/,//,*,[]\}}$, and z be a label not present in $P_1, \ldots, P_n$. Let $w = max\{L_{P_1}^*, \ldots, L_{P_n}^*\}$. If $mod_{w+1}^z(P) \subseteq mod(P_1) \cup \cdots \cup mod(P_n)$, then $P \subseteq P_1 \cup \cdots \cup P_n$.*

Proof. The proof is similar to that of Proposition 3 of [11]. It is a proof by contradiction. The idea is to show that if $P \not\subseteq P_1 \cup \cdots \cup P_n$, then there is $t_1 \in mod_k^z(P)$ for some sufficiently large integer k, such that $P_i(t_1) =$FALSE, for all $i \in [1, n]$ (Step 1). From t_1, we can construct $t' \in mod_{w+1}^z(P)$ such that $P_i(t') =$FALSE for all $i \in [1, n]$, contradicting the assumption that $mod_{w+1}^z(P) \subseteq mod(P_1) \cup \cdots \cup mod(P_n)$ (Step 2).

Step 1: Suppose $P \not\subseteq P_1 \cup \cdots \cup P_n$, then there exists $t \in mod(P)$ such that $P_i(t)=$FALSE for all $i = 1, \ldots, n$. Since $P(t)$ is TRUE, there is a matching e of P in t. By Lemma 1, there exists an extension P^0 of P and there exists a matching e' of P^0 in t such that $e'(v) = e(v)$ for all $v \in N(P)$. We can now construct a data tree t_1 from P^0 by relabeling each *-node with a distinct new label z which is not present in $P_1, \ldots, P_n$. Clearly $P(t_1)$ is TRUE. We can show $P_i(t_1)$ is FALSE for all $i = 1, \ldots, n$. Indeed, suppose $P_i(t_1)$ were TRUE, then there is a matching e'' of P_i in t_1, thus we can construct a matching f of P_i in t, $f \equiv e''e'$ (note this is possible because the nodes in t_1 and P^0 are identical except the *-nodes in P^0 are relabeled with z, and that e'' cannot map any non-* node in P_i into a z-node in t_1). This contradicts the fact that $P_i(t)$ is FALSE.

Step 2: From t_1, we can construct a tree $t' \in mod_{w+1}^z(P)$ using Lemma 2 of [11], in exactly the same way as in the proof of Proposition 3 of [11], such that $P_i(t')$ is FALSE for all $i \in [1, n]$. The details are omitted since they are identical to that in [11].

Lemma 3 provides a way to check $P \subseteq \bigcup_{i=1}^n P_i$ for Boolean patterns. Due to the correspondence between tree pattern containment and Boolean pattern containment [11], we can use the similar way to check the containment of tree patterns in $P^{\{/,//,*,[]\}}$. In the following, we will use $mod_m(P)$ to denote the set of tree patterns obtained from P by replacing the i-th $//$-edge (u, v) with a *-chain between u and v of length l_i, where $0 \leq l_i \leq m$. In other words, $mod_m(P)$ consists of extensions of P where each $//$-edge is replaced with a *-chain of maximum length m. Note that the only difference between $mod_m(P)$ and $mod_m^z(P)$ is that the *-nodes in $mod_m(P)$ are relabeled with z in $mod_m^z(P)$.

Lemma 4. *Let $P, P_1, \ldots, P_n$ be tree patterns in $P^{\{/,//,*,[]\}}$. Let $w = max\{L^*_{P_1}, \ldots, L^*_{P_n}\}$. If every pattern in $mod_{w+1}(P)$ is contained in $P_1 \cup \cdots \cup P_n$, then $P \subseteq P_1 \cup \cdots \cup P_n$.*

Proof. Choose a label z that is not present in $P, P_1, \ldots, P_n$. Denote P^z the Boolean pattern obtained by adding a /-child node labeled z under the selection node of P. Similar to Proposition 1 of [11], we can prove that $P \subseteq P_1 \cup \cdots \cup P_n$ iff $P^z \subseteq P^z_1 \cup \cdots \cup P^z_n$ (the proof is very similar thus omitted). Now if every pattern Q in $mod_{w+1}(P)$ is contained in $P_1 \cup \cdots \cup P_n$, then Q^z is contained in $P^z_1 \cup \cdots \cup P^z_n$, that is, every pattern in $mod^z_{w+1}(P^z)$ is contained in $P^z_1 \cup \cdots \cup P^z_k$. Thus by Lemma 3, $P^z \subseteq P^z_1 \cup \cdots \cup P^z_k$. Hence $P \subseteq P_1 \cup \cdots \cup P_n$.

We can now prove the following theorem.

Theorem 1. *Let Q be a pattern in $P^{\{/,//,*,[]\}}$ and V be a pattern in $P^{\{/,[\ ],*\}}$. For every contained rewriting Q' of Q using V, there are useful embeddings $h_1, \ldots, h_k$ from Q to V such that $Q' \subseteq \mathtt{CAT}_{h_1} \cup \cdots \cup \mathtt{CAT}_{h_k}$.*

Proof. Let $w = L^*_Q$, and $mod_{w+1}(Q') = \{Q_1, \ldots, Q_n\}$. Let $h_1, \ldots, h_k$ from Q to V be all of the useful embeddings from Q to V. Recall that $Q_1, \ldots, Q_n$ are obtained from Q' by replacing each //-edge with a *-chain of length no more than $w + 1$. Clearly $label(\mathtt{rt}(Q_i)) = label(\mathtt{rt}(Q'))$. Since $Q' \circ V \subseteq Q$, and $Q_i \subseteq Q'$, we know $Q_i \circ V \subseteq Q$. Since $Q_i \circ V$ does not have //-edges, we know there is a containment mapping δ_i from Q to $Q_i \circ V$ (recall from Section 2). This containment mapping, if restricted to nodes of Q whose image under δ_i is in the V-part of $Q_i \circ V$, is a useful embedding from Q to V (Note that we are using the modified definition of useful embedding, so that any node that can be mapped into the merged node of $\mathtt{rt}(Q')$ and $\mathtt{sn}(V)$ by δ_i can be mapped into $\mathtt{sn}(V)$ by the useful embedding). Let us denote this useful embedding by h_i. The remaining nodes in Q (i.e., those nodes that are not mapped into the V-part of $Q' \circ V$ by δ_i) together with an artificial root node r constitute the corresponding clip-away tree $\mathtt{CAT}_{h_i}$. Note that $label(r)$ is either * or $label(\mathtt{rt}(Q'))$: if $label(\mathtt{sn}(V)) \neq *$, then by definition $label(r) = label(\mathtt{sn}(V)) = label(\mathtt{rt}(Q'))$ (note the condition (3) we added to the definition of contained rewriting); otherwise $label(\mathtt{sn}(V)) = *$, hence either $label(r) = *$ (when δ_i maps a *-node to the merged node of $\mathtt{sn}(V)$ and $\mathtt{rt}(Q')$) or $label(r) = label(\mathtt{rt}(Q'))$ (when $label(\mathtt{rt}(Q')) \neq *$, and δ_i maps a $label(\mathtt{rt}(Q'))$-node to the merged node of $\mathtt{sn}(V)$ and $\mathtt{rt}(Q').$) Therefore, there exists a containment mapping from $\mathtt{CAT}_{h_i}$ to Q_i. In effect we have proved that for every $Q_i \in mod_{w+1}(Q')$ there is a useful embedding h_i (from Q to V) such that $Q_i \subseteq \mathtt{CAT}_{h_i}$. Therefore, $Q_i \subseteq \mathtt{CAT}_{h_1} \cup \cdots \cup \mathtt{CAT}_{h_n}$. Note that the *-length of $\mathtt{CAT}_{h_i}$ is no more than w. Hence by Lemma 4, $Q' \subseteq \mathtt{CAT}_{h_1} \cup \cdots \cup \mathtt{CAT}_{h_n}$.

The above theorem implies that, if V has no //-edges, then to find the MCR of Q using V, we only need to find all useful embeddings from Q to V and then union the corresponding CATs.

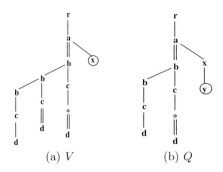

(a) V   (b) Q

Fig. 2. V, Q in Example 1

3.2   The General Case

In the general case where both Q and V may be any pattern in $P^{\{/,//,*,[]\}}$, the MCR of Q using V may not be found by finding all useful embeddings from Q to V. In other words, we can find some of the contained rewritings using the useful embeddings, but there is no guarantee we will find all of them. The following example demonstrates this point.

Example 1. Consider the view V, query Q and data tree t shown in Fig. 2, where V and Q are modified from the examples in [11]. It is easy to verify that x/y is a contained rewriting of Q using V, but this rewriting cannot be found using the approach of useful embeddings.

Next we provide a method to find $\mathtt{MCR}(Q,V)$ for the general case. We can prove the following theorem.

Theorem 2. *Let V,Q be patterns in $P^{\{/,//,*,[]\}}$. Let $w = L_Q^*$, i.e, w be the *-length of Q, and $mod_{w+1}(V) = \{V_1,\ldots,V_n\}$. Then Q' is a contained rewriting of Q using V iff it is a contained rewriting of Q using V_i for all $i = 1,\ldots,n$.*

Proof. (Only if) Let Q' be a contained rewriting of Q using V, that is, $Q' \circ V \subseteq Q$. Since $V_i \subseteq V$, we know $Q' \circ V_i \subseteq Q$ (for $i \in [1,n]$). Hence Q' is a contained rewriting of Q using V_i.

(if) Suppose Q' is a contained rewriting of Q using V_i for all $i = 1,\ldots,n$, then $Q' \circ V_i \subseteq Q$. Suppose Q_j is a pattern in $mod_{w+1}(Q')$. Then $Q_j \subseteq Q'$, thus $Q_j \circ V_i \subseteq Q$. It is easy to verify that $mod_{w+1}(Q' \circ V) = \{Q_j \circ V_i \mid Q_j \in mod_{w+1}(Q'), V_i \in mod_{w+1}(V)\}$. Therefore, by Lemma 4, $Q' \circ V \subseteq Q$. That is, Q' is a contained rewriting of Q using V.

The following corollary follows immediately from Theorem 2.

Corollary 1. *Let V,Q be tree patterns in $P^{\{/,//,*,[]\}}$, $w = L_Q^*$ (i.e., w be the *-length of Q), and $mod_{w+1}(V) = \{V_1,\ldots,V_n\}$. Then*

$$\mathtt{MCR}(Q,V) = \mathtt{MCR}(Q,V_1) \cap \cdots \cap \mathtt{MCR}(Q,V_n).$$

Theorem 1 and Corollary 1 together imply a method for finding $\mathtt{MCR}(Q, V)$ for the general case, which consists of the following steps:

1. Find the set $mod_{L_Q^*+1}(V)$ of patterns (call them *sub-views* of V).
2. Find the MCR of Q using each of the sub-views, using the approach of (modified) useful embeddings.
3. Intersect these MCRs to obtain $\mathtt{MCR}(Q, V)$, the MCR of Q using V.

As an example, one can easily verify that the rewriting x/y in Example 1 can be found by first splitting the view into eight sub-views with only $/$-edges, then finding the contained rewriting x/y of Q using each of the sub-views, and then intersecting the eight identical contained rewritings.

Complexity. The above method for finding the MCR of Q using V requires us to find the MCRs of Q using each sub-view in $mod_{L_Q^*+1}(V)$. We do not need to actually do the intersection of the MCRs, i.e., to transform $\mathtt{MCR}(Q, V_1) \cap \cdots \cap \mathtt{MCR}(Q, V_n)$ into a union of tree patterns. Instead, we can take the intersection $\mathtt{MCR}(Q, V_1)(V(t)) \cap \cdots \cap \mathtt{MCR}(Q, V_n)(V(t))$ when evaluating it over $V(t)$. The cost for finding each of $\mathtt{MCR}(Q, V_i)$ has been discussed in [9]. The cost of finding $\mathtt{MCR}(Q, V)$ would be the total cost of finding all of $\mathtt{MCR}(Q, V_1), \ldots, \mathtt{MCR}(Q, V_n)$ (if we ignore the cost to find $mod_{L_Q^*+1}(V)$).

3.3   An Improvement

Consider the view V and query Q in Example 1 again. Using the approach discussed above, we need to split the view into eight (2^3) sub-views and then find the MCR using each of them. However, as we will show shortly, we do not really need all eight sub-views, we only need to replace the edge $c//d$ in V with a *-chain of 0 or 1 to obtain two sub-views, and then find the contained rewriting x/y using each of these two views. Generally, we only need to consider a subset of the sub-views in $mod_{w+1}(V)$. We will discuss how to find such a subset next.

Let $x, y \in \Sigma \cup \{*\}$ be two labels. Given pattern $Q \in P^{\{/,//,*,[]\}}$, we define $Q(x, y)$ to be the set of *-chains from an x-node or *-node to a y-node or *-node:

$$Q(x, y) = \{u \rightsquigarrow v \in Q \mid label(u) \in \{x, *\}, label(v) \in \{y, *\}\}$$

Define the function $L_Q(x, y)$ as follows:

$$L_Q(x, y) = \begin{cases} -1 & \text{if } Q(x, y) = \emptyset; \\ max\{|u \rightsquigarrow v| \mid u \rightsquigarrow v \in Q(x, y)\} & \text{otherwise.} \end{cases}$$

For example, for the pattern Q in Fig. 2, $L_Q(a, b) = -1$, $L_Q(c, d) = 0$, and $L_Q(*, d) = -1$.

We can prove the following extension to Theorem 2.

Theorem 3. *Let V, Q be patterns in $P^{\{/,//,*,[]\}}$. Let $\mathtt{Mod}_Q(V)$ be the set of patterns obtained from V by replacing each $//$-edge $e = (u, v)$ with a *-chain $u \rightsquigarrow v$ of length l_e, where l_e is an integer in $[0, L_Q(label(u), label(v)) + 1]$. Suppose $\mathtt{Mod}_Q(V) = \{V_1, \ldots, V_n\}$. Then Q' is a contained rewriting of Q using V if and only if it is a contained rewriting of Q using V_i for all $i = 1, \ldots, n$.*

The proof of the above theorem can be done in a similar way to that of Theorem 2, using the following result on tree pattern query containment:

Theorem 4. *Let* $P, P_1, \ldots, P_n$ *be patterns in* $P^{\{/,//,*,[]\}}$. *For any pair of labels* x, y, *define*

$$w(x, y) = max\{L_{P_1}(x, y), \ldots, L_{P_n}(x, y)\}.$$

Let $\text{Mod}(P)$ *be the set of patterns obtained from* P *by replacing each* //-*edge* $e = (u, v)$ *with a* *-chain $u \rightsquigarrow v$ *of length* w_e, *where* $0 \leq w_e \leq w(label(u), label(v)) + 1$. *If every pattern in* $\text{Mod}(P)$ *is contained in* $P_1 \cup \cdots \cup P_n$, *then* $P \subseteq P_1 \cup \cdots \cup P_n$.

The proof of Theorem 4 is omitted due to page limit.

With Theorem 3, we can replace the set $mod_{L_Q^*+1}(V)$ with the set $\text{Mod}_Q(V)$ in our method for finding the MCR of Q using V without affecting its correctness. It is easy to see that for any pair of labels x, y, $L_Q(x, y)$ is less or equal to the *-length L_Q^* of Q, hence $\text{Mod}_Q(V) \subseteq mod_{L_Q^*+1}(V)$.

As an example, consider the query Q and view V in Example 1 again. Since $L_Q(a, v) = L_Q(*, d) = -1$, the //-edges (a, b) and $(*, d)$ in V may each be replaced with a /-edge, and the //-edge (c, d) may be replaced with a *-chain of 0 or 1, resulting only two sub-views in $\text{Mod}_Q(V)$.

Discussion. If $\text{Mod}_Q(V)$ is large, then finding the MCR of Q using the sub-views in $\text{Mod}_Q(V)$ one by one using the approach of useful embeddings and then intersect the MCRs can still be very expensive. Since the sub-views in $\text{Mod}_Q(V)$ are structurally similar to each other, we may design a more efficient algorithm that can find all useful embeddings of Q in all sub-views all at once, and thus saving time in finding $\text{MCR}(Q, V)$. Developing such an algorithm is our future work.

3.4   Minimizing the Size of $\text{Mod}_Q(V)$

Obviously the cost to find $\text{MCR}(Q, V)$ in the general case depends on the size of $\text{Mod}_Q(V)$. Obviously the size $\text{Mod}_Q(V)$ depends on the number of //-edges in V, M, and lengths of *-chains in Q. To make $\text{Mod}_Q(V)$ smaller we should try to minimize both M and the lengths of *-chains. To minimize M we can use tree pattern minimization techniques [2] to remove redundant //-edges. To minimize the length of *-chains in Q we can *normalize* Q into an equivalent pattern with fewer /-edges. The normalization of Q can be done as follows:

(A) Any leaf *-node, if it is not the selection node, can be reconnected to its parent with a //-edge. For example, $a[*/*]/b$ can be changed to $a[*//*]/b$.

(B) If there is a path $v_0, v_1, \ldots, v_n, v_{n+1}$ such that
  (1) $v_1, \ldots, v_n$ are labeled * and they are not the selection node
  (2) v_{i+1} is the unique child of v_i for $i = 1, \ldots, n$, and
  (3) one of the edges in the path is a //-edge,
  then we change all edges on the path to //-edges. For instance, the pattern $a[*//*]/b$ can be normalized into $a[// * //*]/b$, and the pattern $a/ * // * /b$ can be normalized to $a// * // * //b$. However, $a/*[// * /b]$ cannot be normalized into $a// * [// * //b]$, nor can the pattern $a/ * [c]//b$ be normalized into $a// * [c]//b$.

4   Related Work

Over the last few years there have been a significant number of papers published on rewriting tree pattern queries using views. In this section we only compare with a few of them which are the mostly closely related to this work. As mentioned in Section 1, apparently [9] is the first paper on MCR of tree pattern queries using views, and it proposed the technique of useful embeddings for queries and views in $P^{\{/,//,[]\}}$. The paper also proposed an algorithm to find the MCR under a non-recursive and non-disjunctive DTD which can be represented as an acyclic graph. The basic idea is to reduce the original problem into one without DTDs by chasing the tree patterns repeatedly using constraints that can be derived from the DTD. Recently, [12] proposed a method for finding contained rewritings for queries and views in $P^{\{/,//,*,[]\}}$, based on the concepts of *trap embedding* (and its *induced pattern*) and *trap relay*, where a trap embedding is a mapping a tree pattern to a tree, and a trap relay is a mapping from a pattern to another pattern. It can be shown that the induced pattern of a trap embedding from Q to a pattern V_i in $mod_{L_Q^*+1}(V)$ is the same as a CAT if the *attach point* (see [12]) of V_i is $\text{sn}(V_i)$. Thus Theorem 3.3 of [12] and Theorem 2 of this paper are mutually *derivable* from each other, although they look quite different and are proved independently using different techniques. The authors of [12] did not consider how to find the MCR in the general case. Instead, they gave a method for the special case where the query Q has no *-nodes connected to a //-edge and no leaf node u such that u and the parent of u are both labeled *, which is an improvement to a result in [15]. In addition, [14] studied XPath rewriting using multiple views for queries and views in $P^{\{/,//,[]\}}$ and gave algorithms for finding MCRs using an intersection of views, and [13] gave an algorithm for identifying redundant contained rewritings. There have also been works on *equivalent rewritings (ER)*, where an ER is a special contained rewriting Q' which satisfies the condition $Q' \circ V = Q$. Among them [18] showed that for V and Q in $P^{\{/,//,[]\}}$, $P^{\{/,[\ ],*\}}$ and normalized $P^{\{/,//,*\}}$, there is an ER of Q using V iff Q_k is an ER, where k is the position of $\text{sn}(V)$ on the selection path of V, and Q_k is the subtree of Q rooted at the k-th node on the selection path of Q. [1] extended the above result to $Q, V \in P^{\{/,//,*,[]\}}$ and showed that for many common special cases, there is an ER of Q using V iff either Q_k or $Q_{k//}$ is an ER, where $Q_{k//}$ is the pattern obtained from Q_k by changing all edges connected to the root to //-edges. Thus in those special cases, to find an ER we only need to test whether Q_k or $Q_{k//}$ is an ER. Since an ER is also a contained rewriting, it follows that any ER is contained in the MCR. It is interesting to compare the ER found above (i.e., $Q_{k//}$ or Q_k) with the MCR found using our method. For instance, consider the example view and query in [1]: $V = a[//e]/*[//d]$ and $Q = a//*/b[e]/d$. Using the techniques in [1] we find $*//b[e]/d$ to be an ER. Using our method we find $\text{MCR}(Q, V_1)$ to be the union of $*//*/b[e]/d$ and $*/b[e]/d$, which is equivalent to $*//b[e]/d$. [10] assumed both Q and V are minimized, and reduced tree pattern matching to string matching, allowing a more efficient algorithm for finding equivalent rewritings. They further proposed a way to organize the materialized views in Cache to enable efficient view selection and cache look-up. [6] investigated equivalent rewritings using a single view or multiple views for queries and views in $P^{\{/,//,[]\}}$, where the set of views is represented by grammar-like rules called *query set specifications*. The work can be seen as an extension to an earlier work [5] which investigated the same problem for a set of explicit views.

There are also works on answering twig pattern queries using views, where a twig pattern has /-edges and //-edges, but no *-nodes. In particular, [3] studied answering twig patterns using multiple views in the presence of *structural summaries* and integrity constraints, where the answers to the new query are obtained by combining answers to the views through a number of algebraic operations. [17] investigated the problem of answering a twig pattern using a materialized view, where for each node v in the view, a list of data tree nodes (which correspond to the matching nodes of v) are materialized. The authors proposed stack-based algorithms based on holistic twig-join [4] for the computation of answers to the new query. [7] proposed a storage structure for the materialized view answers, which enables an efficient join algorithms for the computation of answers to the new twig query.

5   Conclusion

We showed that the MCR of a tree pattern query Q using a view V can be found by finding all useful embeddings and the corresponding CATs in the special cases where Q has no /-edges or V has no //-edges. We also gave a method for finding the MCR in the general case where both Q and V may be any pattern in $P^{\{/,//,*,[]\}}$, which is to divide view V into a finite set of sub-views (according to Q) and then finding the MCRs using the sub-views. We showed how the number of sub-views we have to consider can be reduced.

The method for the general case can still be expensive, however. It is worthwhile to find more efficient algorithms.

Acknowledgement. This work is supported by the Australian Research Council Discovery Grant DP1093404.

References

1. Afrati, F.N., Chirkova, R., Gergatsoulis, M., Kimelfeld, B., Pavlaki, V., Sagiv, Y.: On rewriting XPath queries using views. In: EDBT (2009)
2. Amer-Yahia, S., Cho, S., Lakshmanan, L.V.S., Srivastava, D.: Minimization of tree pattern queries. In: SIGMOD (2001)
3. Arion, A., Benzaken, V., Manolescu, I., Papakonstantinou, Y.: Structured materialized views for XML queries. In: VLDB (2007)
4. Bruno, N., Koudas, N., Srivastava, D.: Holistic twig joins: optimal XML pattern matching. In: SIGMOD Conference, pp. 310–321 (2002)
5. Cautis, B., Deutsch, A., Onose, N.: XPath Rewriting using multiple views: Achieving completeness and efficiency. In: WebDB (2008)

6. Cautis, B., Deutsch, A., Onose, N., Vassalos, V.: Efficient rewriting of XPath queries using query set specifications. PVLDB 2(1), 301–312 (2009)
7. Chen, D., Chan, C.-Y.: Viewjoin: Efficient view-based evaluation of tree pattern queries. In: ICDE, pp. 816–827 (2010)
8. Halevy, A.Y.: Answering queries using views: A survey. VLDB J. 10(4) (2001)
9. Lakshmanan, L.V.S., Wang, H., Zhao, Z.J.: Answering tree pattern queries using views. In: VLDB (2006)
10. Mandhani, B., Suciu, D.: Query caching and view selection for XML databases. In: VLDB (2005)
11. Miklau, G., Suciu, D.: Containment and equivalence for a fragment of XPath. J. ACM 51(1) (2004)
12. Tang, J., Fu, A.W.-C.: Query rewritings using views for XPath queries, framework, and methodologies. Inf. Syst. 35(3), 315–334 (2010)
13. Wang, J., Wang, K., Li, J.: Finding irredundant contained rewritings of tree pattern queries using views. In: Li, Q., Feng, L., Pei, J., Wang, S.X., Zhou, X., Zhu, Q.-M. (eds.) APWeb/WAIM 2009. LNCS, vol. 5446, pp. 113–125. Springer, Heidelberg (2009)
14. Wang, J., Yu, J.X.: XPath rewriting using multiple views. In: Bhowmick, S.S., Küng, J., Wagner, R. (eds.) DEXA 2008. LNCS, vol. 5181, pp. 493–507. Springer, Heidelberg (2008)
15. Wang, J., Yu, J.X., Liu, C.: Contained rewritings of XPath queries using views revisited. In: Bailey, J., Maier, D., Schewe, K.-D., Thalheim, B., Wang, X.S. (eds.) WISE 2008. LNCS, vol. 5175, pp. 410–425. Springer, Heidelberg (2008)
16. Wang, J., Yu, J.X., Liu, C.: Independence of containing patterns property and its application in tree pattern query rewriting using views. World Wide Web 12(1), 87–105 (2009)
17. Wu, X., Theodoratos, D., Wang, W.H.: Answering XML queries using materialized views revisited. In: CIKM, pp. 475–484 (2009)
18. Xu, W., Özsoyoglu, Z.M.: Rewriting XPath queries using materialized views. In: VLDB (2005)

Implementing Automatic Error Recovery Support for Rich Web Clients

Manuel Quintela-Pumares, Daniel Fernández-Lanvin,
Raúl Izquierdo, and Alberto-Manuel Fernández-Álvarez

University of Oviedo, Computer Sciences Department, c/Calvo Sotelo s/n Oviedo, Spain
{UO156108,dflanvin,raul,alb}@uniovi.es

Abstract. The way developers usually implement recoverability in object oriented applications is by delegating the backward error recovery logic to the ever-present database transactions, discarding the in-memory object graphwhen something goes wrong and reconstructing its previous version from the repository. This is not elegant from the point of view of design,but a cheap and efficient way to recover the system from an error. In some architectures like RIA, the domain logic is managed in the client without that resource, and the error prone and complex recoverability logic must be implemented manually, leading to a tangled and obfuscated code. An automatic recovery mechanism is adapted to that architecture by means of a JavaScript implementation. We developed several benchmarks representing common scenarios to measure the benefits and costs of this approach, evidencing the feasibility of the automatic recovery logic but an unexpected overhead of the chosen implementation of AOP for JavaScript.

Keywords: Recoverability, error recovery, exception, JavaScript, RIA, AOP.

1 Introduction

The architecture of web applications has been continuously evolving since the creation of the primitive transactional script based systems. The following architectural models concentrated the whole domain logic on the server side, while the web client was reduced to some simple presentation logic that was usually implemented using a scripting language. These are called *lightweight clients*and were specially justified when the customers access the application using the primitive low-band connections. However, the last evolution in web architecture has changed thiswith the advent of the*Rich Internet Applications*. Due to the popularization of Ajax and other similar technologies, the way we manage the information during a session has changed.In this new scenario, the user can get from the server not just the required data, but also the interactive logic that allows its full manipulation without communicating with the server except when needed. Some well-known web applications like Google Docs or Google Calendar are good examples of this approach.The RIA approach involves many points that we must consider when we design an application. The most important one is the fact that what we move to the web client are not simple validation rules and presentation logic anymore, but part of

L. Chen, P. Triantafillou, and T. Suel (Eds.): WISE 2010, LNCS 6488, pp. 630–638, 2010.

the domain logic and data that determines the consistency of the object model. This is a closer approach to the classical client-server architectural model. The customer gets a copy of part of the domain model[1](data and business logic), interacts with that information changing it locally, and once he/she finishes, the new state of that model is sent to the server in order to update its version with the new state generated in the web client. That process involves high complexity rules that are difficult to implement and maintain, especially if that must be done with not strongly-typed languages like JavaScript.

Facing the Problem: What Happens with The Model if Something goes Wrong?

Managing the object model involves some serious difficulties we must deal with wherever we do it, either in the web client or in the server. Changes in the state can generate transitional inconsistencies during an operation that, in case it was interrupted, would lead the system to an inconsistent state and, therefore, generate a failure. The most popular way to solve that problem is to recover the last consistent state of the model, a kind of recoverability. We can provide recoverability by programming it by ourselves or by delegating to any external tool or library. In fact, there were several approaches and solutions to face it in different ways and for different contexts: based in new languages (or extensions of existing ones) [1][2][3]; based on specific-purpose design patterns [4][5][6] ; or based in in-memory transactions [7][8]. All these solutions solve at least part the problem (with their own benefits and drawbacks), but the fact is that none of them are usually applied when developing common information systems. Why? The way developers usually implement recoverability in object oriented applications is by delegating the backward error recovery logic to the ever-present database. Transactions guarantee the consistency of the database, and on the other side, most of the applications are strongly database oriented, updating any change occurred in the model in order to preserve the information in case of a system crash. In that scenario, the easiest way to return to the previous consistent state when something wrong happens is just discarding the in-memory object graph and reconstructing its previous version from the information stored in the database. Even when this is not fair neither elegant from the point of view of object-oriented design, it is the cheapest and more efficient way to recover the system from an error. But… what happens when we do not have the database so "close" to the domain logic? In RIA architecture, part of the model is being moved to the web client, and several changes can be done on it before it's updated back in the server. So, there is no updated version of the previous state of the model available for each operation anymore. Once we cannot delegate recoverability to the database, the developer has to implement it manually in the web client. Otherwise the user will lose the work done from the beginning of the interaction until the failing operation crashes because the only checkpoint of the model is the one stored in the server. Implementing recoverability logic is a complex, tedious and error prone task, especially if that must be done using a not strongly-typed language like JavaScript. Furthermore, recoverability logic tends to result tangled with the business

---

[1] In this document we consider the "model" as the object graph that represents the domain state. Parts of this graph are transferred to the RIA via web for their local update. Then, this modified sub-graph is sent back to the server.

logic, obfuscating even more the code and making maintenance more expensive. Unluckily, none of the existing studied solutions for recoverability are applicable to rich Internet clients. An automatic error recovery support would simplify the task of developing web clients, making code simpler and easier to maintain.

Objective

From our point of view, it is possible to provideto the technologies usually applied in the construction of rich web clients an automatic error recovery mechanism that:(i)Would save the developer from the responsibility of implementing and maintaining this error recovery logic that can be managed automatically by the system; (ii)Could be applied to the platforms that are used in the industry, respecting their specification; (iii)Would not spoil the user experience due to unfeasible CPU or memory overheads; (iv)Would avoid the tangling of the hand-made recoverability logic.In this paper we propose and evaluate an implementation of an automatic error-recovery support for RIA clients to help the developer to implement consistent and feasible web clients.

2 Our Approach

Recoverability in object oriented programming has been faced before by our team. In [9] a new semantic construction called *Reconstructor* is proposed and integrated with OO languages using attribute oriented programming in Java. An evolution of this implementation that reduced the recoverability logic tangling by means of Aspect Oriented Programming (AOP) is now under development and evaluation.Basically, the proposal is based on the fact that most of the information needed to recover the system from the occurrence of an exception can be automatically managed by supporting processes, leading to the developer just the task of determining the scope of the recovery. We think that the same mechanism can be applied to rich Internet clients, providing support for recoverability to those scenarios in which the developer is forced to implement it manually in order to avoid the losing of user's work in case of abnormal termination.None of platforms that cover the development of rich web clients support automatic error recovery, and given that JavaScript is supported by all modern web browsers, this work will focus on implementing the Reconstructors based mechanism on this language.

2.1 Reconstructors on JavaScript

A *Reconstructor* is *an element specialized in restoring the consistency of a specific part of the model.* Initially, while a *constructor* must initialize a component to its initial state, the *reconstructor* must restore it to a consistent state. As with Hansel and Gretel's bread pieces, each reconstructor can disallow one specific change that has happened during an operation execution. We consider that any action (internal change of the model or external effect) whose effect can be disallowed is a *reconstructable action.* So, a **reconstructor will be created each time a reconstructable action is executed.**As they are created, the reconstructors are stacked in the *reconstruction path*, a sequence of reconstructors that stores the disallowing actions that the system needs to restore the consistency. Reconstruction paths are organized in *reconstructioncontexts*. Once a *reconstructable operation* is started, a new context is created, becoming the *current context*. Each time a reconstructor is generated, it is added to the *current context*, and thus each context knows exactly the way it can restore the state of the system just to the point it was at before the current operation started. There are two ways a reconstruction context can be closed: by *reconstructing* it (*reconstruct* command) or *discarding* it (*discard* command). Both operations are equivalent to the *commit* and *abort* operations in transactions. The reconstruct command should be invoked when something incorrect happens during the execution. When an error arises, the system should disallow all the reconstructable actions that were executed since the beginning of the current operation. The way to do that is by running the reconstruction path in the reverse order from that in which it was created, and executing the disallowing action of each reconstructor it finds there. On the other hand, the *discard* command assumes that everything was right during the execution of the current context and, consequently, the reconstructors can be discarded and the context destroyed. The developer has to determine which elements of the model are reconstructable. In fact, this is the keystone of our approach, given that we give the developer the opportunity to specify which the important objects in the model are. A developer can declare*Reconstructable attributes, compensable methods,* and *reconstruction contexts.* In the Java implementation, this was made by means of annotations. In JavaScript, a slightly different approach is used in order to respect the different features of this language. The code decoration technique used in the Java implementation has been replaced by a method API that the developer can use to identify the reconstructable elements of the system. These declarations are all the developer has to do in order to allow his system to recover from errors.This does not replace the try/catch mechanism, but complements it. Thrown exceptions still should be handled, but the developer won't need to implement clean up actions in the catch block [23].An object can have different reconstructable characteristics. For each one of these different characteristics, there is a method in the API that, among other parameters, receives a reference to an object, adding that concrete characteristic to it. As JavaScript is a dynamic language, this is done by adding to that object a new property with the information of the applied reconstructable characteristic.In addition, the API allows the developer to define more advanced options over these reconstructable characteristics that are oriented to control even more the scope of the ever expensive automatic recovery system in order to reach feasible rates of over heading.The execution of certain actions involving those objects signaled by the developer (modifications of some of their properties, execution of some of their methods, …) will fire automatic actions such as the creation of the appropriate reconstructors, the managing of the contexts, or the activation of the reconstruction process in case of error.Fig.1shows an example of an API method. The properties 'name' and 'age' of the 'myPerson' object will go back to the values they had before that operation started.

```
reconstructable.object.theseProperties(myPerson,['name', 'age']);
```

Fig. 1. Using the API to define a reconstructable attribute

The next question is how is this information that the API adds to the objects used to recover the system? Our approach is the use of AOP.

Using AOP for Automating Actions

Several previous works [10],[11] suggests that the exceptional behaviour of a system is a cross-cutting concern that can be better modularized using the AOP approach.As error recovery logic is completely independent of the domain model, its tangling with the domain specific logic does not provide any useful information for the developer, but obfuscates the code. Consequently, we consider that this concern is a perfect candidate to be "aspectized".The information that the API methods add to the objects is used in combination with AOP to discriminate during runtime when and how automatic actions are needed.

Several lightweight AOP implementations for JavaScriptlike Ajaxpect or AspectJS among others were evaluated, but the main drawback with these kind of implementations is that they only allow to attach advices to functions, when we also need attaching them to operations like assignations. A more complex AspectScript implementation for JavaScript covered all our needs [20]. Its approach allows attaching extra logic to certain actions. Using AspectScript it is possible to add the proper automatic actions depending on how the developer has declared the way the elements of the system should be recoverable.

3 Evaluation

The solution has been applied to the construction of the prototypes of the SHUBAI project [22]. They are based on augmented accessibility for handicapped users, some of them by means of rich clients where our implementation has been tested. The architecture applied in SHUBAI (Nikko) is based on remote smart sensors in mobile devices that are plugged to a communication bus. Communication errors are frequent in thisarchitecture, leading the device to an inconsistent state, a perfect scenario to apply the solution.On the other hand, we developed some very simple scripts that could show the functionality and purpose of the reconstructors in common scenarios.In the first scenario we implemented a script for sorting a table by any column selected by the user. Without implementing the appropriate clean up actions, if there was an error during the ordering of the elements, the data shown to the user would lose its consistency.The second scenario consists on making several requests to the Google translation API for translating a page. If any of them was not successful, the page would resultonly partially translated.In both cases, if the action cannot be fully completed, the system should go back to its last consistent state. These both examples will be used to evaluate different aspects of our implementation[2].

---

[2] The experiments described in the following sections were run on an Intel core 2 quad q6600 2,40GHz with 4 GB RAM, running Windows Vista SP 2, using the Google chrome (5.0.375.70) browser.

3.1 Code Reduction

For the first example, we developed various implementations with two different ordering algorithms, bubble sort and quick sort. For each of them, we developed two kinds of error recovery behavior, having scripts with different degrees of complexity in their business and error recovery logic, and each one of them will have two versions for taking the clean up actions, one with reconstructors and the other with design patterns based handmade implementation.We applied eLOC to measure the code dedicated to clean up actions in each version[3].

Version	Handmade	Reconstructors
Bubble sort without retry	6,53%	2,05%
Bubble sort with retry	10,62%	2,05%
Quick sort without retry	8,67%	1,85%
Quick sort with retry	26,85%	1,85%
Translator	13,21%	1,30%

Fig. 2. % of eLOC used for recoverabilityComparative of results

Fig. 2 shows that the handmade versionsneeded between 3.18 and 14.51 times more code dedicated to the clean up actions than the versions that used reconstructors, and the average increaseinthehandmade versions is 7.54 times.

3.2 Execution Time and Memory Usage

In order to measure the overhead, we took the bubble sort and quicksort examples with retry functionality. The scenario of the test is a table of 20 elements, ordered in the worst case for each of the algorithms. After ordering the table, an artificial exception is thrown, so the original order is reestablished and the algorithm is retried, repeating this process 50 times. Each experiment is executed 10 times for obtaining the average values.For each of the algorithms, 4 different versions were made. Three of them are the same handmade solution but running under different circumstances: (i) using plain JavaScript (*hand-plain*); (ii) with the transformed code needed to run AspectScript but without using it, not weaving any aspect (*hand-as-off*); (iii) with our complete solution working and ready to be used, but not using it(*hand-rec-off*). And finally, (iv) the fourth version is the one using reconstructors (*rec*). For all these cases, we measured the execution time, and the memory usage using the Google chrome task manager.

Just the hand-as-off version is 159,5 and 337,72 times slower than the hand-plain version respectively, being that the overhead produced only by code transformation of AspectScript (Fig.3). The hand-rec-off and rec versions are between 16.27 and 21.99 times slower than the hand-as-off version. In the bubble sort algorithm the rec version

---

[3] The eLOC(*effective Lines Of Code*) metric is widely discussed as a programming effort metric [24], in our scenario, however, similar pieces of code with commonpurposes are being compared, so we considered it as a valid metric tomeasure the work that the reconstructor based solution is doing onbehalf of the developer.

is a 15.11% slower than the hand-rec-off, and in the quick sort case, its 0.82% slower.The increase of memory usage between the hand-plain and hand-as-off versions is around the 72% in both cases, and the rec and hand-rec-off versions increased the memory usage between 10.29 and 11.14 times from the hand-plain version (Fig.3).In the bubble sort algorithm, the hand-off-version takes a 0.4% more memory than the rec version. In the quick sort algorithm the rec version takes a 4.05% more memory than the hand-off-version.

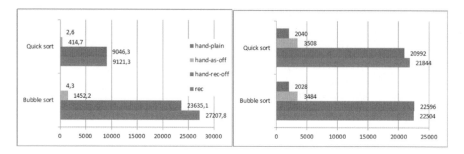

Fig. 3. Execution time (ms) and memory (KB) usage results

3.3 Interpreting the Results

The first thing to notice is the great overhead in both memory usage and execution time that the use of AspectScript implies even when there are not aspects defined, growing even more when the aspects needed for using reconstructors are defined, even if they are never executed.However, once the needed infrastructure is working, the overhead that the reconstructors involve in comparison with the handmade solution is feasible. These results are similar to the ones we observed in [9] with the Java implementation of reconstructors.As we expected, the use of reconstructors can reduce the amount of code used to develop a robust system. The different benchmarks developed show that as they complexity in both business logic and error recovery is increased, the percentage of handmade code needed for recovering from errors grows. Using reconstructors the percentage of error recovery code is greatlyreduced, and its size keeps maintained even when the complexity of the scenario grows. We think that this leads to a cleaner and simpler code that should make easier the development and maintenance of robust web clients.

4   Conclusions and Future Work

The results we obtainedevidence that this implementation of the reconstructors for the JavaScript language is far from being the best, mainly because of our use of AspectScript for weaving aspects.AspectScript helped us to solve quickly the integration problem without the need of implementing our own solution, allowing us to center on finding out the best way for adapting the reconstructor concept to the concrete characteristics of the JavaScript language.If the overhead produced by our use of AspectScript could be enclosed, the implementationcould beefficient enough to be applied in real projects.Different AOP implementations are now under evaluationin order to obtain a more efficient alternative than AspectScript. These approaches imply the use of lightweight AOP implementations for weaving aspects on methods. Because of the limitations of these alternatives, techniques for complementing them are being evaluated, including the use of some of the characteristics recently added to the 5th edition of the ECMAScript standard, or preprocessing the code, but only transforming certainoperations into functions in order to be able to weave aspects on them.Anyway, these other more efficient implementations would have little impact in the present architecture, since only the AspectScript related code should be substituted.

Acknowledgements

The SHUBAI Project: Augmented Accessibility for Handicapped Users inAmbient Intelligence and in Urban computing environments(TIN2009-12132) is developed thanks to the support of the MCYT(Spanish Ministry of Science and Technology).

References

1. Randerll, B., Campbell, R.H., Anderson, T.: Practical Fault Tolerance Software for Asynchronous Systems. IPAC Safecom, Cambridge (1983)
2. Oki, B.M.: Reliable object storage to support atomic actions. MSc thesis,: MIT Dept. EE and CS (May 1983)
3. Cristian, F.: A recovery mechanism for modular software. In: Proc. 4th Int. Conf. Software Engineering, pp. 42–50. IEEE Press, Piscataway (1979)
4. Tikhomirova, N.V., Shturtz, I.V., Romanovsky, A.: Object-oriented approach to state restoration by reversion in fault tolerant systems, Computing Science. University of Newcastle upon Tyne, Newcastle (1997)
5. Silva, A., Pereira, J., Marques, J.: Customizable Object Recovery Pattern. Pattern Languages of Program Design. Addison-Wesley, Reading (1997)
6. Blanck Lisboa, M.L., Fernandes, A.P.: Reflective implementation of an object recovery design pattern. In: VII CongressoArgentino de Ciencias de la Computación (2001)
7. Plank, J.S., et al.: Memory Exclusion: Optimizing the Performance of Checkpointing Systems. Tennessee, USA: University of Tennessee Technical Report UT-CS-96-335 (August 1996)
8. Garthwaite, A., Nettles, S.: Transactions for Java. In: Int. Conf. Computer Languages, pp. 16–27 (1998)
9. Fernández Lanvin, D., et al.: Extending Object-Oriented Languages with Backward Error Recovery Integrated Support. Computer Languages, Systems and Structures. Elsevier, Amsterdam (July 2010)
10. Kiczalez, G., et al.: Aspect-oriented programming. In: Liu, Y., Auletta, V., et al. (eds.) ECOOP 1997. LNCS, vol. 1241, Springer, Heidelberg (1997)
11. Lipper, M., Videira Lopes, C.: A study on exception detection and handling using aspect-oriented programming. In: Proc. 22nd International Conference on Software Engineering, pp. 418–427. ACM Press, New York (2000)

12. Toledo, R., Leger, P., Tanter, É.: AspectScript: expressive aspects for the web. In: Proceedings of the 9th International Conference on Aspect-Oriented Software Development, Rennes and Saint-Malo, France, pp. 13–24. ACM, New York (2010)
13. Cristian, F.: A recovery mechanism for modular software. In: Proc. 4th Int. Conf. Software Engineering, pp. 42–50. IEEE Press, Piscataway (1979)
14. http://martinweb.zobyhost.com/hcirg/en-projects.htm#SHUBAI (June 2010)
15. García, A.F., Rubira, C.M.F., Romanovsky, A., Xu, J.: A comparative study of exception handling mechanisms for building dependable object-oriented software. J. Systems and Software, 197–222 (2001)
16. Rosenberg, J.: Some Misconceptions About Lines of Code. In: Proceedings of the 4th International Symposium on Software Metrics. IEEE Computer Society, Los Alamitos (1997) ISBN:0-8186-8093-8

Author Index

Amagasa, Toshiyuki   240
Amann, Bernd   262
Arase, Yuki   225
Arellano, Cristóbal   294
Artières, Thierry   262
Asano, Yasuhito   480

Bain, Michael   342
Bayir, Murat Ali   166
Bertok, Peter   23
Bhuiyan, Touhid   357
Bobed, Carlos   190
Bratsberg, Svein Erik   587
Bressan, Stephane   456
Burella, Juan   564

Cai, Xiongcai   342
Cai, Yi   510
Cao, Jinli   216
Chang, Edward Y.   20
Chen, Chun   1
Chen, Gang   1
Chen, Junquan   175
Chen, Li   128, 365
Chen, Lijiang   578
Chi, Chi-Hung   322
Ciglan, Marek   91
Comerio, Marco   52
Compton, Paul   342
Cong, Gao   105
Cox, Clive   357
Cui, Bin   105

De Paoli, Flavio   52
de Spindler, Alexandre   411
Díaz, Oscar   271, 294
Ding, Chen   322
Dolog, Peter   120
Dustdar, Schahram   38, 52

Ester, Martin   442

Fang, Chuanyun   204
Fernández-Álvarez, Alberto
   Manuel   630

Fernández-Lanvin, Daniel   630
Fujimura, Ko   328

Gamini Abhaya, Vidura   23
Gedikli, Fatih   157
Ghafoor, Arif   428
Grigera, Julián   564
Grossniklaus, Michael   411, 471
Gu, Xiaoyan   204
Guo, Junxia   602

Haller, Armin   400
Han, Hao   602
Han, Jun   67
Hara, Takahiro   225
Hausenblas, Michael   400
Hemayati, Reza Taghizadeh   254
Heward, Garth   67
Horincar, Roxana   262
Huang, Hai   376
Huang, Sijia   578
Huang, Yi   578
Huang, Yuxin   105

Ilarri, Sergio   190
Iturrioz, Jon   294
Iwata, Mayu   225
Iwata, Tomoharu   328
Izquierdo, Raúl   630

Jannach, Dietmar   157
Jensen, Christian S.   21
Jiang, Dawei   1
Jiang, Huifu   204
Jin, Wei   175
Jonassen, Simon   587
Jøsang, Audun   357

Kabutoya, Yutaka   328
Kannengiesser, Udo   534
Kim, Yang Sok   342
Kitagawa, Hiroyuki   240
Kou, Yue   496
Krzywicki, Alfred   342

Le, Dung Xuan Thi   456
Leone, Stefania   411, 471
Li, Qing   510
Li, Shijun   204
Li, Xuhui   428
Liang, Huizhi   357
Liu, Chengfei   376, 419
Liu, Cong   322
Liu, Mengchi   428
Liu, Renjin   120
Liu, Wei   524
Liu, Yan   534
Lu, Zhihui   578

Mahidadia, Ashesh   342
MahmoudiNasab, Hooran   390
Maurino, Andrea   52
Medina, Nuria Medina   564
Mena, Eduardo   190
Meng, Weiyi   254
Mlýnková, Irena   279
Moser, Oliver   38

Nebeling, Michael   471
Nečaský, Martin   279
Ng, Yiu-Kai   142
Nguyen, Khanh   216
Nie, Tiezheng   496
Nishio, Shojiro   225
Norrie, Moira C.   411, 471
Nørvåg, Kjetil   91

Ooi, Beng Chin   1
Ozonat, Kivanc   308

Panziera, Luca   52
Pardede, Eric   456
Pera, Maria Soledad   142
Pérez, Sandy   271
Phan, Tan   67

Qi, Luole   128
Qian, Tieyun   77
Qian, Weining   442
Quintela-Pumares, Manuel   630
Qumsiyeh, Rani   142

Rahayu, Wenny   456
Robles Luna, Esteban   564

Rosenberg, Florian   38
Rossi, Gustavo   564

Sakr, Sherif   390
Shen, Derong   496
Sheng, Quan Z.   548
Sheu, Philip C-Y.   428
Singhal, Sharad   308
Sóztutar, Enis   166
Swee, Joshua K.Y.   548

Takahashi, Tsubasa   240
Taniar, David   456
Tari, Zahir   23
Tokuda, Takehiro   602
Toroslu, Ismail H.   166
Trillo, Raquel   190
Truong, Hong-Linh   52

Umbrich, Jürgen   400

Versteeg, Steve   67
Vo, Hoang Tam   1

Wang, Haofen   175
Wang, Junhu   616
Wang, Shuo   77
Wang, Xiaoling   442
Whang, Kyu-Young   22
Wobcke, Wayne   342
Wu, BangYu   322
Wu, Jie   578
Wu, Sai   1

Xiao, Jianguo   524
Xie, Haoran   510
Xie, ZhiHeng   322
Xu, Guandong   120
Xu, Jiajie   419
Xu, Kaifeng   175
Xu, Quanqing   1
Xu, Xiwei   534
Xu, Yue   357

Yamaguchi, Yuto   240
Yang, Yang   77
Yao, Junjie   105
Yongchareon, Sira   419
Yoshikawa, Masatoshi   480
Yu, Clement   254

Yu, Ge   496
Yu, Jeffrey Xu   616
Yu, Jian   548
Yu, Lijuan   510
Yu, Wei   204
Yu, Yong   175

Zhang, Xinpeng   480
Zhang, Yanchun   120
Zhao, Xiaohui   419
Zhou, Aoying   442
Zhu, Liming   534
Zong, Yu   120